REDRESS SCHEMES FOR PERSONAL INJURIES

This ground-breaking book takes a fresh look at potential non-litigation solutions to providing personal injury compensation. It is the first systematic comparative study of such a large number—over forty—of personal injury compensation schemes. It covers the drivers for their creation, the frameworks under which they operate, the criteria and thresholds used, the compensation offered, the claims process, statistics on throughput and costs, and analysis of financial costings. It also considers and compares the successes and failings of these schemes.

Many different types of redress providers are studied. These include the comprehensive no-blame coverage offered by the New Zealand Accident Compensation Corporation; the widely used Patient, Pharmaceutical, Motor Accident and Workers Compensation Insurance systems of the Nordic states; the far smaller issue-focused schemes like the UK Thalidomide and vCJD Trusts; vaccine damage schemes that exist in many countries; as well as motor vehicle schemes from the USA.

Conclusions are drawn about the functions, essential requirements, architecture, scope, operation and performance of personal injury compensation systems. The relationships between such schemes, the courts and regulators are also discussed, and both calls and need for reforms are noted. Noting the wide calls for reform of NHS medical negligence litigation within the UK, and its replacement with a no-blame approach, the authors' findings outline options for future policy in this area. This major contribution builds on general shifts from courts to ADR, and from blame to no blame in regulation, and is a work that has the potential to have a major impact on the field of personal injury redress.

With contributions from Claire Bright, Raymond Byrne, Matti Urho, Mary Walker and Herbert Woopen.

Volume 5 in the series Civil Justice Systems

Civil Justice Systems

Series General Editor: Christopher Hodges, Director, Swiss Re/CMS Research Programme, Centre for Socio-Legal Studies, University of Oxford

This series covers new theoretical and empirical research on the mechanisms for resolution of civil disputes, including courts, tribunals, arbitration, compensation schemes, ombudsmen, codes of practice, complaint mechanisms, mediation, and various forms of Alternative Dispute Resolution. It examines frameworks for dispute resolution that comprise combinations of the above mechanisms, and the parameters and conditions for selecting certain types of techniques and procedures rather than others. It also evaluates individual techniques, against parameters such as cost, duration, accessibility, and delivery of desired outcomes, and illuminates how legal rights and obligations are operated in practice.

Volume 1: *The Costs and Funding of Civil Litigation: A Comparative Perspective* edited by Christopher Hodges, Stefan Vogenauer and Magdalena Tulibacka

Volume 2: *Consumer ADR in Europe* by Christopher Hodges, Iris Benöhr and Naomi Creutzfeldt-Banda

Volume 3: *Law and Corporate Behaviour: Integrating Theories of Regulation, Enforcement, Compliance and Ethics*
Christopher Hodges

Volume 4: *A Comparative Examination of Multi-Party Actions*
Joanne Blennerhassett

Redress Schemes for Personal Injuries

Sonia Macleod
Researcher, Centre for Socio-Legal Studies, University of Oxford

Christopher Hodges
Professor of Justice Systems and Supernumerary Fellow of Wolfson College, Oxford
Head of the CMS/Swiss Re Research Programme on Civil Justice Systems,
Centre for Socio-Legal Studies, University of Oxford
Solicitor (non-practising)

With Contributors

Raymond Byrne
Law Commissioner, Ireland

Claire Bright
Senior Lecturer in Law, London School of Business and Management

Shuna Mason
Partner, CMS Cameron McKenna Nabarro Olswang LLP

Magdalena Tulibacka
Adjunct Professor, Emory University School of Law, Atlanta

Matti Urho
University of Turku

Mary Walker
Barrister, New South Wales

Herbert Woopen
Lawyer, Cologne

• HART •
PUBLISHING
OXFORD AND PORTLAND, OREGON
2017

Hart Publishing
An imprint of Bloomsbury Publishing Plc

Hart Publishing Ltd
Kemp House
Chawley Park
Cumnor Hill
Oxford OX2 9PH
UK

Bloomsbury Publishing Plc
50 Bedford Square
London
WC1B 3DP
UK

www.hartpub.co.uk
www.bloomsbury.com

Published in North America (US and Canada) by
Hart Publishing
c/o International Specialized Book Services
920 NE 58th Avenue, Suite 300
Portland, OR 97213-3786
USA

www.isbs.com

HART PUBLISHING, the Hart/Stag logo, BLOOMSBURY and the Diana logo are trademarks of Bloomsbury Publishing Plc

First published 2017

British Library Cataloguing-in-Publication Data
A catalogue record for this book is available from the British Library.

ISBN: HB: 978-1-50991-661-0 (Hart Publishing)
HB: 978-3-406-71910-3 (Verlag CH Beck)
ePDF: 978-1-50991-662-7
ePub: 978-1-50991-663-4

Library of Congress Cataloging-in-Publication Data

Names: Macleod, Sonia, author. | Hodges, Christopher, author.

Title: Redress schemes for personal injuries / Sonia Macleod, Research Fellow, Centre for Socio-Legal Studies, University of Oxford; Christopher Hodges, Professor of Justice Systems and Supernumerary Fellow of Wolfson College, Oxford Head of the CMS/Swiss Re Research Programme on Civil Justice Systems, Centre for Socio-Legal Studies, University of Oxford Solicitor (non-practising) ; with contributors, Raymond Bryne, Law Commissioner, Ireland [and six others].

Description: Oxford [UK] ; Portland, Oregon : Hart Publishing, 2017. | Series: Civil justice systems ; v. 5 | Includes bibliographical references and index.

Identifiers: LCCN 2017032005 (print) | LCCN 2017035616 (ebook) | ISBN 9781509916634 (Epub) | ISBN 9781509916610 (hardback : alk. paper)

Subjects: LCSH: Personal injuries. | Industrial accidents—Law and legislation. | Workers' compensation—Law and legislation.

Classification: LCC K925 (ebook) | LCC K925 .M33 2017 (print) | DDC 346.03/23—dc23

LC record available at https://lccn.loc.gov/2017032005

Typeset by Compuscript Ltd, Shannon
Printed and bound in Great Britain by CPI Group (UK) Ltd, Croydon CR0 4YY

ACKNOWLEDGEMENTS

We owe enormous thanks to a large number of people across the globe who have given us considerable help in understanding their systems. The first category is obviously our small group who have contributed text for chapters: Claire Bright, Raymond Byrne, Shuna Mason, Magdalena Tulibacka, Matti Urho, Mary Walker and Herbert Woopen. Special thanks are due to Naomi Creutzfeldt for research at the start of this project.

The second group are officials and contacts, listed below, who have given generously of their time in answering our detailed questions about their national systems:

Mike Playle, Policy Manager, New Zealand Accident Compensation Corporation

Dr Pelle Gustafsson, Medical Director of the Swedish Patient Insurance (Landstingens Ömsesidiga Försäkringsbolag (LÖF)), Sweden

Anders Öhlén and Anders Hultman of the Swedish Drug Insurance (LFF Service AB, Läkemedelsförsäkringen), Sweden

Agneta Karlbom of the Swedish Patient Insurance Association, Sweden

Rolf Gunnar Jørstad, Director General of the Norwegian Patient Insurance (NPE), Norway

Ragnar Wiik, Director of ECC-Norway

Peter Jakobsen and Kristina Askjaer, both Chefkonsulent, of the Patient Compensation Association (Patienterstatningen), Denmark

Sive Jensen, Martin Solomon, Jakob Steanstrup of the Danish Consumer Council, Denmark

Lone Rasmussen and Bo Dalsgaard of the Danish Chamber of Commerce (Dansk Erhverv), Denmark

Leif Rasmussen, ASK Denmark

Tiina Hellgren, CEO, and Liisa Jahnukainen, Pharmaceutical Injury Insurance, Helsinki

Elina Muukkonen, Claims Manager, Patient Insurance Centre, Finland

Antti Tuulensuu, Claims Manager, Insurance Centre, Motor Insurers Centre, Finland

Asko Nio, Head of Legal, Patient Insurance Centre, Finland

Finnish Workers' Compensation Center (TVK), Finland

Kauma Jussi, TVL, Finland

Kirsi Pohjolainen, Director, TVK, Finland

Hanna Eskola, Actuary, TVK, Finland

Kirsi Salo, Lawyer, TVK, Finland

Gunnar Soerli, Partner, Advokatfirmae BA-HR. Norway

Walther Neuhaus, Actuary, Zabler-Neuhaus AS, Norway

Dr Johannes Ried, Actuary, Zabler-Neuhaus AS

Haslum Norwayel, former President of the Oberlandesgericht Köln (Higher Court of Cologne), and president of the Gutachterkommission Ärztekammer Nordrhein, Düsseldorf

Dr Johannes Riedel, formerly President of the Oberlandesgericht Köln (Higher Court of Cologne), and President of the Gutachterkommission Ärztekammer Nordrhein, Düsseldorf

Connor O'Brien, Injuries Board, Ireland

Maurice Priestley, Injuries Board, Ireland

Jeff Varnell MD, Physician Risk Manager, COPIC

Kenney Shipley, Executive Director, NICA Florida

Deborah Jack, Executive Director, The Thalidomide Trust

Martin Johnson, Formerly Executive Director, The Thalidomide Trust

Yoshihiro Nakagawa, Corporate Officer and Global General Counsel, Takeda

Jeremey Fletcher, Senior Associate, Nabarro; Coal Industry Pneumoconiosis Compensation Scheme

Rose-Anna Lidiard, Solicitor, Field Fisher Waterhouse LLP

Paul Balen, Consultant Freeths LLP

Paul Sim, S&P Medical Devices Knowledge Manager, BSI Group

Jeremy Fletcher, Senior Associate, Nabarro LLP

Continental practice is to show decimal numbers as 1,2 whereas we have standardised in this work to the English 1.2.

TABLE OF CONTENTS AND AUTHORS

Part V: Vaccines and Pharmaceuticals

Part VI: Patient Injuries

Part VII: UK Schemes

Part VIII: Corporate Schemes for Medical Devices

Sonia Macleod

Part IX: Conclusions

Sonia Macleod and Christopher Hodges

DETAILED TABLE OF CONTENTS

Part I: Introduction

Part III: Nordic States

Part VI: Patient Injuries

Part VIII: Corporate Schemes for Medical Devices

Part IX: Conclusions

LIST OF TABLES

LIST OF FIGURES

Part I

Introduction

1

An Introduction to the Schemes

SONIA MACLEOD AND CHRISTOPHER HODGES

Background to No-Fault Schemes

This book sets out details of various leading approaches to compensation of personal injuries that are 'alternative' to the courts.[1] The inquiry grew from a series of key ideas. Firstly, for some 15 years from the mid-1980s on, one of us (Hodges) was involved as a practitioner in, or close to, a series of large product liability cases, the majority of which failed to establish liability on the part of any defendant but left both patients and manufacturers unsatisfied, whilst conferring substantial benefits on the intermediaries in the legal transactions, the lawyers. This prompted the notion that there must be a better way of processing compensation claims, screening those that are valid and invalid, and generally caring for those who suffer misfortune.[2]

Researching that idea led, in the early 2000s, to work on the Nordic drug schemes,[3] taken further in a 2013 doctoral thesis by Matti Urho, and then to investigating a range of 'no blame' or administrative injury schemes across the world. The 'no blame' focus also led us to widen our inquiry beyond patient and drug schemes to include examples of road traffic and medical schemes. The analysis here was also informed by extensive research on the related and expanding phenomenon of consumer alternative dispute resolution (ADR) in Europe.[4]

The current default position for personal injury is tort litigation on the basis of negligence. For some years this has been a source of considerable disquiet; leading academics in both the United States and Europe have identified that the two principal justifications for tort law and the tort system—deterrence and compensation—fall far short of being satisfactorily realised, and that no blame compensation schemes or other insurance-based arrangements presented significantly better approaches. By contrast, research into ADR schemes has found that complaint handling procedures can be designed to collate and

[1] This work draws in particular on the following previous research, supplemented by further currently unpublished research: J Dute, M Faure and H Koziol (eds), *No-fault Compensation in the Health Care Sector* (Wien, New York, Springer, 2004); C Hodges, 'Nordic Compensation Schemes for Drug Injuries' (2006) 29 *J Consumer Policy* 143–75; WH van Boom and M Faure (eds), *Shifts in Compensation between Private and Public Systems* (Wien, New York, Springer, 2007); C Hodges, I Benöhr and N Creutzfeldt-Banda, *Consumer ADR in Europe* (Oxford, Hart Publishing, 2012).

[2] The cases are detailed in C Hodges, *Multi-Party Actions* (Oxford, Oxford University Press, 2001).

[3] See Hodges (n 1).

[4] Reproduced from Hodges, Benöhr and Creutzfeldt-Banda (n 1).

feedback aggregated data that can support learning and behavioural change in a systemic way that litigation cannot.

These ideas suggested that 'no blame' compensation schemes may be an idea that offers considerable advantages to solving the problem of how contemporary society responds to undesired injuries of its citizens. Indeed, some states have adopted this approach decades ago, and for others it may be an idea whose time has come—in Europe, at least. It is possible to conceive of a no blame framework in which different types of complaints are dealt with in ways that are more effective than litigation in terms of both providing appropriate redress and opportunities for learning and improving behaviour.

We therefore decided to examine how many 'no blame' or ADR-like schemes exist for personal injuries. This book aims to identify a number of different compensation schemes, to see how they work, and what can be learned from these existing schemes about best practice in designing such schemes. We have identified far more administrative schemes for personal injuries than we, and we suspect many people, were aware of. Both their existence and how they work are matters of considerable interest.

Some of the schemes occupy positions in their national contexts that are far from alternative, and are, in fact, mainstream. The schemes were established on a 'no-fault' model, meaning that a claimant would not be required to prove that a particular defendant had been at fault (negligent). The consequence of the basis of 'liability' for payment of compensation for injury in a 'no-fault' scheme is not just that the legal basis changes from negligence or strict liability to some form of 'no-fault' trigger, but that the administrative architecture of the claim-processing arrangement is able to change from an adversarial-based litigation model to, in almost all cases, a 'softer' procedure such as an administrative procedure that is closer to an inquisitorial process than an adversarial process.

Our inquiries have taken place against growing political interest in injury schemes from the UK Department of Health and the Scottish Government, both of which have been working on proposals. We decided to exclude coverage of those proposals from this book, since it deals only with extant schemes. We intend to look at the position in the UK, in its particular context of changes in the landscape of litigation, courts and lawyers, in a further volume.

This book essentially describes schemes. We do not here delve into the normative or theoretical issues that inevitably arise. However, we have found that such matters repay detailed consideration in the light of the facts that we set out here, as we show that a no blame approach has spread widely, possibly more so than many realise, and therefore pre-existing theories need to be reviewed.

We draw a series of conclusions about the design and operation of injury schemes in the Conclusions chapter at the end of the book, followed by a chapter that sets out in list form the main parameters for the design of a contemporary compensation scheme, and how it should operate, drawn from those Conclusions on the information and learning set out in this book.

Summary of the Schemes

By way of introduction we provide below a short summary of the main mechanistic elements of each of the schemes in this book. Inevitably, this summary is selective, which is why serious students of the mechanics should consult the detail in the individual chapters.

We group the summaries here by type of injury: general person injury, workers, motor vehicle, patients, birth, vaccines, medicines, and other social schemes.

A. General Injuries

The Irish Injuries Board

Ireland continues to rely on court litigation for medical injuries, requiring a claimant to prove negligence. However, it has introduced a single national personal injury damages assessment mechanism *before* claimants are permitted to start court proceedings for personal injuries (but not, for the present at least, for clinical negligence cases). This in effect reverses the normal litigation sequence (party investigation, court claim, decision on liability, assessment of damages) by placing the assessment of damages as the first and compulsory step. Since many claims settle before a full trial, this adjustment facilitates early settlement.

The Injuries Board was established with state funding, which was repaid after six years, and is now self-funding from modest user fees. Overhead costs have been successfully reduced comparatively quickly. Application fees are charged, currently €600 per case.

The system replaces *part* of the litigation process with an independent non-adversarial assessment of the quantification of damages. This mechanism has assisted the speedy and economic resolution of claims by negotiation between the claimant's lawyer and defendant's insurer. The liability trigger, of course, theoretically remains negligence.

The New Zealand Accident Compensation Scheme

The New Zealand Accident Compensation Scheme (ACS) has formed the central mechanism for compensation of injuries, originally workplace and motor vehicle accidents, for 40 years. Since a financial shortfall in 1986 it has remained economically stable and in recent years has virtually achieved full funding for accrued claims. The single national administration runs five discrete funds covering the injuries arising in the following contexts: work, motor vehicle, earners, non-earners and treatment. Each of the five accounts is funded by levies on defined constituent groups (eg employers, the self-employed, vehicle owners and health professionals), priced separately. Thus, the risk-pooling is done in five discrete groups, limiting cross-subsidies to members of each group. The scheme pays 'real' but not 'full' compensation. There is a strong focus on rehabilitation.

Despite the word 'accident' still being part of the scheme's title, the trigger is no longer the occurrence of an 'accident': that term gave rise to historical difficulties of interpretation at the boundaries, and was dropped. Broadly, The ACS pays for conditions caused by injury but not by disease. The essential trigger is existence of a personal injury. Different triggers are now used for individual injury types, and triggers have become highly specific and more fact-based. The scheme covers all qualifying injuries, including treatment injury and prosthesis failure that is not due to wear and tear.

Court proceedings may not be brought for personal injuries covered by the scheme. The scheme has in practice replaced litigation, which would have to be based on allegations of negligence. In 2015/16, the average time to make a coverage decision was 1.1 days, average time to commence payment was 8.3 days, and the 'return to work' rate was 67.6% after 70 days and 92.8% after 273 days. The administrative cost is around 8% of total costs. This is very low, and is partly a consequence that screening of claims is limited. As at 30 June 2016 scheme is in deficit, NZ$3.367 billion, for the first time since June 2012. However, levies were reduced in 2015/16, and could be adjusted to compensate for the deficit.

The Nordic Insurance Schemes for Motor, Medical and Drug Injuries

The Nordic states (Sweden, Denmark, Finland and Norway) all have similar injury compensation insurance schemes for injuries caused by *motor vehicles, medical treatment and medicinal products*. These schemes have operated for many years and should be seen as part of an integrated single approach to compensation for injury of whatever type.

They also form part of a comprehensive Nordic approach to dispute resolution that involves ADR bodies for consumer trading disputes and results in very low usage of lawyers and courts. An important structural feature is that the injury insurance compensation schemes form part of a unified model, so that they are either administered by a single body for a country or can refer claims to another sister fund if appropriate. They operate on a document-based inquisitorial model, with no hearings or oral submissions, so are efficient in cost terms and notably speedy. The details of each scheme differ, and the triggering criteria differ; they can also be to some extent administratively adjusted, so as to be more or less generous to applicants.

The cost of schemes is notably low, but this is because of the existence of a high level of social security funding in Nordic states, which results in the medical and especially pharmaceutical insurance funds functioning only as a 'top up' to other sources of support for injured people. The pharmaceutical schemes are funded by levies on all manufacturers, save in Denmark where the state provides funding. In Denmark, the fund may reclaim against a manufacturer who is liable under the EU product liability directive, but this has never occurred. This lack of indemnity/recourse is an important economic factor in relation to the cost of the schemes.

The existence of the drug scheme enabled a flexible ad hoc solution to be applied in compensating victims of blood products contaminated with hepatitis C, in which the pharmaceutical industry made ex gratia payments that were administered through the scheme arrangements.

The Nordic states have recently begun to use the increasing database of claims information so as to support improvements in frontline medical practice. Outstanding results have been achieved in Sweden, where the incidence of claims for babies damaged at birth has fallen dramatically. Harm reduction has had dual benefits: firstly it reduces the number of lives of individuals and their families blighted; and secondly it has significantly reduced the overall costs, since such injuries are very expensive. In some states, however, inertia or impediments impede this recycling of information. The lesson is that the whole healthcare system has to be designed so as to collect as much data as possible, and the data then be used to enable an entity to make use of it. Merely collating or passing on information may be insufficient.

UK Public Schemes and Trusts

The UK has a large number of small individual schemes. They have been created piecemeal since the 1960s, and are all different. This leads to confusion over their differing individual scope, procedures and entitlements, and even a lack of knowledge of their existence. It can also lead to overlaps.

The general *Industrial Injuries Disablement Benefit* is paid to employees who suffer industrial injuries as a result of an accident, or from prescribed diseases, as a weekly sum dependent on the extent of disablement. A related scheme covers payments for those suffering mesothelioma from asbestos.

The *Armed Forces Compensation Scheme* applies to injuries to military personnel 'caused by service'. However, although the parameters and procedure are formalised, compensation is only payable at the discretion of the Ministry of Defence. All awards include a lump sum element for pain and suffering. Notionally, the burden of proof (which is a concept that is illogical in relation to a wholly discretionary scheme) rests on the applicant, who must show that the injury was caused by military service, and there are various listed exceptions, so reasonable clarity over that trigger. The process is designed to be straightforward and inquisitorial, and can include 'fast payment' and temporary sums. Proposals are being consulted upon to make this scheme a compulsory ADR mechanism before litigation can commence, which seems inconsistent with the discretionary nature of the scheme.

The *Criminal Injuries Compensation Scheme* pays awards to victims of violent crime of between £1,000 and £250,000 on a tariff basis, and in severe cases plus non-tariff expenses such as loss of earnings and medical expenses, subject to a total cap of £500,000. It functions as a last resort after other sources of compensation, and is subject to a series of exclusions and deductibles, although the application procedure is relatively straightforward. The administrative body, the Criminal Injuries Compensation Authority, is a public body.

B. Workers' Compensation

Workers' compensation arrangements for injuries sustained while at work exist in many jurisdictions across the world—although not in the UK.[5] They are, by definition, 'no blame' schemes. Our inquiries into the holistic systems in the Nordic states led us to look at the industrial accident schemes in Denmark and Finland (as representative examples). As a comparison with a series of schemes from another part of the world, we have included the schemes in six of the eight states and territories in Australia.

Denmark and Finland Industrial Accident Compensation

For those who are injured at work, employer's liability insurance is compulsory, and provides the second tier after the initial tier of state free healthcare provision. It is noted in

[5] G Wagner (ed), *Tort Law and Liability Insurance* (Berlin, Springer, 2005).

chapters 6 and 7 that the details of what benefits, and how much, vary considerably, depending on the details of insurance policies.

Australian Workers' Compensation

Workers' compensation arrangements have existed in Australia for over 100 years. They are not federal, but are based on regional territories, and there are considerable differences in the way different areas deal with claims.

All of the six schemes, Western Australia, the Northern Territory, the Australian Capital Territory, South Australia, Tasmania and New South Wales, detailed in chapter 3 share some similarities. All operate on the basis of insurance defined by a statutory framework. All define what it is to be a worker and the types of injuries that will be compensated, but the definitions used vary the regions. All schemes define how much compensation is due in different circumstances. The procedure and interactions with tort also vary between regions.

In addition to workers' compensation, all of the states studied also offer dust disease compensation.

Western Australia

In Western Australia workers' compensation is governed by the Workers' Compensation and Injury Management Act 1981 as amended, and WorkCover WA is charged with ensuring the efficient and effective operation of the workers' compensation scheme. Insurance is offered by approved insurers. In the event of an accident or injury the employer must be given notice. Claims are made to the employer in a prescribed form, and must include a First Medical Certificate from the treating practitioner. This information is then forwarded to the insurer. If the insurer accepts the claim it will be paid.

The First Medical Certificate contains information on the diagnosis, the likely impact of the injury on the worker's ability to work and details of forthcoming medical treatment. If further information is needed on the condition of the injured worker a Medical Assessment Panel, or a Specialist Medical Assessment Panel, can be convened at any point in proceedings. Thus the scheme relies on external medical expertise, and does not determine injury severity or impact in-house. The independence and impartiality of the Medical Assessment Panels is valued.

If the insurer disputes the claim, then the parties must make an attempted to resolve the claim by negotiation. If it cannot be resolved it will be sent to the Workers' Compensation Conciliation Service, and, if accepted, it will be dealt with by a conciliation officer. Should this fail to resolve the issues it will be sent for arbitration at the Workers' Compensation Arbitration Service. Again the Workers' Compensation Arbitration Service decide whether to accept the claim or not. Decisions of an arbitrator are binding and cannot usually be challenged by the court.

Northern Territory

The Northern Territory Workers' Compensation arrangements are governed by the Return to Work Act 2016 and administered by NT WorkSafe. Interestingly, the Northern

Territory Workers' Compensation arrangements are an exclusive remedy and prevent the recovery of damages in tort for injuries which are eligible for compensation under the 2016 Act.

In order to make a claim notice must be given to the employer of the injury. A claim is then initiated in a prescribed form; again it must include a medical certificate from the treating practitioner. The claim is then passed to the employer's insurer. If the insurer accepts the claim it is paid.

If there is a dispute over the claim, then the dispute is passed to mediation, and if still unresolved to the Health Courts for determination.

Australian Capital Territory

The Workers' Compensation Act 1951 governs the provision of workers' compensation in ACT and is administered by WorkSafe ACT. There is recourse to common law damages for injured workers in certain, limited circumstances. Workers' Compensation Insurance must be obtained by employers from approved insurers.

Notice of the injury must be given to the employer. A claim must be made to the employer and, depending upon the compensation sought, it may need to be accompanied by a medical certificate. This is then passed to the insurer.

Disputes over claims must go to conciliation, and if this fails claims may be sent to arbitration.

South Australia

ReturnToWorkSA administers the Worker's Compensation scheme created under the statutory framework of the Return to Work Act 2014 and subordinate regulations. Seriously injured workers can claim in tort, but for less seriously injured workers the right to litigate in tort has been extinguished. Employers may either purchase insurance from ReturnToWorkSA or, if they fulfil the requirements, they can opt to be self-insured.

In order to make a claim notification of the injury has to be given to the employer. A claim is then lodged with Return; this does not require a claim form as it can be dealt with by ReturnToWorkSA claims handlers by phone or online. Depending upon the compensation sought, a certificate from a medical practitioner may be required. ReturnToWorkSA will then investigate the claim; this can involve sending the claimant for medical assessments. If a claim is accepted compensation will be paid and a return to work/recovery plan will be initiated if appropriate.

If a claim is disputed the claim must be redetermined by a new case worker at ReturnToWorkSA. If this does not resolved the situation, then claims can be reviewed by the South Australian Employment Tribunal.

Tasmania

The principal statute for workers' compensation in Tasmania is the Workers' Rehabilitation and Compensation Act 1988. This is administered by WorkCover Tasmania. Limited rights to common law recourse for workers' compensation claims remain in force in Tasmania. Employers are required either to take out a workers' compensation insurance policy or to apply to WorkCover Tasmania for a certificate of self-insurance.

Injured workers are required to notify their employer of the injury and if seeking compensation to fill in a claim form, which is submitted to the employer with a Workers' Compensation Medical Certificate from a WorkCover-accredited doctor. The employer then has to commence 'without prejudice' payments of medical expenses up to $5,000 and to commence additional weekly payments. These payments must begin regardless of any dispute over liability. The employer then informs their insurer of the claim and the insurer will accept or dispute the claims. A full range of compensation payments is available for accepted claims.

Should a claim be disputed it can be referred to the Workers' Rehabilitation and Compensation Tribunal. This first step to resolve the claim will be conciliation, and if this is unsuccessful, then arbitration will follow. In certain circumstances the Tribunal has the right to refer questions on medical matters to a Medical Panel.

New South Wales

Two principal Acts, the Workers' Compensation Act 1987 and the Workplace Injury Management and Workers' Compensation Act 1998, govern the Workers' Compensation Scheme. Following recent reorganisations, the workers' compensation scheme is overseen by icare. Since March 2017 icare has been the sole provider of workers' compensation policies; employers have either to insure with icare or certify with icare that they are self-insurers. The right to seek redress at common law for workplace injuries is restricted to claimants with more serious injuries.

The claims process requires that the worker notifies the employer, who then notifies the insurer. Notification must contain certain specified details, but there is no obligation to provide a medical certificate. Provisional payments must (with certain exceptions) be made by the insurer after notification. Provisional payments cover up to 12 weeks' wages and medical expense of up to AU$7,500 for reasonably necessary treatment. If provisional payments are refused or the claim is likely to persist for more than 12 weeks a claim form must be submitted. Once a claim is submitted the insurer can accept liability and pay, decline liability, or state that they cannot find a policy in place for that employer and forward the claim to the nominal insurer. Accepted claims are paid according to the statute.

If the extent of a claimant's ability to work is in question, then a Work Capacity Decision may be required. As the insurer, icare can request that injured workers attend a Work Capacity Assessment, which is used to aid the formulation of the Work Capacity Decision. Failure to attend a Work Capacity Assessment may result in reduced compensation payments. Work Capacity Decisions can be reviewed, both internally and externally.

Disputed claims follow a designated pathway, starting with internal review by the insurer. If resolution is not obtained, then the dispute can be heard by the Workers' Compensation Commission. If the dispute is over the degree of permanent impairment suffered by the claimant, then an Approved Medical Specialist will assess the claimant. All other disputes are handled using conciliation and/or arbitration. There is the potential for legal representation for the claimant while the dispute is being resolved.

While not based on fault it is interesting to note that the workers' compensation processes seen in Australia are akin to the litigation process, so much so that legal representation is often allowed through the conciliation and arbitration processes. The use of negotiation and conciliation is widespread in Australian workers' compensation, but is highly unusual in context of no blame compensation schemes. These ADR techniques tend to coexist with litigation, but are not commonly seen within ADR mechanics. This may be because the Australian workers' compensation schemes have been in existence for far longer than the other schemes studied here. They developed in a time when common law remedies served as the dispute resolution mechanism and they often sat alongside tort, rather than replacing it entirely. The fact that they pre-date the widespread development of ADR mechanics is reflected in their procedures.

UK Industrial Injuries Disablement

The UK arrangements for industrial injuries are not uniform or comprehensive and are formed of both private and state provision.

Employers' liability insurance is obligatory for most UK employers and covers injuries at work.

Industrial Injuries Disablement Benefit is a part of the state-provided social security which provides employed earners with benefits in the event of an injury arising out of and in the course of employment or suffering from an occupational disease prescribed in relation to that employment. Certain categories of workers are excluded, such as the self-employed and members of the armed forces.

Cases of mesothelioma arising from employment can receive compensation via various different pathways. The appropriate compensating body will depend upon whether an applicable employer's liability insurance policy can be found, the nature of any asbestos exposure and the ability to trace an employer responsible for the exposure. Compensation can be made under the Industrial Injuries Disability Benefit Scheme, the Diffuse Mesothelioma Payment scheme of 2008, the Diffuse Mesothelioma Payment scheme of 2014, the Pneumoconiosis Act 1979 or via civil claim. This is hardly a clear or easily navigable situation for potential claimants, who are usually quite unwell by the time they commence a claim.

A separate scheme, the Coal Workers Pneumoconiosis Scheme 1974, also exists for pneumoconiosis suffers who had previously worked as miners for the British Coal Corporation for at least 10 years and who were in receipt of benefits under the Industrial Injuries Disability Benefits scheme.

The coal health schemes for chronic obstructive pulmonary disease (COPD) and vibration white finger (VWF) were two of the largest administrative compensation schemes in British history. Litigation in the late 1990s found British Coal liable in negligence for employees' and previous employees' respiratory diseases and VWF. The liability was the absorbed by the Department of Trade and Industry (as was) and an administrative scheme was created to assess the health impacts on past employees and to provide compensation where appropriate. The objective was to make claiming simpler for victims and to cut costs. The basis was again court-overseen schemes that were designed to operate like litigation. A Claims Handling Agreement was drawn up for each scheme to define how claims would be dealt with, including the levels of recompense for solicitors and claims handlers.

The volume of claims was substantial: the COPD scheme received 592,000 claims; and the VWF scheme received 170,000 claims. One of the major objectives was to keep these claims out of the courts as it was felt they would block the civil litigation system for years.

In that regard the schemes were successful. However, the administrative costs of the scheme were huge; overall for every £2 spent on compensation, £1 was spent on administering the scheme. While this is not out of line with what would have been expected if each case had been litigated, it is vastly disproportionate when compared to the expected cost of processing claims though an administrative scheme. Solicitors fees were paid by the scheme and made up over half of the administrative costs and were often felt to be excessively high given the work undertaken.

To add insult to injury, some solicitors' firms were found to have behaved in highly inappropriate ways, and were disciplined by the Law Society.

The miners' scheme turned out to have spectacularly unsuccessful consequences and avoidable costs. There was hardly any need for any legal representation of individual claimants, since the process was an inquisitorial one undertaken by the administrator, and by private-sector claims management firms.

Conclusions on Workers' Injury Compensation

Workers' compensation schemes exist within both civil and common law jurisdictions, both as a general injury compensation mechanism and in the more piecemeal manner seen in the UK in relation to specific injury types. Workers' compensation schemes take a variety of forms, from quasi-litigation procedures seen in Australia to the more administrative system seen in Denmark.

The majority of schemes tend to require external medical expertise to aid with eligibility decisions, though the coal health schemes included the provision of medical testing for specific diseases.

All of the schemes here have clear formulas as the basis for calculating compensation, with compensation only available for defined needs, such as medical costs or loss of earnings. With the exception of the coal health schemes, general damage for pain, suffering, etc, are not seen in workers' compensation awards.

The UK coal health schemes demonstrate that the design of a scheme is key to overall efficiency and effectiveness. The Nordic schemes are defined by statutes such that there is very little discretion for the administering schemes. The Australian schemes go a step further and create a statutory body charged with overseeing the efficient and effective running of the scheme. Any oversight option needs to be independent and flexible, the coal health schemes' Claims Handling Agreements were negotiated between solicitors and did not have the flexibility to be amended easily to regulate the excessive profits made by some of the solicitors handling the claims.

C. Motor Vehicle Injuries

US Motor Vehicle No-Fault Coverage

Although the economic cost of motor accidents rose hugely in the first quarter of the 20th century, and prompted calls for no-fault schemes, it was not until the 1970s (when fatality and accident rates were in fact declining) that no-fault injury schemes were adopted by around 20 US states. The current insurance arrangements across the 50 states vary enormously. Nine of them still have a no-fault model, but subject to a threshold over which a claim *can/must* be brought in tort. The evidence suggests that verbal/textual thresholds are preferable to monetary thresholds, since the latter can affect fraudulent claiming, at last in some communities, or at least a 'loading' of claims to overcome the required value. Prioritisation of payer sources is also subject to huge differences.

The detailed rules of the no-fault scheme in Florida have been amended many times, mainly to address what is perceived to be a significant battle against fraudulent claims. North Dakota's scheme has also been amended many times, albeit relate to relatively minor points. Fraud does not seem to be a particular problem in North Dakota, possibly attributable to a range of factors including different population spreads (it has no major cities). It limits the sum paid by the employer to the first $10,000, after which healthcare and other insurers have to step in.

The arrangements in Colorado were similar to those in North Dakota (including a $25,000 threshold), but Colorado repealed its no-fault scheme in 2003—although there have been attempts to reintroduce it. Certain differences between those two states are thought potentially to have made one stable and the other problematic. For instance, the total benefits available under personal injury protection in North Dakota have a cap of $30,000, whereas it was in excess of $130,000 in Colorado. Further, North Dakota allows coordination of medical payments, with motor insurers only covering the first $10,000. This was not the case in Colorado, where motor insurers were primary payers (only workers' compensation took priority) and there was no coordination.

The combination of a very high available limit with a low threshold was considered responsible for the demise of no-fault in Colorado. The impact of these limits on both consumer premiums and insurer profitability was substantial. However, Florida has a relatively low cap on total available compensation, but the scheme limps along, rather than functioning well. Coverage limits, while potentially useful to moderate a scheme, are not sufficient to ensure a healthy scheme.

Usage of medical services impacts on the costs of any no-fault scheme. It is difficult to determine whether the increased uptake and use of medical services seen in no-fault states represents 'over-claiming' in these states or whether the more limited provision seen in tort states is inadequate. The impact of the limitations in the 2012 reforms in Florida on the types of medical services is yet to be evaluated. Limiting available services to treatments with proven benefits does not seem to be unreasonable.

US motor claims are processed by the injured person's insurer, but the procedural rules are laid out in the relevant state legislation. There is therefore a shift away from the courts as the determiners and towards the insurers as the decision maker. In order to prevent the insurers from being partisan and acting in their own (profit-making) interests a set of minimum eligibility criteria, procedure, entitlements and timeframes are set out in statute. This arrangement can be effective, as is seen in North Dakota. However, the statutory requirements can also hamper insurers if they are seen as too generous, as was the case in Florida before the 2012 reforms. US no-fault motor insurance shows that personal injury compensation can be dispensed outside of a court setting, but that clear guidance is required on eligibility, procedure, entitlements and timeframes. This guidance must be adaptable and capable of change with input from the insurers/decision makers, ideally with any changes being overseen by a body that is independent of the insurers/decision makers. A right to appeal insurer decisions, such as happened in *Lasky v State Farm Insurance Company*,[6] provides an additional safeguard ensuring the correct interpretation of the legislation by insurers.

Nordic Motor Vehicle No-Fault Coverage

Nordic states operate a no-blame motor insurance in certain defined situations. The vast majority of motor insurance is offered in the standard way; individuals take out insurance and in the event of an incident they make a claim to their insurer. If there is a second party involved in the accident, then the insurers attribute fault between the parties and compensation duly follows the fault. However, in certain circumstances this cannot happen—usually if the innocent individual cannot identify the at-fault party or if the at-fault party is uninsured. In these situations compensation may be offered by a pooled redress mechanism, and not by the individual's own insurer. The pooled redress mechanism varies slightly country by country, but is essentially similar in that it is funded by insurers, but independent of them, and with clear guidelines on eligibility, procedure and quantum.

These countries provide compensation to those who were blamelessly injured if the standard insurance compensation mechanisms are unavailable. Payment into the system by insurers is mandatory. Although the system is funded by insurers, the contributions are made up from payment of premiums by citizens; hence this can be viewed as an example of collective responsibility for compensation. Clearly this system requires a certain level of civil obedience; it is only functional where a sufficient proportion of the population hold motor vehicle insurance and where hit-and-run incidents remain at a fairly low level.

A similar system is found in the UK and is run by the Motor Insurers' Bureau.[7]

D. Patient Injury Schemes

ADR for German Medical Claims

Claims for injuries caused by medical negligence in Germany theoretically have to be brought in court and to prove fault, but are in practice handled by specialist medical ADR bodies (*Schlichtungsstellen*) in each region (Land). Each Land's scheme operates somewhat differently, on rules established by their local medical chambers. All the schemes work on the basis of three principles: participation is *voluntary*, *non-binding* and *free of charge* as such. Costs outside of the procedure, particularly those of lawyers, are to be borne by the parties.

It has been found that the asymmetry of expertise and status between patients and doctors does not support a negotiation or mediated discussion in relation to liability, but mediation can be helpful at a subsequent stage of agreeing quantification of damages. Around 12,000 cases are handled by the *Schlichtungsstellen* nationally each year. It is thought that

[6] *Lasky v State Farm Insurance Company* 296 So 2d 9 (1974).
[7] www.mib.org.uk/.

somewhere between another 12,000 and 28,000 cases progress in courts rather than in the *Schlichtungsstellen* system. *Schlichtungsstellen* decisions are accepted by the parties in around 85% of cases, avoiding further recourse to court. Interestingly, when cases that have been through the *Schlichtungsstellen* are subsequently litigated the *Schlichtungsstellen* conclusions are overturned in 42% of cases.

The *Schlichtungsstellen* essentially make voluntary non-binding determinations on whether there was fault, and whether that had adverse consequences for the patient. The process requires the consent of both patient and doctor, either of whom can withdraw consent at any time. The processes are in writing, so akin to non-binding arbitration. The *Schlichtungsstellen* generally have full access to all records, and consult relevant medical specialists. The *Schlichtungsstellen* operate independently, often chaired by a senior judge. Decisions are made by panels of at least one doctor and one lawyer, sometimes with a review procedure.

After a determination of fault, the patient would have to launch a court case, but in practice cases are generally settled quickly by insurers. Indeed, insurers are involved in the process during the liability phase, and are then able to respond quickly to settle claims.

In recent years, cases have been entered into a Medical Error Reporting System (MERS), coordinated by the federal Doctors' Chamber (Ärztekammer), which provides data to assist in disseminating information on risk and expertise, as well as on the nature of legal decisions on what does or does not constitute negligence.

The drawback of this mechanism is that it applies the legal standard of fault. The nature of the decision-making tribunal and to some extent the procedure may have changed, but the trigger to compensation remains the legal test of fault.

German Medical Insurance Ombudsman

Citizens may purchase extra insurance cover to supplement the services provided by German state healthcare, and, any employers provide medical insurance for their staff. Disputes under medical insurance policies are determined by two ombudsmen; one the ombudsman of the Association of Private Healthcare Insurers (PKV-Verband), and the other the general insurance ombudsman. The procedure is similar to that of all ombudsmen in Germany: a case file is assembled by staff and, if the case is not settled by mediated negotiation, a non-binding decision will be made by the ombudsman.

US Medical Schemes

Several schemes have arisen in the United States to deliver fairer redress following clinical injuries. They range from the large-scale provision afforded by the Department for Veterans' Affairs under the US code to much smaller, more localised initiatives such as COPIC, a Denver-based medical insurer's approach. COPIC initiated a fair patient compensation approach called '3Rs' in 2000 partly to reduce liability risk, but it was equally an enlightened patient-focused response to assist physicians to respond to their patients and to learn from mistakes in a no-blame environment. These schemes represent a change in outlook from the usual tort and blame-focused approach traditionally seen in US medical negligence.

The Netherlands

We do not mention the arrangements outlined in this paragraph further in this book, but it is interesting to note some similarities with, and some differences from, Germany above.[8] For some years, disputes between consumers and health insurance companies in the Netherlands were largely resolved through an out-of-court ADR process operated by the Complaints and Dispute Foundation for Health Insurances (Stichting Klachten en Geschillen Zorgverzekeringen, SKGZ).[9] SKGZ was established in 2006 by the Association of Dutch Health Insurers (Zorgverzekeraars Nederland) and the Federation of Patients and Consumer Organizations in the Netherlands (NPCF), and is fully financed by the Dutch association of health insurers.

After claims have been raised with the relevant health insurer, they may be filed at SKGZ. SKGZ then offers an ombudsman process as a first stage, which is free of charge. Cases are managed by a team of case officers but every decision is made by the health insurance ombudsman himself. The ombudsman tries to seek a settlement between parties and writes a (non-binding) recommendation/decision. If the first stage is unsuccessful, the second stage is for the complaint to be considered by the Complaint Commission for Health Insurances (Geschillencommissie Zorgverzekeringen). The consumer has to pay a modest fee (€37) at this stage, which will be reimbursed if the consumer wins the case. The Commission resolves disputes on the 'binding advice' (binding outcome) basis that is common to the large body of consumer ADR entities (Geschillencommissie) in the Netherlands.[10]

The Dutch legislator recently passed a new Act on the (overall improvement of the) quality of healthcare and the out-of-court ADR of complaints and disputes between patients and healthcare *providers* (doctors and other healthcare professionals).[11] Regarding the out-of-court resolution of healthcare complaints and disputes, the Act provides that:

- All Dutch healthcare providers maintain an in-house complaints officer (funded by the healthcare provider), for first-tier resolution of complaints and disputes.[12]
- It is mandatory for all healthcare providers to (establish and) adhere to an ADR scheme (approved by the Minister of Health, Welfare and Sport)[13] for the out-of-court resolution of healthcare complaints and disputes.[14] The ADR procedure is binding advice, as noted above, and patients can be compensated up to (at least) €25,000.[15]

[8] We are grateful to Eline Verhage of Leiden University for the following information.

[9] www.skgz.nl/.

[10] See Hodges, Benöhr and Creutzfeldt-Banda (n 1) ch 6; E Verhage, 'The Implementation of the Consumer ADR Directive in the Netherlands' in C Cortes (ed), *The New Regulatory Framework for Consumer Dispute Resolution* (Oxford, Oxford University Press, 2016); M Knigge and E Verhage, 'The Impact of the ADR Directive on Article 7:904 par 1 DCC Explored' in *BW Krant Jaarboek* (2016) 85, https://openaccess.leidenuniv.nl/handle/1887/43410.

[11] Healthcare Quality, Complaints and Disputes Act (Wet kwaliteit, klachten en geschillen zorg (Wkkgz)) *Stb* 2015, 407.

[12] See Arts 13–17 Wkkgz.

[13] See Art 18(2) Wkkgz.

[14] See Arts 18–22 Wkkgz.

[15] See Art 20 Wkkgz.

The Act came into effect on 1 January 2016. Healthcare providers were obliged to adhere to an ADR entity by 1 January 2017 at the latest.[16] One striking consequence has been that 26 different ADR entities have been established and accredited under the Act.[17]

The Nordic Patient Compensation Schemes

Each of the Nordic states included in this study has well-established and stable patient compensation arrangements (see Figure 1.1). They are considered to function efficiently and effectively. They have replaced tort liability-based court actions almost completely for medical compensation claims. The shift away from fault liability was a conscious policy move, underpinned by a series of conclusions. First, fault liability is a difficult triggering criterion for an injured person to establish. Second, a fault-based trigger leads to a significant restriction in the number of injured people who are awarded compensation, and withholds compensation from a significant number of people who may be widely considered to deserve financial social support. Third, having to fund and prove a fault-based investigation and law suit is a highly unattractive task for an injured person to have to confront, at a time when he or she needs social support rather than difficult challenges. Fourth, the fault and litigation approach causes a major division between patients and carers, and amongst carers themselves. In particular, it does not support self-reporting of information that is essential for the continuous monitoring and improvement of safety performance. Thus, fifth, the tort system is intended to regulate behaviour by creating deterrence and a climate of fear, since the prevailing view in those involved with all of the Nordic schemes is that in fact deterrence does not improve medical performance, and instead it decreases it. Accordingly, sixth, if the purpose of the tort/fault system is merely to provide compensation, there is a cheaper, fairer and more inclusionary way of achieving that, namely by a compensation scheme.

The schemes are based on the fact that virtually all healthcare in each state is provided by the state (over 90% in Sweden, Denmark, Finland and Norway). In Denmark, Finland and Norway, the (usually regional) authorities responsible for providing public healthcare pay the cost of compensation awarded to patients, annually in arrears. In Sweden, the public providers use a common mutual insurer, LÖF. In all four states, the private healthcare providers either pay through commercial insurance arrangements or, in Norway, pay into a separate fund maintained by the single national claims administration body.

In Finland, Norway and Denmark, patient compensation functions as a top-up secondary payer behind state benefits, with limited or no recourse being available. In Sweden both patient and drug compensation schemes are secondary payers to state benefits: the total compensation package is provided by the state, with LOF paying elements that state benefits do not pay, such as financial compensation for loss of income or for pain and suffering.

Sweden applies both a deductible from compensation (which is unpopular) and a ceiling on total compensation (which can lead to injustice in high-value cases).

In all of the states, administration of the patient compensation system is centralised in one body, which also processes private claims in Denmark, Finland and Norway, although

[16] www.government.nl/topics/quality-of-healthcare/contents/healthcare-quality-complaints-and-disputes-act-wkkgz.

[17] www.geschilleninstantieszorg.nl/erkende-instanties.

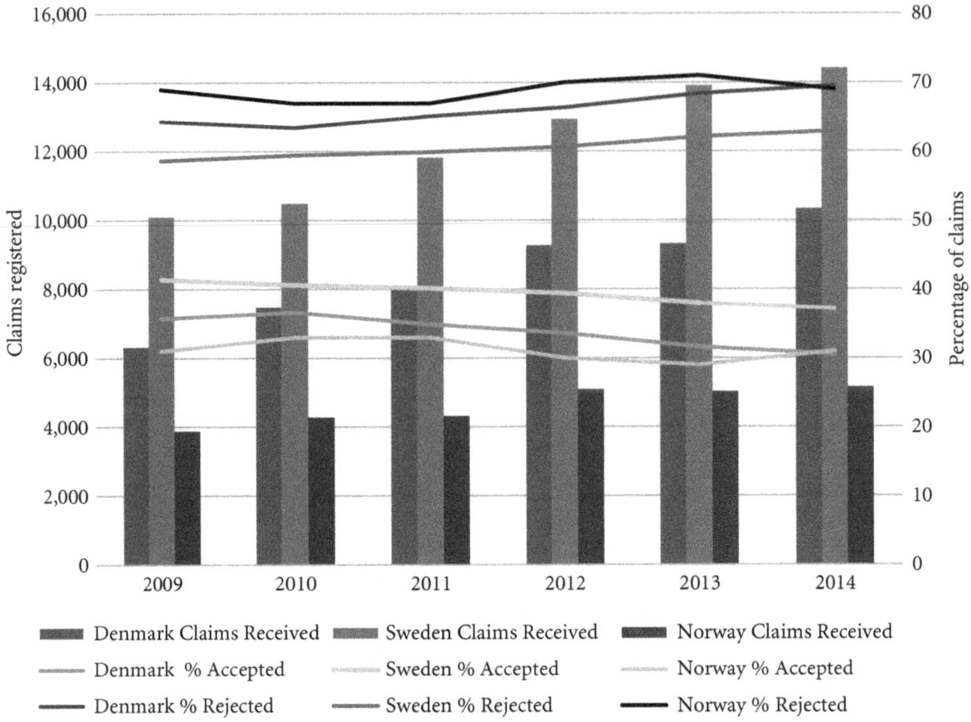

Figure 1.1: Patient injury compensation trends for Denmark, Sweden and Norway, 2009–14

private claims are handled by a separate body in Sweden. Such centralised administration bodies all adopt an inquisitorial approach to investigating and deciding claims.

A major advantage of having a unified, central administrative system lies in the ability to combine all data on claims so as to be able to recycle it with the aim of improving standards of practice. All the Nordic states are involved in this. Impressive results from Sweden are quoted above; the fact that the regional authorities who pay the costs of both hospitals and compensation own the claims administration body may have led to the ability to achieve not only the speedy recycling of safety data but also its application in practice, such as in retraining staff, and in focusing on reducing adverse and costly outcomes.

There are some differences between the triggers applied in the four Nordic countries. In Sweden and Denmark a 'specialist standard of care' is applied retrospectively. In Finland the criterion is that the care has been given by an 'experienced professional person in the healthcare sector'. Norway applies a main rule concerning failure in connection with the providing of health services, although it is not necessary that anyone is at fault.

The French ONIAM Scheme for Medical Injuries

The ONIAM scheme covers compensation for injuries caused by medical accidents and a number of specific medical procedures, such as vaccination and blood transfusion. It is possible to add on other types, such as occurred with the addition of claims arising from administration of the drug Benfluorex.

The ONIAM scheme is free to claimants and has an inquisitorial model. Claims, accompanied by medical records, are lodged with regional committees Commissions Régionale de Conciliation et d'Indemnisation (CRCI) of medical and legal experts, who undertake initial investigation, verify the injury and causation, may request a medical examination of the claimant, and decide whether the injury was due to negligent medical treatment or resulted from a serious non-negligent medical adverse event. The investigation process should not exceed 6 months. Based on that finding, claims are allocated to the relevant public or private insurance for payment.

The CRCI have strong national central administration provided by ONIAM from its Paris headquarters. A finding by a CRCI on whether or not the medical service provider was at fault remains critical, since if there was fault the provider's liability insurer is required to compensate, whereas where fault is not established the state body ONIAM pays. It can be seen, therefore, that the scheme is expressly based on the principle of solidarity. ONIAM also compensates accepted cases from various vaccines from its own state funds.

The reason why the fault trigger was retained was that it was the trigger used in widely held pre-existing insurance policies. Hence, such policies could continue to apply once the scheme was introduced: the change was that the state stepped up to provide compensation for those who qualified where fault had not existed. That was a pragmatic way of extending cover and compensation, whilst continuing to rely on privately funded insurance and thereby limiting the exposure of the state. However, the continuation of finding fault in the system would have other disadvantages, namely criticising individual doctors and healthcare providers.

Insurers must make a settlement offer of an acceptable amount within 4 months or be subject to a 15% punitive surcharge. That punitive mechanism appears to work well in incentivising insurers to make a single fair offer of the right sum of money, rather than enter into negotiation. Tariffs of payments apply. A claimant may reject an offer and start a court claim, although this is very rare. Decisions related to Benfluorex are made by a specific expert panel rather than the normal CRCI.

The throughput has been stable for some years, at around 4,300 applications of all types of claims, of which very roughly half are accepted. The number of claims for medical accidents has been roughly 750 a year, with average annual payouts of around €85,000–99,000.

The ONIAM model is efficient in providing a single inquisitorial investigation, which allocates valid claims to be settled by the relevant insurance fund, and includes a mechanism.

Uncertainty over the definition of 'medical accident' remains a concern.

The Polish Medical Liability Scheme

This has strong similarities to the French scheme, but differs in certain details. It is based on decisions by regional investigative committees, composed of equal numbers of doctors and lawyers, that decide causation and liability but not quantum. A committee may seek an expert opinion. Making decisions by committee can slow decisions that could have been made by experienced case-handlers.

Various design issues reveal the influence of civil procedure, rather than a clean shift to an administrative system. First, the claimant must produce evidence that the injury was caused by a 'medical incident', namely something that takes place in a manner 'contrary to current medical knowledge'. The 'fault' of any individual doctor or worker is irrelevant. The concept of 'medical incident' may be expected to give rise to difficulties, and it appears from the statistics that a 'medical incident' is established in only a relatively small percentage of claims (see chapter 19). The Committee makes its decision on the basis of the evidence presented by the parties and, where necessary, an expert opinion that it obtains. This is still essentially an adversarial model, rather than an inquisitorial model. It is unclear whether claimants have found the burden of producing evidence that would establish whether there was a 'medical accident' has presented an impossible barrier and defeated otherwise good claims. Certainly, the number of claims in the system appears to be modest, and falling (432 in 2012, 1310 in 2013, 1197 in 2014, and 1030 in 2015).

Magdalena Tuilibacka notes in her chapter academic criticism of the fact that doctors (or other healthcare professionals) are not represented on the committee that makes the decisions on whether the patient's injury was caused by their fault (see chapter 19). That view is, we suggest, not relevant. Such involvement would be relevant if the healthcare professional's actions were being examined on the basis of whether there was, or was not, fault, and such fault would then trigger compensation. That analysis would be the typical judicial analysis of law in a (quasi-adversarial) court procedure. But, as various Polish authorities that Tulibacka identifies have rightly said, that approach is simply not relevant where the analysis is not based on the fault of an individual or institution, and money is being allocated to a patient on the basis of criteria that are not related to any such fault. The professional or institution should not need to 'defend' or represent itself in such a system. The Swedish Patient Compensation Scheme has made that point clearly.

The second strong influence of a civil procedure influence is in relation to costs and loser pays. The claimant must pay a small access fee (about €50). Other costs are allowable against tariffs set by the Ministry of Health, and again involve modest sums. All these fees and costs are allocated at the end of the procedure on a loser-pays basis. The proceedings are quite informal. The committee must make a decision within 4 months of receiving the complaint (although in practice delays are reported), and a party may request that a decision be reconsidered by different committee members.

Within 30 days of receiving the final decision of the committee, the hospital or insurer must propose an amount of compensation to the claimant, which the claimant has 7 days to accept. If accepted, the contract is binding on both parties, bars further court action and is enforceable as if it were a court judgment. The claimant is free to reject the offer, and commence court proceedings if he or she wishes.

The scheme initially made it compulsory for all hospitals to hold insurance, but this requirement did not work and had to be made voluntary. There were not enough insurers in the country, and their premiums were too great. Hospitals may now self-insure, ie pay claims out of their own resources.

The major difference from the French system, of course, is the absence in Poland of a mechanism requiring the hospital or insurer to make a *fair* offer. Tariffs set out payments for particular types of injuries and levels of incapacity, and those sums may not be exceeded. However, there must be a strong risk that offers are made that are too low.

Overall, this scheme could be considerably improved.

UK Injury-Specific Schemes

The standard approach to injuries caused by medical treatment or products is through litigation based on establishing negligence, strict product liability, or, in case of private but not NHS care, breach of contract. Medical litigation comprises a significant industry, but is expensive. Individual compensation schemes have been created for injuries caused by specific situations. This has sometimes been a consequence of the high cost of litigation, or difficulties in establishing liability, and sometimes after an initial trial that established liability, following which administrative arrangements were put in place to handle individual applications.

The leading example was the Thalidomide Trust, which was a ground-breaking event in British legal history, involving settlement of claims in circumstances in which legal liability was unclear but social responsibility was evident. In 1968, the court oversaw and approved the establishment of a initial settlement for a limited number of children, and subsequently in 1973 a second settlement for the remaining children lead to the creation of the Thalidomide Trust to supervise the distribution of annual payments to what was almost a fixed class of victims (although the class was subdivided into three categories and some others were later allowed to join). A notable feature over time was that, despite the initial settlement being reached on a 'full and final' basis, later campaigning led to additional sums being contributed to the Trust both by the commercial enterprises involved and by government. The additional funding and change of tax status campaigns were driven by the beneficiaries of the trust, with the trust providing a communication hub and a focus of expertise for them to draw upon.

The Coal Workers' Pneumoconiosis Scheme, established in 1974, has made lump sum payments averaging £2,700 to over 90,000 workers as a top up to Industrial Injuries Disablement Benefit and other state benefits.

The leading examples of compensation schemes created after public bodies have been found liable to mine workers are those relating respectively to COPD and VWF.

Payments for infection with hepatitis C and HIV from blood or blood products supplied through the NHS are made on an ex gratia basis. Two private companies and three trusts make payments. The companies—the Skipton Fund for hepatitis C and the MacFarlane and Eileen Trust Limited for HIV—have no discretion and must pay a one-off lump sum (£12,800 for the former and on a scale basis for the latter up to £60,500); ongoing payments are specified at set values when eligibility criteria are met. The three charities give discretionary support in terms of lump sums, grants and ongoing payments. They have eligibility criteria, but may change them. The claimant has the burden of proof. In practice, the applicant's doctor verifies the medical details to the Skipton Fund, which then undertakes its own verification. However, patients have faced difficulties in identifying records going back some decades. The idea of replacing the five bodies with a single body has rightly been floated.

Litigation by victims of variant Creutzfeldt–Jakob disease (vCJD), in which the government was held to have been negligent in its arrangements for producing safe human growth hormone, prompted the establishment of a trust fund by the government in 2001 with no admission of liability. The potential number of claimants may be around 33,000. Claimants need a certification that they probably suffer from vCJD from the specialist National CJD Surveillance Unit at the University of Edinburgh and must submit claims to a firm of solicitors that represents the trust. Payments were initially made on an interim basis, and were revised from 2010 to include a basic sum (£120,000–125,000) and further items in relation to adverse family experiences, expenditure (not legal costs) and dependency. Payments may also be made at the trustees' discretion from a discretionary fund. One of the hallmarks of this scheme is the inclusion of classes of beneficiaries who would not be eligible for damages at common law, including families. This highlights that the parameters of compensation schemes are flexible and can be more accommodating than settlements achieved through the courts.

A more recent development, not described below further, is the establishment in 2006 of the Dental Complaints Service (DCS) by the dentists' regulatory body, the General Dental Council. The DCS operates in a similar way to a growing number of other consumer ADR complaint bodies that have been created in recent decades,[18] but, unlike most others, can intrinsically resolve a personal injury problem in the context of resolving a problem about service level. Many outcomes involve free remedial treatment, a refund, an apology or an explanation. Two aspects are particularly interesting. Firstly, it appears that payments for 'pain and suffering' can be made but are rare. Secondly, the consequence of using this scheme is that most patients who complain through the DCS remain with the practice of the dentist they complained about, and even with the individual dentist.

E. Medicines

German Pharmaceutical Claims Arrangements

In Germany, the response to the Contergan (Thalidomide) tragedy was threefold. Firstly, the German legislator established a foundation to pay pensions to handicapped children, based on the degree of handicap. The foundation was jointly financed by the drug company and the state, but since 1987 has been entirely funded by the state. Secondly, the newly introduced medicines regulatory law was the first in Europe to introduce strict liability for injuries caused by medicinal products (with no development risk defence).[19] Thirdly, the industry and insurers established a pooled high layer of reinsurance coverage (the Pharmapool) to provide cover for a situation where the level of claims turned out to be very high. The Pharmapool facility has not been called on save in making payments to victims of contaminated blood products, which were made on an ex gratia basis.

Vaccine Damage Schemes

Many advanced states have introduced vaccine damage schemes since the 1980s, as a means of supporting public confidence in vaccination programmes in view of their

[18] Hodges, Benöhr and Creutzfeldt-Banda (n 1).

[19] This is unusual: most Member States have included a development risk defence in implementing this optional provision of Directive 85/374/EC, and whereas Germany has included it in general it has excluded it for pharmaceutical products.

importance to public health. It is accepted that rare and unpreventable, but sometimes serious, injuries can be caused by vaccination, without fault of the manufacturer. Manufacturers, especially in the United States, are loath to make and sell vaccines without either indemnity from the state or a compensation scheme. The schemes are typically administered by the state.

An initial attempt at a protective scheme was the US 1976 Swine Flu Vaccination Program, which offered claimants the option of an administrative scheme or litigation. The US government voluntarily removed the need for these patients to prove negligence, because that was considered 'unfair'. The unpredictability of pharmaceutical science was shown when only a small number of swine flu cases appeared, but instead over 400 cases of Guillain–Barré syndrome unexpectedly occurred.

The US Vaccine Scheme (VICP), introduced in 1988, is funded by a surcharge on all doses, collected from manufacturers and reflected in prices charged. The VICP process operates on a quasi-adversarial basis, with a claimant lodging a claim (with a filing fee of $400), which is then reviewed by a government physician who makes a recommendation to the Court of Claims, where the government position is presented by an attorney at a hearing at which the claimant or an attorney may appear (whose fees can be claimed), before a decision is made by a Special Master. The process assumes that claimants will seek legal advice, and there are provisions for legal costs to be paid. After a determination on eligibility, the level of award is usually negotiated between the Department of Justice and the claimant's attorney, and is otherwise determined by the specialist court. The whole process takes on average 2–3 years, but can take 10 years. A claimant is required by law to apply through the VICP before starting a court claim.

The VICP has simplified the hurdles for certain types of claimants, by providing that if certain injuries and conditions are found to have occurred within a prescribed period of time after administration of a vaccine that is listed in the Vaccine Injury Table, they will be subject to a rebuttable presumption of causation. In other cases, the claimant has to demonstrate causation through traditional tort law principles, relying on scientific studies and expert medical evidence. This traditional approach, even with the option of using either a specialist court or the normal courts, remains as a considerable barrier to be overcome by claimants.

In response to the 2014 outbreak of Ebola virus, the US government enacted a statutory immunity for Ebola vaccine manufacturers from litigation in US courts.

Between 1988 and 2013, 14,730 vaccine claims were filed in the VICP, of which 3,368 were compensated and 9,642 rejected (a 35% success rate).

The UK Vaccine Damage Payments Scheme (VDPS) was introduced in 1979 and provides a single one-off payment lump sum (currently £120,000). It is regarded as support and not compensation, and payment does not bar a legal claim. Claimants must be 'severely disabled' (at least 60%) and involve a vaccine that is listed. Applications are made to an administrative unit, which obtains medical evidence from the doctors or hospitals involved. Funding is from general taxation, and is administered by the government, which obtains medical evidence on an investigative basis.

In the UK between 1977 and 2000 just under 21% of claims were compensated, leading to total payments of £9,760,000. However, for the period 2006/07–20011/12 a total of 38 awards were made and 407 rejected, a success rate of 8.5%. Hence, UK success rates are consistently low.

Although broadly similar and shaped by the same rationale there are clear differences between the UK and US vaccine compensation schemes. The US scheme has a substantially higher level of lawyer involvement, including the payment of considerable sums in lawyers' fees. The most striking differential for victims is the contrasting success rates of applications: up to 2012 approximately 35% of US applications have been compensated, contrasting with a success rate of 8.5% between 2006 and 2012 in the UK.

The Nordic Drug Compensation Schemes

The compensation schemes for injuries caused by medicines in the Nordic states can be viewed as logical adjuncts to the larger patient schemes; indeed, they share administration in Denmark and Norway. The schemes are voluntary in Sweden and Finland, and compulsory by law in Norway and Denmark. The pharmaceutical schemes pay top-up benefits above social security cover. They are financed by contributions levied from the pharmaceutical industry in Sweden, Finland and Norway, and from general taxation in Denmark. The levy rates can range from 0.1% to 0.4% of sales. The administration costs of the schemes are low in each state.

The claim success rates are around 20% in Sweden, Denmark and Norway, but 50% in Finland (possibly influenced by a single drug, Pandemrix). In general, the number of claims is low when compared with the number of patient claims, and the schemes resolve pharmaceutical cases with almost no recourse to litigation or class actions. The schemes cover *unforeseeable injuries*, ie injuries that are either not warned about or are more severe than expected, and in a number of respects, the drug schemes have wider coverage than under product liability law.

The schemes are administered centrally in each country. Claims are made to and investigated by either the commercial insurer who operates the national scheme for the time being (as originally in Sweden), or a body representing the pool of insurers or producers (as in Finland, and from 2007 in Sweden), or a body that handles both medical and pharmaceutical claims (the statutory Danish Patient Insurance Association and the Norwegian System of Patient Injury Compensation, NPE). Such bodies make decisions on the facts, on causation and on the appropriate level of award. There may be an appeal by a dissatisfied claimant to another body, save in Norway, thus providing an element of scrutiny and standardisation of decisions. The systems are, therefore, inquisitorial, rather than adversarial, at least at the primary stage, and provide for access to patients' medical information, and controls on maintaining the confidentiality of such information.

In the past decade, the major change has been a switch to captive insurance companies, which now exist for each country except under the state-funded scheme in Denmark. The industry switched in Norway from 2004 after it was unable to obtain an acceptable quote from the insurance market. In Finland the transfer to a captive occurred in 2012.

The Japanese Pharmaceutical Compensation Scheme

A pharmaceutical compensation scheme was introduced in Japan in 1980 after a tragic epidemic, starting in the 1960s, of subacute myeloptic neuropathy (SMON) attributed to a drug for protozoal diarrhoea. The scheme provides a platform for five different compensation schemes: SMON, HIV and AIDS, hepatitis C and, most unusually, *any* adverse drug reaction (ADR) requiring hospitalisation caused by medicines prescribed at a hospital or clinic or purchased at a primacy, even if such drugs were properly used.

Unusually, the scheme is administered by the national Pharmaceutical and Medical Devices Agency (PMDA), which also distributes funding for specific issues, such as SMON patients.

The funds for the five schemes are kept separate, without cross-subsidy. Funding for the specific injury types comes from various sources. SMON funding comes from a court settlement with some elements of state funding, whereas the HIV scheme is state funded. The general drug ADR scheme is funded, similarly to the Nordic drug schemes, by an annual levy on all the companies who hold a national marketing authorisation for a medicine (in 2014 at a rate of 0.00027% of sales), and contributions from pharmacies who compound medicines. The PMDA has the ability to charge individual companies whose products are found to have caused injuries, but this apparently has never occurred. The rate of claims and contributions has generally been historically stable, and the fund has built up a contingency fund.

This arrangement provides a seamless single point of contact for claimants regardless of the source of the funding they are claiming from. The claimant technically has the burden of proof, but in practice only has to submit a doctor's certificate. The PMDA evaluates claims and allocates and distributes funds when appropriate from the correct source. The Japanese scheme is successful and stable.

Product Manufacturer Schemes

Manufacturers of medicines and medical devices have created and funded compensation schemes at various times, usually to avoid litigation (in the UK) or as a second stage after litigation (especially in the USA). These schemes cover various products including: medicines, such as the Thalidomide Trust (UK) and the PMDA ADR and SMON streams (Japan), ICI's Eraldin scheme (UK); and medical devices, such as breast implants (Dow Corning, Trilucent) and orthopaedic hips (ASR and Stryker's ABG II Modular-Neck System) and the US Vaccine Scheme.

Whilst all of these schemes are funded by manufacturers, the way they are administered varies, some being administered by the state or a state agency, while others are run by private bodies of various legal forms. State-administered schemes arise in a wide range of contexts and can be seen administering court settlements (SMON), or in response to the needs of a particular class of claimant (the industrial disease schemes in the UK, the Skipton Fund) or in response to the needs of the whole population (the US and UK Vaccine Schemes). Schemes arising from litigation tend to favour private companies or, in common law jurisdictions, trust funds as administrators. Voluntary compensation schemes are usually administered by a private body, often a third-party claims handler or a company set up specifically for that purpose.

In the case of implants, the primary benefit is not to pay damages as pain and suffering, but to fund an investigation into whether an explant and possible new implant may be necessary (a revision, in medical terminology).

The Dow Corning scheme involved a compensation scheme agreed after extensive US litigation to avoid even further litigation, and was then taken over by the US bankruptcy protection and reorganisation procedure. This origin is unusual to European eyes in a number of respects. Important features were the need to draw up detailed parameters for different categories of claimants to qualify, including limitation rules, and tariffs of payment entitlements. The scheme has processed around 80,000 claims, of which 12% were rejected. The Dow Corning scheme, established in 1994 with the continuing involvement of attorneys, and subject to court oversight, has been widely criticised for being slow and difficult to access.

The UK Trilucent scheme, another breast implant scheme, was improved in a number of respects compared with the Dow Corning approach. It was established in UK around 2000 contemporaneously with regulatory action that drew attention to the medical hazard (recommending that patients were assessed and the implant removed), and to a specialist care facility that had been set up to provide the associated care package. It was a considerable practical advantage that the establishment of those three elements (safety action, care package and compensation scheme) was announced at the same time, with the regulatory agency playing a key role. The principal focus was on providing advice, verification and care (including explantation) to women. The primary process of 'application' was very straightforward: women contacted a trained nurse at the Trilucent Care Centre, and were referred to a consultant for a full examination, including ultrasound and MRI imaging, explantation and, if desired, replacement with another product. This process delivered what the women wanted, which was swift access to medical advice and reassurance, followed by action[20] that was automatically paid for. Provision of compensation was a secondary priority, but was facilitated by providing for a very simple two-part structure, one with fixed modest cost (which was taken up by the overwhelming majority) and the other for more complex injuries, involving a more traditional procedure for lodging relevant evidence and review by lawyers.

The ASR reimbursement programme of 2010 developed the ideas further. It is not a traditional no-fault compensation scheme, but offers reimbursement of medical and out-of-pocket expenses to patients implanted with potentially faulty ASR hip systems (around 93,000 people worldwide). As an initial step, surgeons were urged to contact patients, as some patients might not know of the arrangements and might not seek advice until after joint damage had already occurred. Patients register with an independent claims administration company, which assists them through subsequent tests and revisions with their relevant hospitals worldwide. Interestingly, it does not include payments for pain, suffering and loss of amenity, and this has led to litigation in some jurisdictions. As the scheme is ongoing, it is too early to assess its impact, but it is interesting to note that similar schemes have been created by other manufacturers.

A major challenge that exists with any product that is widely used is how to identify and contact all users, especially if this has to be done in a hurry. In the Trilucent situation, this seems to have been managed relatively well. The ASR reimbursement programme allocated

[20] There was considerable disquiet as the amounts offered by the Care Scheme were often deemed insufficient by the hospitals carrying out the removal and new implant procedures, meaning some women were left in a position where the revision surgery was not automatically paid for.

funds specifically for this purpose and used a variety of strategies in different parts of the world depending on the local requirements. A major part of the solution with implants is to require technical traceability as a regulatory requirement.

F. Industry Arrangements in the UK for Research Injuries

In a separate category are the arrangements for compensating those who suffer injury in *clinical trials on medicinal products or clinical evaluation of medical devices.* The British approach was to sidestep the need for litigation by providing an alternative to the claimant having the uncertainty of whether there would be a successful legal basis for a claim. Instead, healthy volunteers in drug trials were given a contractual right to compensation, and patients in clinical research were given the benefit of a moral obligation that the sponsor of the research would pay compensation (backed by insurance where needed), in both cases based on sums that would be awarded by a court if liability were to be accepted. The arrangements were standardised in documents issued by the manufacturers' trade associations around 1990, and have since become required by medical authorities as matters of ethical and administrative approval of clinical research. There have fortunately been very few instances in which research damages have occurred, so compensation scheme arrangements have not been established.

G. Birth Injuries

Birth-Related Neurological Injury Compensation

The Virginia Birth-Related Neurological Injury Compensation Program, established by statute in 1987, is a payer of last resort, which pays only the expenses remaining after all available insurance, government programmes and any other sources have contributed their share. It can only make an award relating to medical expenses, loss of earnings, reasonable legal expenses, and for death of a child. It is administered by the state's Virginia Workers' Compensation Commission. The Program is entirely funded from participating healthcare insurers, hospitals ($55 per live birth, subject to an annual maximum of $200,000) and providers (annually $6,200 for participating physicians, $300 for non-participants as a condition of being licensed in Virginia). The Program is not fully financially secure, so in recent years a statutory fallback provision has required annual contributions by liability insurers in Virginia of up to 0.25% of all 'net direct premiums written' by liability insurers.

The Virginia Programme handles around 10 cases a year. Claimants are required to pay a $15 filing fee. The procedure seems somewhat court-like on paper, with potential provisions for hearings, serving interrogatories, holding depositions and cross-examination of witnesses. However, in practice cases are investigated by panels of three impartial doctors, who submit a report with recommendations to the Commission. Various appeal stages are available.

The Virginia Program, quite similar to the Florida Program below, has a number of striking features. First, administrative expenses form only 5% of the costs. Over 70% of payments in recent years go towards nursing costs, the next largest category being lost wages, at around 6–7% per cent. Second, there is a deliberately joined-up relationship between various different official bodies. Circuit Courts are required to send potentially qualifying claims to the Program, and the Program (through its host the Workers' Compensation Commission) is required to report to the state medical regulatory bodies, namely the Virginia Department of Health and the Virginia Board of Medicine.

The Florida Birth-Related Neurological Injury Compensation Association (NICA), established by statute in 1988, is a payer of last resort after other sources. It can only make an award relating to medical expenses, a periodic or lump sum award (note the difference from Virginia here), reasonable legal expenses, and for death of a child. NICA is entirely funded from participating healthcare insurers, hospitals ($55 per live birth, subject to an annual maximum of $200,000) and providers (annually $5,000 for participating physicians, $2,500 for participating midwives, $250 for non-participating physicians as a condition of being licensed in Florida). NICA is currently actuarially sound. If it were not, extra funding could be obtained, first, from a transfer of up to $20 million from the Insurance Regulatory Trust Fund, and secondly, from casualty insurance carriers, who can be required to pay up to 0.25% of net premiums.

In recent years, NICA has handled between 12 and 47 cases a year, of which the acceptance rates have varied between 28% and 50%. Claimants are only permitted to file civil actions in very limited circumstances. Claimants are required to pay a $15 filing fee to file a claim with the state's Division of Administrative Hearings, and within 10 days provide all medical and other specified evidence, although third parties are required to provide medical records to the parents/guardian, and NICA serves all relevant parties. If NICA decides to accept a claim, it is passed for approval to an Administrative Law Judge (ALJ), who then approves the compensation to be paid. If NICA rejects the claim, it passes to a hearing before the ALJ unless it is withdrawn. In either case, the ALJ holds a hearing to determine the facts and eligibilities.

Parallel regulatory investigations are also held, and the Administrative Court is required to report to the regulatory bodies, namely the Division of Medical Quality Assurance and the Agency for Healthcare Administration.

The largest category of payments in recent years is nursing costs (around 62%), with 'other', parental awards and legal expenses all roughly around or below 10%. General and administrative expenses were 4.7% in 2015.

H. Social Schemes

The Magdalen Laundries

The Irish government established a statutory scheme in 2015 to assess and administer ex gratia payments for now elderly women who when young had been admitted to institutions known as 'Magdalen Laundries'. Although individual women are required to come forward to claim, full assistance is given, in terms of, for example, access to records, in order to establish the length of time they were resident in relevant institutions. The process was designed to be a healing and reconciliation process, rather than an adversarial process.

Restorative justice principles were applied to provide a significant element of monetary compensation, with specific sums payable based on length of stay, capped at €100,000. In order to protect these now elderly and potentially vulnerable women, any ex gratia payments in excess of €50,000 are payable as tax-free weekly income for the remainder of their lives. ADR principles were applied to assist eligible women to obtain health, monetary and other benefits. These include pension top-up payments (since many would have been denied an education and thus a job that would have triggered pension contributions).

The scheme is notable for its 'light touch' bureaucracy and procedures, and its basis in social solidarity.

September 11 Fund

Two Victim Compensation Funds were created in response to the terrorist attacks in 2001. The first fund (VCF1) was created within an astoundingly short period of 11 days, and was administered by a respected individual, Ken Feinberg, who was appointed Special Master. He and his team paid over $7 billion in compensation to 2,680 individuals who suffered physical harm, were present on site, or sustained injury at the time or in the immediate aftermath, or their families—in total, 5,560 eligible claimants. Details of proof were not unduly difficult to provide. Two procedural tracks were made available, one involving notification of both eligibility and the amount of presumed award (after which the claimant could accept the proposal or request a hearing), and the other involving a written notification of eligibility determination followed by a hearing (which could be by phone or otherwise informal). Claims on behalf of deceased victims were fairly evenly split between use of the two procedures, whereas personal injury claims predominantly used the former track (89%). The average award for a fatality was over $2 million, and the average personal injury award was just under $393,000. The administrative costs of VCF1 were $87 million, which represents proportionately an incredibly low 1.2%. One element of contention was the decision to base compensation of personalised loss of earning, under which bankers were worth more than rescue workers. Feinberg argued that a more equitable basis would not have been politically acceptable either.

A new fund (VCF2) was created in 2011, with expanded definitions of eligibility and physical harm, so as to encompass emergency workers or those harmed in a geographically wider area. Again, two tracks have been used.

Part II

New Zealand and Australia

2

New Zealand: The Accident Compensation Scheme

CHRISTOPHER HODGES AND SONIA MACLEOD

A. General Description

New Zealand operates a comprehensive national no-fault personal injury insurance scheme, which is available to all New Zealanders and visitors to New Zealand. The Accident Compensation Scheme[1] came into operation in 1974 following the influential 1967 Woodehouse Report, which aimed to create a coordinated approach to the diverse and differing mechanisms of tort, workers' compensation, social security and the 1964 criminal injuries compensation scheme.[2] The original 1974 scheme had five guiding principles: community responsibility, comprehensive entitlement, complete rehabilitation, real compensation and administrative efficiency.[3] The scheme has been amended on various occasions, often depending on whether the National government wished to reduce costs on employers or the Labour government wished to extend coverage for workers.[4] It is currently governed by the Accident Compensation Act 2001 (ACA),[5] and administered by the Accident Compensation Corporation (ACC).[6] The ACC is a Crown Entity of the New Zealand Government, whose governance structure is prescribed under the Crown Entities' Act 2004, under which

[1] See www.acc.co.nz.

[2] *Compensation for Personal Injury in New Zealand: Report of the Royal Commission of Inquiry* (Royal Commission of Inquiry into Compensation for Personal Injury, 1967).

[3] R Tobin and E Schoeman, 'The New Zealand Accident Compensation Scheme: The Statutory Bar and the Conflict of Laws' (2005) 53 *Am J Comp L* 493.

[4] ibid. The previous legislation was: the Accident Compensation Act 1972, the Accident Compensation Act 1982, the Accident Rehabilitation and Compensation Insurance Act 1992, the Accident Insurance Act 1998, and the Accident Insurance (Transitional Provisions) Act 2000. See T Ison, *Accident Compensation: A Commentary on the New Zealand Scheme* (London, Croom Helm, 1980); special edition of *Victoria University of Wellington Law Review* (2003) 34(2); K Oliphant, 'Beyond Woodehouse: Devising New Principles for Determining ACC Boundary Issues' (2004) 35 *Wellington Law Review* 915–36; K Oliphant, 'Landmarks of No-Fault in the Common Law' in WH van Boom and M Faure (eds), *Shifts in Compensation between Private and Public Systems* (Berlin, Springer, 2007).

[5] see www.legislation.govt.nz/act/public/2001/0049/latest/DLM99494.html; it was initially called the Injury Prevention, Rehabilitation and Compensation (IPRC) 2001; the name was changed in 2010.

[6] ACA, ss 259ff.

the Government appoints a Minister for ACC who appoints a Board that is responsible for the financial management and operational direction of the ACC

The ACA sets out a clear statement of the scheme's purpose, involving what would in European terminology be referred to as social solidarity and cost efficiency:

> The purpose of this Act is to enhance the public good and reinforce the social contract represented by the first accident compensation scheme by providing for a fair and sustainable scheme for managing personal injury that has, as its overriding goals, minimizing both the overall incidence of injury in the community, and the impact of injury on the community (including economic, social and personal costs).[7]

The core functions of the ACC are to provide cover for *rehabilitation and compensation* to the injured, through managing claims and purchasing services, and to reduce the incidence and severity of injuries through injury prevention and by providing levy payers with a range of insurance products that encourage them to manage their injury risks more effectively. The ACC publishes a Statement of Service Performance including outcome and output measures in its *Statement of Intent 2011–2014* and *Statement of Intent 2015–2019*, and annual achievements against the targets set. The three principal objectives are: delivering levy stability by achieving long-term financial sustainability; rehabilitating injured people in New Zealand more effectively; and reducing the incidence and severity of injury where it is cost effective to do so. It also states a series of performance management targets and outputs.

B. Replacement of Tort Liability

Court proceedings may not be brought for personal injuries covered by the ACA, although proceedings are permitted for damage to property, breach of contract, unjustifiable dismissal or exemplary damages.[8] Thus, the scheme therefore replaced tort liability as a means of compensation for injury, since a common law claim for compensatory damages cannot be brought for an injury that comes within the scope of the scheme.[9] Claims for exemplary damages are not precluded, and may be brought for conduct that is malicious, misuse of power or a cynical disregard for the effects of the defendant's actions on the person who has suffered harm.[10] Exemplary damages in common law jurisdictions are intended to be a civil mechanism to punish defendants rather than compensate the victim. It is very rare for litigation to be brought claiming exemplary damages for personal injury in New Zealand, and tends to be where there has been gross injustice or where the ACC contribution does not cover the full costs of an injured individual. However, as will appear below, litigation over decisions made by the ACC is still a significant occurrence.

7 ACA, s 3.

8 ACA, ss 317, 318, 319.

9 PDG Skegg, 'Compensation in the New Zealand Health Care Sector' in J Dute, M Faure and H Koziol (eds), *No-Fault Compensation in the Health Care Sector* (Wien, New York, Springer, 2004).

10 ACA, s 319.

C. The Five Accounts and their Funding

All taxpayers, employers, self-employed people and motor vehicle licence-holders contribute to the funding of the scheme by paying levies.[11] The Government also funds part of the Scheme for people who are not in the paid workforce. Cover is managed under five accounts, which the ACC is required to maintain, keeping levies and payments of various types separate from each other. This facilitates appropriate risk-pricing and limits cross-subsidy. The accounts are:

— the Work Account, funded from levies paid by employers[12] and self-employed people,[13] covering all work-related injuries;[14]
— the Motor Vehicle Account, funded by motor vehicle licence fees, a motorcycle safety levy and petrol levy, financing entitlements paid for personal injuries suffered as a result of motor vehicle injuries on public roads;[15]
— the Earners' Account, financed by levies paid by earners (paid through PAYE), plus self-employed levies based on earnings, covering entitlements for non-work personal injuries (including at home, and during sport and recreation);[16]
— the Non-Earners' Account, funded by appropriations made by Parliament,[17] covering entitlements to non-earners, including students, social security beneficiaries, older people and children, other than motor vehicle injury and treatment injuries;
— the Treatment Injury Account, funded roughly equally from the Earners' and Non-Earners' Accounts (via the Government). The Account was originally set up with a view to levying treatment providers for the cost of treatment injuries, but such levy has never been implemented because of ongoing debate over the relative benefits and costs.

Large employers may choose to enter the Accredited Employer Programme (AEP), which allows them to self-insure some risks in return for significantly lower levies. The goal of the AEP is to improve workplace safety and rehabilitation performance by providing employers with a financial incentive.[18] Entry is subject to satisfactory workplace safety standards, claims management ability and the financial backing to carry the self-insurance risk.

[11] The Accident Compensation (Financial Responsibility and Transparency) Amendments Act 2015 set out to improve the framework for determining how the ACC's levied accounts are funded so it is more transparent, has a longer-term focus and supports more stable levies. The Act also included provisions to ensure that residual levies are not over-collected. The Accident Compensation (Work Account Levies) Regulations 2016, the Accident Compensation (Motor Vehicle Account Levies) Regulations 2016 and the Accident Compensation (Earners' Levy) Regulations set out the provisions for these accounts. For a more detailed summary see the ACC's Reports on rates of levies prescribed in the Accident Compensation (Work Account Levies) Regulations 2016 and the Accident Compensation (Earners' Levy) Regulations 2016.

[12] ACA, s 167.

[13] The self-employed were included in this fund from 2007: ACA, s 168B.

[14] The 2016/17 levy will be NZ$0.80 per NZ$100 liable earnings: in 2015/16 it was NZ$0.90 per NZ$100 of liable earnings.

[15] ACA, s 213. The 2016/17 the levy per vehicle is NZ$130.26 down from NZ$194.25 in 2015/16. The motorcycle safety levy was introduced by s 24(3) of the Accident Compensation Amendments Bill 2010.

[16] ACA, s 218. The 2016/17 levy is NZ$1.21 per NZ$100 of liable earnings: in 2015/16 it was NZ$1.26 per NZ$100 of liable earnings.

[17] In 2015/16 funding provided was NZ$1.092 billion.

[18] See *Financial Condition Report 2013* (Accident Compensation Corporation, 2014) paras 1.6.3 and 1.6.4. There are two product options. Under Full Self Cover, employers take the full financial risks of all claims apart

All three of the levied Accounts are currently fully funded.[19] This is a considerable achievement as in recent years there have been deficits running into billions of dollars. The Motor Vehicle Account carries a high proportion of serious injury claims caused by motor accidents, and was therefore the last to be fully funded.[20] It has achieved full funding faster than was predicted. Initial estimates were that it would not reach full funding until 2019. That said, complacency would be ill advised. Initial figures for 2016 suggest that there is a funding deficit, rather than the surplus predicted for 2016.[21]

Difficulties have sometimes been encountered in matching liabilities and funds. A major shortfall in funds in 1986 led to a large rise in levies. Since then the scheme has been generally stable, but has experienced volatilities. Claiming behaviour is influenced by government policy (different parties lean more towards generosity or austerity), global and national economic performance, and public perceptions of the stability of the ACC. There was a steep increase in the number of claims from 2005 to 2008, in a period of economic growth. The economic crash after 2008 led to a perception that the ACC was in trouble as workers' compensation funds generally suffered with falling industrial performance, and claim volumes fell from 2008 to 2012,[22] but are now rising again.

Entitlements to be paid are established by the Government in legislation but the Board has a considerable amount of discretion around decisions on *eligibility*. It can—and does—change its policy regarding eligibility and access, for example on whether particular situations constitute incapacity or not, and whether individuals are capable of working.

The ACC has achieved full funding—and therefore no longer collects a residual levy to fund past claims[23]—and future levy rates are projected to decrease and stabilise.[24] There is concern that the volume and cost of claims will increase in future as demographic change increases the elderly population, who suffer more falls and recover more slowly. In addition,

from serious injury and sensitive claims, and manage claims for up to 5 years, usually through an external claims administrator. On expiry of the period, when cover is handed back to the ACC, employers pay the ACC the estimated remaining lifetime costs of open claims and remain liable for unreported or reopened claims. Under the Partnership Discount Plan, employers self-insure claim payments for 2 or 3 years.

19 See *Financial Condition Report 2015* (Accident Compensation Corporation, 2016).

20 Amongst reasons put forward for this are excessive speed, alcohol and the fact that there relatively few dual carriageways in New Zealand, and therefore few central barriers between carriageways. This situation leads to a significant number of head-on collisions, producing serious and expensive injuries. Given advances in medical response and treatment, the numbers of fatalities have dropped over time but survivors can be expensive.

21 ACC *Third Quarterly Report 2015–16*, available at: www.acc.co.nz/PRD_EXT_CSMP/groups/external_communications/documents/reports_results/wpc139147.pdf. The funding deficit is primarily attributed to the decreasing single-effective discount rate.

22 See *Financial Condition Report 2015* (Accident Compensation Corporation, 2016) para 3.2.3.

23 Until 2016/17 the Earners' and the Motor Vehicle Levy contained two components: a current portion and a residual portion. The residual portion was to cover the ongoing costs of claims that occurred before 1 July 1999, when the Scheme was funded on a pay-as-you-go basis. Up to 2009 the remaining cost of these claims was assessed as generally increasing each year, which meant that these claims were putting a greater burden on levy-payers over time. Therefore, in 2009 the government fixed the total amount to be collected via residual levies. Since then the estimate of the pre-1999 claim costs has decreased and it is now substantially lower than the amount fixed in 2009. The amount of the residual levy that will have been collected as at 1 April 2016 has been assessed as being sufficient to fund the remaining costs of pre-1999 claims. See the *Financial Condition Report 2015* (Accident Compensation Corporation, 2016) para 5.6

24 *Annual Report 2015*, (Accident Compensation Corporation, 2016), 26. The Accident Compensation (Financial Responsibility and Transparency) Amendments Act 2015 sets out to stabilise levies and to provide a clearer, more transparent mechanism for the funding of the levied accounts.

growing co-morbidities (eg obesity, diabetes and mental health) in the general population will continue to impact claim rates and costs.[25]

D. Eligibility

The scheme provides cover for personal injury, as defined in relation to a number of specific categories. In order to qualify for cover the client must meet what is in effect a two-stage test of having an injury and that injury being caused by either a specific event, series of events, or gradual process, disease or infection.[26] The eligibility provides comprehensive injury cover regardless of what the claimant was doing when he or she was injured, how the injury happened (including whether the claimant contributed to the injury) and irrespective of the age or employment status of the claimant.

The ACC has to be satisfied that the claimant has suffered a *personal injury*, which can mean any of:

— physical injury;[27]
— mental injury[28] suffered due to a physical injury;
— mental injury caused by certain criminal acts;[29]
— damage (other than wear or tear) to dentures or prostheses[30] that replace part of the human body;
— death due to a physical injury.

The legislation specifies rules relating to the accident[31] or event that caused the personal injury in relation to mental injury, treatment injury and gradual process. In relation to

25 *Annual Report 2015* (n 24) 18.

26 ACA, s 20.

27 ACC considers that a physical injury has not occurred unless there is actual damage to the body from the injury. Neither is a diagnosis of pain sufficient to establish there has been a physical injury; an actual diagnosis of the injury is required. Physical injuries can include: wounds, lacerations, contusions, burns, sprains and strains, fractures, amputations or dislocations, damage to dentures or prostheses, work-related gradual process injuries, such as tendonitis, and deafness caused by noise at work, infections or diseases caused at work by performing a particular task or being exposed to a particular environment (this excludes any conditions you may have had since birth), blindness, poisoning, choking, loss of consciousness, a foreign body in the eye.

28 Defined in the Accident Compensation Act 2001 as 'a clinically significant behavioural or psychological dysfunction'. A mental disorder with a specific diagnosis that requires treatment.

29 Listed in ACA, Sch 3.

30 Defined on the ACC website as '[a]n artificial substitute for a missing body part, such as an arm, leg or eye, which is used for functional or cosmetic reasons, or both. Also includes a device used to improve function, such as a hearing aid.' See also s 26(1)E and s 26(5) ACA 2001.

31 ACA, s 26. The definition of 'accident' is lengthy, and includes specified occurrences:

(a) a specific event or a series of events, other than a gradual process, that—(i) involves the application of a force (including gravity), or resistance, external to the human body; or (ii) involves the sudden movement of the body to avoid a force (including gravity), or resistance, external to the body; or (iii) involves a twisting movement of the body:

(b) the inhalation of any solid, liquid, gas, or foreign object on a specific occasion, which kind of occurrence does not include the inhalation of a virus, bacterium, protozoan, or fungus, unless that inhalation is the result of the criminal act of a person other than the injured person:

(ba) the oral ingestion of any solid, liquid, gas, fungus, or foreign object on a specific occasion, which kind of occurrence does not include the ingestion of a virus, bacterium, or protozoan, unless that ingestion is the result of the criminal act of a person other than the injured person:

cover for drugs, for example, cover is in general available for oral ingestion or inhalation that causes personal injury, and potentially as a treatment injury for adverse side effects.

Specific rules apply to *treatment injuries*.[32, 33] Treatment is widely defined, and includes the failure of equipment or a device,[34] and personal injury suffered by a person or certain others as part of a clinical trial.[35] The scheme does not appear to include injuries caused by licensed medicinal products.[36] This Account therefore provides free insurance for healthcare costs, paid for by salaried and other wage earners and the Government.

(c) a burn, or exposure to radiation or rays of any kind, on a specific occasion, which kind of occurrence does not include a burn or exposure caused by exposure to the elements:
(d) the absorption of any chemical through the skin within a defined period of time not exceeding 1 month:
(e) any exposure to the elements, or to extremes of temperature or environment, within a defined period of time not exceeding 1 month, that,— (i) for a continuous period exceeding 1 month, results in any restriction or lack of ability that prevents the person from performing an activity in the manner or within the range considered normal for the person; or (ii) causes death.

[32] Defined by ACA s32, a personal injury that is:

(1) (a) suffered by a person—
(i) seeking treatment from 1 or more registered health professionals; or
(ii) receiving treatment from, or at the direction of, 1 or more registered health professionals; or
(iii) referred to in subsection (7); and
(b) caused by treatment; and
(c) not a necessary part, or ordinary consequence, of the treatment, taking into account all the circumstances of the treatment, including—
(i) the person's underlying health condition at the time of the treatment; and
(ii) the clinical knowledge at the time of the treatment.
(2) Treatment injury does not include the following kinds of personal injury:
(a) personal injury that is wholly or substantially caused by a person's underlying health condition:
(b) personal injury that is solely attributable to a resource allocation decision:
(c) personal injury that is a result of a person unreasonably withholding or delaying their consent to undergo treatment.
(3) The fact that the treatment did not achieve a desired result does not, of itself, constitute treatment injury.

[33] ACA, s 228. See K Oliphant, 'Defining "Medical Misadventure": Lessons from New Zealand' (1996) 4 *Medical Law Review* 1–31.

[34] ACA, s 33(1). For the purposes of determining whether a treatment injury has occurred, or when that injury occurred, treatment includes—

(a) the giving of treatment:
(b) a diagnosis of a person's medical condition:
(c) a decision on the treatment to be provided (including a decision not to provide treatment):
(d) a failure to provide treatment, or to provide treatment in a timely manner:
(e) obtaining, or failing to obtain, a person's consent to undergo treatment, including any information provided to the person (or other person legally entitled to consent on their behalf if the person does not have legal capacity) to enable the person to make an informed decision on whether to accept treatment:
(f) the provision of prophylaxis:
(g) the failure of any equipment, device, or tool used as part of the treatment process, including the failure of any implant or prosthesis (except where the failure of the implant or prosthesis is caused by an intervening act or by fair wear and tear), whether at the time of giving treatment or subsequently:
(h) the application of any support systems, including policies, processes, practices, and administrative systems, that—
(i) are used by the organisation or person providing the treatment; and
(ii) directly support the treatment.

[35] ACA, s 26(4)–(6).

[36] But it may be speculated whether the courts would permit this if a situation arose, which does not seem to have happened.

The scheme provides compensation for those ordinarily resident in New Zealand for personal injury suffered in New Zealand,[37] mental injury or nervous shock caused by certain criminal acts,[38] or work-related mental injury,[39] suffered since April 2002. It also covers treatment costs for personal injury suffered outside New Zealand before April 1974.[40]

E. Compensation

The scheme aims to provide 'real' but not 'full' compensation. A claimant who has suffered a personal injury is entitled to the statutory entitlements if he or she has cover for the personal injury and is eligible under the ACA.[41] The main entitlements are:[42]

(a) rehabilitation, comprising treatment, social rehabilitation and vocational rehabilitation;[43]
(b) first week compensation (but see below);
(c) weekly compensation after the first week;
(d) lump sum compensation for permanent impairment: this is intended to cover non-economic loss and is paid for permanent impairment assessed at 10% or greater;
(e) funeral grants, survivors' grants, weekly compensation for the spouse or partner, children or other dependants of a deceased claimant, and childcare payments.

Where a claim is accepted, the ACC must determine within 13 weeks whether the claimant is likely to need social or vocational rehabilitation, and if so make an appropriate plan, which must be subsequently updated.[44] Where the ACC decides that it is reasonably practicable for the claimant to return to the same employment, an employer must take all reasonably practicable steps to assist with the claimant's vocational rehabilitation and plan.[45] Social rehabilitation includes aids and appliances; attendant care; childcare; education support; home help; modifications to the home; training for independence; and transport for independence.[46]

It is the duty of an employer to pay compensation for the first week of incapacity resulting from a work-related injury, at 80% of earnings.[47] ACC cover of 'weekly compensation' of 80% of the claimant's weekly earnings applies after the first week, capped at NZ$1,341.31 but subject to indexation.[48] The ACC may suspend or cancel an entitlement if it is not

[37] ACA, s 20.
[38] ACA, ss 21 and 21A.
[39] ACA, s 21B.
[40] ACA, s 22.
[41] ACA, s 67.
[42] ACA, s 69, see also Parts 10 and 11.
[43] The objective is to return a person to a state of being capable of resuming employment or other prior activities. Rehabilitation rates measure the percentage of clients who stop receiving weekly compensation under the ACS and move to benefits provided by the Ministry of Social Development (MSD) if they are work-ready with no job to return to. In 2012 the rate moving to MSD benefits within 3 months of ceasing weekly compensation was around 4%: *Financial Condition Report 2013* (Accident Compensation Corporation, 2014) para 1.4.30.
[44] ACA, ss 75 and 77.
[45] ACA, s 71.
[46] ACA, s 81.
[47] ACA, s 97–99.
[48] ACA, ss 100 and 115, and Sch 1, cl 32, 46.

satisfied that a claimant is entitled to continue to receive the entitlement.[49] Certain disentitlements apply, such as for willfully self-inflicted injuries or where the claimant has been convicted of murder or is imprisoned.[50]

F. Procedure

A claimant must lodge a claim with the ACC, in a manner prescribed by the ACC.[51] A claim form is lodged through a treatment provider, who certifies that the injury has been sustained and was thus not caused by disease. A claim must be lodged within specified time limits, primarily within 12 months after the date of injury or the date on which entitlement arose.[52] The ACC must decide in a timely manner whether or not it accepts that the person has cover, and if so provide information about entitlements and facilitate the claimant's access to those entitlements,[53] and give notice of its decision.[54] It may require the applicant to provide a certificate by a registered health professional, or any other relevant information, or to make a written statement about any relevant specified matters, or to authorise the ACC to obtain medical and any other records that may be relevant, or undergo a medical or any other kind of assessment at the ACC's expense.[55] Claims are screened by the ACC's computerised programme, which automatically processes most claims and filters complicated claims for manual assessment.

The ACC is subject to a statutory Code of Claimants' Rights, the purpose of which is 'to meet the reasonable expectations of claimants (including the highest practicable standards of service and fairness) about how the [ACC] should deal with them'.[56] The Code is subject to ministerial approval and confers legal rights on claimants and obligations on the ACC.

The ACC operates through 25 branches and service centres across the country, comprising 3,300 staff (including 1,942 frontline staff) in 2016.

A claimant may apply to the ACC for independent review of its decisions. The reviewer's costs are paid by the ACC and decisions are binding on both sides (the process is therefore a form of arbitration).[57] In 2015–16 only 2.5% of entitlement cases were subject to formal review, and 84.2% of reviews were upheld in the ACC's favour. While this is a low percentage of reviews the sheer volume of claims handled means that over 80 entitlement claims are reviewed each year. Certain subsequent rights of appeal to a District Court are given to claimants, the ACC and employers,[58] again with all costs being paid by the ACC. A significant level of litigation is brought by this means over cover and entitlements, especially in relation to boundary issues. ADR is encouraged.[59] Over time, the courts have clearly

[49] ACA, s 117.
[50] ACA, ss 118–22A.
[51] ACA, ss 48, 49, 52.
[52] ACA, s 53.
[53] ACA, ss 50, 54.
[54] ACA, s 64.
[55] ACA, s 54.
[56] ACA, s 40.
[57] ACA, ss 133–48, esp 147.
[58] ACA, ss 149–61.
[59] ACA, s 328A.

extended the scheme,[60] appearing to base decisions on the basis that the scheme comprises the national injury insurance mechanism and that it should generally therefore be inclusionary as a matter of social policy.[61] The Government is considering whether such disputes can be processed by shorter and cheaper means than through the District Courts.

G. Statistics

Details of the claims numbers and finance for 2015–16 are shown in Table 2.1.[62] In 2015–16, the ACC accepted 1.9 million new claims and had a total claims cost (including outstanding claims liabilities) of NZ$6.1 billion, with NZ$3.2 billion paid out on claims. Category totals were $669 million on medical treatment, $322 million on hospital treatment and elective, $590 million on social rehabilitation, $469 million on public health acute services and $1,045 million on non-fatal weekly compensation. Total funds under management were $35.5 billion. Over 1,022,000 clients visited their GP, over 491,000 clients received physiotherapy, and over 37,500 clients had surgery.

Table 2.1: Summary of the ACC and accounts at end of June 2016

	Motor Vehicle Account	Non-Earners' Account	Earners' Account	Work Account	Treatment Injury Account	Total
Total assets (NZ$ millions)	11,272	3,389	10,206	9,804	3,2616	38,275[a]
Total liabilities (NZ$ millions)	10,460	7,901	9,258	8,520	5,245	41,383
Funding ratio (%)[b]	107.8	41.9	112.8	117.2	68.3	94.9
New claims registered in 2015–16[c]	28,841	859,562	653,224	204,165	7,849	1,753,641
New entitlement claims in 2015–16[d]	5,329	27,451	58,198	34,912	3,378	129,268

[a] Net assets before outstanding claims liability.

[b] Funding ratio comprises the (total assets)—(payables, accrued liabilities and provisions)—(unearned levy liability)—(unexpired risk liability)/(outstanding claims liability); for a full breakdown see p 56 of ACC's Annual Report 2016.

[c] Data taken from ACC's online statistics tool www.acc.co.nz/about-acc/statistics/injury-statistics-tool/index.htm on 04/04/2017.

[d] ibid. Entitlement claims are those which receive one or more of the entitlements provided under the AC Act, including weekly compensation.

[60] The ACA is considered to be broadly written, affording scope for differences in interpretation, and for litigation. Extensions in cover have largely occurred through court decisions. See *Financial Condition Report 2013* (Accident Compensation Corporation, 2014) para 1.3.2.

[61] Decisions in recent years have upheld cover for failed screening for spina bifida and workplace mental trauma. There is pressure to include situations where the outcome of treatment was not as desired.

[62] *Annual Report 2016* (Accident Compensation Corporation, 2016) and see *Financial Condition Report 2013* (Accident Compensation Corporation, 2016).

The claims-handling expenses for the five accounts as a proportion of the claims payments, which gives a crude measure of efficiency, are shown in Table 2.2. These figures are impressive, particularly when compared to the tort system that the ACC replaced four decades ago.

Table 2.2: Claims handling expenses as a proportion of claims paid in 2014/15[a]

	Motor Vehicle Account	Non-Earners' Account	Earners' Account	Work Account	Treatment Injury Account	Total
Claims paid (NZ$ millions)	477	964	1,190	707	164	3,502
Claims-handling costs (NZ$ millions)	40	100	153	99	23	415
Claims-handling costs as a percentage of claims paid (%)	8.4	10.4	12.9	14.0	14.0	11.9

[a] *Annual Report 2016* (Accident Compensation Corporation, 2016) 77.

The income and expenditure statements for 2016 are shown at Table 2.3. The number of new claim registrations and new entitlement claims over recent years is shown in Figure 2.1, and the payment totals are given in Figure 2.2. The picture shows broad stability.

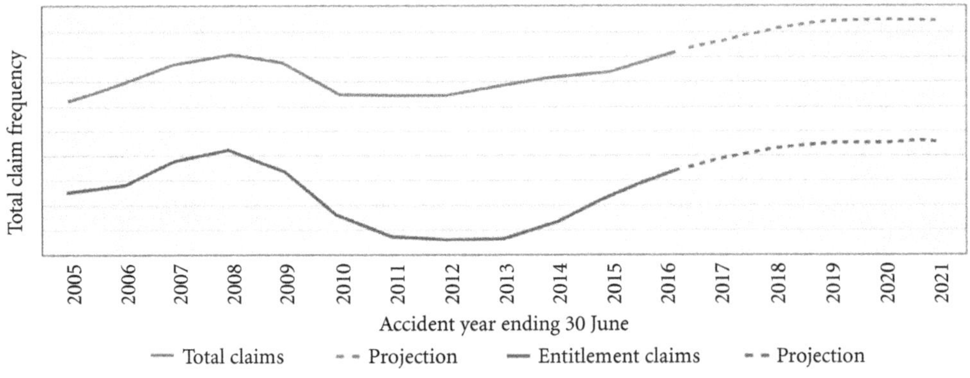

Figure 2.1: New estimated ultimate reported total and entitlement claims per 1,000 people, 2005–21

Table 2.3: ACC statement of consolidated revenue and expense for 2016[a]

$M	Notes	Actual 2016	Actual 2015	Budget 2016
Net levy revenue	4	3,926	4,313	4,146
Other revenue		1	2	–
Total net levy and other revenue		**3,927**	**4,315**	**4,146**
Investment revenue		3,273	3,997	1,628
Less investment costs		63	61	65
Net investment revenue	5	**3,210**	**3,936**	**1,563**
Claims paid		3,502	3,219	3,427
Increase in outstanding claims liability	19	6,334	2,632	1,538
Movement in unexpired risk liability	18	103	265	48
Total claim costs		**9,939**	**6,116**	**5,013**
Injury prevention costs	7	50	30	51
Operating costs	7	515	494	516
Net (deficit] surplus		**(3,367)**	**1,611**	**129**
Total comprehensive revenue and expense for the year		**(3,367)**	**1,611**	**129**

[a] *Annual Report 2016* (Accident Compensation Corporation, 2016) 58.

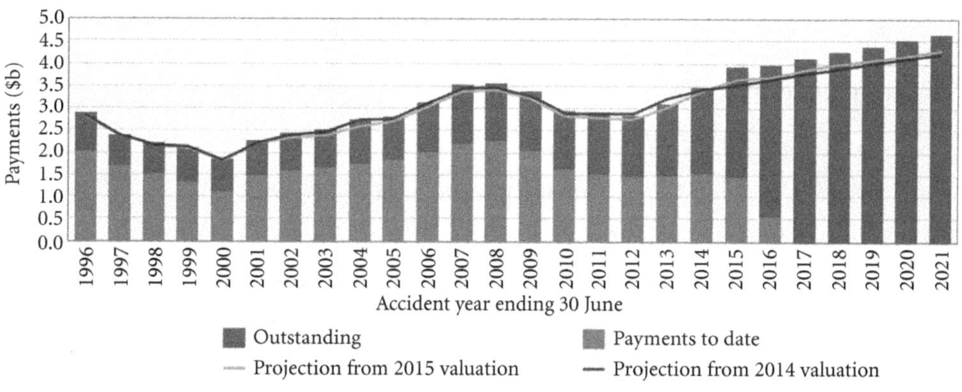

Figure 2.2: ACC total payments, actual and predicted 1996–2021. Source: *Financial Condition Report 2016* (Accident Compensation Corporation, 2016)

The average time taken to make a cover decision on a claim in 2015–16 was 1.1 days for all claims, with an average time to commence payment of 8.3 days.[63] In 2015–16, the 'return to work rate' after 70 days was 67.6% and after 273 days was 92.8%.[64]

[63] *Annual Report 2016*, ibid.
[64] ibid.

H. Improving Safety

The ACC has available comprehensive national data on the types, nature, incidence, causes and locations of personal injury accidents in New Zealand, going back to 1974. Historically it has used this data to undertake a number of injury prevention initiatives. Particular campaigns in 2011–12 were to reduce claims involving drivers aged 15–19 through a young driver programme, and to reduce falls in the elderly by increasing vitamin D take up.

In 2013 the Cabinet replaced the New Zealand Injury Prevention Strategy (NZIPS) governance group with a Cross-Government Injury Prevention Work Plan. Following this, in 2014, the ACC introduced a new approach to injury prevention. The aim of this was threefold:

— to improve collaboration with a range of stakeholders;
— the improve the use of data to design ACC programmes;
— to target ACC programmes' areas of focus.

The intention is that the Cross-Government Injury Prevention Work Plan will establish an outcome-based framework that will enhance and support the injury prevention activities of government agencies in particular, and other organisations and groups in the wider community. It will focus cross-government and ACC activity on 'what works' in priority areas, and what provides tangible benefits to both the Scheme and New Zealand. In its initial phase, the Work Plan will address four key priority areas for ACC and its partner agencies:

— vulnerable children;
— sexual violence;
— family violence;
— the role of dementia in falls.

I. Workplace Safety

In 2013–14 the ACC entered into a partnership with the Ministry of Business, Innovation and Employment (MBIE) to effect safer workplaces. The ACC and WorkSafe New Zealand (the workplace health and safety regulator) have drawn up a Reducing Harm in New Zealand Workplaces Action Plan.[65] This is looking at how to improve safety, including the role of economic incentives for employers.

The ACC currently offers two incentive programmes designed to encourage workplace safety, both of which give discounts on levies in return for businesses meeting certain health and safety standards. The Workplace Safety Management Practices programme is available to all employers, and involves three levels of accreditation, providing discounts of 10%, 15% and 20%. This tends to attract around 30% of the eligible larger employers. The Workplace

[65] *Reducing Harm in New Zealand Workplaces Action Plan 2013–19* (Accident Compensation Corporation, 2016), available at www.acc.co.nz/PRD_EXT_CSMP/groups/external_ip/documents/reference_tools/wpc139179.pdf.

Safety Discounts programme is targeted at businesses (including the self-employed) with 10 or fewer employees, especially in high-risk industries. It is administered by industry bodies and gives a discount of 10%. It currently attracts under 2% of eligible businesses. The efficacy of these two programmes has been regarded as 'unclear', hence the evaluation under the current Action Plan.[66] An incentive scheme for operators of commercial fleet vehicles was launched in December 2013 and now has 71 members, representing 5.7% of the heavy vehicles fleet.

The ACC publishes an Annual Financial Condition Report by independent actuaries, which includes a significant amount of information on aggregated accident data, but it does not publish data that would enable individual employers or hospitals to be identified or their safety performance to be assessed or ranked. Such information is sent to the Government and to Worksafe New Zealand, the workplace safety regulatory authority, which was established only in December 2013 after a major mining incident. Information on a hospital's own performance is sent confidentially to all the District Health Boards. The ACC does not regard its responsibilities to stretch to seeking improvements in workplace or hospital safety outcomes, which would be regulatory functions. It considers that publication of data that would enable public comparisons to be drawn would be a highly contentious development that would require political approval.

J. Conclusions

In establishing the ACC, New Zealand took the choice of preferring a national personal injury insurance scheme to providing compensation through liability litigation. As a result, litigation for personal injury falling within the scope of the scheme effectively does not exist in the country. The ACC is the largest insurer in New Zealand. Approximately one-third of the national population (30.8% in 2016) receives compensation every year, almost all of which are for small injuries and sums. The scheme is not intended to provide any deterrent or regulatory effect.

The scheme is set up to deliver the fast processing of volume claims. It is imperative that the system should perform at least as well (and preferably better) than the courts. This it does in terms of general speed of administration and certainty of cover and payment for most injuries. The five funds provide comprehensive coverage of all types of injury and enable appropriate risk-pricing while avoiding cross-subsidisation. The scheme is structured with a single easily identifiable point of access for claimants, regardless of the cause of their injury.

Inevitably, some concomitant drawbacks exist. The 'tyranny of volume' means that individual claims are not closely screened, the scope of cover is not clear and/or cover is not provided for non-injury conditions. No particular concern appears to exist that claimants make fraudulent or exaggerated claims, since a treatment provider is necessarily involved as a first stage in giving a professional opinion on the fact and cause of the injury. Further, statistical comparison of the rates for workers' returning to work (rehabilitation) shows a

[66] *Financial Condition Report 2013* (Accident Compensation Corporation, 2014) para 1.6.13.

consistently better performance by New Zealand over Australia since at least 2000, albeit less so since 2004.[67]

There is an inherent and perennial boundary issue over whether a person's condition was caused by injury or disease, the former triggering insurance cover under the ACC but the latter not. An example frequently occurs with knees: a claim for a contusion on the knee would be immediately accepted but a couple of years later the request might be for a new knee as an entitlement. Another example would be a lumbar sprain accepted by a GP as having been caused by lifting a washing machine, but which might later be diagnosed by a consultant as a disc prolapse. These situations illustrate the difficulty of placing clear boundaries on what constitutes injury, and the problems of applying a rule at an initial point in time when subsequent medical information may indicate a different result.

A person is clearly better off if their adverse condition is attributed to injury rather than to illness. A worker whose condition is attributed to a spinal cord injury, for example, will receive significant benefits, such as 80% of earnings, wheelchairs, vehicle and house adaptation, a case manager, vocational and rehabilitation placements, and private surgery. By contrast, if the condition is diagnosed as cancer of the spine, benefits are fewer and capped. This 'two-tier' approach causes some debate and divisive disputes. It produces conflict in the case of co-morbidities, such as diabetes and obesity.[68]

A criticism, particularly made by those outside the country, is that the level of payments is modest and not the 'full compensation' that would be awarded by courts. However, the scheme forms an essential part of the national landscape of injury support,[69] is well established and widely accepted. New Zealanders value the breadth, certainty and speed of this insurance scheme over higher monetary value. It appears that only high earners take out additional first-party insurance that will provide higher levels of income replacement.

The matching of liabilities to funding is a perennial issue, as with all insurance schemes. In its initial years funding for the ACC was somewhat precarious, but it has been essentially financially stable for over 25 years. Current-year funding of present and future claims has not been achieved in 2015/16, when a $3.3 billion deficit was recorded. Overall the funding deficit has been decreasing in the last decade, with 2014–15 in surplus, and the deficit reduction is a trend that is expected to continue. This longevity demonstrates that a no-fault scheme can achieve long-term sustainability, while still adapting to changing social and political backdrops. Nevertheless, the scheme operates for a relatively small population (4 million) on a relatively limited scale, currently with $40 billion in liabilities and annual weekly compensation payments annually totalling $10 billion. Future sustainability as the population ages and health conditions produce further impact on injury remains a concern.

Economic analysis holds that the cost of accidents should be charged to those who cause them so that they will be incentivised to improve safety performance and reduce the number and severity of injuries that are under their control.[70] The New Zealand system pays little heed to this theory. The long-term trends in claims figures suggest that claim rates are principally affected by public perceptions of the financial strength of the fund. Incentives

[67] See *Financial Condition Report 2013* (Accident Compensation Corporation, 2014) graph 1, 20.

[68] *Financial Condition Report 2013* (Accident Compensation Corporation, 2014) para 9.8.12.

[69] K Oliphant, 'Landmarks of No-Fault in the Common Law' in WH van Boom and M Faure (eds), *Shifts in Compensation between Private and Public Systems* (Berlin, Springer, 2007), 73.

[70] This classic theory and its absence under the ACS is recognised in *Financial Condition Report 2013* (Accident Compensation Corporation, 2014) 8.

to improve safety are not made by differentiating between the premiums paid by individual hospitals or drivers—a single, simple national levy rate applies to all. The low take up of the two discount schemes available to employers, especially one for smaller businesses, suggests that these payers afford little value to such incentives. Further, it appears that the ACC is not persuaded that publishing safety data of individual firms would lead to significant improvements in safety. The system for feedback is limited, and not publicly transparent. Indeed a workplace safety regulator has only recently been created. We suspect that closer attention to the function of behavioural oversight (regulation by public authorities and through public transparent scrutiny) based on the rich seam of accident and injury data that is held by the ACC would lead to improvements in safety culture and outcomes. Examination of the New Zealand data should be able to identify which hospitals, business activities and drivers produce higher risks, and then enable an inquiry into whether economic or other incentives in fact affect relevant behaviour.

Litigation is not altogether removed. It has been removed as the mechanism by which most injured people establish a right to cover, but it remains as a residual mechanism to resolve some disputes over coverage and entitlements. It appears that a more efficient process could be adopted than currently applies. However, it is interesting that extension of the scheme has been made over time by the courts rather than through legislative channels.

3

Australia: Workers' Compensation

MARY WALKER

Workers' Compensation: A Global Approach

Workers' compensation schemes exist in many jurisdictions (both common law and civil) as an alternative to the traditional tort system for injuries at work. Generally, these schemes do not offer full compensation; they provide assured payments for medical provision, economic losses, lost wages, disability payments and sometimes death benefits, in the event of a workplace accident on a no-fault basis. Upon joining the scheme, the worker agrees that the scheme is the exclusive remedy, and therefore waives his/her right to sue in the tort of negligence. Schemes vary somewhat, but damages for pain, suffering, loss of amenity and punitive damages against an employer for negligence are generally absent, or highly restricted, under workers' compensation programmes.

Workers' compensation schemes have advantages and disadvantages for both parties, often referred to as the compensation bargain.[1] A substantial body of literature exists on the impacts of, and arguments for and against the compensation bargain, though this area of discussion is sadly outside the scope of this volume.

Briefly, claimants benefit from a reduced burden of proof; instead of the difficult of proving negligent breach of duty of care on the part of their employer they merely have to demonstrate that the injury occurred out of or in the course of employment. Claimants also benefit from the certainty of known, usually statutorily fixed, benefits. The trade-off for this certainty and reduced burden of proof is the relinquishing of the right to undertake higher-risk, but potentially more lucrative litigation. For an employer a no-fault arrangement removes the defences that would have been available under litigation, and they will potentially pay a greater proportion of claims. However, they gain far more predictable, lower payments; damages are fixed at lower levels than would be awarded in negligence, and the exclusiveness of the remedy ensures no surprise outlays. The heads and values of damages paid are often capped or restricted under workers' compensation schemes; non-economic losses, such as pain, suffering and loss of amenity, are simply not available, and other heads, such as lost wages, are capped under workers' compensation schemes, where they would be available in litigation.

[1] For a detailed description, see RA Epstein, 'The Historical Origins and Economic Structure of Workers' Compensation Law' (1982) 16 *Georgia Law Review* 775, 800–01.

Workers' compensation schemes first appeared in Europe in the 1880s, beginning with the inception of the German scheme in 1884.[2] Unofficial and sector-specific schemes often pre-dated statutory schemes—there are reports of an English pirates' workers' compensation scheme to compensate injured or disabled crew members, which pre-dated the official statute of 1897.[3]

This movement was primarily in civil law jurisdictions, but spread to common law jurisdictions. The largest common law jurisdiction to have adopted workers' compensation is the USA, but Australia, Canada and many other common law jurisdictions have also incorporated workers' compensation into their legal frameworks. A detailed description of the workers' compensation arrangements in six Australian territories will follow.

In the USA in the decade following 1910 nearly all states relocated workplace accident claims from tort systems into administrative schemes. The advent of workers' compensation in the USA has been widely written about, and will be only briefly detailed here. In his incisive account Robert L Rabin has suggested that workers' compensation schemes took effect in the USA on the back of a prevailing ideological driver.[4] Several major components have been identified.

Workers' compensation arose during the Progressive Era, an epoch that saw substantial health and safety reforms, an expansion of the protection for child workers, the enshrining of women's rights, increased democratic accountability, a reconsideration of the state's power and obligations coupled to a widespread faith in administrative systems.[5]

These developments occurred when there were scant opportunities for recompense for injured workers, there was little first-party insurance available to workers[6] and the 'unholy trinity'[7] of employer defences often prevented workers from recovering damages, leading to a climate of dissatisfaction with the existing tortious arrangements.[8]

Additionally, these schemes arose during a time when workplace injuries were on the rise, and media portrayals increased awareness both of this upsurge in injuries and of the consequences for affected individuals, leading to the perception of a burgeoning crisis by the US public.

Finally, there was a critical shift in attitudes towards accidents among Americans. Previously accidents were viewed as the unpredictable unhappy consequences of the actions

[2] Starting with Germany in 1871 and 1884, Austria in 1887, Norway in 1894, Finland in 1895, Denmark, France and Italy in 1898, Greece in 1901, Belgium and Russia in 1903; taken from SB Horovitz, 'Worldwide Workmen's Compensation Trends' (1970) 59 *Kentucky Law Journal* 37, 40–41.

[3] S Talty, *Empire of Blue Water: Captain Morgan's Great Pirate Army, the Epic Battle for the Americas, and the Catastrophe that Ended the Outlaws' Bloody Reign* (Three Rivers Press, 2008).

[4] RL Rabin, 'Some Reflections on the Process of Tort Reform' (1988) 25 *San Diego Law Review* 13, 18.

[5] Ibid, 17–18.

[6] KS Abraham, *The Liability Century: Insurance and Tort Law from the Progressive Era to 9/11* (Cambridge, MA, Harvard University Press, 2008) 44–45.

[7] These were contributory negligence, voluntary assumption of risk and the fellow servant defence (that an employer could not be found liable for damages arising from the actions of a fellow servant). The impact of these defences at common law was slowly being eroded, but they were still a substantial hurdle to claimants. See PV Fishback and SE Kantor, *A Prelude to the Welfare State: The Origins of Workers' Compensation* (Chicago, University of Chicago Press, 2000).

[8] Ibid, 23.

of individuals. During the Progressive Era, however, there was a change towards viewing workplace accidents as predictable and inevitable, increasing the rationale for collective restitution rather compensation that relied upon individual blame.[9]

Workers' compensation schemes are widespread in the USA, where they are organised on a state-by-state basis. Every US state has some form of workers' compensation scheme.[10] As there is considerable writing on workers' compensation arrangements in the USA it will not be covered in any further detail in this volume.

* * *

Compensation schemes for personal injury are 'prominent features of the legal landscape in many developed countries'.[11] As would be expected, Australia has a well-established system of workers' compensation. Australian workers' compensation arrangements are organised separately in eight states and territories rather than on a national basis. Descriptions of workers' compensation arrangements in six of these states and territories follow. All eight jurisdictions have a workers' compensation scheme, and there are an additional three Commonwealth schemes. Each jurisdiction operates under local legislation.[12] The three Commonwealth schemes cover the following groups

- Australian Government employees, Australian Defence Force personnel who were in service before 1 July 2004, and the employees of licensed self-insurers under the Safety Rehabilitation and Compensation Act 1992;
- Certain seafarers under the Seafarers Rehabilitation and Compensation Act 1992; and
- Australian Defence Force personnel in service on or after 1 July 2004 under the Military Rehabilitation and Compensation Act 2004.

A detailed comparison and summary of these schemes can be found in the comparison of workers' compensation arrangements in Australia and New Zealand (August 2014) produced by Safe Work Australia.

Table 3.1 summarises the key features of the six schemes described below as at June 2013.[13]

As can be clearly seen these schemes have considerable differences, some would say inconsistencies, in their organisation, funding, access to legal recourse and solvency. These discrepancies arise from the historical origins of these schemes, which developed on a state-by-state rather than federal basis. The impact of these differences will be discussed in the Conclusions.

[9] Abrahams (n 6) 52–53 and JF Witt, *The Accidental Republic: Crippled Workingmen, Destitute Widows, and the Remaking of American Law* (Cambridge, MA, Harvard University Press, 2006).

[10] Texas is the only state where workers' compensation insurance has never been mandatory.

[11] G Grant and D Studdert, 'Poisoned Chalice? A Critical Analysis of the Evidence Linking Personal Injury Compensation Processes with Adverse Health Outcomes' (2009) 33 *Melbourne University Law Review* 2.

[12] Safe Work Australia, *Comparison of Workers' Compensation Arrangements in Australia and New Zealand* (2015), available at www.safeworkaustralia.gov.au/sites/swa/about/publications/pages/comparison2015.

[13] Ibid, 15.

Table 3.1: Key features of the workers' compensation schemes in Western Australia, South Australia, Tasmania, the Northern Territory, the Australian Capital Territory and New South Wales

	Western Australia	South Australia[a]	Tasmania	Northern Territory	Australian Capital Territory	New South Wales
Employees covered for workers' compensation (2012–13)	1,223,700	740,000	209,700	120,000	122,859 private-sector workforce	3,320,800
Number of serious claims with 1 week or more incapacity (2012–13)	12,950	8,840	2,720	1,130	1,700	37,580
Number of serious claims per 1,000 employees (2012–13)	10.6	11.9	13.0	9.4	11.1	11.7
Compensated deaths per 100,000 employees (2012–13)	2.0	0.7	1.4	3.3	0.8 (ACT private sector only)	1.4
Scheme funding	Private insurers	Central	Private insurers	Private insurers	Private insurers	Managed
Standardised average premium rate (% of payroll)	1.28 (2012–13) 1.22 (2011–12)	2.48 (2012–13) 2.51 (2011–12)	2.11 (2012–13) 1.96 (2011–12)	1.95 (2012–13) 1.81 (2011–12)	1.99 (2012–13) 2.11 (2011–12)	1.67 (2012–13) 1.70 (2011–12)
Standardised funding ratio (%) (2012–13)	132	66	105	91	N/A	118
Excess/unfunded (30 June 2014)	NA	$1,132 m unfunded	$2 m unfunded	NA	NA	$2.6 bn funded
Access to common law	Yes	No	Yes—limited	No	Yes	Yes

[a] These figures refer to the old scheme; the new scheme has only been running from 1 July 2015, so only has limited figures are available. It allows access to the common law in limited circumstances, as will be detailed below.

A. Western Australia Workers' Compensation

Introduction: The Compensation Scheme

The primary mechanism for workers' compensation in Western Australia is under the Workers' Compensation and Injury Management Act 1981 (WA).[14] Originally, workers' compensation was governed by the Workers' Compensation Act 1902 (WA), introduced at the onset of the 20th century. This legislation was implemented to provide no-fault compensation with regard to work-related injury (and subsequently disease), based in very large part on what was then the British model.[15] Over the course of the next 79 years, the Act was subject to 'frequent and complex amendments'.[16] Therefore, workers' compensation law was consolidated in the Workers' Compensation Act 1981 (WA). Further changes were to come through the Workers' Compensation Reform Bill 2004 (WA) which introduced changes to statutory benefits, injury management, access to common law, employer incentives in relation to return to work for disabled workers, and fairness in dispute resolution.[17] As part of the reforms the Act was renamed the Workers' Compensation and Injury Management Act 1981, which reflects an emphasis on injury management within the workers' compensation scheme in Western Australia. Finally, in 2011 the Act was amended once again by the Workers' Compensation and Injury Management Amendment Act 2011 (WA).[18] These amendments had the effect of restructuring and reconstituting the entire dispute-resolution process provided by the Act.

In its current form, two primary purposes of the Act have been emphasized by WorkCover Western Australia. These are:

- ensuring workplace injuries are managed in a manner that enables a worker's prompt and safe return to work following an injury;[19] and
- ensuring injured workers are compensated for lost wages, medical expenses and associated costs while they are unable to work.[20]

WorkCover Western Australia Authority

The purpose of WorkCover is to ensure the 'efficient and effective operation of the workers' compensation scheme'.[21] This extends to functions such as monitoring the extent to which employers and insurers are complying with the scheme,[22] promoting awareness and

[14] Workers' Compensation and Injury Management Act 1981 (WA). Unless otherwise advised, all references to sections refer to this Act.

[15] Neil Gunningham, 'Asbestos Related Diseases and Workers Compensation' (2012) 34 *Sydney Law Review* 271.

[16] Safe Work Australia, 'Comparison of Workers Compensation Arrangements in Australia and New Zealand' (Report, 2010) 181–86.

[17] ibid, 12.

[18] Workers' Compensation and Injury Management Amendment Act 2011 (WA).

[19] WorkCover WA, 'Workers Compensation & Injury Management: A Guide for Workers' (Report, 2016) 5.

[20] ibid.

[21] s 100.

[22] s 100(a).

disseminating information regarding workers compensation,[23] and resolving or assisting in the resolution of disputes through conciliation and arbitration.[24]

Eligibility Criteria for Compensation

There are two essential prerequisites for eligibility. The first is being a worker within the meaning of the Workers' Compensation and Injury Management Act (the Act). Under the Act, workers are defined as 'any person who has entered into or works under a contract of service or apprenticeship with an employer, whether by way of manual labour, clerical work, or otherwise and whether the contract is expressed or implied, is oral or in writing'.[25] It also applies to certain 'deemed workers'.[26] These are workers who perform work in circumstances that fall outside of the general statutory definition of worker, yet are still considered workers for the purposes of receiving legislation. Deemed workers include members of the clergy,[27] jockeys[28] and Crown workers.[29] Furthermore, the 2011 Amendment had the effect of removing age-based entitlement limitations. As a result, injured workers aged 65 years or older are able to access weekly income payments on the same terms as all other injured workers.[30]

The second prerequisite for eligibility is the suffering of an injury within the meaning of the Act. Accordingly, an injury is 'a personal injury by accident arising out of or in the course of the employment'.[31] Accordingly, this extends to a disease contracted by a worker in the course of his employment at or away from his place of employment, where the workers' employment was a significant contributing factor in contracting the disease.[32]

Note that in April 2016 a new regulation was effected to expand the definition of an 'instrument' which amounts to an 'industrial award' under section 5(1)(d) of the Act.[33] The prescribed instruments are a fair work instrument as defined in the Fair Work Act 2009 (Cth) and an award-based transitional instrument as defined in the Fair Work (Transitional Provisions and Consequential Amendement) Regulations 2009.

Application Procedure for Workers' Compensation

A necessary formal requirement before any application is the provision of written notice to the employer within 12 months of the alleged injury. Notice requires the name and address of the person injured,[34] the cause of the injury expressed in 'ordinary language',[35] the date and place at which the injury occurred,[36] as well as any other information prescribed by

[23] s 100(h).
[24] s 100(e).
[25] s 5(1).
[26] Part II of the Act and Division II of the Act.
[27] s 8.
[28] s 11A.
[29] s 14(2).
[30] s 56(2).
[31] s 5.
[32] s 5.
[33] Workers' Compensation and Injury Management Regulations 1982, reg 3A.
[34] s 178(2)(a).
[35] s 178(2)(b).
[36] s 178(2)(c).

regulations. Claims must be made within 12 months from the occurrence of the injury or, in case of death, within 12 months from the time of death.[37] However, defects or inaccuracy in the notice will not be fatal to the claim, provided that such a defect does not prejudice the employer or was occasioned by mistake, absence from the state or some other reasonable cause. Additionally, failure to provide notice within 12 months will not be fatal for the same reasons. Furthermore, a worker must also attach a copy of their First Medical Certificate (FMC). The FMC is a correspondence from the treating doctor to the employer and his or her insurer outlining his diagnosis of the injury, the likely impact of the injury on the worker's ability to work, and a plan for future appointments and treatment.[38]

The employer then has 5 days to lodge this claim with his or her insurer.[39] The insurer has up to 14 days from the date the claim is lodged to reach a decision on liability.[40] Should the insurer accept liability, the provision of compensation entitlements will commence. Should liability be disputed, recourse to the dispute resolution process will be necessary. The insurer must inform the Director of the Workers Compensation Conciliation Service that the claim is being disputed.[41] Failure by the employer to lodge the claim with his insurer and failure by the insurer to provide notice within the designated time frame will result in penalties.[42]

The Dispute Resolution Scheme

The purpose of the Dispute Resolution Scheme is to ensure a fair and cost-effective system for the resolution of disputes that is timely, accessible, approachable and professional.[43] The scheme was introduced under the Workers Compensation and Injury Management Amendment Act 2011 (WA).[44] The amendments contained within this Act implemented certain recommendations arising from WorkCover WA's 2009 legislative review,[45] and drastically restructured the pre-existing dispute resolution framework.[46]

Workers' Compensation Conciliation Service: Principles, Power and Procedure

An application to the Workers' Compensation Conciliation Service (WCCS) must demonstrate that the parties have made a reasonable attempt to resolve the matter through negotiation.[47] Only once negotiation is completed can a party make an application to the Director of the WCCS. The Director may accept the application or reject it. Rejection occurs if the matter at hand is unsuitable for the WCCS in that it is unlikely to be resolved

37 s 178 (1)(b).
38 WorkCover WA (n 20) 8.
39 s 57A(2A).
40 s 57A(3).
41 s 57A(3a)(b).
42 s 57A(3).
43 s 177.
44 Workers' Compensation and Injury Management Amendment Act 2011 (WA).
45 Explanatory Memorandum, Workers' Compensation and Injury Management Amendment Bill 2011 (WA) 1.
46 ibid.
47 s 182F (1)(b).

through conciliation.[48] If a claim is rejected the Director provides both parties with a certificate to this end, allowing the parties to make an application for the arbitration service.[49]

Accepted applications will be heard before a Conciliation Officer (CO). In accordance with the emphasis on efficiency, fairness and effectiveness that is embedded within the Act, the CO has a duty to act 'fairly, economically, informally and quickly'[50] and 'according to the substantial merits of the case without regard to technicalities and legal forms'.[51]

During the hearing, the CO has the power to require parties to answer questions or provide documents to assist with conciliation.[52] The Act increased accessibility by providing a range of ways in which the matter can be heard; in person, by telephone or video links or solely through the provision of documents.[53] This flexibility provides a more convenient and more accessible service for all parties.

Legal representation may also be utilised by both parties throughout the conciliation conference.[54] However, in an instance where the worker is not legally represented, the CO has the power to refuse the employer his own representation.[55] This potentially reduces imbalances of power between a wealthy employer and a vulnerable worker. Additionally, for parties that face difficulties with the English language, interpreters can be utilised.[56] Finally, and in the same vein, the CO may also direct that compensation payments be made while a dispute is being resolved or in relation to minor matters.[57] If the parties are disputing the nature of the injury suffered, the CO may refer the matter to a medical assessment panel or an approved medical specialist panel (this process will be discussed more below).[58]

The conciliation process will come to a conclusion when the parties have reached an agreement, or where it becomes evident that further discourse would not lead to any further agreement beyond what has already been agreed to (if anything).[59] A time limit of 56 days beginning from acceptance of an application is also imposed, in effect focusing the process and preventing drawn-out discussions where no real conciliation is achieved.[60] At the cessation of the conciliation, a Certificate of Outcome (the certificate) is issued.[61] The certificate outlines the matters that were in dispute at the start of the process, the outcome of the conciliation and the terms of any directions provided during the conciliation. Should the conciliation end in agreement, under section 182N the CO can make a binding order that gives effect to the agreement if the parties consent to such an order.[62] Accordingly, this requires a memorandum of consent.[63]

48 s 182H.
49 s 182H.
50 s 182I(2)(a).
51 s 182I(2)(b).
52 s 182J.
53 s 182V.
54 s 182S(1).
55 s 182S(2).
56 s 182U.
57 s 182K.
58 s 182ZD.
59 s 182O(1).
60 s 182O(1)(c).
61 s 182O(2).
62 s 182N.
63 ibid.

Workers' Compensation Arbitration Service: Principles, Powers and Procedure

An application for arbitration must be made within 28 days of receiving the Certificate of Outcome.[64] The application is considered by the Registrar for the Workers' Compensation Arbitration Service (WCAS), who will only allow such services to be utilised where attempts at conciliation have failed, or the matter was originally determined unsuitable for conciliation.[65]

Arbitration is heard by an arbitrator chosen by the Chief Executive Officer (CEO) of WorkCover.[66] Arbitrators must be a legal practitioner and may be an officer for WorkCover.[67]

When conducting the arbitration, the arbitrator is to act according 'to equity, good conscience and the substantial merits of the case without regard to technicalities and legal forms'.[68] The arbitration process itself is designed to follow principles of 'natural justice'.[69] Whilst this raises questions as to whether such vague terms give too much power to the arbitrator himself, these provisions reflect the Scheme's emphasis on fairness and efficiency. Additionally, in reaching his decision the arbitrator may draw from established common law precedent if the arbitrator thinks it to be 'relevant to the proceeding'. Furthermore, in accordance with its deeply embedded principles of fairness, the arbitrator may treat as legal illegal contracts under which the worker claiming compensation was employed.[70] Such a provision is evidently designed to ensure a fair and equitable outcome for the worker. If the worker was unaware as to the illegality of the contract, it would be a gross injustice to deny his claim due to the invalidity of the contract.

The power of the arbitrator to give directions is especially wide reaching, with the scope of this power extending to 'whatever is necessary for the speedy and fair conduct of the proceeding'.[71] The arbitrator also has the power to confirm, revoke or vary a payment direction made by a CO.[72] Additionally, many of the powers of the arbitrator are similar to those held by the CO. For instance, the arbitrator has the power to request the parties to provide any documentation, material and information to the other party and to the Registrar of the WCAS.[73] If a party fails to do so, they face a penalty of $2,000.[74]

As with conciliation conferences, the arbiter has the power to refuse representation to the employer when such representation is not enjoyed by the worker. Arbitration proceedings also allow for the use of interpreters,[75] and may be conducted in a variety of ways.[76] Throughout the course of the arbitration the arbitrator may also call on witnesses to provide evidence as to the matter at hand.[77]

[64] s 182ZU.
[65] s 182ZU.
[66] s 182ZQ.
[67] s 182ZQ (3).
[68] s 188(2)(b).
[69] s 188(1).
[70] s 192.
[71] s 190(1).
[72] s 190.
[73] s 193.
[74] s 193.
[75] s 197.
[76] s 198.
[77] s 202.

Where new evidence comes to light the arbitrator himself may review the decision. Otherwise the decisions of the arbitrator are final and binding, subject only to appeal to a District Court in certain circumstances. Generally, appeals will only be allowed when a dispute concerns a question of law, ensuring that the court need not deal with vexatious complaints regarding the amount of compensation awarded. Whether an appeal is allowed or not will depend on an examination of the written determination of the arbitrator, with applications for an appeal being made within 28 days from the receipt of written determination by the appealing party. Additionally, the court is restricted with regards to any new evidence it is to consider, which may only be accepted with the leave of the court. Whilst the appeal is being deliberated, the District Court may decide to stay the decision of the arbitrator.[78]

Medical Assessment Panels

Throughout the course of the dispute, if contention arises as to the extent of an injury or the effect of that injury on the worker's capacity to work, the CO or arbitrator may refer the matter to a Medical Assessment Panel (MAP).[79] The MAP can determine the degree of any alleged permanent *loss* to the worker's use of any body part contained within column 1 of Part 1 of the table set out in Schedule 2,[80] and whether or not the employee has contracted AIDS.[81]

The MAP is obliged to act 'speedily and informally, and in accordance with good conscience, without regard to technicalities or legal forms'.[82] This further reflects the emphasis that this system places on efficiency and effectiveness. The MAP is selected by either the Director or the Registrar, depending on how the matter is being heard.[83] The MAP is generally comprised of three members, with at least one of those members specialising in the area of medicine that the injury relates to.[84] A person who has previously treated the worker cannot be a member of the MAP.[85] Once before the MAP, a worker must answer questions and produce any relevant documents for the MAP,[86] as well as submitting to a medical examination.[87] Failure to comply with these requirements can lead to a suspension of the right to compensation.[88]

The MAP has 28 days to make its determination.[89] The determination of the panel is final and binding on all parties and on any court or tribunal.[90]

[78] s 250.
[79] s 182ZD, 210.
[80] s 145A(2) This includes total loss of sight, loss of hands, loss of legs or any permanent loss of the full efficient use of the back.
[81] s 31F.
[82] s 145D(1).
[83] s 145C(1).
[84] s 145C(2).
[85] s 145C(3).
[86] s 145D(2)(b)(c).
[87] s 145D(2)(d).
[88] s 145D(5).
[89] s 145E(2).
[90] s 145E(6).

Approved Medical Specialist Panel

The role of the Approved Medical Specialist Panel (AMSP) is similar to that of the MAP, in that it resolves contention between a worker and an employer where the extent of an injury is being disputed. However, its jurisdiction is over more difficult areas of medicine which thus require a more refined opinion. A matter will be referred where there are concerns of *impairment* (permanent or otherwise) of any body part contained within column 1 of Part 2 of the table in Schedule 2,[91] the degree of an alleged disability, the degree of a workers' permanent whole person impairment (WPI) for the purposes of determining eligibility for a specialised retraining programme,[92] or determining whether an exceptional circumstance exists and a worker is entitled to an additional amount of compensation beyond the amount already awarded.[93]

In these circumstances the matter is determined by an arbitrator rather than a CO.[94] The AMSP is selected by the Registrar, and must contain at least two Approved Medical Specialists (AMSs) who are experts in the relevant field.[95] As with the MAP, the members of the AMSP must not have previously treated the worker. Like the MAP, the powers of the AMSP extend to requiring the worker to answer any question about the injury.[96] Additionally, the worker, the employer and the employer's insurer are obliged to produce any relevant documentation or information should the AMSP deem it necessary. The worker must also submit to an examination.[97] Failure to comply with these requirements will result in financial penalties as well as a suspension of the worker's right to an assessment of his or her permanent impairment.[98]

In making its determination, the AMSP is to act in accordance with the relevant WorkCover guides, namely the WorkCover WA *Guides for the Evaluation of Permanent Impairment*. These guides are largely based on the American Medical Association's *Guides to the Evaluation of Permanent Impairment*.[99] Instructions regarding how to decide these questions are contained within the relevant guides, such as consideration of whether the worker has refused medical treatment,[100] or whether the worker has a pre-existing condition.[101] Accordingly, an AMS must be trained in the use of these guides.[102] In addition to these guides, WorkCover may provide the AMSP with any information relevant to the matter at hand, provided that the worker consents to such a release of information.[103]

A determination requires a unanimous decision by the AMSP.[104] Upon reaching this determination, the AMSP must inform the Registrar as soon as practically possible.

[91] s 31D(4). This includes impairment of sight, impairment of hands, legs and genitals. Accordingly, a worker who suffers such an injury will be eligible for a lump sum payment under s 31C(1). Whenever a worker is seeking a lump sum under that section the matter must be referred to an AMSP.

[92] s 158C(2).

[93] Sch 1, cl 18A(2aa)(a).

[94] s 146F.

[95] s 146K.

[96] s 146M.

[97] s 146M.

[98] s 146M.

[99] Work Cover, Guides for the Evaluation of Permanent Impairment (Report, 2013) 4.

[100] ibid, 9.

[101] ibid, 3.

[102] s 146R.

[103] s 146I.

[104] s 146P.

The Registrar then passes on the determination to the parties. The determination is final and binding.

Benefit of Medical Panels

The use of panels such as the MAP and AMSP, as well as panels such as the Industrial Diseases Medical Panel, has been praised as a way of ensuring impartial outcomes. Niselle notes that the adversarial process inherent within the court system encourages the use of so-called medical 'gladiators' or 'rent-a-quacks'.[105] These are specialists who are selected because their opinion advances the client's case, rather than for their expertise in the particular area.[106]

Industrial Diseases Compensation

Whilst still contained within the general framework of workers' compensation, a different procedure is involved when a worker is applying for compensation for an 'industrial disease'.[107] Under sections 33 and 34 of the Act, these diseases include pneumoconiosis, mesothelioma, lung cancer and diffuse pleural fibrosis. In order for a claim to be successful, two basic elements must be established:

(i) That the worker is less able to earn full wages, or has died from, one of the prescribed industrial diseases.[108]
(ii) That the disease is or was caused by any employment which the worker was employed at any time previous to the date of being so rendered.[109]

Unlike other diseases, there is no time limit for claiming compensation for one of these prescribed industrial diseases. However, the worker claiming compensation must not have left the state for a period of 6 months or been employed interstate for a period of 6 months or more.[110]

This area of compensation law is particularly significant for Western Australia, given its history regarding outbreaks of asbestos-related diseases. This is largely attributed to mining industry practices throughout the 20th century, such as the mining of blue asbestos in Wittenoon in the Pilbara.[111] The death toll is quite significant; between 1960 and 2008, about 1,400 men and 220 women in Western Australia were diagnosed with mesothelioma.[112] The endemic nature of such diseases remains relevant today. Given the long latency for these diseases, it is predicted that the death toll will rise in the coming years.[113] Estimates

[105] P Niselle, 'Stress, Colon Cancer and Workers' Compensation' (2004) 180 *Australian Medical Journal* 341.
[106] ibid, 342.
[107] WorkCover Western Australia, *Industrial Diseases* (20 December 2013) www.workcover.wa.gov.au/Workers/Industrial+Diseases/Default.htm.
[108] Workers' Compensation and Injury Management Act 1981 (WA), s 33(a)(b)(c)(d). Certain dates apply to claims for these diseases: for mesothelioma, claims can only be made for diseases contracted on or after 8 May 1970; for lung cancer, on or after 1993; for diffuse pleural fibrosis, 19 September 2009.
[109] s 33.
[110] s 33(a)(b).
[111] L Nicholson, 'Push to Close WA's Asbestos Compensation Loophole' *WA News* (Perth) 14 April 2014.
[112] ibid.
[113] Gunningham (n 16) 270.

suggest that the incidence of mesothelioma in Australia will peak between 2015 and 2020.[114] This creates a need for knowledge in this area, including understanding the rationale behind any reforms.

Making a Claim

As with all claims for compensation, the worker must first approach a medical practitioner of his own choosing to diagnose his illness.[115] Upon diagnosis the medical practitioner will provide the worker with a signed certificate stating that the worker is unfit for work.[116]

Theoretically an employer can request its own diagnosis by sending the worker to a medical practitioner of the employer or their insurer's choice; unlike other jurisdictions this practice is exceptionally rare in Western Australia. Generally once a worker is diagnosed by a practitioner and has notified the employer of their claim, the matter is referred to a medical panel administered by WorkCover.[117] Diagnosis by a medical panel is compulsory for any claim regarding industrial diseases.[118]

The employer has 14 days from when the worker first lodged the claim to send particulars of the claim to WorkCover WA. The WorkCover WA chief executive officer shall then refer the question of the worker's condition and fitness for employment to a medical panel.[119] Importantly there is no time limit for the prescribed diseases.[120] However, a person is not eligible for compensation if his disease arises from employment in any mine or mining operation.[121] Accordingly, employers who have workers engaged in any mine or mining operation are required to effect cover for industrial disease with the Insurance Commission of Western Australia.

Pre-emptive action may also be taken upon exposure to asbestos through private schemes. For instance, certain law firms allow exposed workers to register their exposure. This can assist in protecting any future compensation claims, particularly in terms of discerning the facts of any claim and attributing liability to a particular employer.[122] The registration form may require the name of the employer, the period in which exposure occurred, the product that the worker was exposed to and whether any preventative measures were taken by the employer to prevent exposure.

Medical Panel

Workers' compensation claims for industrial diseases are determined by an independent medical panel. This entity has complete jurisdiction over all matters regarding the diagnosis of the prescribed industrial diseases as contained within sections 33 and 34 of the Act.[123]

[114] Asbestos and Mesothelioma Association of Australia, 'Mesothelioma', www.asbestosassociation.com.au/about-asbestos/mesothelioma/.
[115] s 57A.
[116] s 57.
[117] s 36.
[118] s 36(1).
[119] s 36.
[120] s 33.
[121] s 47.
[122] Turner Freeman Lawyers, 'Asbestos Disease Compensation Western Australia' (Report, 2011) 18.
[123] s 36.

This medical panel is not to be confused with the aforementioned AMSP or MAP, which have no jurisdiction over matters concerning these prescribed industrial diseases.

The medical panel is generally comprised of two or three physicians who are chosen by the CEO of WorkCover WA.[124] Those chosen are physicians who specialise in diseases of the chest or occupational disease.[125] The operation of the panels is largely independent, with WorkCover WA providing administrative support and funding only.

The panel determines whether the worker is or was suffering from a prescribed disease and, if so, whether the worker is now less able to earn full wages,[126] whether the worker's capacity to exert physical effort has been adversely affected,[127] and if so, to what extent,[128] whether the worker is fit for work, and if so at what level,[129] whether any other disease or condition has led to the worker's injury or death and a number of other related questions.[130] The medical panel may meet with a worker privately, ask the worker questions, require the worker to produce copies of relevant documents, and require him to submit to a medical examination.[131] A medical panel may also require any medical practitioner who has examined the worker to meet with the panel to answer questions and supply any relevant documents

Under section 37, a medical practitioner who has previously examined the worker may come before the panel and present oral submissions.[132] The practitioner is to be paid for as a witness.[133] Such inducement suggests a desire for all relevant information to be provided, and serves to disrupt an imbalance of power that may exist between a worker and his employer. Both parties may also tender any relevant certificates or documentation that will help the panel come to its decision.

Determination of Panels

Following the examination of the medical evidence provided and a review of the worker, the medical panel makes a determination as to whether one of the four listed diseases is present and to what extent this disease impairs the ability of the worker to work. The medical panel does not consider or determine what caused the worker's incapacity.

The medical panel's determination is provided to the worker and the employer. The determination of the medical panel is final, conclusive and binding on the worker and employer.[134] There is no restriction on the number of appearances an applicant may make before the medical panel.

If the medical panel determines the worker is not suffering from one of these four diseases, it does not imply that there is nothing wrong, nor does it prevent the worker from

[124] s 36(1).

[125] s 38(1)(a)(b). At least one physician must specialise in diseases of the chest.

[126] s 38(1)(a)(b).

[127] s 38(1)(d).

[128] s 38(1)(e).

[129] s 38(1)(e).

[130] This contrasts to the system in place for medical panels in NSW, where no guidelines are provided by the relevant authority.

[131] s 38.

[132] s 37.

[133] s 37.

[134] s 38(4).

making a claim for other conditions which may be compensatable under the provisions of the Act.

In 2014/15, the panel convened on 26 occasions and made 49 determinations that indicated the presence of a prescribed industrial disease that rendered the worker less able to earn full wages. Only 6 determinations were made against such a finding.[135]

Employer Liability

The medical panel does not (and should not) make a decision as to whether the employer actually caused the impairment.[136] Causation may be determined by an arbitrator, but more often by court proceedings.[137] The operation of the presumption is dependent upon the disease being listed within Schedule 3 of the Act. If an asbestos-related disease is not contained within Schedule 3, normal causation applies, requiring the worker to prove his employment was a significant contributing factor to their contraction of the disease.[138]

Whilst the operation of the presumption would seemingly make the matter of claiming compensation relatively simple, significant complexities arise due to the long latency periods that asbestos-related diseases have.[139] This leads to situations where a worker has been exposed to asbestos when working for multiple employers, often in different jurisdictions. This leads to immense difficulties in attempting to define the cause of the disease after such a long period.[140] Accordingly, the worker's employment history must be provided to the employer, allowing the employer to dispute the claim by asserting that the exposure to asbestos occurred whilst the worker was employed elsewhere.[141] The persistence of such disputes can literally be fatal to these claims, as the time period between contraction and death for a person with mesothelioma is particularly short.[142] A worker can easily die before the matter is settled and compensation awarded.

Limits of Compensation and the Common Law Loophole

Claiming compensation through the statutory scheme often proves to be a more viable option for workers than any common law option. However, it is evident that seeking compensation through the court system can lead to substantially greater monetary compensation through damages than through the compensation scheme. The current scheme allows workers to redeem weekly wage payments, often in the form of a lump sum. In addition, a worker will be able to claim for all reasonably incurred expenses relating to medical treatment.[143]

135 WorkCover Western Australia, 'Annual Report 2014/15' (Report, 2015) 81.
136 *Ansett Australia Ltd v Medical Assessment Panel* (1998) 19 WAR 395.
137 s 33.
138 ibid.
139 Gunningham (n 16) 279. These periods can range from 15 to 40 years.
140 Ibid, 270.
141 s 41(2).
142 Gunningham (n 16) 270.
143 s 99. Entitlement to medical and like services ceases 52 weeks after the entitlement to weekly payments ceases or, if compensation is payable only for medical and like services, 52 weeks after this entitlement commences—s 99AD(1) and (2). In certain circumstances, entitlement does not cease (eg if the worker's health or

In contrast, at common law a person may seek damages for pain and suffering, damages for loss of expectation of life, damages for economic loss (where a person can no longer work because of their dust disease), and damages for past and future out-of-pocket expenses.

Given that common law claims are more attractive in this regard, it makes it more difficult to attract workers to rely on the scheme rather than entering into the cumbersome court process. A number of limitations exist with regard to common law claims. For instance, as this is a common law matter, compensation will only be provided when it is determined that the employer was negligent. In contrast, the workers' compensation scheme operates on a no-fault basis. Additionally, in order to be eligible for a common law claim, the medical panel must have determined that the worke has suffered a level of impairment of 15%.[144] Furthermore, even if this is accepted, damages will not be unrestricted unless it is established that a person has more than 25% impairment.[145]

Finally, there is the so-called common law 'compensation loophole',[146] which arises because a worker claiming compensation may only claim compensation under the common law once.[147] Accordingly, the claim process operates on a 'once and for all' basis.[148] The issue with this is that the inhalation of asbestos or other dusts causes not one, but a number of latent diseases that will not manifest within the subject until many years after the initial exposure.[149] Such diseases can differ in terms of severity, and will not always manifest at the same time.[150] It follows that a person can claim compensation for a disease such as asbestosis, yet years later if he develops a related (and perhaps more severe) disease such as mesothelioma he can have no recourse to common law.[151] This can lead to situations where a worker is encouraged by his employer to claim damages at common law for a less severe disease (eg asbestosis, chronic bronchitis). Should this occur, this will ensure that no action can be taken against the employer at common law in the instance where the worker contracts a more debilitating, or potentially fatal, disease.[152]

Attempts have been made to rectify this situation, including the introduction of the Asbestos Diseases Compensation Bill 2013. The Bill was to allow a court to 'award damages on a provisional basis for an asbestos disease sufferer's first illness and a further award of damages in the unfortunate event that the person goes on to develop a more serious asbestos related illness'.[153] The Bill was never enacted.

Whilst providing greater access to common law claims is certainly beneficial, perhaps the better solution appears (at a glance) to be to improve the compensation available to workers under an expanded statutory scheme. This will ensure that workers, employers and insurers can diverge from seeking court claims altogether, and have the matter settled quickly, inexpensively and effectively.

ability to undertake activities of daily living or remain at work or surgery is required)—s 99AC(5). If injury is a severe injury or results in death, counselling services will be provided to family members, not exceeding $5,410—s 99(1)(b). Injured workers also have an entitlement to attendant care—s99(1).

[144] ibid.
[145] ibid.
[146] Nicholson (n 112).
[147] s 46(1).
[148] Explanatory Memorandum, Asbestos Diseases Compensation Bill 2013 (WA) 3.
[149] ibid.
[150] ibid.
[151] ibid.
[152] Explanatory Memorandum, Workers' Compensation and Injury Management Amendment Bill 2011 (WA) 1.
[153] Explanatory Memorandum, Asbestos Diseases Compensation Bill 2013 (WA) 3.

Statistics[154]

Claims numbers have been steadily declining for the past 5 years. Figure 3.1 shows the number of claims, broken down by lost time or no lost time, between 2011/12 and 2014/15. This shows a 15% decrease in the total number of claims lodged per year, with a 22% decrease in the number of claims without lost time and an 8% drop in claims involving lost time.

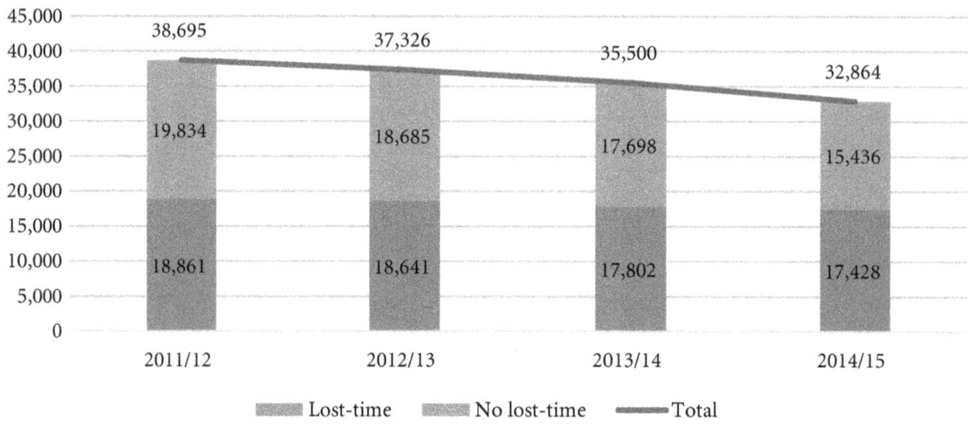

Figure 3.1: Number of claims lodged between 2011/12 and 2014/15 broken down by lost time and no lost time

The majority of lost-time claims are for shorter-duration, 1–59 day, claims. There has been a steady decline in the proportion of shorter-duration lost-time claims, with a corresponding increase in longer-duration claims.

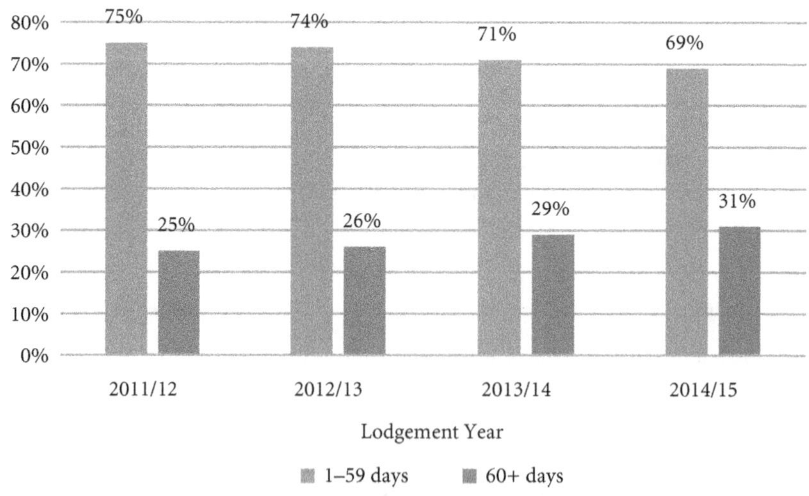

[154] Unless otherwise stated figures in this section are taken from WorkCover WA's Workers' Compensation Scheme Trends October 2015, available at www.workcover.wa.gov.au/content/uploads/2014/09/October-Scheme-Trends.pdf.

Of the lost-time claimants in 2014/15, 80% returned to work either full or part time, with 13% not returning to work due to injury-related reasons. See Figure 3.2. This is an increase in return-to-work rates, up from 76% in 2011/12.

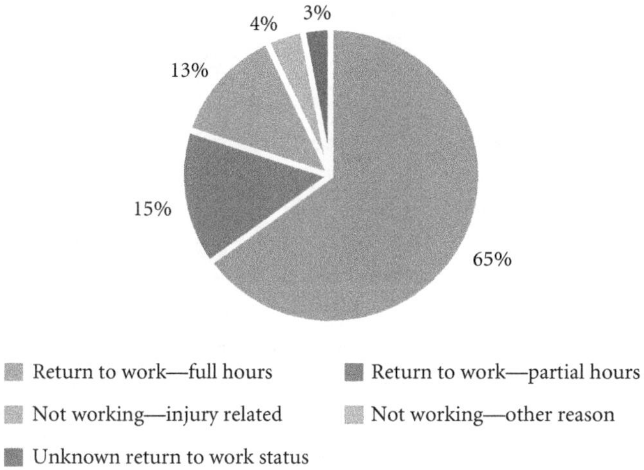

Of the compensation paid out, two-thirds is paid as direct compensation (eg income payments and lump sums) with the remaining third paid to service providers. These proportions have been stable for several years (see Figure 3).

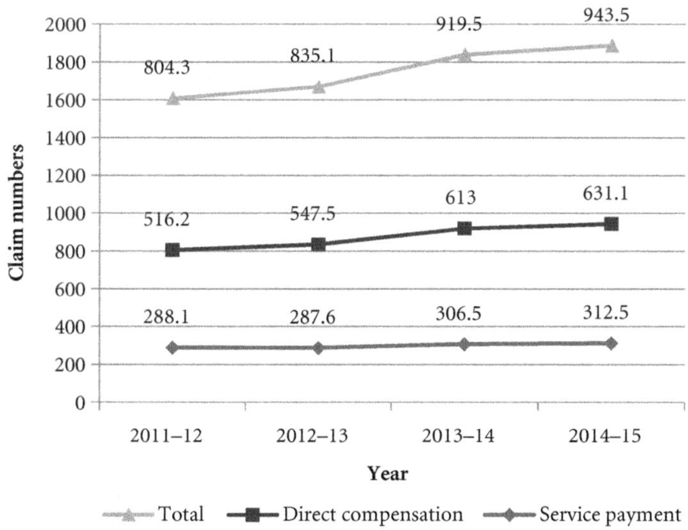

Figure 3.2: Claims payments (adjusted) by recipient type 2011/12–2014/15

The payments to service providers can be further broken down, and again show a relatively stable position over the last 5 years (Figure 3.3).

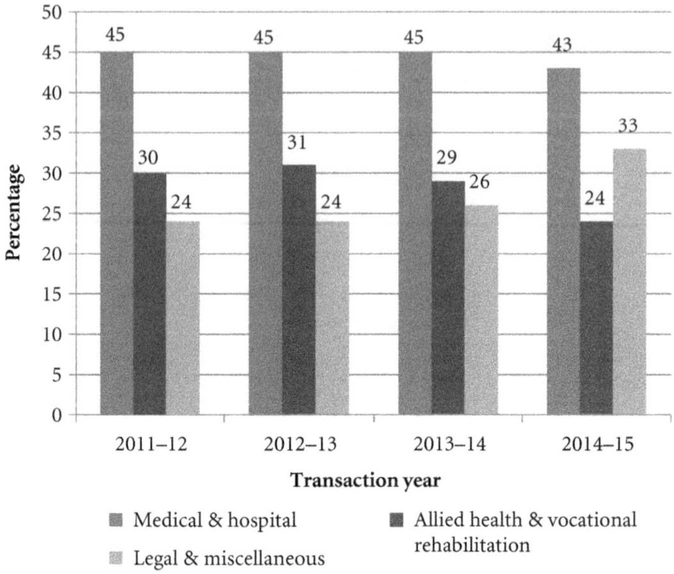

Figure 3.3: Proportions of service claim payments by type 2011/12–2014/15

Dispute Resolution

A claim is considered disputed if it progresses to conciliation and is then resolved or if it progresses to conciliation and then onto arbitration. Disputation rates are the number of 'disputes' lodged per year as a proportion of active claims (defined as a claim on which any payment type has been made in the financial year). Disputation rates are low, but have been increasing since 2011/12 (see Figure 3.4).

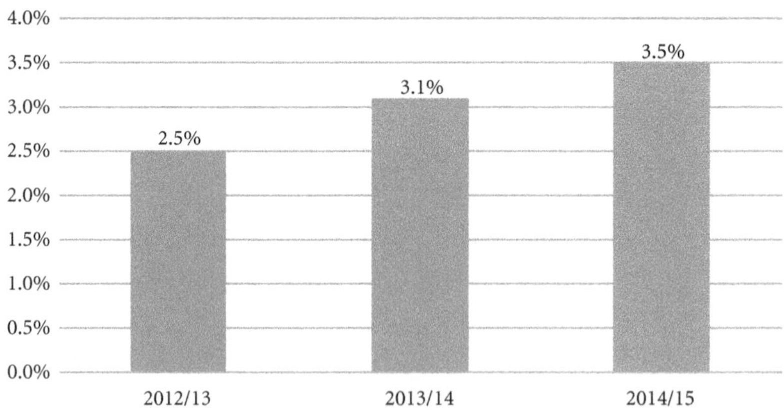

Figure 3.4: Disputation rates from 2011/12 to 2014/15

The majority of claims which became disputes were rapidly resolved in 2014/15, with 95% of disputes resolved within 9 months (see Figure 3.5).

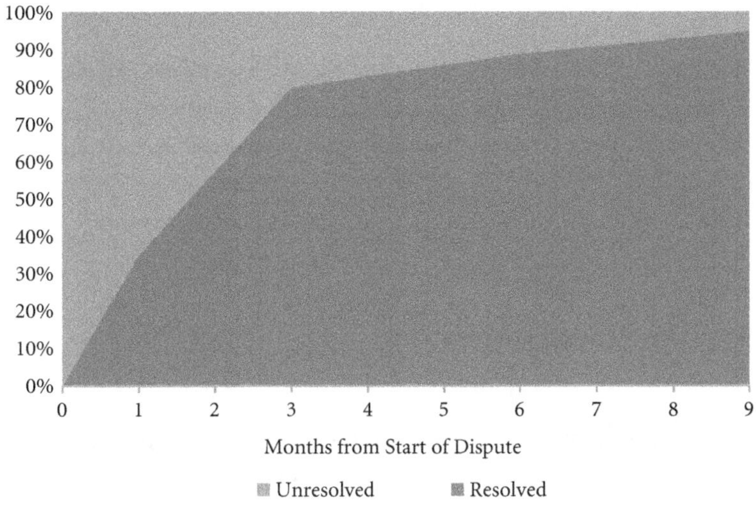

Figure 3.5: Dispute resolution percentages within 9 months of the dispute arising for 2014/15

Financial Information

WorkCover WA is a not-for-profit entity, which has the ability to generate income via payment for services and levies. The financial section of the 2015 Annual Report[155] shows that WorkCover WA is financially solvent, with no reason for concerns over its long-term financial stability. The revenue for 2014/15 was AU$34.854 million with an expenditure of AU$21.566 million. The majority of the income comes from insurer contributions in the form of annual levies on workers' compensation premiums paid by employers and by direct levy on self-insuring employers. See Figure 3.6 for the income breakdown and Figure 3.7 for the breakdown of expenditure.

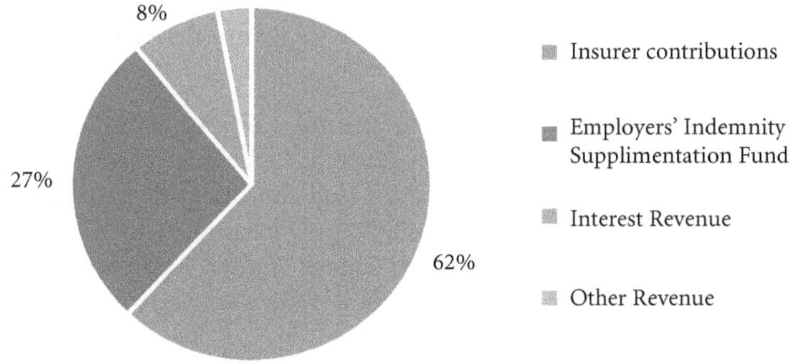

Figure 3.6: WorkCover WA revenue sources 2014/15

[155] Workcover WA, *Annual Report* (2015) 107, available at www.workcover.wa.gov.au/content/uploads/2014/09/Annual-Report-Web-version-Interactive-Section-7-p.104-148.pdf.

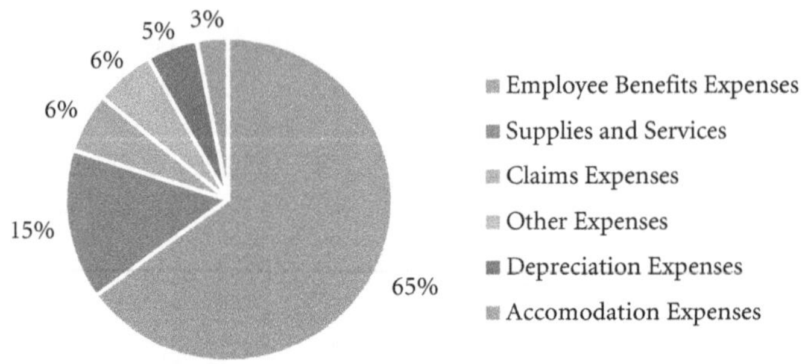

Figure 3.7: WorkCover WA expenditure 2014/15

B. Workers' Compensation Arrangements in the Northern Territory

An Overview

The Return to Work Act (NT) (formally the Workers' Rehabilitation and Compensation Act 2008 (NT)) is the primary instrument which manages the no-fault workers' compensation and rehabilitation scheme in the Northern Territory.

The scheme is administered by NT WorkSafe. NT WorkSafe is the administrative and regulatory arm of the Northern Territory Work Health Authority, the statutory body established under the Workplace Health and Safety Act 2011. The Work Health Authority has the tasks of:

— administering the Act;
— monitoring compliance;
— furthering the Act's objects;
— encouraging rehabilitation of injured workers; and
— collecting and providing information about incidents and compliance under the Act.[156]

The Act establishes the Workers' Rehabilitation and Compensation Advisory Council, whose function is to maintain review of the Act, provide recommendations and advice, and carry out investigations relating to compensation and administration of the Act.[157]

The Act also establishes the Scheme Monitoring Committee, which monitors the viability and performance of the compensation scheme, monitors premium rates, receives submissions and reports on premium rates, and advises the Minister on the basis of the

[156] Return to Work Act (NT) s 6(1). Unless otherwise advised, all references to sections refer to this Act.
[157] s 9.

information it obtains.[158] The Committee is provided with information from the Work Health Authority to assist in performing its functions.[159]

Amending Acts

Following a review of the legislation in 2014, significant amendments were made to the Workers' Rehabilitation and Compensation Act to reduce costs and to ensure the longevity of the scheme.[160] The review made 58 recommendations in its final report to the Government, all of which were accepted with minor variations.[161]

The Workers' Rehabilitation and Compensation Legislation Amendment Bill 2015 came into effect on 1 July 2015. It renamed the legislation in line with the objectives of the Act—to provide effective rehabilitation and compensation for injured workers and to provide for prompt and effective management of workplace injuries in a manner that promotes and assists the return to work of injured workers as soon as practicable.[162] It also:

— recognised that fire fighters are at greater risks of developing certain types of cancers as a result of exposure to hazardous substances while performing firefighting activities;[163]
— aligned the definition of a 'worker' with the Australian Taxation Office's PAYG guidelines;[164]
— extended the 26-week period of entitlement to compensation for workers over the age of 67 to a period of 104 weeks;[165]
— increased the death benefit for dependants of a deceased worker from 260 times to 364 times the average weekly earnings;[166] and
— capped the calculation for normal weekly earnings after 26 weeks.

The second set of amendments was made effective on 1 October 2015 by way of the Return to Work Legislation Amendment Bill 2015. Included were changes to:

— provide for family counselling to assist with rehabilitation;[167]
— ensure reasonable payment for medical and rehabilitation costs where a decision is made to defer liability;[168]
— replace 'reasonable administrative action' with 'management action taken on reasonable grounds and in a reasonable manner' for a defence to a mental injury claim;[169]

[158] s 145.
[159] a 146.
[160] NT WorkSafe, *Changes to the Workers' Compensation Legislation*, www.worksafe.nt.gov.au/LawsAndCompliance/WorkersCompensationLaws/Pages/Changes-to-Workers-Compensation-Legislation.aspx.
[161] ibid.
[162] s 2.
[163] s 50A.
[164] s 3.
[165] s 65(1)(d).
[166] s 62.
[167] s 73.
[168] s 85.
[169] s 3A.

- provide for payment for legal advice incidental to mediation[170] and the involvement of support persons during mediation;[171] and
- permit finalisation of claims by way of a negotiated settlement and lump sum.[172]

Interaction with Common Law Liability

Damages cannot be awarded for a worker or dependant of a worker against an employer, a fellow employee or the nominal insurer for any injuries sustained by the worker.[173] The purpose of this provision is to preclude the bringing of an action for damages for personal injury covered by the Act.[174] When a personal injury arising from employment accrued before commencement of the Act, then a claim can be made.[175] If the injury compensatable under the Act is caused in circumstances which appear to create legal liability in a person other than the employer, the claimant can both take proceedings against that person to recover damages and also pursue a claim for compensation against the employer.[176] If both claims succeed the claimant must repay the employer the compensation that he has paid out up to the value of the damages recovered from the other party.[177]

Coverage and Eligibility

Where a worker[178] suffers an injury[179] which results in or materially contributes to death, impairment or incapacity, the employer must pay a prescribed amount of compensation.[180] The employment that gives rise to the injury must be 'connected with the Territory';[181] although the injury can still be compensatable if the worker is outside the Northern Territory.[182]

The Act also provides that where a person (principal subcontractor) contracts with another person (the subcontractor) for work in the course of the principal contractor's business or trade, the principal contractor is liable to pay compensation as if the contractor were their employee.[183]

[170] s 103FA.
[171] s 103F.
[172] s 78A, 78B.
[173] s 52(1).
[174] s 52(2).
[175] s 189(1).
[176] s 176.
[177] s 176(2).
[178] A 'worker' is a natural person pursuant to s 3 of the Return to Work Act (NT), where a worker performs work or a service of any kind for another person under a contract; and is, in relation to the contract, an employee for the purpose of assessment for PAYG withholding under the Taxation Administration Act 1953(Cth), Sch 1, Parts 2–5; or who is a person, or a member of a class of persons, prescribed for this definition as a worker; note that s 3 provides 8 exceptions to the above.
[179] Injury for the purpose of coverage means a physical or mental injury arising out of or in the course of employment and includes disease and aggravation or recurrence of a pre-existing condition: s 3.
[180] s 53(1).
[181] s 53AA. The principal determinant is whether the worker works in that jurisdiction; is usually based in the jurisdiction; or the principal place of business of the employer is in the jurisdiction.
[182] s 53(2), (3).
[183] s 127.

To be covered, the injury must also arise 'out of or in connection with the employment'.[184] This means that the injury has occurred in circumstances where the worker is at the workplace, while the worker was making a journey for work purposes by the shortest convenient route (including journeys to and from work), or while the worker was taking a break either onsite or offsite.[185] For diseases, a more stringent threshold applies. The employment must have *materially contributed* to the worker's contraction of the disease, or its aggravation.[186] This requires that the employment was the 'real, proximate or effective cause' of the injury or disease.[187]

Specific provisions are made for compensation in respect of contact with bodily fluids. If a worker in the course of his employment comes into contact with another's bodily fluids and has a *reasonable* apprehension he has contracted a disease, the worker is entitled to the reasonable costs of medical testing, and for wages lost due to any absence while being tested from his employer.[188]

Deliberately self-inflicted injuries or injuries attributable to the worker's own serious and wilful misconduct will not be compensated (unless it resulted in death or long-term incapacity).[189] The Act specifies that a worker is not entitled to compensation where the worker was driving while affected by alcohol where the alcohol materially contributed to the injury.[190]

What Benefits are Recoverable under the Scheme?

The Act divides the benefits recoverable under the Act into:

— death benefits;
— total incapacity and loss of earning capacity;
— compensation for permanent impairment;
— medical treatment, rehabilitation and other compensation

Death Benefits

The person responsible for the cost of the funeral is paid a funeral benefit (the lesser of either the cost of the funeral or a prescribed amount[191]).[192] The worker's dependants are paid an amount equal to 364 times average weekly earnings at the time the payment is made.[193]

184 s 3.
185 s 4(1).
186 s 4(6A).
187 s 4(8).
188 s 53A.
189 s 57.
190 s 60(1).
191 The prescribed amount is 10% of the annual equivalent of average weekly earnings: Reg 6, Workers Rehabilitation and Compensation Regulations (NT).
192 s 62(1)(a).
193 s 62(1)(b). Note also that the proportion of distribution among dependants varies depending on the number of dependants. Alternatively, proportions can be determined on application to the Court having regard to the relative needs and dependency of the dependants.

In addition to this amount, each child[194] of the worker is paid 10% of weekly average earnings, capped at 100% for 10 or more children.[195]

Total or Partial Incapacity

Where a worker has suffered from a work injury causing incapacity, he or she is entitled to compensation. The compensation is the difference between what the injured worker actually earned in a week and his/her normal weekly earnings before the injury.[196] This provision covers the worker's loss of earning capacity resulting from the injury for the first 26 weeks of incapacity.

After these first 26 weeks, the injured worker is entitled to compensation amounting to the lesser of either:

— 75% of the worker's loss of earning capacity; or
— 150% of average weekly earnings.[197]

The calculation of an individual's average weekly earnings is capped at 250% of the average weekly earnings at the time.[198]

Permanent Impairment

Compensation will only be provided where the impairment affects at least 5% of the person.[199] This is assessed by a medical practitioner, but can be reassessed on application to the Work Health Authority, who may refer the application to a panel of 3 medical practitioners.[200] The process presupposes that the permanent impairment being assessed was caused by a compensatable injury for which either the employer has accepted liability or the Work Health Court has made a determination.[201]

The compensation amount is expressed as a percentage of 208 times the worker's average weekly earnings. Where a worker's permanent impairment is assessed at 15% or more, the worker will be paid compensation equal to that assessed percentage of 208 times average weekly earnings. For impairments of 85% or above, compensation will be 100% of 208 times average weekly earnings. For permanent impairment assessed between 5% and 14%, the following table applies:[202]

Column 1	**Column 2**
Degree of permanent impairment	Percentage of 208 × average weekly wage
≥5% and <10%	2
10%	3

(continued)

[194] s 63(3). Child means a child of the deceased worker who has not attained 16 years; or has attained that age but is not yet 21, and is studying or handicapped.
[195] s 63.
[196] s 64.
[197] s 65(1)(b).
[198] s 65(3).
[199] s 70.
[200] S 72.
[201] *Taylor Enterprises v Pointon & Work Health Authority* [2009] NTMC 029, [29].
[202] s 71.

Column 1	Column 2
11%	4
12%	6
13%	8
14%	12

Medical Treatment, Rehabilitation and Other Costs

There is no maximum amount recoverable for the *reasonable* costs of medical treatment following a work-related injury. An employer is liable to pay the costs *reasonably* incurred by the worker as a result of the injury for:

- medical, surgical and rehabilitation treatment;
- hospitalisation and hospital treatment;
- travel for the purposes of the above;
- necessary nursing or personal attendance; and
- accommodation for the purpose of medical, surgical or rehabilitation treatment.[203]

An employer must take reasonable steps to provide the injured worker with suitable employment and, so far as is practicable, to participate in efforts to retrain the worker.[204]

How are Claims Made?

Notice of the relevant injury must have been given to or served on the worker's employer as soon as practicable.[205] Compensation claims must be made in the approved form, and then be served on the employer.[206] Unless it is a claim in respect of compensation for death or under section 73 it must be accompanied by a certificate from a medical practitioner. Defects, omissions or irregularity in the claim will not affect its validity unless it relates to information that is not known by or ascertainable by the employer.[207]

The employer can choose to accept, defer or dispute liability for the compensation and has 10 days from receipt of the claim to notify the claimant of their decision.[208]

Regardless of their decision on liability the employer must request the written opinion of a medical practitioner who is treating or has treated the worker.[209] An employer can require a claimant to attend a medical practitioner (provided and paid for by the employer) for examination at reasonable intervals.[210] If the worker unreasonably refuses, the employer may reduce or cancel the compensation until the examination takes place.[211]

[203] s 73.

[204] s 75A. Note also the maximum penalty for non-compliance is 25 penalty units in the case of a body corporate, and 8 penalty units or imprisonment for 3 months for a natural person.

[205] s 80.

[206] s 82(1). Note that if the claim is against a Nominal Insurer, the form must be served on the Nominal Insurer: s 168.

[207] s 82(3).

[208] s 85(1).

[209] s 90B.

[210] s 91.

[211] s 91(2).

Mechanisms for Disputes

Where there is a dispute as to entitlements, liability or payment under the scheme, the parties[212] can first go to mediation,[213] and then to the Work Health Court.[214] Appeals can be made to the Supreme Court, but only on questions of law.[215] Before resorting to the courts, the claimant must show that an attempt at resolution was made through mediation.[216]

Insurance Arrangements under the Scheme

Insurers who wish to become approved insurers under the Act, or employers who wish to self-insure, may apply for approval in writing (for a specified fee)[217] to the Work Health Authority.[218] In determining whether to approve such applications, the Authority will take into consideration:[219]

— the insurer's/employer's ability to provide the necessary service;
— the financial viability of the insurer/employer'
— the insurer's/employer's ability to provide information required or likely to be required under the Act; and
— such other matters that are prescribed.

Every employer (who is not an approved self-insurer) must obtain an insurance policy from an approved insurer for all its workers.[220]

If a liable employer defaults on payment for over a month, the entitled claimant can claim directly against the employer's approved insurer.[221]

If an employer has committed a misrepresentation, or breached a term of the policy, a provision of the Act, or a law in force in the Norther Territory relating to employment, the insurer may recover the amount they paid to the claimant and any reasonable costs they incurred from the employer.[222]

[212] s 103F: parties is defined as the claimant, the employer and the employer's insurer.

[213] s 103B–E. Also note that unless the Court orders otherwise, the parties must bear their own costs of mediation: s 103G.

[214] ss 93–110B.

[215] ss 115, 116.

[216] s 103J(1).

[217] s 121A. This fee is determined by the Authority, and contributes towards the administration costs of proceedings in the Work Health Court and the Supreme Court or in providing a mediation service.

[218] There are five active Approved Insurers in the Scheme: Allianz, TIO, CGU (IAL), QBE and GIO (Suncorp). There are four active Self Insurers: Westpac, Catholic Church, Woolworths and Coles Myer: see NT WorkSafe, 'Actuarial Review of the NT Workers' Compensation Scheme as at 30 June 2012' (2013) 4.

[219] ss 119 and 120.

[220] s 126(1). The maximum penalty for non-compliance is 255 penalty units. Reg 15 specifies that the employer's insurance policy must be to the amount of no less than AU$2,000,000. Self Insurers approved under the Act are Catholic Church Insurance Ltd, Wesfarmers, Westpac Group and Woolworths: see NT Worksafe (n 219).

[221] ss 132 and 133.

[222] s 136(1).

The Act establishes the Nominal Insurer Fund[223] (administered by the Nominal Insurer):

- to protect injured workers whose employers do not have a workers insurance policy; and
- to protect employers and injured workers where the insurers default in payment of compensation under the scheme.[224]

Where the Nominal Insurer pays compensation arising from the employer's failure to obtain adequate insurance, the employer must pay any amount required under the Act and pay to the Nominal Insurer an amount equal to the highest premium payment for the period where there was no cover.[225]

Where an approved insurer is unable to make a payment for compensation under the Act, the Nominal Insurer shall take on the same rights, powers, duties and liabilities as if they were the approved insurer.[226]

Where a claim has been made against a liable employer, the person entitled to compensation may make a claim against the Nominal Insurer where:

- the employer defaults in payment of compensation; or
- the employer is dead, cannot be located or, in the case of a company, has been wound up, and the approved insurer cannot be identified.[227]

When such claims are made against the Nominal Insurer, the claim is determined as if the insurer were the employer. If the claim is successful the liability of the employer to pay compensation is discharged, except in respect of any amounts already paid.[228] Henceforth, the amount paid by the Nominal Insurer (including the costs of conducting the claim) becomes a debt due to the Nominal Insurer from the employer.[229]

A claim for compensation against the Nominal Insurer cannot be made by a director of a body corporate in respect of an injury arising in the course of employment, or by an immediate family member of the injured worker.[230]

Funding of the Nominal Insurance Fund

The Fund consists of:

- the balance of the former Fund;
- contributions paid by approved insurers, self-insurers and employers;
- moneys recovered by the Nominal Insurer in the course of its functions and powers;
- fines or pecuniary penalties paid under Part 7 (Insurance) of the Act;[231] and
- such other moneys lawfully paid into it.[232]

223 ss 150 and 162.
224 s 150.
225 s 172(3).
226 s 137.
227 s 167.
228 s 170.
229 s 172(2).
230 s 173.
231 s 177.
232 s 162.

The Nominal Insurer may also borrow funds, on approval by the Minister, to be paid into the Fund.[233]

The Nominal Insurer determines, subject to ministerial approval, an amount that each approved insurer and self-insurer must contribute to the Fund.[234] The amount is a percentage of the premium income in the previous financial year (in the case of an approved insurer); or is a percentage of the premium that would have been payable by a self-insurer if they had obtained an insurance policy (in the case of a self-insurer).[235] The Minister may also determine that all employers are required to contribute to the Fund.[236]

Payments are made out of the Fund to cover the amount from claims against the Nominal Insurer, the costs of the Fund's administration and such other moneys as may be lawfully paid out of it.[237] Where the amount of the Fund is insufficient to meet a payment, the Treasurer may make temporary advances to the Fund out of the Public Account (within the meaning of the Financial Management Act).[238] The amount borrowed shall be paid back to the Public Account with interest (determined by the Treasurer) progressively as money is paid into the Fund.[239]

Statistics

Table 3.2 was compiled as part of an actuarial review (prepared by PWC) of the Northern Territory Workers' Compensation Scheme (the Scheme) as at 30 June 2015.[240] The table shows the funding ratio for the Scheme for Approved Insurers and Self-Insurers both separately and combined.

Table 3.2: The funding ratio of the Scheme (Total, Insurers and Self-Insurers) as at 30 June 2015

	Actual provisions, AU$000s (a)	**PwC central estimate, AU$000s (b)**	**Difference, AU$000s (b)–(a)**	**Funding ratio, % (a)/(b)**
Insurers	352,159	322,687	–29,471	109
Self-Insurers	6,922	4,464	–2,457	155
Total	359,080	327,152	–31,928	110

The funding ratio represents a measure of the ability of the Scheme in aggregate to meet its benefit commitments. However, an aggregate funding ratio less than 100% in this context does not necessarily mean that individual insurers may not be able to meet their commitments to the Northern Territory Scheme. As shown in Table 3.2, the funding ratio

233 s 162(4).
234 s 164A(1).
235 s 164A(1).
236 s 164B.
237 s 163.
238 s 166.
239 s 166(3).
240 PWC, *NT Worksafe: Actuarial review of Northern Territory Workers' Compensation Scheme as at 30 June 2015*, available at www.worksafe.nt.gov.au/AboutUs/Documents/workers-comp-actuarial-review-20150630.pdf.

for Approved Insurers is marginally more than 100%. This indicates the Scheme is fully funded. This is an improvement on the position in 2012 when the total funding ratio was a just 79.3% with an approved insurer rate of 78.5%. These figures indicate that there is some volatility in scheme funding ratios, but provided the current trend remains, the scheme is adequately funded.

Reinsurance

In order to mitigate the risk of large claims, individual insurance providers will take out reinsurance. The exact nature of these reinsurance contracts is known only to the parties to the contract; it is merely noted here that reinsurance exists as a mechanism to reduce the risk to an approved insurer of not being able to meet their obligations. Self-insurers do not take out reinsurance.

Approved Insurer Statistics

Claim Count by Accident Year

Figure 3.8 shows the claim count by accident year for approved insurers.[241] It can be seen that while there is some year-to-year volatility, the overall trend is fairly stable.

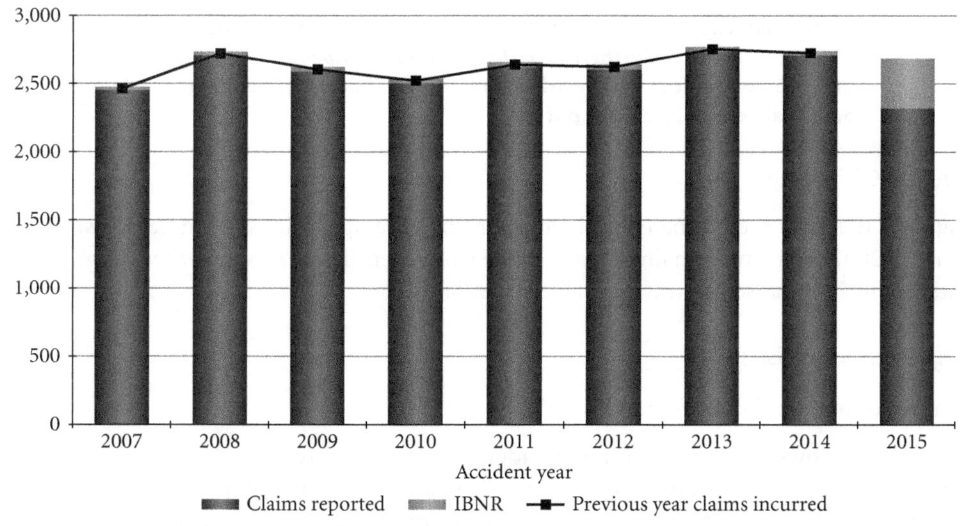

Figure 3.8: Claim numbers incurred by year for insurers

Claim numbers are a relatively crude indicator as they do not incorporate any element of employment levels, wage increases, etc. The claim frequency provides a more sensitive measure. It is calculated as the number of claims incurred divided by the estimated number of full-time equivalent employees. To estimate the number of employees the actual wages paid are inflated to 2015 levels and divided by the average Northern Territory weekly wage.

241 ibid, 10.

Figure 3.9 shows the claim frequency per $77,419 of real wages, and demonstrates the continuation of a downward trend in claim frequency.

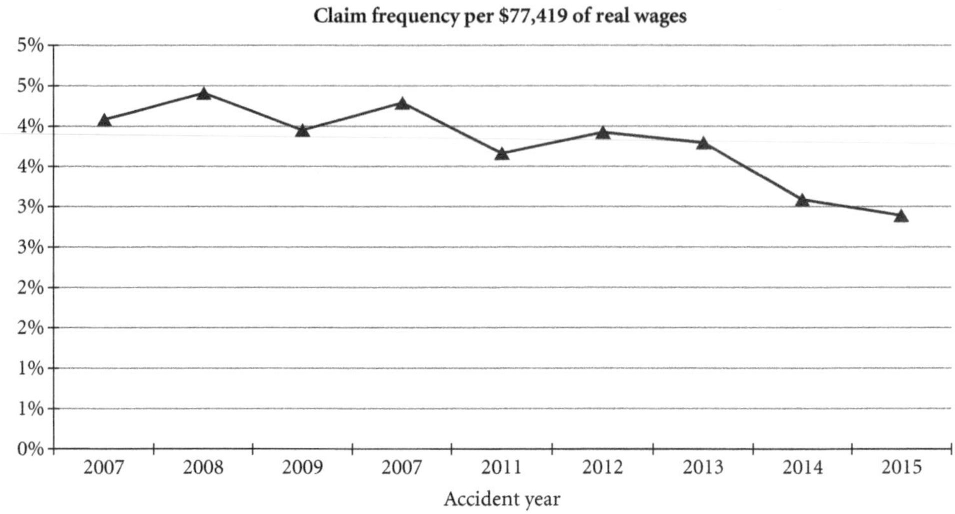

Figure 3.9: Claim frequency per $78,862 of real wages 2006–14. Source: PWC, *NT Worksafe: Actuarial review of Northern Territory workers' compensation scheme as at 30 June 2015*, 11. Available at www.worksafe.nt.gov.au/AboutUs/Documents/workers-comp-actuarial-review-20150630.pdf

Average Claim Size

Figure 3.10 shows the estimated average claim size. Average claim sizes increased substantially in 2007, and have remained fairly stable since then. There was, however, a significant increase in 2015 caused by unfavourable large claims.[242]

Self-Insurer Statistics

Claim Count by Accident Year

Figure 3.11 shows the claim count by accident year for self-insurers. This shows that there is some year-to-year volatility, with a peak in 2011 and a decline since then. Obviously there is some difficulty in estimating claim numbers for recent years due to the incurred, but not reported (IBNR) claims. However, the 2015 figures indicate a slight increase compared to previous years.[243]

Claim numbers are a relatively crude indicator as they do not incorporate any element of employment levels, wage increases, etc. However, for self-insurers the low numbers of claims per year would make any modelling of claim frequency less robust.

[242] ibid, 12.
[243] ibid, 16.

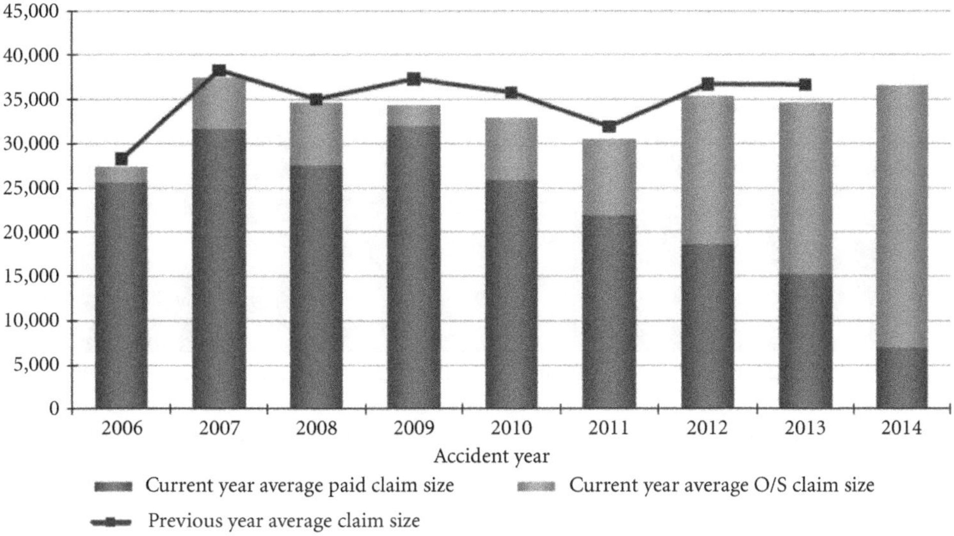

Figure 3.10: Gross average insured claim size (inflated to 30 June 2014 values) for 2006–14

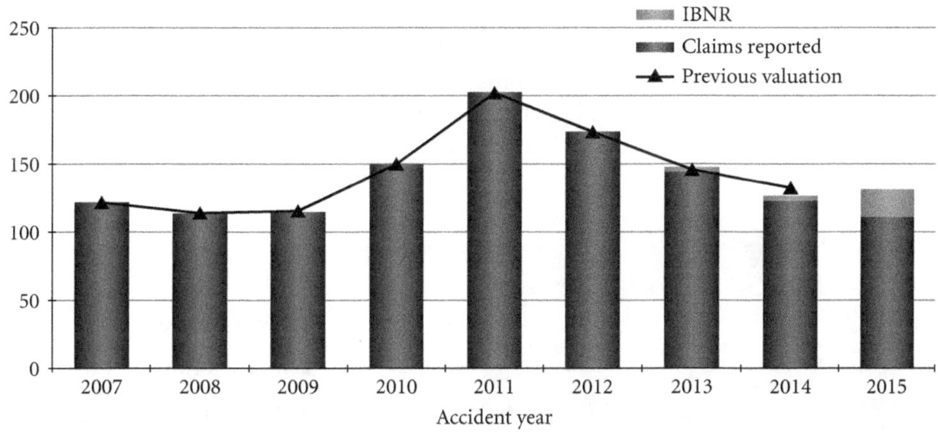

Figure 3.11: Claim numbers incurred by year for self-insurers

Average Claim Size

Figure 3.12 shows the estimated average claim size. Average claim sizes are more variable for self-insurers, and have peaked in 2008 and in 2012, with troughs between. Note that the 2015 average claim size is $15,800 which is a $2,700 (21%) increase on the previous year.[244]

[244] ibid, 17.

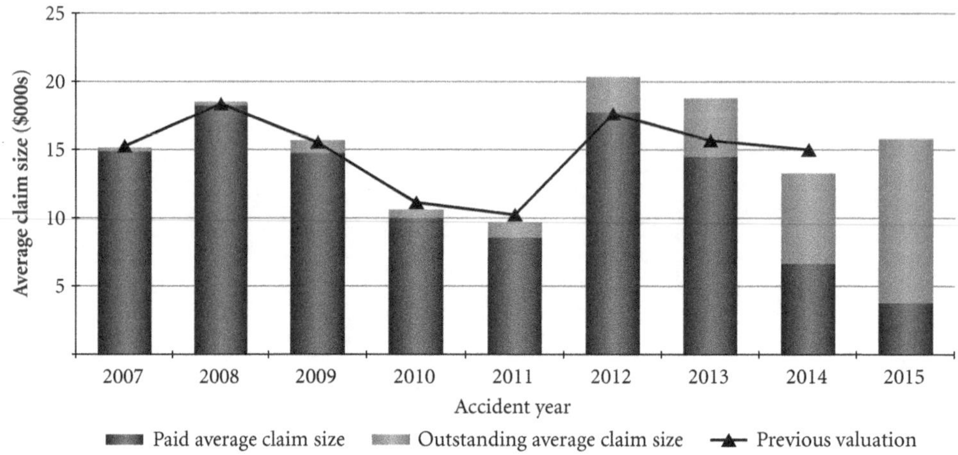

Figure 3.12: Gross average self-insured claim size (inflated to 30 June 2014 values) for 2006–14

Overall Statistics

There are substantial differences between the insured and the self-insured claims, most notably in the claims size, with an average self-insured claim for 2014 being for just under $14,000 compared to around $36,000 for insured claims.

The overall breakdown of costs for claims from 2012 to 2014 is shown in Figure 3.13; this demonstrates that around a third of the incurred costs are spent on weekly benefits, with another third spent on redemptions and non-economic lump sums. The remaining

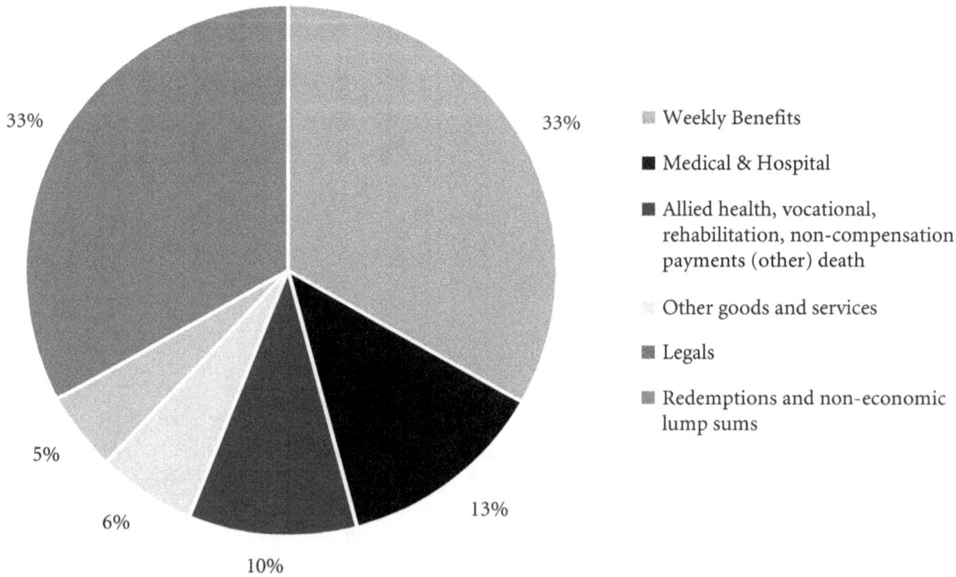

Figure 3.13: Average breakdown of incurred costs by service type 2012–14

third is divided between medical/hospital, allied health/vocational/rehabilitation/non-compensation payments, other goods and services and legal costs.[245]

Table 3.3 indicates that there was a shift in the balance of average incurred costs between 2006–08 and 2012–14, with a slight decrease in the percentage paid as weekly benefits and an increase in redemptions and non-economic lump sums between these points. This data may indicate that period payments were dropping out of favour, with a corresponding increase in lump sum payments, which may prove an important cost driver for the scheme. However, with the addition of the 2015 data,[246] there is no evidence to support either of these trends.

Table 3.3: Breakdown of incurred costs (%) by service provision 2006–08 and 2012–14

	2006–08	**2012–14**	**2015**
Weekly benefits	36.8	33.2	37
Medical and hospital	13.6	12.6	15
Allied health, vocational, rehabilitation, non-compensation payments (other) death	11.7	10.3	12
Other goods and services	5.4	6.0	5
Legal costs	5.7	4.8	5
Redemptions and non-economic lump sums	26.8	33.0	25
Total	100.0	100.0	100

Conclusions

The funding ratio indicates that the scheme is solvent for both approved insurers and self-insurers. Claims-handling expenses are low in the schemes with assumptions for future claims-handling expenses set at 6% for insured claims and 7% for self-insured claims. All indicators point to the scheme being efficient, functional and sustainable in the long term.

C. Worker's Compensation Arrangements in the Australian Capital Territory[247]

Overview

The Workers' Compensation Act 1951 (the Act) governs the workers' compensation scheme in the Australian Capital Territory. The Act aims to provide timely, safe and durable return

245 ibid, 88.

246 ibid.

247 Unless otherwise advised, all references to sections under this heading are to the Workers' Compensation Act 1951 (ACT).

to work through effective injury management and income support for injured workers.[248] It works in conjunction with occupational health and safety legislation which seeks to reduce the human and economic cost of work-related injury through improvements in the management of occupational health and safety in the workplace.

WorkSafe ACT is the government agency responsible for the administration of the workers compensation legislation. WorkSafe ACT provides an advisory and inspectorate function on a range of workers' compensation and occupational health and safety issues.[249]

Workers' compensation in the ACT is a privately underwritten and market-driven scheme. Insurers apply to WorkSafe ACT to become 'Approved Insurers' to offer workers' compensation coverage in the ACT. As at 1 July 2016, there were seven approved insurers and six self-insurers operating under the legislation.[250]

The Workers Compensation (Cross-border Workers) Amendment Act 2014 (ACT) recently ensured that the ACT legislation is in conformity with nationally agreed cross-border procedures and sufficiently displaces contrary common law precedents.[251]

Interaction with Common Law Liability

A person entitled to compensation under the Act may also take proceedings to recover common law damages, in certain circumstances. If the injury creates liability in someone other than the employer, the person may take a proceeding against the person to recover damages and may also make a claim against the employer under the Act.[252] However, if the worker receives both amounts, the amount of the damages recovered must be repaid to the employer.[253]

Compensation under the Act is not payable to the extent that a judgment or agreement for damages has been obtained in relation to the injury.[254]

Insurance Arrangements under the Scheme

Employers must maintain a compulsory insurance policy[255] with an approved insurer.[256] The director-general of WorkSafe ACT may give the employer a notice requiring the employer to obtain a compulsory insurance policy within 10 business days after the notice is given (the 'compliance period').[257] If the employer still does not hold compulsory insurance

248 Workers' Compensation Act 1951 (ACT).

249 www.accesscanberra.act.gov.au/app/answers/detail/a_id/2198.

250 www.accesscanberra.act.gov.au/app/answers/detail/a_id/3170#Current Approved Insurers.

251 Safe Work Australia, *Comparison of Workers' Compensation Arrangements in Australia and New* Zealand (2015) available at www.safeworkaustralia.gov.au/sites/swa/about/publications/pages/comparison2015.

252 s 183.

253 s 183.

254 s 184.

255 'Compulsory insurance policy' means an insurance policy for an unlimited amount for any liability of the employer for an injury or death of each territory worker employed by the employer: s 144.

256 s 147. Note the maximum penalty for non-compliance is 50 penalty units (s 147A).

257 s 147(3). The maximum penalty for non-compliance with this notice is 50 penalty units: s 147A(3).

at the end of the compliance period, the director-general may again require the employer to obtain insurance within 10 days.[258] If the employer still does not comply after this time expires, the director-general may give notice directing the employer to stop conducting its business.[259]

Where an employer fails to maintain a compulsory insurance policy, the director-general must determine the premium that would have been payable to an approved insurer if the employer had maintained a compulsory insurance policy for the period the employer was not insured (up to a maximum of 5 years).[260] The director-general may determine an amount up to double the avoided premium amount,[261] which may be recovered as a debt owing by the employer to the Default Insurance (DI) Fund. The Act establishes the DI Fund.[262] The DI Fund is managed by the DI fund manager[263] and money paid into it is taken to be trust money under the Financial Management Act 1996.[264] The purpose of the Fund is to pay compensation to an injured worker if the injured worker's employer does not have a compulsory insurance policy covering the worker for the claim.[265] This cannot be done if the employer is a self-insurer.[266]

An approved insurer must not refuse to issue a compulsory insurance policy required by an employer unless the employer has not paid for the policy; or the employer has not given the insurer information reasonably requested by the insurer in relation to the policy.[267]

Coverage and Eligibility

The Act provides for compensation to be paid to a worker[268] by his employer if the worker suffers personal injury[269] resulting from his employment.[270] However, for injuries caused by an industrial disease the requisite causality is only established if the employment substantially contributes to the injury.[271,272] Cases where the injury was suffered partly or

[258] s 147(4). The maximum penalty for non-compliance with this notice is 250 penalty units: s 147A(4).

[259] s 147(5). The maximum penalty for non-compliance with this notice is 250 penalty units: s 147A(5).

[260] s 149(3).

[261] In determining the amount, the director-general has regard to various factors including: whether payment of the premium would cause the employer financial hardship; whether the director-general is likely to recover the amount; the employer's history of compliance with the Act; and steps the employer has taken to obtain a compulsory insurance policy: s 149(4).

[262] s 166.

[263] Appointed by the director-general: s 166C.

[264] s 166(3).

[265] s 166A.

[266] s 149. An employer can become a self-insurer once certain requirements are met: see Part 10 of the Regulations, in particular, Reg 87 Workers Compensation Regulation 2002.

[267] s 152. The maximum penalty for non-compliance is 100 penalty units.

[268] A 'worker' means an individual under a contract of service, whether the contract is express or implied, oral or written: s 8(1). However, public servants are not 'workers' under the Act and are therefore not covered: s 9(1).

[269] 'Injury' under the Act means a physical or mental injury (including stress), and includes aggravation, acceleration or recurrence of a pre-existing injury: s 4.

[270] s 31(1).

[271] s 31(2).

[272] s 27.

completely because the worker had a specified pre-existing condition also require that the employment substantially contributed to the injury.[273]

Injuries sustained on an 'employment-related journey' are considered sufficiently connected to the worker's employment.[274] Employment-related journeys include a journey between the worker's home and workplace; and between the worker's home and an educational institution being attended for work-related purposes.[275]

Compensation will only be payable under the Act if the employment is connected with the ACT.[276] This is essentially that the worker usually works or is based in the ACT or the employer's principal place of business is in the ACT.[277] If the worker has already received compensation for the injury under the worker's compensation law in another State or Territory, or outside Australia, they may not double-dip and recover compensation under the ACT Scheme.[278]

Compensation is *not* payable if the injury or death of the worker is caused by an intentionally self-inflicted injury.[279] Compensation is not payable if the injury occurred because of the worker's serious and wilful misconduct, unless the injury results in death or serious and permanent disablement.[280] Compensation is not payable for a period that the worker is imprisoned because they have been convicted of an offence.[281]

What Benefits are Recoverable under the Scheme?

The Act divides the benefits recoverable into:

- weekly payments (Part 4.3);
- compensation for permanent injuries (Part 4.4);
- compensation for medical treatment, damage and other costs (Part 4.5); and
- compensation for death (Part 4.6).

Weekly Payments

If a compensatable injury leaves a worker partly or wholly incapacitated, the worker is entitled to weekly compensation for up to 26 weeks[282] from the date of the injury.[283] For total incapacitation, compensation is paid at the worker's pre-injury weekly earnings.[284]

[273] s 31(3). Pre-existing conditions include diseased heart valve, coronary artery disease, aortic aneurism, cerebral aneurism or any other condition prescribed by regulation under the section.

[274] s 36(1).

[275] s 36(2).

[276] s 36B, 36A.

[277] For other indications of connection where the Territory connection remains unclear, see s 36B.

[278] s 36F.

[279] s 82(2).

[280] s 82(3). This includes where the worker was under the influence of alcohol or a prescribed drug and the alcohol or drug contributed to the injury: s 82(4).

[281] s 83.

[282] A different period applies if the injured worker is over 63 years old: see s 39(3).

[283] ss 39 and 38.

[284] s 39(4)(a).

For partial incapacitation the amount is the difference between the pre-injury weekly earnings and the average weekly amount the worker is earning or could reasonably earn.[285] After the initial 26 weeks the worker is entitled to a proportion of the compensation amount they were receiving within the period.[286]

Entitlement to weekly compensation for total incapacity stops when the worker stops being totally incapacitated, returns to work or dies.[287] Partial incapacity weekly compensation ceases when the worker stops being partially incapacitated or dies.[288] The weekly compensation can resume if the worker becomes totally or partially incapacitated again.[289]

Permanent Injuries

The worker is entitled to a percentage of AU$100,000.[290] The percentage of compensation for permanent injuries varies according to categories of (and subcategories of) speech loss, sensory loss, hearing loss, loss of vision, arm injuries, hand injuries, leg injuries, foot injuries, bowel injury, loss of sexual organs, brain damage, permanent impairment of back, neck or pelvis, disfigurement and disease.[291] A worker who has suffered 2 of the above loss types is not entitled to receive more than $150,000.[292] Where a worker has suffered loss of a proportion of an above function, the worker is entitled to that proportion of the total loss.[293] If a proportion of the total loss is attributable to a previous injury or a pre-existing condition a deduction will be made from the compensation to account for this.[294] If the injury causes the death of the worker within 3 months of sustaining the injury some type of loss cannot be claimed for.[295]

Medical Treatment

The costs of reasonable medical treatment, of resulting reasonable alterations to the worker's residence, of rehabilitation services, and the reasonable cost of replacing or repairing damage to clothing sustained from the injury are available from the employer.[296] The worker must give notice to the employer of the amount and reasonable details of compensation sought.[297] However, failure to do so will not affect a claim if the failure was reasonable or if the error did not affect the employer's position.[298]

Before making payment, the employer may require the worker to be assessed by a doctor chosen by the employer.[299]

[285] s 39(4)(b).
[286] For total impairment proportions, see s 41; for partial impairment proportions, see s 42.
[287] s 43(1).
[288] s 43A(1).
[289] s 43(2) and s 43A(2).
[290] ss 49 and 51.
[291] s 48, Sch 1 of the Act.
[292] ss 50 and 54.
[293] s 59.
[294] s 61(1).
[295] s 68. These losses are loss of sense of taste, smell, loss of sexual organs, loss of breasts, loss of capacity to engage in sexual intercourse, severe facial/ bodily disfigurement.
[296] s 70.
[297] s 71(1).
[298] s 71(2).
[299] s 72.

Death

With fatal work-related injuries, the dependants are entitled to the following:

— a single lump sum payment of $150,000 to be divided between the dependants;[300]
— weekly compensation of $50 for each child dependant;
— the funeral expenses of the worker to a maximum of $4,000.[301]

Other Requirements on Workers, Employers and Insurers

The Act also specifies that an insurer[302] must establish, maintain and give effect to an injury management programme,[303] which aims to achieve the best results for a timely, safe and durable return to work.[304]

An employer has obligations under the injury management programme to give notice of injury to the insurer,[305] to take part and cooperate in establishing a personal injury plan for a worker,[306] to provide suitable work for an injured worker,[307] to establish a return-to-work programme,[308] and to provide the worker with vocational rehabilitation.[309]

However, the above requirements do not apply to the DI Fund.[310] An employer may face a maximum penalty of 10 penalty units for non-compliance with the above requirements.[311]

The injured worker has obligations to take part and cooperate in the establishment of the plan,[312] to nominate a doctor who will take part in their plan,[313] and to take all reasonable efforts to return to work with the worker's pre-incapacity employer.[314] If the worker fails to comply with the above requirements, or fails to attend a medical or employment prospect assessment, or fails to take up an offer of suitable work, the worker's weekly compensation may be suspended.[315] The resumption of weekly compensation does not entitle the worker to recover the compensation they missed during the suspension.[316]

[300] The lump sum payment need not be distributed evenly, but rather may be divided as a magistrates court considers appropriate: s 77(4). The lump sum payment must be paid into a magistrates court, and until the lump sum is paid to the person entitled to it, the court must invest or deal with the payment in a way for that person's benefit: s 78(2).

[301] s 77.

[302] Meaning an approved insurer or self-insurer: s 86A.

[303] s 88, 89.

[304] s 86.

[305] s 93(2).

[306] s 100.

[307] ss 105 and 106.

[308] s 109. Note: a return-to-work programme must provide policies and procedures for the injured worker's rehabilitation: s 109.

[309] s 142. 'Vocational rehabilitation' means the assessment of the needs of the worker for paragraph and the provision of appropriate, adequate and timely services for the worker aimed at maintaining the worker in suitable employment or returning the worker to suitable employment: s 140.

[310] s 88(2).

[311] ss 90 and 91.

[312] s 101.

[313] s 102.

[314] s 104.

[315] s 113.

[316] s 113(4).

If it appears to an insurer, who is liable to pay compensation on behalf of the employer, that the worker has suffered a significant injury,[317] the insurer must establish and give effect to a personal injury plan for the worker.[318]

The worker's personal injury plan may provide for the insurer to pay costs of medical treatment and rehabilitation services for the injured worker, whether or not the worker has made a claim for compensation, or whether the insurer has accepted liability.[319]

How are Claims Made?

Against Employers and Insurers

A worker can make a claim for compensation.[320] A claim is considered made on the day it is given to the employer or insurer.[321] An injury notice[322] must be served on the employer.[323]

Notice of the injury must be given to the employer as soon as practicable, and before the worker voluntarily left the employment in which they were injured.[324] The claim for compensation must be made:

- — within 3 years after the injury, or
- — if the worker was not aware of the injury, within 3 years after the worker became aware; or
- — if the worker dies, within 3 years after the claimant became aware of the death.[325]

If the proceeding is out of time, it may be allowed by a magistrates court.[326] In coming to a decision, the court may have regard to all the circumstances of the case.[327]

After notice is given to the employer, the employer must give the insurer notice of the injury (in an injury notice)[328] within 48 hours of becoming aware of the injury, unless the employer is a self-insurer.[329] If the employer does not give notice in time, the employer is liable to pay weekly compensation from the date of injury until the employer gives the insurer notice, and the employer must not be indemnified by the insurer for these payments.[330] Once the insurer is given notice, they have 3 days to take action under their injury management programme.[331]

[317] 'Significant injury' means a workplace injury likely to result in the worker being partially or totally incapacitated for a continuous period of longer than 7 days: s 97.

[318] s 97.

[319] s 107.

[320] s 116(1).

[321] s 122(1).

[322] An injury notice must contain the name and address of the injured worker, the cause of the injury, and the date and time the injury happened: s 123(1).

[323] s 123.

[324] s 120(1), 93.

[325] s 120(1).

[326] s 120A.

[327] s 120A(4). These include, for example, the length and reasons for the delay, the extent to which the delay will prejudice the employer, and the conduct of the employer and the insurer.

[328] For requirements of contents of notice, see s 94.

[329] s 93.

[330] s 95.

[331] s 96.

Any claim for weekly compensation must be made with a medical certificate from a doctor.[332] The medical certificate must comply with the requirements under the regulation and must include a statement of the doctor's assessment of whether the worker's employment was a substantial contributing factor to the injury.[333]

If a claim is being made for permanent injuries, the claimant must wait 2 years to commence the claim, unless the injury has stabilised[334] or a magistrates court allows it.[335]

If the insurer has not rejected the claim within 28 days of receiving the claim, the insurer is taken to have accepted the claim, and any payment made for the claim by the insurer is not recoverable.[336] Payment may still be recoverable if the insurer is the DI Fund manager.[337]

If an insurer rejects a claim for compensation under the Act they must provide written notice, with reasons for the rejection, to the worker and to the employer. If the insurer is a self-insurer the requirement to give notice to the employer is not applicable.[338] The claim is taken to be rejected when the worker and the employer receive the notice; whichever is the last of them.[339] The insurer is liable to pay compensation until it rejects or settles the claim.[340] Such compensation is not recoverable by the insurer.[341]

Against the DI Fund

The following persons may make a claim for payment against the DI Fund:

- the injured worker;
- a representative of the injured worker;
- if the injured worker has died leaving a dependant—their dependants; or
- if the injured worker has died leaving no dependants—a person acting on behalf of the estate of the injured worker.[342]

The claimant must also be unable to make a claim for compensation or damages against a person other than the employer, and be eligible under the Act.[343] A person is eligible when:

- his employer is liable for compensation but the liability is not covered by a compulsory insurance policy;[344]
- the employer has agreed to pay an amount to discharge its liability, but its liability is not covered by a compulsory insurance policy;[345]

332 s 116(2).
333 s 118(2).
334 For meaning of 'stabilised', see s 121(4).
335 s 121.
336 s 128(1).
337 s 128(2).
338 s 129.
339 s 129.
340 s 134(1).
341 s 134(2).
342 s 170(1).
343 s 170(1).
344 s 170E, F.
345 s 170G.

— the employer is covered by a compulsory insurance policy but the insurer has been wound up or cannot provide the indemnity required;[346]
— the employer is a self-insurer and is unable to pay the compensation;[347]

There is a 1-month time limit for making the claim, starting when the claimant became eligible to make the claim; however, the DI Fund manager has the discretion to extend this deadline following a formal application.[348]

All DI Fund settlements must be approved.[349] Lump sum payments require approval from a magistrates court or by the DI Fund manager in all other cases.[350]

When a compulsory insurance policy is not sufficient to cover the employer's liability the DI Fund pays the amount due under the approved settlement directly to the claimant.[351] When the insurer has been wound up or cannot provide the indemnity; or where the employer is a self-insurer unable to cover liability, the DI Fund must pay the liquidator the amount necessary to satisfy the claim, and the liquidator must pass that amount to the claimant.[352]

If the DI Fund pays a claimant for the liability of an employer, that payment discharges the liability of the DI Fund and of the employer in relation to the claim, and the DI Fund acquires the right of subrogation for any action that the employer has in relation to that claim.[353] Upon the DI Fund making a payment to the claimant the employer automatically owes the DI Fund 3 times the amount that the DI Fund paid the claimant.

Mechanisms for Disputes

Where there is a dispute as to entitlements under the scheme, the parties must make a genuine effort to reach agreement in conciliation. Failing this, they may go to arbitration.[354] Questions of law may be referred to a magistrates court.[355] If a magistrates court requires a decision to be settled by arbitration under the Act, any party to the arbitration may also appeal the decision to the Supreme Court.[356]

Funding

The DI fund consists of:

— amounts received or recovered by the DI Fund manager under the Act;
— income from the investment of amounts of the Fund;

[346] s 170H, I.
[347] S 170HA, HB.
[348] s 170A.
[349] s 171(1).
[350] s 171(2).
[351] s 171D.
[352] s 171E.
[353] s 171I.
[354] s 195, Reg 38, 42 (Part 7) Workers' Compensation Regulation 2*002*. The arbitration may be held by a magistrates court: Regs 51 and 53.
[355] s 195, Reg 52 (Part 7) of the Workers' Compensation Regulation 2002.
[356] s 197.

— amounts borrowed by the Territory for the Fund; and
— other amounts lawfully paid into the Fund.[357]

Each year, the DI Fund manager determines the yearly contribution approved insurers and self-insurers must make to the Fund based on the Fund's existing and expected liabilities and assets; the total premiums for each approved insurer for the last policy period; and the amount required to be paid into the DI Fund to ensure its sustainable functioning.[358] If, after making a determination, an unusual claim that was not reasonably foreseeable is brought against the DI Fund, and the Fund manager believes the Fund cannot meet the unexpected claim, the DI Fund manager must determine a supplementary annual contribution for approved insurers and self-insurers to allow the Fund to meet the claim.[359]

Statistics

Data and graphs taken from the Chief Minister, Treasury and Economic Development publication *ACT Workers Compensation Review of Scheme Performance to 30 June 2014 Directorate* (April 2015)[360] unless otherwise stated.

Gross Average Claim Size (per Non-Nil Claim) by Payment Type

Figure 3.14 summarises the uninflated gross average payment by type for each past accident year. It shows a 16% increase in claim size in 2013/14 over 2012/13 levels. The increase in payments since 2011/12 is due to an increase in common law payments. In 2014/15,

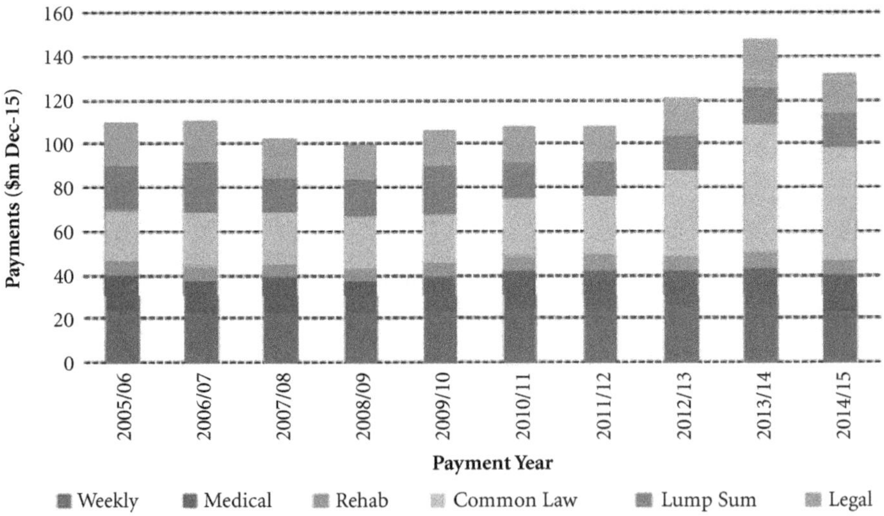

Figure 3.14: Gross payments by type 2005/06–2014/15 (actual uninflated values)

[357] s 166(2).
[358] s 168A.
[359] s 168AA.
[360] Available at www.cmd.act.gov.au/__data/assets/pdf_file/0004/912199/R_2015_ACT-Scheme-Performance_Final.pdf.

there was a AU$14 million (9%) decrease in payments on the year prior. Both the average settlement size and the total number of claims decreased over this period; however, they both remain higher than levels prior to 2013/14.

Lump Sum payments were increased to nearly 300 in 2013/14 from around 200 in 2012/13, see Figure 3.15.[361] This is thought to be due to several factors, including new ACT Court Practice Directives aimed at finalising cases within a year, a push by courts to reduce the number of cases in the court docket, an increased impetus by insurers to settle difficult cases pre-trial and increased advertising by plaintiff lawyers. Aside from the latter, all the other factors indicate that cases are being concluded more quickly, rather than there having been an increase in case numbers.

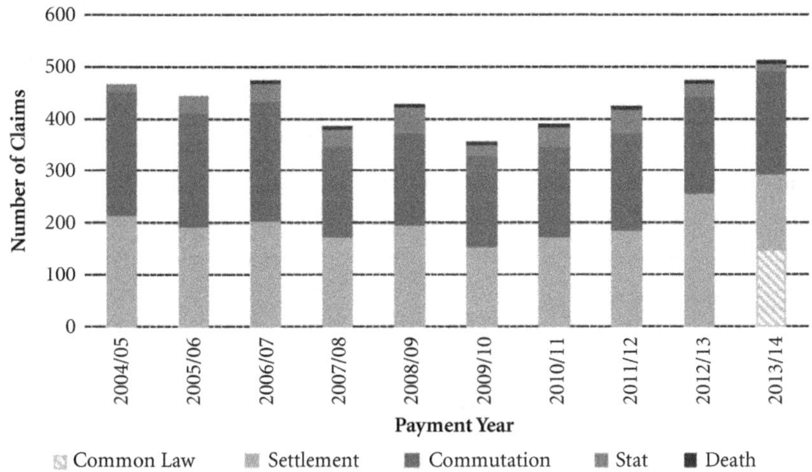

Figure 3.15: Number of claims receiving a lump sum payment 2004/05–2013/14. (Note: prior to 2013/14 there was no way to distinguish between common law awards and negotiated settlements, so all were deemed negotiated settlements.)

Number of Claims Made

Figure 3.16 shows the number of non-nil claims reported in each year. This shows a drop in the number of claims; it is proposed that this drop is related to a safety review of the construction industry undertaken in 2012/13.

Similarly, an 8% drop in lost-time claims was seen between 2013/14 and 2012/13, and then a 1% drop between 2014/15 and 2013/14 (see Figure 3.17). This has also been attributed to the same safety review of the construction industry.

Claim frequency provides a more sensitive measure than simple claim numbers. Figure 3.18 shows the claim frequency both per AU$million wages and per 100 employees. This shows that there has actually been a decline in claim frequency since 2013.

[361] There is some difficulty in interpreting lump sum numbers as different insurers use different classifications for lump sum payments; for a more detailed analysis, see www.cmd.act.gov.au/__data/assets/pdf_file/0004/912199/R_2015_ACT-Scheme-Performance_Final.pdf, 27.

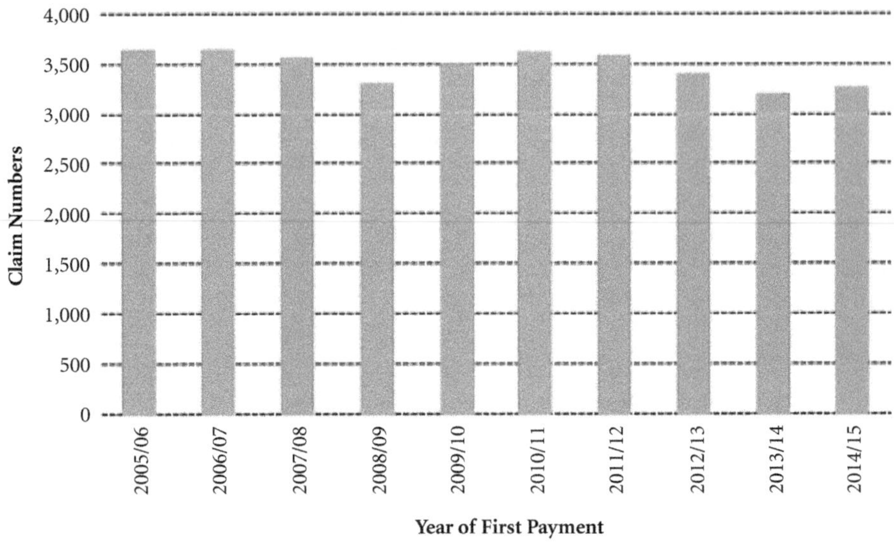

Figure 3.16: Numbers of non-nil claims 2005/06–2014/15

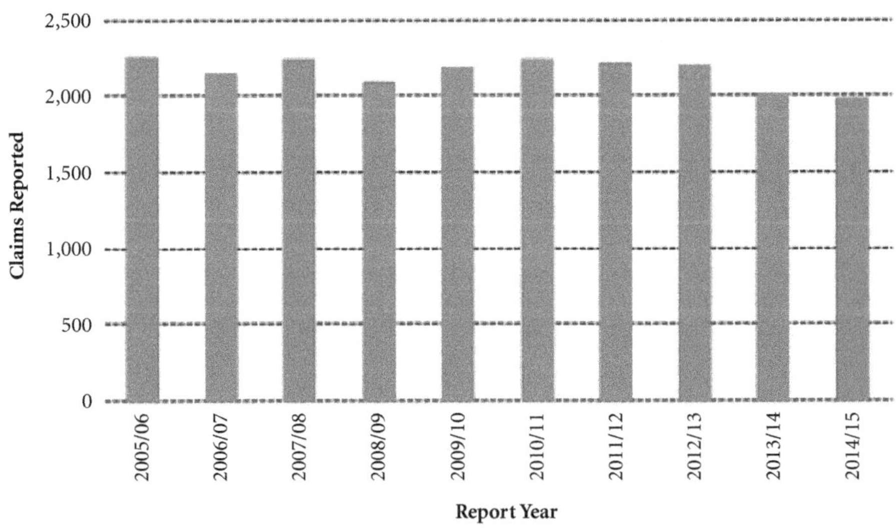

Figure 3.17: Numbers of lost-time claims 2005/06–2014/15

Financial Solvency and Premium Rates

For the scheme to be functional, the premium charged by insurers must be greater than the risk premium paid out to enable them to make a profit. Figure 3.19 shows that the achieved premium rate has consistently been higher than the risk premium, enabling insurers to profit and ensuring the financial stability of the scheme. Clearly these rates are harder to estimate in more recent years, due to greater uncertainty over the estimated risk premium.

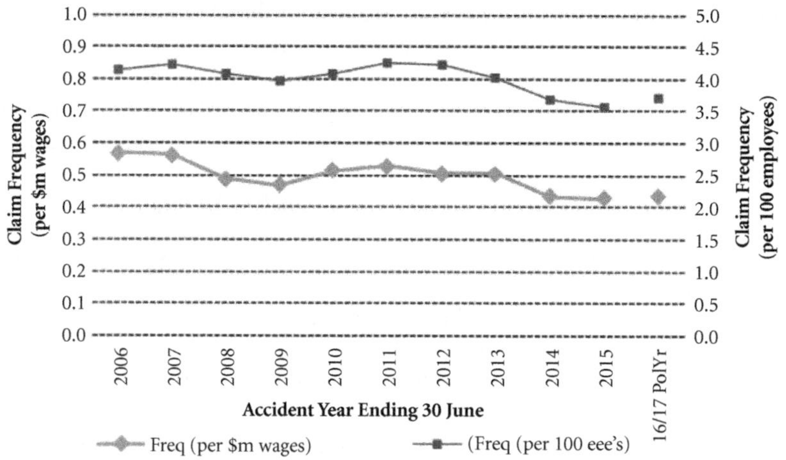

Figure 3.18: Claim frequency 2006–15

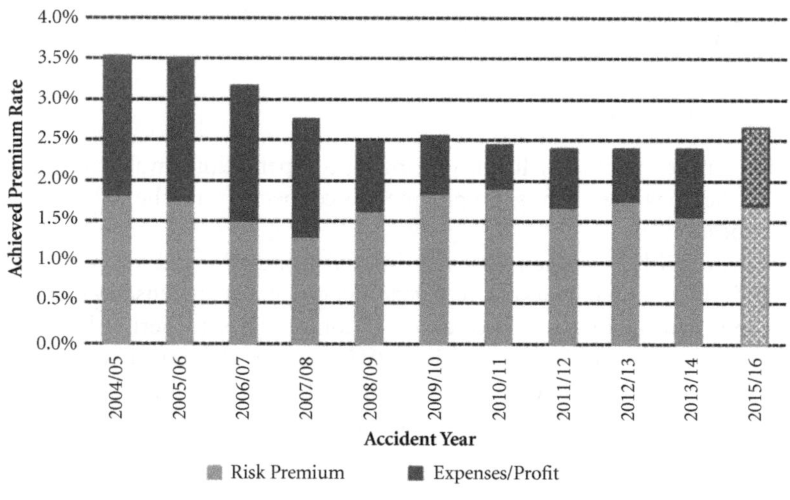

Figure 3.19: Estimated ultimate achieved premium rates 2004/05–2015/16

D. South Australia's Workers' Rehabilitation and Compensation Scheme[362]

General Description

South Australia's Workers' Rehabilitation and Compensation Scheme covers employees who sustain injuries arising from performance of their work.[363] The scheme aims to

[362] Return to Work SA, available at: www.rtwsa.com/.
[363] Return to Work SA, available at: www.rtwsa.com/about-us/returntoworksa.

balance the interests of workers and employers; provide rehabilitation to aid an early return to work; provide fair compensation for injuries; and reduce the social cost to the community caused by work related injuries.[364] ReturnToWorkSA manages the scheme, being a statutory authority under the Return to Work Corporation of South Australia Act 1994[365] to administer the Return to Work Act 2014[366] and the subordinate Return to Work Regulations 2015.[367]

The scheme has undergone several reforms, most notably the following:

- the now-repealed Workers Compensation Act 1971 (SA), which largely resulted from the Byrne Report;[368]
- the now-repealed Workers Rehabilitation and Compensation Act 1986 (SA) ('WRCA');
- in 1992 common law litigation was abolished;[369]
- in 1994 psychiatric injuries were included (and the substantial cause test was implemented for them);
- the expiration and review of the WRCA regulations in 2008, with new ones implemented in 2010; and
- most recently the implementation of the Return to Work Act 2014 (the Act) and the Return to Work Regulations 2015 (the Regulations) which established the ReturnToWorkSA Scheme.

Interaction with Tort Liability

The Act extinguishes employer's liability, bringing compensation payable to injured workers under the Act to workers whose employment is connected with the state, irrespective of the injury's location.[370]

The Act distinguishes between seriously injured workers[371] and less seriously injured workers. Seriously injured workers received higher-value compensation and have the additional ability to sue in tort.[372] Access to the common law for seriously injured workers was reintroduced by the Act. From 1 July 2015 damages claims can be made in tort for injured workers who have died or sustained a whole-body impairment of 30% or more.[373] However, the only permissible damages are for economic loss;[374] claims for treatment, care

[364] Return to Work SA, *Return to Work Scheme*, available at www.rtwsa.com/about-us/return-to-work-scheme.

[365] Return to Work Corporation of South Australia Act 1994 (SA).

[366] Return to Work Act 2014 (SA).

[367] Return to Work Regulations 2015 (SA).

[368] 'A Workers' Rehabilitation and Compensation Board for South Australia—The Key to Rapid Rehabilitation and Equitable Compensation for those Injured at Work' ('Byrne Report').

[369] Braction Consulting Services Pty Ltd and PricewaterhouseCoopers, 'Review of the South Australian Workers' Compensation System Report' (Parliament of South Australia, House of Assembly, December 2007).

[370] Return to Work Act 2014 (SA) s 10. The location where the injury took place is not the most pertinent of criteria—compensation will be potentially payable when the work is in connection with the state, see s 10(1), rather than occurring in the state.

[371] ibid, s 21. Seriously injured workers are those with 30% or more whole-body impairment.

[372] This is a change from the previous position in South Australia, whereby there was highly restricted access to the common law for injured workers claiming against their employer.

[373] Return to Work Act 2014, s 72.

[374] ibid, s 72(2)(c).

and support are not permitted,[375] and the claimant must elect either for weekly payments calculated under Part 4, Division 5 of the Act (releasing the employer from liability) or to claim damages for future economic loss caused by reduction in (including loss of) earning capacity from his employer.[376] A worker must receive legal advice from a practising lawyer before making this choice,[377] with mandatory funding available to the worker to cover the cost of this advice.[378]

Co-worker liability for negligent acts is also excluded when there is no wilful misconduct[379] that gave rise to the negligence of the other worker which caused the injury.[380]

The Act includes the right to litigate for compensatable injuries arising out of third-party actions.[381] However, if compensation arises, there are rules governing the interactions between the rights of parties to recover and to claim any damages awarded.

Injured parties are also precluded from double-claiming compensation in other jurisdictions with the legislation giving employers who have paid compensation the ability to recover the payments made if a double-claim occurs.[382]

Funding

The scheme is funded by premiums paid by employers and revenue generated by the investment of premiums collected by ReturnToWorkSA.[383] The premiums paid by employers are indexed depending on the business's size, the risk levels in their industry and, for medium to large employers, their claims' history.[384] This system takes into account the claims' performance of businesses relative to their competitors and so premiums are proportionate to the likelihood of drawing on funds from the scheme. The premiums are reviewed annually with assistance from independent actuaries.

Eligibility

Employees are eligible for compensation (and are covered) provided that:

- the injury arose from employment,[385] (provision exists for self-employed workers[386] or the dependants of a deceased worker[387] to claim);
- their employment is connected with the State of South Australia;[388] and

[375] ibid, s 72(2)(a).
[376] ibid, s 72(2)(b).
[377] ibid, s 72(6).
[378] ibid, s 72(7).
[379] ibid, s 66(3)(b)(ii).
[380] ibid, s 66(3)(b).
[381] ibid, s 66.
[382] ibid, s 67.
[383] ibid, Division 4 of the Act, ss 136–45; for a breakdown of income figures, see the ReturnToWorkSA Annual Report; the report for 2014/15 is available at www.rtwsa.com/__data/assets/pdf_file/0006/29328/Annual-report-2014-15.pdf.
[384] ibid, Div 4.
[385] ibid, s 7.
[386] ibid, s 175.
[387] Div 8, ss 59–63.
[388] ibid, Div 2, ss 10–12; it is notable that the scheme will cover workers whether the injury occurs within the state or not so long as the worker usually works in South Australia or is based in South Australia for employment or the employer's principal place of business is located in South Australia.

— their injury was significantly caused by their duties or the course of their work[389] or an existing condition was significantly contributed to by their employment.[390]

Cover is applicable irrespective of the worker's employment status (full-time permanent, part-time, casual or fixed-term contract), and injury, in relation to a worker, is defined as:[391]

(a) any physical or mental injury including—
 (i) loss, deterioration or impairment of a limb, organ or part of the body, or of a physical, mental or sensory faculty; or
 (ii) a disease; or
 (iii) disfigurement; or
(b) where the context admits—the death of the worker,
 and includes an injury that is, or results from, the aggravation, acceleration, exacerbation, deterioration or recurrence of a prior injury.

A seriously injured person has two options; either to claim under the Act or to pursue a claim at common law. The procedures to determine severity of the injuries and whether a person is classified as seriously injured are detailed in Divisions 4 and 5 of the Act. The person is to be assessed by a legally qualified medical practitioner and the assessment must be carried out in accordance with the Impairment Assessment Guidelines SA.[392] An assessment can be recommended by the injured individual themselves or by their case handler.

ReturnToWorkSA Procedure

The aim of the Act is to provide a single, integrated process for recovery, rehabilitation and compensation for injured workers.

Application

Injured workers must notify their employer of a compensatable injury, ideally within 24 hours of its occurrence (or as soon as practicable). Notice to an employer can be oral or in writing; if given orally the employer can mandate that the injured employee subsequently give notice in writing.[393] Self-employed individuals must give notice to ReturnToWorkSA.[394] The employer must notify ReturnToWorkSA of the injury within 5 days of being notified.[395]

If a claimant requires compensation for lost wages, then he will require a Work Capacity Certificate from his doctor. This details the level at which an individual is capable of working given the injury he has sustained.

[389] ibid, s 7; this includes work-related travel, on-site breaks and may also include journeys to and from work, though injuries which occur while commuting are not automatically included: see s 7(8).

[390] ibid, s 7(3).

[391] ibid, s 4.

[392] ibid, s 22. Impairment Assessment Guidelines are available at www.rtwsa.com/media/documents/Impairment-assessment-guidelines.pdf.

[393] ibid, s 16(2), (3) and (5).

[394] ibid, s 16(1)(b).

[395] ibid, s 16(6).

Claims can be completed by the employee or his employer, ideally both parties will complete the claim form together. Claim forms are available online.[396] Claim forms include a declaration allowing ReturnToWorkSA access to the individual's medical records. Claim forms also require a Work Capacity Certificate.[397] This certificate must include

(i) the nature of the injury;
(ii) the probable cause of the injury so far as that is ascertainable by the designated person;
(iii) if the claimant claims to be incapacitated for work—the claimant's current and likely future capacity;
(iv) any other matter specified by the Corporation with the approval of the Minister.

ReturnToWorkSA will then investigate the claim using such investigations as are deemed necessary to achieve an evidence-based determination of the claim.[398] Generally ReturnToWorkSA has 10 days to determine the claim, with some exceptions.[399] Workers may be required to submit to an examination by a medical professional selected by ReturnToWorkSA or to answer questions so that specific medical questions can be determined.[400] ReturnToWorkSA has an obligation to give written notice of the determination to the claimant and the employer as soon as is practicable.[401]

If a claim is rejected, then notice must include the legal grounds on which the claim is rejected and information on the claimant's rights to have the claim redetermined.[402]

'Return To Work/Recovery Plans'

Part 3 of the Act establishes a system that seeks to ensure that those injured at work—

(a) achieve the best practicable levels of physical and mental recovery; and
(b) return to their pre-injury work or, if that is not reasonably practicable, is in any event restored to the workforce and the community in a timely, safe and durable way.[403]

To facilitate this, ReturnToWorkSA's early intervention, recovery and return-to-work services have the power to do one or more of the following:[404]

(a) provide for the physical, mental or vocational assessment of a worker;
(b) provide advisory services to a worker, members of the family of a worker, an employer and others;
(c) assist a worker in retaining, seeking or obtaining employment;
(d) assist in the training or retraining of a worker;
(e) assist a worker to find or establish appropriate accommodation;

[396] See www.rtwsa.com/media/documents/Claim-form.pdf.
[397] The Return to Work Act 2014, s 31(1)(c).
[398] ibid, s 31(1).
[399] ibid, s 31(4).
[400] ibid, s 31(2).
[401] ibid, s 31(7).
[402] ibid, s 31(8).
[403] ibid, s 23.
[404] ibid, s 24(1).

(f) provide equipment, facilities and services to assist a worker to cope with any injury at home or in the workplace;
(g) provide assistance to a person who may be in a position to help a worker to overcome or cope with an injury;
(h) provide necessary and reasonable costs (including costs of travel, accommodation and childcare) incurred by a worker in order to receive or participate in any services; and/or
(i) provide anything else that may assist in achieving the objects of this Part.

Where it appears that the worker is or is likely to be incapacitated from work for 4 or more weeks, then ReturnToWorkSA must prepare a return to work plan or a recovery plan.[405]

Subject to certain conditions, employers are mandated by the Act to provide suitable employment, in so far as is reasonably practicable, for the injured employee.[406]

Compensation

Injured workers are entitled to the following benefits:

Interim Payments[407]

Interim payments to the claimant by ReturnToWorkSA are available pending the final determination. If ReturnToWorkSA has not determined a claim within the allowed 10 business days, then payments are mandatory.[408] If, when the claim has been determined, it appears that the claimant has been paid interim compensation payments to which he or she was not entitled, then ReturnToWorkSA can recover the amount of the interim payments as a debt.[409]

Weekly Payments of Income Maintenance or Incapacity Benefits

These differ for seriously injured and non-seriously injured workers (see Table 3.4) and must equate to at least the Federal minimum wage.[410]

Medical and Associated Costs (Rehabilitation) Reasonably Incurred as a Consequence of a Compensatable Injury or Illness[411]

Benefits depend on the circumstances of each case but can include:

(a) medical services;
(b) hospitalisation and all associated medical, surgical and nursing services;

[405] ibid, s 25(1). These plans are to be drawn up in situations where it would be beneficial to the worker and/or the community to return to work; if these criteria are not fulfilled, then it is not mandatory to draw up a plan: see s 25(2) for more detail.
[406] ibid, s 18.
[407] ibid, s 32.
[408] ibid, s 32(2).
[409] ibid, s 32(3).
[410] ibid, s 42.
[411] ibid, s 33.

Table 3.4: Payment Types and Time Frames for Seriously and Non-Seriously Injured Workers

Weeks after the start of the incapacity	Seriously injured workers[a]	Non-seriously injured workers[b]
0–52	Either 100% of the worker's notional weekly earnings for total incapacity, or for partial incapacity if his/her actual earnings are less than his/her pre-incapacity notional weekly earnings, then the difference between these sums	Either 100% of the worker's notional weekly earnings for total incapacity, or for partial incapacity if his/her actual earnings are less than his/her pre-incapacity notional weekly earnings, then the difference between these sums
52–104 weeks	NA	Either 80% of the worker's notional weekly earnings for total incapacity, or for partial incapacity if his/her actual earnings are less than 80% of his/her pre-incapacity notional weekly earnings, then the difference between these sums
104 weeks onwards	NA	Nothing
52 weeks onwards	Either 80% of the worker's notional weekly earnings for total incapacity, or for partial incapacity if his/her actual earnings are less than 80% of his/her pre-incapacity notional weekly earnings, then the difference between these sums	NA

[a] The Return to Work Act 2014, s 41.
[b] The Return to Work Act 2014, s 39.

(c) approved recovery/return to work services;
(d) travelling, or being transported, to and from any place for the purpose of receiving medical services, hospitalisation or approved recovery/return-to-work services (but not where the worker travels in a private vehicle);
(e) where it is necessary for the worker to be accommodated away from home for the purpose of receiving medical services or approved recovery/return-to-work services—such accommodation (but not exceeding limits prescribed by regulation);
(f) attendance by a registered or enrolled nurse, or by some other person approved by the Corporation or of a class approved by the Corporation, where the injury is such that the worker must have nursing or personal attendance;
(g) the provision, maintenance, replacement or repair of therapeutic appliances;
(h) medicines and other material purchased on the prescription or recommendation of a health practitioner;
(i) other services (or classes of services) authorised by the Corporation.

These costs can either be paid directly to the worker or to the person to whom the worker is liable for these costs. It is notable there is no explicit statutory limit for these expenses, but the Minister has the power to publish a scale of charges under this section[412] and healthcare providers must not knowingly charge an injured worker more than the value allocated on the scale of charges for that procedure.[413]

Lump Sum Compensation for Non-economic Loss (Permanent Injuries or Impairment)

The maximum payout is $472,000 (indexed) and weekly income payments and benefits are still payable.[414] This is calculable using the Whole Person Impairment Values and is not available for injuries of less than 5% of the Whole Person Impairment[415] or for psychiatric injury.

In the event of compensatable injuries causing death, four types of payment are potentially available to eligible claimants: payment for approved counselling services; funeral expenses; weekly payments; and lump sum payments.[416]

Payment for approved counselling services is available for family members[417] according to a set cost tariff, and can be paid either directly to the family member or to the professional providing the counselling.[418]

Funeral expenses are available to cover the lesser of either the actual funeral costs or an indexed prescribed amount. These are paid either to the person who paid, or is liable to pay for, for the funeral or to the person who conducted the funeral.[419]

Details for weekly payments[420] are set out in Table 3.5.

Table 3.5: Weekly payments for dependants of deceased workers

Dependent spouse or domestic partner	Dependent child (not orphaned)	Dependent child (orphaned)	Dependent relative (not any of the previous groups)
Total dependency gets *50%* of the notional earnings of the deceased, *partial dependency* gets a *variable* lesser percentage of the notional earnings, determined by the Corporation	*Total dependency* gets *12.5%* of the notional earnings of the deceased, *partial dependency* gets a *variable* lesser percentage of the notional earnings, determined by the Corporation	*Total dependency* gets *25%* of the notional earnings of the deceased, *partial dependency* gets a *variable* lesser percentage of the notional earnings, determined by the Corporation	*Variable*, determined by the Corporation having regard to; the extent of their dependency; their earning capacity and means; and the extent of any other benefits provided under this Act for them

412 ibid, s 33(12).
413 ibid, s 33(16).
414 ibid, s 57(1). This payment is indexed or subject to Regulations specifying a greater amount.
415 ibid, s 58.
416 ibid, ss 59–62.
417 ibid, s 63(5). Family member for the purposes of this section of the act is a spouse, domestic partner, parent, sibling or child of the worker or of the worker's spouse or domestic partner.
418 ibid, s 63(1)–(4).
419 ibid, s 62.
420 ibid, s 59(1).

Details for Lump Sum Payments[421]

The prescribed lump sum is the amount for non-economic loss specified under Division 7 of the Act, less any amounts already paid to the worker. Lump sum payments are available to a partner and/or a child. Under the relevant section of the Act a partner is a spouse or domestic partner, and a child is someone who is either under 18 years of age, or a full-time student under the age of 26 who is attending an educational institution approved by the Corporation or someone who is incapable of earning a living by reason of disability, where disability includes any physical, mental or intellectual disability. Table 3.6 details the permutations of lump sum awards.

Table 3.6: Lump sum payments for dependants of deceased workers

At least one partner and no children	No partner and at least one child	At least one partner and only one child	At least one partner and between two and five children	At least one partner and more than five children
100% of the prescribed sum (divided equally between the partners if more than one)	100% of the prescribed sum (divided equally between the children if more than one)	90% of the prescribed sum for the partner(s) (divided equally between them if more than one) 10% of the prescribed sum for the child	5% of the prescribed sum for each child The remainder for the partner(s) (divided equally between them if more than one)	75% of the prescribed sum for the partner(s) (divided equally between them if more than one) 25% of the prescribed sum to be divided equally between the children

The following are grounds for suspension of benefits under the Scheme, if:

- the worker is guilty of misconduct or acts in contravention of instructions from the employer while commuting;
- on the balance of probabilities the injury is wholly or predominantly attributable to
  - serious and wilful misconduct on the part of the worker;[422] or
  - the influence of alcohol or a drug voluntarily consumed by the worker (other than a drug lawfully obtained and consumed in a reasonable quantity by the worker).[423]
- Breaches an obligation of mutuality[424] or refuses to provide information—including:
  - failure to submit to a medical examination where required by notice;[425]
  - failure to supply a medical certificate where required by notice;
  - fefusal or failure to submit to proper medical treatment;
  - refusal or failure to participate in a rehabilitation programme or frustrates the objectives of the programme;

[421] ibid, s 61.
[422] ibid, s 8(2)(b)(i) and (3); this does not apply in a case of death or permanent total incapacity for work.
[423] ibid, s 8(2)(b)(ii) and (3); this does not apply in the case of death or serious and permanent injury.
[424] ibid, s 48(3).
[425] ibid, s 31(3).

  - failure to comply with obligation under rehabilitation or return-to-work plan;
  - refusal or failure to undertake work offered where the employee is capable of doing the work; refusal or failure to take reasonable steps to find or obtain employment; or the unreasonable discontinuation of the employment;
  - refusal or failure to participate in assessments of worker's capacity, rehabilitation progress or future employment prospects; or
  - anything else recognised as a breach of the obligation of mutuality.
- Is imprisoned.[426]
- Is absent from Australia for a prolonged period and the Corporation deems fit.[427]

Review and Dispute Mechanisms

Claimants and employers are able to have their claim determinations reviewed by the South Australian Employment Tribunal.[428] Claims for review must be made within 1 month of notice of the decision for which a review is being sought.[429] As a first stage in the dispute resolution process ReturnToWorkSA must appoint a suitable person who did not make the initial decision to reconsider the claim decision.[430] This must take place within 10 business days of receiving notice of the dispute.[431] If this determination is not resolved through reconsideration, it must be referred to a conciliation conference where a conciliation officer will identify the issues in the dispute and investigate the possibility of resolving the dispute with agreement from both parties.[432] Should the dispute remain contentious it will be heard by the South Australian Employment Tribunal in accordance with the rules laid out in Part Three of the South Australian Employment Tribunal Act 2014.[433] Briefly, the Tribunal consists of not more than three members,[434] composed of presidential members (the president and deputy presidents of the South Australian Employment Tribunal), magistrates and conciliation officers.[435] A full bench consists of three presidential members.[436] Matters which contain a point of law can be referred to the full bench who may then refer it to the Supreme Court of South Australia for Determination.[437]

Note that the Statute Amendment (South Australian Employment Tribunal) Bill 2016[438] is currently open for public consultation. If enacted, the Bill will amend the jurisdiction of the South Australian Employment Tribunal by replacing the current Industrial Court with a new South Australian Employment Court. The new court will have the ability to hear a claim arising for damages for breach of a contract for employment[439] and an action for the

[426] ibid, s 193; though there is provision to continue payments to the dependants of the worker if the Corporation sees fit.
[427] ibid, s 51.
[428] ibid, s 99(1) and (2).
[429] ibid, s 100. The Tribunal has the discretion to extend the notice period where there is good reason to do so.
[430] ibid, s 102(1).
[431] ibid, s 102(5).
[432] ibid, s 104.
[433] South Australian Employment Tribunal Act 2014 (SA).
[434] ibid, s 19(2).
[435] ibid, s 9.
[436] ibid, s 19(3).
[437] ibid, s 22.
[438] Statute Amendment (South Australian Employment Tribunal) Bill 2016, Draft Bill.
[439] Statute Amendment (South Australian Employment Tribunal) Bill 2016, s 10.

grant of an injunction or specific performance.[440] The Bill was read for the first time on 4 August, 2016.

Should the parties be unhappy with the decision of the Employment Tribunal there is a right of review. Appeals against the decisions made by a magistrate, a conciliation officer or two or three non-presidential members of the Tribunal can be reviewed by a presidential Member of the Tribunal.[441] The review powers are wide ranging: the presidential member can affirm, set aside or vary the decision or send it back for reconsideration (with directions or recommendations) or substitute his or her own decision.

Appeals on a point of law decided by a single presidential member of the Worker's Compensation Tribunal or a tribunal other than a full bench are heard by a full bench.[442]

For further appeals on questions of law, the matter can be referred from the judicial determination of the full bench of the Tribunal to the full court of the Supreme Court of South Australia and heard with permission of a single judge of the Supreme Court.[443] The Supreme Court can decide the question of law, refer the matter back to the Tribunal with appropriate directions or make a consequential or related order (including an order for costs)

Costs

The rules governing costs are set out in sections 106–108 of the Return to Work Act 2014. Essentially capped (according to a scale set by the Minister) costs are available to cover the cost of representation and other applicable costs arising from the review of a decision by the South Australia Employment Tribunal, where it is reasonable to do so.

Specific Diseases

Schedule 2 of the Return to Work 2014 Act includes a list of diseases presumed to arise from general employment. Schedule 3 details diseases presumed to arise from employment as a firefighter. Where a disease is listed in the schedule and the conditions detailed in the Description of the Work given in Schedule 2, it is presumed that the disease arose from the work. For example, for mesothelioma and asbestosis there is a presumption that any work involving the inhalation of asbestos fibres would have caused these conditions. For workers who have been exposed to inhalation of asbestos fibres and have contracted asbestosis (or mesothelioma), disability is presumed (in the absence of proof otherwise) to have arisen in the course of that employment.[444] This presumption eases the burden that can be associated with proving causation in long-latency disease cases.

Statistics

The scheme has only been in place since July 2015, and therefore there are currently no annual statistics available.

[440] ibid, s 26A.
[441] Return to Work Act 2014, s 66.
[442] ibid, s 67.
[443] ibid, s 68.
[444] ibid, s 9.

E. The Tasmanian Workers' Compensation Scheme

General Description

A worker in the State of Tasmania who has suffered an injury or disease arising from or during his/her employment is entitled to pursue a Workers' Compensation Claim. The principal piece of legislation setting out the laws relating to workers' rehabilitation and compensation in Tasmania is the Workers' Rehabilitation and Compensation Act 1988 (the Act). The Act is supported by the Workers' Rehabilitation and Compensation Regulations 2001. The object of the Act is to establish a no-fault rehabilitation and compensation scheme for injured workers that, inter alia:

— provides for the prompt and effective management of workplace injuries in a manner that promotes and assists the return to work of injured workers as soon as possible;
— provides fair and appropriate compensation to workers and their dependants for workplace injuries; and
— provides an effective and economical mechanism for resolving disputes relating to the treatment and management of, and compensation in relation to, workplace injuries.[445]

The scheme is administered by WorkCover Tasmania, and is privately underwritten by licensed insurers. Under the Act, employers are required to take out a workers' compensation insurance policy with a licensed insurer. Alternatively, employers may apply to the WorkCover Board to be granted a permit to self-insure. Court proceedings may still be brought for personal injuries covered by the scheme, although access to the common law is limited and requires an injury which affects at least 20% of the individual.[446]

In July 2006 Alan Clayton conducted a review at the request of the Tasmanian Government, with a subsequent opportunity for public comment.[447] In response the Workers' Rehabilitation and Compensation Amendment Act 2009 (the Amendments) was passed by Parliament in late 2009 and commenced on 1 July 2010. These Amendments modernised the existing system.[448]

[445] Workers' Rehabilitation and Compensation Act 1988 (TAS) s 2A.

[446] Safe Work Australia, *Comparison of Workers' Compensation Arrangements in Australia and New Zealand* (July 2013) 21.

[447] Review of the Tasmanian Workers' Compensation System—Report (Alan Clayton, September 2007) available at www.lawcouncil.asn.au/lawcouncil/images/LCA-PDF/a-z-docs/clayton_report.pdf. The review focused on the adequacy of compensation for workers who could not establish negligence or meet the 30% whole-person impairment threshold. The Report included a number of recommendations, including ones on improving the level of compensation for more seriously injured workers.

[448] The major impacts of the Amendments Act were threefold:

1. it implemented the government's response to the Clayton report;
2. it amended the timing and level of weekly payment step-downs; and
3. it reduced the common law threshold from 30% whole-person impairment to 20%.

Eligibility

To be entitled to compensation under the Act, a person must be a worker. A worker is a person who has entered into, or works under, a contract of service or training agreement.[449] Where a worker has died, the term 'worker' includes the legal personal representatives or dependants of the deceased worker.[450] Some persons are deemed or taken to be workers under the Act. They include, inter alia:

- police volunteers, volunteer firefighters, and volunteer ambulance workers and other prescribed volunteers (if any) while they are engaged in their volunteer duties;[451]
- taxi drivers while they are driving a taxi or performing any associated activity, such as loading, unloading or cleaning the taxi (except where the driver is the responsible operator of the taxi);[452]
- jockeys and apprentice jockeys;[453] and
- salespeople, canvassers and collectors paid by commission.[454]

Some people are specifically excluded from the Act.[455] They are not entitled to workers' compensation if injured while working. These excluded people are:

- people employed on a casual basis for a purpose other than the employer's trade or business;
- uutworkers;
- people employed as domestic servants with a private family, who have done less than 48 hours employment with their employer when they are injured (people in casual domestic employment, eg cleaners, would usually be covered by an extension of a household insurance policy);
- crewmembers of a fishing boat who are paid wholly or mainly on the basis of a share of the profits or gross earnings of the boat;
- people taking part in approved programmes of work for unemployment payment (work-for-the-dole schemes);
- people employed on ships covered by the Commonwealth Seafarers Rehabilitation and Compensation Act 1992; and
- sportspeople.

A worker is entitled to workers' compensation if he suffer an injury, not being a disease, arising out of or in the course of his employment.[456] However, some injuries suffered in

[449] Workers' Rehabilitation and Compensation Act 1988 (TAS), s 3(1).
[450] s 3(1).
[451] s 6A.
[452] s 4DB.
[453] s 4DC.
[454] s 4C.
[455] ss 4(5) and 7.
[456] s 25(1)(a).

the following situations or circumstances are specifically excluded from coverage under the Act:

— any injury that occurs while a worker is travelling between his home and work (unless the injury occurs during a deviation from his normal route that his employer tells, asks or authorises him to make);[457]
— any injury that occurs during an absence from the workplace that was not authorised, directed or requested by his employer;[458]
— any injury that is caused by a worker's serious or wilful misconduct (unless the injury results in his death, or serious and permanent incapacity);[459] or
— any injury that was intentionally self-inflicted.[460]

A worker is entitled to compensation under the Act for a disease that their employment contributed to by a substantial degree. To have contributed to a substantial degree, employment must be the *major* or *most significant factor*.[461] With respect to industrial deafness, a worker must have suffered more than 5% binaural hearing impairment since 16 August 1995.[462]

Compensation

Benefits provided on a no-fault basis under the scheme include:

Weekly Payments

A worker who is incapacitated (either totally or partially) for work as a result of a work-related injury or disease is entitled to weekly payments at the highest amount of these two options:[463]

(i) the worker's ordinary time rate of pay for the employment (as set by an award or other industrial instrument such as an enterprise agreement) that the worker was engaged in immediately before the incapacity began; or
(ii) the normal weekly earnings of the worker averaged over the relevant period of employment (the 'relevant period' depends on how long the worker has been employed by the employer).[464]

For the first 26 weeks of incapacity, the worker receives weekly payments at 100% of his normal weekly earnings. After these 26 weeks, there are two reductions (or 'step-downs') in weekly payments:

(i) If the worker is incapacitated for more than 26 weeks, weekly payments are paid at 90% of his normal weekly earnings. However, if the worker is able to return to some

457 s 25(6).
458 s 25(6).
459 s 25(2)(a)(i).
460 s 25(2)(a)(ii).
461 s 3(2A).
462 s 72A(2) and (3).
463 s 69(1).
464 s 69(14).

form of work, but his employer is unwilling or unable to provide suitable alternative duties, then the worker will receive 95% of normal weekly earnings.[465]

(ii) If the worker's incapacity exceeds 78 weeks, weekly payments are reduced to 80%. However, if the worker is able to return to some form of work, but his employer is unwilling or unable to provide suitable alternative duties, then the worker will receive 85% of normal weekly earnings.[466]

The step-downs do not apply (ie the worker will continue to be paid 100% of normal weekly earnings) if the worker is back at work for 50% or more of his normal weekly hours.[467] The maximum period that weekly payments can be paid depends on the worker's level of whole person impairment (WPI):[468]

— A worker with a WPI of less than 15% is entitled to weekly payments for up to nine years.
— A worker with a WPI of at least 15% but less than 20% is entitled to weekly payments for up to 12 years.
— A worker with a WPI of at least 20% but less than 30% is entitled to weekly payments for up to 20 years.
— A worker with a WPI of 30% or more is entitled to weekly payments until the worker reaches the age of 65.

Entitlement to weekly payments stops at the age of 65.[469] Section 86(1) of the Act enables an employer to reduce or end weekly payments in certain circumstances, for instance where the weekly payments the worker receives relate to total incapacity and the worker has returned to work.

Medical and Other Expenses

The employer must pay the cost of all reasonable expenses the worker necessarily incurs for medical services, hospital services, household services (eg cleaning, laundry and gardening), nursing services, constant attendant services (eg to shower, dress or feed the worker), rehabilitation services and ambulance services.[470]

The employer must pay reasonable expenses for the worker to travel to any medical, hospital or rehabilitation service or to attend any medical examination organised by the employer.[471]

The amount payable for using a private vehicle is calculated based on the 'occasional user' rates set out in the Tasmanian State Service Award.[472]

Lump Sum Compensation for Permanent Impairment

To be entitled to lump sum permanent impairment compensation, the worker must meet the appropriate WPI threshold. These thresholds are:

(i) for the loss of part, or all, of a finger or toe: no threshold applies.[473]

465 s 69B(1)(a).
466 s 69B(1)(c) and 69B(2).
467 s 69B(2A).
468 s 69B(1).
469 s 87(1).
470 ss 74 and 75.
471 s 76.
472 s 76.
473 s 71.

(ii) for any other permanent physical impairment: a threshold of 5% WPI applies.[474]
(iii) for permanent psychological impairment: a threshold of 10% WPI applies.[475]
(iv) for industrial deafness: a threshold of 5% binaural hearing loss, suffered since 16 August 1995, applies.[476]

A worker who suffers a WPI of 20% for an injury that occurred on or after 1 July 2010 may be entitled to pursue common law damages.[477]

Compensation to Dependants of Deceased Workers ('Death Benefits')

Where a worker dies from his work-related injury or disease, his dependants may be entitled to compensation.[478] This may include weekly payments; a lump sum payment; compensation for the worker's medical expenses; compensation for funeral costs; and compensation for counselling costs.

A dependant spouse or caring partner is entitled to weekly payments, paid as:[479]

(i) 100% of the deceased worker's normal weekly earnings/ordinary time rate of pay for the first 26 weeks following the date of death; then
(ii) 90% of the deceased worker's normal weekly earnings/ordinary time rate of pay for the period over 26 weeks and up to 78 weeks from the date of death; then
(iii) 80% of the deceased worker's normal weekly earnings/ordinary time rate of pay for the period over 78 weeks and up to 2 years from the date of death.

If the worker dies more than 78 weeks after sustaining the work-related injuries that caused his death, his dependent spouse or caring partner will be entitled to 80% of the deceased worker's normal weekly earnings or ordinary time rate of pay, from the date of death, and up to 2 years from the date of death.[480]

Dependent children are entitled to weekly payments paid on a different basis. Instead of being paid a proportion of the deceased worker's normal weekly earnings or ordinary time rate of pay, they are entitled to weekly payments of 15% of the basic salary.[481] These weekly payments start 13 weeks from the date of the worker's death and continue until the child reaches 16 years of age (or 21 years of age if a full-time student).[482]

If members of a deceased worker's family require counselling services following the worker's death, the Act allows for the payment of reasonable costs up to AU$4,000.[483]

Procedure

A worker must notify his employer *as soon as practicable* after suffering a workplace injury (including a disease) and before voluntarily leaving the employment where the

[474] s 71.
[475] s 71.
[476] s 72A.
[477] s 138AB.
[478] s 67A.
[479] s 67A(2).
[480] s 67A(4).
[481] s 67A(1)(b).
[482] s 67A(1)(b).
[483] s 75(1) and (10).

injury occurred.[484] On receiving notice of an injury from a worker, an employer is required to serve on the worker a written notice as prescribed in the regulations within 14 days.[485] The prescribed written notice provides information about making a claim for compensation, including the time limits in which a claim must be made. In most cases, a claim for compensation must be made within 6 months of the date of the worker's injury or death (an exception is made for industrial deafness cases).

If an employer receives a workers' compensation claim from one of its workers, it must notify its insurer of the claim within 3 working days.[486] An employer must then:

— start making weekly payments of compensation if the worker has been certified as totally or partially incapacitated for work; and
— start paying for medical and associated expenses up to $5,000.[487]

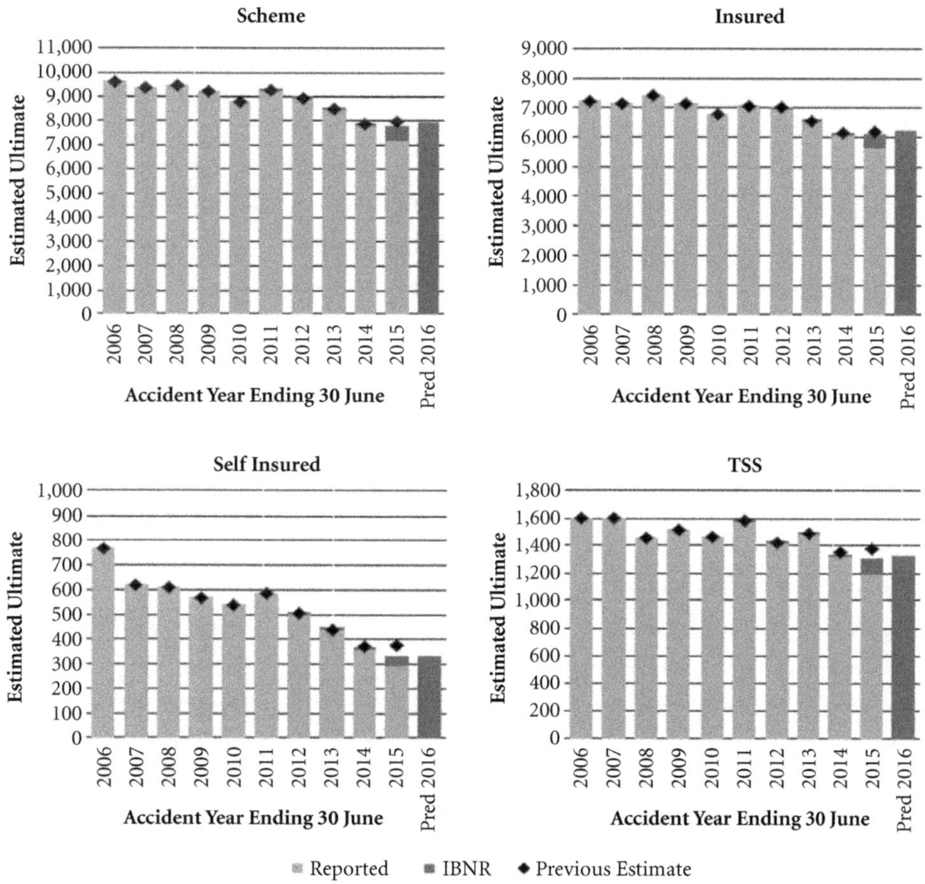

Figure 3.20: Estimated ultimate claim numbers 2006–16, total scheme, insured, self-insured and Tasmanian state services

[484] ss 32 and 33(10).
[485] s 33A.
[486] ss 36(1AA) or 143A.
[487] s 77AB.

These payments are to start regardless of whether the employer disputes liability for the worker's claim. They cannot be claimed from the employer's insurer. If the employer does not dispute liability to pay compensation within 84 days, the employer is taken to have accepted liability for the claim and the Workers' Compensation Tribunal can order the employer to pay compensation.[488]

Statistics

The *WorkCover Tasmania Scheme Review to 30 June 2014 (April 2015)*[489] is the most comprehensive source of data on workers' compensation claims in Tasmania. There were 7,930 new reported claims in 2013–14, a drop of 1% on 2012–13 (see Figure 3.20).

The claim frequency also shows a 2% decline from the previous year (see Figure 3.21). The claim frequency is a more sensitive measure of claiming behaviour, which is a fairly crude marker.

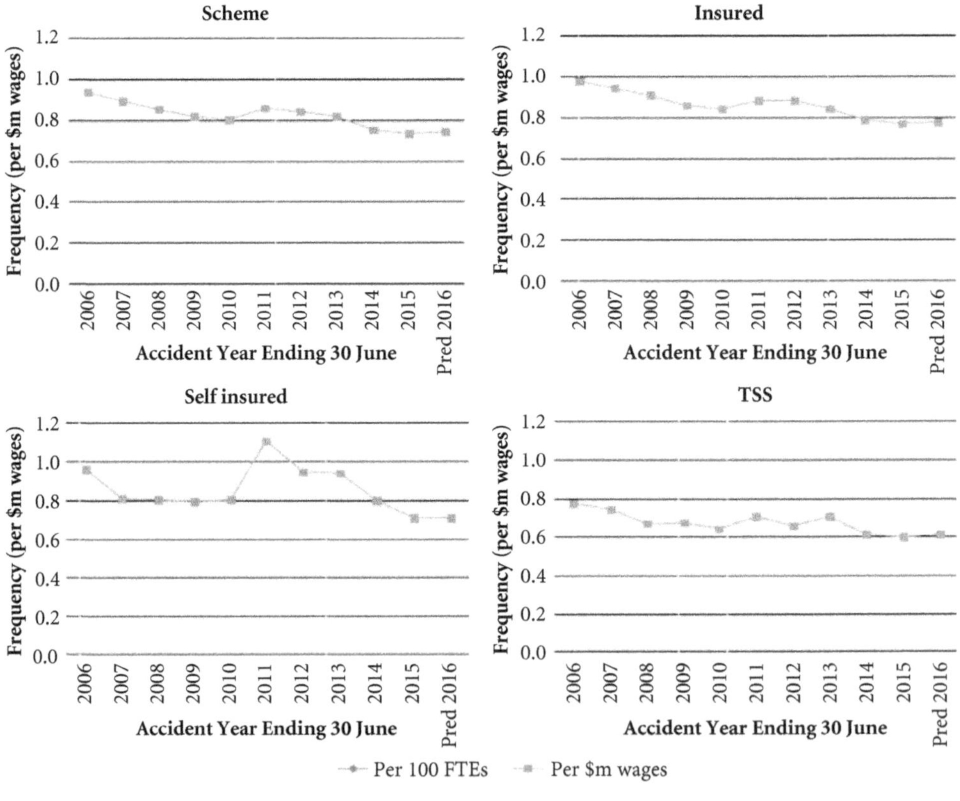

Figure 3.21: Claim frequencies per AU$million in wages for total scheme, insured, self-insured and Tasmanian state services

[488] s 81AB.
[489] Available at www.workcover.tas.gov.au/__data/assets/pdf_file/0003/331860/Scheme_Review_to_30_June_2015.pdf.

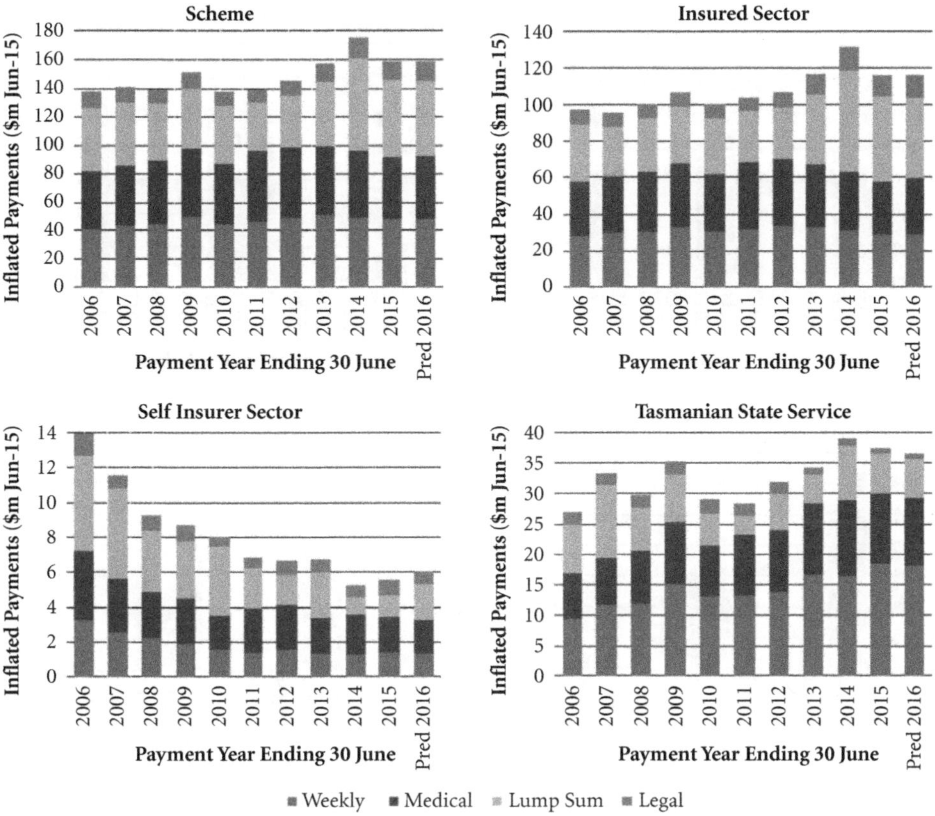

Figure 3.22: Breakdown of inflated claims payments by category for total scheme, insured, self-insured and Tasmanian state services

The breakdown of inflated claim payments is shown in Figure 3.22. Compared to the previous year there have been reductions in payments across all the different categories—weekly, medical, lump sum and legal.

Lost-Time Claims

The number of claims which include a lost-time element has been decreasing across the scheme since 2011, with a slight drop in the proportion of total claims that have a lost-time component (Figure 3.23).

Ultimate Costs of the Scheme

Figure 3.24 shows the ultimate claims cost as a percentage of wages for the scheme as a whole and the insured, self-insured and Tasmanian state service elements. This indicates a relatively stable position, which bodes well for future scheme sustainability.

This is particularly important for this scheme as the insured premium rate has increased by almost 5% per year from 2010/11 to 2013/14 in order to allow insurers to make a profit (see Figure 3.25). It is estimated that insurers have been running at a loss for the 5 years

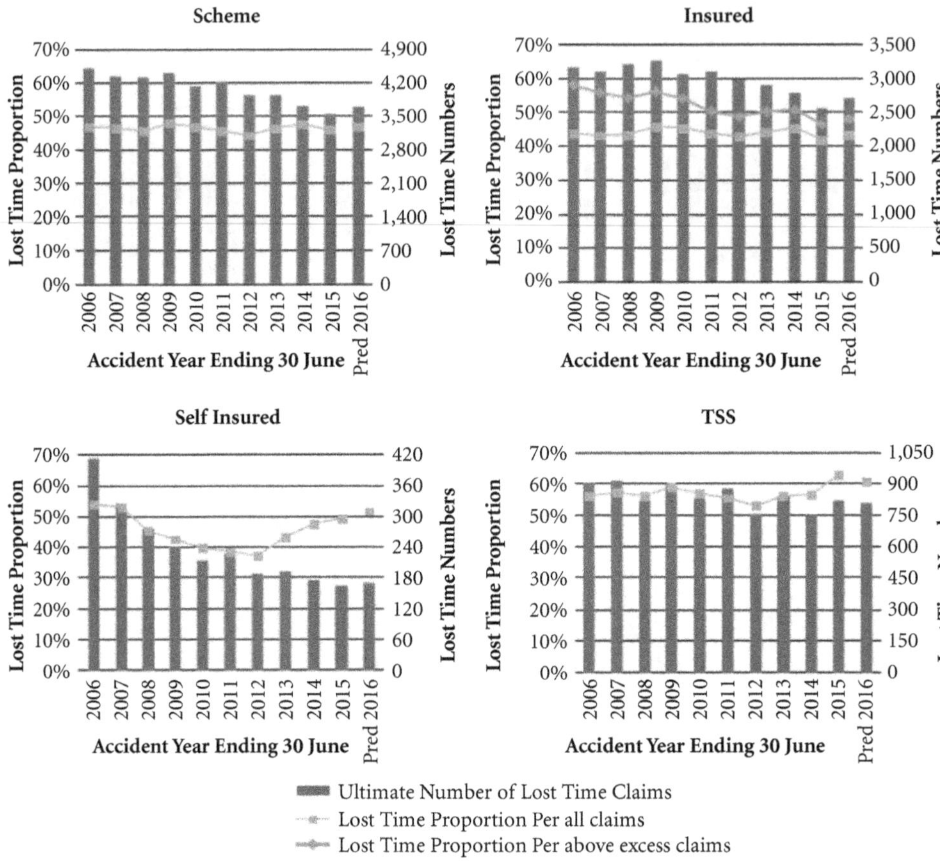

Figure 3.23: Lost-time claims, claim numbers and proportions for total scheme, insured, self-insured and Tasmanian state services

from 2008 to 2012, clearly an unsustainable position in the long term. The increases in premium rate of the preceding years, combined with an increase in 2015/16, should enable insurers to make sufficient profit to meet their target returns on capital. This ought to ensure that the scheme is financially sustainable in the longer term.

The Nominal Insurer

As would be expected of an insurer of last resort, the nominal insurer deals with a very limited number of claims, usually between 3 and 5 per year (see Figure 3.26).

The value of claims to the nominal insurer is variable; Figure 3.27 shows the payments made per year in thousands of dollars.

As expected, when there are lower claim numbers the average incurred claim size under the nominal fund is more volatile than that of the other schemes (see Figure 3.28).

The administration costs of the nominal insurer are shown in Figure 3.29. This shows that the administration costs are variable. The administration expenses as a percentage of the claims payments have been on average 16% for the period 2005–14. This is fairly efficient when you consider the very low claims numbers.

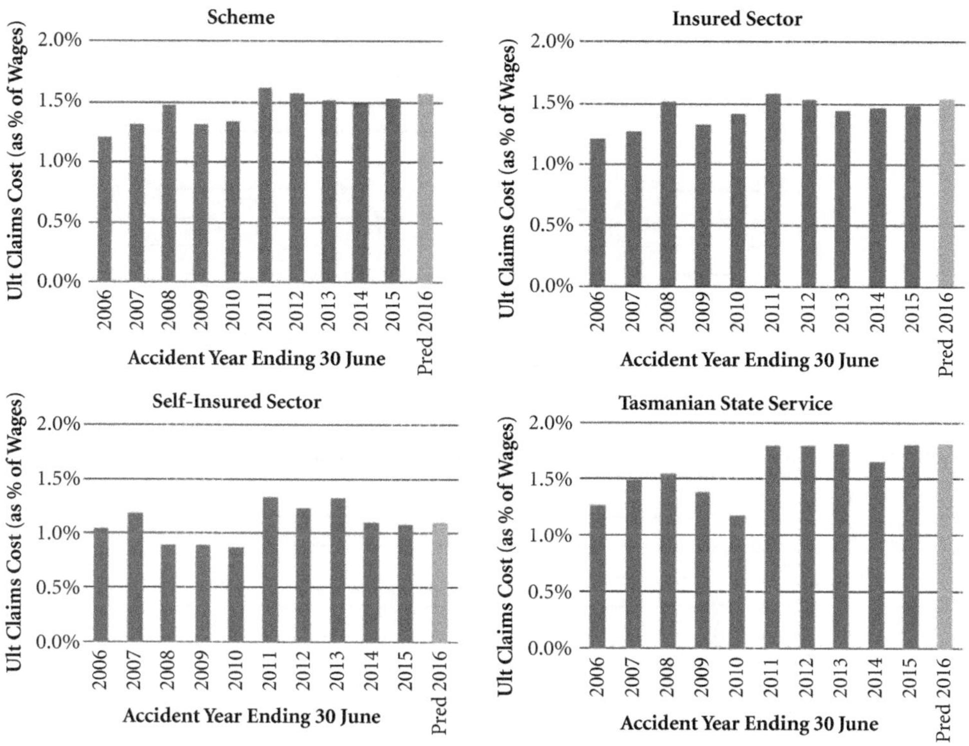

Figure 3.24: Ultimate claims cost as a percentage of wages for the total scheme, insured, self-insured and Tasmanian state services

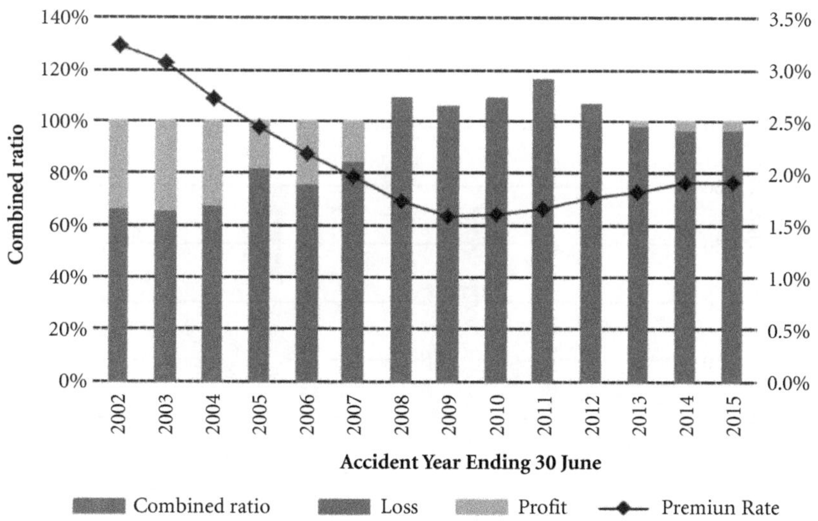

Figure 3.25: Licensed insurer profitability

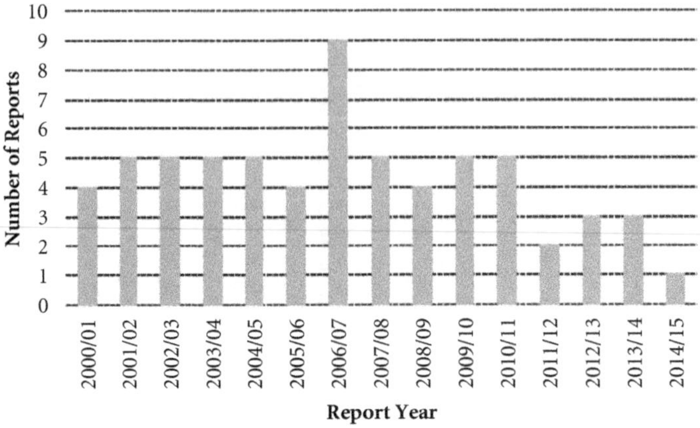

Figure 3.26: Annual claims numbers to the nominal insurer for 2000/01–2014/15

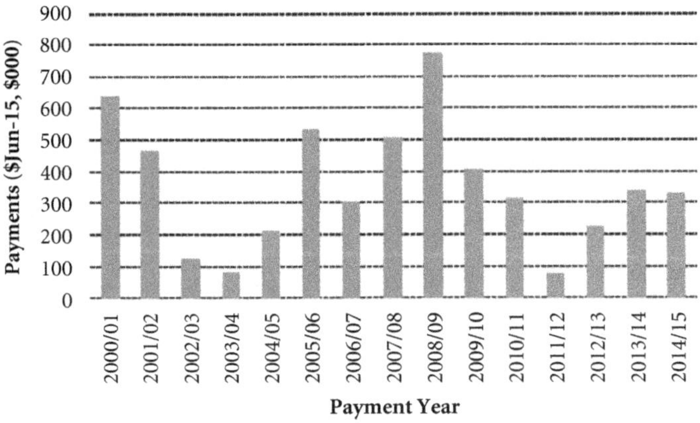

Figure 3.27: Payments made per year by the nominal insurer (in AU$1000s) 2000/01–2014/15

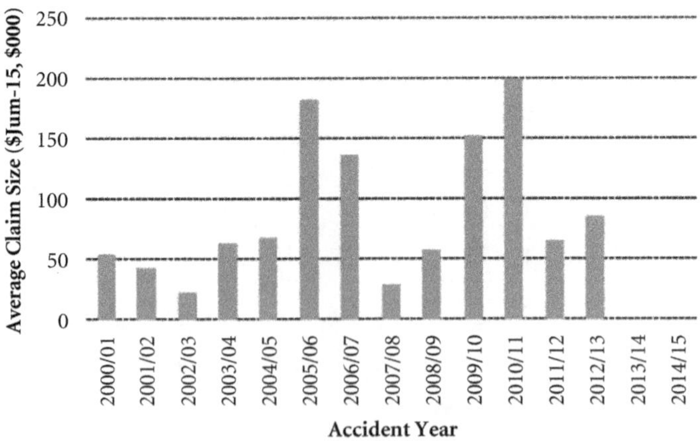

Figure 3.28: Average nominal fund incurred claim size 2000/01–2014/15

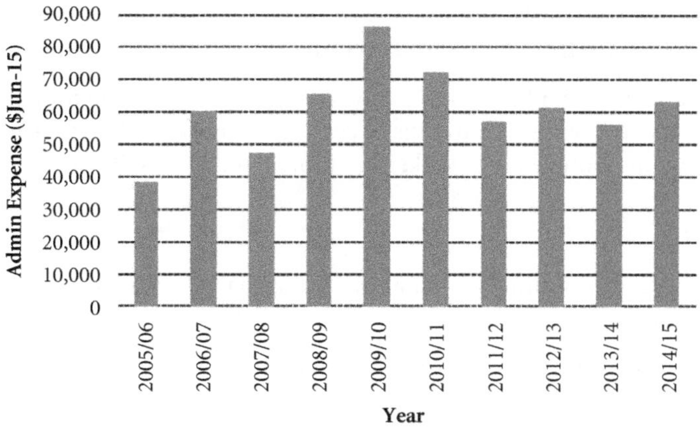

Figure 3.29: Nominal fund administration expenses 2005/06–2014/15

F. The Tasmanian Asbestos Compensation Scheme

General Description

The asbestos compensation scheme provides compensation on a no-fault basis to a worker who develops an asbestos-related disease as a result of exposure to asbestos during the course of his work in Tasmania. The scheme is governed by the Asbestos-Related Diseases (Occupational Exposure) Compensation Act 2011 (the Act). The Act was introduced in response to confusion over the applicability of the Workers Compensation Act 1927 (TAS) and the Workers Rehabilitation and Compensation Act 1988 (TAS). The scheme is designed to have a low burden of proof and process claims quickly. It is funded through a 4% levy applied to all workers' compensation insurance premiums.[490]

Eligibility

To be entitled to compensation under the scheme, a person must be, or must have been, a worker as defined under the legislation at the time of exposure to asbestos.[491] The term 'worker' is taken to apply to a wide range of persons and includes jockeys and apprentice jockeys, ministers of religion, volunteers engaged in firefighting and police volunteers.[492] A worker may be eligible for compensation if:

(i) they have been diagnosed with an asbestos-related disease (ARD); and
(ii) they were exposed to asbestos fibres whilst employed in Tasmania; and

[490] Asbestos-Related Diseases (Occupational Exposure) Compensation Act 2011 (TAS) s 166.
[491] s 12. A worker is a person who has or had entered into or works or worked under a contract of service or training agreement with an employer.
[492] See Div 2.

(iii) they have not previously received any compensation, including through a common law settlement, for the asbestos-related disease they are applying for.[493]

A worker with an ARD who has less than 2 years life expectancy due to that disease must apply for compensation within 12 months of correct diagnosis.[494] There are no time frames to apply for compensation where a worker has more than 2 years life expectancy.[495]

Compensation

Where the worker has an imminently fatal ARD (less than 2 years life expectancy from the date of correct diagnosis):

— The worker is entitled to a lump sum compensation of 360 compensation units (as at 1 January 2012, 1 compensation unit is equal to AU$737.77) plus a further age-based payment (if under 80 years of age).[496]
— The worker is also entitled to have their reasonable medical expenses paid for by the scheme. However, when total medical expenses reach AU$87,000 a review is to be held to ensure that the worker is receiving correct treatment.[497]

Where a worker has a non-imminently fatal ARD (more than 2 years life expectancy from the date of correct diagnosis):

— A worker with a non-imminently fatal ARD must undergo an impairment assessment. Compensation is only payable when there is a whole-person impairment of 10% or more.[498]
— Three lump sum payments are payable to the worker depending on the degree of impairment up to the total of 360 compensation units. However, if the worker is assessed at 51% or more whole person at his first assessment, he will receive all three lump sums at the same time (360 compensation units).[499]
— The worker is also entitled to the payment of reasonable medical expenses. There is no dollar cap on the payment of these expenses.[500]
— Where the worker is still employed, weekly payments are payable based on incapacity.[501]
— Where the worker is subsequently diagnosed as being imminently fatal, or he develops a different imminently fatal ARD, he will be paid any remaining lump sum compensation up to 360 compensation units. He will also receive the age-based payment if eligible.[502]

Members of the family:

— Where a worker has died from an ARD, the members of the worker's family are entitled to the same amount of lump sum compensation (not including payment of

[493] s 6.
[494] ss 36 and 37.
[495] s 38.
[496] s 72(4).
[497] s 118.
[498] s 73(1).
[499] s 73(2).
[500] s 119.
[501] s 85(2).
[502] s 37.

medical expenses or weekly payments) that the worker would have received had he not died.[503] They may also be entitled to funeral expenses in relation to the deceased worker.[504]

— Members of the family include a spouse and a child who is less than 22 years of age (natural child, adopted child and, in some circumstances, a stepchild).[505]

Procedure

All persons suffering from an ARD must obtain a medical certificate with a diagnosis from a specialist (including a determination regarding life expectancy).[506] Once lodged with the Asbestos Compensation Commissioner (ACC), this information is given to the Medical Panel.[507] The Medical Panel is not bound by the rules or practices as to evidence and may act informally, without regard to technicalities or legal forms.[508] The Medical Panel determines a number of questions including:

(i) Is the disease reasonably attributable to exposure during the course of employment?
(ii) If the disease is non-imminently fatal, what is the worker's impairment and incapacity for work if applicable?
(iii) Is the worker likely to recover from the disease?

The Medical Panel then refers its determination to the ACC. The ACC cannot overturn this determination. A determination on whether compensation is payable is made within 28 working days of all evidence being provided to the Medical Panel.[509]

Statistics

Given that the scheme is in the early stages of implementation, the ACC and WorkCover Tasmania have been progressing the establishment of effective monitoring and reporting systems.

As a result, nothing was published during 2011–13. Annual reports of the ACC have been published for 2013/14[510] and 2014/15[511] (see Table 3.7).

Financial Information

In March 2015 the asbestos levy was reduced from 4% to 3.5% as claim levels were lower than expected. This indicates that the fund is financially solvent (see Table 3.8).

503 s 31.
504 s 117.
505 s 5.
506 s 55.
507 s 47.
508 s 65.
509 s 70.
510 Asbestos Compensation Commissioner, *Annual Report 2013–14*, available at www.justice.tas.gov.au/__data/assets/pdf_file/0004/298651/Asbestos_Ann_Rep_2013-14T.pdf.
511 Asbestos Compensation Commissioner, *Annual Report 2014–15*, available at www.justice.tas.gov.au/__data/assets/pdf_file/0009/329481/ACC_Annual_report_2014-15.pdf.

Table 3.7: Claims to the Asbestos Compensation Commissioner 2013–15

	2014/15	2013/14
Worker applications	17	24
Family member applications	1	3
Total applications	18	27
Successful claims	10 paid	13 paid (5 pending at 30 June 2014)

Table 3.8: Financial position of Asbestos Compensation Fund 2013–15

	2014/15	2013/14
Claim payments	$4.95 million	$6.836 million
Reasonable medical expenses	$240,000	$150,000
Impairment assessor and medical panel costs	$78,000	$107,800
Remaining Asbestos Compensation Scheme funds (as at 30 June)	$8.121 million	$6.139 million

Conclusions

Disease claims involve very different presumptions of causation. In Western Australia if a worker suffers from an asbestos-related disease, and his or her employment involves or involved exposure to and inhalation of asbestos fibres, there will be a presumption that this employment has substantially contributed to the contraction of this disease.[512] Accordingly, the onus is on the employer to rebut this presumption through the provision of evidence to the contrary.[513]

This can be contrasted to the New South Wales and Queensland schemes, in which it must be established that the worker's employment was 'reasonably attributable' or a 'significant contributing factor' to the worker's contraction of the disease.[514] Self-evidently the process of seeking compensation in Western Australia is substantially easier than in New South Wales or Queensland.

G. New South Wales

Overview

Workers' compensation is only one component within a comprehensive set of personal injury compensation systems and schemes in New South Wales. In New South

[512] Workers' Compensation and Injury Management Act 1981, s 32.
[513] s 44.
[514] Table 2.4d of Comparison of Workers' Compensation arrangements in Australia and New Zealand. Safe Work Australia October 2016.

Wales provision exists for personal injury compensation across a variety of injury circumstances—from work-related injuries, transport accidents, to sporting mishaps and healthcare-related injuries. The majority of these schemes incorporate common law elements with 'no-fault' models, or restrict common law rights in preference to these alternatives to court.

Recent changes have attempted to harmonise and streamline compensation processes. Most recently the State Insurance and Care Governance Act 2015 reorganised and restructured the compensation landscape. Under this Act from 1 September 2015 responsibility for workers' compensation in New South Wales was assumed by three organisations:[515]

- The State Insurance Regulatory Authority (SIRA) is responsible for the regulation of workers' compensation insurance.
- SafeWork NSW is responsible for workplace health and safety regulation.
- Insurance and Care NSW (icare) is responsible for workers' compensation insurance.

icare's remit includes workers' compensation, the Sporting Injuries Insurance Scheme, Dust Disease Care and Compensation, no-fault lifetime care for those seriously injured in motor vehicle accidents, as well as the Home Building Compensation Fund. The moves towards a more unified system of no-fault compensation in New South Wales is interesting to note, but is unfortunately outside the scope of this chapter, which will focus on workers' compensation arrangements.

Interaction with Common Law Liability

Workers' compensation provides protection to workers and their employers in the event of a work-related injury or disease.[516] Injured workers in New South Wales have two options to obtain compensation: (i) a claim under the statutory Workers' Compensation Scheme;[517] or (ii) alternatively an injured worker can take a common law action against a negligent or at-fault employer. Under the common law damages are only available when a worker sustains an injury that causes a WPI[518] of 15% or more.

Under the Workers' Compensation Act 1987 there are provisions to prevent double recovery.[519]

[515] The websites for these new bodies are still under construction, so while there is some information on the new websites some links will revert to websites for the organisations that they have superseded.

[516] icare (www.icare.nsw.gov.au) was created by the State Insurance Governance and Care Act 2015 No 19 and replaced WorkCover Authority of NSW (www.workcover.nsw.gov.au/) which was abolished under Schedule 4 of this Act. icare's website is, however, currently under construction (June 2016) so may link through to the WorkCoverNSW site.

[517] Framework created by the Workers' Compensation Act 1987, the Workplace Injury Management and Workers' Compensation Act 1988, the Workers' Compensation Regulations 2010 and the Workers' Compensation (Bush Fire, Emergency and Rescue Services) Act 1987 as amended by subsequent legislation.

[518] Under s 151H of the Workers' Compensation Act 1987, compensation will only be awarded if this can be 15% WPI due to physical injury or 15% WPI due to primary psychological/psychiatric injury.

[519] s 9AC of the Workers' Compensation Act 1987.

Insurance Arrangements under the Scheme

Two principal Acts—the Workers Compensation Act 1987 and the Workplace Injury Management and Workers Compensation Act 1998—govern the Workers' Compensation Scheme. The Scheme is funded through insurance premiums paid by employers, which cover financial and medical support to injured workers and the costs of dispute management and scheme administration.[520]

Coverage and Eligibility

A person is eligible for compensation if he or she is a worker[521] who suffers an injury[522] during the course of employment.[523] Schedule 1 of the 1988 Act provides guidance on

[520] Parliament of New South Wales, New South Wales Workers' Compensation Scheme 6, Report 1 (June 2012).

[521] Defined at s 4 of the Workplace Injury Management and Workers Compensation Act 1988:

> *Worker* means a person who has entered into or works under a contract of service or a training contract with an employer (whether by way of manual labour, clerical work or otherwise, and whether the contract is expressed or implied, and whether the contract is oral or in writing). However, it does not include:
>
> (a) a member of the NSW Police Force who is a contributor to the Police Superannuation Fund under the Police Regulation (Superannuation) Act 1906, or
> (b) a person whose employment is casual (that is for 1 period only of not more than 5 working days) and who is employed otherwise than for the purposes of the employer's trade or business, or
> (c) an officer of a religious or other voluntary association who is employed upon duties for the association outside the officer's ordinary working hours, so far as the employment on those duties is concerned, if the officer's remuneration from the association does not exceed $700 per year, or
> (d) except as provided by Schedule 1, a registered participant of a sporting organisation (within the meaning of the Sporting Injuries Insurance Act 1978) while:
>   (i) participating in an authorised activity (within the meaning of that Act) of that organisation, or
>   (ii) engaged in training or preparing himself or herself with a view to so participating, or
>   (iii) engaged on any daily or periodic journey or other journey in connection with the registered participant so participating or the registered participant being so engaged,
>
> if, under the contract pursuant to which the registered participant does any of the things referred to above in this paragraph, the registered participant is not entitled to remuneration other than for the doing of those things.
>
> This definition is extended so that any reference in this Act to a worker who has been injured includes, if the worker is dead, a reference to the worker's legal personal representative, or the worker's dependants, or any other person to whom or for whose benefit compensation is payable.

[522] Defined at s 4 of the Workers' Compensation Act 1987:

> *injury*:
> (a) means a personal injury arising out of or in the course of employment, and
> (b) includes:
>   (i) a disease contracted by a worker in the course of employment, where the employment was a contributing factor to the disease, or
>   (ii) the aggravation, acceleration, exacerbation or deterioration of any disease, where the employment was a contributing factor to the aggravation, acceleration, exacerbation or deterioration, but
> (c) does not include (except in the case of a worker employed in or about a mine):
>   (i) a dust disease, or
>   (ii) the aggravation, acceleration, exacerbation or deterioration of a dust disease.

[523] Sch 1 of the Workplace Injury Management and Workers' Compensation Act 1998 defines deemed employment.

whether a worker is a deemed worker for the purposes of the Act and whom the Act regards as his or her employer. It includes provisions on; workers lent or on hire (including outworkers); other contractors; rural workers; timbergetters; salespersons; canvassers; collectors and others; tributers; mine employees; mine rescue personnel; jockeys and harness racing drivers; drivers of hire vehicles or hire vessels; caddies and others employed in golf; shearers' cooks and others; firefighters in fire districts; workers at place of pick up; boxers; wrestlers; referees and entertainers; voluntary ambulance workers; ministers of religion covered by policies; and participants in training programmes.

The employment that gives rise to the injury must also be 'connected with the State',[524] although the injury can still be compensatable if the worker is outside the State.[525]

To be covered, the injury must also arise 'out of or in the course of employment'.[526] This means that the injury must have occurred while the worker is at the workplace, while the worker was making a journey for work purposes via a direct route (including journeys to and from work),[527] or while the worker was taking a break either onsite or offsite.[528]

The employment must be a 'substantial contributing factor' to the injury[529] (or in the case of a disease injury the employment must be the 'main contributing factor').[530] No payments are made for psychological injuries where the injury results from the reasonable actions of the employer.[531] No payments are made for heart attacks and strokes unless the nature of the work substantially increases the risk.[532]

Deliberately self-inflicted injuries or injuries attributable to the worker's own serious and wilful misconduct will not be compensated (unless these resulted in death or serious and permanent disablement).[533] The Act specifies that a worker is not entitled to compensation where the worker was driving having knowingly consumed alcohol, and where the alcohol materially contributed to the injury.[534]

What Benefits are Available under the Scheme

Workers' compensation provides protection to workers and their employers in the event of a work-related injury or disease or death.[535] As of 1 August 2016 the SIRA Guidelines for claiming workers' compensation apply.[536]

[524] Workers' Compensation Act 1987, s 9AA(1) and (3).
[525] s 9AA(2).
[526] s 4.
[527] s 10.
[528] s 11.
[529] s 9A.
[530] s 4(b).
[531] s 11A.
[532] s 9B.
[533] s 14(2).
[534] s 11(1)(b).
[535] Part 3 of the Workers' Compensation Act 1987.
[536] Under s 376(1)(c) of the Workplace Injury Management and Workers' Compensation Act 1998 these guidelines operate by force of law as delegated legislation. Within these guidelines are both mandatory expectations, defined by the term 'must', and best-practice expectations, defined by the term 'should'. The guidelines are available at www.workcover.nsw.gov.au/__data/assets/pdf_file/0004/84064/Guidelines-for-claiming-workers-compensation-8084.pdf.

Through the workers' compensation system, injured workers may have an entitlement to:

- — weekly payments;
- — permanent impairment payments (including pain and suffering where applicable);
- — medical, hospital and rehabilitative expenses and domestic assistance payments;
- — property damage expenses;
- — Death benefits (available benefits include a lump sum, weekly payments for dependants and other associated expenses such as funeral costs).

When a worker is injured at work, the employer, injured worker, insurer and treatment provider have responsibilities to ensure that the injured worker is provided with benefits and assistance to recover and return to safe, durable work.

Weekly Payments

Tables 3.9 and 3.10, both taken from the SIRA Guidelines, describe the amount of weekly payments due under different circumstances and the definitions used to calculate them.

Permanent Impairment Payments

Compensation for permanent impairment is a lump sum payment. The rules for determining the value of lump sums vary according to whether the injury was sustained before 1 January 2002 or on or after that date (see Table 3.11).

Table 3.9: Weekly workers' compensation benefits available in New South Wales

Entitlement period	Section of the 1987 Act	No current work capacity (inability to return to work in suitable or pre-injury employment)	Has current work capacity (able to return to suitable employment but not pre-injury employment)	
0–13 weeks: 1st entitlement period	36	The lesser of: (AWE × 95%) – D or MAX – D	The lesser of: (AWE × 95%) – (E + D) or MAX – (E + D)	
14–130 weeks: 2nd entitlement period	37	The lesser of: (AWE × 80%) – D or MAX – D	Working less than 15 hours per week	Working 15 hours or more per week
			The lesser of: (AWE × 80%) – (E + D) or MAX – (E + D)	The lesser of: (AWE × 95%) – (E + D) or MAX – (E + D)
53 weeks onwards	44C	Exclude overtime and shift allowance from AWE		

(continued)

Table 3.9: (***Continued***)

Entitlement period	Section of the 1987 Act	No current work capacity (inability to return to work in suitable or pre-injury employment)	Has current work capacity (able to return to suitable employment but not pre-injury employment)
131–260 weeks: special requirements after second entitlement period	38	Cessation of weekly benefits after 130 weeks of weekly payments unless:	
		Worker is assessed as having no current work capacity which is likely to continue indefinitely. The lesser of: (AWE × 80%) – D or MAX – D	Worker has applied to the insurer after 78 weeks for continuation of weekly payments beyond 130 weeks, and is working 15 hours or more and has current weekly earnings of at least the amount specified in s 38(3)(b) (as indexed) per week, and the insurer has assessed the worker is, and is likely to continue indefinitely to be, incapable of undertaking additional employment that would increase the worker's current weekly earnings *or* worker has applied to the insurer after 78 weeks for continuation of weekly payments beyond 130 weeks, and worker is a worker with high needs (as defined in s 32A of the 1987 Act). The lesser of: (AWE × 80%) – (E + D) or MAX – (E + D)
Certain circumstances do not affect payments after 2nd entitlement period	40		Worker continues to have an entitlement under s 38 even if the worker, for up to 4 weeks in the first 12 consecutive week period (or any subsequent consecutive period of 12 weeks of s 38 payments), has: worked more or less hours (even if less than 15 hours) than the hours worked at the time of making the s 38 application, or received higher or lower current weekly earnings (even if less than the amount specified in s 38(3)(b) (as indexed) per week).
Special provision for workers with highest needs	38A	If the determination of weekly payments for a worker with highest needs is less than a minimum amount, the amount payable is to be treated as the minimum amount.	

(continued)

Table 3.9: ***(Continued)***

Entitlement period	Section of the 1987 Act	No current work capacity (inability to return to work in suitable or pre-injury employment)	Has current work capacity (able to return to suitable employment but not pre-injury employment)
Special compensation because of surgery after the 2nd entitlement period	41	Special compensation for incapacity resulting from injury-related surgery is payable at s 37 rate: incapacity must not occur during the first 13 consecutive weeks after the end of the 2nd entitlement period; payable for a maximum period of 13 weeks following surgery; and only if the worker is not otherwise entitled to payment under s 38 after 130 weeks. To be eligible the worker must have received weekly payments and had current work capacity prior to incapacity from injury-related surgery; and must have returned to work after the initial injury for 15 hours or more and have current weekly earnings of at least the amount specified in s 38(3)(b) (as indexed) per week.	
After 5 years (260 weeks) of weekly payments	39	Weekly payments of compensation cease to be payable unless the worker's injury results in a permanent impairment greater than 20%. A worker with high needs may continue after 260 weeks but requires a work capacity assessment to be done at least once every 2 years. Workers with highest needs continue to be entitled to weekly payments without the requirement for a work capacity assessment. A worker's entitlement after 260 weeks is still subject to the requirements of s 38.	

Table 3.10: Definitions used to calculate weekly workers' compensation benefits in New South Wales

Factors	Definition
MAX (section 34 of the 1987 Act)	The maximum weekly compensation amount applicable as prescribed in section 34 and indexed. Refer to www.sira.nsw.gov.au for maximum amount.
D (section 35 of the 1987 Act)	Deductible amount, sum of the value of each non-pecuniary benefit (eg residential accommodation, use of a motor vehicle, health insurance, education fees) provided by the employer to a worker (for the benefit of the worker or their family) in respect of that week.
E (section 35 of the 1987 Act)	Worker's earnings after the injury and is the greater of the amount the worker is able to earn in suitable employment or the worker's current weekly earnings.
AWE (section 44C of the 1987 Act)	Also known as pre-injury average weekly earnings (PIAWE). Average of worker's ordinary earnings during the relevant period plus overtime and shift allowance. Overtime and shift allowance only to be included in the PIAWE calculation in the first 52 weeks.

(continued)

Table 3.10: ***(Continued)***

Factors	Definition
Workers with high needs (section 32A of the 1987 Act)	A worker whose injury has resulted in permanent impairment and: (a) the degree of permanent impairment has been assessed for the purposes of Division 4 to be more than 20%; or (b) an assessment of the degree of permanent impairment is pending and has not been made because an approved medical specialist has declined to make the assessment on the basis that maximum medical improvement has not been reached and the degree of permanent impairment is not fully ascertainable; or Note: paragraph (b) no longer applies once the degree of permanent impairment has been assessed. (c) the insurer is satisfied that the degree of permanent impairment is likely to be more than 20% or (d) includes a worker with highest needs.
Workers with highest needs (section 32A of the 1987 Act)	A worker whose injury has resulted in permanent impairment and: (a) the degree of permanent impairment has been assessed for the purposes of Division 4 to be more than 30%; or (b) an assessment of the degree of permanent impairment is pending and has not been made because an approved medical specialist has declined to make the assessment on the basis that maximum medical improvement has not been reached and the degree of permanent impairment is not fully ascertainable; or Note: paragraph (b) no longer applies once the degree of permanent impairment has been assessed. (c) the insurer is satisfied that the degree of permanent impairment is likely to be more than 30%.

Table 3.11: Rules for determining the amount of lump sum compensation for permanent impairment

For an injury received on or after 1 January 2002	For an injury received before 1 January 2002
Compensation must be based on an assessment of the impairment contained in a medical report from a permanent impairment assessor listed on the SIRA website, as trained in the assessment of the part or body system being assessed	Compensation may be agreed based upon medical reports or negotiated between the parties

Medical, Hospital and Rehabilitative Expenses

Medical, hospital and rehabilitation expenses will be paid where the treatment or service:

— meets the definitions described in section 59 of the 1987 Act;[537]

537 Workers' Compensation Act 1987, s 59.

- takes place while the worker is entitled to receive compensation (the compensation period, see below) for the medical, hospital and rehabilitation expenses;
- is reasonably necessary because of the injury;
- is pre-approved by the insurer (unless the treatment or service is exempt from pre-approval—see below).

Injured workers (and if required, an escort) are entitled to recover necessary and reasonably incurred fares, travel costs and maintenance to cover the cost of travelling for an approved treatment or service. The worker must obtain prior approval from the insurer for the incurred travel costs (unless the travel is for treatment exempt from prior approval). There is no entitlement to travel expenses for a treatment or service that is provided at a location that necessitates more travel than is reasonably necessary.

Definitions under Section 59 of the 1987 Act

Under this section the following definitions are found:

> ***hospital treatment*** means treatment (including treatment by way of rehabilitation) at any hospital or at any rehabilitation centre conducted by a hospital and includes:
> (a) the maintenance of the worker as a patient at the hospital or rehabilitation centre,
> (b) the provision or supply by the hospital, at the hospital or rehabilitation centre, of nursing attendance, medicines, medical or surgical supplies, or other curative apparatus, and
> (c) any other ancillary service,
> but does not include ambulance service.
>
> ***medical or related treatment*** includes:
> (a) treatment by a medical practitioner, a registered dentist, a dental prosthetist, a registered physiotherapist, a chiropractor, an osteopath, a masseur, a remedial medical gymnast or a speech therapist,
> (b) therapeutic treatment given by direction of a medical practitioner,
> (c) (Repealed)
> (d) the provision of crutches, artificial members, eyes or teeth and other artificial aids or spectacles,
> (e) any nursing, medicines, medical or surgical supplies or curative apparatus, supplied or provided for the worker otherwise than as hospital treatment,
> (f) care (other than nursing care) of a worker in the worker's home directed by a medical practitioner having regard to the nature of the worker's incapacity,
> (f1) domestic assistance services,
> (g) the modification of a worker's home or vehicle directed by a medical practitioner having regard to the nature of the worker's incapacity, and
> (h) treatment or other thing prescribed by the regulations as medical or related treatment,
> but does not include ambulance service, hospital treatment or workplace rehabilitation service.

The Compensation Period[538]

Table 3.12 defines the relevant compensation periods.

[538] s 59A.

Table 3.12: The compensation period for different injury types

Criteria	Compensation period
Workers with no permanent impairment or a permanent impairment assessed as 1–10%	Two years from: — when weekly payments stop; or — from the date of claim if no weekly payments made
Workers with a permanent impairment assessed as 11–20%	Five years from: — when weekly payments stop; or — from the date of claim if no weekly payments made
Workers with high needs. This refers to workers: — with a permanent impairment assessed as greater than 20% — where an approved medical specialist has declined to make an assessment as the worker has not reached maximum medical improvement — whose insurer is satisfied that the worker is likely to have a permanent impairment of greater than 20%	For life

Determining What is Reasonably Necessary

For each case a determination will be reached by the insurer on whether a treatment is reasonably necessary. When considering the facts of the case, the insurer should understand that:

- what is determined as reasonably necessary for one worker may not be reasonably necessary for another worker with a similar injury;
- reasonably necessary does not mean absolutely necessary;
- although evidence may show that the similar outcome could be achieved by an alternative treatment, it does not mean that the treatment recommended is not reasonably necessary.

The above points should be sufficient in most cases for an insurer to determine reasonably necessary. If the insurer remains unclear on whether a treatment is reasonably necessary, then the following factors may be considered:

- the appropriateness of the particular treatment;
- the availability of alternative treatment;
- the cost of the treatment;
- the actual or potential effectiveness of the treatment;
- the acceptance of the treatment by medical experts.

Approved Treatment and Exemptions from the Need for Pre-approval[539]

The majority of treatments require pre-approval by an insurer. Table 3.13 details treatments and services (including transport to and from) which can be reimbursed if accessed without the prior approval of the insurer.

[539] s 60.

Table 3.13: Treatment costs which can be reimbursed if accessed without prior approval of the insurer

Treatment	Expense
Initial treatment	Any treatment within 48 hours of the injury happening
Nominated treating doctor	Any consultation or case conferencing for the injury, apart from telehealth and home visits Any treatment during consultation for the injury, within 1 month of the date of injury
Public hospital	Any services provided in the emergency department, for the injury Any services after receiving treatment at the emergency department for the injury, within 1 month of the date of injury
Medical specialists	If referred by the nominated treating doctor, any consultation and treatment during consultations for the injury (apart from telehealth) within 3 months of the date of injury Note: Medical specialist means a medical practitioner recognised as a specialist in accordance with the Schedule 4 of Part 1 of the Health Insurance Regulations 1975 who is remunerated at specialist rates under Medicare.
Diagnostic investigations	If referred by the nominated treating doctor for the injury: — any plain X-rays, within 2 weeks of the date of injury — ultrasounds, CT scans or MRIs within 3 months of the date of injury, where the worker has been referred to a medical specialist for further injury management. On referral by the medical specialist for the injury, any diagnostic investigations within 3 months of the date of injury Note: A general practitioner's MRI referral must meet the Medicare Benefits Schedule criteria
Pharmacy	Prescription and over-the-counter pharmacy items prescribed by the nominated treating doctor or medical specialist for the injury and dispensed: — within 1 month of the date of injury; or — after 1 month of the date of injury if prescribed through the Pharmaceutical Benefits Scheme.
SIRA-approved physical treatment practitioners (physiotherapist, osteopath, chiropractor, accredited exercise physiologist)	Up to eight consultations if the injury was not previously treated and treatment starts within 3 months of the date of injury Up to three consultations if the injury was not previously treated and treatment starts over 3 months after the date of injury Up to eight consultations per allied health recovery request (AHRR) if the same practitioner is continuing treatment within 3 months of the date of injury and: — the practitioner sent an AHRR to the insurer; and — the insurer did not respond within 5 working days of receiving the AHRR.

(continued)

Table 3.13: *(Continued)*

Treatment	Expense
	One consultation with the same practitioner if the practitioner previously treated the injury over 3 months ago. This is a new episode of care One consultation with a different practitioner if the injury was previously treated. Up to 2 hours per practitioner for case conferencing that complies with the applicable Fees Order Up to $100 per claim for reasonable incidental expenses for items the worker uses independently (such as strapping tape, TheraBand, exercise putty, disposable electrodes and walking sticks) Notes: — Consultations with an accredited exercise physiologist require a referral from a medical practitioner — All treatments exclude home visits, telehealth and practitioner travel — A list of SIRA-approved practitioners can be found at www.sira.nsw.gov.au See the SIRA Workers' Compensation Guideline for the approval of treating health practitioners for more on practitioner approval.
SIRA-approved psychologist or counsellor	Up to eight consultations if a psychologist or counsellor has not previously treated the injury and treatment starts within 3 months of the date of injury Up to three consultations if a psychologist or counsellor has not previously treated the injury and treatment starts over 3 months after the date of injury. Up to eight consultations per allied health recovery request (AHRR) if the same practitioner is continuing treatment within 3 months of the date of injury and: — the practitioner sent an AHRR to the insurer; and — the insurer did not respond within 5 working days of receiving the AHRR. One consultation with the same psychologist or counsellor if the practitioner previously treated the injury over 3 months ago. This is a new episode of care One consultation with a different psychologist or counsellor if the injury was previously treated Up to 2 hours per practitioner for case conferencing that complies with the applicable Fees Order Up to $100 per claim for reasonable incidental expenses for items the worker uses independently (such as relaxation CDs and self-help books) Notes: — These consultations require a referral from a medical practitioner — All treatments exclude home visits, telehealth and practitioner travel — A list of SIRA-approved practitioners can be found at www.sira.nsw.gov.au See the SIRA Workers' Compensation Guideline for the approval of treating health practitioners for more on practitioner approval.

Domestic Assistance Payments[540]

These payments are intended to cover the cost for assistance with domestic tasks, such as gardening or laundry, which the worker is unable to do due to the injury. This section also covers some travel expenses which would not fall within the scope of the medical, hospital and rehabilitative compensation offered under the Act.

Standard domestic assistance payments are calculated by drawing up a care plan and calculating the need for domestic assistance and a reasonable rate of pay for that assistance. There is provision in the Act for paying gratuitous domestic assistance, defined as domestic assistance provided to an injured worker for which the injured worker has not paid and is not liable to pay. Gratuitous domestic assistance is only paid where the assistance provider has suffered a loss of income or foregone employment as a result of providing the assistance.

Rates of pay for domestic assistance are capped.[541] Gratuitous domestic assistance is also capped at a maximum of 35 hours per week. The hourly rate is calculated by taking the Australian Bureau of Statistics' full-time adult average weekly (ordinary time) earnings of all NSW employees and dividing this number by 35.

Property Damage[542]

A worker can claim for the cost of repair or replacement or certain items of property if the property is damaged because of a work-related accident. Claims can be made for crutches; artificial limbs; eyes or teeth; other artificial aids; spectacles; clothes; and any fees paid for or wages lost by the worker due to attending a consultation, examination or prescription to replace the property. The worker does not have to be injured to claim for property damage.

Reasonable costs are paid for the damaged property, up to $600 for spectacles and up to $2,000 for all other items.

Death Benefits[543]

Available death benefits include lump sum payment and weekly payments to dependants, and other expenses (including funeral expenses and the cost of transporting the deceased's body).

— *Lump sum.* In the event that an employer is liable for the death of a worker a lump sum payment of $750,000 is due.[544] This is either paid to the worker's dependant(s) (and apportioned between them if there is more than one according to the provisions of the 1987 Act[545]); if there are no dependants, then it is paid to the worker's legal representative.

— *Weekly payments.* In addition to the lump sum payment, an amount of $66.60 per week is paid in respect of each dependent child under the age of 16 years and each dependent child being a student over the age of 16 years but under the age of 21 years.[546]

540 s 60AA(4).
541 ss 60AA and 61(2).
542 Division 5 of the Workers' Compensation Act 1987.
543 Division 1 of the Workers' Compensation Act 1987.
544 Workers' Compensation Act 1987, s 25.
545 s 29.
546 s 29.

— *Funeral expenses.*[547] These are capped at $15,000. If required, the cost of transporting the deceased's body is calculated separately.

Commutations

In addition, commutations are also available. A commutation is an agreement between the injured worker, employer and Scheme Agent or insurer to pay all of the injured worker's entitlements to weekly benefits, medical, hospital and rehabilitation expenses as a lump sum. By agreeing to a commutation, the injured worker's entitlements to weekly payments and all other expenses will no longer be paid and the Scheme Agent or insurer will not be liable for further claims with regards to the injury. SIRA must also certify the commutation meets all the criteria set out in section 87EA of the Workers' Compensation Act 1987.[548]

A commutation is only available when the following preconditions have been met:

— the injured worker must have a permanent impairment that is at least a 15% WPI;
— compensation for permanent impairment and pain and suffering has been paid;
— it is 2 or more years since the worker first received weekly payments for the injury;
— all opportunities for injury management and return-to-work have been fully exhausted;
— the worker has received weekly benefits regularly and periodically throughout the previous 6 months;
— the worker must be entitled to existing and ongoing weekly benefits;
— weekly benefits have not been stopped or reduced as a result of the worker failing to meet return-to-work obligations.

Prior to receiving a commutation:

— the worker must receive independent legal and financial advice;
— the Scheme Agent or insurer, employer and worker must agree with the commutation;
— SIRA must approve the commutation;
— all agreements must be registered with the Workers' Compensation Commission.

The legislation provides different amounts of benefits depending on the period the worker has received entitlements, the pre-injury average weekly earnings, the date of injury and the date the claim was lodged.

Other Requirements on Workers, Employers and Insurers

Financial provision is available to assist workers with returning to work.[549]

When an injury prevents a worker from returning to his or her pre-injury employment and the worker finds work with a new employer, the pre-injury employer is liable to pay return-to-work costs. These costs include the provision of education or training, transport, childcare, clothing, equipment and any other similar service or assistance. The amount payable for such claims is capped at $1,000.[550]

[547] ss 26 and 28.
[548] ss 26 and 28.
[549] ss 68B and 68C.
[550] s 68B(5).

When a worker has an injury which has resulted in a degree of permanent impairment or over 20% and has been in receipt of weekly benefits for an aggregate period of more than 78 weeks, then the employer is liable to pay compensation for the cost of education and training of the worker to assist the worker to return to work. Payments under this section are capped at $8,000.

How Claims Are Made

Notification

The first step of all claims is that the employer and insurer are notified. The worker is required to give notice (written or verbal) to the employer that an injury has occurred, usually as soon as possible after the incident. Notice must include the name and address of the injured, how the injury happened, and the date upon which the injury happened.[551] The employer is required to keep a readily accessible record of all workplace injuries.[552]

Employers have 48 hours from the time they become aware of the injury to notify their insurer or the Nominal Insurer (initial notification). Initial notification can also be made by the worker or a representative of the worker or the employer. Initial notification can be in writing or verbal.[553]

Notification must include the information detailed in Table 3.14.

Table 3.14: Details required in the initial notification

Worker	Name Contact details (including phone number and postal address)
Employer	Business name Contact details
Treating doctor (where relevant)	Name Name of medical centre or hospital
Injury	Date of the injury or period over which the injury emerged Time of the injury Description of how the injury happened Description of the injury
Notifier	Name Relationship to the worker or employer Contact details (including phone number and postal address)

551 Workplace Injury Management and Workers Compensation Act 1998, ss 244 and 245.
552 s 246.
553 ss 44 and 246.

If initial notice is given to the relevant insurer, then they must forward the initial notice to the Nominal Insurer. If initial notice is given to the Nominal Insurer, then the Nominal Insurer must forward that notice to the relevant insurer as soon as is practical and must give notice to SIRA as soon as is practical.[554]

An insurer has 3 days from initial notification to identify the policy covering the injury. If the insurer cannot find a policy that covers the injury, the insurer should contact the employer and the notifier to learn more and should also:

- tell the worker, employer and notifier that it is not the current insurer;
- refer the matter to the SIRA Customer Service Centre;
- if the current insurer can be identified, then the notification should be immediately passed to the current employer and the worker, employer and notifier should be advised in writing of this.

If the initial notification is incomplete, then the insurer should reply to the notifier (and the worker if possible) within 3 business days, specifying the information required to complete the notification.

The date on which a completed notification is received is the 'initial notification date' for the purposes of provisional payments.[555] Upon receipt of a completed initial notification the insurer should review the information and start to determine liability. The insurer must also either start provisional payments or delay starting weekly provisional payments due to a reasonable excuse.

Provisional Payments

These are made before an insurer determines liability and cover:

- up to 12 weeks' loss of income; and
- up to $7,500 of reasonably necessary medical treatment costs.[556]

Provisional payments must be started within 7 days of the initial notification date unless there is a reasonable excuse not to pay them (reasonable medical treatment costs cannot be withheld).[557] Insurers must notify the worker in writing when they start provisional payments and should notify the employer. This notice must explain: that the payments have started on a provisional basis; how long the payments are expected to last for; that an injury management plan will be drawn up if the worker is unable to return to their pre-injury employment for 7 continuous days; and that the worker can make a claim and details of how to claim.

The notice should also include: the worker's pre-injury average weekly earnings; the amount of weekly payment and how that amount was calculated (including a copy of the completed PIAWE form if one has been provided); who will pay the worker (the insurer or the employer); information on what to do if the worker disagrees with the amount or does not receive payment; what information (if any) the worker needs to supply to the insurer

[554] s 44.
[555] s 266.
[556] ss 267, 275 and 280.
[557] s 279.

to enable payments to continue. A claim form and the Information for Injured Workers Brochure should also be supplied.[558]

If the notice includes information about a work capacity decision, then this must be communicated as per the legislative requirements.

Provisional payments are calculated on the same basis as weekly payments.

Reasonable Excuses for Delaying Provisional Payments[559]

Table 3.15, taken from the Guidelines, details the acceptable reasons for delaying provisional payments. Failing to commence provisional payments without a reasonable excuse is an offence with a maximum fine of 50 penalty units.

Table 3.15: Reasonable excuses for delaying provisional payments

There is insufficient medical information	The insurer does not have enough medical information to establish that there is an injury, as a workers' compensation certificate of capacity, or other medical information certifying that an injury has occurred, has not been provided. Note: use discretion for workers in remote areas if access to medical treatment is not readily available.
The injured person is unlikely to be a worker	The person cannot verify they are a worker or the employer can verify that they are not a worker. If there is any doubt that someone is a worker under the workers compensation law, the insurer must verify that person's status. Information that confirms this may include but is not limited to: — the employer agreeing to the worker's status; — the worker's payroll number; — a current payslip or a bank statement with regular employer payments; — a contract of employment or services. Sections 4 and 5, and Schedule 1, of the 1998 Act
The insurer is unable to contact the worker	The insurer has not been able to contact the worker after at least: — two attempts by phone (made at least a day apart); — one attempt in writing (which may include an attempt by email).
The worker refuses access to information	The worker will not agree to the release or collection of personal or health information relevant to the injury to help determine their entitlement to compensation.
The injury is not work related	The insurer has information that: — the worker did not receive an injury which is compensatable under the NSW workers' compensation law; or — strongly indicates that compensation for an injury may not be payable under the workers' compensation law.

(continued)

[558] s 269.
[559] s 267.

Table 3.15: ***(Continued)***

	If the employer believes the injury is not work related, it should provide to the insurer supporting information, such as: — medical information that the condition already existed and has not been aggravated by work; — factual information that the injury did not arise from or during employment. Note: suspicion, innuendo, anecdotes or unsupported information from any source, including the employer, is not acceptable. Section 4 of the 1998 Act and sections 9A, 9B and 10 of the 1987 Act
There is no requirement for weekly payments	A notification of injury has been received; however, the insurer has verbal or written confirmation from the worker and employer that they do not anticipate that weekly payments will be claimed.
The injury is notified after 2 months	The notice of injury is not given to the employer within 2 months of the date of injury. Note: The insurer should disregard this excuse if liability is likely to exist and it believes provisional payments will be an effective way to manage the injury

Insurers with a reasonable excuse to delay provisional payments must give the worker written notice within 7 days of receiving the initial notification and should also tell the employer in writing as soon as possible.[560]

The notice to the worker must set out:

- — the excuse(s), and should include copies of all relevant information considered in the decision;
- — that the worker can still make a claim for compensation, which the insurer will determine within 21 days of receipt;
- — how the worker can make that claim.

Insurers should supply a claim form and explain:

- — how the excuse can be resolved;
- — that the worker can talk to the insurer for further information;
- — that the worker can seek help from their union, SIRA's Customer Service Centre or a Workers' Compensation Independent Review Officer;
- — that the worker has a right to seek an expedited assessment by application to the Registrar of the Workers' Compensation Commission.

Submitting a Claim Form[561]

Workers can submit claim forms to insurers at any time. Forms are available online from the SIRA website. There are two circumstances where a claim form is mandatory:

- — if a reasonable excuse notice has been issued and the excuse continues to exist; or

560 s 268.
561 ss 260 and 261.

— if compensation is likely to be claimed beyond provisional payments and the insurer determines that there is insufficient information to determine ongoing liability

In other circumstances a claim form is not required; if an insurer feels that they have sufficient information to determine liability they can forego the requirement on the worker to provide one.

Ordinarily a claim must be made within 6 months of the injury, with some exceptions for certain circumstances.

Workers making a claim must supply sufficient information to demonstrate the following:

— they were employed;[562]
— they received an injury from or during the employment;
— that they have lost income, need(ed) medical treatment or may incur other expenses due to the injury.

Employers must forward a claim or any other details relating to a claim to the insurer within 7 days, and have 7 days to respond to questions from an insurer about a claim.[563]

Upon receipt of a claim an insurer has three options available:

— to inform the other parties that a policy cannot be found, as detailed above;
— to accept liability; or
— to decline liability.

Determining Liability

Liability is determined on the basis of the facts. There are different timescales for the different claim types (see Table 3.16).

Table 3.16: Timescales for establishing liability for different claim types

Compensation	Legislation	Outcome	Timescale
Weekly payment	S 275 of the 1998 Act	Accept liability and start weekly payments or dispute liability	Within 21 days, unless provisional weekly payments have commenced, if so then before the provisional payments finish (max 12 weeks)
Permanent impairment payments	S 259 of the 1998 Act	Accept liability and make a reasonable offer of settlement or dispute liability	Within 1 month of an injury being 'fully ascertainable', otherwise within 2 months of the worker providing all relevant information Fully ascertainable means the degree of injury has been either agreed by all parties or determined by an approved medical specialist and not appealed

(continued)

[562] As per ss 4, 5 and Schedule 1 of the Workplace Injury Management and Workers Compensation Act 1998.
[563] Workplace Injury Management and Workers' Compensation Act 1998, s 264.

Table 3.16: (*Continued*)

Compensation	Legislation	Outcome	Timescale
Medical, rehabilitative, etc	S 279 and 280 of the 1998 Act	Accept liability and start payments or dispute liability	Within 21 days, unless provisional medical expenses payments have commenced, if so then before the provisional payments end (max $7,500) If the insurer has approved the expense, then it is liable unless: — the entitlement stops due to s 59A of the 1987 Act; or — the insurer informs the worker that it is disputing liability for the service before the service is provided
Domestic assistance	S 279 and 280 of the 1998 Act	Accept liability and start payments or dispute liability	Within 21 days
Property samage	S 289 of the 1998 Act	Accept liability and start payments or dispute liability	Within 28 days
Death denefits		Accept liability and start payments or dispute liability	The insurer has a duty to act promptly

Different types of claims require different information to allow the insurer to establish liability as outlined in Table 3.17 and further detailed below.

Table 3.17: Information required by insurers to establish liability

Compensation	Eligibility	Information Required
Weekly payment	Incapacity due to an injury causing loss of income	Sufficient information to determine if the injury was due to work A Worker's Compensation Capacity Certificate supplied by the worker and the treating medical professional If required by the insurer capacity to work is determined by a Work Capacity Assessment Insurers should ask the employer and the worker for information to calculate the worker's pre-injury average weekly earning
Permanent impairment payments	Physical injury causing a permanent impairment of 10% or more for physical injury or 15% or more for primary psychological injury	Sufficient information to determine if the injury was due to work A report from a Permanent Injury Assessor

(continued)

Table 3.17: (*Continued*)

Compensation	Eligibility	Information Required
Medical, rehabilitative, etc	Medical treatment as defined by s 59 of the 1987 Act	The treatment must be: — undertaken within the compensation period; — 'reasonably necessary'; and — either pre-approved by the insurer or exempt from the need for approval
Domestic assistance	Domestic assistance as defined by s 60A of the 1987 Act	Must be certified by a medical practitioner, based on a functional assessment that the assistance is: — 'reasonably necessary', and the necessity arise directly from the worker's injury; — the worker did the tasks before the injury; — the injury has caused either a permanent impairment of 15% or more or if the assistance is temporary—up to 6 hours per week is available for a total period of 3 months (not necessarily consecutive); and — is in accordance with a domestic assistance care plan set up by the insurer
Property damage	Cost of repairing or replacing property as defined in Div 5 of Part 3 of the 1987 Act	Sufficient information for the insurer to determine: — the damage was due to employment; — the type and value of the damage items; and — how they were damaged
Death benefits	Section 4 of the 1998 Act	Sufficient information for the insurer to determine liability, such as information from the employer and witnesses, any factual investigations, the death certificate, treating medical records, coroner's or autopsy reports, police reports, etc Sufficient information for the insurer to determine dependency, such as factual investigations, birth or death certificates, marriage certificates, statutory declarations from possible dependants, family and those close to the deceased, financial records, etc

Weekly Payments

To be eligible for weekly payments a worker must supply sufficient information to determine if the injury was due to work. Insurers should ask the employer and the worker for information to calculate the worker's pre-injury average weekly earnings.

Workers must also provide a completed Worker's Compensation Certificate of Capacity[564] (supplied by the treating medical practitioner).

Parts A and C of the certificate are completed by the worker. Part B of the certificate must be completed by the medical practitioner and must:

— specify a period, usually of no more than 28 days;

[564] Workers' Compensation Act 1987, s 44B.

— certify the worker's capacity for work during this period (which must be no more than 90 days before the date the certificate is provided);
— state the expected length of the incapacity.

The worker can give the certificate to his insurer or to his employer, who must forward it to the insurer within 7 days.[565]

Work Capacity Assessments[566]

Insurers are entitled to ask the claimant for information on his or her capacity to work in the form of a Work Capacity Assessment. Work Capacity Assessments are performed by the insurer with the intention of establishing whether a worker has either any current work capacity or no current work capacity.[567]

— Current work capacity—the worker's injury is such that the worker is not able to return to his or her pre-injury employment but is able to return to work in suitable employment
— No current work capacity—the worker's injury is such that the worker is currently unable to return to work, either in the worker's pre-injury employment or in other suitable employment

Work Capacity Assessments should consider two questions:

1. Does the worker have a present ability to return to his or her pre-injury employment?
2. Does the worker have a present ability to return to suitable employment?

The complexity of a work capacity assessment will depend upon the circumstances of the case. Insurers should be governed by the principles of procedural fairness: insurers will need to determine what the principles of procedural fairness require on a case-by-case basis.

A worker is able to provide any information to the insurer that he or she wishes to be considered in a work capacity assessment (eg certificate of capacity, treating specialist reports, job description).

The insurer must keep a record of a Work Capacity Assessment in the worker's file, which must include: the work capacity assessment date; if applicable, dates of contact with the worker and case notes of discussion points; the details and dates of any other assessment the worker had to attend; the assessor's identity; the outcome of the assessment (eg whether a work capacity decision is required).

If it is determined that the worker cannot return to his or her pre-injury employment, then the insurer must assess if the worker can instead work in other suitable employment. If so, then the insurer must also identify the type(s) of employment which the worker is currently suited to.[568]

The timings of Work Capacity Assessments are shown in Figure 3.30.[569] If after 78 weeks of weekly payment it is likely that the worker's entitlement to weekly payments

[565] Workplace Injury Management and Workers' Compensation Act 1998, s 264.
[566] Workers Compensation Act 1987, s 44A.
[567] ss 32A and 44A.
[568] Suitable work is defined at s 32A of the Workers' Compensation Act 1987.
[569] Workers Compensation Act 1987, ss 32A and 38.

will continue beyond 130 weeks of weekly payments, then the insurer must perform a Work Capacity Assessment. This assessment must be completed prior to the worker accumulating 130 weeks of weekly payments. Before the assessment the insurer must notify the worker in writing, including; the purpose of the assessment; what information the worker can provide to the insurer for consideration; the expected completion date; the matters it may decide on.

If a worker is assessed as having a current work capacity, the insurer must provide the worker with an application form for continued weekly payments after 130 weeks and instructions that he or she needs to complete this form for weekly payments to continue.

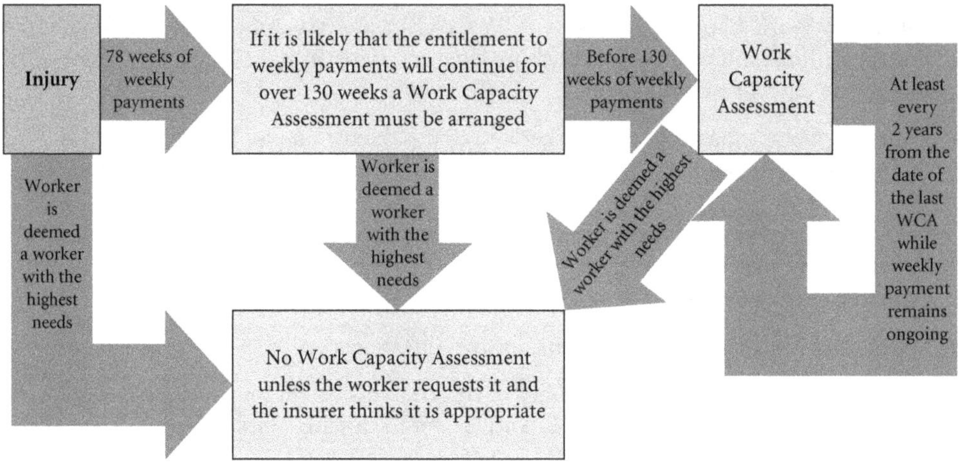

Figure 3.30: Timelines for work capacity assessments

Should an insurer require more information to be able to assess work capacity they can require the worker to attend assessment appointments, providing the assessment appointment is reasonably necessary. There is a maximum of four assessment appointments per Work Capacity Assessment, with no more than one appointment per medical speciality or healthcare professional speciality.[570]

If an assessment appointment with an independent medical examiner is requested it must be conducted in accordance with the guidelines on independent medical examinations and reports. The insurer must advise the worker of the date and time of each appointment at least 10 working days before the appointment occurs.[571] The advice must include:

- the location of the appointment;
- the purpose of the appointment and how it may inform the Work Capacity Assessment;
- that refusing to attend, or failing to properly participate so that the assessment at the appointment cannot take place, may result in the insurer suspending weekly payments until the assessment appointment is completed.[572]

If a worker has refused to attend an assessment or the assessment did not take place due to the failure of the worker to properly participate in the assessment, the insurer may suspend

[570] s 44A(5).

[571] If the worker agrees, less than 10 days' notice can be given for the assessment; the insurer should keep a record of any such agreement.

[572] Workers' Compensation Act 1987, s 44A(6).

the worker's weekly payments. Before suspending the payments, the insurer should be satisfied that it possesses sufficient information to confirm that the failure of the assessment was due to the worker. The insurer should advise the worker that weekly payments will remain suspended until the assessment appointment has taken place. Where suspension has occurred, the insurer should expedite the new assessment appointment and advise the worker of the details.[573]

Work Capacity Decisions[574]

Work capacity decisions can encompass a variety of issues, including, but not limited to:

- the worker's current work capacity;
- what constitutes suitable employment for the worker;
- how much the worker can earn in suitable employment;
- the worker's PIAWE or current weekly earnings;
- whether the injury means that the worker is unable to engage in a particular type of employment because the nature of that employment involves substantial risk of further injury;
- any other decision of an insurer that affects a worker's entitlement to weekly payments of compensation, including a decision to suspend, discontinue or reduce the amount of the weekly payments of compensation payable to a worker on the basis of any decision referred to in the above bullet points.

Work capacity decisions may be made regularly throughout the life of a claim. Insurers should consider the principles of procedural fairness, including fair notice, when making any decision that may affect a worker's rights or interests. The potential impacts on the worker determine the degree of complexity and information gathering required to formulate a work capacity decision. Decisions which impact upon payments will usually require the insurer to provide the worker with all the information and reasons used to reach the decisions, and should include information for the worker on the right to request a review if he or she does not agree with the decision.

When a worker has received at least 12 continuous weeks of weekly payments any work capacity decision that reduces or terminates the worker's weekly payments must be preceded by a 3-month notice period before the changes takes effect. This provides an opportunity for workers to seek a review of the decision and to submit additional information to be considered in the review.

Reviewing a Work Capacity Decision

A worker can ask for a work capacity decision to be reviewed. There are three levels of administrative review:

- an internal review, where the insurer undertakes the review and informs the worker of the review decision;
- a merit review, where SIRA undertakes the review and informs the worker and insurer of its findings and recommendations;

[573] s 44A(6).
[574] s 43.

- — a procedural review, where the Workers' Compensation Independent Review Officer (WIRO) undertakes the review and informs the worker, insurer and SIRA of its findings.

Additionally, a worker can seek judicial review of a decision by the Supreme Court of NSW.

A worker can apply to the insurer to perform an internal review of the work capacity decision at any time. The worker must apply for an internal review within 30 days of receiving the work capacity decision advice for a stay of the decision to apply during the internal review process. The application must be made by supplying a completed Work Capacity—Application for Internal Review by Insurer form to their insurer. The form is available from www.sira.nsw.gov.au or the insurer. The application form must identify the decision that the worker is requesting be reviewed and include the worker's reasons for seeking the review of the decision. It can also include additional relevant information for the insurer to consider (eg medical or employment information). The insurer cannot refuse to perform an internal review after receiving a review application.

Where the work capacity decision involves a decrease or loss of payments, an application for review may stay the operation of the decision. A stay is only applicable if the application for review is made within a set timeframe (see Table 3.18). A stay temporarily prevents the insurer taking action on the decision for the period between the application for the review and the notification of the decision or findings of the review. The purpose of the stay is to provide protection to the worker by maintaining their weekly payments while the review is being undertaken. A stay can only prevent a decision taking effect. It cannot reinstate what has already occurred. A stay does not alter the required period of notice contained in the work capacity decision. If the required period of notice expires at a time when a review is not being undertaken, the worker's weekly payment of compensation will be reduced or discontinued in accordance with the work capacity decision. If the worker then subsequently applies for a review the worker's weekly compensation rate during the review will continue at the rate shown in the work capacity decision.

A stay will no longer apply if a worker withdraws a review application.

Table 3.18: Required timeframes for a review application to create a stay

Review type	**A stay applies where the worker:**
Internal review	Applies to the insurer within 30 days of receiving the work capacity decision
Merit review	Applies to SIRA: — within 30 days of receiving the insurer's internal review decision (or after 30 days from making an application to the insurer for an internal review, where the insurer has failed to conduct the internal review and notify the worker within 30 days); and — before the work capacity decision has taken effect after the required notice period
Procedural review	Applies to WIRO: — within 30 days of receiving the merit review findings; and — before the work capacity decision has taken effect after the required notice period

Internal Reviews

The worker must complete an application form, which must identify the decision that the worker is requesting be reviewed and include the worker's reasons for seeking the review of the decision. It can also include additional relevant information for the insurer to consider (eg medical or employment information). The insurer cannot refuse to perform an internal review after receiving a review application.

The insurer must respond to an application for review in writing (posted to the worker within 5 business days of receipt of the application). The communication must include:

— details on whether a stay is in place or not (dependent upon the relevant timeframe) and the implications of this for the date of the work capacity decision taking effect on compensation payments;
— that the review will be completed within 30 days of the application being made by the worker;
— the date by which the worker will be notified (with consideration of the postal rule);
— how to apply for a merit review, and that this option is available if the worker does not receive the decision within 30 days of making the application for internal review;
— the Work Capacity—Application for Merit Review by the Authority form;
— confirmation of receipt of any new information the worker has supplied.

Insurers must complete the review must be completed within 30 days of the worker's application, and no one involved in the original decision may conduct the review. The reviewer must be identified by name. The reviewer must consider any new information obtained or provided and can ask the worker for more information. The purpose of the internal review is to make the most correct decision based on all the available information; this includes both information that was available when the original work capacity decision was made and any new information.[575]

An internal review decision will either affirm the original decision or deviate from it. The insurer must notify the worker of the review decision as soon as practicable. This must be in writing on the Work Capacity—Notice of Decision of the Insurer Following an Internal Review of a Work Capacity Decision form. The following information should be included:[576]

— the review decision;
— its consequences, including any effects on the worker's entitlement to weekly payments and future medical, hospital and rehabilitation services under the 1987 Act;
— the reasons for the decision;
— the information considered;
— the process and timeframe for requesting a merit review of the decision;
— whether a stay of the operation of the decision applies and, if so, the required timeframe for applying for a merit review;
— a copy of the Work Capacity—Application for Merit Review by the Authority form;
— the review process after a merit review;

[575] s 44BB.
[576] s 44BB.

— that the worker can seek help from their insurer, SIRA's Customer Service Centre, the Workers' Compensation Independent Review Officer (WIRO), or their trade union.

Merit Reviews

If the worker is still not satisfied with the insurer's work capacity decision following an internal review (or if the insurer has not completed the review within 30 days of the application), the worker can seek a merit review of the insurer's work capacity decision by SIRA.

The worker has 30 days to apply for a merit review, starting either upon receiving the internal review decision, or if the insurer has not notified the worker of its decision the timing starts to run from the date that the internal review decision was due. Application must be by using the Work Capacity—Application for Merit Review by the Authority form. The worker must also notify the insurer of the merit review application by providing them with a copy of the application form.

The SIRA reviewer can decline to conduct a merit reviews for three reasons:

— it determines that the application is frivolous or vexatious;
— the worker does not provide information that it has requested;
— the application is made outside the 30-day timeframe outlined above.

The reviewer's findings and recommendations are binding upon the insurer, who must give effect to them. This should happen immediately.

Procedural Reviews

If the worker is still not satisfied with the SIRA reviewer's decision following a merit review, he or she can seek a procedural review of the work capacity decision by the WIRO.[577] The worker has 30 days to apply for a procedural review, starting upon the date of receiving the merit review decision. Application must be by using the WIRO Application for a Procedural Review form. The worker must also notify the insurer of the procedural review application by providing them with a copy of the application form.

A procedural review examines the insurer's procedures in making the work capacity decision. It does not assess the merits of the decision.

Procedural reviews can be declined for the same reasons as merit reviews.

The insurer is bound by the reviewer's findings and recommendations and must give effect to them. This should happen immediately

Judicial Review

If a worker is still unhappy with the outcome he or she can seek judicial review of a decision by the Supreme Court of NSW.

[577] Previously titled the WorkCover Independent Review Office or WIRO. WIRO is still used as the abbreviation despite the change of name.

Mechanism for Disputes

A dispute occurs when the insurer decides that a person is not entitled to workers' compensation benefits because the legal requirements for compensation are not met. There are several reasons why an insurer may dispute liability, the most common of which are shown in Table 3.19.

Table 3.19: Reasons for insurers to dispute liability and the relevant statutory provisions

Reason to dispute liability	Statutory provision(s)
The worker has not sustained an injury	Section 4 of the 1998 Act
The person is not a worker	Sections 4 and 5 and Schedule 1 of the 1998 Act
Employment was not a substantial contributing factor to the injury	Section 9A of the 1978 Act
The psychological injury was wholly or predominantly caused by the employer's reasonable actions	Section 11A of the 1987 Act
Claimed medical, hospital and rehabilitation expenses are not reasonably necessary because of the injury	Section 60 of the 1987 Act
The claim for property damage covers items the Act does not	Sections 74 and 75 of the 1987 Act
There is no total or partial incapacity for work	Section 33 of the 1987 Act
The degree of permanent impairment does not reach the required thresholds for a lump sum payment	Sections 65A and 66 of the 1987 Act
The worker was injured on a journey with no real and substantial connection between his/her employment and the accident that caused the injury	Section 10 of the 1987 Act

If liability is disputed by the insurer there are strict protocols in place that govern the process. Prior to notifying the worker of the dispute, the insurer should make sure an appropriately qualified person reviews all the information it has considered in the decision and the reasons for the decision. This should be someone other than the original decision maker. Once the insurer is satisfied its decision is correct, the insurer must then issue a notice of dispute to the worker. Timeframes vary depending on the type of compensation claimed, and are detailed above in the sections on the types of available benefits.

The notice of dispute must include the items described in Table 3.20.

Understanding Dispute Timeframes

The insurer must give the worker notice by post or in person.[578] When issued by post, the notice cannot take effect until 4 working days after it was posted (or 4 working days after any relevant period the notice specifies).[579]

[578] Workplace Injury Management and Workers' Compensation Act 1998, s 236.
[579] Interpretation Act 1987, s 76.

Table 3.20: Items to be included with the Notice of Dispute

Reasons for and issues in disputing liability	A concise, easy-to-understand statement of: — why the insurer disputes liability — the issues relevant to the decision — any legislative provision the insurer is relying on — any required period of notice before the dispute takes effect
Documents the worker has submitted	A list of reports and documents the worker has submitted and relied on in making the claim
Documents the insurer has considered	A statement identifying the reports of the type referred to in Clause 46 of the 2010 Regulations that are relevant to the decision, whether or not they support the reasons for the decision — a copy of all these reports and a statement detailing them If the insurer believes that giving the worker a report would pose a serious threat to anyone's life or health, the insurer can instead: — give a medical report to the medical practitioner the worker has nominated for that purpose — give any other report to the worker's legal practitioner — seek SIRA's direction on another approach when these options are not appropriate
How to request a review	The procedure the worker should follow to request a review of the decision
Where to seek assistance	Statements that the worker: — can seek further information from the insurer — can seek help from a trade union, a lawyer, SIRA's Customer Service Centre or Workers' Compensation Independent Review Officer
Where to refer an application for determination of a dispute	The physical and email addresses of the Registrar of the Workers' Compensation Commission or the Registrar of the District Court, as appropriate. If the notice is about a claim for work injury damages, it must also include statements that the claimant: — must serve a pre-filing statement before starting court proceedings — cannot raise matters in court that are materially different from the pre-filing statement, except with the court's leave

If the dispute includes a decision on weekly payments, the insurer must also follow the required period of notice specified in section 54 of the 1987 Act.

If a defect is identified in the notice of dispute, the insurer should correct the defect and reissue the notice. The notice period for the dispute restarts on the date the corrected notice is issued.

Reviews of the Insurer's Decision

A worker can request that the insurer reviews the decision to dispute the claim at any time before an application for dispute resolution is lodged with the Workers' Compensation

Commission. A dispute can be referred for review either to the insurer, or to the Workers Compensation Commission. A worker can ask the insurer to review the decision to dispute a claim at any time before an application for dispute resolution is lodged with the Workers' Compensation Commission.[580] However, the request for a review does not delay the timeframe for the dispute to take effect.

When the insurer receives a request, it must review the claim and respond within 14 days. The review might lead the insurer either to accept liability or to continue to dispute liability. The insurer should ensure that the person undertaking the review is appropriately qualified and was not involved in making the original decision.

If the insurer continues to dispute the claim a further notice of dispute must be issued.

The Workers' Compensation Commission[581]

The Workers' Compensation Commission aims to help with dispute resolution over claims. It is usually free to claimants (on some occasions legal costs may be payable). The objective is to enable the parties to agree an outcome, but if this is not possible, then the Commission will issue a decision.

To be accepted a claimant will have to have already submitted a claim to his insurer. An application to resolve a dispute can be made when:

— an insurer has not made a decision on a claim within the appropriate timeframe;
— the insurer has given notice that the claim has been refused;
— the claimant disagrees with the quantum of an accepted claim.

The Workers' Compensation Commission claims process varies according to the type of claim; a schematic of the process is shown in Figure 3.31. Disputes about the degree of permanent impairment follow a slightly different path to almost all other dispute types.

If a dispute is resolved by an arbitrator, then a Certificate of Determination will be issued to all parties. This records how the dispute was resolved and officially closes the matter.

Legal Representation

There are some circumstances in which WIRO will award a claimant a grant towards the cost of legal representation to enable him to pursue his claim. The grant can only be awarded to WIRO-approved lawyers and only in certain circumstances. The claimant must seek a WIRO-approved lawyer, who will then send an application detailing the issues at stake to the Independent Legal Assistance and Review Service (ILARS). ILARS assesses the application against its grant funding criteria. If the application is successful a grant will be awarded and paid by WIRO. This provides assistance for claimants if it is needed and it is felt there is a chance of the claim succeeding.

[580] Workplace Injury Management and Workers' Compensation Act 1998, s 287A.
[581] www.wcc.nsw.gov.au.

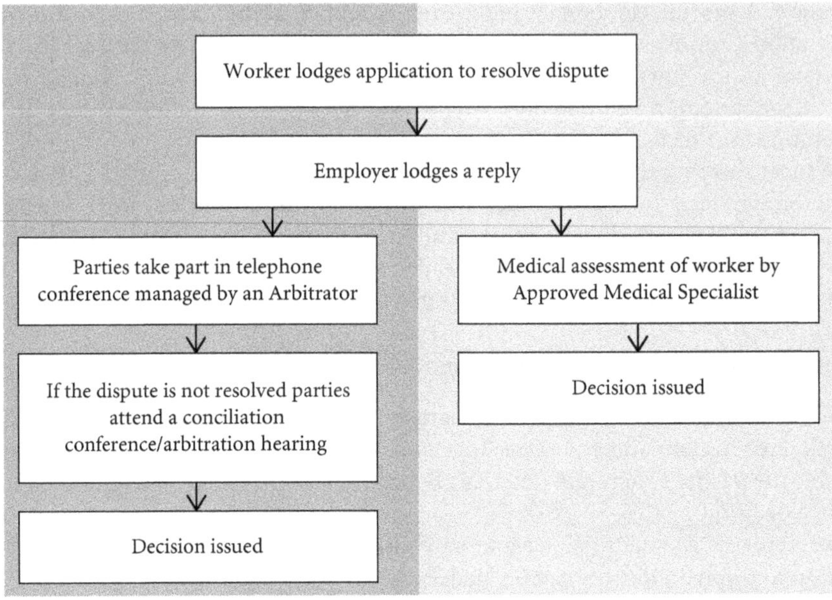

Figure 3.31: Schematic of the Workers' Compensation Commission dispute resolution process

Insurance Arrangements under the Scheme

Workers' compensation insurers[582] fall into one of four categories in New South Wales:

(a) The NSW Workers' Compensation Scheme, consisting of the Nominal Insurer and its agents.[583] The Scheme is administered by Insurance and Care NSW (icare),[584] which acts on behalf of the Nominal Insurer—the independent non-governmental legal entity responsible for the performance of the Scheme.[585]
(b) SICorp (through the Treasury-Managed Fund), which manages workers' compensation, administration and financial liability for public sector employers (except those who self-insure).
(c) Self-insurers, which are organisations with enough capital to underwrite, pay and manage their own claims.
(d) Specialised insurers, which hold restricted licences to underwrite workers' compensation insurance risk for a specific industry or class of business or employers.

Since September 2015 SIRA, a NSW Government agency, has regulated and managed the NSW workers' compensation system, including the licensing of self-insurers and specialised insurers and oversight of service providers.[586]

582 As defined in the 1987 and 1998 Acts.
583 As at June 2016 there were five insurance agents: Allianz Australia Workers' Compensation (NSW) Ltd, CGU Workers' Compensation (NSW) Ltd, Employers Mutual NSW Ltd, GIO General Ltd and QBE Workers' Compensation (NSW) Ltd. Employers can utilise a broker to help them to obtain workers' compensation insurance, brokers must be licenced by the Australian Securities and Investments Commission.
584 State Insurance Governance and Care Act 2015, s 10(1)(a).
585 Workers Compensation Act 1987, as amended, ss 154(A) to 154(C)(A).
586 Part 3 of the State Insurance Governance and Care Act 2015.

All NSW employers must have a workers' compensation policy if they pay more than $7,500 in wages per annum, employ an apprentice or trainee, or are part of a group for premium purposes.

Funding

The Workers' Compensation Nominal Insurer is a not-for-profit legal entity established in 2005 to issue policies of insurance and to manage workers' compensation claims for NSW employers. All premiums received are paid into the Workers' Compensation Insurance Fund (WCIF) to meet the cost of claims and the administration of the Scheme. The assets of the WCIF are owned by the employers of NSW, as a class, who are also responsible for meeting any shortfall. (The scheme is currently in surplus.)

As at 2014/15 approximately 272,000 employers and over 3 million workers were covered by the NSW Workers' Compensation Scheme.[587]

The Nominal Insurer has earned premiums of over AU$2.2 billion income (excluding GST) and assets of AU$17 billion in 2014/15, making it one of Australia's largest insurance businesses.[588]

The 2015 reforms have, so far, resulted in premium rate decreases, particularly where companies are able to demonstrate improved safety, harm reduction and a better claims experience. The longer-term intention behind these reforms is to continue with the dual trends of decreased harm and decreased premiums.

Statistics

According to the most recent annual report 2014/15,[589] the incidence rate for workplace injury in 2013/14 was 7.2 claims per 1,000 employees. This drop in incidence rate is part of a long-term historical trend. The changes to workers' compensation definitions of major workplace injuries introduced in 2012 make comparing the situation before and after these reforms difficult. Figure 3.32, taken from the NSW WorkCover Statistical Bulletin 2011/12,[590] shows the number, frequency and incidence rates of employment injuries from 2002/03 to 2011/12, illustrating the long-term trends.

Alongside injury claims, the number of fatality claims has also been dropping; see Figure 3.33, taken from the NSW WorkCover Statistical Bulletin 2013/14.[591]

As individual insurance companies handle claims, there is little centralised data on claims handling, so no statistics on average claim time frames, etc.

There is some information about the Merit Review Service (MRS), but this is limited as it was only established in January 2013. In 2014/15 there were 748 review applications lodged

[587] NSW Government, *Workcover Authority Annual Report 2014–15* (2015) available at www.workcover.nsw.gov.au/about-us/annual-reports.

[588] ibid.

[589] Figures taken from ibid.

[590] www.workcover.nsw.gov.au/about-us/statistical-bulletins.

[591] www.workcover.nsw.gov.au/about-us/statistical-bulletins.

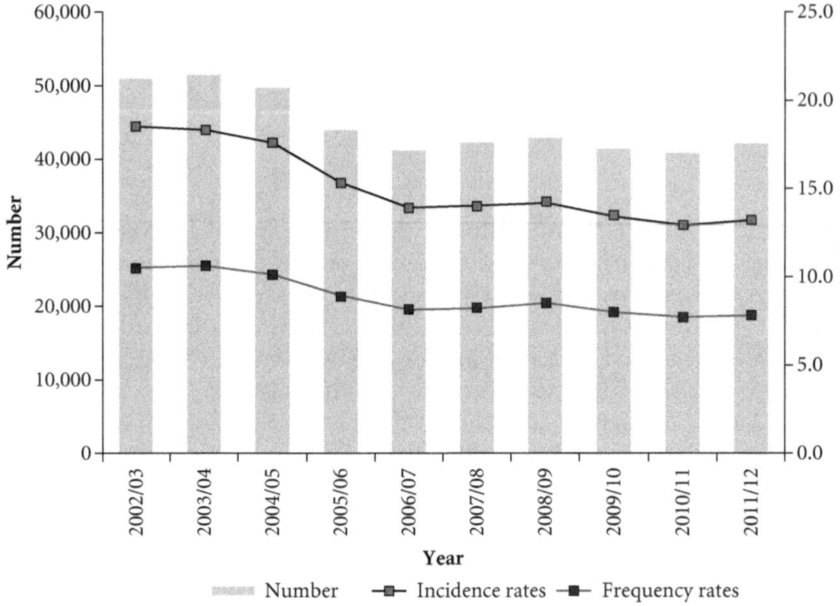

Figure 3.32: The number, frequency and incidence rates of employment injuries from 2002/03 to 2011/12

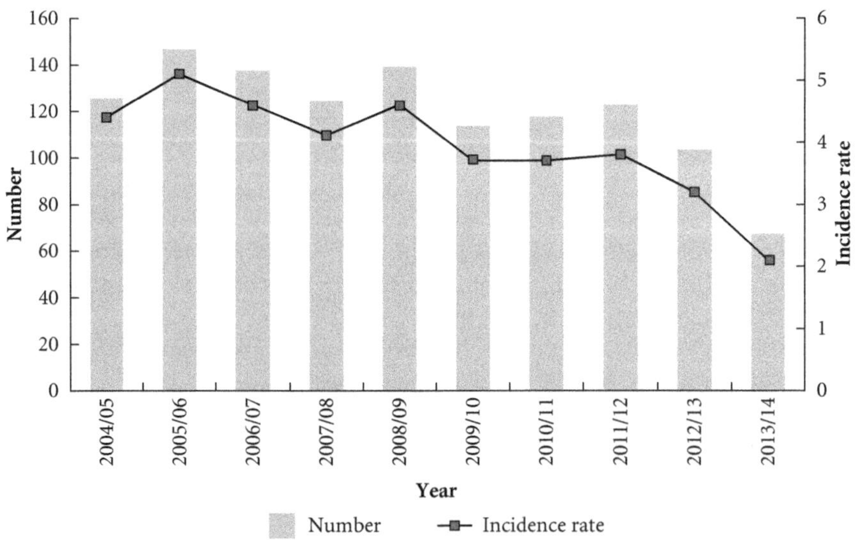

Figure 3.33: Fatality claim numbers and incidence rate (the number of fatalities per 100,000 workers) between 2004/05 and 2013/14

(1,136 in 2013/14) and 956 reviews were finalised (910 in 2013/14) with a median time to finalisation of 30 days (67 days in 2013/14).

Figures from WIRO show that the vast majority of applications for procedural review are accepted, with the majority closed within 1 week.

H. New South Wales Work (Dust Diseases) Related Injuries

The Workers' Compensation (Dust Diseases) Authority (DDA) provides a no-fault workers' compensation scheme to past and present workers who have developed a compensatable dust disease from occupational exposure to dust as a worker in New South Wales.[592] The DDA is an agency within icare.[593] The scheme is governed by the Workers' Compensation (Dust Diseases) Act 1942 and provides compensation to people who have developed one of the following diseases (or a pathological condition of the lungs, pleura or peritoneum that may cause one of the following dust diseases)[594] through occupational exposure to the relevant dust as a worker in New South Wales:

- aluminosis;
- asbestosis;
- asbestos-induced carcinoma;
- asbestos-related pleural disease (ARPD);
- bagassosis;
- berylliosis;
- byssinosis;
- farmers' lung;
- hard metal pneumoconiosis;
- mesothelioma;
- silicosis;
- silico-tuberculosis.

Other diseases affecting the lungs, pleura or peritoneum may be covered if these conditions are related to an exposure to dust and the condition may lead to one of the specified diseases listed above.[595]

It is important to note that the DDA is a separate entity from the Dust Diseases Tribunal of NSW. The DDA can examine, hear and determine claims for compensation and medical expenses (not damages) by NSW workers. Claimants cannot be legally represented. In contrast, the Dust Diseases Tribunal of NSW is a court that has exclusive jurisdiction (under the Dust Diseases Tribunal Act 1989) to hear and determine damages claims for those who have been affected by dust diseases including diseases related to asbestos exposure and claims made by the dependants of those who have died from dust diseases.[596] While the Tribunal appears similar to a court process seeking damages there are some key differences. The emphasis on the Tribunal is to provide a quick resolution of the real issues in dispute, given the fact that many people who lodge a claim are in the advanced stages of their disease and may be dying from their illness. Certain claims can be prioritised so that plaintiffs whose life expectancy is shortest are dealt with more swiftly. Bedside hearings are also available as well as video link-ups to ensure hearings can be held at any time and in most locations.

[592] www.icare.nsw.gov.au/our-services/dust-diseases-care.

[593] As with other agencies the new website is not yet up and running, so information is taken from the Dust Diseases Board website: www.ddb.nsw.gov.au/.

[594] Defined at Schedule 1 of the Workers' Compensation (Dust Diseases Act) 1942.

[595] Workers Compensation (Dust Diseases Act) 1942, s 3.

[596] Information on the Dust Diseases Tribunal is taken from the Tribunal website: www.dustdiseasestribunal.justice.nsw.gov.au/.

Having noted the existence of the Tribunal, the remainder of this chapter will detail the process and outcomes of the statutory DDA scheme.

Process

Application to the DDA is free to the claimant. A claim form can be downloaded from the website and filled in. The form is relatively simple and has been designed so it can be filled in without recourse to professional help.[597]

If a claimant is uncertain whether he or she is suffering from a dust disease the DDA can arrange for a free medical to establish this. The claimant needs to fill out a medical examination application form.

To apply, the claimant must send in a claim form (available online), which contains details of all other required information (such as payslips), any supporting medical tests results/information, a letter from their treating medical practitioner, and documents proving his identity and that of any dependants. If the claimant is a spouse or dependant of a deceased worker he must provide proof of his relationship to the deceased, and if the worker was not receiving benefits from the DDA when he died a supplementary application form should also be included.

Upon receipt of an application the DDA will begin to establish if the person is suffering a dust disease caused by exposure while working in NSW. The initial stage of this may involve an Industrial History Officer from the DDA interviewing the claimant about his work history. The Industrial History Officer will then write a report, which forms part of the claimant's application.

The Application will then be considered by a Medical Assessment Panel. This is a panel of three legally qualified medical practitioners appointed by the Minister. The Chairperson is nominated by the Minister, with one panel member nominated by employers whose employment exposes the worker to the possibility of contracting a dust disease, and the third panel member nominated by such workers.[598] Each member can have a deputy who is entitled to act in their place. Service on the Medical Assessment Panel is usually on 3-year terms, which are renewable.[599] A Medical Assessment Panel can be comprised of 2 or 3 members, and the decision of the Medical Assessment Panel is based on the agreement of at least 2 of the panel members. Interestingly, the Act provides immunity from prosecution for Panel Members in respect of carrying out their duties under the Act.[600]

The Medical Assessment Panel must establish that a partial or total disablement from work due to a dust disease is reasonably attributable to asbestos (or other dust) inhalation in dust disease occupation while the person was a worker in that occupation. In the case of a deceased worker, the Medical Authority determines if the worker's death was attributable to a dust disease. This does not rule out other causes, such as smoking, and does not seek to apportion. The DDA then calculates any compensation entitlement payable in accordance with Workers' Compensation (Dust Diseases) Act 1942 and Workers' Compensation Act 1987.

[597] Available at: www.ddb.nsw.gov.au/making-a-claim/how-to-apply.
[598] Workers' Compensation (Dust Diseases Act) 1942, s 7.
[599] s 7.
[600] s 7(6).

If a person is in receipt of compensation from any other Act for the dust damage the eligibility for compensation from the DDA will be proportionally reduced.[601] Other government benefits, such as a pension from Centrelink or the Department of Veterans Affairs, may be reduced upon receipt of an award from the DDA.

Awards

The following types of benefits are available:[602]

— weekly compensation benefit;
— medical and related treatment expenses;
— funeral expenses;
— allowances for a dependent spouse and any dependent children aged up to 16 years or until 21 years if enrolled as a full-time student; and
— interim payments are also available in some circumstances.

Weekly Benefits

The weekly amount paid varies, according to the degree of disability, as certified by the Medical Authority and the proportion of the individual's total exposure to the disease-causing dust that occurred while that person was a worker in New South Wales.

The rates of payment are calculated according to the Workers' Compensation Act 1987, as detailed above.[603]

Briefly, for the first 26 weeks the worker is paid either his or her pre-injury earnings or the maximum, currently set at $480.50 per week. After 26 weeks, if the claimant is 100% disabled due to a compensatable dust injury and the dust injury occurred exclusively while the individual was a worker in New South Wales, then the claimant will receive the maximum weekly payment. Compensation will be proportionally reduced if the claimant is

— not 100% disabled due to a compensatable dust injury; or
— the dust exposure did not occur exclusively while the individual was a worker in New South Wales; or
— the claimant is in receipt of compensation for the dust injury from another source.

For malignant cases, compensation may be payable from the date of diagnosis or initial investigation of the disease. For non-malignant cases, compensation is paid from the date the application is received by the DDA.

Dependants (spouses, children up to 16 years and/or full-time students up to 21 years) may attract an allowance of between $90.50 and $335.30 per week. There is also an allowance of $9,000 for funeral expenses.[604]

[601] s 8AA(2).
[602] s 8.
[603] s 8.
[604] s 8.

Medical Expenses

Medical expenses are described in the Workers' Compensation Act 1987 and can include the following:[605]

- Ambulance services to transport a worker to a doctor or hospital.
- Treatment for a dust disease at a hospital, nursing home, palliative care or rehabilitation centre (either public or private and with Government accreditation).
- Therapeutic treatments at the direction of a doctor such as exercise, massage, herbal remedies and aromatherapy.
- Wheelchairs and other mobility aids such as walking frames and motorised scooters.
- Therapeutic apparatus such as reclining chairs, beds, pressure relief cushions, commodes and shower chairs.
- Home care including nursing care, cleaning, and lawn and garden maintenance.
- Respite care (out of home) or rehabilitation centre admissions at a designated healthcare facility (either public or private with Government accreditation).
- Pharmaceuticals, domiciliary oxygen and nebulisers as prescribed by the treating doctor for the treatment of a dust disease.
- Reasonable travelling costs to and from treatment of the dust disease.

All medical and ancillary care must be deemed reasonably necessary and related to the dust disease.

Division 3 of the Workers' Compensation Act 1987 sets the maximum amounts that the DDA can spend on medical and hospital costs for each worker:

- $50,000 limit for medical expenses;
- $50,000 limit for hospital expenses;
- $10,000 limit for ambulance expenses.

In exceptional circumstances, the DDA may approve the payment of medical, hospital or ambulance expenses that are greater than the limits listed above if the additional treatment required is necessary as a result of the claimant's dust disease.

Funeral Expenses

Funeral expenses are available in some circumstances, and are limited to the value prescribed in the Workers' Compensation Act 1987.[606] Up to $9,000 is available to cover funeral expenses of individuals who had been in receipt of benefits from the DDA. This can be used to cover the costs of a casket and flowers for the casket, travel to go to the funeral, a single cemetery site or the single component of a double site, newspaper announcements, the issue of a death certificate, clergy and funeral director fees, and transfer fees. It cannot be used for catering and costs associated with a wake, a headstone or memorial plaque, memorial books, media presentations or tributes, a double cemetery plot or limousines. Costs

[605] s 8.
[606] s 8.

can be paid direct to the funeral director, or if they have already been paid the executor can reclaim them.

A completed claim form is required, along with a copy of the death certificate and an itemised invoice from the funeral director. Claims are processed within 21 days of receipt of all of the information required.

Interim payments

Interim payments are also possible under section 8B of the Act, until the DDA has made the first payment.

Funding

The DDA is funded by its investments and by insurance contributions from insurers, self-insurers and the Nominal Insurer under the 1987 Act, as determined by SIRA.[607] As at 30 June 2015, the total estimated claims liability was AU$1,854.4 million; broken down as AU$773.2 million in known claims and AU$1,072.2 million in estimated future claims.

Statistics

As at 30 June 2015,[608] the DDA provided compensation benefits to 4,013 customers: 1,088 workers and 2,925 dependants. AU$81.219 million was paid in compensation benefits during 2014/15 and AU$9.050 million in healthcare and funeral benefits.

The DDA received a total of 672 applications in 2014/15: 345 applications were made by people who had been exposed as a worker, while 210 applications were made by the dependants of a deceased worker. A further 110 applications were for medical screening, while the remainder of applications did not proceed.

In 2014/15, 276 new awards of compensation were approved to workers who had been certified by the Medical Authority as having a disability as a result of their dust disease. The award consists of a fortnightly compensation benefit, payment of all medical and related treatment expenses, and payment of funeral expenses. Allowances are paid for a dependent spouse and any dependent children aged up to 16 years, or until 21 years if enrolled as a full-time student.

The DDA also made 12 new awards for the payment of medical expenses for workers who were in paid employment and ineligible to receive weekly compensation benefits.

An additional 199 awards were made to dependants of deceased workers. Awards to dependants of a deceased worker generally consist of a lump sum payment together with fortnightly compensation benefits.

These figures are broadly consistent with the figures from previous years (see Table 3.21). Clearly the long latency of dust diseases can make predicting the number of cases per year

[607] s 6(7).

[608] NSW Government, *Workers' Compensation (Dust Diseases) Board Annual Report*, 2014/15 available at www.opengov.nsw.gov.au/publications/15224.

more difficult, adding to the complexity of forecasting both future claim numbers and future claim values.

Table 3.21: Compensation awards granted by type by the DDA 2012/12–2014/15

Compensation awards granted	**2012/13**	**2013/14**	**2014/15**
New worker awards	295	220	276
New medical expenses awards	22	26	12
New dependant awards	183	176	199
Total awards granted	**500**	**422**	**487**

Conclusion

The NSW Dust Compensation Scheme has been running since 1942, making it one of the longest running no-fault compensation schemes. This is a considerable achievement given the difficulty in predicting dust disease cases, due to their long latency. The relatively low levels of dust disease claims as a proportion of the total population of New South Wales, around 500 new dust disease claims in a state with a population of over 7.5 million in 2015, may go some way to explaining its success. Although each case can be expensive the combination of capping of awards and the high population means the relative cost per head remains affordable.

GRAPHS AND DIAGRAMS

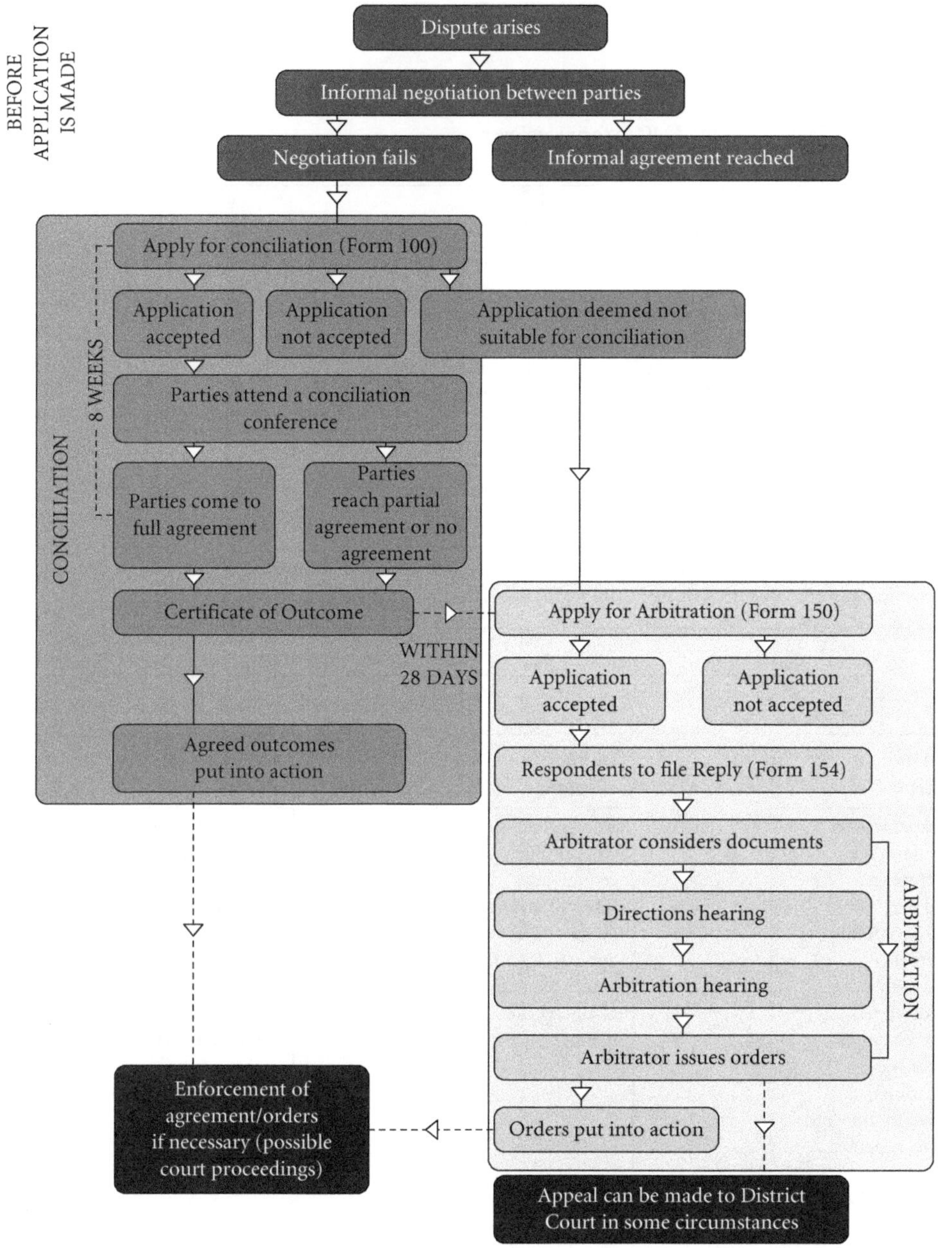

Managing Your Claim

Worker

- Seeks first aid
- Reports incident
- Obtains a First Medical Certificate (FMC) from a doctor of their choice
- Completes worker section of the Workers' Compensation Claim Form
- Submits the Claim Form and FMC to their employer

Employer

- Completes an employer report form provided by the insurer
- Completes employer section of Claim Form
- Provides copy of Claim Form and FMC to worker for their records
- Sends Claim Form and FMC to Insurer within 5 working days of receipt fom the injured worker

Insurer

Has 14 days after the claim documents are submitted to make a decision on liability and advise the worker and employer in writing if liability for the claims is accepted, disputed or undecided.

Liability Accepted

Injured worker's entitlements commence, and may include weekly payments reasonable medical and related expenses, vocational rehabilitation and travel expenses.

Liability Disputed

The injured worker may request an internal review by the insurer or apply to CAS for assistance with resolving the dispute.

Decision Pended

The insurer has a further 10 days to decide the claim. If no decision is made after this time, the worker or employer may apply to CAS for assistance with resolving the dispute.

The Claims Process - Roles and Responsibilities

Part III

Nordic States

SONIA MACLEOD, MATTI URHO AND CHRISTOPHER HODGES

4

Nordic Injury Compensation Schemes

A. Introduction

Each Nordic state has national injury compensation schemes that comprise separate but related arrangements covering workplace, road traffic, medical and drug injuries. Sweden was usually the first to introduce national arrangements, which then inspired schemes in Denmark, Finland and Norway. The arrangements in these countries are similar but not identical. Collectively, they form a distinct, coherent and integrated set of national approaches towards compensation for injuries. The generic approach can be referred to as a 'Nordic model'.

This part of the book sets out the individual arrangements for patient and medicines injuries compensation in all four of the major Nordic states. It also covers industrial injuries in Denmark, workers' compensation in Finland, and motor vehicle injuries in Sweden and Finland. (This does not mean that workers or motor schemes in the other countries do not exist: the authors were simply unable to undertake a wider analysis of all areas in the time available.) A wealth of detail has been assembled, and there is inevitably some repetition, but this is necessary in order to identify some critical differences that might otherwise be missed. Short conclusions are included at the end of most countries or major sections, but a more detailed comparative analysis of the overview and of the different injury types (especially patients and medicines) is given in the last chapter of the book.

All Nordic states have relatively small populations compared with some of the larger European states. Population figures in millions are: Sweden 9.8, Finland 5.5, Denmark 5.7, and Norway 5.2.[1] The modest size has an important effect on the structure and affordability of the systems discussed here.

B. Integration of Coverage and Entitlements

Various points should be noted at the start, which aid understanding of the context and rationale for the national arrangements.

[1] *World Population Prospects*, vol 1: *Comprehensive Tables* (United Nations, 2015).

Firstly, the national architecture encompasses separate but linked compensation arrangements for injuries from four main sources of personal injuries (workplace, road traffic, medical treatment and medicinal products).[2] Thus, none of the individual schemes should be viewed in isolation, since the design and effectiveness of each sectoral scheme depends on the existence of other sources of insurance and compensation. Together, they form part of a *unified and comprehensive system* for compensating personal injuries in each country.

Second, the basic structure is one of *tiers* or *waves* of sources of financial support. In other words, particular individual sources of financial support are placed in a particular prioritised sequence, which results in payers being categorised into primary and secondary tiers. This produces a coordinated effect between the various sources, spreads the cost amongst different funding pools, and reduces the financial impact on any one such pool.

Thus, the first tier is the national insurance system operated by the state, which covers the cost of medical care for all injuries, with limited co-payments by individuals up to a modest limit. Medical care is generally provided free (and generally covers between 92% and 98% of the population). The second tier of cover is provided by the compulsory employer's liability system, which covers certain benefits, although the extent and degree of cover will vary between employers. A third tier comprises social security, covering gaps in employment injury insurance, and giving certain extensions in cover and benefits. The fourth tier is motor vehicle, patient or pharmaceutical compensation schemes, which provide further benefits, but generally do not overlap with the sources of cover mentioned earlier. The national arrangements differ in their details but all produce the same result in practice, namely that injured individuals obtain payments from several possible sources.

The final tier of schemes, for example of patient and pharmaceutical compensation, are, therefore, not the primary sources of compensation, but make top-up payments for, firstly, pain and suffering and loss of amenity, and, secondly, any shortfall in the provision from other sources of loss of income. Overall, the principle of providing full compensation is observed, but the pharmaceutical schemes, for example, merely top-up other extensive sources of compensation, in order to provide what is overall comprehensive cover. Importantly, payments received from social security or other schemes will be deducted from the sum assessed under the patient or pharmaceutical schemes as being the notional sum for full compensation.

In Norway, the same compensation rules apply under the Act on Damages for compensation for all kinds of personal injury (motor vehicle, patient injury, pharma injury) *except* workforce injuries, which are regulated not under that Act but through standardised subsidiary rules under the Act relating to industrial injury insurance, which still applies the principle of deduction. Workers' compensation in Denmark has its own compensation tariff.[3] In Finland, the patient injury scheme is primary and the pharmaceutical injury scheme is secondary, ie it activates only to the extent that paid compensation cannot be reduced under another law.[4]

The Nordic system, therefore, contrasts with a typical tort-based system, in which successful establishment of liability triggers a right to full compensation, and typically also a

[2] In Norway there is a related scheme that compensates for wrongs not eligible under other compensation schemes, www.sivilrett.no.

[3] Patient Injury Law, ss 11, 12, 15–23.

[4] Patient Injury Law s 8.

right of recourse by other payers of benefits against the liability insurer. Instead, under the Nordic system, different tiers of cover respond to particular types of benefits, but there is in practice no cross-indemnity. This leads to the economic result that all of the schemes' funds—and especially the patient and pharmaceutical schemes—are of modest size in their particular national contexts, since their function is limited and secondary, and they do not individually have to cover full quantum as might be awarded in a tort or product liability claim. This system results in the individual schemes being more affordable.

C. Holistic Systems and Nordic Philosophy

A third overview point is that all of these compensation schemes form part of a *wider national architecture on dispute resolution* that also includes a series of alternative dispute resolution (ADR) bodies for consumer-to-trader (C2B) disputes. In Sweden, for example, that architecture comprises a national body, the Allmänna reklamationsnämnden (ARN), which can handle any type of C2B dispute (and has 11 specialist departments), together with a number of separate specialist bodies that can handle disputes in particular sectors, such as financial services, communications, etc.[5] The combined effect of all these arrangements is that consumers and injured persons rarely rely on courts to bring legal claims, since what may be viewed as 'alternative' dispute resolution arrangements in other countries are mainstream in Nordic states. The normal response to a 'legal problem' is to contact these specialist dispute resolution bodies and compensation schemes, rather than to consult lawyers or use courts.[6]

Fourthly, these ADR and compensation schemes together comprise a particularly *Nordic philosophy* towards social security and risk protection. The various component parts of the national systems were built sequentially but so as to provide a coherent national model based on mutual shared support, supported by a series of insurance pools. Adversarial approaches to resolving problems, such as litigation, do not feature in the model. A general timeline of some of the principal historical developments is shown in Table 4.1. The Nordic approach has some parallels with the rise of European consumer protection policy since the 1960s. By 2015 the Nordic approach no longer seems unique but merely a vanguard of a wide-ranging European shift towards *social solidarity and shared protection*. It is, therefore, important not to look at any one of the component parts without recognising how it fits into the comprehensive national structure. That point is explained further below.

[5] These bodies are described in detail in C Hodges, I Benöhr and N Creutzfeldt-Banda, *Consumer ADR in Europe* (Oxford, Hart Publishing, 2012).

[6] This statement, although generally true, should be slightly qualified for Norway, where there remains some activity by lawyers and court cases on personal injury cases, especially road traffic incidents, and around 200 patient claims a year (3%): comment to the authors from Rolf Gunnar Jørstad, who surmises that the increased difference in cultural willingness to consult lawyers in Norway in comparison with other neighbouring states may be influenced by a higher standard of living.

Table 4.1: Historical development of the solidarity components in Nordic states

Sweden	Denmark	Finland	Norway
National Insurance Act 1962 Occupational safety insurance 1972. Patient insurance voluntary scheme 1975 Traffic Damage and Injury Act 1976 Drug Compensation Voluntary scheme 1978 Comprehensive medical injury insurance 1997 from Patient Injury Act 1996.	Patient Compensation Association Scheme 1992; Act 2011 Pharma compensation 1996; Patient and Pharma Act 2011 Labour Market Occupational Diseases Fund 1988; Consolidated Workers' Compensation Act 2013	Workers' Compensation 1895; Act 1948, most recent Act 2015[a] Patient scheme 1987, Act 1998 Pharma voluntary scheme 1984	Patient insurance voluntary scheme 1988; statutory 2001 Act from 2003; extended to private providers 2009 Pharma insurance in Product Liability Act 1988

[a] Came into force 1 January 2016.

Fifthly, the origin of the Nordic approach was a firmly held belief that a compensation scheme based on *tort liability* (or, when it was later introduced after the 1985 Directive, strict product liability)[7] was *unsatisfactory* for several reasons. It was—and remains—thought to be difficult, expensive and take time to prove fault or other liability, and many people who suffer injuries would not be able to do so.[8] Accordingly, an approach was needed that was *more efficient* in investigating and analysing the facts, and which extended compensation to more deserving people, whilst controlling overall cost and still being robustly based on proof of causation.

Furthermore, it was noted that one of the rationales of tort liability, and one which profoundly underpinned the liability system in the United States, which was closely studied by Nordic scholars and policy-makers, was the idea that tort liability controlled behaviour as a regulatory mechanism. In the Nordic medical community, that idea has always been rejected. Indeed, it was regarded as a positively dangerous approach, and as not supporting an open relationship either between healthcare professionals and patients, or amongst the professionals themselves. A fault-based system is not considered conducive to encouraging a high level of voluntary reporting of adverse events.[9] Indeed, there is a strong view that looking for scapegoats through a legal system based on fault is 'a very efficient way of killing

[7] Directive 85/374/EEC.

[8] In the early 1970s only 10 patients per year received compensation for medical malpractice through the tort system (C Oldertz, 'Security Insurance, Patient Insurance and Pharmaceutical Insurance in Sweden' (1986) 34 *American Journal of Comparative Law* 635).

[9] H Johansson, 'The Swedish system for compensation of patient injuries' (2010) 115(2) *Uppsala Journal of Medical Sciences* 80. That study estimated that about 30,000 patients out of a population of 9 million (0.33%) are treated annually for avoidable patient injuries, as contrasted with official figures of 0.2% of hospital admissions in Sweden resulting in patient injury.

more patients'.[10] Before the insurance arrangements were put in place in Sweden, around 100 patients were compensated annually through the tort system, whereas the insurance system currently compensates between 5,000 and 6,000 patients. In Finland it is estimated that only 10% of the total number of people injured come to light and inform the Patient Insurance Centre (PVK).[11]

Instead, the philosophy behind these schemes is that mistakes happen, and that it is important to be honest, open, put things right by providing appropriate care, and move forward by learning from the experience. Thus, the patient insurance systems provide 'no-blame' compensation. Policy-makers believe that this system has created a basis for increased confidence and openness between medical staff and patients. This is illustrated by the fact that in 60–80% of cases reported to the Swedish patient insurance scheme, the physician, nurse or social worker involved helped the patient report the matter—rather than trying to cover it up or refusing to accept that it happened. Some outstanding examples are given below of how data from the complaint systems is now used to inform and drive improvements in healthcare—although some barriers to achieving this exist, and effectiveness varies between countries.

D. Levels of Damages

In each state, the level of damages that are paid under the workplace, patient and medicines schemes are based on the level of damages that would be awarded by a court in a tort or product liability claim.[12] Such levels are, in turn, historically based on insurers' tariffs for road traffic injuries. It can, therefore, be said that no claimant is disadvantaged by claiming under a patient or medicines scheme rather than by asserting a legal claim. The general standard of damages is set by the law or courts, and decisions should not, in theory, be different in any given case if they are made by the various compensation bodies or courts. Thus, there is no particular advantage in claimants attempting to game between different systems so as to recover more under one than another, and there is a general sense of equity about all of the schemes.

However, in practice there is some circularity in how the situation operates. It is obviously in the interests of insurers and the insureds if levels of damages generally are prevented from escalating, and insurers may wield ultimate influence by setting tariffs for road traffic injuries. Judges do have power to increase awards in such cases that come before them, but the existence of comprehensive compensation arrangements leads to the fact that there are few unsettled claims that reach the courts, so few judges have limited opportunities to affect the situation, and may do so generally only over an extended timescale. They will be also aware that if they increase the level of damages awarded, that will have an inflationary

[10] Statement by members of the Swedish Patient Insurance to C Hodges in 2012. It would be interesting to know if there is any empirical evidence to show higher rates of reporting for patient or workers' compensation (which is very high in Denmark, as they have three ways of reporting) maybe compared to 'fault'-based systems, but we suspect that any such comparison cannot be made.

[11] Potilasvakuutuskeskuksen.

[12] For Sweden, see the Undertaking by the pharmaceutical companies, cl 8; discussed in Ch 5.

effect on a complex, interlinked system of insurance, premium levels and prices for goods and services. The result is that the general levels of compensation available, whether under insurance arrangements or through legal channels, have not risen much since the compensation schemes were introduced, stretching back in some cases over 40 years, and are low by European standards[13] and certainly in comparison with the USA. Systemic increases in levels of damages have to be made by statute, which Denmark did from 1 July 2002.

E. Patient Complaints

All the Nordic countries have systems for handling patient complaints. In Norway this is done by Statens helsetilsyn (Norwegian Board of Health Supervision). By way of expansion, the position in Denmark is as follows.

A patient will first be offered an opportunity to meet the healthcare provider that is the subject of the complaint in a 'dialogue meeting'. The meeting does not involve a mediator. The patient may bring a relative or friend. The meeting is run on slightly different ways by each of the five regional authorities, so some achieve greater or lesser success with this technique. A patient may then make a complaint with the Disciplinary Centre of the Patient Ombuds. The way in which such complaints are being processed is gradually changing. The traditional approach is that complaints are heard by a Board of five people (two doctors or relevant specialists, two consumer representatives and one judge). The Board staff assemble the papers, and both sides are given an opportunity to be heard and comment on the allegations and evidence. The process can, therefore, take some months. The Board hears around 30 cases at each meeting, of which between 5 and 10 are decided in favour of upholding the complaint. Professionals who are found to have made a serious error or to have scored 3 upheld complaints within 2 years are 'named and shamed' on a website. All information from the Board's cases is passed to the Health and Medicines Authority: there are no visible consequences from this. The above Board system, focusing on individual actors, is gradually being replaced by an administrative inquiry into the entire systemic aspects of treatment, undertaken by Patient Ombuds staff, first introduced in 2000. In 2015, the Board handled around 40% of cases and the systemic treatment investigation system handled 60%. Overall, the Consumer Council considers that it is difficult for patients to criticise doctors and many other healthcare professionals, and that the current systems do not assist people to resolve their problems by raising issues in a low-key but expert way, so that lessons can be learned and fed back to improve practice.

[13] D McIntosh and M Holmes, *Personal Injury Awards in EU and EFTA Countries*, 2nd edn (The Hague, Kluwer Law International, 2003).

5

Sweden

A. General Benefits and Arrangements

A person injured in Sweden would be able to obtain compensation from the following sources. First, state healthcare insurance covers medical costs and loss of income. In relation to loss of income, an employee who suffers injury would first claim social security payments from the government. The amount that the government would pay is amended from year to year, and is currently 80% of a prescribed notional salary, the latter currently being SEK 300,000 (€32,610) pa. Individual employees or employers may voluntarily take out extra first-party insurance. If there were long-term disability, the loss of income payments would gradually be turned into an early-retirement pension.

However, where the injury was caused by, for example, a medicine, the remaining 20% (or more in the case of employees who are paid more than SEK 300,000) would be topped up by payments from the Pharmaceutical Insurance. The Pharmaceutical Insurance would also make some extra payments for medical expenses that were not covered under the medical insurance, such as for providing a wheelchair or adapting a home for specific disability arrangements. The state does not cover payments for pain and suffering or disability, such as scars. Both of these would be covered by the Pharmaceutical Insurance, to a level of 100% of values prescribed in tables maintained by the insurer of the scheme, which are on a par with levels of damages that would be awarded by the courts. Such tables and values originated with insurers' tables for motor insurance injury compensation payments. It is recognised that such levels are relatively low. For the first 10 or 15 years of the Pharmaceutical Insurance, the levels under that scheme were rather more favourable than the courts would have awarded, in order to make the former more attractive, but since the scheme has been so widely used, the amounts have equalised. In the case of injuries to patients, however, the principal driver in the switch to an insurance system was the fact that it was much easier to obtain compensation because negligence liability did not have to be proved, rather than any differential in the financial arrangements.

Extensive information and advice on insurance issues is provided to consumers by the Swedish Consumers' Insurance Bureau (KFB).[1] Disputes with insurance companies are handled through in-house arrangements, such as committees or ombudsmen, or may be referred free of charge to consumers to one of several ADR Boards. General complaints against traders can be made to the public National Board for Consumer Disputes (Allmänna reklamationsnämnden, ARN). In relation to disputes about insurance,

[1] Konsumenternas Försäkringsbyrå.

the Swedish Insurance Association operates five ADR Boards: the Board for Insurance of Persons (PFN), the Board for Bodily Injury Liability Insurance (APN), the Board for Legal Protection Issues (FNR), the Insurance Appeal procedure (POL, which concerns an attorney's suitability), and the Attorney Cost Board (OKN). The Boards issue non-binding recommendations, almost all of which (in practice 100%) are followed by the insurance company members. Cases are considered in writing, by five people: the Chair, two members representing customer interests and two insurance experts. Expert external medical experts may also be asked to contribute. Together, the five voluntary Boards handle a total of around 1,800 cases a year.

The Board for Bodily Injury Liability Insurance reviews cases of serious bodily injuries. Under its rules, it is compulsory for subscribing companies to consult the APN (unless the claimant has a legal representative who agrees with the insurance company that the process is unnecessary) before making a decision, whenever a case involves medical disability of 10% or more, a specified yearly loss of income (SEK 22,000 in 2012), compensation for loss of financial support, re-evaluations of agreed cases where facts have changed, or if the claimant so requests on any other issue. The APN also provides opinions to courts, the ARN or authorities where requested to do so. It handles around 200 cases a year, and recommends a change in about 20% of the decisions made by insurance companies.

B. Swedish Motor Vehicle Insurance

Overview

Motor insurance is compulsory in Sweden, under the Motor Traffic Damage Act.[2] Owners, purchasers under a hire purchase agreement and those leasing a vehicle for a year or more are required to take out motor third-party liability insurance (MTPL) from the first day of taking ownership,[3] and the requirement for insurance is ongoing. Insurance premiums are calculated according to a number of factors including the type of vehicle, vehicle make and model, engine size, usage, domicile, claims history and method of payment. MTPL insurance covers personal injuries and property damage up to SEK 300 million.[4]

Third-party insurance can be supplemented by voluntary insurance that covers theft of the car or damage to it. Comprehensive insurance can be taken out if additional protection is required. Most insurance companies offer different types of supplementary insurance and these can vary between companies.

Standard claims under MTPL policies are handled by the insurance companies. The traffic injury insurance is essentially a primary payer, and social security payments and

[2] s 2 of the Motor Traffic Damage Act (SFS 1975:1410).

[3] Penalty premiums are liable if a vehicle is not insured in a timely manner, see www.tff.se/en/Motor-insurance-charges/Unpaid-insurance-premiums/. The penalty is considerably higher than the cost of the unpaid premium would have been, and the penalty premium is used to compensate those injured by uninsured drivers. Penalties are collected by the Swedish Motor Insurers (TTF), a trade body to which all Swedish motor insurance providers are legally obliged to belong. See www.tff.se/en/.

[4] www.lansforsakringar.se/globalassets/aa-global/dokument/information/01523-broschyr-personbil-eng.pdf.

workers' compensation are secondary compensation systems but with a right to recover its payouts from the traffic injury insurance. Motor liability insurance consists of personal injury and property damage elements.

Motor insurance covers all parties for all personal injuries. As a rule, damage is compensated in full, even if it was not due to anyone's carelessness. Under the Swedish system all personal injury insurance is on a first-party basis. Damages paid for personal injury are paid by the driver's policy, and there is no apportionment of personal injury damages in the event that the driver is not legally liable for the accident. There are limited circumstances in which personal injury compensation can be reduced, namely if there is intentional or gross negligence on the part of the driver or if the driver was over the drink drive limit or had consumed drugs. In these circumstances the compensation for the driver will be diminished, but major medical costs are not reduced and loss of income is never reduced below 90% of the driver's pre-injury earning level.[5]

Passengers', cyclists' and pedestrians' claims for personal injury are compensated by the MTPL policy of the vehicle on a no-fault basis.

MTPL insurance does not cover damage to the policyholder's own vehicle or his/her property contained in the vehicle. MTPL insurance provides compensation to third parties for any property damage sustained. Damage caused to an innocent party's car by a collision will be compensated. Claims for property can be apportioned according to legal liability. The claimant will seek compensation from the insurer of the vehicle he or she was travelling in, and the insurer will then subrogate and claim back the compensation relating to the 'at-fault' portion.

In most collision situations, the liability for compensation can also be split between the parties pro rata according to their degree of guilt. In situations where vehicles have collided and the damage was caused by the driver's or the owner's carelessness, driving in breach of traffic rules or by the improper condition of the vehicle, then the liability for compensation generally falls upon the party who failed to uphold these requirements.

The system is essentially one of strict liability based on the balance of probabilities, where the right to compensation will be determined on objective grounds.[6]

Since 1 January 2015 a limitation period has applied of 10 years from the day on which the accident occurred. This applies to insurance claims and court cases.

All MTPL policies have a deductible. For road traffic damage, the deductible is SEK 1,000 for a private individual, and SEK 3,500 for legal entities. An extra deductible of SEK 1,000 applies to drivers of 24 years of age or younger. In certain cases, for example driving without a valid licence, an additional deductible for criminal damage may also be added.

In Sweden litigation over personal injury compensation from road traffic accidents is very rare. This is because a special body, the Swedish Road Traffic Injuries Commission (Trafikskadenämnd, TSN) exists to settle such claims.[7] The Commission hears major

[5] See the guidance on the Transport Styrelsen website www.transportstyrelsen.se/en/road/Vehicles/Compulsory-road-traffic-insurance1/ and the Compensation for Personal Injury in Road Traffic Accidents in Sweden (Compensation based on the Swedish Law of Damages) produced by TTF the Swedish Motor Insurers available at www.tff.se/Documents/Skada/Skadeers%C3%A4ttning_eng.pdf.

[6] See the guidance on the Transport Styrelsen website http://www.transportstyrelsen.se/en/road/Vehicles/Compulsory-road-traffic-insurance1/.

[7] Trafikskadenämnd, TSN (the Swedish Road Traffic Injuries Commission): www.trafikskadenamnden.se/Information-in-English/.

compensation cases and determines the compensation payable to the claimant. TSN is funded by levies; all MTPL insurance providers and the TFF are legally required to maintain and fund TSN.

The Road Traffic Injuries Commission

The Road Traffic Injuries Commission and the Patient Injuries Commission (see below) are established by law and their role is mandatory for all insurance companies. The Road Traffic Injuries Commission was created in 1936 in order to bring consistency to awards of compensation.[8] Under the Traffic Insurance Decree,[9] all insurance companies that provide traffic insurance must 'maintain and pay for' a Road Traffic Injuries Commission. The regulations require that the Commission's task is to work for 'fair and uniform' compensation,[10] that the Chairman is appointed by the government, and that its regulations must be approved by the government. The Commission comprises 68 people, in three groups: 6 Deputy Chairmen and their substitutes (all judges) appointed by the Financial Supervisory Authority; 13 representatives of insurance companies and substitutes, appointed by the Financial Supervisory Authority on the recommendation of the Swedish Association of Traffic Insurance; 13 lay representatives and substitutes, appointed by the Financial Supervisory on the recommendation of labour market organisations.

Every case is decided by a panel of 6, comprising 2 from each of the above 3 groups. The secretariat has 20 employees, 17 of whom are lawyers, who present cases to the meetings of the Commission and write up the recommendations

An injured person has a choice of going to court or to the Commission. If the case is not heard in court, it is a requirement that the Commission hear certain cases before the victim and insurance company reach agreement, namely:

(a) Where there is medical disability of 10% or more; or annual loss of income of at least half of the base amount (an index-linked annually adjusted figure set by the Swedish Social Insurance System), approximately SEK 22,000; loss of support; or retrial (on certain conditions).[11]
(b) Where the victim asks for the Commission's recommendation.[12]

The Commission issues about 5,000 recommendations a year. It also annually publishes compensation tables, which are used by courts and the Lawyers' Association, and recommendations on special matters. It applies tort law, and therefore the principle of full compensation for relevant losses. The Commission applies decisions of the Supreme Court. Any court may require the Commission to issue a recommendation on a compensation issue.[13] The Commission takes 3–5 months to process claims.

[8] A similar scheme exists in Finland.
[9] Trafikförsäkringsförordningen.
[10] Traffic Injuries Commission Regulations, s 1.
[11] ibid, s 3.
[12] ibid, s 4.
[13] ibid, s 2.

Decisions are not binding, but are accepted in 99% of cases. Companies are required to inform the Commission if they do not follow a decision, but it is not known how many do not tell the Commission this.

Available Compensation

For personal injury, the claims process is divided into two stages: the first period of medical emergency; then any period of disability. The period of medical emergency is the period from the accident to the stage where the claimant's medical condition has stabilised. If the claimant has fully recovered, then there will be no subsequent period of disability. It may be that at the end of the period of medical emergency the claimant's remaining disability is recognised as temporary, meaning that from the end of the period of medical emergency a measurable disability stated as a percentage will exist for a specific period, eg 2 years, after which a new medical assessment has to be carried out. The degree of disability is considered definitive when the person's medical condition has stabilised to a degree that any residual adverse effects are permanent and foreseeable. (The situation is then said to be static.)

Types of damage for personal injury are categorised as shown in Tables 5.1 and 5.2.

Table 5.1: Categories of compensation available during the medical emergency period

Damage type	Available compensation
Expenses, including any medical and rehabilitation costs	Costs of medication, nursing, tests and treatment and travel expenses to and from medical appointments as well as any other non-medical out-of-pocket expenses
Loss of income	Actual temporary loss of income. This is calculated based on the loss of annual income that is attributable to the injury. Any income compensation received in the form of sick pay, occupational injury compensation or other comparable social insurance benefit is deducted from this amount. Also deducted are any employer's sick pay and any other compensation paid under a collective bargaining agreement, or paid by the employer under another scheme. Once these deductions have been made, what remains is the indemnifiable loss as defined in the Swedish Law of Damages
Pain and suffering	Paid based on a tariff system until the injury has healed or the state of health has stabilised. Compensation levels are determined by the type and severity of injuries, the type of care required and the duration of recovery. Tables to help calculate the sum due are published by TSN[a]

[a] Trafikskadenämnd, TSN (the Swedish Road Traffic Injuries Commission): www.trafikskadenamnden.se/Information-in-English/.

Compensation awarded during the medical emergency period is paid by the vehicle's MTPL insurer. If there is a dispute or question over the compensation, then this can be referred to TSN, which will then supply an opinion.

If the claimant has not made a full recovery after the initial medical emergency period, then compensation for a period of disability will follow.

Table 5.2: Categories of compensation available during the period of disability

Damage type	Available Compensation
Expenses, including any medical and rehabilitation costs	Costs of medication, nursing, tests and treatment and travel expenses to and from medical appointments as well as any other non-medical out of pocket expenses
Loss of income	Compensation quantum is determined by the loss of income that is attributable to the incapacity and that may be foreseen for the future. The calculations are made in annualised amounts For loss-of-income calculations, pension and similar payments are deducted. The loss of income is usually in the form of an annually calculated annuity, payable until normal pensionable age, and reduced by a certain amount thereafter
Pain and suffering	Paid based on a tariff system until the injury has healed or the state of health has stabilised. Compensation levels are determined by the type and severity of injuries, the type of care required and the duration of recovery. Tables to help calculate the sum due are published by TSN[a]
Disadvantage (permanent cosmetic defect) and incapacity	Compensation is payable for a permanent impairment to a person's appearance, such as a scar formation, caused by the injury. When settling a claim photographs of the claimant's scars are compared to reference photographs of scars from past cases settled by TSN. The level of compensation is stated in a separate table produced by TSN. In addition to cosmetic defects, other factors are taken into account, such as impairment of movement and/or functional impairment of sight, speech, hearing, smell, taste, etc. The age of the victim will be taken into account as a younger person will have to live with the injury for longer
Special inconveniences	This category covers any major difficulty or strain at work, over and above that which is covered by the disadvantage and incapacity table. For example, ongoing pain or tiredness may force the claimant to use what would previously have been leisure time to rest. Special inconveniences compensate non-financial damage or loss by subjectively taking into account of the victim's situation and circumstances at work, etc, as distinct from the categories of compensation 'pain and suffering' and 'disadvantage and incapacity', where the objectively measured consequences of injury dominate As for disadvantage and incapacity a younger person will receive more compensation than an older one as the younger claimant will have longer to live with his or her incapacity

(continued)

Table 5.2: ***(Continued)***

Damage type	Available Compensation
Death benefits	Funeral costs and other reasonable expenses in connection with the death are paid Loss of maintenance compensation will be paid to Individuals who were dependent on the support of a person killed by in a traffic accident. The payment is intended to put them in the position they would have been in if the individual had not died.

[a] Trafikskadenämnd, TSN (the Swedish Road Traffic Injuries Commission): www.trafikskadenamnden.se/Information-in-English/.

Sweden's Law of Damages states that compensation must be paid in the form of an annuity if the compensation is of essential importance as a means of support to the claimant. However, compensation may be paid in the form of a capitalised lump sum amount if circumstances justify it. For example, if the claimant is a foreign national and is resident outside Sweden, then a capitalized sum may be used should that form of payment be standard in the claimant's jurisdiction and because an annual payment could be subject to currency fluctuations.

Once a settlement has been finalised there are limited rights to review. Under Chapter 5, §5 of the Swedish Law of Damages, compensation may be reviewed if the conditions on which the compensation was determined have materially changed. Prior to 2002 this was strictly restricted to compensation for loss of income and maintenance; since 2002 the right to review of injury, loss or damage has been extended to cover all forms of compensation including costs and non-financial loss. Major increases in injury-related costs—nursing, medication, etc—can now be compensated via a 'cost annuity', which in such circumstances may include compensation for other inconveniences.

Any right to additional compensation is time limited. The claimant has to contact the company settling the claim within 3 years from the time at which the claimant becomes aware of a possible right to additional compensation, and absolutely no later than 10 years from the case being definitively settled.

For property damage, essentially any type of damage can potentially be covered, with certain derogations. Property, including a car, belonging to the owner of the at-fault vehicle will not be covered by his or her MTPL insurance.

Claims under the Swedish Motor Insurers

The Swedish Motor Insurers (TTF) has a broad remit including safeguarding the rights of individuals who have been involved in a road accident and overseeing aspects of the statutory motor liability insurance. TTF promotes traffic safety by collecting fines for uninsured vehicles. Additionally, TTF has a claims-handling role for specific categories of claims that are not handled by insurers, thereby providing a 'safety net' for claimants in relation to specific claim types. Specifically, TTF handles claims for compensation caused by uninsured vehicles, unknown vehicles and foreign vehicle accidents. The arrangements in relation to each of these categories are shown in Table 5.3.

Table 5.3: Motor claims handled by the TTF

Claim type	Available compensation
Uninsured vehicles	The Motor Liability Insurance Act requires TTF to compensate for traffic damage caused by an uninsured vehicle. As a rule, TTF's liability for damage caused by an uninsured vehicle is similar to an insurance company's liability for damage caused by an insured vehicle
Unknown vehicles	Personal injury and property damaged caused by an unknown vehicle is compensated by TFF
Foreign vehicles	TTF will assist with claims against foreign vehicles, regardless of whether the accident occurred in Sweden or abroad If the Swedish vehicle was insured the claim will be handled by the vehicle insurer. If the foreign vehicle was uninsured or unknown, then the claim will be handled by TTF
Animals	TTF pays compensation for the financial loss suffered by the owner of the animal as a result of a traffic accident with an uninsured or unknown vehicle. TTF receives around 1,500 claims per year, mainly referring to dogs, reindeer, cats and horses.

TFF Claims Process

Claims are made using a specified claims form, which is available online from TTF. The form is usually submitted by the driver causing the damage, or by the injured party if an unknown party causes the damage. The report must contain details of the vehicles involved in the incident, together with a detailed description of the incident. All documents which clarify the damage (eg the police investigation report, possible photos, a copy of the Green Card) should be appended to the damage report. Once completed and signed, a damage report must be sent to TTF.

C. Swedish Patient Insurance

Background[14]

Local government in Sweden is divided into 21 counties and regions, which operate relatively autonomously. Over 92% of health and dental care is publicly funded, with the remaining 8% split roughly equally between private medical and dental services. There are about 40,000 physicians, 11,500 dentists, 120,000 nurses, 8,000 midwives and 300,000

[14] The authors are indebted to Dr Pelle Gustafson, Medical Director of LÖF, for assistance on this section over several years.

assistant nurses, working in around 75 hospitals and 1,400 primary care centres. The Health and Social Care Inspectorate is responsible for supervision of professional standards and discipline.[15]

Before 1970, compensation for all medical injuries had to be claimed through the legal tort system, with court judgments or compromise being the only resolutions. Around 100 patients a year received some kind of compensation. All parties were dissatisfied with the results and with the system, and concern rose at that time when hospital-acquired diseases started to spread. Possible alternative solutions were investigated through an international search. In 1972, the Tort Damages Act was created, which highlighted the theoretical view prevalent in the United States that the purposes of the tort system were not just to deliver compensation but also to create a deterrent effect on behaviour. Swedish debate crystallised around the principles that injured people should be compensated for their actual losses, but the compensation system should not seek to deter or punish. As the number of medical treatments available increased, experiments began with ex gratia, compromise and insurance solutions.

In 1975, a distinctive voluntary insurance system for investigating and compensating medical accidents was created, which covered 99% of claims.[16] The insurance was administered by a consortium of companies, which offered cover to both private and public healthcare providers. This was felt to have worked well, and trust and confidence was felt to have risen both in the arrangements and in relations between patients and healthcare providers. A 1992 review concluded that care-givers' responsibilities to compensate should be delineated and they should be subject to an obligation to compensate. Increased familiarity with EU law raised the objection that the Patient Insurance Consortium was illegal under the Competition Act. Finland introduced an insurance solution by law in 1987, and Denmark in 1992.[17]

On 1 January 1997 the voluntary arrangements in Sweden were replaced by a statutory scheme under the Patient Injury Act (1996:799). The Act reads similarly to the previous insurance contracts, and provides the right of patients to injury compensation and the obligation of care-givers to be insured. The decision to award compensation is taken independently of the opinion of the care-giver, and is paid where an injury was avoidable. Payments under Patient Injury Insurance are secondary to those made under the Social Insurance System, the latter covering ongoing care needs, such as a live-in carer or house modifications. The two payment systems are mutually dependent.

The Insurance Arrangements

The Patient Injury Act requires care-providers to carry patient insurance which pays compensation for injuries. Most medical services in Sweden (perhaps 96%) are provided

[15] This body (the IVO, see www.ivo.se) took over the role from the National Board of Health and Welfare in 2013. Until 2011, this was the Health and Medical Care Liability Board. In 2010, 4,500 complaints were received, and 400 reprimands issued.

[16] A voluntary scheme was introduced in 1975, and was extended by law to full coverage in under the Patient Injury Act (1996:799). See www.patientforsakring.se/Articles-in-English.html. See generally U Hellbacher, C Espersson and H Johansson, *Patient Injury Compensation for Healthcare-Related Injuries* (Visby, 2007).

[17] Finland 1987; Norway provisional rules in 1988, law in 2003; Denmark 1992; and Iceland in 2001.

by county councils and regions, which use a common mutual insurer, Landstingens Ömsesidiga Försäkringsbolag (LÖF, County Councils' Mutual Insurance Company),[18] which is jointly owned by the county councils and regions. Thus, LÖF processes 90–95% of all patient injury claims in Sweden. Patients pay only SEK 100 a day to public providers for any kind of inpatient treatment, from a minor procedure to a triple heart bypass, and SEK 300 for an outpatient visit. Councils pay insurance premiums of SEK 105 per inhabitant.

The remaining claims not covered by LÖF, arising from privately funded care, are managed by the Patient Insurance Association, to which all insurance companies providing relevant cover belong, and under whose rules all patients shall have the same coverage, regardless of care-provider. If a care-provider does not have insurance, the Patient Insurance Association will itself investigate and compensate the injury, thereby guaranteeing cover, and reclaim sums paid from the caregiver.

The purpose of the arrangement is purely to provide compensation, and not to involve any disciplinary, supervisory or regulatory function. Introduction of the statutory scheme was judged to have made it easier for patients to receive compensation, and to have 'considerably improved relations between patients and medical professionals'.[19] Patients can receive compensation without having to search for any error or negligence during treatment, and

> can more easily understand that healthcare-related injuries can occur without doctors or other healthcare workers showing negligence or making a mistake, while healthcare workers do not have to feel their actions will be scrutinized every time the issue of compensation for patient injury arises … medical professionals can feel free to help patients receive compensation.[20]

The Patient Injury Act covers patient injuries that occur in connection with health and medical care services in Sweden.[21] It does not cover medical care delivered abroad, unless this care has been approved by Swedish authorities under certain circumstances. The word 'patient' is not defined in the Patient Injury Act, but preliminary materials for the Act referred to all persons who have established contact with medical personnel with respect to their health. 'Health and medical care services' are defined[22] by reference to the Swedish Health and Medical Care Services Act (1982:763) as medical measures to prevent, examine and treat diseases and injuries. Also included are ambulance transport services and similar forms of patient transport, dental care under the Swedish Dental Services Act (1985:125), and 'other similar medical activities', which includes in vivo studies for medical purposes, forensic psychiatric examinations and medical research on humans. The Act also covers retail sales of pharmaceuticals.

To be covered, the activity must be carried out by personnel subject to the Act on professional activity in health and medical care (1998:531), whether care is public or private. Further details are specified in Regulations.

[18] www.patientforsakring.se. At one stage, claims were outsourced to be handled by an external company, Personskadereglering AB (PSR).

[19] Hellbacher, Espersson and Johansson (n 16) 7.

[20] ibid.

[21] Patient Injury Act, s 3.

[22] Patient Injury Act, s 5.

Compensatable Injuries and Causation

Compensatable injuries include both physical and mental injuries. To be compensatable, the injury must have been caused by the healthcare service provided. Patients are not entitled to compensation if the injury is the consequence of a necessary procedure to diagnose or treat an illness or an injury which, if left untreated, would be directly life threatening or lead to severe disability. Injuries caused by medications are not eligible for compensation unless the injury is due to a prescription or administration error. The Patient Injury Act specifies that compensation is paid if the injury is caused by one of six categories of compensatable injuries:[23]

1. *Avoidable treatment injury*: '[A]n injury caused by examination, care, treatment, or similar measure, provided that the injury could have been avoided either by a different performance of the chosen procedure or by choice of a different manner of performance, or by a different procedure with the same effect, but with a lower total risk.'[24]

   In determining whether or not a treatment or diagnostic injury was avoidable, the Act specifies that the action is to be assessed against that of an experienced specialist or other experienced practitioner in the field. The standard is not that of the most experienced expert, and this test is applied in relation to each relevant area of practice. The evaluation is retrospective, and this subsequent independent viewpoint avoids attributing fault to the professional who dealt with the situation at the time.

   Direct consequences of an underlying disease and injuries that would have occurred or developed anyway, regardless of medical care, are not eligible for compensation. Compensation is not paid if, inter alia, 'the injury is a consequence of a necessary procedure for the diagnosis or treatment of an illness or injury which without treatment is downright life-threatening or entails severe disability'.[25]

   A body of learning and practice has developed from experience of particular situations. In relation to birth injuries, the injury must be caused by a direct medical procedure in conjunction with childbirth, such as the use of vacuum extraction, or by incorrect diagnosis during childbirth.

   The severity of an injury, or its rarity, has no relevance on the right to compensation. An alternative treatment method must be available at the time of treatment, meet the patient's care needs, and offer an overall lower level of risk.
2. *Material-related injury*:[26] '[A]n injury caused by a defect in, or the defective use of, a medico-technical product or hospital equipment used for examination, care, treatment or similar procedure.'[27]
3. *Diagnostic injury*: '[A]n injury caused by incorrect diagnosis.'[28]
4. *Infection injury*: '[A]n injury caused by transmission to a patient of an infectious agent during examination, care, treatment or any similar procedure, which leads to an

[23] Patient Injury Act, s 6.

[24] Patient Injury Act, s 6.1.

[25] Patient Injury Act, s 7.1.

[26] This covers an injury caused by a defect in, or the defective use of, a medical device or hospital equipment used for examination, care, treatment or any similar measure. Such injuries constitute fewer than 1% of all claims.

[27] Patient Injury Act, s 6.2.

[28] Patient Injury Act, s 6.3.

infection.'[29] No compensation is paid if the infection must be reasonably tolerated, or if no transmission has occurred, such that the causative bacteria or virus came from the patient.[30]

5. *Accident-related injury*: '[A]n injury caused by an accident occurring in connection with examination, care, treatment or any similar procedure.'[31] This would include during a patient's transport or in connection with a fire or other damage to healthcare premises or equipment.
6. *Medication injury*: '[A]n injury caused by prescription or administration of medication in contravention of regulations or guidelines.'[32] This would include an injury that arises because the medication is used incorrectly, with an incorrect dosage, or if contraindicated.[33]

A provider's resource constraints may occasionally be an explanation for why something has occurred, but are not taken into consideration in deciding on entitlement to compensation. The patient has the burden of proof of showing that the causal relationship exists. The required standard of proof is that the causal relationship can be established with preponderant probability. In practice, the insurer investigates each case, and makes a determination. If the insurer fails to find a causal relationship between the action and the injury, the patient's burden comes into play at that stage.

The amount of compensation awarded is determined under the personal injury compensation rules of the Tort Liability Act.[34] Under that Act, economic damages (loss of income and costs incurred due to the injury) and non-economic damages (compensation for pain and suffering, disability and disfigurement, and inconvenience) are awarded. In the event of death, compensation is paid for funeral costs, loss of support and psychological problems due to the death of a close relative.

Cases have both a ceiling for the amount of compensation and a deductible. A deduction is made from every patient injury compensation equal to 1/20th of the base amount under the National Insurance Act (1962:381) applicable when the compensation is determined.[35] The deduction is currently roughly €250. This deductible was intended to deter the scheme being flooded by a large number of small and insignificant claims, and mirrors the same technique used in the product liability Directive for property damage.[36] However, there is a general feeling now that the deductible is unhelpful, since it reduces many small sums paid to an amount that patients can consider to be insulting, and involves unnecessary administration cost. One Swedish county decided in 2015 to compensate the patient for the deductible where a case is settled. There is speculation as to whether more counties will follow, or whether the deductible will in time disappear from the Act.

Patient injury compensation for each event is limited to 1,000 times the base amount under the National Insurance Act (1962:381) applicable when the compensation is determined,

[29] Patient Injury Act, s 6.4.

[30] This can lead to difficult decisions in determining what type of germ caused the injury, or whether it was transferred: Finland and Denmark have rules that are somewhat easier to apply, but attempts to change the law in Sweden have so far been unsuccessful.

[31] Patient Injury Act, s 6.5.

[32] Patient Injury Act, s 6.6.

[33] Injuries that are caused by the medicine itself are passed to the Drug Insurance.

[34] Patient Injury Act, s 8.

[35] Patient Injury Act, s 9. The deduction was SEK 2,220 in 2014.

[36] Directive 85/374/EC, Art 9(b); deduction of €500.

which is roughly equivalent to €1 million.[37] For each injury event the patient injury compensation is limited for each injured patient to at most 200 times this base amount.[38] The amounts stated in the previous two sentences do not include interest or compensation for litigation costs.[39] The ceiling is felt by LÖF to produce injustice in a number of high-value cases, especially babies injured at birth. If such a claimant were to succeed in court, proving negligence or culpa, the compensation would not be subject to such a ceiling. In around 10 cases a year, LÖF applies some discretion in voluntarily treating these types of claims as tort claims so as to award extra amounts.

Process

Independent Patient Advisory Committees exist in every region that help patients who run into problems by offering information and advice. They receive 25,000 contacts a year.[40] Patients send their claims to LÖF (for publicly provided care) or to the Patient Insurance (for private providers). A claim to the LÖF or Patient Insurance is completely free to the patient. The claim form is available from the LÖF website, although patients usually register a claim with help from a social worker in each clinic. It is only the patient, relative or legal help who can file a claim. LÖF has around 150 employees and works with 170 medical advisors.

For private claims, a claim adjuster (usually a lawyer or nurse) employed by the insurer will investigate the facts from the records, usually obtaining an opinion from a relevant medical expert, who is a member of a panel of around 150 specialists. The patient merely has to say that he or she has an injury: unlike a legal claim, it is not necessary to make any allegation of causation or fault, or to provide evidence. The same approach applies to claims made to LÖF, where the medical expert is called a Medical Advisor, and thus formally gives advice to LÖF, which makes the decision.

A patient who is unsatisfied with the insurance company's response may refer the claim to the Patient Claims Panel, which is established under the Act, as an alternative to going to court. The Panel consists of a Chair and six other members. The Chair must be, or must have served as, a regular judge. Three of the other members represent the interests of the public/patients—one must be a medical expert, one must be familiar with the personal injury settlement process of the insurers, and one must have special knowledge of healthcare activities (in practice, this results in two doctors being members). The reference to the Panel is free. The Panel's mandate is to promote fair and consistent application of the Patient Injury Act and to issue opinions at the request of a patient or other claimant, care-provider, insurer or court. Opinions may relate to either decisions or quantum of compensation, or both. The Panel's opinions are advisory but insurers always comply with them.

Around 10% of all insurance decisions are referred to the Panel. The Panel's opinion differs from that of the insurer in about 10% of cases. This mechanism is considered by the insurers to provide both an expert appeal mechanism and effective oversight of their

[37] Patient Injury Act, s 10.

[38] ibid.

[39] ibid.

[40] From 2010, the Patient Safety Act requires hospitals to inform patients of the system. The intention has been to make the system better known and open; a similar change in Denmark, where notifications can be made anonymously, resulted in a dramatic increase in claims.

activities and decisions. The questions considered by the Panel are predominantly highly technical, and are resolved by applying combined medicolegal expertise. Some grey areas arise, in which case the Panel tends to err on the side of a patient.

A patient may receive compensation under the Patient Injury Act considered by a court, either instead of going through the insurance route or within a certain time after an insurer's decision. Perhaps 15–20 cases are commenced in court a year, of which 1 or 2 are successful.

Statistics

In 2015 patients reported approximately 15,700 injuries or treatment complications to LÖF. Compensation is paid in about 40% of these cases[41] to around 6,300 patients. The total compensation cost paid per year was estimated at about SEK 400 million in 2007, increasing to SEK 630 million in 2015. The trend in annual cases is shown in Figure 5.1. The number of claims was stable for most of the 2000s but increased after the new Patient Safety Act was passed in 2010, which led to increased visibility of the system, owing to requirements that healthcare providers identify an injury as such to patients and inform them of their right to seek compensation, and that providers break the law if they try to cover things up. The notification numbers for 2015 (15,700) were 9% higher than the 2014 notification level. This increase is thought to be partially due to the increased disclosure requirements of the 2010 Act and also to the county councils and regions taking a more active role in engaging with patients. LÖF interpret the increase in notification numbers as an indication of higher levels of awareness of the compensation system, rather than an increase in adverse events.

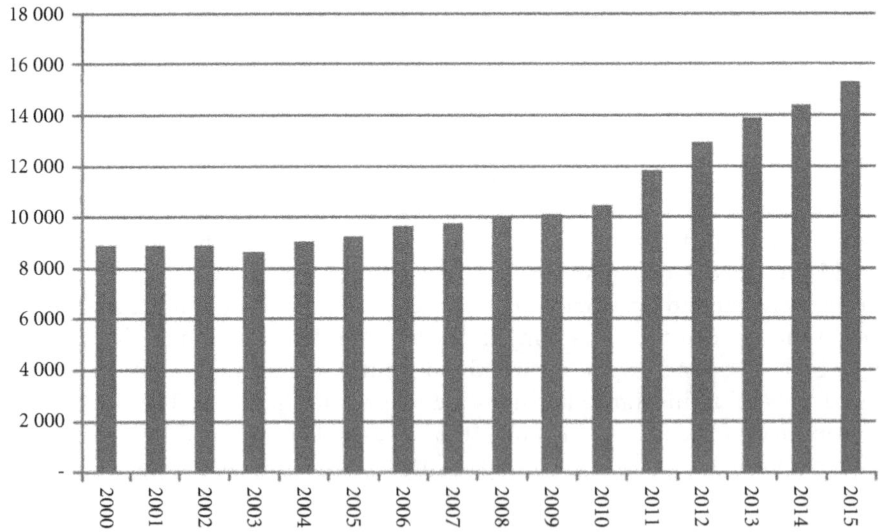

Figure 5.1: Number of annual claims made to LÖF 2000–15

[41] In 2004 these figures were 10,000 reports and 45% compensated.

The distribution of claims and costs by specialty for 2014 is summarised in Table 5.4. The most reported injuries are unclassified, orthopaedic surgery, dental care and general surgery. Obstetrics and gynaecology account for about 7% of claims, but over a quarter of compensation costs because of the serious and ongoing nature of injuries suffered by children harmed at birth, and their need for expensive long-term care. The most expensive claims are those involving brain-damaged infants: around a decade ago, of 100,000 babies born each year, approximately 30 claims annually were settled for birth-related brain injuries. Today, this figure is considerably lower, as discussed further below.

Table 5.4: Distribution of LÖF claims and costs in 2016

Specialty	Claims %	Costs %
Orthopaedic surgery	22	22
General surgery	11	8
Obstetrics and gynaecology	8	28
Primary care	10	8
Dental care	12	2
Other	31	32

Source: Pelle Gustafson, Chief Medical Officer LÖF, personal communication.

The most common reasons for rejection of claims in decisions taken in the period 1997–2004 are summarised at Table 5.5. The most common reasons were, first, that the injury could not have been avoided through another procedure and, second, that no causal relationship was present between the medical care and the reported injury.

Table 5.5: Five most common reasons for rejection in decisions 1997–2004

Reason for rejection	%
Unavoidable consequence	26.3
Not related to treatment, etc	17.6
Not a diagnostic injury	17.0
Not a personal injury	5.5
Applicable insurance not carried at time of injury	4.5
Other reasons	29.1

Source: Taken from Statistics for LÖF 1997–2004. More recent data is not available.

The Patient Claims Panel currently receives somewhat over 1,500 claims a year. This has risen from around 1,100 claims in 2004. For many years, the percentage of claimants awarded compensation has been stable at about 10%. The courts receive around 10 cases a year of which perhaps 2 succeed. The Patient Insurance system resolves 50% of claims within 6 months, and 80% within 12 months.

The premium cost is around €12 per inhabitant per year, and the total premium is around 0.3% of the total national healthcare expenditure of SEK 300 billion (€3.08 billion). The cost per case is around €900 per claim, compared with an estimated €22,000 per court case. LÖF's annual turnover was around €110 million in 2015. It has 140 employees and 150 part-time medical advisors. The total paid in compensation in 2015 was around €66 million. LÖF is involved in around 20 cases a year where patients sue in court; it has two in-house lawyers who handle these. It tries to settle these court claims before the court process begins.

A decision on approval or rejection of a claim is made by LÖF in 75% of cases within 6 months: the aim is to reduce this to 3 months. Decisions on calculation of sums paid take a further 2–3 months. A reference to the Patient Claims Panel takes a further 4–5 months. If a patient's prognosis remains unclear, LÖF tries to pay part of the claim and finalise it when the evidence later becomes available.

Advantages and Disadvantages

The system has a number of significant advantages. The system is transparent. It delivers low-cost full insurance payments to those who qualify, without deduction for cost. It involves no cost or risk for the injured person. It has increased the number of patients who receive compensation who are thought rightfully to deserve it. Individual rights to go to court at any time during the process are preserved. But under the court process, the claimant faces the burden of proving negligence, funding the costs of the action, and the risk of liability for opponents' costs if the case fails. Healthcare workers and patients appreciate that the system supports openness, trust and confidence in their relationships, instead of confrontation and mistrust. The Swedish system also offers the ability for the insurance body LÖF to engage with healthcare bodies and workers in injury prevention, discussed below.

LÖF considers that the scheme is objective and effective to operate. Particular problems identified with the current system by LÖF are as follows. First, the deductible causes more problems than it avoids. Second, the requirement for 'transmission of an infectious agent' is a difficult concept for patients to understand, and can give rise to complex investigations and fine decisions. The Finnish or Norwegian approaches—in which the decision to compensate an infection injury is based on more objective and understandable grounds—are thought to be better,. Third, public knowledge of the insurance arrangements and availability of compensation is not as high as the operators or healthcare community would wish: visibility could be higher.

LÖF considers that it responds to what people want after a medical mishap: first, an explanation of what happened; second, the ability to excuse the healthcare professional (forgiveness); and third, knowledge that the information is used within a system aimed at ensuring that same thing does not happen again (monitoring, learning, improving).

Using Data to Monitor and Improve Practice

The data arising from injury claims has in recent years been mined as a valuable source of information on improving patient safety. It is owned by the employers, and is authorised to be used in various ways. All of the claims received and paid are registered on a database by LÖF. This database currently comprises well over 150,000 claims, including the 14,500 claims that are current on an annual basis. It comprises a major record of the undesirable outcomes flowing from the nation's public healthcare system. This data has been deployed to improve outcomes, in cooperation with the owners of the facilities and (especially) the professionals involved. LÖF's work on safety improvement is unique in the country, and has produced outstanding successes.

The work falls into four main categories: large projects; smaller projects; education, courses and conferences; and research. The large projects are reactive to the main areas of claim types, such as safe deliveries, reducing prosthesis infections, safe abdominal surgery, safe use of drugs in primary care, and safe trauma care. LÖF works with a wide range of professional organisations on these initiatives, investing some of its resources in administrative and financial support. The project on safe deliveries has achieved outstanding success within a short time. LÖF originally adopted an approach to reducing birth injuries from practice in the Netherlands. A Web-based programme and online learning courses were created, focusing particularly on interpreting cardiograms. As shown in Figure 5.2, after the period of intensive outreach of working closely with midwives, obstetricians and paediatricians (marked 'safe delivery and care'), the incidence of birth injuries has been steadily reduced, and in some quarters has been zero. In consequence, not only are fewer children condemned to live difficult lives with significant mental and physical deformities, and the disruption and emotional damage caused to families has been avoided, but the insurance costs of the entire compensation system have been reduced significantly. The graph shows the number of settled claims for serious birth injuries on a quarterly basis, per 100,000 babies born during the period. As far as LÖF is able to assess, the rules governing compensation and their application did not change during the time period covered by this graph.

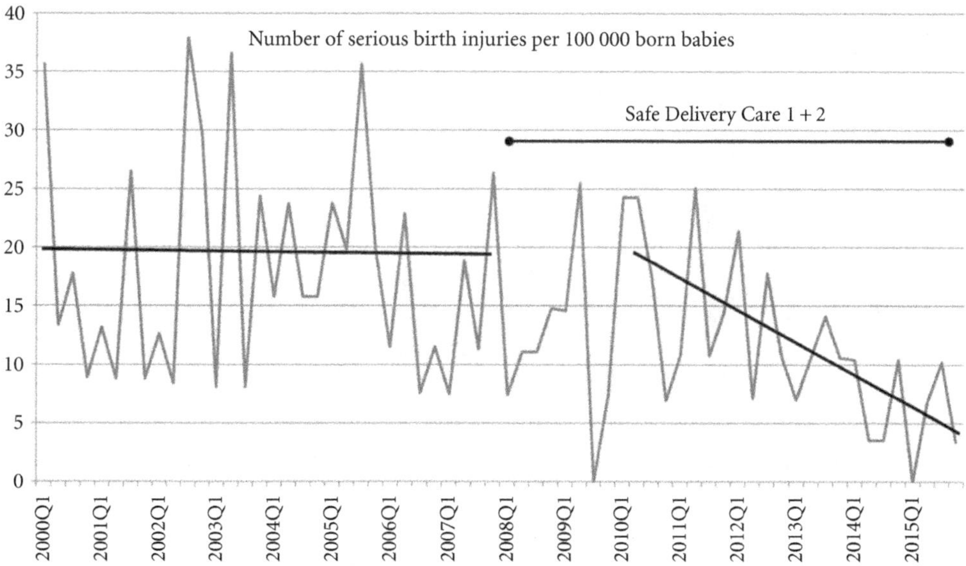

Figure 5.2: LÖF data on settled claims involving serious birth injuries 2000–15, per quarter and per date of birth

A further example of improvement in practice, coupled with reduced need for compensation, is in wound complications after hip and knee arthroplasty. The project used the same approach as the baby delivery project. The aim was to reduce joint infection after joint prosthesis surgery. Figures 5.3 and 5.4 show the reduction in infections for hip and knee arthroplasties after a large orthopaedic department went through the retraining project in 2010 and 2011. A dramatic decrease in both deep and superficial infections can be noted, drastically decreasing the number of injured patients, and the need to pay out compensation. The project has been extended to other departments in Sweden, where similar results have been reported.

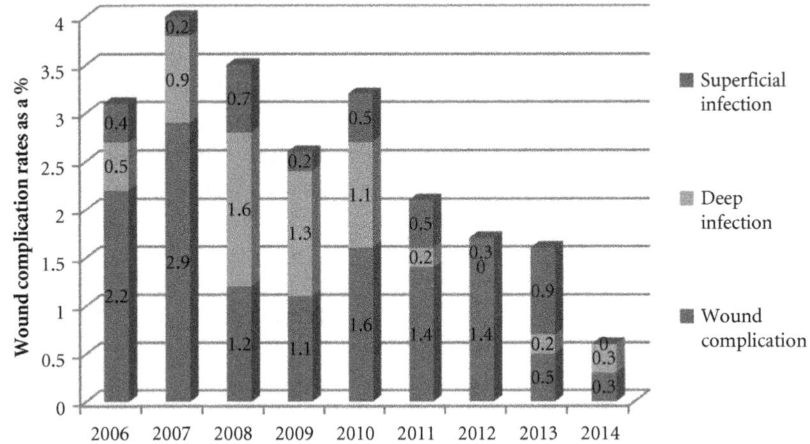

Figure 5.3: Data on wound complications after hip arthroplasty in a large orthopaedic department 2006–14

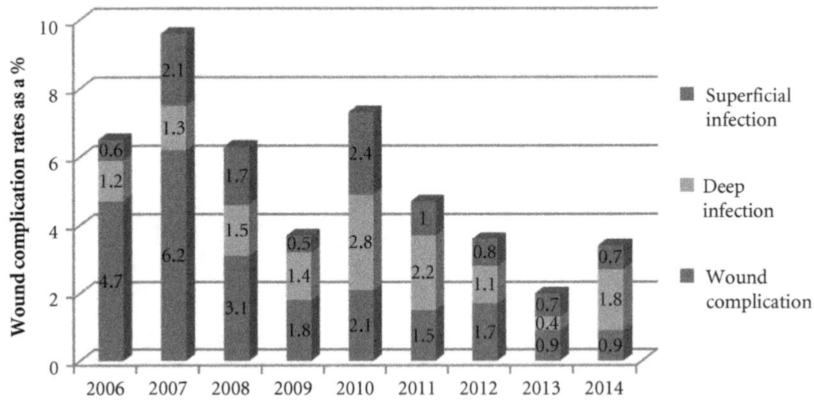

Figure 5.4: Data on wound complications after knee arthroplasty in a large orthopaedic department 2006–14

Examples of smaller projects have been: smoke-free operation, providing claims statistics to caregivers, marking of medication packaging, providing 100-point surgery checklists (SURPASS, invented in the Netherlands), crew resource management, and structured communications (SBAR).

There have been multiple ways in which LÖF has disseminated the information that it holds, and engaged in education, courses and conferences. Authorised officers of employers can be authorised to access the data down to the level of individual patients. LÖF encourages self-analysis of data by caregiver organisations and individuals. It is introducing an online log-in facility. Data is not provided to the media below county level. Individual doctors are notified of a decision on award on compensation, but not of what sum is paid.

Some years ago, systems were in place to provide regular updated information to senior officials.[42] However, it was found to be unnecessary to continue this, since the initiative produced the desired incentives in focusing on behaviour, and under the Patient Safety Act of 2010 root cause analysis became more or less mandatory. Certain legal restrictions on disseminating data have also been introduced, which have prompted investigations into new ways of spreading data.[43]

National patient safety conferences are arranged together with the Hospital Federation, the National Board of Health and Welfare and the medical professions. Facilitation of relationships between Chief Medical Officers has been particularly effective. It is not the function of LÖF to adopt any authoritarian or regulatory position in relation to, say, whether there were problems with a particular hospital or individual, but issues can be flagged with the local responsible person. Thus, LÖF's Medical Director might phone a hospital's Chief Medical Officer to ask if he or she was aware of a trend or concern that had been identified from the claims data. LÖF would, however, not see its function as to inform the Inspectorate.

The claims data are also used for scientific studies. For example, all 500 claims between 1990 and 2007 for brain-damaged babies were analysed, in particular 177 avoidable brain damages from 'normal deliveries'.[44] The causes for brain injuries identified were: cardio-tocography-interpretation, oxytocin prescription, non-optimal mode of delivery, and resuscitation. LÖF uses its resources to provide pump-priming funding for initial stage research that other bodies would not fund.

It is not, of course, known how many incidences of accidents have not been reported. It would therefore be difficult to use the database to measure prevalence or incidence of adverse events.

[42] Every third month the Chief Medical Officer of all participating hospitals received an overview of the claims for the hospital in the previous month, on the basis of which a root cause analysis was undertaken for severe claims, so as to identify why the injury happened but not who did it. The idea was that the root cause of an injury could be identified by asking 'why?' in many similar cases. It was usually found that routines or systems needed to be changed in order to avoid the same kind of injury happening again. LÖF used economic incentives to promote root cause analysis, by paying $1,000 to hospitals who did a root cause analysis of any medical injury that was reported to LÖF.

[43] Earlier practice was that every year the Chief Medical Officers and the heads of Orthopaedics, General Surgery and Obstetrics received full information of medical details for all claims arising from their hospital in the previous 5 years. They also received comparisons of similar medical procedures in all hospitals, such as a diagram covering all hospitals that perform hip replacements. The diagram showed per hospital both the number of claims and the number of performed hip replacements, so readers could compare their results other comparable departments. Diagrams were produced for 60 procedures per hospital. Since 2009 some of that information was also published by the LÖF so as to assist learning by hospital staff.

[44] S Berglund, C Grunewald, H Pettersson and S Cnattingius, 'Severe Asphyxia due to Delivery-Related Malpractice in Sweden 1990–2005' (2008) 115 *BJOG* 316–23.

Conclusion

A major reason why the Patient Insurance system is economic is that its awards are subsidiary to benefits paid by the national welfare system. Thus, the general welfare system provides 80% of salary during sick leave and 64% of salary if early retirement occurs because of illness for incomes of up to €32,000 a year. The Patient Insurance only pays amounts over those levels on the basis of the damages that would be awarded in court under the Tort Liability Act. The cost is €7 per claim per inhabitant per year, with administration costs taking 20% of the total.

D. Swedish Drug Insurance

General Description

The Swedish Drug Compensation Scheme[45] was introduced voluntarily in 1978, against the background of government proposals to introduce a statutory scheme. It was influenced by the Patient Insurance Scheme (voluntary from 1975 and under the Patient Insurance Law that came into force on 1 January 1977) and also the traffic damage model.

Rising costs provoked a series of changes. Before 2007, the Pharmaceutical Insurance Association (LFF) collected premiums and bought insurance on the market, from Zurich Insurance, negotiating the premium annually based on the claims record, where the insurer handled all claims. In 2007 a holding company was established (LFF Service),[46] in which every pharmaceutical company owns one share, regardless of company size, and into which insurance payments are made according to company sales. Further, a captive insurance company was established (SLF),[47] fully owned by LFF. Between 2007 and 2011, claims handling was outsourced to Zurich Insurance, but in 2011 claims handling was brought into SLF. This immediately avoided payment of 25% VAT on outsourced services.

All manufacturers or importers of pharmaceuticals subscribe to an undertaking that agrees to pay compensation and to take out insurance, and sets out the rules under which compensation will be paid for injuries.[48]

A drug-related injury is defined as meaning personal injury that has been caused, on the balance of probabilities, by medication with or any other use of a pharmaceutical for purposes treating disease or other healthcare issue, or for clinical trials of new pharmaceuticals.[49]

[45] www.lakemedelsforsakringen.se/default.aspx?id=16720. See C Hodges, 'Nordic Compensation Schemes for Drug Injuries' (2006) 29 *Journal of Consumer Policy* 143–75; J Dute, MG Faure and H Koziol (eds), *Tort and Insurance Law: No-Fault compensation in the Health Care Sector*, vol 8 (New York, Springer, 2004). The drug scheme was introduced in 1978.

[46] LFF Service AB. The company has two employees, the Managing Director and the secretary.

[47] Svenska Läkemedelsförsäkringen AB.

[48] The Undertaking is amended periodically, the latest applying from 1 January 2015.

[49] Undertaking, cl 3. The following are excluded: a disease or other injury that is due to lack of or reduced effect by the pharmaceutical; dependency on a pharmaceutical; injury that has occurred when an activity that is inappropriate having regard to the intended or predicted effect of the pharmaceutical; injury that further to what is stated above and in cl 7 has occurred through the use of the pharmaceutical in a way that is obviously wrong or inappropriate. A drug-related injury is not indemnified if the injury with preponderant probability was caused due to the ordination or supply of the pharmaceutical in contravention of regulations or instructions: cl 4.

A drug-related injury is indemnified only to the extent that:

(1) the injury suffered is disproportionate to the expected benefit from the treatment; and
(2) the type of injury, or the severity of the injury could not reasonably be predicted.

The consideration according to principle 1 must comprise a weighing in between the scope/extent of the injury on the one hand, and on the other hand the nature and severity of what the treatment was aimed to indicate, cure, mitigate or prevent, had not the treatment been started.

The consideration according to principle 2 shall be made with reference to what an experienced professional should have predicted. In the case of non-prescription drugs, the reference is the text on the packaging and the package leaflet of the drug. The injured person's general health condition shall also be taken into consideration.

Under normal circumstances, a drug-related injury is not indemnified if the period of injury is less than one month.[50]

Compensation for a drug-related injury is to be determined in accordance with the Tort Liability Act,[51] with the following exceptions:[52]

8.1 Compensation for bodily defect or other permanent harm is paid in accordance with the norms determined in respect of each calendar year by The Company.

8.2 When determining the amount of the indemnity, a deduction shall be made for, beside the benefits mentioned in Chapter 5, Section 3, of the Tort Liability Act, compensation that can obviously be received through social security insurance, road traffic insurance or patient injury insurance.

8.3 Compensation is not paid for such additional costs that may arise as a result of benefits provided by the State, county council or municipality being debited in a greater amount or lapsing because the injured person is entitled to reimbursement for costs under this undertaking.

8.4 In the case of death owing to a drug-related injury, indemnity can also be paid to someone who was particularly close to the deceased in accordance with Chapter 5, Section 2, third item of the Tort Liability Act (1972:207).

8.5 Compensation is not paid for costs of representation in conjunction with consideration of a claim for compensation against the insurer.

The following limits on indemnity apply:[53]

— SEK 10 million for each injured person, including the value of annuities capitalised, at the date they are fixed, in accordance with Trafikförsäkringsföreningens Cirkulär A 5/2005.
— SEK 250 million for all injuries that are reported during one and the same calendar year, of which at most SEK 150 million for all series injuries referable to one and the same calendar year.
— An injury shall be referred to the calendar year in which the report was made to the insurer. However, in the case of a series injury all injuries that form part of a series loss shall be considered as reported during the calendar year when the first report of an injury in the series is submitted to the insurer.

[50] Undertaking, cl 5.
[51] Tort Liability Act 1972:207, ch 5, ss 1–5 and ch 6, s 3.
[52] Undertaking, cl 8.
[53] Undertaking, cl 9.

- Series injury means drug-related injuries that have afflicted several persons as a result of the same injurious effect of one or more pharmaceuticals within the same therapeutic field of application, provided the pharmaceutical has been withdrawn from registration or its use restricted due to injurious effects, or the injurious effect was caused by a manufacturing defect.
- However, a series injury comprise only injuries caused by the use of a pharmaceutical that has been distributed for consumption before information has been given to the Swedish prescribers of pharmaceuticals.

Under the rules of the undertaking, a claimant must give notice of injury to the insurer. SLF examines each case and negotiates with the claimant a resolution to the claim. The claimant must give consent to the insurer for the latter to investigate the matter by having access to the patient's records. The investigation is a retrospective, documentary enquiry and does not involve physical examination of the patient. Questions of principle and disputed claims are referred, at the request of the claimant or insurer or LFF, to a Pharmaceutical Injury Panel. The Panel comprises 7 members, of which the Chairman and 4 other members are nominated by the government (of whom 2 members represent medical science and 2 members represent the interest of patients), and 2 are nominated by the insurer (of whom 1 shall have extensive experience from claims handling at an insurer).[54] Since 2007, a claimant who disputes a decision of the Pharmaceutical Injury Panel has the right to go to court.[55] A very small number of such claims have been brought in court, but as at 2015 all such claims had subsequently been dropped by claimants before completion of the court process.

The annual premiums paid by pharmaceutical companies in Sweden totalled around SEK 80 million (€8.88 million) in 2013–15, giving an average cost per head of population of €1.21. The cost fell over the preceding decade, despite rising claims, as a result of efficiencies achieved by the Drug Insurance Association and the creation of the captive insurance company that has brought claims handling in-house. For example, around 2004 the total cost was SEK 100 million (€10.87 million).

Statistics

Table 5.6 shows statistics from 1994. In the 36.5 years up to 31 December 2004 in which the Swedish scheme operated, 10,171 injuries were reported. Of the total number of claims reported, 6,187 were rejected, and 3,462 were compensated. Thus, an average of 384 claims per year were reported, of which 35.9% were compensated. The total compensation paid from 1978 to 2004 was SEK 560.2 million, ie an average of SEK 161,814 per injured person (€17,979). The numbers of claims per year, and the percentage compensated, are reported to have been similar for many years, apart from a rise after a television programme and other media interest in 2001 relating to hepatitis C infection (although for some reason the number of claims paid remained at the normal level of one-third).

[54] LFF Undertaking, cl 14. This was changed from 8 members, as a result of creation of the captive insurance company in 2007.

[55] There was previously an arbitration option.

Claim numbers increased from 2009 to 2012 as a result of various effects. First, there were over 700 cases relating to one drug, Pandemrix. Second, the 2011 Patient Safety Law mandated information to be given out about patient insurance, which led to a 30% increase in patient claims, and this had a knock-on effect on the profile of drug insurance. Thirdly, patients were enabled to self-report side effects to the regulatory authority, and the response letter from the authority includes reference to the drug insurance. It may be that the numbers have now stabilised.

Table 5.6: Swedish Pharmaceutical Insurance statistics as at 1 October 2015. Claims made include claims for serial injuries from Gammagard, Vioxx, Acomplia, Pandemrix (narcolepsy claims) and other injuries

Reported year of injury	Number of claims received	Paid (SEK)	Reserved for future payments on claims assessed (SEK)	Total costs (SEK)
1994	388	34,605,583	1,100,000	35,705,583
1995	380	32,596,766	8,454,552	41,051,318
1996	435	38,090,122	13,882,389	51,972,511
1997	492	39,223,893	3,098	39,226,991
1998	411	15,199,748	6,011,612	21,211,360
1999	352	14,061,161	3,270,689	17,331,850
2000	620	16,898,274	200,812	17,099,086
2001	1,208	53,288,453	4,698,890	57,987,343
2002	850	32,191,185	9,155,004	41,346,189
2003	590	23,026,808	10,792,753	33,819,561
2004	637	33,657,219	8,949,251	42,606,470
2005	690	24,158,664	3,518,600	27,677,264
2006	534	18,519,153	17,374,656	35,893,809
2007	567	22,196,477	15,710,079	37,906,556
2008	556	21,635,106	31,679,001	53,314,107
2009	516	16,019,748	5,140,320	21,160,068
2010	647	22,383,382	6,546,218	28,929,600
2011	755	27,803,999	52,050,281	79,854,280
2012	932	28,737,822	33,895,659	62,633,481
2013	906	21,460,177	20,866,730	42,326,907
2014	748	11,857,123	11,086,831	22,943,954
2015	558	3,408,346	8,118,769	11,527,116

6

Denmark

A. General Benefits

In Denmark, the first line of support for injury for most people is public healthcare, under which most treatment is free. Individuals or certain employees may have private healthcare options. Another line of support comes from employee schemes; details of these vary, depending on what scope and level of cover the employer has agreed to fund. A further line of support is state social security benefits.

B. Danish Industrial Injuries

Overview

Claims for benefits in relation to industrial injuries in Denmark are managed by Labour Market Insurance (AES). AES was established on 1 July 2016 and is a merger of the previous Arbejdsskadestyrelsen (National Board of Industrial Injuries) and parts of the previous Arbejdsmarkedets Erhvervssygdomssikring (Labour Market Occupational Diseases Fund).[1] In addition to workers' compensation claims, the National Board of Industrial Injuries manages claims relating to victims of the Second World War, and can play a role in private insurance claims. Essentially two parallel schemes run in Denmark for accidents at work and injuries from industrial diseases.

First, it is mandatory for almost all employers in Denmark to hold workers' compensation insurance from private insurance companies against industrial accidents and diseases. There are some exemptions for state institutions, municipalities and the self-employed. If an employer does not take out workers' compensation insurance cover the employer will be liable to pay any compensation owed to an injured worker. Details of the benefits for employees are laid down in the Consolidated Workers' Compensation Act 2013[2] rather than in individual policies. Self-employed workers may take out workers' insurance under the Act,[3] but very few individuals use this option.

Second, all employers (including public bodies) must contribute to the Labour Market Occupational Diseases Fund. This is a semi-public institution founded in 1999 which

1 See http://aes.dk/English/About-us.aspx.
2 Consolidated Workers' Compensation Act. Consolidated Act No 278 of 14 March 2013.
3 ibid, s 48(2).

operates on a not-for-profit basis outside the competitive insurance market. Since 2009 occupational diseases coverage has been on a pay-as-you-go basis. Under this system, employers contribute an annual payment to cover the expenditure of the previous year. The contribution rate is calculated using the average injury costs for the relevant sector for the preceding 5 years. Sixteen sectors are specified, each with sector-specific contribution rates, and there is an additional catch-all residual pool and rate. The sector-specific pools mean that there is no cross-subsidisation between the different categories of employment. However, the additional pool acts to equalise surpluses and deficits in the trade pools. The pay-as-you-go contribution covers both the costs of administration and compensation. Contributions are currently viewed as being fairly low, but this is due to the relative youth of the scheme, and it is anticipated that the value of serial payments will increase over coming years. The Fund currently holds reserves in excess of US$300 million in stocks and shares. The Labour Market Occupational Diseases Fund can give an employer a rebate where it considers that the employer has improved the safety of the working environment it offers.[4]

In order to qualify for compensation or other benefits, the applicant's injury or disease must be recognised as an industrial injury under the Danish Workers' Compensation Act.[5] The two primary requirements for such recognition are that the injury must have occurred while the applicant worked for an employer domiciled in Denmark, and the injury must have been due to the applicant's work or working conditions. Thus, there has to be causality (which requires more than merely a correlation) between the work and the cause of the injury. A number of exceptions apply to the two requirements, for example that employment is not a precondition for an injury in connection with saving human life, doing friends a favour or working abroad.

An industrial injury covers two different concepts: accidents at work and occupational diseases (personal injuries caused by either short- or long-term exposure to disease causing agents or conditions).

Accident at Work

An accident at work[6] is a physical or mental injury that follows an incident or exposure that occurred suddenly or within 5 days.[7]

Occupational Disease

An occupational disease[8] is a disease or disorder that is caused by the work or working conditions. The disease may develop due to short- or long-term exposures. Examples of occupational diseases are tennis elbow, allergy, hearing loss and asthma. Exposures in the workplace that may cause some of those diseases are stated as: repetitive work movements, work with

[4] ibid, s 58.
[5] ibid, s 5.
[6] ibid, s 6.
[7] This definition was subjected to scrutiny in the latest round of reforms (effective 1 January 2004 for accidents and 1 January 2005 for industrial diseases). Under the old scheme an accident was conceptualised as 'a sudden, external and unexpected impact on the body', the definition traditionally used in private insurance policies, whereas the new concept of 'a personal injury cause by an incident or exposure that occurs suddenly or within five days' was specifically developed for the workman's compensation scheme. This broader conceptualisation is closer to citizens' expectations.
[8] See s 8 of the Consolidated Workers Compensation Act. Consolidated Act No 278 of 14 March 2013.

arms lifted above shoulder height, heavy lifting work, work in a very noisy environment and work with hazardous substances.

In order to establish that the disease developed due to exposure in the workplace, a documented correlation and mechanism of action between exposure and the disease must exist in the medical literature. In other words, the workplace exposure must be, beyond reasonable doubt, the most likely factor that caused the disease. If there is another more likely cause of the disease, compensation can be either reduced or denied, depending upon circumstances.

Two categories of disease are considered, those included in the list of occupational diseases and those not listed. A disease is included on the list of occupational diseases when there is adequate medical documentation that a disease is caused by a certain exposure. That listing provides a presumption of recognition that a disease is an industrial injury if a person has had specific exposure to a disease-causing agent in the workplace to a higher level than persons not undertaking that type of work. The list is constantly updated by the Occupational Diseases Committee to cover the most recent research. That Committee is composed of representatives from the Danish Board of Health (Sundhedsstyrelsen), the Working Environment Authority (Arbejdstilsynet), two members from the Danish Confederation of Trade Unions (LO), two from the Danish Employers' Confederation (DA), one from the Public Employees, one from the Salaried Employees' and Civil Servants' Confederation, and one from the National Board of Industrial Injuries, who chairs the Committee.

If a disease is not on the list of occupational diseases, it is still possible to recognise it as an occupational disease. Recognition occurs when the disease is solely or mainly caused by the special nature of the work. The Occupational Diseases Committee must, in its recommendation, state that it is beyond reasonable doubt that the disease was caused by special factors in the workplace.

There is a 5-year limitation on compensation claims, which runs from the point at which the employee became aware of the cause of action, or 5 years from when the employee became aware that the claim could be attributed to gross negligence on the part of the employer.

From 1993 the National Board of Industrial Injuries has been self-funded by fees (prior to that it was state funded). Each report is opened as a case, and a fee is due from the insurer for each completed case. The case fee is currently set at DKK 4,000 for industrial accidents and DKK 5,000 for occupational disease claims. These fees are relatively low and there is cross-subsidisation between easy cases (approximately 40% of occupational disease cases are rejected after an initial assessment, but a fee is still payable) and those that require more complex investigation. A Commission report in 2014 recommended limiting the number of 'baseless' cases, possibly by amending instructions to GPs on the reporting of cases to the Board. AES publish detailed Recognition of Accidents guidelines on their website, which give specific examples of what would and would not constitute an occupational injury for commonly occurring claim types.[9]

Claims Process

There are several different routes by which a workers' compensation claim can be made. First, individual citizens can report their own industrial injury/disease, using claim forms

[9] See the Guide to Recognition of Accidents available at http://aes.dk/da/English/Publications/Guides.aspx.

available from the Board website.[10] Second, employers have an online reporting system, EASY. Third, doctors and dentists have a separate online reporting system, ESS.

Employers, general medical practitioners and dentists have a duty to report industrial injuries, so reporting by an applicant personally is generally unnecessary. An application is made to the National Board of Industrial Injuries, which assesses whether the claim should be handled as an accident claim or an occupational disease claim. At that stage, the Board requires: the applicant's employer's signature that the incident or exposure occurred, and any additional information; information about the employer's insurance company; and a medical certificate describing the injury and confirming that it was due to the incident or exposure. The Board then gathers the necessary information from the applicant, the applicant's employer and the applicant's doctor in order to examine what happened and what may be the cause of the injury. The Board will not consider a claim where the applicant or employer might have done something to avoid the incident happening.

The Board employs approximately 200 case-workers organised into teams. All case handling takes place on an open digital platform, so that all parties to the case can log in and see documents, medical reports and the progression of the case. Cases are initially screened against eligibility criteria, and information is gathered from employers and medical treatment providers. If there are any doubts as to the incident or the exposure, the Board may hear other witnesses. The Board may consult a medical consultant from its panel, especially on causation issues. Such consultants are not employed in-house, thereby saving overheads.

The Board usually makes a comprehensive decision, so its decision letter usually covers both whether the injury has been recognised and whether the applicant is entitled to compensation. An applicant who disagrees with the decision may submit the decision to the National Social Appeals Board.

The maximum timeframes for processing cases are set out in the Act and are 3 months to reject an accident claim, 6 months to reject a claim for a listed occupational disease, and 2 years to reject a non-listed occupational disease. Compensation must be awarded within 1 year of the decision. In recent years, approximately 90% of all claims have been decided within 1 year.

Complaints against decisions of the National Board of Industrial Injuries, made within 4 weeks after the decision letter, are considered by the National Social Appeals Board (Ankestyrelsen). Approximately 12% of decisions are appealed, of which 4–5% are amended. Compensation can be either increased or decreased at appeal, so there is a risk for the claimant. If a claimant is dissatisfied with the decision of the Appeals Board there is a right to appeal to the courts. Less than 0.5% of cases are appealed to the courts, of which less than 1% result in an amended decision. Again there is the possibility that the courts will reduce compensation levels.

Compensation

The Workers' Compensation Act offers several different types of compensation, notably:

— reimbursement of expenses for treatment, medicine and aids;

[10] http://aes.dk/da/English/Industrial-injuries/How-to-report-an-Industrial-Injury.aspx.

— compensation for loss of earning capacity;
— compensation for permanent injury; and
— compensation for surviving dependants.

The Workers' Compensation Act does not cover compensation for lost earnings, compensation for pain and suffering, and compensation for tort.

Reimbursement of Expenses for Treatment, Medicines and Aids

Reimbursement for medical care is secondary to public health insurance and is based on three principles:

1. The injured person should have no expenses arising from the work injury.
2. It is in the insurance company's best interest to pay for treatment that can help heal the injured person or lessen the consequences of the injury.
3. The Board cannot make a decision on compensation until the medical treatment is finished.

Reimbursement of aids and appliances is primary to public health insurance and based on a single principle, namely that aids and appliances are considered parts of the injured person.

Compensation for Loss of Earning Capacity

Compensation for loss of earning capacity is the first line of compensation, before an invalidity pension from the state. State invalidity pensions may be reduced in cases where the worker's compensation payment is granted. There is an ongoing discussion on whether this payment priority should be reversed.

Compensation can be taken as either a serial payment or a lump sum. If the loss of earning capacity is less than 50%, then the payment is made as a lump sum. If more than 50% of the individual's earning capacity is lost, then monthly payments are the norm, but claimants can (and usually do) request to have this capitalised into a lump sum. For example, if a claimant had suffered an 80% loss of earning capacity and requested to have this capitalised into a lump sum payment, then 50% would be paid as a lump sum and the remaining 30% would be paid monthly as a serial payment. This mechanism helps to reduce reserves. Monthly payments are subject to taxation, whereas lump sums are not. Loss of earning capacity is calculated as 83% of the actual annual loss of income, multiplied by the percentage incapacity, multiplied by the number of years until retirement. Loss of earning capacity is subject to certain conditions, is capped at a maximum annual income of DKK 459,000, and will only be paid if there is a minimum of 15% incapacity.

Compensation for Permanent Injury

Compensation for permanent injury is a lump sum payment which is not subject to taxation. It is not dependent on loss of earning capacity. The maximum payment in cases of 100% permanent injury is DKK 766,000, but in cases where permanent care is needed, then up to DKK 919,200 (equivalent to 120%) is available, but it will only be paid if there is a minimum of 5% invalidity.

Compensation for Surviving Dependants

The categories of individuals potentially eligible for survivor payments are a spouse or cohabiters and children. There are three types of compensation available:

- A transitional allowance at death for a spouse/cohabiter. This is a lump sum of DKK 172,000 which is not subject to taxation.
- Compensation for loss of a provider for a spouse/cohabiter. This is a monthly benefit of 30% of the deceased's annual income which is available for up to 10 years and is subject to taxation.
- Compensation for surviving children. Each child receives 10% of the annual earned income of the deceased, available as a monthly benefit and subject to taxation.

Statistics[11]

In 2015 approximately 53,000 decisions were made by case handlers. Breakdowns of cases made and accepted for the period 2007–15 are provided in Tables 6.1 and 6.2, respectively.

Table 6.1: Numbers of cases reported to the Board 2007–15

	2007	2008	2009	2010	2011	2012	2013	2014	2015
Accidents	25,881	27,945	36,384	31,284	31,434	31,807	32,684	32,776	25,947
Occupational diseases	21,808	21,771	23,786	20,946	22,727	25,579	26,672	26,110	23,825
Eyeglass injuries	81	82	90	73	76	84	47	42	43
Sudden lifting injuries	106	69	134	67	45	32	25	39	10
Total	**47,876**	**49,867**	**60,694**	**52,370**	**54,282**	**57,502**	**59,428**	**58,967**	**49,825**

Table 6.1 shows that approximately 52% of the workers' compensation cases in 2015 related to accidents, while 47% concerned occupational diseases. This pattern has been fairly consistent, with slight variations, throughout the period 2007–15. There has been a steady increase in case numbers throughout this period. There was a blip in case numbers in 2009, which occurred because 8,500 cases were reopened following a Supreme Court decision.

[11] Taken from the Arbejdsskadestatistik, available at www.ask.dk/da/Statistik/Arbejdsskadestatistik.aspx.

Table 6.2: Percentage of cases recognised (accepted and compensated) by the Board 2007–15

	2007	2008	2009	2010	2011	2012	2013	2014	2015
Number of recognised accidents	16,664	17,167	16,754	16,223	13,800	13,659	14,986	10,099	10,039
Percentage of accident claims recognised	79.8	76.2	78.3	79.4	79.8	78.3	75.7	60.2	50.6
Number of recognised occupational diseases	4,787	4,685	4,810	5,171	3,994	4,748	4,926	3,236	4,660
Percentage of occupational disease claims recognised	24.7	24.4	24.7	27.4	26.4	27.0	24.4	19.3	21.6

The percentage of accidents recognised in 2014 and 2015 is substantially less than in previous years. This is due to a Supreme Court Decision in 2015 on the concept of what constitutes an injury. However, a higher proportion of the cases which are recognised since the Supreme Court Decision receive compensation. (See Tables 6.3 and 6.4.)

Table 6.3: Number of decisions made by the Board and average degree of injury by year for 2007–15

	2007	2008	2009	2010	2011	2012	2013	2014	2015
Number of decisions made	11,752	10,952	11,985	12,489	10,209	10,901	11,194	8,193	9,677
Average degree of permanent injury	10	10	10	10	10	11	10	11	11

Table 6.4: Total spend on compensation by category by year in DKK millions (2015 level) 2009–15

	2009	2010	2011	2012	2013	2014	2015
Accidents	2,984	3,324	2,745	3,061	2,437	1,147	1,211
Occupational diseases	1,077	1,283	1,135	1,367	1,115	1,197	1,138
Sudden lifting injuries	2	4	4	–1[a]	3	1	0
Total	**4,063**	**4,611**	**3,884**	**4,428**	**3,555**	**2,345**	**2,349**
Index	100	113	96	109	87	58	58

[a] Removal of an ongoing service can lead to a negative compensation payment calculation.

Private Compensation Claims

The Board has a small team devoted to private compensation claims. Insurance companies and other boards, such as the Danish Patient Injury Board, and injured persons can obtain an impartial non-binding statement on the cost implications of health effects and occupational consequences of a personal injury, such as injuries sustained in road accidents or in a person's leisure time. This is essentially a quantification service provided by a trusted not-for-profit body with considerable expertise. In 2015 the Board dealt with approximately 5,000 such enquiries.

The statement may include an assessment of:

— the degree of permanent injury—the direct health effects of the injury;
— the loss of earning capacity—the permanent loss of earnings caused by the injury;
— maximum medical improvement—the time when there is no longer any substantial and permanent improvement of the health complaints (only applies to injuries sustained before 1 July 2002).

Fees for this service vary according to the type and complexity of the statement required, but are not considered to be excessive since the Board is a not-for-profit organisation. If the parties do not agree with the statement made by the National Board of Industrial Injuries, the case may be taken to court.

C. Danish Patient Insurance

The Danish landscape on complaints and compensation contains a number of bodies. The Danish Health and Medicines Authority is the public regulatory authority. Various professional societies exist for different healthcare specialists. A move to introduce a national system of accreditation for healthcare professionals was stopped by the government in 2015. Healthcare services are almost all (98%) provided by public hospitals).

Patientombuddet (National Agency for Patients' Rights and Complaints) plays a central role in relation to overseeing complaints and the compensation system, with a number of different functions. The Patientombuddet also deals with complaints from the Patient Compensation Association. Matters concerning complaints and compensation will be outlined below, but one thing to note at this point is that the Patient Ombuds hosts the Patient Safety Database, which records notifications received on unexpected hazards that are reported by professionals. Such reports can be made anonymously.

There are also various private membership bodies, such as the Danish Patient Safety Society,[12] the Danish Patient Association[13] and the Handicap Society. The Danish Consumer

[12] The Danish Society for Patient Safety is a private not-for-profit body established in 2001, supported by both public and private funding, to promote patient safety. It has provided a national focus for a number of national initiatives, such as 'Unintended Actions' (encouraging professionals to report unintended events), 'Strengthen Contact to Citizens' (a Netherlands-inspired campaign to support healthcare professionals in facing complaints proactively, which followed 'Ask Questions', an initiative that was initially resisted by doctors but has helped both sides to get more out of the encounters) and 'Safe Birth' (a programme to educate midwives and obstetricians), all of which are credited with improving practice.

[13] This was founded in 2008 and claims to represent 800,000 patients.

Council has a healthcare function and its representative sits on various sectoral and official committees and boards.

A distinction applies between patients who wish to make a complaint and those who wish to seek compensation. In relation to complaints, the system operates as follows. Advice for patients is widely available from Patient Guidance staff employed by the five Regions, who are readily accessible in major hospitals. These advisers are rated by the Consumer Council as performing well for patients.

Patient Compensation

All claims for compensation must go through the administrative system, namely the Patient Compensation Association and the Patient Ombuds' Board. A patient can only go to court if he or she is not satisfied with the decision from the Patient Ombuds' Board. A court claim requires proof of fault, and is rare. The Patient Ombuds's Board requires proof of injury and a mistake. The Patient Compensation Association (Patienterstatningen)[14] operates on an administrative inquiry basis and is required to be established jointly by public regional authorities and private health insurance companies.[15] Patient injury compensation is paid by the regional authority if the injury occurred in a private hospital, by a private insurance carrier if the injury happened in a private hospital or by the Minister of Health if the claim is for a pharmaceutical injury. An appeal can be made from a decision of the Patient Compensation Association to the Patient Ombuds' Board, which is comprised of 10 members, including consumer representatives. In 2015, 14.5% of such appeals were changed by the Board.

The Patient Compensation Association scheme was introduced in 1992. In the beginning only patient injuries were covered but pharmaceutical injuries have also been covered since 1996. The arrangements have been governed by the same Act since 2007, the current Act dating from 2011.[16] The Association handles claims for vaccines,[17] medical injuries and medicinal products,[18] thereby promoting efficiency.

The Law provides as follows:

> Compensation is paid if, on a preponderance of the evidence, the injury was caused in one of the following ways:
>
> (1) If it may be assumed that an experienced specialist in the field in question would in the given circumstances have acted differently during examination, treatment or the like, thereby avoiding the injury;
> (2) if the injury is due to the malfunction or failure of technical apparatus, instruments or other equipment used for or in connection with examination, treatment or the like;
> (3) if, on the basis of a subsequent evaluation, the injury might have been avoided using another available treatment technique or treatment method that would have been just as effective in treating the patient's illness from a medical point of view; or

[14] www.patienterstatningen.dk/en.aspx. The name was changed from Danish Patient Insurance Association (Patientforsikringen) to reflect the move from an insurance model to a general state compensation model.

[15] Act on the Right to Complain and Receive Compensation within the Health Service, No 1113 of 7 November 2011, s 32.

[16] ibid.

[17] Scheme introduced in 1978. Several vaccine claims are received every year.

[18] Act on the Right to Complain and Receive Compensation within the Health Service, No 1113 of 7 November 2011, s 55. Formerly Act on Damages for Pharmaceutical Injuries No 1120 of 20 December 1995 amended by Act No 1228 of 27 December 1996, s 18.

(4) if injury occurs as the result of examination, including diagnostic procedures, or treatment in the form of infections or other complications that are more extensive than the patient should reasonably have to endure. Account must be taken in this respect of the severity of the injury, the patient's illness and general state of health, the unusualness of the injury and the general possibility of taking the risk of its occurrence into consideration.[19]

These rules are referred to as the specialist, the equipment, the fairness and toleration rules. Specific provisions govern compensation for clinical trial subjects, volunteers and blood donors.[20]

The Act places a notable duty on healthcare professionals to inform and help patients where the latter have become injured:

> It shall be incumbent on every authorised healthcare professional who, in the course of his work, becomes aware of injuries that may be assumed to entitle the patient to compensation under this Part to inform the injured person thereof and to help to the necessary extent with notification of the Patient Insurance Association or the private institutions to which the processing of claims under this Part has been assigned pursuant to section 19(5).[21]

The statutory limitation on treatment injury and drug injury claims is twofold. Injuries occuring from 1 January 2008 have a 3-year limitation from the point at which the claimant became aware of the injury, with an absolute limitation period of 10 years from injury.

Claims may be made to the Patient Compensation Association by telephone or email but most are made through filling in the form on the website. A new feature was introduced in 2015 by which a person may test whether a claim may be valid: this assists in informing patients and weeds out cases that would obviously not succeed. Some people then email the Association to check the position.

The Association operates on an inquisitorial basis, considering all relevant documents. The investigation is based on documentary evidence and oral evidence is not required since it is not viewed as relevant. The Association has power to require all municipal authorities, hospitals, pharmacies, physicians and others to disclose all relevant information that is of importance in relation to handling cases.[22] Cases are processed by individual case handlers, from the Association's staff of around 200, which contains medically and legally qualified personnel. Some of the doctors also work part time in hospitals, so the level of internal expertise is high. Where necessary, the best external expertise in the country can be sought from outside specialists.

Approximately one-third of cases are accepted, the remaining two-thirds are rejected. All decisions—both recognised claims and rejections—are given in detailed formal letters, which set out the specific reasons for the decision, and are usually 3–4 pages long, so patients can be given a clear and complete explanation. The Association is developing an electronic case-handling system, so as to be as efficient as possible. The file contains all relevant information and the evidence that has been collected. A patient will be able to access his or her own file, with an individual identifier. Patients are sent emails to notify them when a development has occurred, such as that particular evidence has been obtained, so that they have an opportunity to log in and comment on the file. This approach is intended

[19] ibid, s 20.
[20] ibid, ss 21, 22.
[21] ibid, s 23.
[22] ibid, s 55.

to avoid the delay and cost of sending every piece of material to each claimant, and allowing them an extended period to respond. When this system was introduced, it immediately reduced the volume of phone calls received from patients, such as asking if there were any developments or when these might be expected. A similar system is operated in the Work Injuries Compensation system.

The average time to process a case to the approval or rejection stage is 200 days (2015 data). It has been falling for some years (average 288 days in 2009). The Association is aiming to reduce it further. It takes longer to calculate the compensation sum, as it may be necessary to wait to ascertain the exact level of impairment of the claimant. Interim payment can be made in such cases.

Compensation is fixed in accordance with the Danish Liability for Damages Act.[23] In calculating compensation sums, the Association applies the rules, working through the specified categories to arrive at a total sum. Payment is generally made as a single lump sum, not as a series of annual payments, so the payers do not have to carry reserves. Interim payments are available where the extent of the loss cannot be quantified at that point. The same entitlements apply under both the patient and pharmaceutical schemes, namely:

— medical expenses and other losses;
— lost wages;
— pain and suffering, subject to a maximum of approximately €10,000;
— permanent injury, subject to a maximum of approximately €100,000;
— loss of ability to work, subject to a maximum of approximately €1,000,000;
— loss of breadwinner, in the range €120,000–350,000;
— burial expenses.

Compensation is not paid to cover recourse claims.[24] Compensation can be reduced or invalidated if the patient contributed to the injury wilfully or through gross negligence.[25]

An appeal may be made from the Patient Compensation Association to the Patient Ombuds' Board, which is comprised of 10 members, including consumer representatives.[26] There are roughly 200 such appeals a year. Appeals can be made by the patient, the Region/Ministry or the private health insurer. In approximately 25% of rejected claims the patient will go on to appeal. In 2015, a different decision was made by the Patient Ombuds' Board in 14.5% of cases. It is possible for the Patient Ombuds' Board to award a patient *less* in compensation than the sum awarded by the Association, and even to overturn the entire award. The Association informs successful patients of this risk in the award letter. Regional Authorities may also appeal, although they have only recently begun to do so, spurred by the need to conserve public funds. It is possible for a Regional Authority to reclaim an award that has been paid to a patient, and this has caused considerable public unrest as being unfair. The time limit for appealing is 3 months, but the Supreme Court has allowed a Regional Authoirty to open up an initial acceptance decision that was taken some years before, where the compensation sum was only awarded much later.

[23] ibid, s 24, referring to the Damages Assessment Act, No 266 of 21 March, 1986, previously the Damages Assessment Act, No 599 of September 8, 1986. Computations are specified on compensation for individual economic losses in case of personal injury or loss of family support following death.

[24] ibid, s 24(4).

[25] ibid, s 25.

[26] ibid, s 34.

The Association informs the relevant payer what sum is to be paid to the approved patient: the payer is one of the five Regional Authorities in relation to public hospitals and care, and private insurers in relation to their insureds.[27] This system avoids the need for the Association to make any payments itself, or keep any reserves. The non-state healthcare providers are subject to an obligation to hold insurance in amounts specified annually by the Minister.[28]

In 2015, the Association's overall administrative cost was DKK 137,525,060 million. The Association believes that this cost level is well below what the total transactional costs would be of the equivalent number of law suits. It processed around 10,000 claims, and awarded DKK 691.0 million in patient compensation, and DKK 55.7 million in drug compensation. Recently these figures have been declining—see Table 6.5.

Table 6.5: Total compensation paid under the Danish patient and drug compensation schemes 2013–15

Category	2013		2014		2015	
	DKK millions	%	**DKK millions**	%	**DKK millions**	%
Treatment injury compensation	747.7	94.2	723.6	94.2	691.0	92.5
Drug injury compensation	46.5	5.8	44.3	5.8	55.7	7.5
Total	796.4	100.0	767.9	100	746.7	100

Table 6.6 shows the number of patient reports (both treatment and pharmaceutical injuries), detailed by the year of the report and the number of reviews carried out. Interestingly there are more reviews carried out than reports made; this is because a single patient report can include more than one healthcare provider if the patient treatment occurred in more than one setting. Table 6.7 indicates that the numbers of notifications and reviews are relatively stable. Tables 6.8–6.10 show the number of claims, numbers of approvals, and average handling times in recent years.

Table 6.6: Numbers of notifications to and reviews by Patienterstatningeg 2011–15

Activity	2011	2012	2013	2014	2015
Patient notifications	7,309	8,455	8,477	9,079	9,014
Annual percentage change in notification numbers	8.4	15.7	0.3	8.2	–0.7
Reviews conducted	8,365	9,705	9,801	10,913	10,617
Annual percentage change in review numbers	6.9	16.0	1.0	11.2	–2.7

[27] ibid, s 29.
[28] ibid, s 30.

Table 6.7: Patient injury claims by healthcare provider 2012, 2013 and 2014

	2012				2013				2014			
	Total decisions	Accepted cases	% Accepted	Compensation DKK millions	Total decisions	Accepted cases	% Accepted	Compensation DKK millions	Total decisions	Accepted cases	% Accepted	Compensation DKK millions
Public hospitals	6,310	2,228	35.3	549.6	7,169	2,436	34.0	538.8	7,034	2,349	33.4	561.2
General practitioners	694	182	26.2	49.3	847	235	27.7	76.5	842	201	23.9	57.9
Private hospitals	681	272	39.9	92.2	705	262	37.2	73.7	576	185	321	52.9
Private specialist practitioners	461	141	30.6	26.4	497	136	27.4	32.3	486	125	25.7	27.6
Emergency doctors	172	36	20.9	13.5	190	45	23.7	20.4	199	34	17.1	10.8
Ambulance service	19	1	5.3	0.2	23	2	8.7	0	32	5	15.6	1.9
Private physiotherapists and occupational therapists	36	8	22.2	0.8	43	4	9.3	0.9	55	10	18.2	4.3
Private chiropracters	55	6	10.9	0.2	34	3	8.8	0.4	49	6	12.2	0.2
Municipal health systems	34	3	8.8	1.1	45	5	11.1	0.5	33	4	12.1	0.6
Emergency phone service					6				10	0	0	0
Clinical dental technicians	3		0		5		0		6	0	0	0
Private psychologists	2		0		1		0		5	0	0	0
Educational institutions									3	1	33.3	0
Radiographers	1	1	100.0	4.0	1		0	0.0	1	0	0	0

(continued)

Table 6.7: *(Continued)*

	2012				2013				2014			
	Total decisions	Accepted cases	% Accepted	Compensation DKK millions	Total decisions	Accepted cases	% Accepted	Compensation DKK millions	Total decisions	Accepted cases	% Accepted	Compensation DKK millions
Bandagers	4		0		1		0		1	0	0	0
Private podiatrists	3	2	66.7	0.0	4	1	25.0	0.1				0.0
Private midwives					3							
Private dentists	3		0									
Hygenists					1							
Other	75	6	8.0	0.9	90	13	14.4	1.5	105	14	13.3	0.6
Undisclosed	35	1	2.9	0.0	52	2	3.8	3.7	254	19	7.5	4.3
Total	**8,588**	**2,887**	**33.6**	**738.2**	**9,717**	**3,144**	**32.4**	**748.8**	**9,693**	**2,953**	**30.5**	**722.3**

Table 6.8: Total claims under the Danish patient and drug compensation schemes 2009–15

	Administration cost (DKK)	Reviews	Cost per review (DKK)	Cost per review in real terms (DKK)
2009	79,224,851	6,664	11,888	11,739
2010	90,016,428	7,809	11,527	11,131
2011	99,502,871	8,363	11,898	11,161
2012	113,347,429	9,699	11,687	10,716
2013	121,382,922	9,804	12,381	11,282
2014	129,568,789	10,887	11,884	10,746
2015	137,525,060	10,617	12,953	12,392

Table 6.9: Total approved decisions under the Danish patient and drug compensation scheme 2009–15

	Administration cost (DKK)	Number of rulings KEL and EAL	Price per opinion (DKK)	Price per opinion in real terms (DKK)
2009	79,224,851	11,340	6,986	6,898
2010	90,016,428	12,505	7,198	6,951
2011	99,502,871	13,844	7,187	6,742
2012	113,347,429	15,060	7,545	6,918
2013	121,382,922	16,928	7,171	6,534
2014	129,568,789	17,236	7,517	6,798
2015	137,525,060	17,127	8,025	7,677

Table 6.10: The average handling time in days from receipt of application to compensation decions for patient injury claims 2012–15

2012	2013	2014	2015
211	216	200	211

Pharmaceutical Compensation

The Danish pharmaceutical compensation scheme was introduced in 1995.[29] The Minister of Health has responsibility for deciding on claims, and has delegated the investigation and administration of claims to the Danish Patient Compensation Association, which operates as discussed above. Appeals from the Association's decisions may be made within 3 months to the Patient Compensation Appeals Board, appointed by the Minister and comprising various representatives chaired by a judge.[30]

Under the pharmaceuticals arrangements, the Act provides that damages are paid

> to patients who sustain physical injury as a result of properties in pharmaceuticals which are used in connection with examination, treatment or the like (pharmaceutical injury). The same shall apply to the dependents of such persons.[31]

The pharmaceutical must have been dispensed commercially in Denmark for consumption or a clinical trial.[32] Damages shall be paid for mental injury to trial persons and donors.[33] Damages are only paid to the extent to which the injury is not covered by the Danish Act on Insurance against Consequences of Occupational Injuries.[34] A pharmaceutical injury shall not include any disease or disorder or other injury which (i) is due to the pharmaceutical

[29] Act on Damages for Pharmaceutical Injuries No 1120 of 20 December 1955, amended by Act No 1228 of 27 December 1996. The law is now the Act on the Right to Complain and Receive Compensation within the Health Service, No 1113 of 7 November 2011.

[30] ibid, s 58.

[31] ibid, s 38.

[32] ibid, s 39.

[33] ibid, s 38(2).

[34] ibid, s 38(3).

not having had the intended effect on the patient in question or (ii) is due to errors or negligence in connection with prescription or dispensing of the pharmaceutical to the patient.[35]

Damages are only paid for a pharmaceutical injury that has occurred as a result of adverse drug reactions if the nature or extent of the adverse drug reactions exceeds that which the injured person should reasonably accept.[36] The provision comprises both known and unknown adverse drug reactions as well as concretely foreseeable and unforeseeable adverse drug reactions.[37] In applying the test, the following points must be taken into special consideration:

1. the nature and degree of severity of the disease or disorder at which the treatment was aimed;
2. the injured person's state of health;
3. the extent of the injury; and
4. the general possibilities of taking the risk of the occurrence of the injury into consideration.[38]

Damages are only paid if there is a preponderance of evidence that the injury has been caused by the use of pharmaceuticals.[39] There is a threshold for payment of DKK 3,000.[40] The amount of damages is limited to DKK 150 million per year, save that it is DKK 100 million in relation to each 'serial injury',[41] which is defined as pharmaceutical injuries sustained by several persons and caused by the same property in the same substance in one or more pharmaceuticals and which are the result of adverse drug reactions that were not included in the approved literature, or due to a defect as a result of instruction errors, defects in manufacture or distribution errors.[42] A claim must be filed within 3 years after the person entitled has acquired or should have acquired knowledge of the injury, and is barred 10 years after the day on which the injury was caused.[43]

Compensation and administration in Denmark are funded entirely by the state[44] out of general taxation, as a matter of political policy. The pharmaceutical scheme was initially financed by reducing state support for the purchase of medicines (75%) by 0.1%, so compensation was therefore indirectly financed by patients themselves, who paid proportionately increased prescription or purchase fees. However, the scheme is now funded exclusively from general taxation.

Various consequences flow from the fact that the scheme is statutory, and funded by the state. First, damages are not paid to cover any derived claims of recourse,[45] ie an employer or his insurer cannot recover sums paid to a patient. Secondly, the state is subrogated to the patient's claim against the relevant pharmaceutical company if the product is defective,

[35] ibid, s 42.
[36] ibid, s 43.
[37] ibid.
[38] ibid, s 43(2).
[39] ibid, s 44.
[40] ibid, s 46.
[41] ibid, s 50.
[42] ibid, s 49.
[43] ibid, s 59.
[44] ibid, s 54.
[45] Act on the Right to Complain and Receive Compensation within the Health Service, No 1113 of 7 November 2011, s 47; formerly Act on Damages for Pharmaceutical Injuries, s 10.

as defined by the Product Liability Law, and may claim indemnity on that basis.[46] In the 20 years in which the scheme has operated, only one case has been brought under this provision, which the state lost.[47] This data is of interest in the context of considering the level of safety of medicines, and the legal standard of strict liability set by the EU product liability Directive 85/374/EC. It is predictable that such cases would be very rare, since product liability, although strict, is limited to where the drug is 'defective', so it would have to be proved that the product did not provide the level of safety that should be expected. Given that the claims under the schemes are mostly 'system injuries', the products involved would not be 'defective' under the product liability legislation. Tables 6.11–6.18 give relevant statistics from recent years.

Table 6.11: Claims approved and denied, Danish pharmaceutical compensation 2011–13

Category	2011		2012		2013		2014	
	No	%	No	%	No	%	No	%
Approved claims	79	21.9	115	29.7	121	22.3	175	30.7
Denied claims	281	78.1	272	70.3	357	74.7	395	69.3
— Outside scope of the law (eg not an injury or no causation)	52	14.4	67	17.3	107	22.4	147	25.8
— Limitation	26	7.2	27	7.0	21	4.4	33	5.8
— Conditions not fulfilled (eg not rare or serious)	185	51.4	169	43.7	213	44.6	205	36.0
— Compensation under the minimum threshold of DKK 3,000	14	3.9	6	1.6	10	2.1	9	1.6
— Miscellaneous	4	1.1	3	0.8	6	1.3	1	0.2
Total	**360**	**100.0**	**387**	**100.0**	**478**	**100.0**	**570**	**100.0**

Table 6.12: Total compensation paid by category of healthcare provider, Danish pharmaceutical compensation 2012–14, DKK millions

Category	2011	2012	2013	2014
Public hospitals	25.3	30.7	29.1	27.1
General practitioners	14.6	12.7	16.8	15.8
Others	3.3	0.0	0.2	0.0
Private specialist doctors	4.9	0.2	0.2	0.3
Emergency	0.0	0.0	0.2	0.2
Private hospitals	0.0	0.0	0.1	0.5
Municipal health systems	0.0	0.0	0.0	0.1
Total	**48.0**	**43.7**	**46.5**	**44**

[46] Act 371 on Product Liability. The procedure is that the state would send a letter before action, and institute proceedings if the claim were not settled.

[47] The case was against Servier Danmark A/S in 2011 in relation to patients who suffered heart disease that was associated with use of the weight-loss product isomeride. The case failed on the basis that the product was not defective since the risk was warned about when the product was placed into circulation.

Table 6.13: Breakdown of claim decisions by healthcare provider 2012–14

	2012				2013				2014			
	Total decisions	Accepted cases	% Accepted	Compensation DKK millions	Total decisions	Accepted cases	% Accepted	Compensation DKK millions	Total decisions	Accepted cases	% Accepted	Compensation DKK millions
Public hospitals	195	42	21.5	30.7	263	43	16.3	29.1	261	48	18.4	27.1
General practitioners	148	68	45.9	12.7	157	65	41.4	16.8	237	112	47.3	15.8
Private specialist practitioners	28	3	10.7	0.2	28	6	21.4	0.2	46	8	17.4	0.3
Private psychologists	2		0						1	0	0	0
Municipal health systems	1		0						1	1	100	0.1
Private hospitals	5	1	20.0	0.0	4	1	25.0	0.1	6	4	66.7	0.5
Private dentists	3		0		10		0		1	0	0	0
Emergency doctors	1	1	100.0	0	2		0	0.2	0	0	0	0.2
Other	4		0	0	9	3	33.3	0.2	12	1	8.3	0.0
Unknown					5	1	20.0		5	1	20	0
Total	**387**	**115**	**21.5**	**43.7**	**478**	**119**	**24.9**	**46.5**	**570**	**175**	**30.7**	**44**

The number of claims for drug compensation has been stable in Denmark for some years at around 300 cases, of which around 36% were awarded compensation. That contrasts with a rise in claims related to medical practice in recent years (6,673 claims in 2009, of which 34% were approved for compensation).[48]

An issue arises whether patients have or can obtain sufficient understanding of the availability of the schemes. Doctors hand out information on the medical treatment scheme where injuries occur, but not on the drug injuries scheme. Industry does not give out information about the compensation system. The Association forwards all information to the drug companies and the medical agency, so the compensation data are put to use in assessing safety and practice issues.

Table 6.14: Drug injury compensation in DKK millions by award category 2012–14

	2012	**2013**	**2014**
Permanent loss of ability to work	24.4	23.1	19.3
Permanent injury	8.8	11.5	13.4
Pain and suffering	1.9	2.7	3.7
Death benefits to spouses	3.0	3.8	3.4
Loss of income	4.4	3.3	2.8
Death payments (to help with transitions)	0.7	0.6	0.7
Death benefits to children	0.1	0.6	0.7
Medical expenses	0.2	0.5	0.5
Funeral expenses	0.1	0.1	0.0
Death payments to other dependants	0.0	0.0	0.0
Total due under the new law (since 2002)	43.5	46.2	44.5
Total due under the old law (pre 2002)	0.2	0.2	–0.2
Total	**43.7**	**46.4**	**44.3**

Table 6.15: The average handling time in days from receipt of application to compensation decions for drug injury claims 2012–14

2012	**2013**	**2014**
207	217	200

[48] *Annual Reports 2009–2015* (Patientforsikringen, 2009, 2010, 2011, 2012, 2013, 2014 and 2015), available at www.patientforsikringen.dk.

Table 6.16: Claims reported to the Danish Pharmaceutical Insurance 1996–2012

Year	Number of claims received	Number of claims determined	Total payments, excluding interest, DKK	Average compensation per claim, DKK
1996	41	21	0	—
1997	60	49	91,105	1,859
1998	68	56	1,348,845	24,087
1999	79	68	620,575	9,126
2000	81	70	4,650,167	66,431
2001	128	76	4,245,651	55,864
2002	132	131	10,983,520	83,844
2003	182	198	17,021,917	85,969
2004	220	176 (54 successful)	c14,500,000	c82,386
2005	352	261	15,867,716	60,795
2006	255	321	30,480,027	94,953
2007	334	243	32,009,204	131,725
2008	312	264	32,982,493	124,933
2009	356	404	53,771,277	133,097
2010	347	425	51,126,822	120,298
2011	378	353	48,040,334	136,091
2012	430	387	43,800,000	115,873
2013	476	478	46,500,000	
2014	570	570	44,300,000	

Table 6.17: Decisions under the Danish Act on Damages for Pharmaceutical Injuries, by type from 1996 to 2010 with data for 2006–10 shown by year[49]

	1996–2005		2006		2007		2008		2009		2010	
	No	%	No	%	No	%	No	%	No	%	No	%
Recognised claims	**343**	**30**	**57**	**18**	**55**	**23**	**94**	**36**	**144**	**6**	**106**	**25**
Rejected as the injury occurred before 1 January 1996	0	0	5	2	4	2	6	2	4	1	7	2
Rejected as it is not a physical injury	276	24	47	15	44	18	48	18	69	17	68	16

(continued)

[49] Report on the Act on Damages for Pharmaceutical Injuries, 2003; Patientforsikringen *Annual Reports*, http://pebl.dk/Viden-om-skader/Tal-og-statistik/2015-2010.

Table 6.17: *(Continued)*

	1996–2005		2006		2007		2008		2009		2010	
	No	%	No	%	No	%	No	%	No	%	No	%
Rejected as it did not involve a medicine	0	0	9	3	2	1	3	1	2	0	0	0
Rejected as the injury is not sufficiently serious	277	24	58	18	51	21	53	20	77	19	138	33
Rejected as there is no causal relationship	56	5	52	16	21	9	30	11	61	15	72	17
Rejected because limitation period expired	0	0	91	28	66	27	22	8	25	6	15	4
Rejected as compensation is under DKK 3,000	0	0					4	2	12	3	13	3
Miscellaneous	190	17	2	1	0	0	4	2	10	2	6	1
Total	**1142**	**100**	**321**	**101**	**243**	**101**	**264**	**100**	**404**	**100**	**319**	**101**

Table 6.18: Comparison of the national pharmaceutical insurance schemes 2007–12

	2007	2008	2009	2010	2011	2012
Number of annual claims	334	312	356	347	378	412
Percentage of claims per head of population	0.000059	0.000055	0.000063	0.000062	0.000067	0.000073
Percentage of successful claims	23	36	36	25	19	27
Number of claims paid per year	55	94	144	113	68	103
Percentage of claims paid per head of population	0.0000098	0.000016	0.000025	0.000020	0.000012	0.000018
Compensation per successful claim, €	77,598	46,783	49,788	60,326	94,196	56,699
Total cost of compensation (excluding administration costs) paid per year, € million	4.2	4.3	7.1	6.8	6.4	5.8
Total contributions as percentage of national sales turnover	0.02	0.02	0.03	0.03	0.03	—
Cost of administration of the scheme per year, € million	0.17	0.3	0.3	0.3	0.3	0.3

Conclusions

Danish commentators regard the system as stable and satisfactory. There are no plans for change. New Acts were passed in Denmark in 2007[50] and 2011.[51] These were consolidations and Parliament did not take the opportunity to make substantive changes to the scheme. The method of funding through the state is stable. Industry made a successful defence on the basis that it would be contrary to EU law to pay if legal liability (strict liability for a defective product) were not proved, and this has not been established in any case since the scheme started. The industry position was later supported by the ECJ *Sanchez* case,[52] in the light of which private funding in Sweden and Finland (and Norway) looks curious, rather than the other way round. The Danish scheme always considers whether it can reclaim from the manufacturer. It sends each decision to the Ministry of Health, which pays the claimant within 1 month (previously 14 days).

The Association sends a great deal of information to hospitals and Regional Authorities, such as reports on their cases. The Association does not, however, know how the Regional Authorities react, and how they put the information to use, and with what results. The Association's staff regularly write articles on aspects of interest. It has an in-house doctor who works full time on the statistics, writing articles for the medical press and journals, and working on outreach.

The Association's data revealed that small hospitals tend to have more and more serious birth injuries than larger hospitals, which was attributed to the availability of resources and expertise. However, when this was announced in 2015, the Patient Safety Association responded that the data was in fact from 2012 and out of date since practice and results had since improved as a result of the 'Safe Birth' project, aimed at reducing the incidence of babies injured at birth. All midwives and gynaecologists in the country have since been required to take a special course and pass a test based on cases processed by the Association. The incidence of such problems appears to have begun to fall.

[50] Danish Act on the Right to Complain.

[51] Act on the Right to Complain and Receive Compensation within the Health Service, No 1113 of 7 November 2011.

[52] Case C-183/00 *María Victoria González Sánchez v Medicina Asturiana SA*.

7

Finland

A. General Benefits

An injured employee's salary in Finland usually continues to be paid in full by the employer for 10 days, although the duration may vary depending on the terms of collective agreements. Thereafter, the employee could claim 'daily money' from the state social security fund. The level of daily money varies depending on the employee's salary level (usually around 60%), and how many dependants he has. There is no separate payment for mortgage or housing expenses.

If the employee suffers an injury as a patient or from a drug, he could claim the remaining 40% from either the Patient Compensation or Pharmaceutical Insurance Fund (and his employer could also reclaim 40% from that fund in respect of the initial period of 2 months). In the unlikely event that he was to make a legal claim (based on culpa, ie negligence), and he succeeded, he would in addition be awarded a specific sum for pain and suffering. The Pharmaceutical Insurance Scheme payments also include an element for pain and suffering, based on extensive tariffs maintained by the insurance companies that include such factors as how long or how many times the claimant was hospitalised.

B. Finland's Workers' Compensation System

Finland was a particularly early adopter of workers' compensation; legislation governing workers' compensation has been on the statute books since 1895. Workers' compensation insurance is compulsory in Finland, under the Workers' Compensation Act, 459/2015, repealing from 1 January 2016 the Employment Accident Insurance Act[1] of 1948 and other applicable Acts.[2] Amendments to the current 1948 law were made most recently in 2011 and in early 2015 (implemented from 1 January 2016) The new Workers' Compensation Act entered into force at the beginning of 2016. The new Act will apply to claim events that occur on 1 January 2016 or later. The three previous Acts will continue to apply to

[1] Employment Accident Insurance Act 608/1948. Available in English at www.finlex.fi/en/laki/kaannokset/1948/en19480608.

[2] Including the Occupational Diseases Act (1989), the Farmers' Employment Accidents Insurance Act (1981) covers the employment accident security for rural self-employed persons and the Government Employees' Accident Compensation Act (1990) (which defines the accident coverage given to government employees).

occupational accidents and diseases that occurred or manifested earlier.[3] The basic structure remains the same under the 2016 reforms, which essentially codify the current system and define more precisely the circumstances of what is classified as an accident at work. The provisions on what constitutes a claim event compensable by the workers' compensation insurance are more specific in the new Act than in the previous Employment Accidents Insurance Act.

Workers' compensation is the primary source of funding for injured persons, taking priority to other social security payments and any other secondary sources.

The Obligation to Insure

An employer is obliged to insure its employees against occupational accidents and diseases, as provided in the Act. An employer does not have an insurance obligation if the wages and salaries paid, or agreed to be paid, by the employer for the work commissioned in a calendar year do not exceed €1,200 (section 3).

The length of a working day is immaterial—even 1 hour of work a day meets the conditions of a person-day. The duty to insure applies regardless of the number of employees or the value of salaries paid. Therefore, employers with the duty to take out insurance include both large industrial companies and individual households or organisations providing employment under an employment contract.

There are two main exemptions to the obligation to insure:

— Small employers, who pay less than €1,200 in wages and salaries to employees per calendar year, are not required to take out insurance. However, an employee always has a right to receive compensation in the event of an accident. The Finnish Workers' Compensation Centre (TVK)[4] pays compensation for accidents at work and occupational diseases that occur while an employee was employed by a small employer.
— The state has no insurance obligation; however, the compensation for occupational accidents or diseases resulting from central government work is paid from state funds, as provided in the Act (section 3).[5]

In both these exemptions the compensation awarded is calculated according to the Workers' Compensation Act, so employees are not disadvantaged by these arrangements.

[3] The Workers' Compensation Act (Työtapaturma- ja ammattitautilaki) (no 459/2015 in the Statutes of Finland, abbreviated TyTAL): unofficial translation at www.tvk.fi/en/about-tyta/Legislation/. TyTAL entered into force on 1 January 2016, combining the provisions of three Acts valid at the time, namely the Employment Accidents Insurance Act (608/1948), the Occupational Diseases Act (1343/1988) and the Act on Rehabilitation Compensable under the Employment Accidents Insurance Act (625/1991, the Rehabilitation Act). TyTAL determines the eligibility for compensation and the rates of compensation for accidents and diseases that occur or manifest after 1 January 2016. The older Acts determine the eligibility for compensation and the rates of compensation for accidents or diseases that occurred or manifested prior to this point. However, all claims lodged after 1 January 2016 will be processed according to the provisions in TyTAL, including the processing of appeals.

[4] www.tvk.fi/en/, prior to 2016 known as Tapaturmavakuutuslaitosten (TVK) or the Federation of Accident Insurance Institutions in English.

[5] Employment Accident Insurance Act 2015, s 1.

Other public bodies, such as municipalities, joint municipal authorities and parishes, are obliged to take out insurance in the same way as normal employers.

Employers shall file an insurance application with the accident insurance institution of their choice. Valid insurance must exist when work begins. The insurance becomes valid on the day when the application was demonstrably submitted to an insurance institution or its representative. The following are the categories of temporary and continuous cover.

- *Continuous insurance policy.* The insurance policy shall be continuous and the policy period is the calendar year (continuous insurance), unless the insurance policy enters into force during the calendar year, in which case the first policy period ends on the last day of the next calendar year.
- *Fixed-term insurance policy.* The insurance policy can also be made for a fixed term for work or a work site not exceeding 1 year (fixed-term insurance). In this case, the policy period is the same as the validity period of the insurance policy. A fixed-term insurance is terminated without notice. However, if the work continues beyond the time limit and the policyholder informs the insurance company about the said continuation before the time limit expires, the validity of a fixed-term insurance will be extended until the new time limit set by the policyholder, though for no more than 12 months. (Section 158)

Transferring a Continuous Insurance to Another Insurance Company

The policyholder can transfer a continuous insurance policy to another insurance company in writing, using a special transfer notice. The insurance policy can be transferred so that it expires on the last day of March, June, September or December, but not before the expiry of the first policy period. The policyholder submits the transfer notice to a new insurance company which shall forward it to the old insurance company no later than 3 calendar months prior to the above-mentioned days.

The insurance policy enters into force in the new insurance company from the moment the previous insurance policy expires. If the policyholder has submitted a transfer notice to two or more insurance companies and indicated the same date of commencement, the first transfer shall remain valid, unless otherwise agreed between the policyholder and the insurance companies.

Further provisions on the insurance policy transfer procedure are given by a Decree of the Ministry of Social Affairs and Health.

Work Performed as an Employee

In general, all persons in employment are entitled to statutory accident insurance

Unless otherwise provided, the Workers' Compensation Act applies to a person who is working:

(a) in an employment relationship referred to in section 1 of the Employment Contracts Act (55/2001);

(b) in an employment relationship referred to in section 1 of the Seafarers' Employment Contracts Act (756/2011);

(c) in a public-service employment relationship under the Public Servants Act (750/1994);
(d) in a public-service employment relationship referred to in the Act on local government officers (304/2003);
(e) in a public-service employment relationship under the Church Act (1054/1993);
(f) in a public-service employment relationship referred to in the Act on parliamentary officers (1197/2003);
(g) as the President of the Republic, Member of the Government, public servant in the President's Office, the Parliamentary Ombudsman, the Deputy Ombudsman or a Member of Parliament;
(h) in other statutory employment relationships governed by public law.[6]

However, the following employee groups are *not* necessarily covered by statutory accident insurance:

Family Members

If a family member is self-employed and he or she has entered into an employment contract with him or her, he or she is usually covered by the compulsory insurance. Since he or she is an employee, he or she cannot be covered by the voluntary working hours insurance. A family member refers to a spouse, a cohabiting partner or a direct descendant or ancestor of the person referred to in subsection 1 and living in the same household. A cohabiting partner refers to a person living in conditions resembling marriage in the same household with the person working in a leading position in the company. (Section 9)

Athletes

This Act does not apply to sports activities. The provisions on athletes' right to compensation for an accident are laid down in the Act on athletes' accident and pension cover (276/2009).

Persons in Leading Positions[7]

Persons who work for remuneration in a limited company or other corporate body in leading positions are comparable to employees; however, they are not in an employment relationship with the limited company or other corporate body if:

1. a shareholder working for a limited company in a leading position personally owns no more than 30% of the company shares or him or her and his or her family members together own no more than 50% of the company shares and of these his or her personal share is no more than 30% or he or she personally owns no more than 30% or him or her and his or her family members together own no more than 50% of the voting rights arising from the ownership of the shares and of these his or her personal voting rights no not exceed 30%; or

[6] S 8 of the Employment Accident Insurance Act 2015 defines categories of people eligible as workers.
[7] S 9 of the Employment Accident Insurance Act 2015 defines work performed by persons in leading positions.

2. a person working in a leading position in another corporate body have control, either alone or together with family members, over the corporate body equivalent to that referred to paragraph 1.

A leading position refers to the managing director, member of a board of directors or equivalent positions or equivalent factual powers in a limited company or other corporate bodies.

A partner in a general partnership or a partner in another corporate body or group who is personally responsible for the liabilities and commitments of the corporate body or group is not comparable to an employee.

Self-Employed Persons

The provisions on the application of this Act on work performed by self-employed persons are laid down in Part VI (section 10). The definition of self-employment was changed to correspond to that provided in the earnings-related pension legislation. If a person has taken out a self-employed person's pension insurance, he or she does not have to take out the compulsory insurance cover. However, he or she can still take out a voluntary self-employed person's accident insurance policy. If he or she has taken out a self-employed person's pension insurance, he or she does not have to take out the compulsory insurance cover. However, he or she can still take out a voluntary self-employed person's accident insurance policy.

Certain categories of worker are not covered by the standard workers' compensation arrangements detailed in Table 7.1: the obligation to insure these special groups is not based on the Workers' Compensation Act.

Section 13, this Act applies to work performed in Finland, unless otherwise provided below.

Table 7.1: Special workers' compensation groups

Category of person	Coverage statutes and provisions
Farmers	*Work performed by farmers and grant recipients* This Act does not apply to the work performed by farmers referred to in the Farmers' Accident Insurance Act (1026/1981) or from 1/1/2016 under Act (873/2015) which repeals the previous Act or work performed by grant recipients. Farmers are covered by statutory accident insurance defined in Farmers' Accident Insurance Act. Compensation is determined according to the Employment Accidents Act, and is similar to that to which employees are entitled. Farmers' accident insurance is obligatory for all farmers with obligatory farmers' pension insurance (MYEL pension insurance). Other people working as self-employed persons in agriculture may take out a voluntary occupational accident insurance. Accident insurance for farmers is provided by the Farmers' Social Insurance Institution.

(continued)

Table 7.1: ***(Continued)***

Category of person	Coverage statutes and provisions
Grants and scholarships	Artists and researchers who receive a grant or a scholarship have been covered by statutory accident insurance since 1 January 2009. Statutory insurance for those receiving a grant or a scholarship is provided by the Farmers' Social Insurance Institution. (992/2008)
Students	Students in grades 7–9 in lower secondary education, upper secondary education, or completing degree studies in vocational training or higher education are insured against accidents at work and occupational diseases under a special Act. A precise definition of the insured student groups is included in the Act on Compensation for Injury or Illness Occurred in Training-Related Conditions Comparable to Work (1318/2002) or from 1/1/2016 under Act (460/2015) which repeals the previous Act. Students' accident insurance only covers accidents that take place in practical training in conditions that are typical for the type of studies in question. An educational institution is comparable to an employer in terms of taking out an insurance policy.
Persons in adult labour market training	Such persons are entitled to be covered by statutory accident insurance under the Employment Accidents Act. The body organising the training is comparable to an employer in terms of taking out an insurance policy.
Family carers	As defined in the Family Carers Act: such persons are entitled to be covered by statutory accident insurance under the Employment Accidents Act. The municipality is comparable to an employer in terms of taking out an insurance policy.
Persons in rehabilitating employment	The municipality organising rehabilitating work is responsible for taking out an insurance policy defined in section 3, paragraph 1 of the Employment Accidents Act in case of an accident occurring in rehabilitating employment (Act on Rehabilitating Work Experience, 189/2001).
Informal carers	The municipality is responsible for taking out an insurance policy defined in section 3, paragraph 1 of the Employment Accidents Act for informal carers (Act on Support for Informal Care).
Persons in specific employment for people with disabilities	The body organising specific work for people with disabilities is responsible for taking out an insurance policy defined in section 3, paragraph 1 of the Employment Accidents Act for persons in such employment (section 27e, paragraph 3 of Social Welfare Act).

Special Groups Receiving Compensation from State Funds

Compensation from State funds is paid under several Acts for accidents at work or similar accidents that have occurred under conditions defined in those Acts. The amount of compensation is determined according to the Employment Accidents Act. The compensation is paid by the State Treasury. These groups are listed in Table 7.2.

Table 7.2: Special groups compensated from state funds

Statute	Persons covered
Military Accidents Act (1211/1990) (Now replaced by new Law 1521/2016: The Law for Replacement of Military Accident and Service Illness)	— conscripts — persons attending refresher courses — women in voluntary military service — persons in non-military service — persons in training for a military career — persons participating in voluntary national defence: Act 556/2007 on Voluntary National Defence (only for training contracted by National Defence Forces)
Rescue Act (379/2011)	— those who have taken independent rescue action under the general duty to act — those belonging to a contract brigade or other contracted organisation with no work contract — persons ordered to assist in rescue operations — persons who have voluntarily participated in rescue operations — persons ordered to participate in civil defence training
Maritime Rescue Act (1145/2001)	— those who have assisted or have been ordered to assist in a maritime rescue operation — those participating in maritime rescue operations as a member of voluntary association or similar organisation
Emergency Powers Act (1080/1991) (now replaced by new Law 1552/2011	— those ordered by the government to undertake certain work
Police Act (493/1995) (Now replaced by new Law 872/2011, which came into force 22 July 2011)	— those obliged to assist due to an order by a policeman those assisting voluntarily
Act on Accident Compensation for Persons Having Assisted in Performing Official Duties in Certain Cases (625/1967)	— those assisting police officers, border guards or customs officers or officers in prison guard duties, or transporting prisoners
Act on Accident Compensation for Persons in Certain Penal and Other Institutions (894/1946)	— prisoners — persons in care in state-owned mental hospital or other institution — persons ordered or placed to work outside the institution — persons performing unpaid work or tasks as part of their monitoring sentence, community service or juvenile punishment — persons performing tasks or undergoing programmes as part of the monitoring of a conditional imprisonment
District Courts Act (673/2016) which came into force on 1/1/2017 and replaced the previous Act (581/1993)	— lay judges

(continued)

Table 7.2: ***(Continued)***

Statute	Persons covered
Act on Integration (1386/2010)	— immigrants receiving labour market subsidy in the form of integration assistance
Act on the Reception of Persons Seeking International Protection (746/2011)	— asylum seekers and other persons seeking or receiving international protection
Act on the Public Employment Service (916/2012) which entered into force on 1/1/2013 and replaces the previous Act (1295/2002)	Job-seekers and other customers of the employment authority, who are — in a work or training try-out — in a work try-out, coaching for working life or training at a workplace — on a visit to an educational institution — in jobseeking training — in job coaching

Regulation of Insurance Companies and Premiums

Companies wishing to underwrite workers' compensation insurance in Finland must belong to TVK. The EU Damage Directives are implemented in the Workers' Compensation Act. The provision of statutory accident insurance is undertaken by 13 private insurance institutions, the State Treasury (for state employees) and the Farmers' Social Insurance Institution (for farmers). Insurance premiums are collected by insurance institutions under sections 166 and 167–76 of the Workers' Compensation Act.

Since joining the EU in 1999, Finnish workers' compensation insurance premiums have been based on free competition, taking into account the provisions of the Employment Accidents Act on insurance premiums. This means that insurance institutions compete against each other and determine the insurance premiums independently. Essentially there are three rules on the pricing of premiums:

- the premium must be based on risk;
- the premium cannot be excessive;
- the premium cannot be too low (such as to risk leaving a pension shortfall).

Various measures are in place to ensure these rules are adhered to. An amendment to the Employment Accidents Act commencing 1 January 2012 introduced further deregulation of insurance premiums. The current insurance premium regulations comprise legal provisions on the key principles of insurance premium calculations and on the public monitoring task assigned to the Financial Supervisory Authority. Additionally, each insurance company is required to have its calculation base for insurance premiums (premium base) approved by the Board of Directors of the company, stating how the insurance premiums are determined and applied. The premium base must be applied uniformly to all policyholders.

The key principles laid out for determining insurance premiums[8] are that there must be correlation between risks and costs, the reasonability and coverage principle, equal treatment of policyholders, and promotion of safety at work.

[8] Employment Accidents Act, s 35; Workers' Compensation Act, ch 21.

Annual profitability reports for all the insurance companies act as a soft market supervisory tool. There have been instances of insurance companies failing, but so far they have always been merged before they enter bankruptcy.

Summary of Workers' Compensation

Two types of claim are recognised: claims for injuries from accidents at work and claims for occupational accidents. Both categories are defined by statute, accidents in sections 20–25 of the Workers' Compensation Act.[9] A brief summary follows, but for a full definition of what comprises an accident at work, the Act should be consulted.

Compensation for either an accident at work or an occupational disease will only be awarded when there is causality between the injury or illness and an accident at work or a work-related exposure agent. In most cases, establishing causality is straightforward. However, sometimes causality is established on the basis of likelihood. In such cases, questions of fact, eg whether the accident was intense enough and had a suitable mechanism to cause the injury or disease concerned, will be considered. Causality assessments are based on medical knowledge and experience on the emergence mechanism of injuries and diseases.

Causality is rarely established unequivocally using medical methods. In practice, decisions on compensating injuries caused by accidents are based on likely medical causality between an incident and an injury. In the case of occupational diseases, this likelihood requirement is included in the legislation. The test is essentially a balance of probabilities, namely that the reasons for causality have to be stronger than reasons against causality. A medical explanation that a certain injury could be caused by the incident may not be sufficient basis for compensation by itself. A new definition of causality is provided in section 16:

> Medical causality assessment
>
> Compensation for the claim event is conditional upon probable medical causality between the claim event and the injury or illness, unless otherwise provided hereinafter. In particular, the causality assessment will take into account medical findings and observations, the type of occurrence and previous injuries and illnesses.

(a) Accidents at Work

An accident is defined as a sudden, unpredictable incident caused by an external factor and leading to an injury or illness. *Unpredictable* means that the incident is beyond the control of the employee, ie unexpected and unintended by the employee; deliberate self-harm is not an accident. *Sudden* means that the incident is rapid and immediate, such as a trip, a fall or a crash. *An external factor* means a factor independent of the injured person that causes the accident, such as a slippery street, a pothole, an object falling on top of a person or a sharp object grabbed accidentally. Accident is defined in section 17. An accident refers to a sudden

[9] Occupational Diseases Act (1343/88) and subsequent amendments (1643/1992, 1315/2002, 1366/2004); te Occupational Diseases Decree (1347/88) and its amendments (1315/2002, 252/2003); Workers' Compensation Act, ch 5.

and unforeseen event arising from an external factor that causes the employee to be injured or develop an illness.

An accident at work takes place at work or in circumstances arising from work. Only accidents that occur at work, ie when the person is working or in circumstances arising from work, are compensated as workplace accidents. Circumstances arising from work can also include various situations that are secondary to the actual job. An accident will only be compensated if it takes place at work, in an area belonging to the place of work, on the way to or from work, or when the employee has been sent by the employer to perform a certain task.

An injury or illness caused by an assault or other deliberate action by another person at work or in circumstances arising from work is treated as an accident at work and compensated accordingly. An accident at work that has occurred during war or armed conflict will also be compensated, provided that prerequisites of the definition of work accidents is fulfilled.

(b) Occupational Diseases

Occupational diseases are defined in the Workers' Compensation Act as a disease likely to have been principally caused by a physical, chemical or biological factor at work. Illnesses due to psychological factors do not qualify as occupational diseases.

An occupational disease is compensated by the insurance company that insured the employee at the time when the occupational disease occurred. If an occupational disease occurs subsequently, the insurance institution responsible for compensating the employee is the one that insured the last job during which the occupational disease could have been caused.

Long-latency occupational disease claims, those with a 10-year or greater gap between exposure to the causal factor and the onset of the disease, are processed normally by insurance companies and funded by the pay-as-you-go system,

The Occupational Disease Decree contains a list of the most common illnesses considered to be occupational diseases and the factors causing them, ie exposure agents. The list is neither exhaustive nor limitative. Illnesses not included in the Decree can be compensated as occupational diseases if there is a causal relationship between the disease and exposure to a factor defined in the legislation and if the presence of that factor at the workplace can be established with sufficient probability.

The most common occupational diseases include acoustic traumas, respiratory allergies, skin diseases, asbestos illnesses and strain injuries in upper limbs. The Institute of Occupational Health in Finland publishes annual statistics on compensated occupational diseases and illnesses suspected to be occupational diseases.

Available Compensation

The intention is to provide full compensation for personal injury. There is no maximum sum, and compensation is paid according to the injury. However, the new Workers' Compensation Act will apply to claim events that occur on January 2016 or later and is only available for certain types of damage, which are summarised in Table 7.3.

Table 7.3: Forms of compensation available under the Employment Accidents Insurance Act 2016

Damage type	Available compensation
Compensation is paid for the following medical treatments (section 37)	(1) emergency care, examination, diagnosis and treatment of the injury or illness provided or prescribed by a physician or a dentist or another licensed healthcare professional or, within his or her competence, a person who is entitled to perform the tasks of a licensed professional; (2) medications and treatment supplies; (3) medical rehabilitation provided as medical treatment; Medical rehabilitation provided as medical treatment includes: (1) rehabilitation counselling and rehabilitation guidance; (2) assessment of functional ability and ability to work and rehabilitation needs; (3) rehabilitation examination to establish the patient's rehabilitation options; (4) therapies aimed at improving and maintaining functional capacity and other necessary measures to promote rehabilitation; (5) medical rehabilitation aids, including fitting, modification, trial, teaching the use of and monitoring of rehabilitation aids, transport of rehabilitation aids on reasonable grounds, and maintenance and repair of rehabilitation aids; (6) adaptation training; (7) episodes of rehabilitation in institutional or outpatient care consisting of the measures referred to in paragraphs (1)–(6). Medical rehabilitation aid refers to a standard-level appliance, device, supply, programme or a similar solution that supports, maintains or improves the injured person's ability to work or functional ability in the activities of daily living or prevents deterioration of the ability to work or functional ability. However, compensation can be paid for the cost of aids that are above the standard level if their use enables the restoration of the ability to work or materially reduces the need for personal assistance. Further provisions on the meaning of measures, services and aids covered by medical treatment in this section may be given by a decree of the Ministry of Social Affairs and Health.
Compensation for certain personal objects (section 54)	The injured person is compensated for spectacles, hearing aids, dentures, orthopaedic dressings, back braces and artificial limbs, joints and organs used by the injured person and damaged in connection with the claim event. In addition, the injured person is compensated for torn clothes or broken rings during medical treatment.
Compensation for additional housekeeping costs (section 53)	An injured person who, as a result of the damage, is unable to perform housekeeping is paid a reasonable compensation for the unavoidable additional costs of housekeeping, but not exceeding a period of 1 year from the occurrence of the claim event. In assessing reasonableness, the starting point is the fee charged for a corresponding service in accordance with the provisions of the Client Fee Act.

(continued)

Table 7.3: ***(Continued)***

Damage type	Available compensation
	Housekeeping refers to house cleaning, laundry, shopping, childcare and other such routine maintenance at home.
Care allowance (section 51)	The injured person is paid a care allowance if he or she is in need of care, assistance, supervision or guidance due to injury or illness. If the need is regular or nearly regular in some everyday activities, a basic care allowance of €8.70 per day is paid. If the need is regular and daily, an increased care allowance of €19.55 per day is paid. If the need is continuous, time-consuming and daily, the maximum care allowance of €23.41 per day is paid. A blind person is paid the maximum care allowance for a period of 2 years from the time blindness occurred and half of this amount after that. The care allowance is, however, not paid for a period during which the injured person is treated in a hospital or other institution.
Clothing supplement (section 52)	A clothing supplement is paid to the injured person if he or she as a result of the damage uses an aid made of soft materials, a support or dressing for the body to be used in a small area or any other aid or appliance that is likely to cause wear and tear of clothing. The clothing supplement amount is €0.58 per day. The injured person is paid an increased clothing supplement if he or she has had, as a result of the damage, double amputation or he or she uses a thigh or leg prosthesis, long support bandage for a lower extremity, body support vest or corset made of a hard material or any other aid or appliance which, due to its structure, purpose or high usage is likely to cause exceptional wear and tear of clothing comparable to those referred to above. The amount of increased clothing supplement is €2.31 per day.
Compensation for loss of earnings (section 55) and right to claim daily allowance (section 56)	Compensation for the loss of earnings under this Act refers to the daily allowance, worker's compensation pension and rehabilitation allowance. The injured person is entitled to claim a daily allowance for 1 year from the date of the claim event if, due to the damage, he or she is incapable of performing his or her work in part or in full. The injured person, however, is not entitled to claim the daily allowance if his or her ability to perform his or her work has deteriorated by less than 10%. The allowance cannot be granted unless the injured person's earnings have reduced by at least a 20th of the minimum annual earnings referred to in section 79. The daily allowance is paid for each calendar day, excluding the date of the claim event. Daily allowance is only paid, however, if the injured person has been partially or completely incapable of performing his or her own work for a period of at least three consecutive days, excluding the date of the claim event.

(continued)

Table 7.3: ***(Continued)***

Damage type	Available compensation
	If at the time of manifestation of the occupational disease the injured person was claiming an old-age or disability pension, and the exposure occurred before retirement, the right to claim a daily allowance shall be assessed on the basis of the work which the injured person performed last before retirement. In this case, by derogation from section 58, the amount of the allowance is based on the annual earnings determined in accordance with section 74.
Rehabilitation allowance (section 69)	The injured person has the right to claim a rehabilitation allowance during vocational rehabilitation pursuant to sections 89 and 90. The injured person is not entitled to claim daily allowance or workers' compensation pension for this period. The rehabilitation allowance corresponds to the full amount of the daily allowance for 1 year from the date of the claim event, irrespective of any deterioration of the ability to work. After that, the rehabilitation allowance shall correspond to the full amount of a workers' compensation pension, irrespective of any deterioration of the ability to work. However, if rehabilitation does not prevent the injured person from engaging in suitable gainful employment, the rehabilitation allowance is determined in accordance with sections 56 and 57 or sections 63 and 64. The amount of the rehabilitation allowance also corresponds to that referred to in subsection 2 during a holiday included in the curriculum of the training pursuant to section 89(3)(4).
Amount of the workers' compensation pension (section 66)	The maximum workers' compensation pension is 85% of the annual earnings in accordance with sections 71–79 until the injured person turns 65 years of age, after which it will be 70% of the annual earnings, though subject to a one-time increase pursuant to section 67. The workers' compensation pension is adjusted each calendar year using the earnings-related pension index referred to in section 98 of the Employees Pensions Act.
Deducting employee payments from the daily allowance (section 62)	60% of the total amount made up by the daily allowance portion of the health insurance referred to in Chapter 18, section 21(1) of the Health Insurance Act (1224/2004), the earnings-related pension contribution for those under 53 years of age referred to in section 153(1) of the Employees Pensions Act, and the unemployment insurance contribution referred to in section 18(1) of the Act on the financing of unemployment benefits (555/1998) is deducted from the daily allowance. No deduction is made from a daily allowance which is: (1) paid to the employer; (2) based on the self-employed person's annual earnings or on work insured under the Farmers' Accident Insurance Act, (3) based on the minimum annual earnings referred to in section 79. The provisions of subsections 1 and 2 on the daily allowance also apply to the rehabilitation allowance which corresponds to the amount of daily allowance.

(continued)

Table 7.3: ***(Continued)***

Damage type	Available compensation
Rehabilitation compensation	Compensation paid to an injured person for rehabilitation is governed by provisions 27.1.1991/627.
Survivors' pension and funeral assistance (section 99)	Survivors' pension and funeral assistance shall be paid if the claim event referred to in this Act has resulted in the injured person's death. The survivors' pension comprises the surviving spouse's pension and child's pension. The survivors' pension is also paid when the injured person has gone missing and has probably encountered a claim event referred to in this Act, which has caused immediate danger to his or her life and there is no reason to believe the injured person survived. In this case, the survivors' pension is granted for a fixed period of 1 year, not to exceed the date when the decision issued by the court under the Act on declaring the missing person legally dead (127/2005) has gained legal force. The death of an injured person in disability category 18 or higher is considered to be caused by the claim event referred to in this Act unless it can be thought probable that the injury or illness caused by the claim event did not significantly contribute to the injured person's death.
Surviving spouse's pension entitlement (section 100) Amount of the survivors' pension (section 104)	The injured person's spouse is entitled to a surviving spouse's pension. A person who permanently cohabited with the deceased injured person in conditions resembling marriage in the same household without entering into matrimony and who has or had a child with the deceased injured person or has a notarised agreement with the deceased injured person on joint maintenance is entitled to a surviving spouse's pension. The pension entitlement also arises in cases where the mutual child was born after the injured person's death. In addition to the provisions of subsection 2, the person referred to in the said subsection is entitled to the survivors' pension provided that the injured person was not married at the time of his or her death. If at the time of death the injured person cohabited in conditions resembling marriage with a person referred to in subsection 2 while being separated from his or her spouse and an application for dissolution of marriage was pending, the person referred to in subsection 2 is entitled to a surviving spouse's pension instead of the spouse. The maximum annual amount of the survivors' pension is 70% of the injured person's annual earnings. The maximum amount of the surviving spouse's pension of the injured person's annual earnings is: (1) 40% if there are no persons entitled to a child's pension; (2) 35% if one of the beneficiaries is also entitled to a child's pension; (3) 30% if two of the beneficiaries are also entitled to a child's pension; (4) 20% if three of the beneficiaries are also entitled to child's pension; and (5) 15% if four or more of the beneficiaries are also entitled to child's pension. 27 459/2015

(continued)

Table 7.3: ***(Continued)***

Damage type	**Available compensation**
	When determining the amount of the surviving spouse's pension, the earnings or pension income of the person entitled to the surviving spouse's pension is taken into account in accordance with section 107.
Child's pension entitlement (section 101) Amount of the survivors' pension (section 104)	A child of the injured person who was under 18 years of age at the time of the injured person's death is entitled to a child's pension. A child of the injured person who has reached the age of 18 is also entitled to a child's pension if he or she studies full time or at the time of the injured person's death was unable to support himself or herself due to disability arising from illness, defect or injury, for as long as the said criteria are met, though not to exceed the date when the child reaches the age of 25. A child of the spouse or person referred to in section 100(2), entitled to a surviving spouse's pension on the injured person's death, is also entitled to a child's pension on the injured person's death under subsection 1 if he or she was supported by the injured person. A child surviving his or her parent is always entitled to a child's pension. However, the child's pension is not paid on the death of more than two persons at the same time. If a child who has survived two deceased persons and is receiving a child's pension is subsequently granted a child's pension on the death of his or her own parent, the child's pension granted first shall discontinue from the date on which the pension granted on the death of his or her own parent begins. The aggregate amount of the child's pension of the injured person's annual earnings is: (1) 25% if one person is entitled to child's pension; (2) 40% if two persons are entitled to child's pension; (3) 50% if three persons are entitled to child's pension; and (4) 55% if four or more persons are entitled to child's pension. The aggregate amount of a child's pension is divided evenly between the beneficiaries. If there are orphans without both parents among the beneficiaries, a further 15% of the injured person's annual earnings is added to the aggregate child's pension, though not exceeding an amount which would make the aggregate amount of a child's and surviving spouse's pension exceed the maximum amount of the survivors' pension laid down in subsection 1. The additional amount is divided evenly between the said fully orphaned beneficiaries.

Information Flow in the Claims Process: Standard Claims under Private Workers' Compensation Insurance Policies

The information flow in standard claims is handled by insurance companies using the procedure outlined in Figure 7.1.[10] All reports are logged as a claim, then investigated, and a decision is reached. A decision must be received by the claimant within 30 days of the insurance company receiving all the necessary documentation. Only a very small percentage of claims that are investigated result in no compensation being paid.

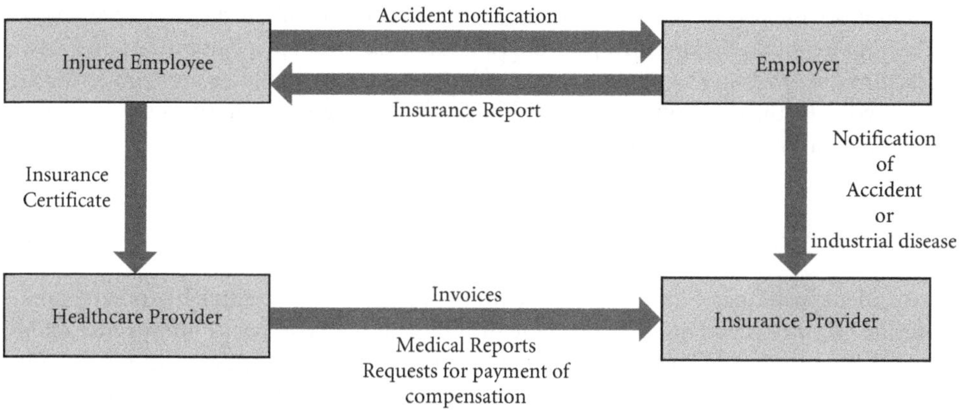

Figure 7.1: Claims process for a Finnish workers' compensation claim handled by an insurance company

The employee generally reports the problem to the employer immediately. The employer issues an insurance certificate for the employee to take to the healthcare professional, who in turn submits reports, invoices and requests of compensation payment to the insurer. Insurers pay the healthcare costs direct to the municipality.

The employer also reports the incident to the insurance company using an accident notification form within 10 working days of the date they were made aware of the matter. This is a specified form for basic information which is available from TVK's website. The form contains relevant employment data (salary, etc), details of the work and the circumstances in which the incident happened.

Any accident which is considered a serious accident must also be reported to the police and the worker protection authorities. Serious accidents are those that led to death or a serious injury (defined as an injury that is likely to be permanent and which complicates normal life).

If a claimant is dissatisfied with the decision of the insurer, he has the right to appeal, free of charge, to the Employment Accident Appeals Board.[11] Information on how to lodge an

[10] The procedure is detailed in ch 4 of the Employment Accidents Insurance Act 1948. Figure taken from www.TVK.fi/en/Claims/Compensation-procedure/.

[11] www.fin-fsa.fi/en/Financial_customer/Customer_protection/Assistance/Appeals/Employment_Accidents_Appeal_Board/Pages/Default.aspx.

appeal is annexed to an insurance company's decision. The Employment Accidents Appeal Board is an independent appellate instance whose operations are based on the Workers' Compensation Appeal Act (1316/2010)

The process of appealing to the Employment Accidents Appeal Board is governed by the Workers' Compensation Act.

The Employment Accidents Appeal Board handles complaints filed on insurance companies' decisions on statutory workers' compensation insurance. These decisions relate mainly to occupational accidents and injuries, occupational diseases, agricultural entrepreneurs' occupational accidents and draftees' military accidents and illnesses. The Board does not handle cases involving private accident insurances, which are dealt with by the Insurance Board.

The Employment Accidents Appeal Board is the first-level appellate instance for cases concerning the statutory workers' compensation insurance. Should an individual still be dissatisfied with the decision of the Board he has a final right of appeal to the Insurance Court, the highest court in accident cases.

Accident Insurance Compensation Board

The Accident Insurance Compensation Board operates within the Finnish Workers' Compensation Centre. Its purpose is to promote the uniformity of compensation practice under this Act by issuing guidelines and statements (section 226). The Accident Insurance Compensation Board comprises (section 227) a chairman, 3 lawyer members, 4 labour market members, at least 5 medical experts, and a necessary number of deputies, all of whom are appointed by the Ministry of Social Affairs and Health for a period of up to 3 years at a time. The members must be familiar with statutory accident insurance. The chairman and the lawyer members shall have a master's degree in law. The medical experts must be physicians licensed in Finland and familiar with insurance medicine.

A representative of the Ministry of Social Affairs and Health is appointed chairman of the Accident Insurance Compensation Board. Lawyer members and at least 2 medical experts as well as their deputies are appointed on a proposal by the Finnish Workers' Compensation Centre. Two of the labour market members are appointed on a proposal by the employers' key central organisations, and 2 on a proposal by the employees' key central organisations. One of the labour market deputies is appointed on a proposal by the employers' key central organisations, and 1 on a proposal by the employees' key central organisations. One of the medical experts is appointed on a proposal by the employers' key central organisations, and 1 on a proposal by the employees' key central organisations. The Board, which may also operate in sections, elects the necessary vice-chairmen from among its lawyer members. If the chairman or a member, medical expert or deputy of the Accident Insurance Compensation Board is permanently prevented from attending to his or her duties, the Ministry of Social Affairs and Health appoints a replacement chairman, member, medical expert or deputy for the remaining period of office, taking into account the provisions of subsections 1 and 2.

Pay-as-You-Go

The Pay-as-You-Go scheme is regulated by the Workers' Compensation Act and participation in the scheme is mandatory. Every insurance company is obliged to participate in and pay into the pay-as-you-go scheme. According to the current Act, the insurance companies cover the costs annually in proportion to their premium income.

Specifically Pay-as-You-Go absorbs the risks of claims for compensation caused by:

— inflation (future increases in pension provision, etc);
— major catastrophe;
— medical costs inflation; and
— long-latency diseases (those with a 10-year or larger gap between exposure to the causal agent and diagnosis).

It also acts to pay-as-you-go scheme to spread the risk (and payments for) workers' compensation for uninsured individuals, for example where an employer falls into the small business exemption or has failed to take out workers' compensation insurance according to their obligations. In this specific category, ie claims concerning uninsured work, the TVK plays a claims-handling role.

In the Act, catastrophe refers to a temporally and spatially limited event or a sequence of events of the same origin for which a total compensation amounting to more than €75,000,000.00 (catastrophe threshold), provided in the Act, is paid to one or more injured person or beneficiary. Compensation for a catastrophe is taken into account only insofar as such a compensation exceeds the catastrophe threshold. The insurance company´s payment share constitutes the compensations which the insurance company pays after the compensations paid by all insurance companies for the same catastrophe in total exceed the catastrophe threshold.[12] There is no ceiling for the size of a catastrophe event.

As the compensation paid is based on actual earnings, the Pay-as-You-Go top-up protects insurance companies in the event of a catastrophic high-value event/claim. The largest claim to date under the Pay-as-You-Go scheme was €4.2 million for a helicopter disaster in 2005.

Each year TVK calculates the costs that have accrued over the past year for these risks/claims. To cover this amount, the insurers transfer their payments (calculated according to their market shares) to te TVK. TVK then apportions the Pay-as-You-Go funds and reimburses the cost of the insurance companies' risks accrued over the past year.

The premiums for 2007–15 are in Table 7.4.

Table 7.4: Pay-as-You-Go premiums 2007–15

Year	Premium income (€)	PAYG income (€)	Gross premium income (€)	PAYG as a percentage
2007	585.7	100.9	686.6	14.7
2008	586.3	93.8	680.6	13.8
2009	554.5	91.8	648.5	14.2
2010	537.6	109.7	656.8	16.7
2011	584.2	143.7	727.9	19.7
2012	619.9	136.4	756.3	18.0
2013	623.1	117.8	740.9	15.9

(continued)

[12] Workers' Compensation Act, s 231.

Table 7.4: *(Continued)*

Year	Premium income (€)	PAYG income (€)	Gross premium income (€)	PAYG as a percentage
2014	589.4	107.3	696.7	15.4
2015	599.7	109.1	708.8	15.4

Statistics

Annual claims numbers for workers' compensation are collated by TVK[13] and are shown in Figure 7.2. The number of workers' compensation claims has remained fairly constant after an increase in 2005 due to legislative changes. The number of compensation claims includes a steady level of claiming for commuting accidents/injuries, with approximately 2,000 such claims per year.

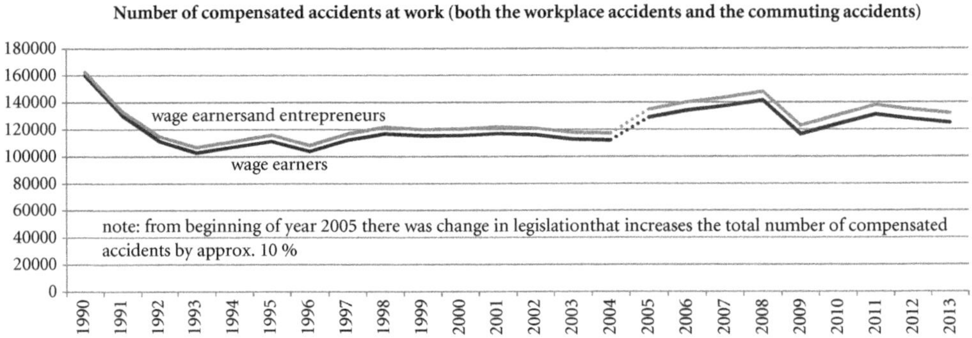

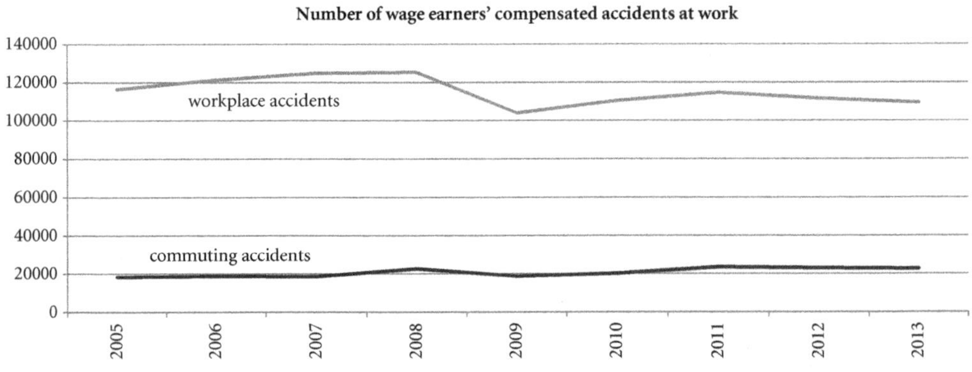

Figure 7.2: The numbers of compensated accidents at work 1990–2013 and the breakdown of wage-earners' compensated accidents at work into workplace accidents and commuting accidents 2005–13

While commuting accidents account for a relatively low proportion of total workers' compensation claims, they form a disproportionately higher proportion of fatal claims, as demonstrated in Figure 7.3.

[13] Annual Workers' Compensation Accident Statistics kindly supplied by Jussi Kauma, CEO of TVK available on the website as 2014 slides: www.tvk.fi/en/Publications/.

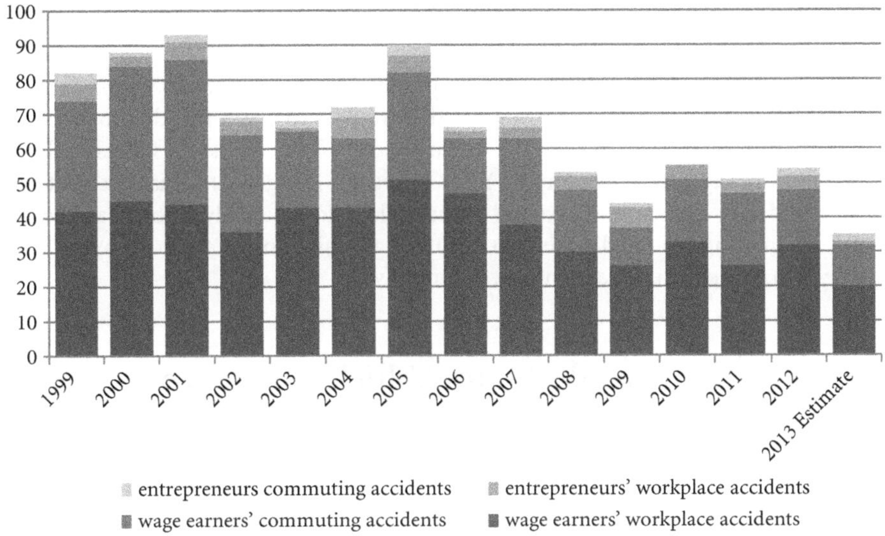

Figure 7.3: Fatal accidents at work 1999–2013 (estimate) by category

As can be seen from Figure 7.4, the majority of claims for workers' compensation are for relatively short-duration disabilities, with over 90% of disabilities resolved within 30 days.

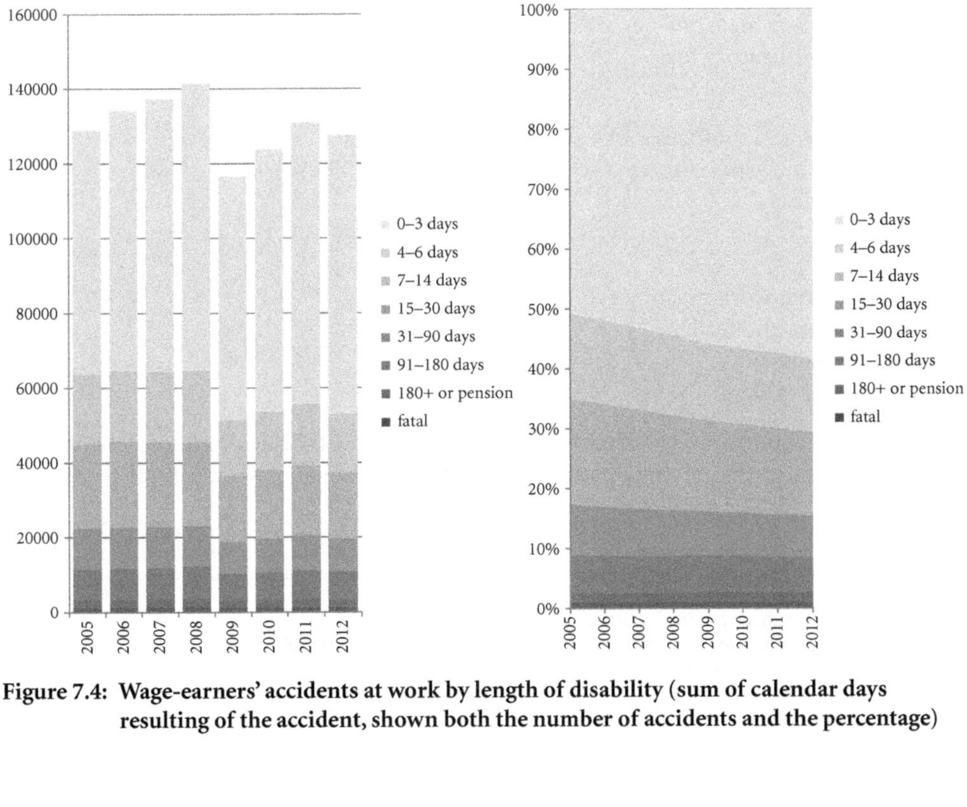

Figure 7.4: Wage-earners' accidents at work by length of disability (sum of calendar days resulting of the accident, shown both the number of accidents and the percentage)

Conclusions

Workers' Compensation in Finland is unusual in that it reflects actual salary loss, and is not capped. The potential implications of this are that injuries to just one individual, for which the standard example is given as the chief executive of Nokia, can have a substantial impact on compensation payments. The Pay-as-You-Go system mitigates the risk associated with the social solidarity principle of paying actual salary loss (locally known as the Nokia problem), by creating a pooled system to ensure high-value losses are born by all insurers collectively, not by an individual insurance company.[14]

C. Finland's Motor Vehicle Insurance

Overview

Motor insurance is compulsory in Finland, under the Act Governing Motor Insurance[15] and the Decree Governing Motor Insurance Liability.[16] Owners and registered keepers of a vehicle are required to insure it within 7 days of taking ownership,[17] the requirement for insurance is then ongoing. Insurance premiums are calculated according to a number of factors including the type of vehicle, vehicle make and model, engine size, usage, domicile, claims history and method of payment. Included in the premium is an insurance tax (24%), a road safety fee (approximately 1%) and some other central costs. Standard claims under motor liability insurance policies are handled by insurance companies.

Insurance is obligatory for:

— cars;
— motorbikes and motocross bikes;
— motorised bicycles (with a power output in excess of 250 W);
— snowmobiles;
— ATVs;
— tractors;
— power-driven work machinery;
— semi-trailers and trailers;
— caravan trailers;
— lightweight trailers.

The traffic injury insurance is essentially a primary payer, and social security payments and workers' compensation are secondary compensation systems but with a right to recover

[14] The authors are grateful to Kirsi Pohjolainen (TVK lawyer) for updated information and to Hanna Eskola (TVK actuary) for the updated statistics above.

[15] Available in Finnish at www.finlex.fi/fi/laki/ajantasa/1959/19590279.

[16] Available in Finnish at www.finlex.fi/fi/laki/ajantasa/1959/19590324?search%5Btype%5D=pika&search%5Bpika%5D=liikennevakuutusasetus.

[17] Penalty premiums are liable if a vehicle is not insured in a timely manner, see www.lvk.fi/en/obligation-to-insure/penalty-premium/ for more detail.

back from the traffic injury insurance. Motor liability insurance consists of personal injury and property damage elements. Motor insurance covers all parties for all personal injuries. There are no minimum or maximum requirements for personal injury coverage as compensation is paid according to the injury. As a rule, damage is compensated in full, even if it was not due to anyone's carelessness. Third parties are compensated for any property damage sustained. Damage caused to an innocent party's car by a collision will be compensated. However, it is a prerequisite for compensation in a situation where vehicles have collided that the damage was caused by the driver's or the owner's carelessness, driving in breach of traffic rules or improper condition of the vehicle. In collision situations, the liability for compensation can also be split between the parties pro rata to their degree of guilt. Damage to the policyholder's vehicle and/or property within that vehicle is not covered by the standard motor liability insurance. Comprehensive insurance can be taken out if additional protection is required.

Under the Finnish system all personal injury is on a first-party basis. Where there is an apportionment of liability, then the 'at-fault' party will pay. The claimant will seek compensation from his own insurer and the insurer will then subrogate and claim back the compensation for the 'at-fault' portion.

The system is essentially one of strict liability based on the balance of probabilities, and only requires showing causation between the use of the vehicle and the injury, with very few exceptions. Where contributory negligence is proved, compensation can only be reduced if there is intentional or gross negligence on the part of the claimant, eg attempted suicide. Another derogation is permitted for drunk and drugged drivers, where compensation for the driver will be diminished, but passengers will get full compensation even if they knowingly got into a car with a drunk/drugged driver. A further derogation applies where the vehicle was stolen, where there is no entitlement to compensation.

Available Compensation

For personal injury, the policy intention is to provide full compensation. There is no maximum sum, with compensation paid according to the injury. Types of damage are categorised as shown in Table 7.5.

Table 7.5: Categories of personal injury damage and compensation

Damage type	Available compensation
Medical and rehabilitation costs	Costs of tests and treatment and travel expenses to and from medical appointments
Loss of income	Actual temporary loss of income, as evidence by — a medical report stating the type of injury, the period of being incapable of working; and — an employer's report on the earnings before the injury, during the period of being incapable of working, and after it

(continued)

Table 7.5: *(Continued)*

Damage type	Available compensation
Delayed earnings	Where an injury results in a verifiable delay to the completion of a claimant's training to qualify for a post or their graduation, then compensation may be paid for the delay in earnings
Aches, pains and other temporary harm	Paid based on a tariff system until the injury has healed or the state of health has stabilised. Compensation levels are determined by the type and severity of injuries, the type of care required and the duration of recovery
Permanent incapacity	Injuries are classified in 20 handicap categories according to the severity, based on Decree 1649/2009 by the Ministry of Social Affairs and Health. The amount of compensation is affected by the type of injury and the resulting category of handicap; the person's age at the moment of the accident, and in some cases a particular deterioration of quality of life due to the injury
Permanent cosmetic defect	Compensation is payable for a permanent impairment to a person's appearance, such as scar formation, caused by the injury. The amount of compensation depends mainly on the type, extent and location of the injuries and the person's age
Motor insurance policy's care allowance	A care allowance may be payable to an injured person who is permanently dependent on another person's care. The amount of care allowance depends on the severity of the condition caused by the traffic accident. The scale of severity is specified in the care allowance classification of the Traffic Accident Board
Vocational rehabilitation	The aim of vocational rehabilitation is to assist the injured person in his/her return to working life and to support coping with work in spite of the limitations caused by the injury. The measures qualifying for compensation may be diverse, such as training for a new vocation or various support actions Rehabilitation can be carried out in cooperation with the Insurance Rehabilitation Association of Finland (VKK)[a]
Disability Pension	When an injury results in a permanent disability the claimant is entitled to a disability pension paid by the motor liability insurance. Before a disability pension is granted the possibility of using vocational rehabilitation to enable the claimant to return to work is investigated
Loss of support (survivor's pension)	Individuals who were dependent on the support of a person killed by in a traffic accident are entitled to reasonable compensation for loss of support (survivors' pension). The amount of survivors' pension covered by a motor insurance policy depends, among other things, on the earnings of the deceased and the widow/widower, and by the size of the family

[a] www.vkk.fi/english/index_eng.html.

For property damage, essentially any type of damage can potentially be covered,[18] and the categories shown in Table 7.6 are commonly used.

[18] See detailed provisions at www.liikennevahinkolautakunta.fi.

Table 7.6: Categories of property damage and compensation

Damage type	Available compensation
Vehicle damage	The owner of a damaged motor vehicle is entitled to receive compensation (ie the repair costs or the corresponding amount of money). If the motor vehicle cannot be repaired at reasonable expense, the insurance company shall pay compensation matching the valid value of the vehicle. In that case, the motor vehicle ownership is assigned to the insurance company. The decrease in value of the damaged vehicle will not be compensated for
Substitute vehicle costs	The rental costs of a substitute vehicle will be covered if the need for a substitute vehicle is considered well grounded, eg due to the person's profession When a car is privately owned motor liability insurance covers 94% of the total rental cost of a substitute vehicle (not including any excess elimination fee paid for the risk of damage). The remaining 6% of the rental costs corresponds to the average savings in the vehicle's operation costs during the vehicle's period of non-use. Aside from the rental costs, no other compensation for period of non-use is paid for the rental period The rental costs of a vehicle substituting a maintenance lease vehicle are paid in full
Compensation for period of non-use	It is a prerequisite for paying compensation that the company has been provided with the repair shop's certificate for the period of non-use Compensation for the time taken to repair the vehicle based on the price of the vehicle. When determining the amount of compensation, the price of the vehicle is deemed to be the cost of acquiring a new vehicle at the time of the accident if the vehicle is no more than 5 years old. If the vehicle is older than 5 years, the fair value of a similar vehicle at the time of the accident is regarded as the price of the vehicle If the vehicle is written off, compensation for non-use is payable for a maximum of 14 days. For heavy equipment, compensation is paid for 30 days or for the time spent on obtaining and equipping a new vehicle
Towing expenses	Motor liability insurance may compensate for the expenses incurred when towing a vehicle so damaged that it no longer is roadworthy

Claims under the Finnish Motor Insurers' Centre

The Finnish Motor Insurers' Centre (LVK) has a broad remit including safeguarding the rights of individuals who have been involved in a road accident and overseeing aspects of the statutory motor liability insurance. LVK promotes traffic safety by coordinating the work of road accident investigation committees and compiling road traffic accident statistics. Additionally, LVK has a claims-handling role for specific categories of claims that are not handled by insurers, thereby providing a 'safety net' for claimants in relation to specific claim types. Specifically, LVK handles claims for compensation caused by uninsured vehicles, unknown vehicles, foreign vehicles, vehicles holding transfer insurance policies, vehicles holding frontier insurance policies, and loss of reindeer. The arrangements in relation to each of these categories are shown in Table 7.7.

Table 7.7: Motor claims handled by the LVK

Claim type	Available compensation
Uninsured vehicles	The Motor Liability Insurance Act requires the LVK to compensate for traffic damage caused by an uninsured vehicle. As a rule, the LVK's liability for damage caused by an uninsured vehicle is similar to an insurance company's liability for damage caused by an insured vehicle.[a] However, the compensation liability does not cover — damage suffered by the owner or keeper of the uninsured vehicle; or — damage suffered by the driver when he or she knew or should have known that no motor liability insurance had been obtained for the vehicle
Unknown vehicles	Personal injury caused by an unknown vehicle is compensated by the LVK in accordance with the Act on Motor Liability Insurance. Property damage caused by an unknown vehicle is only compensated when it has been caused in the context of a major personal injury.
Foreign vehicles	If the damage was caused by a vehicle registered in an EEA country, the claims is subrogated to a green card representative authorised by the LVK[b] Compensation is paid according to the Motor Liability Insurance Act. If a foreign vehicle is insured in a green card system country, its traffic insurance policy is used to compensate for personal and property damage suffered by the other party (including pedestrians' and bicyclists' damage), as well as injuries suffered by the passengers of insured vehicles. However, with the exception of Swedish and Norwegian vehicles, the owner and driver of the vehicle which caused the damage do not receive compensation for their personal injuries If the vehicle causing the damage has a border insurance policy, it is also used to compensate for personal injuries suffered by the guilty vehicle driver and owner In order to establish the degree and cost of personal injuries, the following are required: — medical report; — invoices and receipts for the expenses incurred; — if compensation is sought for loss of income, a report showing the decrease of income In order to quantify the damage to property, the following are required: — inspection report for the damaged vehicle; — repair bill; — a report of the type, condition, age and damage of the damaged article; — purchase price of the damaged article and its fair value (at the time of damage)

(continued)

Table 7.7: ***(Continued)***

Claim type	Available compensation
Vehicles holding transfer insurance[c]	Compensation is paid according to the Act Governing Motor Liability Insurance and the regulations or instructions issued pursuant to it
Vehicles holding frontier insurance[d]	Compensation is paid according to the Act Governing Motor Liability Insurance and the regulations or instructions issued pursuant to it
Vehicles intended for use by the disabled	Compensation for damage of electric wheelchairs and other lightweight unregistered vehicles intended exclusively for use by disabled persons. If the vehicle is insured the insurer pays, but if there is no insurance in place, then LVK will provide benefits in line with insurance
Reindeer	Reindeer damage refers to the financial loss suffered by the owner of the reindeer as a result of losing the reindeer in a traffic accident. The LVK is the centralised place for compensating for reindeer damage caused by motor vehicles. The LVK also compensates for reindeer damage caused by foreign vehicles The LVK does not compensate for any other damage occurring in conjunction with the reindeer collision. Compensation for vehicle damage or personal injuries caused by a reindeer collision is sought from the vehicle's own motor and all-risk vehicle insurance policy

[a] The owner of the uninsured vehicle is not entitled to any compensation, neither is the driver if the driver is not the owner and the driver knew that the vehicle was not insured. The burden of proof falls on the Finnish motor insurer to prove this. Passengers are covered regardless of their awareness.

Liability for collisions between uninsured motorist and other users can be complex. If, for example, an insured at-fault driver collides with an uninsured vehicle, then the driver of the uninsured vehicle can claim against the at-fault driver. There is no claim for recourse by LVK as the vehicle is uninsured (effectively it cannot subrogate). A lack of insurance cover is not a reason for recourse. Driving without insurance is an administrative offence but not a criminal offence. There is a very limited right of recourse if the driver has acted with intent, under gross negligence, or drunk/drugs.

[b] Details on the procedure for claiming can be found at www.lvk.fi/en/compensation-for-accidents/damage-compensated-for-by-the-motor-insurers-centre/.

[c] www.trafi.fi/en. Temporary transfer insurance is available from some Vehicle Inspection Stations and Custom Offices. The premiums are paid to LVK, who then handle any claims. Transfer permits can also be used in conjunction with standard insurance policies, who will then handle and resulting claims.

[d] Frontier insurance must be taken for a foreign-registered vehicle, brought to Finland for temporary use, if (a) the vehicle is registered in a country outside the green card system, or (b) the vehicle does not have a valid green card. Frontier insurance will not be granted for a vehicle registered in an EEA country. The vehicle must be in Finland or at the border on its way to Finland when the insurance period starts. Frontier insurance is granted by the Finnish Motor Insurers' Centre, and is also sold in customs offices. The frontier insurance is valid in the whole EEA and in Switzerland. If the validity of the green card expires before the vehicle leaves the country, frontier insurance must be taken for the remaining time. The minimum period of validity of frontier insurance policies is 30 days and the maximum period of validity is 1 year.

LVK Claims Process

Claims are made using a specified claims form, which is available online from the LVK website. The form is usually submitted by the driver causing the damage, or by the injured party if an unknown party causes the damage. The report must contain details of the vehicles involved in the incident, together with a detailed description of the incident. All documents which clarify the damage (eg the police investigation report, possible photos, a copy of the green card) should be appended to the damage report. Once completed and signed, a damage report can be sent to the LVK by mail, fax or e-mail.

If the damage was caused by a foreign vehicle, the case can be processed faster if the vehicle's insurance company details are provided. Whenever possible, a copy or photograph of the foreign vehicle's international insurance certificate (green card) should be sent to the Finnish Motor Insurers' Centre.

Statistics

Annual claims numbers for the Finnish motor insurance industry are collated.[19] They show that in 2014 there were 93,067 road accidents in Finland, of which 17,733 resulted in personal injuries. In these accidents, 23,084 people died or were injured (136 died, 353 were seriously injured, the remaining 22,595 were slightly injured): see Figures 7.5 and 7.6. In 2014 a total of €483,000,000 was paid out in compensation (an average of €4,800 per accident).

These figures do not include the 5,716 claims made to LVK in 2014, which were:

— 326 claims for personal injuries caused by unknown motor vehicles;
— 638 uninsured vehicle claims;
— 729 foreign transport claims;
— 51 transfer damage claims;
— 160 claims for damage occurring abroad; and
— 3,812 reindeer claims.

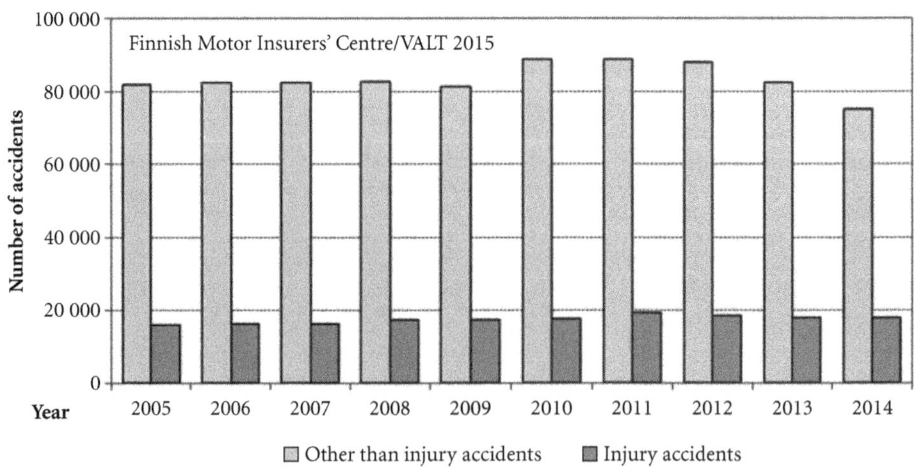

Figure 7.5: The annual number of accidents resulting in injury and accidents not resulting in an injury 2005–14. Source: Data taken from the Annual Road Accident Statistics of Insurance Companies 2014, www.lvk.fi/en/statistics/OTI-road-accident-statistics/

[19] Collated by the Finnish Crash Data Institute (OTI) Annual Road Accident Statistics of Insurance Companies 2014, available at www.lvk.fi/en/statistics/OTI-road-accident-statistics/.

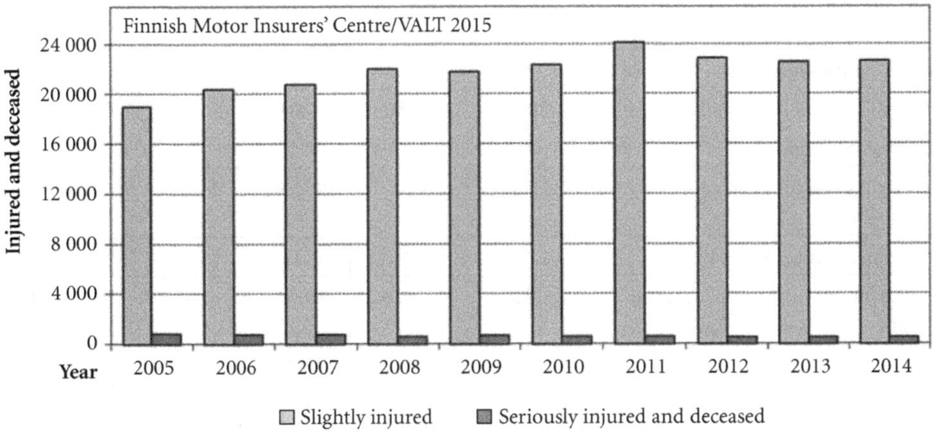

Figure 7.6: The annual number of injuries and deaths from motor vehicle accidents 2005–14. Source: Data taken from the Annual Road Accident Statistics of Insurance Companies 2014, www.lvk.fi/en/statistics/OTI-road-accident-statistics/

Gross annual premium income from insurance is approximately €80 million. Table 7.8 shows the value of claims paid for traffic injury insurance between 2009 and 2013 by category. Payments are generally paid as lump sums, but some sums are paid as serial payments, such as 'loss of income', which is paid on a monthly basis as an ongoing expense.

Table 7.8: Claims paid 2009–13 under motor liability insurance (€1,000)

Type of compensation					
	2009	**2010**	**2011**	**2012**	**2013**
Bodily injuries					
Medical treatment	53,448	50,922	53,603	54,571	58,173
Rehabilitation	25,980	28,266	29,154	27,636	26,059
Compensation for loss of income	40,249	40,575	42,873	40,138	41,027
Continuing benefits	58,440	57,381	63,093	66,425	66,630
Non-recurring compensation for defect or permanent handicap	8,130	8,160	8,392	8,136	9,300
Non-recurring compensation paid instead of continuing benefits	–185	147	236	374	77
Funeral expenses	903	811	839	830	816
Compensation for pain & suffering	9,814	8,888	9,619	9,259	9,432
Other compensation for bodily injuries	5,299	4,994	1,113	4,440	11,788
Index increases	74,911	72,428	73,977	77,337	77,814
Total	276,988	272,572	282,898	289,147	301,115

(continued)

Table 7.8: ***(Continued)***

Type of compensation					
	2009	**2010**	**2011**	**2012**	**2013**
Material damages					
Compensation for damage to motor vehicle	162,285	175,705	188,492	187,914	184,094
Compensation for loss of use	14,445	15,046	15,604	12,780	13,847
Other compensation for material damage	11,520	8,920	18,383	15,559	14,281
Total	188,250	199,670	222,478	216,252	212,222
Claims settled by Motor Insurers' Centre	6,913	7,285	6,785	7,403	11,066
Claims settlement costs	43,799	44,964	49,975	54,211	69,233
Total claims paid	**515,951**	**524,492**	**562,137**	**567,014**	**593,637**

Source: Insurance Supervisory Authority's Statistics.

D. Finnish Patient Insurance

General Overview

The Finnish Patient Injuries Act (Potilasvahinkolaki 585/1986)[20] obliges all healthcare providers to take out patient insurance that provides compensation for injuries specified by this Act.[21] In Finland 85% of the healthcare market share is provided by the state, with around 15% being provided privately. The medical district pays in arrears for damage that occurs in state hospitals, whereas the private hospitals take out private insurance with a private insurance company. State arrangements are not therefore insurance provision. The Finland Patient Injury Compensation Organisation holds reserves to cover claims occurring in the future. Claims under the Act are processed by the Patient Insurance Centre.[22]

Patient insurance is not a 'primary payer', but is 'secondary' to most of the other statutory insurances and benefits. The compensation from the Finnish Patient Insurance Centre is effectively a 'top-up'. It covers situations where there is a shortfall between the costs and loses attributable to the patient injury amount available from other statutory benefits and insurance provisions. In general, therefore, the primary payer is social security, with the exception of the extra costs associated with assisted living, which is provided by municipalities for severely injured patients. Municipalities have a right of recourse against the Patient Insurance Scheme. This right of recourse is not limited in value, but has to be justified.

[20] www.finlex.fi/fi/laki/ajantasa/1986/19860585.

[21] In the event that a healthcare provider has been declined insurance, the Finnish Patient Insurance Centre will insure the healthcare provider pursuant to s 7 of the Act.

[22] Potilasvakuutuskeskuksen, www.pvk.fi/en/.

Patient insurance will only cover any additional losses created by the patient injury. Costs and losses that would have been incurred regardless of the patient injury are not covered. For example, the costs of the original examination and/or treatment of the patient's illness or injury will not be awarded as they would have accrued regardless of the patient injury. When quantifying compensation, any costs and financial losses due to the pre-existing illness or injury will be ineligible for compensation.

Generally, costs and losses arising from an injury will be compensated for in full.[23] However, the injured party is also expected to prevent any subsequent loss whenever possible. Costs incurred in the treatment of patient injuries must be deemed necessary.

Compensation payable under patient insurance is determined by applying the provisions contained in the Tort Liability Act and the guidelines issued by the Traffic Accident Board. The decision policy of the Patient Injuries Board will also be taken into account in the compensation.

The law is currently (March 2017) being examined with a view to updating it and changes may follow.

Types of Damage Compensated

There are seven categories of injury which are potentially compensated under the Act, as noted in Table 7.9.

Table 7.9: Types of patient compensation covered in Finland

Damage type	Explanation
Treatment Injury	A bodily injury caused by an examination, treatment or other similar action (eg a vaccination, and blood or organ donation) performed on the patient, or the failure to do so. An injury caused by an erroneous prescription or dosage of drugs may also be compensated for as a treatment injury A prerequisite for compensation is that an experienced medical professional (taking into account the proficiency required of the professional) could have performed a different procedure in the examination or treatment situation in question, thereby avoiding the injury Compensation will not follow if an injury was unavoidable or if the actions of a healthcare professional do not materially contribute to the injury
Infection injury	Compensation may follow when an infection results from medical treatment, depending on the findings of an infection tolerance evaluation. This evaluation examines the balance of the risks related to the procedure against patient risks, such as pre-existing illnesses, injuries or medications. The evaluation will take into account the severity of the injury caused by the infection, the nature of the illness or injury being treated, and the overall health of the patient

(continued)

[23] The costs incurred must be necessary when claiming for the treatment of a patient injury.

Table 7.9: ***(Continued)***

Damage type	Explanation
	The patient is expected to tolerate more serious consequences of an infection if the original illness was more serious and carried a higher risk of infection than most illnesses/injuries. However, an injury caused by an infection may be compensable if it is rare and sudden
Accidental injury	Compensation is paid for various sudden, unexpected and external events occurring without the will of the injured party and causing injury in connection with a medical examination or treatment. This includes accident risks directly related to medical procedures, but not accident risks that are part of ordinary life
Equipment-related injury	Compensation is paid for an equipment-related bodily injury if the injury was caused to the patient by a defect in the equipment or device that was used for the examination, treatment or other similar action. Cases where the injury was caused by health professional usage of equipment and injury was due to implantable medical devices are not covered
Injury due to incorrect delivery of pharmaceuticals	Compensation is paid for bodily injuries caused by prescription drugs if they were dispensed by the pharmacist contrary either to the prescription or to the law governing the delivery of pharmaceuticals This provision refers to the incorrect delivery of prescription pharmaceuticals from a pharmacist. If pharmaceuticals were administered to the patient in conjunction with an examination or treatment they are processed as treatment injuries Side effects from pharmaceuticals which were appropriately prescribed and dosed are not compensated for under patient insurance. Compensation for side effects may be available from voluntary pharmaceutical injuries insurance provided by the Finnish Mutual Insurance Company for Pharmaceutical Indemnities, discussed below[a]
Unreasonable injury	Compensation is paid where an appropriate examination, treatment or other corresponding processing results in a severe injury which is materially disproportionate to the initial situation, regardless of whether the injury was preventable Compensation for unreasonable injury is only available in cases where the patient has been caused a permanent severe illness, injury or loss of life. An illness or injury is considered to be severe if it falls into Classes 7 or 8 in the Ministry of Social Affairs and Health's classification of injuries Compensation will be decided on the basis of an unreasonableness evaluation. When evaluating the possible unreasonableness of an injury, the following factors will be taken into consideration: — the severity of the additional injury caused by treatment; — the nature and severity of the illness or injury examined or treated; — the patient's overall health; — the rarity of the injury; and — the likelihood of the injury in the individual case in question

(continued)

Table 7.9: ***(Continued)***

Damage type	Explanation
	An injury might be compensable as unreasonable if there is considerable disproportion between the consequence and the patient's illness or injury originally treated and his or her overall health. The greater the disproportion and the greater the negative effect on the patient's health caused, the more evident it will be that the injury might be unreasonable Even severe complications are not compensated for if the risks were foreseeable by medical experts, or, if left untreated, the illness or injury could have led to the death or serious disablement of the patient

[a] www.laakevahinko.fi/in-english/pharmaceutical-injury/.

Exclusions from Coverage

Not all consequences of medical treatment and healthcare are eligible for compensation. Compensation is only paid for the bodily injuries that fulfil the conditions prescribed in the Patient Injuries Act. The following will not be compensated:

(a) Injuries that have occurred outside the geographical area of Finland, even if the patient was referred for treatment in Finland.
(b) Material damage, such as torn clothing or the loss of dentures. In these instances the claimant may claim damages from the tortfeasor under the Tort Liability Act.
(c) 'Pure financial loss', which is financial loss that is not connected to bodily injury or material damage.
(d) Injuries that occurred before 1 May 1987.
(e) Injuries occurring during ordinary assistance services (eg domestic help, or treatment and assistance and monitoring services connected to living in a residential care home); these are not regarded as medical treatment or healthcare.
(f) Insignificant injuries, even if they qualify for compensation on other grounds. An injury is considered to be insignificant if it causes the patient only slight pain and suffering, no permanent functional disability, no aesthetic injury, or the costs incurred do not exceed €200 (as of September 2003).

Types of Loss Compensated

The compensation that is payable in respect of different defined types of loss is summarised in Table 7.10.

Table 7.10: Compensation available for different loss types

Loss type	**Available compensation**
Cost of medical care	The cost of medical care due to the injury, eg extra hospital and/or outpatient treatment, doctor's appointments, home healthcare, laboratory examinations, medicines, physical therapy and rehabilitation. If the body granting healthcare aid devices considers an aid device necessary, the cost of that will be covered Injuries that occurred in public healthcare settings are compensated for according to the rates charged by the public sector, unless there are exceptional medical grounds for using private medical services, eg if the treatment is not available in the public sector
Other expenses due to the injury	Travel expenses incurred by additional treatment may be compensated. In addition, the following compensation may be paid: — care allowance, if the injured party requires help in her own home for 'personal activities', such as dressing up and washing; — increased home care costs, if the injured party requires help with the more strenuous housework and the household does not include other family members; — clothing allowance, if the aid devices necessitated by the injury cause wear or soiling of the claimant's clothes while living at home
Temporary incapacity (pain and suffering)	Acute pain and suffering and other inconvenience sustained by the patient immediately after the injury The amount of compensation is determined by the degree of severity of the illness or injury, the duration of the recovery time, procedures, and the need for hospitalisation. If the pain becomes chronic, it will be classified as permanent incapacity and compensated accordingly.
Permanent functional incapacity	Only permanent functional incapacity (disability) resulting from a patient injury will be compensated. Compensation will not be paid for a permanent incapacity due to the originally treated illness. Compensation is determined by using the classification of permanent disabilities. Depending on the time of receiving the injury, the incapacity will be determined on the basis of either the Decision of the Ministry of Social Affairs and Health (1012/1986) or the Decree (1649/2009). In the classification of permanent disabilities, functional incapacity has been divided into incapacity categories 1–20 Permanent incapacity is established using medical reports. Compensation value is determined by the nature of the injury and the claimant's age
Permanent cosmetic incapacity	If the patient injury causes permanent impairment to a person's appearance, such as a scar, it will be compensated by paying damages for cosmetic incapacity. However, compensation will not be paid for cosmetic incapacity caused by the treatment of the original treatment

(continued)

Table 7.10: ***(Continued)***

Loss type	**Available compensation**
	When assessing the amount of compensation, the following criteria will be taken into account: the extent, form, colour, location and visibility of the scar. The age of the injured party is another factor that affects the amount of compensation payable
Loss of income	Compensation for loss of income of an employee or a self-employed person due to occupational disability caused by the patient injury. Compensation can be paid for partial occupational disability When assessing a person's working capacity, the remaining working capacity and opportunities for occupational rehabilitation will be taken into account. The effects of the possible pre-existing illness or injury are also considered. The patient is obliged to investigate opportunities to enhance their working capacity by medical or occupational rehabilitation. Loss of income will be compensated according to the actual loss. It will be calculated with reference to the estimated income that the injured party would have received had the patient injury not occurred. If the other benefits paid on the basis of other laws do not cover the loss of income in its entirety, the remaining part will be compensated from the patient insurance. When calculating the compensation, daily allowances and pensions received under other statutory insurance schemes, such as Motor Liability Insurance Act (Liikennevakuutuslaki 279/1959), Workers' Compensation Insurance Act (Tapaturmavakuutuslaki 608/1948), various acts of employee's pensions and the National Pensions Act (Kansaneläkelaki 347/1956) will be deducted Loss of income paid from patient insurance is subject to tax
Compensation to other family members and close individuals	Certain family members of the injured party and others who are especially close to the injured party may have the right to receive reasonable compensation for necessary expenses and loss of income as a result of taking care of the injured party during a period of acute convalescence. Family members are entitled compensation for injuries that occurred in connection with treatment given on 1 January 2006 or after
Loss of life	In cases where the injury leads to the loss of life, funeral expenses and other related costs can be covered to an equitable amount. Insurance will cover costs relating to, for example, the burial place, coffin, gravestone, funeral ceremony, flowers, obituary notice and mourning clothes. Funeral expenses will be paid to the estate or the person who paid for the funeral Compensation to cover funeral costs received under other insurance provisions may be deducted from the patient insurance compensation. The costs arising from the estate inventory are not covered, since such costs are not considered compensable according to established legal practice as they are not related to actual funeral costs

(continued)

Table 7.10: ***(Continued)***

Loss type	Available compensation
	In cases where there is a loss of life, necessary maintenance may be granted to those entitled to this compensation. People entitled to this compensation are the husband or wife and under-age children and, in some cases, children under 21 who are studying. Compensation may also be payable to persons who received de facto maintenance from the deceased Compensation is payable under patient insurance to the extent that other sources of income and compensation payable under other statutory schemes do not guarantee sufficient maintenance to the applicant.

The patient claims for the cost of the injury which they have already paid; if it is found that compensation is due, then the injury insurance reimburses the patient the cost of these payments. Alternatively, if liability for the accident has already been determined, the healthcare provider can invoice the patient injury insurance directly for the cost of the patient's care. When providing compensation for the costs of medical care the Patient Insurance Centre is only obliged to pay a proportion of the statutory fixed fee to state hospitals. This 'top-up' payment is for care in the state hospital and is the same amount regardless of whether the patient injury occurred in the public or private sector. There is a presumption that treatment of the patient injury will take place in the public sector unless there is a particular reason why it should be undertaken in the private sector (eg the treatment required is not available in the public sector). This presumption exists regardless of whether the patient injury occurred in the public or private sector, which has led to some disquiet about the public healthcare sector having to correct mistakes which occur in the private sector. Should a patient injury need to be treated in the private sector, then the cost will be paid in full. Table 7.11 shows the statutory fees in euros that are paid for various procedures from 2010 to 2015.

Table 7.11: Statutory fees in euros for various procedures 2010–15

	Inpatient fee per night (amount deducted from the fee when it is paid by PVK)	Outpatient appointment (amount deducted from the fee when it is paid by PVK)	Minor operation (all plus the potential cost of additional inpatient and outpatient treatment if needed
2015	38.10 (17.60)	32.10 (0)	105.50
2014	34.80 (16.10)	29.30 (0)	96.40
2012–13	32.60 (15.10)	27.50 (0)	90.30
2010–11	32.50 (15.00)	27.40 (0)	89.90

Claims Process

Claims follow a two-stage process. The initial stage establishes eligibility and the second quantifies the level of compensation. An injury notification form must first be completed and signed and returned to the Finnish Patient Insurance Centre.[24] Injury notification forms are available online or by phone from the Finnish Patient Insurance Centre, from the Patient Ombudsman or from the social worker at the healthcare provider.

Injury notification forms require information on, first, the time, date and location of the occurrence of the injury; second, the identity of the healthcare provider (including all other healthcare providers where a claimant was treated for the same complaint both before and/or after the injury); and third, a description of the injury and the grounds on which compensation is sought (detailed medical or legal analysis of the grounds is not required, merely a clear, cogent description of why the claimant feels compensation is appropriate).

The Finnish Patient Insurance Centre then requests reports from the healthcare provider on the treatment provided based on the information in the injury notification form.

Claimants generally have 3 years from the date when the injured party first knew, or should have known, of the injury to file the injury notification with the Patient Insurance Centre. The limitation period does not start at the point of treatment, but at the point when the injured party understood or should have understood that the injury resulted from the medical treatment and healthcare received. There are some exceptions to the 3-year limitation, but there are some absolute limitation periods. Claims for treatment given on 1 May 1999 must be made no later than 10 years after the treatment. A 20-year absolute limitation period exists for treatment given prior to 1 May 1999. Absolute limitation periods are from the date of treatment, not from the date when the patient became aware of the injury. Notifications filed after the expiry of the absolute limitation period will not be examined. Effectively if a patient only become aware of an injury after the expiry of an absolute limitation period, their claim will not be examined, even if they are within 3 years of discovering the injury.

Approximately two-thirds of claims are deemed not eligible for compensation. The remaining third are passed to the in-house claims' assessors who handle the procedures for paying compensation. A form for compensation will be attached to the decision that is sent to the patient, which the claimant is required to fill in and return to the Patient Insurance Centre. The form requires claimants to provide further information to substantiate their claims. This includes doctor's prescriptions, receipts, and documentation on other compensation and benefits which the claimant may be received from other sources. The form must be returned within 3 years of the decision that the claimant is eligible for compensation. Once the Finnish Patient Insurance Centre has received the completed form the claims assessor will determine if any additional information is required. When all necessary information has been obtained a claims decision will be made and compensation will be

[24] Forms must be signed and returned; electronic submission is not possible. Usually the form is signed by the adult concerned, but in cases where the patient lacks legal competence a trustee can sign on the patient's behalf. Parents can usually act as trustees for children, and when an individual is deceased a party to the estate is usually able to act.

paid. If there is a lengthy injury/recovery period a number of claims for compensation may be raised against the same initial eligibility decision.

The four steps in the patient insurance claims-handling process, which are shown in Figure 7.7, are as follows:

1. preliminary handling and registering, requesting info (20 staff);
2. decide if the injury is within the patient insurance scheme or not, eg within limitation (18 staff);
3. if eligible, then determining if the injury should be compensated (same 18 staff as above);
4. payment of compensation (total of 25 staff): divided into two teams, lump-sum payments, eg medical costs and intangibles (team of 18 staff) and ongoing periodic payments, eg loss of income (7 staff).

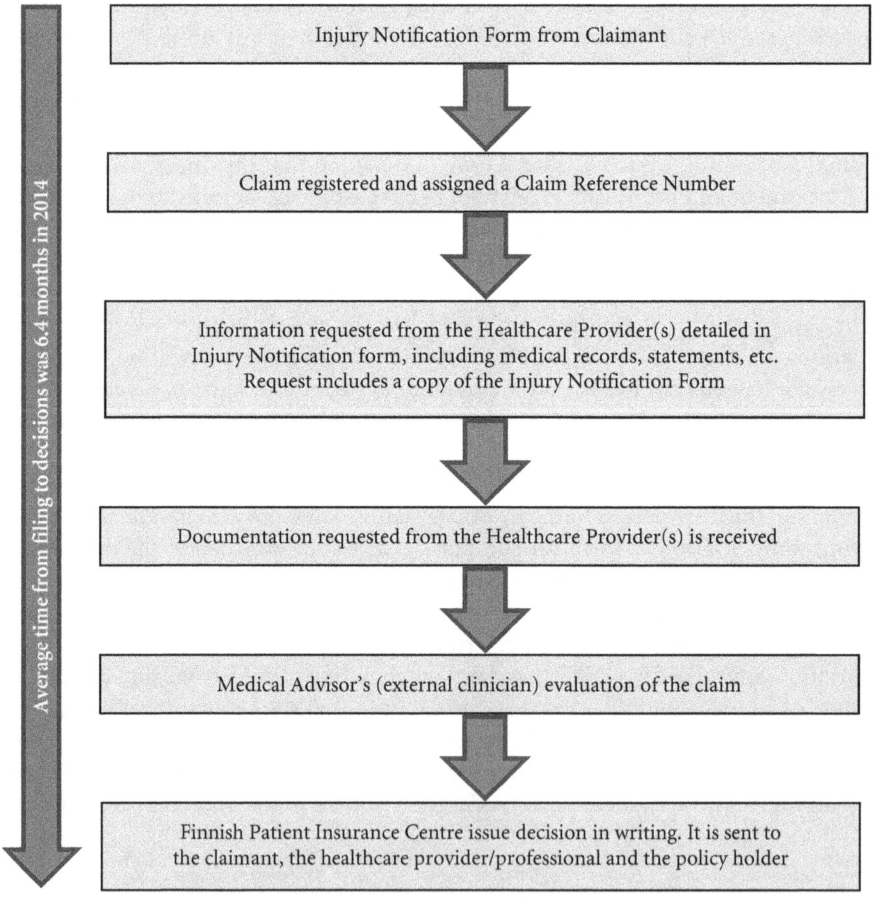

Figure 7.7: Diagram of the process for determining eligibility of claims for compensation

Revisions and Appeals against Claim Decisions

If a claimant is unhappy with the decision she can request an in-house revision. If she is still not satisfied, there is a right of appeal to the Court. These processes are detailed below.

Request for Revision

Revisions are available when a claimant considers that the claims decision has been based on insufficient documentation. Requests for revisions must be sent in writing to the Finnish Patient Insurance Centre. Requests should detail why the decision is considered to be unsatisfactory. Any missing documents should be attached to the request, or the request should contain information on where these documents may be obtained.

Appeal Procedure

Instructions for lodging an appeal are attached to each claims decision. There are two options for individuals who are dissatisfied with the claims decision: first, to refer the issue to the Patient Injuries Board;[25] or second, to the Court of First Instance. The Patient Injuries Board will process the request free of charge. The Patient Injury Centre submits cases on permanent loss of income to the Board.

The task of the Patient Injuries Board is to issue recommendations in patient injury cases in court. A Board recommendation may be requested by the claimant, the party engaged in the health and medical care, or the Finnish Patient Insurance Centre. The Patient Injuries Board may also issue a recommendation during court proceedings following a request by the court or the parties involved.

The Patient Injuries Act does not specify a time limit within which a patient should refer his or her case to the Patient Injuries Board. In practice the Patient Injuries Board considers requests where the patient has not brought the matter to court within the statutory 3 years as being effectively unfounded. Consequently, the Patient Injuries Board will not consider requests where the patient no longer has an opportunity to bring the matter to court.

Where a claimant with an accepted claim requests a Board recommendation solely regarding the quantum of compensation, the Board may evaluate the level of compensation on its own initiative.[26]

The normal composition of the Patient Injury Board is a chairman and secretary, a doctor or nurse or medical professional, and an insurer. Complex decisions can be decided by a larger group of members of the Board. The Board can consult independent experts.

Around 1,100 cases go to the Patient Injuries Board each year. Most appeals are on eligibility for compensation. In 2015 1,099 requests were made to the Board, 844 by patients, 179 by the Patient Insurance Centre, 12 by insurers and 7 opinions were sought. Of these 18.6% gave a favourable outcome to the victim.

The claimant may bring an action against the Finnish Patient Insurance Centre at the District Court of Helsinki or at the local District Court where the claimant is domiciled or

[25] http://pvltk.fi/.
[26] This procedure is based on the Patient Injuries Act, s 11b.

a permanent resident. The action may also be brought at the District Court within whose jurisdiction the injury (or the omission leading to the injury) took place or where the injury became apparent.

An action against Finnish Patient Insurance Centre must be brought within 3 years of the date on which the claimant was informed in writing of the decision (and of the limitation period).[27]

Statistics

In 2015 a total of 8,242 (7,952 in 2014) new claims were made to PVK. In total, 7,792 (8,509 in 2014) claims settlement decisions were made. The median for the time to process claims was 5.9 (5.7 in 2014) months and the mean was 6.4 (6.4 in 2014) months. The processing times of patient injury cases that had been deemed compensable shortened by 2 weeks. In 2015 a total of €40.5 million (€39.6 million in 2014) was paid out in compensation for patient injuries.[28] Statistics on claims, types, decisions and insurance premiums between 2010 and 2015 are shown in Tables 7.12 to 7.14 and Figure 7.8.

Table 7.12: A breakdown of the number of decisions and percentages accepted by injury type 2010–15

Decision	**Year of decision**					
Claim settlement decision	2010	2011	2012	2013	2014	2015
Patient injuries entitled to indemnity						
Treatment injury	2,029	2,059	1,958	2,255	2,236	2,141
Infection injury	125	124	132	125	129	113
Accidental injury	29	30	22	21	22	22
Others	33	31	25	37	36	24
Total	2,216	2,244	2,137	2,438	2,423	2,300
Minor patient injuries	68	67	61	66	62	58
No patient injury						
Injury was unavoidable or tolerable	1,762	2,013	1,675	2,005	2,074	2005
Other claims reported	3,126	3,583	2,823	3,149	3,950	3,429
Total	4,888	5,596	4,498	5,154	6,024	5,434
Total	7,172	7,907	6,696	7,658	8,509	7,792

(continued)

[27] The period for filing a suit is based on the Insurance Contracts Act (Vakuutusopimuslaki 543/1994).
[28] Taken from the 2015 Annual Reports and Statistics, available from www.pvk.fi/en/.

Table 7.12: *(Continued)*

Decision	Year of decision					
Claims decided						
Patient injuries entitled to indemnity	30.9%	28.4%	31.9%	31.8%	28.5%	29.5%
Minor patient injuries	0.9%	0.8%	0.9%	0.9%	0.7%	0.7%
No patient injury	68.2%	70.8%	67.2%	67.3%	70.8%	69.8%
Total	100.0%	100.0%	100.0%	100.0%	100.0%	100.0%
Patient injuries entitled to indemnity						
Treatment injury	91.6%	91.8%	91.6%	92.5%	92.3%	93.1%
Infection injury	5.6%	5.5%	6.2%	5.1%	5.3%	4.9%
Accidental injury	1.3%	1.3%	1.0%	0.9%	0.9%	1.0%
Others	1.5%	1.4%	1.2%	1.5%	1.54%	1.0%
Total	100.0%	100.0%	100.0%	100.0%	100.0%	100.0%
No patient injury						
Injury was unavoidable or tolerable	36.1%	36.0%	37.2%	38.9%	34.4%	36.9%
Other claims reported	63.9%	64.0%	62.8%	61.1%	65.6%	63.1%
Total	100.0%	100.0%	100.0%	100.0%	100.0%	100.0%

Table 7.13: Compensation paid in €100s by compensation category 2010–15

Payment year	2010	2011	2012	2013	2014	2015
Compensation for expenses						
Medical treatment and rehabilitation	2,774	2,939	3,006	3,635	3,801	3,887
Increased costs of living	2,010	1,742	1,535	2,355	1,830	1,548
Funeral expenses	219	264	279	257	298	335
Other expenses	2,677	2,582	2,827	3,695	3,913	5,211
Total	7,680	7,527	7,647	9,941	9,842	10,981
Compensation for loss of income						
Loss of income	8,766	8,321	8,951	10,766	10,665	11,677
Compensation for intangible losses						
Temporary incapacity	4,443	4,541	4,832	4,977	5,377	4,842
Cosmetic impairment	573	545	588	581	540	529
Permanent incapacity	4,513	5,026	5,353	5,057	5,768	4,849
Total	9,529	10,111	10,773	10,615	11,686	10,220

(continued)

Table 7.13: ***(Continued)***

Payment year	**2010**	**2011**	**2012**	**2013**	**2014**	**2015**
Claims settlement expenses						
Costs	6,671	7,012	7,225	7,530	7,756	8,300
Total paid	**32,645**	**32,972**	**34,596**	**38,853**	**39,949**	**41,178**

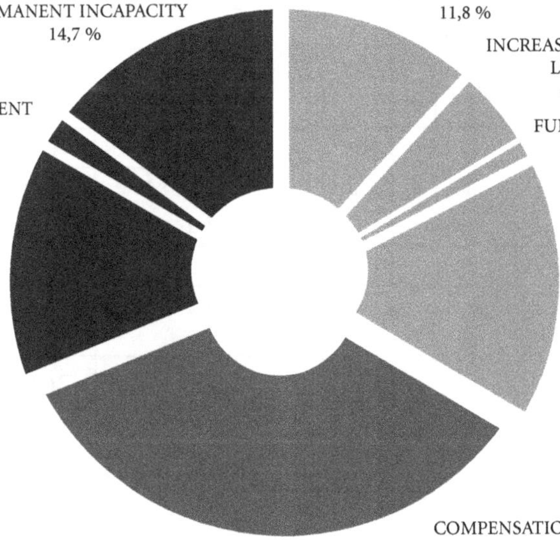

Figure 7.8: Compensated patient injuries by type of compensation in 2015

Table 7.14: Patient insurance premiums and indemnities 2009–15 (including a breakdown of claims incurred into claims paid and changes in provision for claims)

	2010	**2011**	**2012**	**2013**	**2014**	**2015**
1. Premiums						
Self-insurance provided by the Patient Insurance Centre	27,676	37,845	42,200	67,924	210,061	53,785
Paid in cash	28,622	30,056	27,587	31,245	34,996	31,856
Change in receivables	–946	7,789	14,613	36,679	175,065	21,929
Other insurance	6,349	6,082	11,302	14,495	13,123	15,140
Total	**34,025**	**43,927**	**53,502**	**82,419**	**223,184**	**68,925**

(continued)

Table 7.14: ***(Continued)***

	2010	2011	2012	2013	2014	2015
3. Claims incurred total figures						
Self-insurance provided by the Patient Insurance Centre	26,355	36,066	40,351	65,492	204,879	51,515
Other insurance	7,211	6,521	13,829	17,093	28,195	24,823
Total	**33,565**	**42,487**	**54,179**	**82,585**	**233,074**	**76,339**
breakdown of claims incurred						
Claims paid: self-insurance provided by the Patient Insurance Centre [a]	26,510	27,180	28,480	30,719	29,998	30,036
Claims paid: other insurance[a]	6,135	5,793	6,116	8,133	9,951	11,142
Claims paid: total[a]	**32,645**	**32,973**	**34,595**	**38,852**	**39,949**	**41,178**
Changes in provision for claims: provided by the Patient Insurance Centre	–155	8,886	11,871	34,773	174,881	21,479
Changes in provision for claims: other insurance	1,076	728	7,713	8,960	18,244	13,681
Changes in provision for claims: total	920	9,614	19,584	43,733	193,125	25,161
Provision for claims at the end of the year						
Self-insurance provided by the Patient Insurance Centre	164,326	173,211	185,082	219,855	394,736	416,215
Other insurance	20,585	21,314	29,026	37,987	56,231	69,912
Total	**184,911**	**194,525**	**214,108**	**257,842**	**450,967**	**486,127**

[a] Includes claims settlement expenses.

Conclusion

Finnish patient insurance provides a stable compensation system. The processing times of around 6 months combined with the relatively low appeal rate to the Patient Injuries Board, of just under 15%, indicate an efficient system.

E. Finnish Drug Insurance

Evolution of the Pharmaceuticals Insurance Scheme

A voluntary scheme—virtually identical to the Swedish scheme—was introduced in Finland in 1984,[29] and was followed by a statutory medical scheme in May 1987.[30] The original scheme provided for joint insurance of pharmaceutical companies under which the beneficiaries are users of medicines. A pharmaceutical manufacturer, importer, marketer or sponsor of a clinical trial applied to be a member of the Finnish Cooperative for the Indemnification of Medicines-Related Injuries. The Cooperative negotiated premiums on an annual basis with a pool comprising the Finnish insurance companies (the Finnish Pharmaceutical Insurance Pool).[31]

Briefly, claimants were able to submit claims to the Finnish Pharmaceutical Insurance Pool. The same experienced personnel at the injuries department of the Insurance Pool for Medicines-Related Injuries examined injury reports for to claims relating to both the medicines insurance and patients' statutory treatment insurance based on the Patients' Treatment Injuries Act. This arrangement was efficient in allocating claims to the 'right' insurance. An applicant who was dissatisfied with the decision of the insurance institution could request an opinion to be given by the Pharmaceutical Injuries Board, which was comprised of neutral doctors and lawyers or judges appointed by the Pool.[32] Opinions were issued by the Board free of charge. Any dispute between an insurer and an injured person could also have been referred to arbitration[33] if the parties so agreed.[34] An injured person could also have brought an action against the insurer in the former's local court.

As of 1 January 2012, the Pharmaceutical Injury Insurance was transferred to a new insurance company, the Finnish Mutual Insurance Company for Pharmaceutical Injury Indemnities, which is a captive owned by the Finnish Cooperative for Indemnification of Medicine-Related Injuries.[35] At that time, claims handling was outsourced to the British

[29] For details of the previous scheme, see the Finnish Pharmaceutical Insurance Pool's General Terms and Conditions, effective as of 1 January 2004, and applying to injuries that occurred on or after 1 January 1996 and before 1 January 2012.

[30] Act of 1986, later Act of 17 August 1990/694. The existence of the medicines scheme, and its success, was a significant reason why Finland delayed implementing the Product Liability Directive until 1990: there was in practice simply no urgent need for strict liability, since the scheme provided an indemnity for drug claims and there were no significant claims for other products. The Finnish government had originally proposed that the medicines scheme should be statutory, and had drafted an Act, but the pharmaceutical industry pre-empted that by negotiating and establishing voluntary arrangements with the Finnish insurance sector. This development was influenced by economic considerations: it had been calculated that a statutory scheme would have cost the equivalent of €8 or €9 per thousand of population, whereas the cost of the voluntary scheme is around €1 per thousand.

[31] In 2004, there were four such companies in the pool.

[32] Pharmaceutical Injuries Insurance, General Terms and Conditions, effective as of 1 January 2004, cl 11.

[33] Under the Arbitration Proceedings Act (967/92).

[34] Pharmaceutical Injuries Insurance (n 32) cl 12.

[35] For the terms and conditions of insurance, see www.laakevahinko.fi/in-english. The previous Pharmaceutical Injury Pool does not accept new claims after January 2012 except for Pandemrix vaccine-related narcolepsy claims (see below).

loss adjuster Crawford. In 2015, claims handling was brought back in-house as an efficiency measure. In the new company the premiums are better able to be kept at a reasonable level than they were under the previous pool.

When the new company was established it took on all new claims from that point forward, with the exception of Pandemrix vaccine claims, which were still being handled by the old pool as a residual function. There are, therefore, currently two parallel schemes in operation dealing with defined, different claim types.

The old terms and conditions of the scheme (effective 1 January 2007) were replaced by new terms and conditions (effective 1 January 2015). There are some key differences between the old Pool and the new company, as detailed in Table 7.15.

Table 7.15: Major differences between the old and new Finnish pharmaceutical insurance schemes

Old provisions		**New provisions**	
2/2007	Definition of pharmaceutical included intrauterine devices	2/2015	Definition of pharmaceutical includes all intracorporeal products
11/2007 and 12/2007	Defined the composition of the Pharmaceuticals Injuries Board and the appeal mechanism	11.1/2015 and 12/2015	New Board is the Finnish Financial Ombudsman Bureau (FINE) Complaints Board Number 5 (Pharmaceuticals) The composition of the Board and the rules for complaints are different
12.2.1/2007	Right to appeal to Court of Arbitration		Deleted in the new rules
—	No personal claims limit existed	8/2015	Claims limit of €4 million per person or €30 million total claim per year against the company.[a] If this is insufficient to satisfy all those entitled to compensation, then all compensation paid shall be reduced in equal proportion[a]

[a] At present there is, according to paragraph 8 of the terms and conditions, pharmaceutical injuries insurance in force as of 1 January 2017.

Liability for any each injury suffered by a person is limited to €4 million over the individual's lifetime. Once the liability is clear, then the value of the claim is reserved; if it appears that the total cost now and in the future will be over €4 million, then the monthly payments are decreased so that the total lifetime payment does not exceed the € 4 million limit.

The scheme has a total annual payout limit of €30 million for all injuries that are reported during that calendar year. Should the amount of compensation awarded in a calendar year exceed €30 million, then all compensation paid shall be reduced by an equal proportion. This is only likely to occur if a serial injury (multiple claims for the same drug) were to occur.

In order to maintain financial stability the scheme holds reserves and takes out reinsurance. At present the reinsurance is in place for individual claims valued at over €3 million and for serial claims up to €30 million per year. As the company is highly solvent, it may be able to build greater reserves, which would make the reinsurance premium lower. The taking out of reinsurance is a scheme risk management strategy; even if a claim was covered by the reinsurance policy the individual claimant would have their claim processed by the scheme and be paid by the scheme in the usual way and the claimant would never have any dealings with the reinsurer.

Claims under the Finnish Cooperative for the Indemnification of Medicine-Related Injuries

The Finnish pharmaceutical injuries insurance is a top-up or secondary compensation scheme. Any other statutory benefits and compensation from insurance schemes are subtracted from the loss due to the injury and the pharmaceutical injuries insurance pays the claimant the remainder.

Under the current arrangements for handling new pharmaceuticals claims, pharmaceutical injuries insurance is taken out by the Finnish Cooperative for the Indemnification of Medicine-Related Injuries. This covers unexpected adverse effects sustained by users of pharmaceuticals sold or supplied for consumption in Finland by members of the Cooperative. As this is a voluntary scheme, injured individuals can only claim against this scheme if the manufacturer or supplier of the pharmaceutical in question is a member of the Cooperative.

The insurance for pharmaceutical injuries insurance is issued by the Finnish Mutual Insurance Company for Pharmaceutical Injury Indemnities. The insurance company handles the claims and pays accepted claims.[36]

To qualify for compensation under the pharmaceutical injuries insurance, the damage must be due to a pharmaceutical as defined in the Medicines Act, or an intrauterine contraceptive device. Injuries due to traditional herbal and homeopathic preparations are not covered.

Pharmaceutical injuries insurance covers bodily injuries caused by the use of a pharmaceutical as medication or investigational drug. Pharmaceutical injuries are by definition personal injuries caused by a pharmaceutical. There is no prescribed form of injury—any bodily or psychiatric injury caused by a pharmaceutical can be considered. The insurance covers investigational drugs used in clinical trials in Finland providing that the trials are conducted or sponsored by signatories to the insurance contract in accordance with legislation governing medical research. If a comparator used in a trial has been made by another manufacturer, the comparator is covered by the insurance if it is supplied for consumption in Finland by a signatory to the insurance contract.

A likely causal connection between the use of the pharmaceutical and the adverse effect must be demonstrated. The use of a pharmaceutical must be the most likely cause of the injury. Where there are multiple possible causes of the injury, compensation will only be

[36] www.laakevahinko.fi/in-english/.

paid if the pharmaceutical is deemed more likely to have caused the injury than all of the other causes. This causal connection is established with the help of information contained in adverse effects registers (and compiled works published therein) as well as in medical journals. Other factors considered for this purpose include the person's overall health and diseases, treatment received, and any other medication taken.

Since the use of pharmaceuticals invariably involves the possibility of adverse effects, an injury will not be covered if it results from medically justified risk-taking in the treatment of a disease or injury which, if left untreated, would be life-threatening or may lead to severe physical injuries. Insignificant injuries are not covered.

An injury will not be covered if the adverse effect of the pharmaceutical is reasonably tolerable given the nature of the disease or injury that is being treated, the health of the injured person, the extent of the injury, the ability of specialists to predict the effects of the pharmaceutical and other similar circumstances.

The pharmaceutical injuries insurance does not cover failure of the drug to have the intended effect, such as failure to cure the disease or injury that was being treated.

The pharmaceutical injuries insurance only covers adverse effects caused by the medication itself. Damage resulting from incidents when a pharmaceutical has been incorrectly prescribed or administered are not be covered by this insurance, but may be covered under the Finnish Patient Insurance Centre.

To qualify for full compensation under the pharmaceutical injuries insurance, the use of the pharmaceutical must have resulted in a loss of functional ability, lasting 14 days or more without interruption, or in a permanent physical injury or illness or death. If the injury is not severe enough to meet the above criteria, compensation will only be available for actual expenses incurred and loss of income suffered, providing that these together amount to more than €85 after deduction of any amounts that may be due out of public or statutory compensation schemes.

The scheme specifies that compensation shall be paid in accordance with the Damages Act (412/74) but 'in such a way that compensation in respect of pain and suffering and any permanent impediment shall be calculated in accordance with Guidelines applied by the Pharmaceutical Injuries Board'.[37] Standards and guidance issued by the Traffic Accident Board might also apply. All expenses and losses relating to the injury are primarily covered in full. The claimant is expected to take all possible steps to prevent further injuries and losses.

The period of limitation has recently been changed in the terms and conditions. According to the new definition; any claim for compensation in accordance with these terms and conditions shall be submitted to the Insurer within a year of the date on which the person claiming compensation became aware of the validity of the Insurance, the injury caused by the pharmaceutical or blood product involved and the injured event. The claim form for compensation must be submitted no later than within 10 years from the occurrence of the injured event.

The policy conditions also exclude injuries resulting from use of pharmaceuticals acquired illegally or from misuse of pharmaceuticals.

The scheme only covers any necessary additional expenses and losses arising from a pharmaceutical injury. Expenses or losses that would have accrued without the injury are not reimbursed. Compensation is not paid for the treatment costs of the disease or injury

[37] Pharmaceutical Injuries Insurance (n 32) cl 7.

that was originally being examined or treated, or for the cost of the pharmaceutical that caused the injury.

The Finnish Financial Ombudsman Bureau (FINE)[38] provides an advisory service related to the pharmaceutical injuries insurance to private customers. The Bureau also assists the customer in the negotiations and investigations related to the case. The customer will also be provided with information on legislation, contract terms and the past decisions issued by the Insurance Complaints Board. Moreover, the customer will be advised of the different ways to solve the dispute.

The new Board is a private company, in 2011/12 a post-tax profit of €178,341.79 was made, which was used to capitalise the company; in 2012/13 an identical profit was again used to capitalise the company. This arrangement does not have any state funding.

Type of Loss Compensated

The types of loss compensated by the Finnish Drug Insurance Indemnity Scheme are summarised in Table 7.16.

Table 7.16: The types of loss compensated by the Finnish Drug Insurance Indemnity Scheme

Type of loss	Available compensation
Medical treatment expenses	Compensation is paid for any necessary extra medical treatment expenses caused by the injury, eg extra inpatient and/or outpatient charges, doctor's fees, costs for home nursing and laboratory tests arising from the injury, medicinal expenses and physical treatment In general, medical treatment expenses are reimbursed at the cost level of public healthcare. In exceptional cases, where there is a specific reason for using private healthcare services, compensation for that will be granted
Other expenses arising from the injury	Compensation for other expenses attributable to the injury, eg travel expenses for extra treatment, the cost of extra wear and tear on clothing, and extra costs arising from a period of home care These expenses are paid as they are incurred, eg additional clothing compensation is paid if the injured party incurs extra wear and tear on clothing because of treatment due to a prosthesis, forearm or underarm crutches, a wheelchair, different compression bandages Care allowance and compensation for increased home care expenses are awarded to cover the extra costs incurred during home care. No compensation is paid for the duration of institutional care
Temporary handicap (pain and suffering)[a]	Compensation for temporary handicap reimburses the injured party for the pain and suffering and other discomfort caused in the acute phase immediately after the injury. The amount of compensation is affected by a variety of factors, including the severity of the disability or illness, the recovery period, any treatment procedures performed, and the need for hospital care

(continued)

[38] www.fine.fi

Table 7.16: ***(Continued)***

Type of loss	Available compensation
	Should the pain become chronic it will be compensated as a permanent handicap
Permanent handicap	A permanent functional handicap (invalidity) caused by pharmaceutical injury is potentially compensated as a permanent handicap. In accordance with a decision made by the Ministry of Social Affairs and Health (1012/1986), handicaps are divided into 20 handicap categories. The amount of compensation is affected by the nature of the disability and the age of the injured party and, in some special cases, by the injured party's occupation and hobbies The amount of compensation paid for each handicap category is determined on the basis of guidelines of the Traffic Accident Board.
Permanent cosmetic handicap	If a pharmaceutical injury causes a permanent handicap to a person's appearance, eg a scar, compensation may be available for a cosmetic handicap. When determining the compensation, the extent, shape, colour, location and visibility of the scar are taken into account. The compensation quantum is influenced by the age of the injured party: the compensation paid is lower for older parties. The amount of compensation paid is determined on the basis of guidelines of the Traffic Accident Board (Liikennevahinkolautakunta)
Loss of income	If an employed or self-employed individual loses income due to incapacity for work caused by a pharmaceutical injury, the losses can be reimbursed from the pharmaceutical injury insurance. Any assessment of the ability to work considers the injured party's remaining ability to work and possibilities for vocational rehabilitation. When the assessment is made, the effect of the basic illness or disability on the ability to work is also reviewed. Compensation can also be paid on the basis of partial incapacity for work. The injured party must investigate opportunities to enhance their working capacity by medical or occupational rehabilitation This type of compensation is reimbursed according to actual losses. The basis of calculating the amount of compensation is an estimate of the earned income the injured party would have received had the pharmaceutical injury not occurred. Compensation is only paid to top up any reduction in income resulting from the pharmaceutical injury to the pre-injury level. Provision from statutory or other benefits will be deducted from the individual's pre-injury income to determine the amount of compensation due from the pharmaceutical injury insurance. Daily allowances and pensions paid on the basis of, for example, the Sickness Insurance Act, the Traffic Insurance Act, the Employment Accidents Act, Employment Pension Acts and the National Pension Act as well as the Patient Injuries Act are considered when calculating compensation.

(continued)

Table 7.16: ***(Continued)***

Type of loss	Available compensation
Death	Direct funeral expenses and other costs related to burial can be reimbursed up to a reasonable amount. Compensation can be paid for expenses arising from, for example, the burial plot, coffin, headstone, funeral service, flowers, newspaper advertisements and bereavement-related clothing. No compensation is paid for estate inventory costs because, according to established practice in tort liability legislation, such costs are not regarded as compensatable expenses relating to burial.
	Subsidies received from other instances for the covering of burial costs can be deducted from the compensation paid from the pharmaceutical injury insurance. Burial expenses are reimbursed to the deceased's estate or to the person who has paid the costs from personal funds.
	In case of death, required financial support is paid to the person(s) entitled to it; in other words, the surviving spouse and minor children and, in some cases, children under the age of 21 who are students. Compensation is only paid insofar as the recipient's other income and benefits paid out on the basis of other legislation, eg family pensions, do not cover the necessary support.

[a] The term 'compensation for pain and suffering' was changed to 'compensation for temporary handicap' following amendments of the Tort Liability Act in early 2006.

Claims Process

The Claims Unit of the Finnish Mutual Insurance Company for Pharmaceutical Injury Indemnities assesses the notice of injury, investigates, and sends a written decision to the claimant. If the injury is compensatable, the decision is accompanied by information as to what kind of compensation or benefits are due under the scheme and the decision is accompanied by a claim form on which the claimant may detail the compensation being claimed.

The claims-handling process is made up of the following steps:

(a) Notice of Injury

The claim process is started by sending a completed notice of injury form by mail to the Finnish Mutual Insurance Company for Pharmaceutical Injury Indemnities. An electronic Notice of Injury Form is available from the Pharmaceuticals Injury Indemnities website, which must be completed, signed and returned by post, as electronic submission is not yet possible (but will be possible in the near future). No receipts or other documentation evidencing loss are required at this stage.

The notice of injury form is signed by the claimant where she has the capacity to do so. In cases where the claimant lacks capacity, the notice of injury may be signed by the patient's guardian (usually the parents of a child) or, in the case of death, by a beneficiary of the deceased's estate. By signing the notice of injury, the claimant consents to the treating healthcare institution or another body providing the Finnish Mutual Insurance Company for Pharmaceutical Injury Indemnities with the information and documentation needed to handle the claim.

(b) Information Gathering

Upon receipt of a new notice of injury, the Finnish Mutual Insurance Company for Pharmaceutical Injury Indemnities records the case and allocates it a handling code. Once the injury notice is recorded, the claim handler acquires sufficient documentation to enable a thorough investigation of the claim, on the basis of information recorded on the notice of injury. The treating institution is asked for the required documents even if the notice of injury was accompanied by the patient's records. A copy of the notice of injury is sent to the treating institution for their information. This process can be time consuming. It is deemed worthwhile in order to obtain as extensive a set of documents as possible to facilitate the claims-handling process.

(c) Statement from a Medical Adviser

Once the required patient documentation has been obtained, the case is assessed from a medical perspective. This is usually in the form of a specialist opinion from a medical adviser.

(d) Decision

On receipt of the information and the specialist opinions required for deciding the case, the Finnish Mutual Insurance Company for Pharmaceutical Injury Indemnities issues a written decision on the case. The decision records whether the reported injury is compensatable under the terms and conditions governing pharmaceutical injuries insurance. The acceptance or denial is sent to the claimant and to the policyholder (the Finnish Cooperative for the Indemnification of Medicine-Related Injuries). The decision is accompanied by appeal instructions.

(e) Payment of Compensation

If the claim is accepted as eligible for compensation, the decision informs the claimant as to what kind of compensation is paid and attaches a claim form to this acceptance. The claimant is given further details of next steps to be taken. If it is possible on the basis of the information supplied, compensation for pain and suffering is paid at the time the notice of injury is accepted.

(f) Complaints

Anybody dissatisfied with the decision taken by the Finnish Mutual Insurance Company for Pharmaceutical Injury Indemnities on a claim for compensation may seek redress. Appeal instructions are attached to the company's decision. Essentially there are three main options: requesting rectification from the Finnish Mutual Insurance Company; utilizing the Insurance Ombudsman (including its Insurance Complaints Board); or taking the case to court.

The claimant can address the request of rectification to the Finnish Mutual Insurance Company for Pharmaceutical Injury Indemnities if there is significant new information in the case at hand, relevant to the compensatability of the injury. As far as possible, the new information should be included in the request for rectification or the claimant should indicate the way to access such information.

An appeal against the decision of the Finnish Mutual Insurance Company for Pharmaceutical Injury Indemnities can be lodged with the Insurance Complaints Board operating under FINE (Board No 5). The Board will issue a recommended solution for the cases it receives. There is no fixed application form when applying to the Board for a recommendation; a free-form letter is used. It is recommended that a photocopy of the indemnity decision and any other relevant information should be attached to the request.

In 2014, 10 cases were appealed to the Board, with amended decisions (in favour of the claimant) given in 3 cases. In 2013, 9 cases were appealed to the Board, with amended decisions (in favour of the claimant) given in 3 cases.

The final option for complainants is to lodge a case in the courts; this is very seldom used. Since the new scheme was introduced in 2012, only a single case has been brought to courts.

Statistics

The following statistics are separated into two elements: those since the 2012 reforms under the new captive company, and those from the pre-2012 Pool arrangements. Since the post-2012 arrangements have so far only produced limited data, the older Pool data has been included for comparison.

(a) Post-2012: The Current Arrangements

Recent information is given in Table 7.17 and Figure 7.9. The compensation paid out by the new scheme is slightly lower than was previously paid out under the old Pool. This is thought to be due to changes in the way claims are processed, leading to slightly higher rejection rates; the intention is to investigate this and to provide consistency between the decision of the old and new arrangements

Table 7.17: Claims made to the Finnish Pharmaceutical Insurance Pool 2012–15

Year	Number of claims received (and decided)	Percentage of decided claims compensated	Claims incurred (€000s)	Premium income (€000s)	Operating expenses (€000s)
2012	98 (49)	30%	930	4,800	772
2013	164 (141)	40%	1,103	4,800	639
2014	187 (175)	–	1,214	5,120	781
2015	230	–	2,261	5,120	600

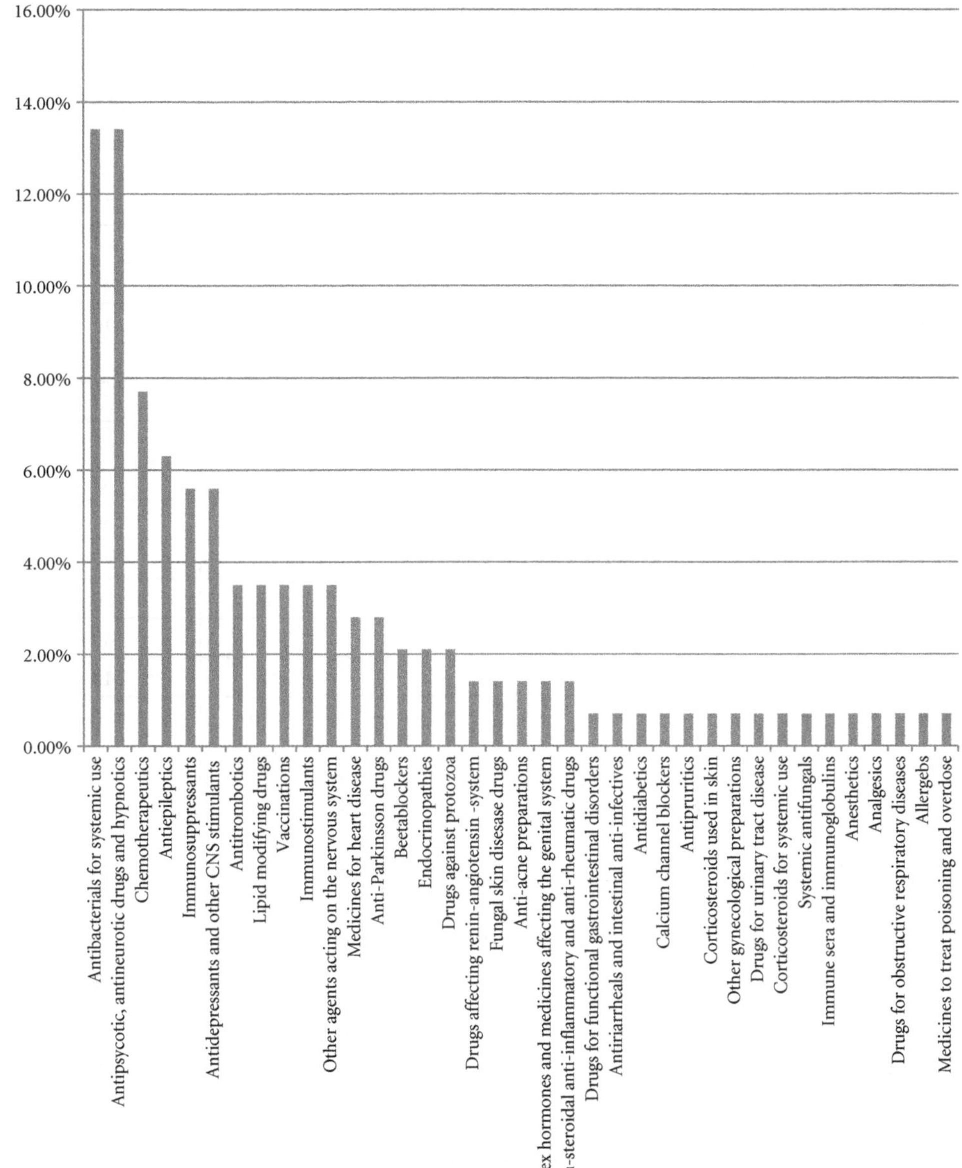

Figure 7.9: Breakdown of applications to the Finnish Pharmacuetical Insurance scheme by drug category for 2014

(b) Pre-2012: The Previous Pool Arrangements

Tables 7.18 and 7.19 show the information from 1995 to 2011. The number of claims received annually, and the success rate, were fairly stable over this period.

Table 7.18: Claims made to the Finnish Pharmaceutical Insurance Pool 1995–2011

Year	Number of claims received	Percentage of claims compensated	Total payments (€000s)	Premium income (€000s)	Administrative expenses (€000s)
1995	222	45%	321	857	56
1996	250	49%	409	620	26
1997	192	47%	549	763	43
1998	189	47%	719	852	27
1999	186	55%	608	694	46
2000	172	54%	1096	856	77
2001	162	58%	890	893	72
2002	206	59%	1042	919	65
2003	224	58%	928	879	55
2004	254	52%	913	1262	54
2005	282	46%	1170	1584	92
2006	227	57%	1035	1802	70
2007	218	56%	1473	2373	87
2008	211	50%	1403	2948	43
2009	167	61%	1404	3600	77
2010	251	46%	1187	4380	104
2011	352	58%	1551	5040	109

Table 7.19: Claims reported to the Finnish Pharmaceutical Pool 1995–2015. Figures shown in parentheses are for Pandemrix vaccine injuries.

	Drug injuries entitled to indemnity	**Minor drugs injuries**	**No drug injury**			**Total claims**	**Proportion of claims made by outcome (%)**				**Outcomes for non-drug injuries (%)**		
			Injury should have been tolerated	Other claims Reported	Total		Drug injuries entitled to indemnity	Minor drugs injuries	No drug injury	Total	Injury should have been tolerated	Other claims reported	Total
1995	100	21	39	62	101	222	45	9	45	100	39	61	100
1996	123	26	27	74	101	250	49	10	40	100	27	73	100

(continued)

Table 7.19: *(Continued)*

	Drug injuries entitled to indemnity	Minor drugs injuries	No drug injury			Total claims	Proportion of claims made by outcome (%)				Outcomes for non-drug injuries (%)		
			Injury should have been tolerated	Other claims Reported	Total		Drug injuries entitled to indemnity	Minor drugs injuries	No drug injury	Total	Injury should have been tolerated	Other claims reported	Total
1997	91	15	24	62	86	192	47	8	45	100	28	72	100
1998	89	9	22	69	91	189	47	5	48	100	24	76	100
1999	103	17	16	50	66	186	55	9	35	100	24	76	100
2000	95	21	8	48	56	172	55	12	33	100	14	86	100
2001	94	12	6	50	56	162	58	7	35	100	11	89	100
2002	122	20	9	55	64	206	59	10	31	100	14	86	100
2003	132	15	15	62	77	224	59	7	34	100	19	81	100
2004	134	19	22	80	102	255	53	7	40	100	22	78	100
2005	131	18	21	112	133	282	46	6	47	100	16	84	100
2006	130	7	22	68	90	227	57	3	40	100	24	76	100
2007	122	21	19	56	75	218	56	10	34	100	25	75	100
2008	106	11	31	63	94	211	50	5	45	100	33	67	100
2009	102 (101)	9 (7)	11 (11)	45 (44)	56 (55)	167 (163)	61 (60)	5 (4)	34 (33)	100 (98)	20	79	98
2010	115 (92)	8 (8)	40 (39)	88 (78)	128 (117)	251 (217)	46 (37)	3 (3)	51 (47)	100 (86)	30	61	91
2011	205 (110)	11 (11)	43 (43)	93 (68)	136 (111)	351 (232)	58 (31)	3 (3)	39 (32)	100 (66)	32	50	82
2012	15 (54)				34 (18)	98 (73)	(74)		(25)				
2013	56 (32)				85 (28)	164 (61)	(52)		(46)				
2014	(30)				(17)	187 (48)	(63)		(35)				
2015	(7)				(22)	230 (31)	(23)		(71)				

Conclusions

The current pharmaceutical compensation arrangements in Finland are in their very early stages, so meaningful analysis on performance is not appropriate. However, the current system derives from a long-standing stable compensation mechanism, and there is no reason to think that this performance will not continue.

8

Norway

A. General Benefits

The first line of support for those injured is the social system (Folketrygden). Mandatory insurance for this is required by an employer. Injured persons have direct claims against insurers under a variety of measures, including those pertaining to motor vehicles, the Act on Industrial Injury Insurance 1989, and the rules on medical (including pharmaceutical) injuries. Some citizens may have additional private accident or health insurances, which are mainly related to ownership of a house, car, boat, etc. It has been estimated that over 95% of all persons who suffer injuries that are eligible for compensation have their loss compensated irrespective of whether the person who caused the injury is legally liable.[1]

B. Norway's Patient Compensation System

Structural Arrangements

The Norwegian Patient Compensation System (NPE) handles injuries caused by both healthcare and medicines.[2] The NPE is a public body under the Ministry of Health and Care Services. The rationale for introducing the scheme in 1988 was as follows. It was well established that few claimants succeeded in getting compensation through the tort law system, and by the 1980s this situation had provoked political unrest. It had been noted that Sweden had introduced a patient insurance scheme in 1975. Norway decided to try a compensation system scheme for 3 years from 1988 on a temporary trial basis in relation to injuries in the main wards of hospitals. The initial intention was to convert the scheme into a law after 3 years, but it in fact expanded over time to include other areas, such as psychiatric wards, general practitioners, and so on. The law was finally passed in 2001[3] and came into force on 1 January 2003, initially for all public healthcare services, which cover 90% of all healthcare in the country. It took further time to include private healthcare, since the insurance companies providing cover for private healthcare were not interested in joining an insurance-based public scheme, so they eventually were cut out and the system was changed to a scheme funded by contributions by private providers in 2009. Since then, the scheme

[1] P Lodrup, 'Norwegian Report', AIDA XIth World Congress, New York, 22–25 October 2002, 117–50). Dansk Selskab for Forsikringsret [Danish Section of the International Association of Insurance Law].

[2] See generally O Molven and J Ferkis (eds), *Healthcare, Welfare and Law* (Oslo, Gyldendal Akademisk, 2011).

[3] https://lovdata.no/dokument/NL/lov/2001-06-15-53/KAPITTEL_5#§19

has covered all healthcare delivered in Norway. The largest area of private care is dental services, plus there are some physiotherapists, chiropractors and certain private medical specialties. Some private hospitals exist, and that sector is in fact increasing as a result of the strong Norwegian economy, enabling people to purchase private care and avoid queues.

When it was introduced in 1988, doctors were somewhat suspicious of the new approach, but that attitude changed fairly quickly and there is a high degree of support for, and trust in, the system. The general consensus is that the system enables patients to have trust in medical professionals and to cooperate in a fruitful manner.

The Norwegian Drug Insurance scheme was included as a chapter in the 1988 Act on product liability.[4] NPE handles drug insurance claims as an assignment on behalf of the Drug Insurance scheme (regulated by an agreement approved by the Ministry of Health and Care Services). The drug insurance side of NPE is comparatively small, with around 30–50 claims a year; these are paid by the Drug Liability Association. Some confusion arises here. Claims for drug injuries can be made to NPE under either the Patient Insurance Act or the Product Liability Act. The trigger for payment is fairly similar under both Acts, but such claims are usually paid under the former scheme as an exception to the normal rule, namely under the reasonability rule. However, NPE is required to process claims made under the Patient Insurance Act at its own (the state's) cost, whereas it can charge the Pharmaceutical Association for its costs for claims under the Product Liability Act, but not where a claim is denied (as most are). Hence denied claims are funded by a public subsidy, which NPE regards as unfair. NPE has urged the industry to provide more information to the public on the existence of the Drug Insurance scheme, so as to raise its profile. Statistics on drug claims are not sent to the Drug Regulatory Authority, although some statistics on claims are available on the Drug Insurance scheme website.[5]

Funding

Public providers make contributions on an annual pay-as-you-go basis. Over 82% of the public funding comes from the four regional Health Authorities that own the public hospitals. The remaining 17% is paid by all of the 428 municipalities, based on number of inhabitants, and the 19 counties, in relation to the small level (0.2%) of public dental services. In contrast, private providers make annual contributions, and NPE is required to maintain reserves sufficient to cover all claims in a year. The accounting required is complex, and providers are grouped into different risk groups.

Procedure and Conditions

In order to claim compensation from NPE, a special injury report form has to be completed and submitted. The form is widely available from the Norwegian health service, or can be downloaded from the NPE website.[6] From April 2015, claims could be submitted online, and within a few months over half of claims were submitted that way. In general, there is no

[4] Produktansvarsloven as LOV-1988-12-23-104 from 1988. Act No 430 of 10 June 2003 disapplied the arrangements for vaccine injuries caused from 1 January 2004, from when vaccine injuries were included in the general medicines arrangements.

[5] www.npe.no/en/.

[6] www.npe.no.

need for legal advice in order to apply for compensation. In normal circumstances, the cost of legal advice will not be covered for the period during which NPE is looking into the case to decide whether there is entitlement to compensation.

NPE has 160 staff, organised as shown in Figure 8.1. On receiving the completed form, claims are first screened and all documentation relevant to the case is collected,[7] including medical records and statements issued by the institution that provided the treatment, before the file is passed to 1 of 4 claims assessment teams[8] to determine if the claimant is entitled to compensation and, subsequently to 1 of 2 compensation assessment teams,[9] to determine the amount of compensation payable. Assessments from experts may be required. Over 90 agreements exist with experts, covering 25 medical specialties, including some dentists and a nurse. These agreements are subject to EU public procurement rules, which entails considerable administration of the public tender process, but has proved to be satisfactory in practice. One advantage is that whilst it may be difficult to terminate a relationship with an unsatisfactory provider under general employment law, the public procurement regime provides for a 3-month termination period. These 90 standing agreements cover 85% of the medical need. The remaining 15% are arranged on an ad hoc basis from experts throughout the country, often selected for their proximity with a claimant, and might involve 500 or 600 people over a 2-year period. Four teams handle liability issues.

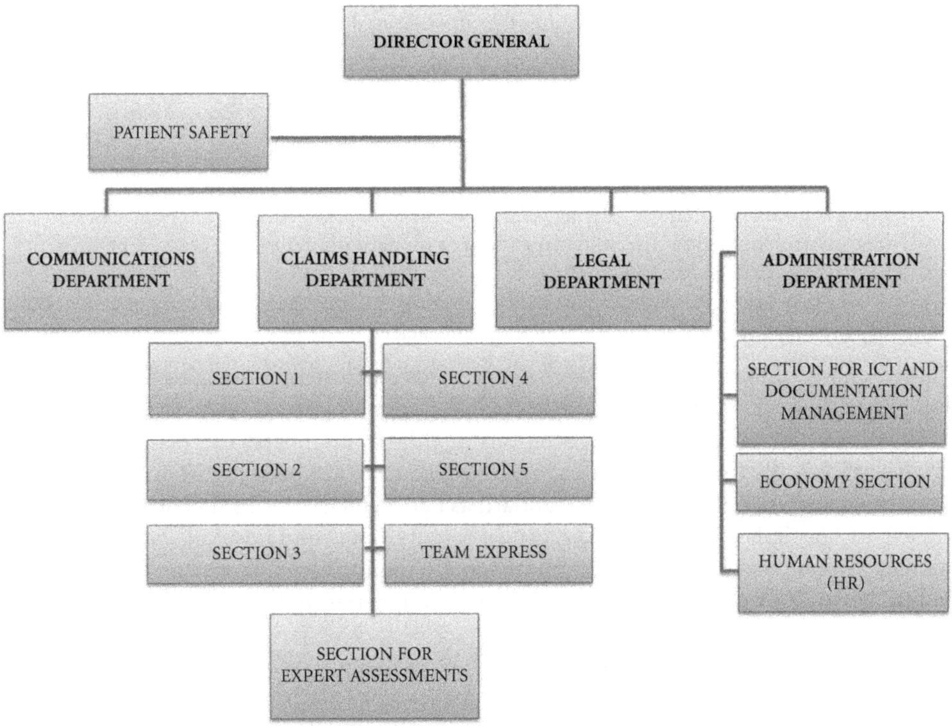

Figure 8.1: Organisation of the Norwegian patient compensation system

[7] The initial claims handling happens in seksjon 5.
[8] Seksjon 1, 2, 3 or 4.
[9] Seksjon 2 (which does both 'liability assessment' and compensation assessment) and seksjon 6.

NPE's published information[10] begins by stating: 'All medical treatment involves a risk, and sometimes things go wrong.' Eligibility for compensation from NPE is based on fulfilling four conditions, which are stated as follows:

1. *The injury must have been caused by the treatment.*
   You must have suffered an injury caused by treatment, examination, diagnosis or follow-up provided. An injury may be temporary or permanent. If the injury was caused by the ailment for which you received treatment, you are not entitled to compensation.
2. *The injury must have been caused by failings in the treatment provided.*
   For you to be entitled to compensation, your injury must have been caused by failings in the treatment you have received. If the injury sustained is not caused by failings in the treatment, you will not normally be entitled to compensation.
   In some extra special cases, it will be possible to claim compensation even if there have been no failings. This happens if the injury is of a particularly significant or unexpected nature.
3. *The injury must have caused financial loss.*
   The injury must have caused financial loss. The intention is for the compensation to cover the increased costs incurred, for medical treatment, medication, transport etc. You may also be entitled to compensation for loss of earnings or the loss of a breadwinner.
   If your financial loss is less than NOK 10,000, you will not be eligible for compensation from NPE. Instead, you should contact the institution where you received the treatment.
   If you have had a bad experience or suffered temporary pain, but no financial loss, you are not entitled to compensation. A compensation award can never be granted as a means of redress. In some cases, you may be eligible for compensation even if you have not incurred a financial loss, e.g. if you have sustained a permanent and significant injury. This is referred to as 'permanent injury compensation'. The permanent injury compensation is intended to compensate for a diminished quality of life.
   There are some exceptions where compensation may be granted where there has been no financial loss, but these will usually be confined to cases where there is a permanent and significant injury. 'Permanent' usually means that the injury lasts for a period of at least ten years. 'Significant' means that the injury must lead to medical disability of at least 15 percent.
4. *The injury must not be too old.*
   You may apply for patient injury compensation no later than three years after you ought to have understood that your treatment, or the lack of treatment, is the cause of your injury.

The primary liability rule requires there to have been a failure in connection with the provision of health services, even if no one is at fault. Something irregular must have occurred, and it is considered whether the expectations that the injured party can reasonably require of the service at the time of delivery of care have or have not been met. This indicates that it can be correct to apply a different standard as between large and small hospitals, general and expert hospitals and, for example, general practitioners and hospitals/experts. The terminology of fault is avoided, and the approach is consciously referred to as 'no blame'. The leading Norwegian academic Professor Aslak Syse has referred to the approach as 'culpa light'.[11]

Liability is strict in relation to technical failure, contagion or infection, and vaccination. Even if there is no basis for liability under the main rule or the 'strict liability rule', compensation may be given in exceptional cases where the patient has suffered especially extensive

[10] www.npe.no/no/Om-NPE/Brosjyrer/Information-in-English/.

[11] A Syse, M Kjelland and RG Jorstad, *Pasientskaderett* (Oslo, Gyldendal,Akadmisk 2011) 597 sider (no 1/2012).

and unanticipated damage. This exceptional 'reasonability' rule is applied in about 3% of approved cases.

As at the end of 2016, it takes an average of 260 days (8.5 months) to determine eligibility and then 370 days (12 months) to assess compensation. The goal is to achieve both segments within 210 days by the end of 2018. One impediment to faster turnaround is that the Public Administration Act gives every claimant the right to know everything that happens affecting his or her case, such as to be informed of the expert assessment and then to have the opportunity of 3 weeks to respond. This 'gold standard' of transparency has its roots in a traditional legal litigation model of safeguarding individual rights in an adjudicative process, but it causes significant delay that is not experienced by the private insurance schemes of other countries.

Claimants may appeal to the Patient Injury Compensation Board (PasientSkadeNemnda, PSN), which is a separate public body, composed of 5 people. The panel normally consists of 3 permanent members. Its leader is a judge and 1 of the other 2 members shall be a medical expert. In difficult cases, the leader can decide that the panel shall consist of 5 members. The panel has delegated to its secretariat to make decisions with no fundamental character. The appeal process is free.

NPE digitally passes its files over to the PSN, and this saves delay in any reinvestigation. Around 40% of NPE's decisions that find against the claimant are appealed, of which around 10% succeed. Around 20–22% of compensation sum decisions are appealed to the PSN, of which 25–30% are awarded an increased compensation sum. NPE views this process as providing useful oversight and feedback, and regards the success rate as a healthy level. One aspect of potential overall inconsistency is that the PSN does not see any of the claims that NPE approves. NPE has responded to general indications set by the PSN, for example where the PSN increased the award for birth injuries and NPE adjusted its general approach accordingly.

Claimants in the public healthcare system are free to start a case in court after having been to the PSN: it is mandatory for them to go to NPE and the PSN first, and the courts reject cases that attempt to go to them directly. In 2016 there were 103 such cases (around 1.5% of the total caseload in the system), but this low number is related to a reduction in capacity within the PSN in 2016 due to reorganisation and relocation of the organisation from Oslo to Bergen. Usually the PSN receives about 200 court cases annually. The PSN decision is upheld in about 70% of such cases. Technically, the claim involves suing the government, and the defence is handled by the PSN. Claimants who have had private care may go straight to court and sue the private provider, but they have to prove negligence and shoulder their own costs and the risk of the opposition's costs. It is very seldom that private healthcare providers are sued. It should be noted that when a claimant in the private healthcare system has had his or her case handled by NPE/PSN, he or she can also sue the government through the PSN.

As a public body, NPE is required to be open and it publishes extensive information. It publishes information down to the level of individual private doctors and dentists.

Contributing to Improving Healthcare Practice

As of the end of 2016, the NPE system has built up a database of over 67,000 claims, of which 24,000 were approved. Patient identities are anonymised. All healthcare workers can create an account and log into the healthcare statistics portal, where they will find patient injury numbers for their hospital, the region and the entire country. The NPE invites hospitals and researchers to use anonymised data (including statistical data) and—subject to obtaining consent from patients—detailed data to learn from the information. NPE has hired a full-time physician to work on making use of the information.

Quite a few research projects on the data have started. However, it may be necessary to obtain consent from patients before using their data, which can prove something of a barrier. An earlier study on birth injuries was undertaken in cooperation with the Norwegian Society of Gynaecologists, which is keen to collaborate with NPE. The findings have been published in three academic articles.[12] A project on gynaecology in cooperation with the Society of Gynaecologists is underway with a PhD-candidate at Trondheim University, in which 55% of the women who were asked to consent did so.

The work done by NPE on the use of data for harm reduction has been recognised by the Norwegian government and included in the White Paper on Patient Quality and Safety 2015, published in autumn 2016. This highlights the shift in position towards the increased use of patient injury data to contribute to increased patient safety.

Cooperation between all the four Nordic patient insurance bodies has commenced recently, giving the prospect of creating a database that is significantly more powerful than any of the four national populations, and covers 25 million people. A joint project is underway on gastroenterological surgery.

Statistics

Table 8.1 shows the number of claims for the period 2011–16. In 2016 there were 5,303 decisions by NPE, of which 1,420 (27%) were accepted. In 2016 5,628 claims were made; 1,295 more than 5 years earlier. In 2014 4,333 claims were made. This increase does not seem to be related to changes in the underlying level of medical safety, but rather to rising visibility of the scheme, fuelled by the focus of the media on healthcare statistics.

In 2016, NPE paid out €110 million in compensation. The highest individual award was €1.4 million. The level of payments is regarded as not exorbitant. It needs to be remembered that social security benefits are the first line of support, and such payments are deducted from compensation awarded by NPE. Total turnover in 2016 was €20 million, or around €3,520 per case. The budget process is undertaken in close cooperation with the supervising ministry, and NPE is under constant pressure to be forward-looking in becoming more efficient and increase productivity. A considerable amount of time is spent on compiling administrative reports under public administration requirements.

[12] S Andreasen, B Backe, RG Jorstad and P Oian, 'A Nationwide Descriptive Study of Obstetric Claims for Compensation in Norway' (2012) 91 *Acta Obstetricia et Gynecologica Scandinavica* 1191–95; S Andreasen, B Backe and P Øian, 'Claims for Compensation after Alleged Birth Asphyxia: A Nationwide Study Covering 15 Years' (2014) 93 *Acta Obstetricia et Gynecologica Scandinavica* 152–58; S Andreasen, B Backe, S Lydersen, K Øvrebø and P Øian, 'The Consistency of Experts' Evaluation of Obstetric Claims for Compensation' (2014) 122 *BJOG* 948, doi:10.1111/1471-0528.12979.

Table 8.1: Norwegian compensation claims 2011–16

	2011	2012	2013	2014	2015	2016
Number of claims	4,333	5,123	5,065	5,217	5,742	5,628
Accepted claims	1,419	1,363	1,384	1,504	1,492	1,420
Denied claims	2,010	3,137	3,416	3,729	3,747	3,883
Total compensation (NOK millions)	847	706	946	1062	933	987
Total turnover (NOK millions)	124	144	161	172	176	179
Average cost claim (NOK)	25,010	28,120	29,410	30,510	30,660	31,750

The main types of causation in claims for 2016 are shown in Table 8.2.

Table 8.2: Norwegian compensation claims 2016 by type of causation

Failure in connection with providing healthcare	83%
Failure in treatment	54%
Failure in diagnosis	26%
Contagion or infection	14%
The reasonability rule	3%

The breakdown of denied claims in 2016 was that no failure was found in 53%, no causal connection in 43%, and no economic loss in 4% of claims. The main types of medical specialties for claims in 2014 are listed at Table 8.3. Birth injuries constitute 2% of claims but 18% of total payments. They also give rise to lifelong human suffering, and affect the whole family.

Table 8.3: Decisions 2012–16 in the 10 largest areas of medical specialisation

Type	Accepted	Denied	Total
Orthopedic surgery	2,505	4,366	6,871
Tumours/cancer	1,131	2,057	3,188
Gynaecology and obstetrics	309	954	1,263
Odontology	470	713	1,183
Gastroenterologic surgery	415	693	1,108
Psychiatry	182	869	1,051
Cardiology	214	621	835
Neurology	188	642	830
Ophthalmology	165	527	692
Infectious diseases and venerology	138	295	433
Others	1,289	3,529	4,818
Total	**7,006**	**15,266**	**22,272**

C. Norwegian Drug Insurance

General Description

All manufacturers and distributors of medicines which are distributed in the country, and persons conducting clinical trials of drugs, are required by law to be members of the Medicines Liability Association (Legemiddelans-varsforeningen).[13] Failure to register as a member of the Association when obliged to do so exposes the company to unlimited liability for any claim for compensation. The Association is required to take out insurance (Medicines Insurance) for the benefit of injured persons.[14] Premiums are negotiated annually, although special rules relate to clinical trials.

Claims are made to NPE, who by agreement with the Drug Insurance scheme is engaged as the claims adjuster. NPE decides on the validity and quantum of claims. The Act does not provide for an appeal against the decision of the adjuster. A claimant who is dissatisfied with the outcome may bring a claim in the courts against the insurance scheme, since the Drug Insurance scheme has stated that it will not adhere to decisions made by the Norwegian Financial Services Complaints Board. The normal legal rules of quantum apply as they would to a standard product liability case.[15]

Norway created a captive insurer in 2004, largely because of difficulties in purchasing reinsurance. The total annual cost for medicines insurance is around NOK31 million (€3.88 million), or €0.85 per capita.

Statistics

Claims made from 1994 are shown in Table 8.4. The number of pharmaceutical injury claims dropped in 2011–13 but rose again in 2014 and 2015. In 2015 there were 107 new pharmaceutical cases reported, up from 64 in 2014. In 2015 93 cases were completed (compared to 50 in 2014). Interestingly the value of compensation decreased from NOK8,073,000 million in 2014 down to NOK7,738,700 million in 2015, indicating a decrease in the average award value.

The Medicines Liability Association believes that the increase in claim numbers may have been caused by a change in the routines of the authority that handles the claims for both pharmaceutical injuries and patient injuries (NPE). It seems that a larger proportion of claims than previously are filed as patient claims first, and only reclassified as pharmaceutical injury claims at the time of settlement. In this way, unsuccessful claims will probably never be reclassified or reported.

There has been a significant reduction in the time to settlement from around the mid-2000s.

[13] As defined in ss 3–4 of the Product Liability Act, No 104/1988 of 23 December 1988. The Medicines Insurance may claim recourse against a person who has not satisfied his obligation to be a member of the Association or to have approved insurance: ss 3–8.

[14] Product Liability Act, ch 3.

[15] Product Liability Act, s 1, para 3, referring to Compensation Act of 13 June 1969, ss 2–5.

Table 8.4: Claims made to the Norwegian Drug Insurance scheme 1989–2014

Year	Number of claims received	Payments made (NOK) as at 31 December 2014	Sums reserved for claims reported but not settled (NOK) as at 31 December 2014	Premium income paid to insurers (NOK millions)	Premium collected from producers: % of turnover
1989 (6 months)	2	14,531	0	3,500,000	0.15
1990	6	2,041,663	0	7,000,000	0.31
1991	29	764,916	0	7,000,000	0.29
1992	19	688,300	0	4,000,000	0.18
1993	16	5,765,540	0	5,750,000	0.20
1994	25	6,332,930	0	4,030,000	0.20
1995	25	3,486,526	0	3,520,000	0.12
1996	52	15,336,949	0	4,880,000	0.20
1997	57	6,073,073	0	8,510,000	0.25
1998	64	19,820,813	0	10,600,000	0.30
1999	50	8,895,382	0	5,300,000	0.30
2000	63	9,265,118	0	5,200,000	0.25
2001	56	8,338,254	0	5,000,000	0.20
2002	43	27,751,424	0	5,900,000	0.20
2003	48	13,244,969	0	8,000,000	0.18
2004	97	15,769,836	0	11,000,000	0.30%
2005	201	31,160,648	0	16,000,000	0.30
2006	116	18,045,390	0	16,000,000	0.35
2007	112	31,548,366	0	16,000,000	0.35%
2008	77	7,898,324	7,000,000	18,000,000	0.35
2009	66	18,614,517	1,415,000	19,000,000	0.30
2010	80	7,003,595	1,108,718	19,000,000	0.25
2011	30	3,363,869	1,891,020	17,000,000	0.20
2012	41	1,951,623	9,624,855	15,000,000	0.15
2013	41	285,899	2,750,147	12,000,000	0.15
2014	66	70,292	5,104,791	6,000,000	0.10
Grand total	**1,482**	**263,532,747**	**28,894,531**	**253,190,000**	

Table 8.5: Spread of Norwegian pharmaceutical claims

Reported Year \ Injury Year	Pre 1989	1989	1990	1991	1992	1993	1994	1995	1996	1997	1998	1999	2000	2001
1989	0	15	0	0	0	0	0	0	0	0	0	0	0	0
1990	0	2025	17	0	0	0	0	0	0	0	0	0	0	0
1991	2	0	753	10	0	0	0	0	0	0	0	0	0	0
1992	0	588	0	100	0	0	0	0	0	0	0	0	0	0
1993	0	0	1566	1034	0	3166	0	0	0	0	0	0	0	0
1994	0	817	1434	2	755	975	2351	0	0	0	0	0	0	0
1995	1	0	1	108	1071	104	1774	428	0	0	0	0	0	0
1996	6	0	0	0	1954	507	7674	200	4996	0	0	0	0	0
1997	0	0	29	0	30	327	3264	857	989	577	0	0	0	0
1998	0	0	0	1502	1185	8	8	7827	1377	2871	5044	0	0	0
1999	0	581	0	1	0	0	2	3	145	586	7527	50	0	0
2000	0	0	0	6	0	23	40	4510	22	241	1163	1942	1318	0
2001	17	0	0	0	11	0	10	0	0	1	8	2983	1922	3387
2002	0	0	0	0	0	0	0	12	0	210	1165	3703	996	20467
2003	148	0	0	0	0	13	0	0	720	3828	0	9	13	2554
2004	0	13	0	8	0	0	0	0	0	0	1463	672	420	2857
2005	22	0	0	0	0	0	0	36	0	242	1	4457	272	2706
2006	0	0	3	0	0	0	0	6	0	7	5	10	1367	254
2007	0	462	8	0	0	4	0	5	0	112	48	2637	51	520
2008	16	0	0	7709	0	4	0	0	7	0	13	2	0	0
2009	0	0	0	0	0	0	0	0	0	9	0	143	12	0
2010	7	0	0	0	0	22	0	153	0	7	0	7	10	53
2011	0	0	0	0	0	0	0	0	0	0	0	0	8	20
2012	0	0	17	0	0	0	0	0	0	0	0	0	0	6
2013	0	0	29	0	0	0	0	0	0	0	0	0	0	0
2014	1865	0	0	0	0	0	0	10	70	0	14	0	140	210
Total Sum	2083	4501	3857	10479	5006	5153	15123	14046	8327	8692	16451	16615	6530	33035
Paid	2083	4501	3857	10479	5006	5153	15123	14046	8327	8692	16451	16615	6530	33035
Reserved	0	0	0	0	0	0	0	0	0	0	0	0	0	0

2002	2003	2004	2005	2006	2007	2008	2009	2010	2011	2012	2013	2014	Total Sum	Paid	Reserved
0	0	0	0	0	0	0	0	0	0	0	0	0	15	15	0
0	0	0	0	0	0	0	0	0	0	0	0	0	2042	2042	0
0	0	0	0	0	0	0	0	0	0	0	0	0	765	765	0
0	0	0	0	0	0	0	0	0	0	0	0	0	688	688	0
0	0	0	0	0	0	0	0	0	0	0	0	0	5766	5766	0
0	0	0	0	0	0	0	0	0	0	0	0	0	6333	6333	0
0	0	0	0	0	0	0	0	0	0	0	0	0	3487	3487	0
0	0	0	0	0	0	0	0	0	0	0	0	0	15337	15337	0
0	0	0	0	0	0	0	0	0	0	0	0	0	6073	6073	0
0	0	0	0	0	0	0	0	0	0	0	0	0	19821	19821	0
0	0	0	0	0	0	0	0	0	0	0	0	0	8895	8895	0
0	0	0	0	0	0	0	0	0	0	0	0	0	9265	9265	0
0	0	0	0	0	0	0	0	0	0	0	0	0	8338	8338	0
1198	0	0	0	0	0	0	0	0	0	0	0	0	27751	27751	0
4765	1196	0	0	0	0	0	0	0	0	0	0	0	13245	13245	0
3988	3483	2864	0	0	0	0	0	0	0	0	0	0	15770	15770	0
2643	3113	15780	1888	0	0	0	0	0	0	0	0	0	31161	31161	0
447	87	3033	11689	1137	0	0	0	0	0	0	0	0	18045	18045	0
1575	195	16599	6555	2733	44	0	0	0	0	0	0	0	31548	31548	0
161	20	266	27	28	5579	1006	0	0	0	0	0	0	14898	7898	7000
8	4949	254	621	9007	1782	3322	23	0	0	0	0	0	20030	18615	1415
9	18	108	17	32	5276	365	42	1985	0	0	0	0	8112	7004	1109
1	0	0	0	0	10	3445	23	69	1679	0	0	0	5255	3364	1891
2	0	0	0	1	1726	42	99	793	8749	142	0	0	11576	1952	9625
0	0	0	447	9	19	1	788	128	25	1380	211	0	3036	286	2750
0	12960	0	72	141	84	210	92	142	150	473	122	390	5175	70	5105
14797	12960	38905	21315	13088	14521	8451	1067	3116	10604	1995	1322	390	292427	263533	28895
14797	12960	38905	21315	11533	13441	7226	237	1787	953	210	52	20	263533		
0	0	0	0	1555	1080	1225	830	1329	9651	1785	1270	370	28894		

Table 8.6: Norway claims statistics 2007–14

	2007	**2008**	**2009**	**2010**	**2011**	**2012**	**2013**	**2014**
Number of annual claims reported	112	77	66	80	30	41	41	66
% of claims per head of population	0.0024	0.0016	0.0014	0.0016	0.0006	0.0008	0.0008	0.0013
% of successful claims	38	28	24	17	22	24	20	25
Number of claims paid per year	44	38	31	18	16	10	11	14
% of claims paid per head of population	0.0009	0.0008	0.0006	0.0004	0.0003	0.0002	0.0002	0.0003
Average compensation per claim (€)	16,861	30,418	30,281	25,467	62,287	27,440	13,255	38,139
Total cost of compensation (excluding admin costs) paid per year (€ million)	2.2	4.4	2.9	3.5	3.4	1.3	0.9	2.2
Total contributions as % of national sales turnover	0.35	0.35	0.30	0.25	0.20	0.15	0.15	0.15
Cost of administration of the scheme per year (€ million)	0.3	0.4	0.3	0.4	0.1	0.2	0.2	0.2

Conclusion

As with its neighbouring Nordic countries, Norway has an effective and firmly established injury compensation system: the elements relating to patient and medicines injuries continue to operate swiftly, efficiently and economically.

Part IV

US Schemes

SONIA MACLEOD

9

No-Fault Schemes in the USA

For a country that is widely perceived as highly litigious, the USA has a potentially surprising number of no-blame compensation schemes for personal injury. These include well-established schemes for

— workers' compensation,
— clinical accident schemes, including birth injury schemes;
— automobile insurance;
— vaccine damage schemes;
— drug and medical device compensation schemes; and
— schemes arising from specific events such as the 9/11 fund.

Inevitably in a federal country, some schemes are national, others are state-wide or operate on more limited eligibility criteria.

At state level, various no-blame systems have been brought in to deal with specific areas of personal injury claims, from the birth-injury schemes in Florida and Virginia to the motor vehicle accident schemes seen in many states. These have had varying degrees of success. At national level, the scheme introduced for victims of the 9/11 terrorist attacks is unique. Before continuing, we also note here two notable schemes. First, there is the integrated approach to treatment, rehabilitation and compensation operated for veterans' healthcare. Second, there is the 'fair compensation' programme operated by Denver medical insurer COPIC that provides a swift no-blame response to injured patients.

It is worth noting the strength of the litigation background in the USA. The architecture of the legal system is strongly based on private enforcement by private actors (dubbed 'private attorneys general') in preference to enforcement by federal public authorities. The latter are widely mistrusted as intruding on personal freedoms, ineffective or captured. It is argued that private law remedies may enforce public norms equally well, based on the idea that the objectives are deterrence and compensation, and either public or private enforcement achieves deterrence.[1]

[1] Amongst many sources, see RA Kagan, *Adversarial Legalism: The American Way of Law* (Cambridge MA, Harvard University Press, 2001); JC Coffee Jr, 'Rescuing the Private Attorney General: Why the Model of the Lawyer as Bounty Hunter is not Working' (1983) 42 *Maryland Law Review* 215, 217; DR Hensler, NM Pace, B Dombey-Moore and B Giddens, *Class Action Dilemmas: Pursuing Public Goals for Private Gain* (Santa Monica CA, RAND, 2000); S Farhang, *The Litigation State: Public Regulation and Private Lawsuits in the United States* (Princeton NJ, Princeton University Press, 2010).

A. Veterans' Healthcare

The Department for Veterans Affairs (VA) provides healthcare for military veterans and has the nation's largest integrated healthcare system, with more than 1,700 hospitals, clinics, community living centres, domiciliaries, readjustment counselling centers, and other facilities. The Department for Veterans affairs also administers a variety of benefits and services that provide financial and other support to service members, veterans and their dependants and survivors. This dual function is interesting; it provides an integrated service which is able to provide both healthcare and also compensation in the event of an adverse event occurring during treatment or rehabilitation.

In the late 1980s, the Veterans Affairs Medical Center, Lexington, Kentucky decided that its strategy for contesting lawsuits was not working.[2] A turning point occurred when a team identified a case where the patient had died as a result of negligence and the patient's family was unaware of the circumstances leading up to the death. There was no risk of litigation, but the hospital staff felt that the case was a moral turning point, since the hospital's role as caregiver extended to the aftermath of harm caused by the hospital itself. In response they designed a 'humanistic risk management' scheme, whereby when a patient was injured, administrators meet the patient or family to describe what happened, offer an apology, accept responsibility, discuss measures to prevent future errors, answer questions, and arrange for the patient to receive financial compensation or further medical treatment, as needed. This approach has now spread to the entire VA healthcare services.[3]

VA has special dispensation under title 38 of the US Code[4] to pay compensation for death or disability 'as if service related'[5] for death or injury occurring during treatment or rehabilitation. To be eligible the claimant must be a veteran, with a disabling condition that

[2] Described in N Berlinger, *After Harm: Medical Error and the Ethics of Forgiveness* (Baltimore, Johns Hopkins University Press, 2005).

[3] Disclosure of adverse events is now mandated in the NHA Handbook, see www.ethics.va.gov/Handbook 1004-08.pdf.

[4] The US Code is a consolidation and codification by subject matter of the general and permanent laws of the United States. It is prepared by the Office of the Law Revision Counsel of the United States House of Representatives and is available at http://uscode.house.gov/.

[5] Eligibility exists where the disability or death was not the result of the veteran's willful misconduct and:

> (1) the disability or death was caused by hospital care, medical or surgical treatment, or examination furnished the veteran under any law administered by the Secretary, either by a Department employee or in a Department facility as defined in section 1701(3)(A) of this title, and the proximate cause of the disability or death was—
>
> (A) carelessness, negligence, lack of proper skill, error in judgment, or similar instance of fault on the part of the Department in furnishing the hospital care, medical or surgical treatment, or examination; or
>
> (B) an event not reasonably foreseeable; or
>
> (2) the disability or death was proximately caused (A) by the provision of training and rehabilitation services by the Secretary (including by a service-provider used by the Secretary for such purpose under section 3115 of this title) as part of an approved rehabilitation program under chapter 31 of this title, or (B) by participation in a program (known as a 'compensated work therapy program') under section 1718 of this title.

is the result of or has been aggravated due to VA-sponsored medical treatment or training. To qualify for compensation the evidence must demonstrate that the individual has either

— an additional disability or disabilities; or
— an aggravation of an existing injury or disease.

This disability must be:

— the direct result of VA fault such as carelessness, negligence, lack of proper skill or error in judgment; or
— not a reasonably expected result or complication of the VA care or treatment; or
— the direct result of participation in a VA Vocational Rehabilitation and Employment or compensated work therapy programme.

Application for these benefits is by online application form available from the VA website or by working with an accredited agent. This policy reflects a shift from a defensive position to a more open approach to information sharing and fair compensation.

B. COPIC

As well as top-down schemes underpinned by legislation, there are some bottom-up initiatives, both from healthcare insurers and healthcare providers. For example, COPIC is a Denver-based insurer that provides malpractice insurance to physicians in Colorado, Iowa and Nebraska, and insures hospitals and provides health plans.[6] Since 2000 COPIC has taken an innovative approach toward patient safety and risk management. It has a fair-compensation programme, called '3Rs': Recognize, Respond to and Resolve patient injury.[7] It originated in 2000, partially as a mechanism for limiting liability and reducing costly settlements and litigation. COPIC's approach focuses on responding to the patient's *injury*, and providing specific types of recompense, without making a determination on negligence.

Physicians are required to report mistakes to COPIC, and then the physician and a COPIC risk manager disclose mistakes to patients within 72 hours of the report. This prompt action is aimed at ensuring that the physician–patient relationship is maintained. This is achieved by facilitating open and honest communication and disclosure, and reimbursement for the patient for related out-of-pocket and medical expenses. Payments are to reimburse lost income and medical expenses resulting from the injury, and for medical care or other services needed by the patient. Payments are capped at $30,000. They do not cover compensation for death or wrong-site surgeries and other major errors, which remain within the traditional litigation process.

Physicians have a financial incentive to report mistakes to COPIC quickly and collaborate in the response. This arises because the programme draws a distinction between 'settlements'

[6] www.callcopic.com/Pages/default.aspx.

[7] www.callcopic.com/copic-services/safety-and-risk/Pages/3rs.aspx.

and 'payments'. Settlements are paid out of malpractice insurance and must be reported to the National Practitioner Data Bank (NPDB). Payments under the 3Rs programme are made from COPIC's own funds and are not required to be reported to the NPDB, and do not affect the physicians' future insurance premiums.

COPIC has a policy of not settling 'nuisance suits' (claims which it regards as without merit but made to try to achieve a payment) to make them go away.

10

Motor Vehicle Coverage in the USA

A. Background to No-Fault

Globally increased vehicle ownership and usage has led to increased numbers of accidents, fatalities and injuries. This has prompted the World Health Organization (WHO) to initiate a decade of action, 2011–20,[1] to reduce accidents and fatalities. This initiative focuses on action in low- and middle-income countries, which account for over 90% of the fatalities, but also includes the USA, which has a good road traffic safety profile and has already made considerable improvements to road traffic safety relative to population size.

Since the early 20th century a seismic shift has occurred in motor vehicle ownership and use in the USA, progressing from a handful of vehicles at the turn of the century to just under 260 million registered vehicles in 2009.[2] Figure 10.1[3] shows the number of fatal road traffic accidents per 100,000 of the population in the USA between 1900 and 2000. The increase in fatalities was driven by increased vehicle ownership and use, and declines were achieved by reduced usage (eg during the Second World War) and later by increased safety requirements, on both vehicles and drivers.

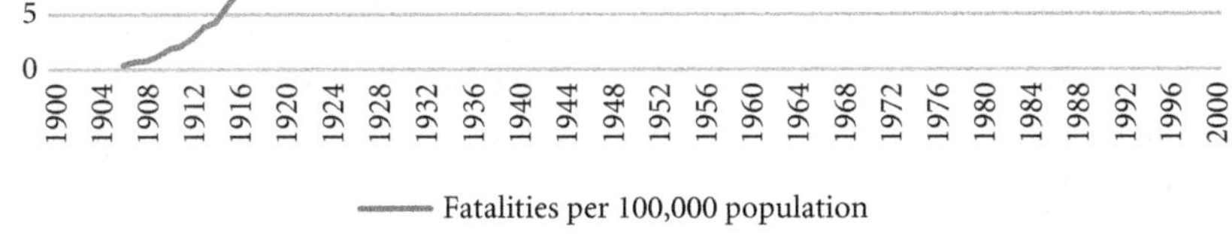

Figure 10.1: The rate of fatal accidents per 100,000 of the US population 1900–2000

[1] www.who.int/roadsafety/decade_of_action/plan/en/.

[2] World Health Organization (WHO), *Global Status Report on Road Safety 2013*, available at www.who.int/violence_injury_prevention/road_safety_status/2013/en/.

[3] Data taken from SB Carter, SS Gartner, MR Haines, AL Olmstead, R Sutch and G Wright (eds), *Death Rate, by Cause 1900–1998 of Historical Statistics of the United States* (2000) table Ab929-951; and from the Fatality Analysis Reporting System (FARS) encyclopaedia at www-fars.nhtsa.dot.gov/Main/index.aspx.

This graph only details fatalities: the number of accidents which resulted in a personal injury would be vastly greater. However, in recent years the number of fatalities per 100,000 of population has been declining. Figure 10.2[4] shows the decline in fatal road traffic accidents from the mid-2000s.

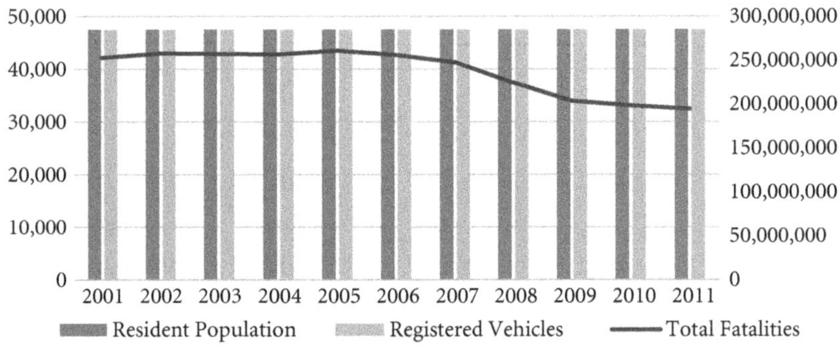

Figure 10.2: Road traffic fatalities, resident population and registered vehicle ownership in the USA 2001–11

Although the overall number of accidents in the USA is relatively low, the cost of compensation is substantial; the economic impact of motor vehicle crashes occurring in 2010 was assessed as US$277.0 billion. To put this in context, it is equivalent to approximately US$897 for every person living in the United States and 1.9% of the US gross domestic product.[5]

This rapid expansion in ownership and accident numbers in the first three-quarters of the twentieth century lead to an examination of the provision for compensating accident victims. There was considerable concern over the ability of tort law adequately to meet the compensation needs of road traffic victims.

Table 10.1 shows the current system adopted by each of the US states, along with the minimum limits for bodily injury, property damage and uninsured motorist coverage (but not PIP). This table illustrates the considerable diversity seen across the US states, both in terms of choice of system, no-fault, choice, add-on or tort based, and in terms of minimum financial responsibility limits. Although not explored in Table 10.1, the way in which financial responsibility can be demonstrated can also vary. Some states will only accept insurance, whereas others will allow bonds or deposits. This table highlights that there is no pan-American approach, and provisions should be considered on a state-by-state basis.[6]

[4] Data taken from the FARS (n 3).

[5] *The Economic and Societal Impact of Motor Vehicle Crashes, 2010* (National Highway Traffic Safety Administration, May 2014), available at www-nrd.nhtsa.dot.gov/Pubs/812013.pdf.

[6] G Wicket, *Automobile Insurance Subrogation: In All 50 States National Treatise* (New York, Juris Publishing, updated annually), ISBN: 978-1-57823-355-7, 2012. This provides an authoritative, regularly updated guide.

Table 10.1: Automobile insurance coverage requirements by US state

State	Minimum bodily injury and property damage (per person/per accident/ property damage, $000)	Uninsured motorist property damage (UMPD)	Uninsured motorist bodily injury (UMBI) (per person/ per accident, $000)
Florida[a]	10/20/10 or 30 combined	No	Mandatory offer
Hawaii	20/40/10	Mandatory offer w/ $50+ deductible	Mandatory offer
Kansas	25/50/10	No	Required 25/50
Massachusetts	20/40/5	No	Required 20/40
Michigan	20/40/10	No	No
Minnesota	30/60/10	No	Required 25/50
New York	25/50/10 (50/100 for death)	No	Required 25/50
North Dakota	25/50/25	No	Required 25/50
Utah	25/65/15 or 80 combined	Required w/ $250 Deductible	Mandatory offer 25/65 or 80
Kentucky	25/50/10 or 60 combined	No	Mandatory offer
New Jersey	15/30/5	No	No
Pennsylvania	15/30/5 or 35 combined	No	Mandatory offer
Arkansas	25/50/25	Mandatory offer (only available with UMBI)	Mandatory offer
Delaware	15/30/10	Mandatory offer, comes with a $250 deductible	Mandatory offer, comes with a $250 deductible
Maryland	30/60/15	Required 15	Required 30/60
Oregon	25/50/20	Required	Required 25/50
South Carolina	25/50/25	Required 25, approx $200 Deductible	Required 25/50, approx $200 Deductible
South Dakota	25/50/25	No	Required
Texas	30/60/25	Mandatory offer	Mandatory offer
Virginia	25/50/20	Required w/ $200 deductible	Required w/ $200 deductible
Washington	25/50/10	Mandatory offer (only available with UMBI)	Mandatory offer
Wisconsin[b]	25/50/10	No	Required
Alabama	25/50/25	No	Mandatory offer

(continued)

Table 10.1: ***(Continued)***

State	Minimum bodily injury and property damage (per person/per accident/ property damage, $000)	Uninsured motorist property damage (UMPD)	Uninsured motorist bodily injury (UMBI) (per person/ per accident, $000)
Alaska	50/100/25	Mandatory offer	Mandatory offer
Arizona	15/30/10 or 40 combined bond	No	Mandatory offer
California	15/30/5	Mandatory offer—upper limit of $3,500	Mandatory offer
Colorado	25/50/15	Mandatory offer w/ request	Mandatory offer
Connecticut	20/40/10	No	Required
Georgia	25/50/25	Required w/ $250+ deductible	Required w/ $250+ deductible
Idaho	25/50/15	No	Mandatory offer
Illinois	20/50/15	No	Required
Indiana	25/50/10	Mandatory offer	Mandatory offer
Iowa	20/40/15	No	Mandatory offer
Louisiana	15/30/25	No	Mandatory Offer
Maine	50/100/25 or 125 combined	No	Required 50/100
Mississippi	10/20/05	Mandatory offer; insured may reject	Mandatory offer; insured cannot reject UMBI cover but keep UMPDI
Missouri	25/50/10	No	Required
Montana	25/50/20	No	Mandatory offer
Nebraska	25/50/25	No	Required
Nevada	15/30/10	No	Mandatory offer
New Hampshire[b]	25/50/25	No	Required
New Mexico	25/50/10	Required w/ $250 deductible	Required
North Carolina	100/300/50	Required	Required
Ohio	12.5/25/7.5	No	No
Oklahoma	25/50/25	No	Mandatory offer
Rhode Island	25/50/25	Mandatory offer	Mandatory offer
Tennessee[b]	25/50/15 or 60 deposit	Mandatory offer if UMBI Purchased	Mandatory offer

(continued)

Table 10.1: ***(Continued)***

State	Minimum bodily injury and property damage (per person/per accident/ property damage, $000)	Uninsured motorist property damage (UMPD)	Uninsured motorist bodily injury (UMBI) (per person/ per accident, $000)
Vermont	25/50/10	Required w/ $150 deductible	Required
West Virginia	20/40/10	Required w/ $300 deductible	Required
Wyoming	25/50/20	No	Mandatory offer

Source: Information from the 2000 Florida Grand Jury report and updated from state websites, correct at 10 December 16.

[a] BI only required if driver is subject to Financial Responsibility Law.
[b] Insurance is not compulsory in this state if driver complies with financial responsibility requirements.

Shading:

No-fault	Choice state	Tort/add-on state	Tort state

B. The Columbia Plan

In 1928 the Committee to Study Compensation for Automobile Accidents was created. It was overseen by Columbia University's Council for Research in Social Sciences, and the resulting output became known as the Columbia plan.[7] Following a 3-year multi-state empirical study, the Committee concluded that accident rates were rising and that the compensation available from insurance and court claims was inadequate. In response, it proposed compulsory liability insurance for all drivers, applying strict liability for any injuries they caused (eg if two cars collided, each driver would be responsible for compensating the other's loss, and they would be jointly and severally liable for any injuries to third parties). The compensation proposed was akin to workers' compensation schemes, which limited claims for non-economic loss (essentially no pain and suffering payments, and restrictions on lost wages).

This plan was discussed in a number of states, but was never fully enacted in the USA. The major reasons that the Columbia plan stalled were general opposition to the ethos of the plan (that it was socialist, anti free enterprise, etc), specific objections from the insurance industry (which was against compulsory insurance), and the social and economic backdrop (the Depression and then the Second World War both reduced mileage, and therefore accidents, and the deaths and injuries caused by motor vehicle accidents seemed far less significant when compared to the scale of deaths and injuries due to the war).[8]

[7] *Report by the Committee to Study Compensation for Automobile Accidents to the Columbia University Council for Research in Social Sciences*, 1 February 1932.

[8] For a discussion on the resistance to the Columbia plan, see KR and J O'Connell, *Basic Protection for the Traffic Victim: A Blueprint for Reforming Automobile Insurance* (1965) 89–102, and S Lewis, 'The Merits of the Automobile Accident Compensation Plan' (1936) 3 *Law and Contemporary Problems* 583–97, available at http://scholarship.law.duke.edu/lcp/vol3/iss4/13.

C. The Keeton–O'Connell Plan

The idea of no-fault motor vehicle coverage regained traction in the USA in the late 1960s and early 1970s, culminating in 1975 in the publication of a no-fault plan by Professors Robert Keeton and Jeffrey O'Connell.[9] This book proposed a comprehensive no-fault plan (with the burden of providing basic protection for traffic victims borne by the motorist as part of the cost of driving), and it was accompanied by a formal actuarial opinion that the plan would reduce costs and even a draft statute (the Motor Vehicle Basic Protection Insurance Act) which the authors said was ready to be enacted.

This plan seemed to provide an equitable, inexpensive workable solution to compensating the victims of road traffic accidents. In essence the plan was not technically 'no-fault', but was a combination of tort and an insurance-backed administrative scheme. Essentially all vehicle owners would be obliged to purchase personal injury protection (PIP), namely first-party insurance against economic loss covering the car owner, passengers and injured third parties. PIP only covered economic losses, and payments were capped (under the original plan a cap of $10,000 was proposed, but levels varied between states). There were no provisions for non-economic losses such as pain and suffering. A more detailed description is given in Nora Freeman Engstrom's incisive and informative appraisal of the Keeton–O'Connell plan.[10]

The timing of the plan coincided with an era where no-fault was perceived as both desirable and achievable. There followed a period between 1970 and 1975 where no-fault schemes were taken up rapidly, starting with Massachusetts and spreading to over 20 states. By way of examples, several schemes are analysed in more detail below.

D. Costs of No-Fault

It is widely acknowledged that the initial cost savings anticipated under no-fault have not materialised. There is a variety of possible reasons for this, some of which are explored below: these are working hypotheses which require further research to confirm or disprove them.

Thresholds

None of the states offer a 'pure' no-fault option: each state has a threshold beyond which a claim can be brought in tort. Thresholds can be either verbal or they can be monetary. Verbal thresholds involve a statutorily defined severity of injury, so that claims which are less severe than the definition are dealt with by the no-fault provisions, whereas if the threshold has been reached or exceeded a tort claim can be issued and no-fault will

[9] Keeton and O'Connell (n 7).

[10] N Freeman Engstrom, 'An Alternative Explanation for No-Fault's 'Demise' (2012) 61 *DePaul Law Review* 303.

not apply. Under a monetary threshold, claims valued below the specified threshold are dealt with under the no-fault system. Those which extend over the specified amount of money can be brought in tort. Thresholds have been criticised as they lead to claim inflation.

Figure 10.3, from an analysis by James M Anderson and colleagues at RAND Corporation, shows the impact of amended thresholds in Massachusetts, a no-fault state with a monetary threshold. There are clear kinks in the chart, showing that here claiming behaviour is moderated by the threshold. In 1987, when the threshold was set at $500, there is little evidence of any claim distortion, but the threshold increased in 1989 to $2,000, and this led to a clear kink in claiming behaviour in subsequent years. There is a relative paucity of claims just below the $2,000 threshold, with a resulting increase in claims just over the threshold, leading to the kink seen in the graph. By 2002, the evidence of this shift of claims is lessened, and by 2007 the kink is minimal, which is likely to be due to the threshold not increasing while medical cost inflation increases that element, so there are simply fewer claims that fall below the threshold. This graph illustrates that the threshold can impact on claiming behaviour, in combination with other factors, including, but not limited to, medical claim inflation and the value of the threshold.

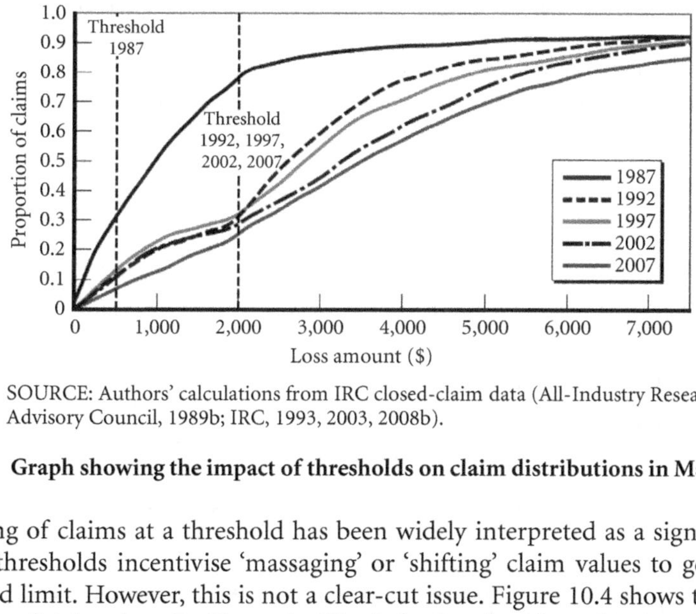

SOURCE: Authors' calculations from IRC closed-claim data (All-Industry Research Advisory Council, 1989b; IRC, 1993, 2003, 2008b).

Figure 10.3: Graph showing the impact of thresholds on claim distributions in Massachusetts*

This shifting of claims at a threshold has been widely interpreted as a sign of fraud, that monetary thresholds incentivise 'massaging' or 'shifting' claim values to get them above the required limit. However, this is not a clear-cut issue. Figure 10.4 shows the cumulative distribution of claimed medical losses in Illinois (again taken from the same RAND report). Illinois has a verbal threshold, so there are no incentives to make a claim reach any specific value. As would be expected, the graphs are smooth, with no kinks.

* JM Anderson, P Heaton and SJ Carroll, *The US Experience with No-Fault Automobile Insurance: A Retrospective* (RAND Corporation, 2010), ISBN 978-0-8330-4916-2, available at http://www.rand.org/content/dam/rand/pubs/monographs/2010/RAND_MG860.pdf

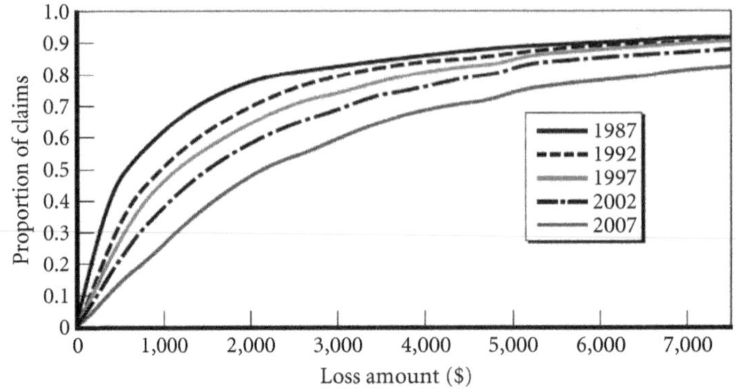

SOURCE: Authors' calculations from IRC closed-claim data (All-Industry Research Advisory Council, 1989b; IRC, 1993, 2003, 2008b).

Figure 10.4: Impact of thresholds on claim distributions in Illinois*

A snapshot view of the above graphs makes a compelling case for verbal thresholds being preferable to monetary thresholds when looking to reduce fraud. However, claiming behaviour is influenced by a wide range of factors, as is demonstrated by Figure 10.5, the graph from Minnesota from the same RAND report. Minnesota has a monetary threshold (currently $4,000), but shows no kinks associated with over-claiming. Claiming behaviour may be impacted by a wide variety of factors. There does seem to be a greater problem with over-claiming in urban areas, and the drivers for this differential are unclear.

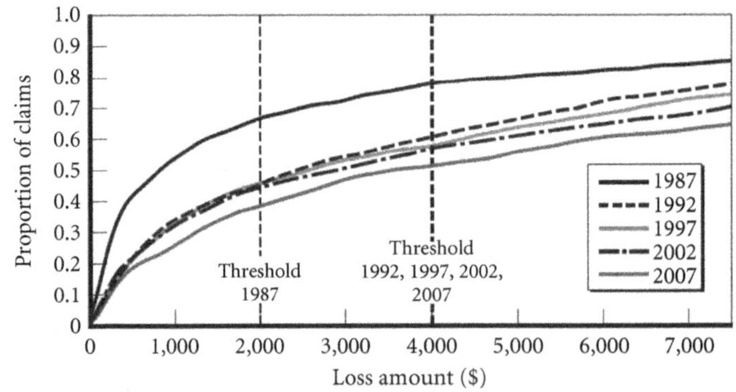

SOURCE: Authors' calculations from IRC closed-claim data (All-Industry Research Advisory Council, 1989b; IRC, 1993, 2003, 2008b).

Figure 10.5: Impact of thresholds on claim distributions in Minnesota*

* JM Anderson, P Heaton and SJ Carroll, *The US Experience with No-Fault Automobile Insurance: A Retrospective* (RAND Corporation, 2010), ISBN 978-0-8330-4916-2, available at www.rand.org/content/dam/rand/pubs/monographs/2010/RAND_MG860.pdf

Use of Medical Services

One of the interesting factors highlighted in the RAND report is the different usage of medical services under no-fault and tort arrangements. No-fault claimants are more likely to claim higher usage of medical services. For example, on average they make more visits to medical service providers than their counterparts under tort regimes. The choice of service also varies, with higher usage of physical therapy and psychotherapy in no-fault states. Arguably this could be a manifestation of better coverage under no-fault than is seen in tort states: no-fault was intended to counteract the inadequate medical care received by victims under tort.

Payment Priority

The other important issue is the order of priority that different forms of insurance take in the payment of personal injury claims from motor vehicle accidents. In the USA there is a variety of potential payees, including, but not limited to, motor insurance (both compulsory and voluntary), healthcare insurance, workers' compensation and some state programmes, such as Medicare and Medicaid. The circumstances of the accident will determine who the potential payees are. The payment priority which PIP or 'no-fault' takes is determined by state legislation, whereas in many 'no-fault' states, such as Florida, motor insurance pays before healthcare insurance. This can lead to a redistribution of costs from one form of insurance to another. This is not without consequence when determining the overall cost of 'no-fault' motor insurance and will be discussed below in the Conclusions. Other states adopt a cooperative policy, whereby the 'no-fault' automobile insurance pays up to a set limit and any costs in excess of this limit are met by the medical insurer.

One of the interesting issues raised by payment priorities is the ability of the different providers to obtain cost-effective healthcare. Healthcare insurers clearly have more specialised expertise in the provision of healthcare than auto-insurers. The ability of no-fault auto-insurers to control treatment cost is often curtailed by statutory provisions, whereas healthcare insurers are not encumbered in this way. This can lead to a relative increase in healthcare costs for auto-insurers.

Fraud, Hard-to-Verify Injuries and Bad-Faith Lawsuits

When no-fault was proposed, opponents argued that the removal of the need to prove fault and the resulting less rigorous investigation procedure would trigger an increase in fraudulent claims. Clearly there are some categories of injury, so-called 'hard injuries', broken bones, lacerations, etc, which are more difficult to fake. However, soft-tissue injuries, in particular sprains and strains, are harder to verify. If no-fault did lead to increased fraudulent claiming, higher levels of claims for hard-to-verify soft-tissue injuries would be expected.

The RAND report gave a detailed analysis. It carried out comparisons of no-fault insurance holders and those with full recourse to tort in two choice states, Pennsylvania and New Jersey. In these states, the claims made by individuals with no-fault coverage were less likely to contain hard-to-verify injuries. This is likely to be due to the limited access to general damages under no-fault provisions. However, hard-to-verify injuries became relatively more common in the no-fault states in the 2000s relative to the 1990s. As is usually the case for claiming behaviour, there is not a simple linear relationship between no-fault and increased hard-to-verify claims.

While it seems obvious that if an auto-insurer suspects a claim it should investigate it, this may not follow. In many states an insured can sue his own insurer for bad-faith failure to pay. This can be an attractive option for a potential claimant, and worrying for potential defendants, as the damages from such claims can be substantial. The threat of bad-faith claims may put insurers off fully investigating suspect claims. Bad-faith claims apply to dealings between an insured and his own insurer, hence this threat does not apply in tort states as the claimant would not be suing his own insurer.

E. Florida

Florida has had a no-fault motor insurance programme since 1 January 1972 (with a brief lapse in 2007). The scheme consists of compulsory liability insurance, currently set at 10/20/10 with no compulsory requirement to take insurance for bodily injury caused by uninsured motorists.[11] Florida uses a verbal threshold, currently defined as:[12]

(a) significant and permanent loss of an important bodily function;
(b) permanent injury, within a reasonable degree of medical probability, other than scarring or disfigurement;
(c) significant and permanent scarring and disfigurement;
(d) death.

Basic no-fault compensation covers economic loss arising from bodily injury, subject to certain caps (Table 10.2). The overall cap is set at $10,000 per claimant for bodily injury, with an additional $5,000 death benefits available where required.[13]

Table 10.2: Basic no-fault motor insurance benefits caps in Florida

Benefit	**Maximum available payment per person**
Disability benefits (work loss or survivors' income loss and other reasonable expenses)	60% of the loss of gross income of the injured/ dead individual
Medical benefits (subject to restrictions on service providers, etc)	80% of reasonable expenses for medical care ≤$10,000 ($2,500 for non-emergency care)
Total disability and/or medical benefits resulting from bodily injury	**≤$10,000 maximum**
Funeral/cremation and burial expenses	≤$5,000

[11] Chs 627.730–627.7405 of the Florida Statutes 2008, cited and known as the 'Florida Motor Vehicle No-Fault Law', available at www.leg.state.fl.us/Statutes/index.cfm?App_mode=Display_Statute&Search_String=&URL=0600-0699/0627/Sections/0627.730.html.

[12] ibid, chs 627.737–(2), available at www.leg.state.fl.us/Statutes/index.cfm?App_mode=Display_Statute&Search_String=&URL=0600-0699/0627/Sections/0627.737.html.

[13] Ch 627.736 of the Florida Statutes 2014, available at www.leg.state.fl.us/Statutes/index.cfm?App_mode=Display_Statute&Search_String=&URL=0600-0699/0627/Sections/0627.736.html.

Priority of Payment

In Florida if there are multiple potential insurers as payees, then the first-resort payee is workers' compensation, after which the no-fault insurers pay, and medical insurers are the insurer payee of last resort. Certain payments made by Medicare or Medicaid are to be repaid by the motor vehicle insurer.[14]

Various limitations on claiming apply. An insurer may exclude benefits on the following grounds:[15]

(a) For injury sustained by the named insured and relatives residing in the same household while occupying another motor vehicle owned by the named insured and not insured under the policy or for injury sustained by any person operating the insured motor vehicle without the express or implied consent of the insured.
(b) To any injured person, if such person's conduct contributed to his or her injury under any of the following circumstances:
    1) Causing injury to himself or herself intentionally; or
    2) Being injured while committing a felony.

Background

Florida was an early adopter of no-fault, enacting laws known as the Florida Automobiles Reparation Reform Act 1971 (hereinafter the Act) to replace traditional tort from 1 January 1972.[16] The purpose of the Act was specified as

> to require medical, surgical, funeral and disability insurance benefits to be provided without regard to fault under motor vehicle policies that provide bodily injury and property damage liability insurance, or other security, for motor vehicles registered in this sates and, with respect to motor vehicle accidents, a limitation on the right to claim damages for pain suffering, mental anguish and inconvenience.[17]

This purpose was later expounded in 1974 in *Lasky v State Farm Insurance Company*,[18] giving the following detailed objectives of the no-fault scheme:

— to ensure that persons injured in vehicular accidents would be directly compensated by their own insurer, even if the injured party was at fault, thus avoiding dire financial circumstances and removing the cost of injuries to uninsured persons from general taxation;
— to reduce court congestion and delays by cutting the number of lawsuits;
— to lower automobile insurance premiums; and
— to end the inequalities of recovery seen with traditional tort remedies.

[14] ibid, ch 627.736(4).
[15] ibid, ch 627.736 (2).
[16] §§627.730–627.741, Florida Statutes 1971, known as the 'Florida Automobile Reparations Reform Act', effective 1 January 1972, available at from the Florida State University College of Law website at http://fall.fsulawrc.com/collection/FlaStat/FlaStat1971/vol2/FlaStat1971v2_OCR_Part29.pdf.
[17] ibid, §§627.731.
[18] *Lasky v State Farm Insurance Company* 296 So2d 9 (Fla 1974).

Under the Act, PIP cover for expenses up to $5,000 was mandatory.[19] One aim of the Act was to remove lower-value claims from the courts. Claimants could still sue in tort if, and only if, (a) the PIP benefits payable for the injury (or payable but for a policy deductible) exceeded $1,000 or (b) the plaintiff either died or suffered an injury or disease consisting in whole or in part of: permanent disfigurement; a fracture to a weight-bearing bone; a compound, comminuted, displaced or compressed fracture; loss of a body member; permanent injury within reasonable medical probability; or permanent loss of bodily function.[20]

In exchange for the limitation of the right to claim under tort law, the Act provided victims with prompt payment of allowable expenses, with expenses considered overdue if they were not paid within 30 days of the insuring being notified of the fact and amount claimed.[21] Medical advisors were only authorised to charge insurers a 'reasonable' amount for the care provided, in line with the amount they would customarily charge for like products, services and accommodations in cases involving no insurance.[22] Provisions were included in the Act for an Independent Mental and Physical Examination (IME) of an injured person covered by PIP when the condition of the claimant was material to the claim.[23,24]

Separate provisions in the Financial Responsibility Law 1955 ensured mandatory coverage for bodily injury and property damage. The requirements were coverage of $10,000 for bodily injury or death of one person in any one accident, rising to $20,000 for death or bodily injury of two or more individuals in one accident, and $5,000 for damage and destruction of another's property. These levels were raised in 1972 from $20,000 to $25,000. A further increase was made in 1973 from $25,000 to $30,000.

In addition, the Act removed the burden of paying for uninsured claimants from general taxation and placed the cost onto motorists directly.

Major Reforms Prior to 2007

Various reforms were made over time; Table 10.3 summarises the major reforms of the key provisions for PIP, bodily injury and property damage liability coverages, PIP deductibles and the legal threshold.

The initial $1,000 limit for claiming medical benefits encouraged 'padding' of medical claims to overcome this threshold. This was addressed in 1976, when the $1,000 'dollar threshold' was replaced by a 'verbal threshold', requiring the claimant to show that the injury met specified severity conditions.[25] In order to be eligible to sue, the injury had

[19] §§627.736(1), Florida Statutes 1971, with subsections covering (a) medical benefits, (b) disability benefits (85% loss of income), and (c) funeral, cremation or burial benefits (not to exceed $1,000).

[20] ibid, §§627.7367(2).

[21] ibid, §§627.736 (4)(a).

[22] ibid, §§627.736 (5).

[23] ibid, §§627.736 (7).

[24] Ss 1, 5, ch 29963, 1995 known as the Financial Responsibility Law of 1955, as amended to §§324.011–324.7251, Florida Statutes 2013, available at www.leg.state.fl.us/Statutes/index.cfm?App_mode=Display_Statute&URL=0300-0399/0324/0324ContentsIndex.html&StatuteYear=2013&Title=-%3E2013-%3EChapter%20324).

[25] Ch 76-266 Laws of Florida, now incorporated (as amended to incidents where the injury or disease consists in whole or in part of: '(a) Significant and permanent loss of an important bodily function. (b) Permanent injury within a reasonable degree of medical probability, other than scarring or disfigurement. (c) Significant and permanent scarring or disfigurement. (d) Death') into s 2, ch 627.737, Florida Statutes 2013, available at www.leg.state.fl.us/Statutes/index.cfm?App_mode=Display_Statute&Search_String=&URL=0600-0699/0627/Sections/0627.737.html.

to consist of death, permanent injury within a reasonable degree of medical probability (other than scarring or disfigurement), significant and permanent scarring or disfigurement, loss of a bodily member, or serious temporary injury that had a material bearing on the claimant's activity and lifestyle during substantially all of the 90 days following the accident.

Table 10.3: History of major statutory provisions for no-fault motor coverage in Florida prior to the 2007 repeal

Year	Chapter law	PIP benefits/ deductibles	Legal threshold	Bodily injury	Property damage
1971	Ch 71-252	$5,000 total PIP benefit; 100% medical; 85% lost wages; $1,000 funeral; PIP deductibles; $250, $500 and $1,000; property damage no-fault coverage	$1,000 monetary threshold or a verbal threshold	$10,000/$20,000	$5,000
1972	Ch 72-297			$25,000	
1973	Ch 73-180			$15,000/$30,000	
1976	Ch 76-266	$2,000 maximum deductible	$1,000 monetary threshold eliminated; verbal threshold revised	$10,000/$20,000	
1977	Ch 77-468	PIP benefit reduced; 80% medical; 60% lost wages; added $3,000 and $4,000 deductibles		Bodily injury repealed	Property damage repealed
1978	Ch 78-374	Increased PIP benefit to $10,000 (effective 1 January 1978) added $6,000 and $8,000 deductibles	Strengthened verbal threshold		
1982	Ch 82-243	Increased funeral benefits to $1,750; reduced maximum deductible to $2,000			

(continued)

Table 10.3: ***(Continued)***

Year	Chapter law	PIP benefits/ deductibles	Legal threshold	Bodily injury	Property damage
1988	Ch 88-370				Reinstated property damage: $10,000
1989	Ch 89-238				$30,000 property damage/ bodily injury
1989	Ch 89-243	Increased funeral benefit: $5,000; renamed 'death' benefit			
2001	Ch 2001-271	Defined benefits/ services to be 'medically necessary'			
2003	Ch 2003-411	Eliminated $2,000 PIP deductible; changed calculation of PIP deductibles			

Source: Data from *Florida's Motor Vehicle No-Fault Law. Report Number 2006-102* (Committee on Banking and Insurance for the Florida Senate, November 2005), available at http://archive.flsenate.gov/data/publications/2006/senate/reports/interim_reports/pdf/2006-102bilong.pdf.

Table 10.3 demonstrates that there has been an evolution of the coverage thresholds, which continues to this day. The thresholds for bodily injury insurance were increased, then reduced back to $10,000/$20,000 and then removed altogether in 1977.

After a Grand Jury was held on the scheme in 2000,[26] in 2001 and 2003 the Act underwent significant reforms. In essence, the problem was fraud.[27] A sunset clause was included in the 2003 legislation, effectively bringing the no-fault provisions to an end on 1 October 2007.[28] These provisions specified that if the legislature took no action prior to 1 October 2007, then the Florida no-fault motor vehicle laws would be repealed.

There followed a period of discussion and in 2006 the Florida legislature passed a law to reprieve the no-fault law from repeal. However, this was vetoed by Governor Jeb Bush,

[26] The 2000 Grand Jury Report, available at http://myfloridalegal.com.

[27] The 2000 Grand Jury Report defined PIP fraud as follows: (1) the illegal solicitation of accident victims for the purpose of filing for PIP benefits and motor vehicle tort claims; (2) brokering patients between doctors, lawyers and diagnostic facilities, as well as attendant fraud, which can include the filing of false claims; (3) billing insurers for treatment not rendered; (4) using phony diagnostic tests or misusing legitimate tests; (5) inflating charges for diagnostic tests or procedures through brokers; and (6) filing fraudulent motor vehicle tort lawsuits. The Grand Jury concluded that 'certain people have turned the $10,000 of personal injury protection coverage into their own personal slush fund'.

[28] Ss 627.730–627.7405, Florida Statutes 2007.

as it was thought the proposed new law did not contain sufficient anti-fraud measures. The legislature had a second opportunity to devise new legislation to save the no-fault law in the 2007 session, but failed to reach agreement in the regular session. It had reached an agreement by the time a special session was called in October 2007 to address Florida's budget shortfall, and the new no-fault motor provisions were signed into law with an effective date of 1 January 2008. During the intervening 3 months between the sunset and the resurrection, Florida motorists were not required to carry PIP, only property damage insurance. However, there were provisions within the 2008 Act to cover accidents occurring in this 'hiatus period'. If both parties had maintained their PIP coverage as was required prior to the sunset clause, then the limited tort immunity and injury threshold provisions in the 2008 Act applied. If either party's PIP coverage had lapsed, then tort law was applied to the claim.[29]

The 2008 No-Fault Motor Provisions[30]

There was surprisingly little change of substance between the pre-sunset and the 2008 no-fault motor provisions. The major changes are given in Table 10.4.

Table 10.4: A comparison of some of the major differences between the pre-sunset, the hiatus period, and the 2008 Florida Act

	Pre-sunset (pre 1 Ocotber 2007)	**Hiatus period (1 Ocotber 2007–31 December 2007)**	**2008 Act (post 1 January 2008)**
PIP/property damage liability coverage	Mandatory \$10/10K	No PIP required \$0/10K	Mandatory \$10/10K
Individual/per crash bodily injury liability	\$10/20K	NA	\$10/20K
Verbal tort threshold	Applied to all cases	Applied if both parties had PIP coverage at pre-sunset levels	Applied to all cases
Reservation of part of the total available of benefits for ER/ in-patient care	NA	NA	Minimum \$5K, must be held for minimum of 30 days after the insurer is notified of the accident
Response time to pre-suit demand letters to avoid litigation	15 days to pay the claim, plus interest and stat. penalty	N/A	30 days to pay the claim, plus interest and stat. Penalty

(continued)

[29] Ch 627-7407(6) of the Florida Statutes 2008, available at www.leg.state.fl.us/Statutes/index.cfm?App_mode=Display_Statute&Search_String=&URL=Ch0627/SEC7407.HTM&Title=-%3E2008-%3ECh0627-%3ESection%207407#0627.7407.

[30] Chs 627.730–627-7405 of the Florida Statutes 2008, cited and known as the 'Florida Motor Vehicle No-Fault Law', available at www.leg.state.fl.us/Statutes/index.cfm?App_mode=Display_Statute&Search_String=&URL=Ch0627/SEC731.HTM&Title=-%3E2008-%3ECh0627-%3ESection%20731#0627.731.

Table 10.4: ***(Continued)***

	Pre-sunset (pre 1 Ocotber 2007)	Hiatus period (1 Ocotber 2007–31 December 2007)	2008 Act (post 1 January 2008)
Costs of medical fees and diagnostic tests	Insurers could introduce limited fee schedules for some diagnostic tests	NA	Insurers can introduce fee schedules limiting charges to 200% of the Medicare amount, for emergency care 75% of the usual charge and community care at the usual rate for that service[31]
Single action restriction	No limit on the number of treatment claims per injured person	NA	All claims from the treatment of an injured person to be brought in the same action

The 2008 Act did not undertake substantive reforms of the pre-sunset Act. This led to further demands for reform. Interestingly in 2009 Florida topped the table for PIP questionable claim (QC) referrals, with the very dubious honour of having twice the number of QCs as New York, the second highest ranking state.[32] From 2008 to 2010 the number of QCs in Florida increased by 34%, with the Florida Department of Financial Services, Division of Insurance Fraud (DIF) reporting that 49% of the referrals they received in the financial year 2010/11 related to PIP fraud. Clearly the issue of fraud had not been resolved by the 2008 Act. A further cause for concern were the costs associated with soft-tissue injuries; 62% of the total PIP costs and 43% of treatment costs were associated with treating these injuries. Approximately 1 in 5 soft-tissue treatments comprised massage therapy. From 2005 to 2010 massage therapists per patient charges increased by 51%. This led to calls for further reforms.

2012 Reforms

In the 2012 legislative session, a two-stage reforming Bill[33] was signed into law on 4 May 2012. The first component came into force on 1 July 2012 and brought in stronger penalties for medical practitioners found to have committed PIP fraud. The second component introduced restrictions on the rights of injured parties to claim for medical treatment and came into force on 1 January 2013.[34] Briefly these were:

— for a claim to be eligible, injured parties had to seek medical treatment within 14 days of the accident;

[31] ibid, 5(a)(2).
[32] C Longino, *ForeCAST Report: Florida PIP* (National Insurance Crime Bureau, 19 March 2014).
[33] House Bill 199 of 2012, available at www.flsenate.gov/Session/Bill/2012/119.
[34] S 627.723 Florida Statutes 2013.

— a decrease in the available PIP benefit for 'non-emergency' treatments to $2,500 (if a proper licensed medical practitioner determines that the injuries are an 'emergency medical condition', then the full $10,000 is still available); and
— treatments by massage therapists and acupuncturists were no longer eligible under PIP claims.

There were legal challenges from acupuncturists and massage therapists, leading to a temporary injunction preventing these provisions from applying between March and October 2013, but on 23 October the injunction was lifted.

Indications to date are that the 2012 revisions are driving a decrease in QCs.[35]

2016 Review of the 2012 Revisions

The Florida Office of Insurance Regulation[36] commissioned a report in 2016 to assess the impact of the 2016 revisions. Pinnacle Actuarial Resources Inc released the report[37] in September 2016. The remit of the report was to:

— estimate any savings that occurred as a result of the changes introduced by the 2012 reforms;
— estimate premium impacts if PIP coverage requirements were repealed and replaced with various combinations of bodily injury liability (BI) and/or medical payments (MP) insurance; and
— estimate premium impacts if the compulsory requirement to purchase automobile insurance was repealed.

The report is very detailed, but the headline findings are that the 2012 reforms produced an estimated aggregate saving in PIP claim costs of 17.5%, and an estimated state-wide average saving in PIP premiums of 15.1%.

If PIP were repealed, the report concluded that an average driver would receive an overall reduction in their automobile insurance of 9.6%, or $81 per year, if no additional BI or MP coverage was taken out. If no-fault first-party MP coverage were selected, the saving would vary according to the level of MP coverage selected: $2,500 coverage would give a 4.9% or $41 saving per year; $5,000 of MP coverage would decrease the liability premium by 1% or $9.

If the requirement to purchase auto insurance was also repealed, the impact on premiums would be between a 0.4% decrease in the savings obtained by repealing PIP alone and a saving of 2.3% more than repealing PIP alone, depending upon the stringency with which the Financial Responsibility Law was enforced.

These headline figures are for auto insurance. Necessarily, the report has detailed analysis of the impact of repealing PIP coverage on other insurance programmes and on the Florida

[35] Longino (n 33) and *Report on Review of the Data Call Pusuant to House Bill 119—Motor Vehicle Personal Injury Protection (PIP) Insurance* (Florida Office of Insurance Regulation, 1 January 2015), available www.floir.com/siteDocuments/HB119DataCallReport.pdf.

[36] See www.floir.com/ for more detail.

[37] *Review of Personal Injury Protection Legislation* (Florida Office of Insurance Regulation, 13 September 2016), available at www.floir.com/siteDocuments/FLOIRReviewPIP20160913.pdf.

healthcare system. Table 33 of the report, is reproduced here as Table 10.5, estimates the impact of repealing PIP on health insurance premiums. This shows that the estimated savings on auto insurance would not be as great as they appear at first, as they would need to be balanced against increases in health insurance premiums.

Table 10.5: Estimated impact of repealing PIP in Florida on health insurance premiums

Healthcare plan	Healthcare insurance premium	Assumed number of cars	PIP medical amount transferred: low estimate	% impact: low estimate	PIP medical amount transferred: high estimate	% impact: high estimate
Single	$6,251	1 car	$42.99	0.7%	£56.44	0.9%
Family	$17,545	1 car	$42.99	0.2%	$56.44	0.3%
	$17,545	2 cars	$85.98	0.5%	$112.88	0.6%
	$17,545	2½ cars	$107.48	0.6%	$141.10	0.8%

Conclusions

While there have been many modifications of the no-fault law, these have been aimed at addressing the problem of fraud rather than ideological or substantive changes. Indeed, it is interesting to note that when the sunset clause was triggered, a 'revival' of no-fault followed very quickly. Florida faces a constant battle against fraud and while there is clearly support for no-fault, there is equally entrenched opposition. The prediction that Florida will soon ditch no-fault has been around for decades. The latest analysis does not present a compelling case to change the status quo. Time will tell if the 2012 revisions will be sufficient to keep Florida as a no-fault state.

F. North Dakota

North Dakota instigated a no-fault motor insurance programme from 1 January 1976.[38] This consists of compulsory liability insurance, currently set at 25/50/25[39] with compulsory insurance for bodily injury caused by uninsured motorists at the same level or higher.[40] North Dakota uses a combined dollar or verbal threshold, current requiring a 'serious injury' (an accidental bodily injury which results in death, dismemberment, serious and permanent disfigurement or disability beyond 60 days) or medical expenses in excess of $2,500.[41] Basic no-fault covers economic loss arising from bodily injury, subject to certain

[38] Currently enacted by the North Dakota Century Code (NDCC) 39-08-20 as amended, available at www.legis.nd.gov/general-information/north-dakota-century-code.

[39] ibid, ch 39-16.1.

[40] ibid, s 26.1-40-15.2.

[41] ibid, s 26.1-41.

caps. The overall cap is set at $30,000 per claimant regardless of the number of claimants per accident, with caps on the different heads of damages. These benefits are detailed in Table 10.6.[42]

Table 10.6: Basic no-fault benefits caps in North Dakota

Benefit	Maximum available payment per person
Work loss or survivors income loss	≤$150 per week (pro-rated for shorter periods)
Funeral/cremation and burial expenses	≤$3,500
Total benefits resulting from bodily injury	**≤$30,000 maximum**

North Dakota no-fault prioritises any workers' compensation payments above basic no-fault motor payments. Healthcare insurers are required to coordinate with no-fault motor insurers, such that the no-fault motor insurer pays the first $10,000, but the health insurer will pay amounts in excess of $10,000 and must coordinate with the no-fault insurer to ensure there is no overpayment. If there are multiple no-fault policies, the occupant's vehicle coverage has priority over the injured passenger's coverage, followed by the assigned claims plan. The assigned claims plan pays no-fault benefits to an individual not otherwise excluded by law or on the financial inability of the basic no-fault insurer.

Basic no-fault coverage is subject to certain behavioural restrictions and is not available to any person who is injured while:

1. Occupying any motor vehicle without the expressed or implied consent of the owner or while not in lawful possession of the motor vehicle.
2. Occupying a motor vehicle owned by such person which is not insured for the benefits required by this chapter unless uninsured solely because the insurance company of the owner has not filed a form pursuant to subsection 2 of section 26.1-41-05 to provide the basic no-fault benefits required by this chapter.
3. During a racing or speed contest, or in practicing or preparing for a racing or speed contest.
4. Intentionally causing or attempting to cause injury to oneself or another person.[43]

Background[44]

After a failed Bill in the 1971 session, during the 1971–72 interim period the North Dakota Legislative Council and Business Committee studied no-fault and in 1973 Senate Bill 2031 was proposed. This was against a backdrop of the possibility that Congress would mandate no-fault insurance by enacting federal legislation in accordance with the Hart–Magnuson Bill. The Hart–Magnuson Bill required states to enact no-fault legislation within a set time period, and if a state failed to act, then more stringent provisions would go into effect.

[42] ibid, s 6.1-41-01.

[43] ibid, s 26.1-41-07.

[44] A memorandum *History of No-Fault Insurance in North Dakota* was drawn up in 2004 for the North Dakota Legislative Council by the Transportation Committee, available at www.legis.nd.gov/files/resource/committee-memorandum/59152.pdf?20141126081807.

As a predominantly rural state with low levels of court congestion, the Council and Committee felt that the stringent provisions of the Hart–Magnuson Bill would be overkill for North Dakota. Therefore Bill 2031 was proposed. It closely followed the Dual Protection Plan drafted by the National Association of Independent Insurers. In summary, the Bill stated that all policies insuring private passenger automobiles must include the following first-party benefits:

— $2,000 per person for medical, hospital, surgical, dental, vocational rehabilitation expenses, etc;
— disability benefits of up to $750 per month up to a maximum of $6,000 for a wage-earner and $4,500 for essential substitute services of non-income producers;
— in addition to the above coverage insurers would have to offer supplemental coverage to an aggregate of not less than $100,000.

However, Bill 2031 was rejected. In the 1975 session, a Bill[45] was put forward bringing no-fault insurance into force from 1 January 1976. This was still against the backdrop of impending federal no-fault legislation, which would have contained exclusions for states which already had no-fault arrangements in place. The Bill included a legislative declaration that the express purpose of the act was:

1. To avoid inadequate compensation to victims of motor vehicle accidents, to require registrants of motor vehicles in this state to procure insurance covering legal liability arising out of ownership or operation of such motor vehicles, and to provide benefits to persons occupying such motor vehicles and to persons injured in accidents involving such motor vehicles; and
2. To limit the right to claim damages for non-economic loss in certain cases and to organise and maintain an assigned claims plan.

The 1975 Bill set a tort threshold of serious injury based on medical expenses in excess of $1,000. The amount payable under a no-fault policy was limited to $15,000 and subject to the caps detailed in Table 10.7.

Table 10.7: Limitations on no-fault benefits in House Bill 1214, the North Dakota Auto Accident Reparations Act 1975

Benefit	**Limit**
Wage loss/survivors' benefit	$150 per week (the average ND weekly wage in 1975)
Death benefits for funeral expenses	$1,000
Replacement services	$15 per day
Total benefits	**£15,000**

Over the intervening years there have been several amendments; the major Bills are detailed in Table 10.8.[46]

45 House Bill No 1214, the North Dakota Auto Accident Reparations Act 1975/.

46 Information obtained from searching www.legis.nd.gov/ and from *History of No-Fault Insurance in North Dakota* (n 45).

Table 10.8: Major amendments to North Dakota no-fault provisions 1979–2014

Year	Bill No	Amendment
1977	Senate Bill 2139	Clarified the definition of 'owner' as to lessees and defined 'disability', 'commercial vehicle' and 'bus'. Amended requirements for no-fault to when use is contemplated. Made an injured party use their own policy in preference to a bus policy when injured on a bus
1977	House Bill 1510	Set a £5,000 limit for a no-fault insurer to coordinate with a health insurer. Allowed the no-fault insurer to subrogate against the health insurer after the first $5,000 of no-fault benefits are paid, allowing more no-fault benefits for non-medical benefits
1979	House Bill 1503	Created an exception under the assigned claims plan for uninsured motorists. If an individual owns a motor vehicle and is uninsured, then that individual cannot collect from the assigned claims plan in the event that he or she is injured in an uninsured motor vehicle accident.
1981	Senate Bill 2061	Included health maintenance organisations within the coordination of benefits provisions
1981	Senate Bill 2051	Set the priority for payment in ridesharing cases to promote ridesharing by lessening the liability of the owner of the vehicle
1981	Senate Bill 2070	Amended the position on ownership after the sale of a motor vehicle. The motor vehicle owner was subsequently defined as the person to whom ownership has been transferred, regardless of title, for the purposes of no-fault
1983	House Bill 1194	Limited the liability of the assigned claims plan to those situations where the claimant would have been eligible for no-fault insurance
1983	House Bill 1195	Prohibited the stacking of insurance coverage for uninsured motorist coverage and no-fault benefits
1985	Senate Bill 2078	Repealed the statutory title and statement of purpose of the 1975 no-fault law as they were non-substantive
1985	House Bill 1528	Increased the maximum level for basic no-fault benefits from $15,000 to $30,000 and the optional excess no-fault benefits for motor vehicle insurance from $40,000 to $80,000. The threshold amount defining serious injury of $1,000 for medical expenses increased to $2,500
1987	Senate Bill 2413	Clarified that a basic no-fault insurer may coordinate any benefits it is obligated to pay for medical expenses in excess of $5,000 which occurred as a result of accidental bodily injury
1989	Senate Bill 2056	Text corrections such as changing workmen's to workers' compensation
1989	House Bill 1409	Mandated that an insured who purchased optional excess no-fault benefits is entitled to optional excess no-fault benefits commencing upon the exhaustion of basic no-fault benefits if the injured person or that person's relative is injured in a motor vehicle not owned by the insured or as a pedestrian

(continued)

Table 10.8: ***(Continued)***

Year	Bill No	Amendment
1989	House Bill 1467	Increased the time for filing a no-fault insurance claim as follows: — survivors' income loss/replacement services loss/funeral expenses—whichever is the earlier of 2 years from the death or 6 years from the fatal accident — further survivors' income loss/replacement services loss benefits—6 years after the last benefits payment
1991	Senate Bill 2089	Clarified the exclusion of basic no-fault insurers from the prohibition from coordinating benefits without providing the purchaser with an equitable reduction or savings in cost. In addition, the bill allows a basic no-fault insurer to recover all no-fault benefits, not solely basic no-fault benefits, from another no-fault insurer when tort law would require recovery.
1991	Senate Bill 2302	Included a motor vehicle owned by a political subdivision and operated as part of a public transit system for which the costs are subsidised by the government in the definition of bus for the purposes of no-fault insurance
1991	Senate Bill 2555	Increased the funeral expense benefit from $1,000 to $3,500
1997	Senate Bill 2046	This bill made technical corrections
1999	Senate Bill 2376	Limits the recoverable damages of a person who is in a motor vehicle accident and does not have liability insurance if that person has at least 2 convictions of operating a motor vehicle without liability insurance Effectively a person with no-fault insurance may not be assessed damages for pain and suffering in favour of a person who has 2 or more convictions of operating a motor vehicle without liability insurance
1999	Senate Bill 2406	The bill requires a person who has been convicted of driving a motor vehicle without liability insurance to provide proof of insurance to the Department of Transportation or else that person's driving privileges are suspended. The proof of insurance must be a certificate from an insurance carrier. The convicted person's license must contain a notation showing that the person must keep proof of liability insurance on file with the department The fee for the notation is $50. The bill requires insurance carriers to notify the director of a cancellation or termination of an insurance policy required for a person convicted of driving without liability insurance
2003	Senate Bill 2275	Increased the amount of no-fault medical expenses a no-fault insurer may coordinate with a health insurer from in excess of $5,000 to $10,000. In short, the no-fault insurer pays the first $10,000 of medical expenses and the health care insurer pays medical expenses after $10,000

(continued)

Table 10.8: ***(Continued)***

Year	Bill No	Amendment
2003	House Bill 1190	Removed the expiration date on the section of law that prohibits a person who had two convictions for driving without liability insurance and was driving without liability insurance from receiving noneconomic loss for serious injury in an action against the insured. In addition, the bill lowers the previous convictions requirement from 2 to 1.
2005	Senate Bill 2047	Changed various definitions as follows: — removed from the definition of 'accidental bodily injury' injury resulting from entering or alighting from a stopped motor vehicle and not caused by another motor vehicle — changed the definition of 'medical expenses' so that the charges must be usual and customary instead of merely reasonable — expressly included diagnostic services as medical expenses and excluded charges for drugs sold without a prescription, experimental treatments and medically unproven treatments — changed the definition of 'occupying' to exclude getting into or out of a motor vehicle — provided for a court to order the insured to reimburse the insurer for an independent medical examination that the insured failed to appear for without good cause. Also repealed NDCC section 26.1-41-17, which provided for equitable allocation of losses among insurers. This section provided for an insurer to recover no-fault benefits paid to an injured person from the motor vehicle liability insurer of a secured person based upon tort law principles
2009	House Bill 1245	Removal of the need for a written request to suspend automobile coverage
2011	House Bill 1308	Establishment of primary and excess automobile liability coverages in certain instances. (mainly those buying, selling, renting and repairing motor vehicles)
2013	House Bill 1263	Ambiguous amendment which seemed to amend the statute concerning driving without liability insurance by reducing a violation from a Class B misdemeanour to an infraction, making the offence a non-criminal moving violation traffic offence, and adding enhanced penalties for a second or subsequent offence. For a repeat violation, the opinion indicated the officer either may advise the driver of the penalties, bond and mandatory impound order, or may take the driver to a magistrate who may then advise the driver of the enhanced charge, bond and mandatory impound order and release the driver on a promise to appear
2013	Senate Bill 2042	Failure to provide security for basic no-fault insurance is reason to refuse a certificate of title for a vehicle

The amendments to the North Dakota no-fault provisions are interesting in that they are relatively minor changes and updates. For example, increases in payments would be expected in line with increased costs. The increased funeral expenses in 1991 under Senate Bill 2555 from $1,000 to $3,000 reflect increased costs. The legislative history reveals that in 1991 a 'no-frills funeral' cost between $3,000 and $4,200.

This lack of change was highlighted again in 2015 when section 1 of House Bill 1073 directed the legislative management to study required motor vehicle insurance. The study included a review of all required motor vehicle insurance limits, specifically limits on no-fault benefits. The study concluded that no changes should be made to the existing provisions.

The smoother progression of the North Dakota provisions, particularly when compared to the upheaval seen in Florida and other states, is interesting. There are some similarities which seem to occur across states, for example the restriction on medical service providers and the limitations on certain types of treatment. Although not known as a state with high levels of fraud, North Dakota tightened up on the allowable treatments in 2005 in Senate Bill 2047.

There are differences in the demographics and other factors (climate, road configurations, vehicle usage patterns, population density, fraud levels, etc) which may explain the success or otherwise of no-fault. North Dakota is not densely populated and does not have the multiple large cities seen in other no-fault states such as Florida and Michigan. One of the very notable differences is the arrangement of payer priority. North Dakota limits the amount which is paid by a no-fault employer to the first $10,000, with healthcare and other insurers paying thereafter. Healthcare insurers' core business involves negotiating with healthcare providers to achieve cost-effective healthcare provision for their customers. By contrast, no-fault insurers have a more diverse profile of activities with potentially lower buying power, so they are unlikely to be able to achieve such competitive rates for healthcare costs. This apportionment arrangement combined with a dollar threshold may be key to the scheme's success in North Dakota.

G. Colorado

Colorado is one of four states that have repealed no-fault (not including the brief repeal seen in Florida). Nevada repealed no-fault in 1980; Georgia in 1991; Connecticut in 1994; and Colorado from 1 July 2003.

Colorado currently has mandatory insurance limits of $25,000 per person for bodily injury, $50,000 per accident for bodily injury, and $15,000 for property damage.[47] Uninsured motorist and underinsured motorist cover must be offered; the minimum coverage allowable must be equal to the bodily injury limits. Uninsured/underinsured coverage can be refused in writing by consumers. Uninsured/underinsured coverage can only be purchased as an addition to a motor vehicle or motorbike policy; it cannot be purchased as a stand-along policy.[48]

[47] Colorado Revised Statutes 42-7-103.
[48] ibid, 10-4-609.

Since 1 January 2009, a mandatory offer of Medical Payment coverage, or 'Med-Pay' coverage, must be made on all policies sold.[49] Med-Pay is a no-fault insurance to cover the cost of immediate medical payments up to the limit purchased by the consumer. The minimum Med-Pay coverage is $5,000. Med-Pay coverage can be refused in writing, but if it is not refused, then it is automatically added to a policy. Med Pay is primary to any healthcare insurance held by the insured, and the coverage and coordination of benefits under Med Pay are subject to rules made by the Commissioner.[50]

Prior to its revocation, Colorado no-fault had a combined verbal/monetary threshold of consisting or either the injuries resulting in death/dismemberment/permanent disability/permanent disfigurement or medical/rehabilitation costs equal to or over $2,500.

Standard coverage was the norm, with a minimum basic cover, which was open to families with annual household income of less than $20,000 from 14 May 1992 until 1 July 1997. Table 10.9 details the maximum benefits coverage available under the standard PIP and the basic policies.

Table 10.9: Colorado's repealed no-fault arrangements

	Standard PIP benefits maximums	Basic policy maximums (annual income ≤$20,000)
Medical benefits	$50,000 (within 5 years)	$25,000
Rehabilitation benefits	$50,000 (within 10 years)	$0
Loss of income benefit (gross)	$400 per week for 52 weeks	$5,000
Death benefit	$1,000 per person	$5,000 per person
Essential services	$25 per day for 52 weeks	$0

In Colorado, auto-insurers were primary, except where benefits were due under workers' compensation. There was provision for coordination of benefits between automobile and accident and health insurers for an appropriately reduced premium.

PIP coverage was required for private motor vehicles with exceptions made for motorcycles, mini-bikes, motorbikes, snowmobiles and other off-road vehicles. Subrogation provisions applied to collisions with non-private motor vehicles, whereby the injurer of the private passenger motor vehicle could seek reimbursement from the responsible owner or operator of the non-private public motor vehicle.

Background

A brief synopsis of Colorado's no-fault automobile insurance legislation will be given. This is not intended to be a comprehensive description of the legislation. The aim is merely to provide an outline of the provisions with a more detailed examination of how various sociolegal factors contributed to the instigation and demise of no-fault in Colorado, including the impact this has had on Coloradans.

[49] ibid, 10-4-635.
[50] ibid, 10-4-641.

Colorado enacted a no-fault auto insurance system under the Colorado Auto Accident Reparations Act 1974[51] (effective 1 April 1974). This followed attempts to introduce no-fault in 1972 with House Bill 1064 and by way of an 'initiative ballot' in November 1972 proposed by common cause.[52] The ballot was rejected at a public vote with a 74% no vote and 26% in favour.[53] While this seems a convincing rejection, there were drafting and policy errors in the Bill proposed at the ballot. The third attempt (a revision of House Bill 1064) proved more successful as the sponsors had listened to requests from the automotive insurance industry that they wished to be the primary payer of healthcare costs, with health insurers as secondary payers, and had modified the bill accordingly. This may seem counterintuitive, but at the time this type of policy was not available to automobile insurers, who saw this as a new business opportunity. Additionally, the rationale was that competition from automobile insurers would drive down healthcare insurers' costs, thereby making this proposal more attractive to consumers. Combined with the increased coverage provided by no-fault and the quicker processing of claims, the scheme appeared to be a win–win. Even at this point there were dissenting voices against no-fault, the public vote in 1972 was unequivocal, and in 1972 4 of the 16-strong Committee on Automobile Insurance tasked with assessing automobile insurance were unconvinced by the no-fault proposals put forward.[54]

No-fault then existed without considerable controversy in Colorado until the mid-1990s. Throughout this period reforms were fairly low level, which is why the threshold remained at a minimal $2,500 (up from $500 in 1985). This level would have been sufficient to keep cases from the courts when the legislation was enacted, but rapid rises in medical care costs meant that the threshold was ineffective by the 1990s. In addition, the law required all 'reasonable' treatments to be covered, including non-traditional treatments. In effect, claimants were able to claim from their own insurance company for any 'reasonable' medical treatment and, because of the low tort liability threshold, they could also go to court to collect for pain and suffering and other non-economic damages. This position was unsustainable and led to increases in premiums.

The Interim Committee Report to the General Assembly Recommendations for 2006 stated:[55]

> Based on data provided by the National Conference of State Legislatures (NCSL), premium costs for Colorado policyholders increased $290.95 during the 10-year period beginning in 1992 and ending in 2002. This marked the largest premium increase for any state during this period, comparing unfavorably to the national average premium increase of only $168.24. Most of the premium increase in Colorado was attributed to a substantial rise in PIP premium costs between 1999 and 2002. The table below compares premium increases in no-fault states between 1992 and 2002.

[51] ibid, 10-4-701–10-4-726 (repealed 1 July 2003).

[52] More detailed background can be found in the Colorado Legislative Council, Research Publ No 190, *Report to the Colorado General Assembly: Committee on Automobile Insurance* (Legislative Council Report, 1972), available from the Colorado State Publications Library Digital Repository at http://cospl.coalliance.org/fedora/repository/co:2872.

[53] Ballot results are available online from state archives at www.leg.state.co.us/lcs/ballothistory.nsf/835d2ada8de735e787256ffe0074333d/79812b464a8787fd87256ffd006a4947?OpenDocument.

[54] See the Colorado Legislative Council (n 53).

[55] Interim Committee on Auto Insurance, Research Publ No 540, *Report to the Colorado General Assembly*, Recommendations for 2006 (December 2005), available at http://cospl.coalliance.org/fedora/repository/co:3022.

Comparison of automobile insurance premium increase in modified no-fault states between 1992 and 2002		
State	**Premium increase**	**Difference between state amount and US average ($168.34)**
Colorado	$290.65	$122.31
New York	$290.38	$122.04
Minnesota	$260.46	$92.12
North Dakota	$259.69	$91.35
Utah	$247.87	$79.53
Kansas	$246.99	$78.65
Kentucky	$231.84	$63.50
Michigan	$206.20	$37.86
Florida	$191.34	$23.00
New Jersey	$163.97	–$4.37
Pennsylvania	$154.04	–$14.30
Massachusetts	$100.09	–$68.25
Hawaii	–$259.65	–$427.99

Insurers also struggled in the later stages of no-fault. There are two main mechanisms by which insurers remain solvent: firstly, by profits on premiums paid by policyholders; secondly, by returns made on the investments insurers make using policyholder premiums. This first income stream was diminishing by the 2000s. It was said:

> In 2002, PIP was costing Colorado insurers an average of $240.47 while actual premiums were only generating $224.03. This meant that for the 2002 policy period, automobile insurers on average paid out $16.44 more in claims than they received premium payments from policyholders.[56]

At the same time the returns from investments made by insurers were also declining, as there was very little available to invest. Therefore, reform was necessary not just for consumers, but for insurers as well.

Repeal

Although not repealed until 2003, the provisions relating to the repeal of Colorado's no-fault automobile laws were contained in section 8 of House Bill 1209 of 1997.[57]

[56] ibid, 6.

[57] House Bill 971209 available at www.leg.state.co.us/preclics/1997/hbills97/HB1209.htm. S 8 reads: 'Repeal. Part 7 of article 4 of title 10, Colorado Revised Statutes, 1994 Repl Vol, as amended, and sections 4241409 and 42-41410, Colorado Revised Statutes, 1993 Repl Vol, as amended, are repealed, effective July 1, 2001, unless the review conducted by the department of regulatory agencies pursuant to section 427608, Colorado Revised Statutes, and reported to the general assembly during the 1999 regular session of the general assembly indicates that the number of uninsured motorist claims reported by insurers declined between July 1, 1997, and the date of the department's review.'

This pre-dated some of the substantial rises in auto insurance premiums. Interestingly the sunset clause on no-fault was not central to the Bill, which primarily concerned the Motor Insurance Identification Database Program. Bill 1209 set a sunset date of 1 July 2001 if the report prepared by the regulatory agencies before 15 October 1999 had not demonstrated a decline in the number (not percentage) of uninsured motorist claims between 1 July 1997 at the date of their report.

From 1 July 2003 Colorado repealed its no-fault provisions. All new policies were founded in tort. Existing no-fault policies continued until they were due for renewal, at which point a new fault-based policy was offered.

The reason cited for the transition was very high auto insurance premiums. Coloradans were paying substantial premiums, which were on a steeply increasing trajectory. Even against this backdrop, the repeal of no-fault without any public discussion is remarkable. It was commented upon in the Department of Regulatory Affairs 2000 Sunset Report into the Colorado Motorist Insurance Identification Database Program.[58] Recommendation Three stated: '[I]t is not likely the General Assembly intended to repeal Colorado's entire No Fault Law based on the success of the MIDB without a public debate. The repeal provision in the No Fault Law should be removed unless the General Assembly determines, after a full hearing, that the law should be repealed.' Regardless of the intention of the intention of the General Assembly, this was what happened. There was never a public discourse on the repeal of no-fault.

The 'no-fault' provisions in Colorado were repealed by section 6 of Senate Bill 01-109, effective 1 July 2002.[59] However, a reprieve was given by Senate Bill 02-90 extending the sunset date to 1 July 2003.[60] The requirements under the review had been altered from a reduction in the number of uninsured motorist claims to a proportional decrease in uninsured motorists between 1 July 2000 and the review date.

A further sunset review in 2002 estimated that there had been a decline in the percentage of uninsured motorist claims, but as there had been an increase in car numbers during the period in question the overall number of uninsured motorist claims had increased.[61] The report concluded that it was not possible to attribute the decline in uninsured motorist claims to the database, and that a further period of study should be undertaken. However, this was not accepted, and no-fault was allowed to sunset.

Alternative compulsory tort insurance provisions were brought into place. As mentioned previously, motor vehicle owners are required to maintain liability insurance in minimum amounts of $25,000 per person for bodily injury, $50,000 per accident and $15,000 for property damage. Motorcycle owners are required to maintain complying policies. Insurance companies must keep both bodily injury liability and property damage liability statistics, which have to be filed annually with the Insurance Commissioner.

[58] Available at http://cdn.colorado.gov/cs/Satellite?blobcol=urldata&blobheadername1=Content-Disposition&blobheadername2=Content-Type&blobheadervalue1=inline%3B+filename%3D%22Motorist+Insurance+Identification+Database+Program%2C+Colorado+-+2000+Sunset+Review.pdf%22&blobheadervalue2=application%2Fpdf&blobkey=id&blobtable=MungoBlobs&blobwhere=1251815541848&ssbinary=true.

[59] http://tornado.state.co.us/gov_dir/leg_dir/olls/sl2001/sl_165.htm.

[60] http://tornado.state.co.us/gov_dir/leg_dir/olls/sl2002a/sl_189.htm.

[61] Available at http://cdn.colorado.gov/cs/Satellite?blobcol=urldata&blobheadername1=Content-Disposition&blobheadername2=Content-Type&blobheadervalue1=inline%3B+filename%3D%22Motorist+Insurance+Identification+Database+Program+Act+-+2002+Sunset+Review.pdf%22&blobheadervalue2=application%2Fpdf&blobkey=id&blobtable=MungoBlobs&blobwhere=1251815535810&ssbinary=true.

There remains a mandatory requirement to offer uninsured/underinsured motorist and collision coverage. Collison coverage must be offered on a no-fault basis against accidental property damage to the insured motor vehicle.

Impact of Repeal

The cost of motor vehicle insurance premiums seen in Colorado since the no-fault provisions sunset is complex, with some sources stating large premium reductions[62] whilst others claim an average 45% increase in pure premium driven by a substantial increase in claim frequency.[63] Some commentators have seized upon the reports of reduced premiums paid by consumers, as was required by Governor Owens, and hailed the repeal as a success. However, to consider only the impact on motor insurance premiums would be to ignore the overall impact on consumers. Given the more limited nature of primary insurance, a substantial proportion of drivers now rely on secondary payment sources to pay for medical treatment in the event of an accident. Healthcare insurance providers have seen increased claims costs since the sunset. Ascertaining both the overall cost to consumers and the level of service provided by these alternative approaches is complex,[64] and the fact that in 2009 and 2012 there were attempts to reinstate no-fault demonstrates that this is not a simple one-dimensional issue.

House Bill 09-1226 would have returned Colorado to a no-fault system, but has been postponed indefinitely. House Bill 09-1226 required the following minimum coverages:

- legal liability coverage of $25,000 to any one person and $50,000 to all persons in one accident;
- property damage coverage of $15,000 per accident;
- PIP regardless of fault up to $25,000 per person for medical care, up to $20,000 per person for rehabilitation services, and up to $15,000 for payment of lost income and essential services; and
- up to $5,000 for death benefits.

These PIP limits are substantially lower than those under the old no-fault regime.

Conclusions

It is interesting to note that the $2,500 threshold under no-fault in Colorado is identical to the $2,500 threshold seen in North Dakota, where no-fault appears to be functioning well. The obvious difference is the amount available under PIP benefits: North Dakota has a total benefits cap of $30,000, whereas in Colorado in excess of $130,000 could potentially have been claimed.

[62] See eg Insurance Research Council Press Release of 7 March 2008, www.insurance-research.org/sites/default/files/downloads/20080307.pdf.

[63] See table 18 of the Florida Office of Insurance Regulation: Review of Personal Injury Protection Legislation 13 September 2016, available at www.floir.com/siteDocuments/FLOIRReviewPIP20160913.pdf.

[64] See eg S Owings-Edwards, 'Hospital Cost Shifts as a Result of Changes in Auto Insurance Regimes' (2005) 392) *Journal of Business and Economics Research*, which uses Colorado as a representative state.

The other major differential is the coordination of payment. North Dakota allows coordination of medical payments, with motor insurers only covering the first $10,000. This was not the case in Colorado, where motor insurers were primary payers (only workers' compensation took priority) and there was no coordination.

There is little doubt that these factors made the enacted version of no-fault unsustainable in Colorado. If the intention was to 'fix' no-fault, then it is surprising that reforms to these elements were not tried, prior to repeal. The repeal of no-fault without either public consultation or attempts at reform serves as a reminder that compensation arrangements are potentially subject to the prevailing political and policy considerations.

The predicted reductions in Colorado auto insurance premiums since no-fault was repealed in 2003 have proved difficult to measure, with both reductions and increases in premiums being claimed, as discussed above. Hence, the impact (and success) of the repeal is difficult to clarify.

H. Conclusions

The effectiveness of no-fault automobile accident programmes in the USA is clearly highly variable on a state-by-state basis. However, it is worth examining the impact of some key factors, including the threshold (type and value), the required coverage limits, the primacy of auto-insurer payers, and the demographic composition of the state (urban, rural, populous, etc).

The impact of a monetary threshold on claiming behaviour is interesting. While some states clearly show a 'loading' of claims to overcome the required value, others do not. Colorado had a relatively low threshold of only $2,500, which was not updated in almost 20 years. An examination of the situation in North Dakota, which has a comparable threshold within a functional no-fault system, indicates that the threshold alone cannot determine the success of a no-fault scheme.

Required coverage limits may have more impact on the functionality of a scheme. The combination of a very high available limit with a low threshold was considered responsible for the demise of no-fault in Colorado. The impact of these limits on both consumer premiums and insurer profitability was substantial. However, Florida has a relatively low cap on total available compensation, but the scheme limps along, rather than functioning well. Coverage limits, while potentially useful to moderate a scheme, are not sufficient to ensure a healthy scheme.

Usage of medical services impacts on the costs of any no-fault scheme. It is difficult to determine whether the increased uptake and use of medical services seen in no-fault states represents 'over-claiming' in these states or whether the more limited provision seen in tort states is inadequate. The impact of the limitations in the 2012 reforms in Florida on the types of medical services is yet to be evaluated. Limiting available services to treatments with proven benefits does not seem to be unreasonable.

The impact of fraud on no-fault schemes is hugely variable, both between states and within states. Urban areas seem to be more affected, but the relationship is not straightforward. It should be remembered that fraud is a criminal activity, with appropriate sanctions should the parties chose to pursue them. The impact of bad faith claims in no-fault states is worthy of further consideration.

One of the major differentials between no-fault states is the priority of payment which the no-fault insurer takes. It was never envisaged in the original Keeton–O'Connell plan that no-fault insurers would be primary payers. Nora Freeman Engstrom argues persuasively that this is a key determinant in the demise of no-fault.[65] In the examples given above, it is striking that North Dakota, which adopts a cooperative policy, has a less contentious, more functional no-fault system than either Florida or Colorado, where the no-fault insurer is/was the primary payer.

[65] Freeman Engstrom (n 9).

11

The Virginia Birth-Related Neurological Injury Compensation Program

This Program was established to provide last-resort exclusive[1] lifelong compensation for those who suffer a qualifying birth-related injury in the state of Virginia. In the later 1980s the state faced a potential collapse in obstetric services; up to one-quarter of the state's obstetricians were considering closing their practices. At this point the charges for medical malpractice insurance were soaring and insurance companies were on the brink of eliminating coverage. In response to this crisis, Virginia worked with all interested parties—including physicians, associations, insurers, lawyers and others—to develop the Virginia Birth-Related Neurological Injury Compensation Program. It was established in 1987 by statute.[2] The Program was the first of its kind in the USA.

The Virginia Birth-Related Neurological Injury Compensation Program is a payer of last resort, effectively a safety net, meaning that it will pay only the expenses remaining after all available insurance, government programmes and any other sources have contributed their share. It provides the exclusive remedy[3] for covered birth-related neurological injuries (but not injuries due to congenital or genetic causes) in Virginia for births on or after 1 January 1988.[4]

All claims for admission to the Program must be made through the Virginia Workers' Compensation Commission.[5] Legal support for the Program is provided by the Virginia Office of the Attorney General.[6]

[1] It is an exclusive remedy against participating hospitals/physicians; if a claim involves only non-participating hospitals/physicians, then a tortious claim can still be initiated in the courts: for more details see section D of 38.2-5002 Code of Virginia at http://law.lis.virginia.gov/vacode/title38.2/chapter50/section38.2-5002/.

[2] The Virginia Birth-Related Neurological Injury Compensation Act (1987, c 540) taken from the Code of Virginia Title 32.8 Insurance Chapter 50 Virginia Birth-Related Neurological Injury Compensation Act, available at http://law.lis.virginia.gov/vacode/title38.2/chapter50/section38.2-5000/.

[3] 38.2-5002 Code of Virginia, http://law.lis.virginia.gov/vacode/title38.2/chapter50/section38.2-5002/.

[4] 38.2-5014 Code of Virginia, http://law.lis.virginia.gov/vacode/title38.2/chapter50/section38.2-5014/.

[5] 38.2-5003 Code of Virginia, http://law.lis.virginia.gov/vacode/title38.2/chapter50/section38.2-5003/.

[6] 38.2-5002.1 Code of Virginia, http://law.lis.virginia.gov/vacode/title38.2/chapter50/section38.2-5002.1/.

A. Organisation and Structure

The organisation and structure of the Program are established by statute, including provisions concerning the auditing, composition of the Board of Directors and structure of the fund. The Program is governed by a nine-member Board of Directors. The Board is appointed by the Governor of Virginia, with six citizen representatives and one representative each of participating physicians, participating hospitals and liability insurers. Some of the citizen representatives also have relevant experience, including legal and financial expertise. The Board's powers and composition are defined by the enabling legislation. The Board hires an Executive Director and a Deputy Director to oversee the day-to-day operations. Additional staff are hired as required to support the Executive Director.

The Program is funded through the Virginia Birth-Related Neurological Injury Compensation Fund (the Fund), which is held as a trust fund for the benefit of claimants and to cover the administrative costs of the Program.[7] The assets of the Fund are administered by the Board of Directors of the Program with an annual audit as per the empowering legislation.[8] The Board is required to retain investment advisors to manage the Program's assets,[9] and these advisors have to produce annual statements for the Board.[10]

B. Program Funding[11]

The Program receives no state funding; it is funded entirely through contributions (termed assessments) from participating healthcare providers, hospitals and insurers. All changes in contribution levels require a legislative action.

Participating physicians are required to pay an assessment. Since 2013 this has been set at $6,200. In addition, all licensed physicians that do not participate in the Program are required to pay a fee of $300 annually as a condition of being licensed in Virginia. A number of exclusions to the assessments apply for physicians with extenuating circumstances.

Hospitals pay an assessment of $55 per live birth to participate, subject to an annual maximum of $200,000 in assessments.

Finally, if, and only if, the Program is determined not to be actuarially sound, an assessment of up to 0.25% of all 'net direct premiums written' by liability insurers in Virginia may be charged. These assessments of liability insurers have been charged at the maximum amount for many years.

Claimants are required to pay a $15 dollar filing fee.

[7] 38.2-5015(A) Code of Virginia, http://law.lis.virginia.gov/vacode/title38.2/chapter50/section38.2-5015/.

[8] 38.2-5015(B) Code of Virginia, http://law.lis.virginia.gov/vacode/title38.2/chapter50/section38.2-5015/.

[9] 38.2-5016(F) Code of Virginia, http://law.lis.virginia.gov/vacode/title38.2/chapter50/section38.2-5016/.

[10] 38.2-5016.1 Code of Virginia, http://law.lis.virginia.gov/vacode/title38.2/chapter50/section38.2-5016.1/.

[11] 38.2-5020 Code of Virginia, http://law.lis.virginia.gov/vacode/title38.2/chapter50/section38.2-5020/. Data taken from the 2014 annual actuarial reports of the Program produced by Pinnacle. Available at http://vabirthinjury.com/wp-content/uploads/2015/01/2014-Actuarial-Scan.pdf.

C. Eligible Injuries

The injury must have resulted from oxygen deprivation or mechanical injury during labour, delivery or resuscitation in the immediate post-delivery period in a hospital. The injury must result in both physical and mental impairment.[12] In addition, either the obstetrical services related to the birth must be provided by a participating physician or they must have occurred in a participating hospital, or both.

Participation is voluntary for physicians, registered nurses, midwives and hospitals.[13] Public awareness of the Program is achieved by providing information about the Program to all parents in the state. All physicians, hospitals and midwives, regardless of whether or not they participate in the Program, must inform their patients of the Program in writing according to the stipulation of the Board of Directors of the Program. All healthcare providers are obliged to indicate whether they participate in the Program or not to parents. A list of participating providers is also available on the Birth Injury Compensation Fund website.[14] Additionally if a baby is hospitalised in a neonatal intensive care unit, then the mother (or other appropriate adult) must be provided with an informational brochure approved by the Board of Directors. The brochure will detail the rights, remedies and limitations under the Program.

If a claim is made to an insurance company licensed to do business in the Commonwealth of Virginia or to a self-insurer, alleging a qualifying injury, the insurer is obliged to report the claim to the Program on a specified form (the making of such a report does not infer any liability). Upon receipt of this information the Program will contact the parents or guardians of the child to inform them of the Program and the eligibility criteria.[15]

D. Limitations

All claims under the Programme must be filed before the child reaches his/her 10th birthday.[16]

E. The Claims Process

Figure 11.1[17] shows a schematic of the potential routes to claiming under the Program.

[12] 38.2-5001 Code of Virginia, http://law.lis.virginia.gov/vacode/title38.2/chapter50/section38.2-5001/.

[13] 38.2-5001 Code of Virginia, http://law.lis.virginia.gov/vacode/title38.2/chapter50/section38.2-5001/.

[14] www.vabirthinjury.com/participating-providers/.

[15] 38.2-5004.1 Code of Virginia, http://law.lis.virginia.gov/vacode/title38.2/chapter50/section38.2-5004.1/.

[16] 38.2-5013 Code of Virginia, http://law.lis.virginia.gov/vacode/title38.2/chapter50/section38.2-5013/.

[17] Taken from the Virginia Birth-Related Neurological Injury Compensation Program Participant Handbook 2011 (with 2016 revisions) http://vabirthinjury.com/wp-content/uploads/2016/08/2016-Handbook-4-2.pdf.

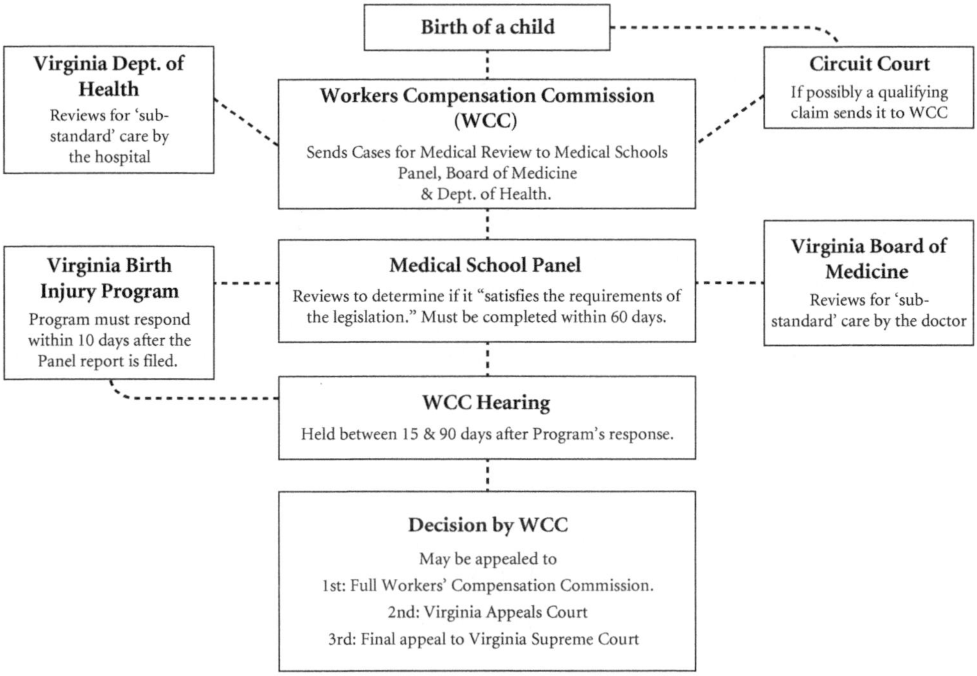

Figure 11.1: Schematic of the claims process

All claims are handled by the Virginia Workers Compensation Commission.[18] A petition[19] is filed (along with a $15 dollar filing fee) and has to include all of the following:[20]

(a) the name and address of the legal representative and the basis for his representation of the injured infant;
(b) the name and address of the injured infant;
(c) the name and address of any physician providing obstetrical services who was present at the birth and the name and address of the hospital at which the birth occurred;
(d) a description of the disability for which claim is made;
(e) the time and place where the birth-related neurological injury occurred;
(f) a brief statement of the facts and circumstances surrounding the birth-related neurological injury and giving rise to the claim;
(g) all available relevant medical records relating to the person who allegedly suffered a birth-related neurological injury and an identification of any unavailable records known to the claimant and the reasons for their unavailability;
(h) appropriate assessments, evaluations and prognoses, and such other records and documents as are reasonably necessary for the determination of the amount of

[18] www.vwc.state.va.us/portal/vwc-website.
[19] This is available online at www.vabirthinjury.com/eligibility-frequently-asked-questions/.
[20] 38.2-5004 Code of Virginia, http://law.lis.virginia.gov/vacode/title38.2/chapter50/section38.2-5004/.

compensation to be paid to, or on behalf of, the injured infant on account of a birth-related neurological injury;
(i) documentation of expenses and services incurred to date, which indicates whether such expenses and services have been paid for, and if so, by whom; and
(j) documentation of any applicable private or governmental source of services or reimbursement relative to the alleged impairments.

The petition itself is fairly straightforward. The additional information requires a third-party provide medical records to the parents/guardian. To ensure this happens there is an obligation on hospitals to release the medical records (including fetal monitoring strips, printed or electronic) to the legal representative of the child once they have been notified of the intention to file a petition.[21] Upon receipt of a petition the Workers' Compensation Commission serves notice on the Program.

Programme Investigations

Investigating each claim is a multistage process, proscribed by statute with set time-limits. The Commission sets out to establish the following on the basis of the evidence provided.

1. *Whether the injury claimed is a birth-related neurological injury rather than a congenital or genetic condition?*

This is decided on the facts of the case, including information detailed in medical records. In determining this, a rebuttable presumption arises that the injury alleged is a birth-related neurological injury where it has been satisfactorily demonstrated that the infant has sustained a brain or spinal cord injury caused by oxygen deprivation or mechanical injury, such that the infant was thereby rendered permanently disabled. This disability must manifest as a motor disability combined with either a developmental disability or for infants sufficiently developed to be cognitively evaluated, a cognitive disability.

Any party disagreeing with this presumption has the burden of proving that the injuries alleged are not birth-related neurological injuries as defined by the legislation.

Should a hospital fail to provide fetal heart monitor traces, then a rebuttable presumption of fetal distress, an element of a birth-related injury, arises.

2. *Whether obstetrical services were delivered by a participating physician at the birth?*

3. *Whether the birth occurred in a participating hospital?*

4. *How much compensation, if any, is to be awarded?*

If the Commission determines either that the injury alleged is not a qualifying injury or that obstetrical services were not delivered by a participating physician at the birth and that the birth did not occur in a participating hospital, it shall dismiss the petition and cause a copy of its order of dismissal to be sent immediately to the parties by registered or certified mail.

[21] F 38.2-5004(E) and (F) Code of Virginia, http://law.lis.virginia.gov/vacode/title38.2/chapter50/section38.2-5004/.

Panel Investigations

To assist the Commission in reaching its decision outside expertise is sought. When a petition is entered, a Panel is convened to investigate the injury and circumstances. The major function of the Panel is checking the facts of the case against the eligibility criteria. The Panel consists of three qualified impartial doctors (from relevant specialities—obstetrics, paediatrics, paediatric neurology, neonatology, physical medicine and rehabilitation—or any other specialty deemed appropriate given the facts of a particular case). There is a limit of one obstetrician per Panel. The Panels are convened by the Deans of the Medical School of Eastern Virginia, the Medical School of the University of Virginia and the Medical College of Virginia of the Virginia Commonwealth University. Each of these medical schools maintains a pool of potential Panel Members, and the Panels rotate from school to school on a case-per-case basis. The remuneration paid to the Medical School is set by statute at $3,000 per claim reviewed.

The Panels create a report for the Commission with a detailed statement of the opinion of the Panel's members on whether or not the child's injuries are eligible injuries addressing all of the elements for eligibility. The report must include the basis on which the Panel reached its conclusion for each individual element. The Panel must file its report within 60 days of the date the Petition was filed with the Commission. The Panel sends copies of its report to the Commission, the Program and to all parties to the proceedings. At the request of the Commission at least one member of the Panel must be available to testify at the hearing.

The Panel report must be considered by the Commission, but is not binding on it.[22] Within 10 days of the Panel report being filed with the Commission, the Program will respond to the petition and submit relevant written information as to whether the injury is an eligible injury (as defined above).

Hearings[23]

If the Program files a response to the petition deeming the injury eligible, then the Commission shall set a date for a hearing and notify the parties. The Claimant and the Program are required parties for the hearing. The Hearing must be between 15 and 90 days from the filing of the Program's response, and shall be held in the city or county where the birth-related neurological injury occurred, or in a contiguous city or county, unless otherwise agreed to by the parties and authorised by the Commission.

All parties to a claim have the right to serve interrogatories or to cause the depositions of witnesses[24] (residing within or without of the Commonwealth of Virginia) and to cross-examine witnesses.[25]

[22] 38.2-5008(C) Code of Virginia, http://law.lis.virginia.gov/vacode/title38.2/chapter50/section38.2-5008/.
[23] 38.2-5006 Code of Virginia, http://law.lis.virginia.gov/vacode/title38.2/chapter50/section38.2-5006/.
[24] 38.2-5007 Code of Virginia, http://law.lis.virginia.gov/vacode/title38.2/chapter50/section38.2-5007/.
[25] 38.2-5008.1 Code of Virginia, http://law.lis.virginia.gov/vacode/title38.2/chapter50/section38.2-5008.1/.

F. Awards

If it has been determined that the child fits the required definition of a qualifying injury and the care was delivered by a participating provider, then the Commission will make an award, with compensation available under the following four heads.[26]

1. Medical Expenses

Funds provided under this head cover actual medically necessary and reasonable expenses of medical and hospital, rehabilitative, therapeutic, nursing, attendant, residential and custodial care and service, medications, supplies, special equipment or facilities, and related travel (these expenses are paid as they are incurred).

Reimbursement may be provided for nursing and attendant care that is provided by a relative or legal guardian of a Program beneficiary when the care given is beyond the scope of childcare duties and services normally and gratuitously provided by family members to uninjured children.

These are calculated at the rate that a similarly injured person would pay for treatment in that locality if the injured person were funding his own treatment. The Program is a payer of last resort. To calculate the value of any award due to a Program beneficiary, his needs will be assessed and the amounts of any of the following which he is due/has received will be deducted:

(a) expenses for items or services that the beneficiary has received, or is entitled to receive, under the laws of any state or the federal government except to the extent prohibited by federal law;
(b) expenses for items or services that the beneficiary has received, or is contractually entitled to receive, from any prepaid health plan, health maintenance organisation, or other private insuring entity;
(c) expenses for which the beneficiary has received reimbursement, or for which the beneficiary is entitled to receive reimbursement, under the laws of any state or federal government except to the extent prohibited by federal law; and
(d) expenses for which the beneficiary has received reimbursement, or for which the beneficiary is contractually entitled to receive reimbursement, pursuant to the provisions of any health or sickness insurance policy or other private insurance programme.

These expenses may be met:

— by directly awarding the sums involved to the family; or
— by requiring the family to purchase private health insurance which provides coverage for such expenses with the Fund paying the cost of the premium and other such expenses; or

[26] 38.2-5009 Code of Virginia, http://law.lis.virginia.gov/vacode/title38.2/chapter50/section38.2-5009/.

— by requiring the claimant to participate in the State Medicaid Program, the Children's Health Insurance Program or other state or federal health insurance programme for which the claimant is eligible, or
— if the Commission determines that it would be unreasonably burdensome to require the claimant to purchase private health insurance and that the infant is ineligible for a health insurance programme described above, to make an award providing compensation for the cost of private accident and sickness insurance for the infant.

2. Loss of Earnings

These are available from the age of 18, and are to be paid in regular instalments beginning on the 18th birthday of the claimant. An infant found to have sustained a birth-related neurological injury shall be conclusively presumed to have been able to earn income from work from the age of 18 until the age of 65, if he had not been injured, in the amount of 50% of the average weekly wage in the Commonwealth of workers in the private, non-farm sector. Payments shall be calculated based on the Commonwealth's reporting period immediately preceding the 18th birthday of the claimant child, and subsequently adjusted based upon the succeeding annual reports of the Commonwealth.

3. Reasonable Legal Expenses

Legal expenses incurred by the claimant in connection with the filing of a claim are available, subject to certain stipulations. Any award for expenses, including attorney's fees, incurred by the claimant in connection with the filing of a claim under this chapter shall be subject to the approval and award of the Commission.[27]

Once an award has been decided a copy of the award shall be sent immediately by registered or certified mail to the parties.

The Commission will not make any payments towards attorney's fees or other expenses incurred by any physician, hospital or nurse midwife that is party to a proceeding, or by a medical malpractice liability insurer of such party. This prohibition does not affect the statutory requirement to reimburse for photocopying costs, or the requirement that the Program compensate the Office of the Attorney General for its provision of legal services to the Program.

4. Death of a Child

If a child does not survive more than 180 days and fulfils the eligibility criteria, the Commission has the discretion to make an award of up to $100,000 to the infant's family.[28]

[27] 38.2-5009(2) Code of Virginia, http://law.lis.virginia.gov/vacode/title38.2/chapter50/section38.2-5009/.

[28] As used in this section, an infant's family means the infant's father, mother, or both, or if neither is a party to the proceeding, the infant's legal guardian. 38.2-5009.1(D) Code of Virginia, http://law.lis.virginia.gov/vacode/title38.2/chapter50/section38.2-5009.1/.

This award is in addition to, and not in lieu of, any other award providing compensation. As for living claimants, prior to making an award the Commission shall conduct a hearing to determine whether such award is appropriate and, if so, the proper amount of such an award and how it should be paid. To enable this process, they will receive evidence pertaining to sorrow, mental anguish, solace, grief associated with the death of the infant and all other material factors that are relevant. This hearing may be conducted as part of a hearing to determine eligibility, with the same procedural requirements as those outlined above.

G. Appeals

There are three levels of appeal; the first is to have the case re-examined by the full Commission, then the Court of Appeal then the Supreme Court.[29]

Appeals for a review by the full Commission.[30] These can re-examine either the determination of eligibility to the Program or the value of the Award made by the Program. An application must be made within 20 days from the date of a determination or 20 days from the date of the award by the Commission. The full Commission (excluding any member of the Commission who made the determination or award, if the first hearing was not held before the full Commission) shall review the evidence. If deemed advisable and as soon as practicable, the Commission instead may hear the parties, their representatives and witnesses and shall make a determination or award, as appropriate. Such review or determination, together with a statement of the findings of fact, rulings of law and other matters pertinent to the questions at issue, shall be filed with the record of the proceedings and shall be sent immediately to the parties.

If a claimant is still dissatisfied with the outcome after a Review by the full Commission he can appeal to the Court of Appeal,[31] under the Supreme Court Rules. Should this fail to resolve the issues, then a final appeal is permitted to the Supreme Court.

H. Parallel Regulatory Investigations into Care

Once a petition/claim has been received by the Program, investigations ensue. The exact nature of these depends upon the nature of the complaint.[32] The investigations include establishing the individual circumstances of the case and potentially proscribing regulatory change, which can range from advice to individual practitioners to mandatory systematic changes for hospitals.

If a complaint is against a physician, it is investigated by the Department of Health Professions, utilising a standardised process for investigating complaints against doctors. If the

[29] 38.2-5011 Code of Virginia, http://law.lis.virginia.gov/vacode/title38.2/chapter50/section38.2-5011/.
[30] 38.2-5010 Code of Virginia, http://law.lis.virginia.gov/vacode/title38.2/chapter50/section38.2-5010/.
[31] 38.2-5011 Code of Virginia, http://law.lis.virginia.gov/vacode/title38.2/chapter50/section38.2-5011/.
[32] 38.2-5004 Code of Virginia, http://law.lis.virginia.gov/vacode/title38.2/chapter50/section38.2-5004/.

Board feels the conduct is sufficient to give rise to disciplinary proceedings it will then refer the case to the Board of Medicine. If a notice or order is made by the Board of Medicine, a copy of it must be mailed to the petitioner or claimant.

If the complaint is against a participating hospital, then the Department of Health will investigate the claim, using standardised procedures laid out in Title 32.1 of the Code of Virginia. Should the Department of Health conclude that substandard care contributed to the injury then it shall take any action it deems appropriate within the limits of its powers.

I. Claim Statistics[33]

The delay between a child's birth and the development and/or confirmation of damage means that at any given point there will be children who have been born, and who fulfil the criteria for the Program, but who have not yet been admitted to the Program. As of 31 December 2013, there were 183 accepted Program participants (Admitted Participants) and it is estimated that there are another 48 children who will eventually be admitted to the Program (Non-Admitted Participants). The number of Admitted Participants combined with the number of Non-Admitted Participants gives the Ultimate Participant Number (231). Table 11.1 provides a breakdown of the Participants in these categories by birth year. Once the 10-year limitation period has elapsed the Admitted Participants number becomes the Ultimate Participant number and it is not possible for any more children born in that year to be admitted to the Program; for the most recent 10 years these figures are based on estimates.

Table 11.1: Admitted, Non-Admitted and Ultimate Participants by birth year 1988–2013

Birth year	Admitted Participants	Non-Admitted Participants	Ultimate Participants
1988	2	0	2
1989	9	0	9
1990	5	0	5
1991	9	0	9
1992	8	0	8
1993	11	0	11
1994	6	0	6
1995	11	0	11
1996	8	0	8
1997	10	0	10

(continued)

[33] Data taken from the 2014 annual actuarial reports of the Program produced by Pinnacle. Available at http://vabirthinjury.com/wp-content/uploads/2015/01/2014-Actuarial-Scan.pdf.

Table 11.1: ***(Continued)***

Birth year	Admitted Participants	Non-Admitted Participants	Ultimate Participants
1998	7	0	7
1999	6	0	6
2000	13	0	13
2001	13	0	13
2002	11	0	11
2003	11	0	11
2004	4	1	5
2005	4	1	5
2006	7	2	9
2007	8	3	11
2008	7	4	11
2009	8	3	11
2010	1	8	9
2011	1	8	9
2012	0	10	10
2013	0	10	10

Breakdown of the Fund's Outgoings by Category

Table 11.2 shows the estimated future financial position, including the expense detailed as administration.

Table 11.2: Estimate of future claim payments and administrative expenses of Admitted and Non-Admitted Claimants as of 31 December 2013 ($ millions, on a present value basis)

	Claimant numbers	Estimate of future claim payments	Estimate of future claim administration expenses	Value of total Assets	Forecasted surplus/ (deficit)
Admitted Claimants	183	286.0	17.6		
Non-Admitted Claimants	48	107.9	4.8		
Grand total	231	393.9	22.4	383.8	(32.6)

On this basis, the administrative expenses form a very small proportion of the total paid out by the fund, approximately 5%. This appears to give the Program a very low level of administrative costs. However, it should be remembered that some of the cost which might be classified as administrative in other systems are included in the benefits payments. Table 11.3 gives a breakdown of the categories which make up the benefits payments. Medical Review/Intake costs and Legal costs make up a relatively small proportion of benefits, (0.17% and 2.57%, respectively, of the total payments made up to 2013) but they still amount to $3,889,935 million up to 2012. The $145,000 increase in legal fees between 2012 and 2013 is notable, particularly as $246,000 of the legal fees paid in 2013 was attributable to just one individual. Clearly increases on this scale are concerning, particularly as the increase is driven by a low number of high-value claims.

Table 11.3: Payment by category for 2010, 2011 and all years up to (and not including) 2012

	Total payments up to 31 December 2013 ($)	% of total payments up to 31 December 2013	2012 Payments ($)	% of 2012 payments	2013 payments ($)	% of 2013 payments
Nursing	90,683,905	69.96%	8,910,682	72.97%	9,762,227	73.72%
Hospital/ physician	2,254,514	1.59%	58,144	0.48%	75,913	0.57%
Physical therapy	3,671,919	2.59%	299,958	2.46%	338,736	2.56%
Medical equipment	2,752,361	1.94%	161,997	1.33%	187,404	1.42%
Prescription drugs	1,488,826	1.05%	89,796	0.74%	83,506	0.63%
Incidental	4,365,891	3.27%	245,832	2.01%	251,591	1.90%
Housing	19,764,980	13.94%	573,908	4.70%	562,719	4.25%
Vans	8,013,173	5.65%	760,684	6.23%	538,732	4.07%
Insurance	1,542,464	1.09%	144,813	1.19%	160,393	1.21%
Lost wages	3,087,417	2.18%	738,391	6.05%	916,958	6.92%
Medical review/ intake	239,398	0.17%	12,425	0.10%	3,575	0.03%
Legal	3,650,537	2.57%	215,190	1.76%	360,583	2.72%
Total	141,785,386	100.00%	12,211,818	100.00%	13,242,387	100.00%

Source: 2014 actuarial report on the Virginia Birth-Injury Neurological Injury Compensation Program available at www.vabirthinjury.com/publications/.

By far the largest cost is the cost of nursing, which is not surprising. Housing is another substantial cost, again given that a high proportion of recipients are either physically disabled or non-ambulatory this is to be expected. Lost wages make up a relatively small proportion of the outgoings. This may seem surprising if you consider that they are payable from aged 18 to 65, but wages are paid at 50% of the average wage in the state and the life expectancy of claimants is such that payments for lost wages, where they are started at all, are likely to be made for a far shorter duration.

J. Actuarial Soundness of the Fund

A deficit of $32.6 million shows the Fund is not actuarially sound as the predicted payments are greater than the predicted income. However, as of 31 December 2013, the Fund has sufficient funding to meet the liabilities of eligible claimants (Admitted and Non-Admitted) for over 30 years. The Fund does not have the ability to generate additional funding from insurer assessments or physician and/or hospital fees without changes to legislation, and its liabilities are likely to increase. The Fund can generate additional income by judicious investment of the Fund assets.

Although currently not actuarially sound, it is worth noting the fact that the expected future annual income generated by the Fund is greater than the current expected annual present lifetime value of new participant liabilities. Therefore, if investment returns continue this predicted trajectory, the Fund deficit should decrease over time.

K. Theft from the Fund

In 2010 Iris Allen, a former claims manager from the Fund, was sentenced to 10 years for defrauding the Fund of around $755,000.[34] She had inflated genuine claims and kept the difference as well as making entirely fraudulent claims, which were paid to companies she owned. This loss highlighted the need, as far as is possible, to protect funds against such criminal actions. A forfeiture money judgment was issued against Allen for almost $1.3 million.

L. Conclusions

The Fund operates on a limited scope as it is a payer of last resort, but it serves a very important function, both as a safety net for individual claimants and as a mechanism for reducing the premiums for liability insurance for healthcare providers.

The Fund has a no-fault approach to eligibility, with an inquisitorial investigation of the facts by doctors. The issues of eligibility and quantum are dealt with at separate stages of the process and by different individuals.

[34] https://archives.fbi.gov/archives/richmond/press-releases/2010/ri052710.htm.

Of particular interest is the interaction between the Fund and other organisations: the Circuit Courts are required to send potentially qualifying claims to the Workers' Compensation Commission and the Commission is required to report to the regulatory bodies, the Virginia Department of Health and the Virginia Board of Medicine. This joined-up approach to compensation and regulation should help to avoid the 'lost data' issues seen in other jurisdictions. Too often information on potentially substandard care progresses through a single channel, usually the courts, and does not always reach regulators.

There is considerable flexibility in the Program for ensuring that claimants receive healthcare, from the mandating of private family healthcare insurance, to the requirement to enrol in state-funded programmes, such as Medicaid. This flexibility enables the Fund both to control its expenditure and provide continuity of care for their beneficiaries.

The funding shortfall is not immediately concerning, and should resolve given time and favourable investment returns. However, if the investments do not materialise as expected or the expenditure is greater than predicted, then the deficit will grow. The Fund is fully codified, with no discretion to raise additional funds, unless there is a legislative change. With the benefit of hindsight, it is easy to say that more adaptable framework or a sliding levy scale established on the basis of need would be more suitable than fixed assessments.

The scheme is generally felt to be a success. A 2002 Joint Legislative and Review Commission stated that: 'In addition to serving more birth-injured children than the tort system, the program provides benefits that exceed the medical malpractice cap for the typical child.'[35]

[35] Page iv of the Review of the Virginia Birth-Related Neurological Injury Compensation Program. The Joint Legislative Audit and Review Commission of the Virginia General Assembly 2003. Available at http://jlarc.virginia.gov/pdfs/reports/Rpt284.pdf.

12

The Florida Birth-Related Neurological Injury Compensation Association

The Florida Birth-Related Neurological Injury Compensation Association (NICA)[1] was established to provide last-resort exclusive[2] lifelong compensation for those who suffer a qualifying birth-related injury in the state of Florida. The reasons behind the Florida and Virginia Birth-Related Neurological Injury Compensation schemes were similar: in the later 1980s both states faced a potential collapse in obstetric services as malpractice insurance prices were soaring and insurance companies were on the brink of eliminating coverage.

NICA was created by statute in 1988.[3] It is a payer of last resort, effectively a safety net, meaning that it will pay only the expenses remaining after all available insurance, government programmes, and any other sources have contributed their share. It provides the exclusive remedy for qualifying birth-related neurological injuries (but not injuries due to congenital or genetic causes) in Florida for births on or after 1 January 1989.[4] NICA replaces the tort system. Claimants are only permitted to file civil actions in very limited circumstances.

All claims for admission to the programme must be made through filing a petition with the Division of Administrative Hearings and are determined by an Administrative Law Judge (ALJ).[5]

A. Organisation and Structure

The organisation and structure of the programme are established by statute, including provisions concerning the auditing, composition of the Board of Directors and structure of

[1] Further information available at www.nica.com/.

[2] The exclusivity of the remedy is defined by s 766.303 of the Florida Statutes, available at www.leg.state.fl.us/Statutes/index.cfm?App_mode=Display_Statute&Search_String=&URL=0700-0799/0766/Sections/0766.303.html.

[3] Ss 766.301–766.316 of the Florida Statutes, available at www.leg.state.fl.us/Statutes/index.cfm?App_mode=Display_Statute&URL=0700-0799/0766/0766ContentsIndex.html&StatuteYear=2013&Title=-%3E2013-%3EChapter%20766.

[4] S 766.303(1) of the Florida Statutes, available at www.leg.state.fl.us/Statutes/index.cfm?App_mode=Display_Statute&Search_String=&URL=0700-0799/0766/Sections/0766.303.html.

[5] S 766.305 of the Florida Statutes, available at www.leg.state.fl.us/Statutes/index.cfm?App_mode=Display_Statute&Search_String=&URL=0700-0799/0766/Sections/0766.305.html.

the Fund.[6] NICA is governed by a five-member Board of Directors, with directors operating on staggered terms of 3 years or until their successor is appointed. The Board is appointed by the Chief Financial Officer of the State of Florida with one representative of each of the following: participating physicians, participating hospitals, liability insurers, a citizen representative and non-participating physicians.[7] The Board's powers and composition are defined by the enabling legislation. The Board hires an Executive Director to oversee the day-to-day operations. Additional staff are hired as required to support the Executive Director.

Funds held on behalf of the plan are funds of the State of Florida.[8] The assets of the Fund are administered by the Board of Directors of the programme with an annual audit, which is to be made available to any participant who requires it, as per the empowering legislation.[9]

B. NICA Funding[10]

NICA currently receives no funding from the state; it is funded entirely through contributions (termed assessments) from participating healthcare providers, hospitals and, if required, insurers. All changes in contribution levels require a legislative action. An initial payment of $20 million was made to NICA from the Florida Department of Financial Service Insurance Regulatory Trust Fund.

Participating physicians are required to pay an assessment set at $5,000 ($2,500 for participating midwives). A number of exclusions to the assessments apply for physicians with extenuating circumstances. In addition, all licensed physicians that do not participate in the NICA are required to pay a fee of $250 annually as a condition of being licensed in Florida.

Hospitals pay an assessment of $50 per live birth to participate.

[6] S 766.315 of the Florida Statutes, available at www.leg.state.fl.us/Statutes/index.cfm?App_mode=Display_Statute&Search_String=&URL=0700-0799/0766/Sections/0766.315.html.

[7] S 766.315(2) of the Florida Statutes, available at www.leg.state.fl.us/Statutes/index.cfm?App_mode=Display_Statute&Search_String=&URL=0700-0799/0766/Sections/0766.315.html. The Chief Financial Officer may select the representative of the participating physicians from a list of at least 3 names to be recommended by the Florida Obstetric and Gynecologic Society; the representative of hospitals from a list of at least 3 names to be recommended by the Florida Hospital Association; the representative of casualty insurers from a list of at least 3 names, one of which is recommended by the American Insurance Association, one by the Alliance of American Insurers, and one by the National Association of Independent Insurers; and the representative of physicians other than participating physicians from a list of 3 names to be recommended by the Florida Medical Association and a list of 3 names to be recommended by the Florida Osteopathic Medical Association. In no case shall the Chief Financial Officer be bound to make any appointment from among the nominees of such respective associations.

[8] S 766.315(5)(e) of the Florida Statutes, available at www.leg.state.fl.us/Statutes/index.cfm?App_mode=Display_Statute&Search_String=&URL=0700-0799/0766/Sections/0766.315.html.

[9] S 766.315(5)(d) of the Florida Statutes, available at www.leg.state.fl.us/Statutes/index.cfm?App_mode=Display_Statute&Search_String=&URL=0700-0799/0766/Sections/0766.315.html.

[10] S 766.314 of the Florida Statutes, available at www.leg.state.fl.us/Statutes/index.cfm?App_mode=Display_Statute&Search_String=&URL=0700-0799/0766/Sections/0766.314.html.

Finally, if, and only if, NICA is determined not to be actuarially sound, a transfer of up to $20 million can be made from the Insurance Regulatory Trust Fund. If further funding is required casualty insurance carriers can be mandated to pay a contribution, which is limited by statute to equal to or less than 0.25% of net premiums. NICA is currently actuarially sound.

Claimants are required to pay a $15 dollar filing fee to the Court.

C. Eligible Injuries

A birth-related neurological injury is defined as an injury to the brain or spinal cord of a live infant weighing at least 2,500 grams for a single gestation or, in the case of a multiple gestations, a live infant weighing at least 2,000 grams at birth caused by oxygen deprivation or mechanical injury occurring in the course of labour, delivery or resuscitation in the immediate post-delivery period in a hospital, which renders the infant permanently and substantially mentally and physically impaired. This definition shall apply to live births only and shall not include disability or death caused by genetic or congenital abnormality.[11] In addition the obstetrical services related to the birth must be provided by a participating physician or they must have occurred in a participating hospital, or both.

Participation is voluntary for physicians, registered nurses, midwives and hospitals.

Public awareness of the programme is achieved by providing information about NICA to parents in Florida. All physicians, hospitals and midwives who participate in the programmemust inform their patients of the programme by providing them with written information on forms furnished by NICA.[12] A list of participating providers is available on the NICA website.[13]

D. Limitations

All claims under NICA must be filed before the child reaches his/her 5th birthday[14] (prior to 1994 this was 7 years).

[11] S 766.302(2) of the Florida Statutes, available at www.leg.state.fl.us/Statutes/index.cfm?App_mode=Display_Statute&Search_String=&URL=0700-0799/0766/Sections/0766.302.html.

[12] S 766.316 of the Florida Statutes, available at www.leg.state.fl.us/Statutes/index.cfm?App_mode=Display_Statute&Search_String=&URL=0700-0799/0766/Sections/0766.316.html. These forms are also available on NICA's website in multiple languages under the heading 'Brochures' at www.nica.com/parents/forms_brochures.html.

[13] www.nica.com/participating_providers.html.

[14] www.leg.state.fl.us/Statutes/index.cfm?App_mode=Display_Statute&Search_String=&URL=0700-0799/0766/Sections/0766.313.html.

E. The Claims Process

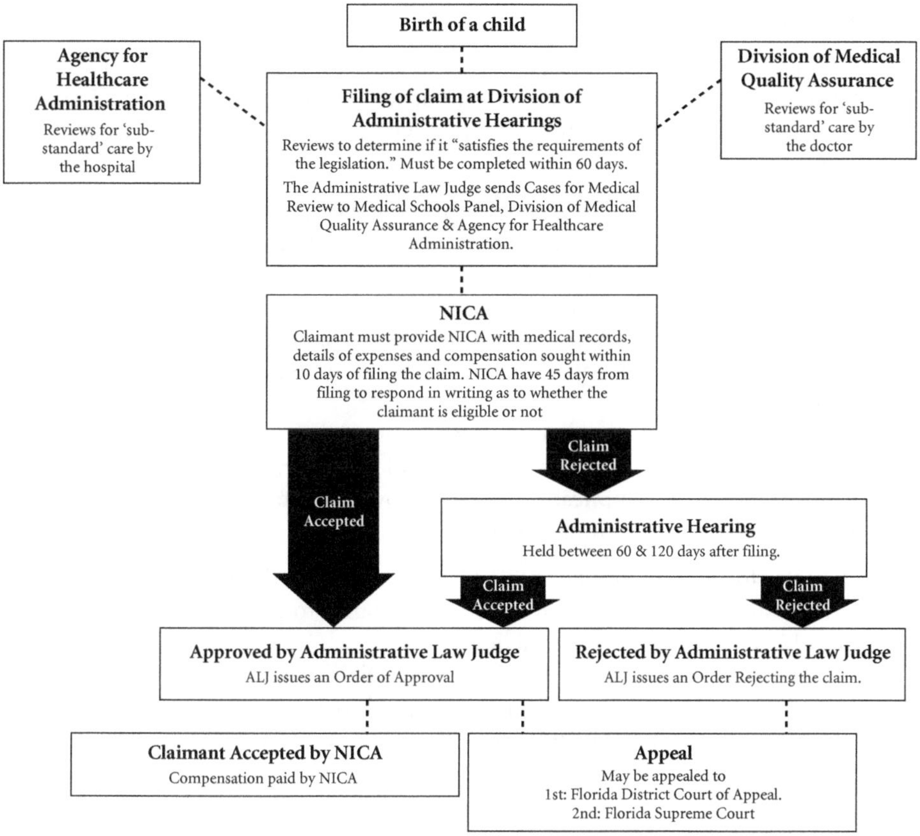

Figure 12.1: Schematic of the NICA claims process

Figure 12.1[15] shows a schematic of the potential routes to claiming under the programme. All claims are filed at the Division of Administrative Hearings.[16] A petition[17] is filed (along with a $15 dollar filing fee) and has to include all of the following:[18]

1. The name and address of the legal representative and the basis for her or his representation of the injured infant.
2. The name and address of the injured infant.
3. The name and address of any physician providing obstetrical services who was present at the birth and the name and address of the hospital at which the birth occurred.

[15] Taken from the NICA Benefit Handbook, available at www.nica.com/parents/index.html.
[16] www.doah.state.fl.us/ALJ/.
[17] This is available online at www.nica.com/parents/index.html.
[18] S 766.305 of the Florida Statutes, available at www.leg.state.fl.us/Statutes/index.cfm?App_mode=Display_Statute&Search_String=&URL=0700-0799/0766/Sections/0766.305.html.

4. A description of the disability for which the claim is made.
5. The time and place the injury occurred.
6. A brief statement of the facts and circumstances surrounding the injury and giving rise to the claim.

The petition itself is fairly straightforward; there are slightly different forms for different circumstances, single parent, two parents, with or without legal assistance, etc. In addition to the petition, the claimant has to furnish NICA with the following within 10 days of filing the petition:

(a) All available relevant medical records relating to the birth-related neurological injury and a list identifying any unavailable records known to the claimant and the reasons for the records' unavailability.
(b) Appropriate assessments, evaluations and prognoses, and such other records and documents as are reasonably necessary for the determination of the amount of compensation to be paid to, or on behalf of, the injured infant on account of the birth-related neurological injury.
(c) Documentation of expenses and services incurred to date which identifies any payment made for such expenses and services and the payer.
(d) Documentation of any applicable private or governmental source of services or reimbursement relative to the impairments.

The additional information requires a third party provide medical records to the parents/guardian. To ensure this happens, NICA offers assistance to parents, including a toll-free number on which parents can obtain advice.

Upon receipt of the petition, the Administrative Division serves the petition on NICA, and on any doctors, healthcare providers and hospitals named in the petition. It also mails the petition to the Division of Medical Quality Assurance and the Agency for Health Care Administration.

NICA Investigations

Investigating each claim is a multistage process, proscribed by statute with set time-limits. NICA (and subsequently the ALJ) set out to establish the following on the basis of the evidence provided:

1. Whether the injury claimed is a birth-related neurological injury. This will be a matter of fact, it is for the claimant to demonstrate that the infant has sustained a brain or spinal cord injury caused by oxygen deprivation or mechanical injury and that the infant was thereby rendered permanently and substantially mentally and physically impaired. If these factors have been established, then a rebuttable presumption shall arise that the injury is a birth-related neurological injury, as defined in the governing legislation.
2. Whether obstetrical services were delivered by a participating physician in the course of labour, delivery, or resuscitation in the immediate post-delivery period in a hospital; or by a certified nurse midwife in a teaching hospital supervised by a participating physician in the course of labour, delivery or resuscitation in the immediate post-delivery period in a hospital.

Administrative Law Judge Approval

If NICA accepts the claimant, the ALJ has to decide whether to accept and approve the order for compensation, and to determine how much compensation, if any, is awardable.[19] The ALJ also has to make a determination on notice if this issue is raised by a party to the proceedings. The ALJ must make a factual determination on whether the notice requirements are satisfied.[20] This is essentially that the participating hospital made the patients aware of NICA according to the statutory provisions. The ALJ has the exclusive jurisdiction to make this factual determination.

If NICA rejects the claim, then the case proceeds to a hearing overseen by the ALJ.

Hearings[21]

The ALJ shall set the date for a hearing (between 60 days and 120 days after the filing of a petition by the claimant). The ALJ shall immediately notify the parties of the time and place of such hearing, which shall be held in the county where the injury occurred unless otherwise agreed to by the parties and authorised by the division.

The parties to the hearing must include the claimant and NICA, and can include other parties as relevant. Any party to the proceedings may, upon application to the ALJ setting forth the materiality of the evidence to be given, serve interrogatories or cause the depositions of witnesses residing within or without the state to be taken (the costs thereof to be taxed as expenses incurred in connection with the filing of a claim). Such depositions shall be taken after giving notice and in the manner prescribed for the taking of depositions in actions at law, except that they shall be directed to the ALJ before whom the proceedings may be pending.

At a hearing the ALJ determines the questions of fact: whether the injury is a birth-related neurological injury; whether it took place in hospital; whether it was provided by a participating healthcare provider; and whether notice had been given. Proceedings may be split, at the discretion of the ALJ, to determine the issues of eligibility, notice and quantum separately.[22]

If the ALJ determines that the injury alleged is not a birth-related neurological injury or that obstetrical services were not delivered by a participating physician at the birth, etc, she or he shall enter an order and shall cause a copy of such order to be sent immediately to the parties by registered or certified mail.

If the ALJ determines that the injury qualifies, then the issue of quantum will be addressed.

[19] Pursuant to s 766.31 of the Florida Statutes, available at www.leg.state.fl.us/Statutes/index.cfm?App_mode=Display_Statute&Search_String=&URL=0700-0799/0766/Sections/0766.31.html.

[20] Pursuant to s 766.316 of the Florida Statutes, available at www.leg.state.fl.us/Statutes/index.cfm?App_mode=Display_Statute&Search_String=&URL=0700-0799/0766/Sections/0766.316.html. A document signed by patients confirming receipt of the NICA brochure provides a rebuttable presumption of notice.

[21] S 766.307 of the Florida Statutes, available at http://www.leg.state.fl.us/Statutes/index.cfm?App_mode=Display_Statute&Search_String=&URL=0700-0799/0766/Sections/0766.307.html.

[22] S 766.309(4) of the Florida Statutes, available at www.leg.state.fl.us/Statutes/index.cfm?App_mode=Display_Statute&Search_String=&URL=0700-0799/0766/Sections/0766.309.html.

F. Awards

If it has been determined that the child fits the required definition of a qualifying injury and the care was delivered by a participating provider and notice was properly given, the Commission will make an award. The award requires the immediate payment of expenses already incurred and requires that future expenses be paid as incurred. A copy of the award shall be sent immediately by registered or certified mail to each person served with a copy of the petition

Compensation is available under the following heads:[23]

1. *Medical expenses*

Actual expenses are paid for the following costs: medically necessary and reasonable medical and hospital costs, habilitation and training, family residential or custodial care, professional residential, and custodial care and service, medically necessary drugs, special equipment, and facilities, and related travel. Expenses paid under this head are limited to reasonable charges prevailing in the same community for similar treatment of injured persons when such treatment is paid for by the injured person.

Awards for medical expenses take into account the following and reduce the award accordingly:

1. Expenses for items or services that the infant has received, or is entitled to receive, under the laws of any state or the federal government, except to the extent such exclusion may be prohibited by federal law.
2. Expenses for items or services that the infant has received, or is contractually entitled to receive, from any prepaid health plan, health maintenance organisation or other private insuring entity.
3. Expenses for which the infant has received reimbursement, or for which the infant is entitled to receive reimbursement, under the laws of any state or the federal government, except to the extent such exclusion may be prohibited by federal law.
4. Expenses for which the infant has received reimbursement, or for which the infant is contractually entitled to receive reimbursement, pursuant to the provisions of any health or sickness insurance policy or other private insurance programme.

2. *Periodic or lump sum award*

An order can be made for payment of up to $100,000 to the parents or legal guardians of the infant found to have sustained a birth-related neurological injury.[24] This is usually made as a periodic payment, but, at the discretion of the ALJ, may be made in a lump sum.

[23] S 766.31 of the Florida Statutes, available at www.leg.state.fl.us/Statutes/index.cfm?App_mode=Display_Statute&Search_String=&URL=0700-0799/0766/Sections/0766.31.html.

[24] S 766.31(1)(b)(1) of the Florida Statutes, available at www.leg.state.fl.us/Statutes/index.cfm?App_mode=Display_Statute&Search_String=&URL=0700-0799/0766/Sections/0766.31.html.

3. *Reasonable legal expenses*

Reasonable legal expenses incurred by the claimant in connection with the filing of a claim are available, subject to certain stipulations.[25] Should the claim succeed and the claimants accept an award, the claimants shall not be liable for any expenses, including attorney's fees, incurred in connection with the filing of the claim, other than the reasonable legal expenses awarded by the ALJ.

4. *Death of a child*

Death benefits of up to $10,000 are available to the infant's family.[26]

G. Appeals

There are two levels of appeal: the first is to have the case re-examined by the Court of Appeal, then the Supreme Court.[27]

H. Parallel Regulatory Investigations into Care

Once a petition/claim has been received by NICA, investigations ensue, the exact nature of which depends upon the nature of the complaint. The investigations include establishing the individual circumstances of the case and potentially proscribing regulatory change, which can range from advice to individual practitioners to mandatory systematic changes for hospitals.

If a complaint is against a physician, it will be reviewed by the Division of Medical Quality Assurance to determine if disciplinary action is required.[28] Such actions utilise a standardised process for investigating complaints against doctors.[29]

If the complaint is against a participating hospital, the Agency for Healthcare Administration will investigate the claim.[30] Should the Agency conclude that substandard care contributed to the injury, then it shall take any action it deems appropriate within the limits of its powers.

[25] S 766.31(1)(c) of the Florida Statutes, available at www.leg.state.fl.us/Statutes/index.cfm?App_mode=Display_Statute&Search_String=&URL=0700-0799/0766/Sections/0766.31.html.

[26] S 766.31(1)(b)(2) of the Florida Statutes, available at www.leg.state.fl.us/Statutes/index.cfm?App_mode=Display_Statute&Search_String=&URL=0700-0799/0766/Sections/0766.31.html.

[27] S 766.311 of the Florida Statutes, available at www.leg.state.fl.us/Statutes/index.cfm?App_mode=Display_Statute&Search_String=&URL=0700-0799/0766/Sections/0766.311.html.

[28] S 766.305(5) of the Florida Statutes, available at www.leg.state.fl.us/Statutes/index.cfm?App_mode=Display_Statute&Search_String=&URL=0700-0799/0766/Sections/0766.305.html.

[29] S 456.073 of the Florida Statutes, available at www.leg.state.fl.us/Statutes/index.cfm?App_mode=Display_Statute&Search_String=&URL=0400-0499/0456/Sections/0456.073.html.

[30] S 766.305(6) of the Florida Statutes, available at www.leg.state.fl.us/Statutes/index.cfm?App_mode=Display_Statute&Search_String=&URL=0700-0799/0766/Sections/0766.305.html.

I. Claims Statistics[31]

The delay between a child's birth and the development and/or confirmation of damage means that at any given point there will be children who have been born, and who fulfil the criteria for the programme, but who have not yet been admitted to the programme. Table 12.1 provides a breakdown of the accepted claims and reported claims by birth year. Once the 5-year limitation period has elapsed it is not possible for a child born in that year to be admitted unless they have an ongoing application; therefore these figures can only be finalised when the limitation period has elapsed and all ongoing claims/appeals have been settled.

Table 12.1: Participants by birth year 1989–2014 (first 9 months only)*

Birth year	Accepted claims	Reported claims	% of claims accepted	Actual/estimated claims accepted
1989	15	32	46.9	15
1990	10	39	25.6	10
1991	8	38	21.1	8
1992	14	48	29.2	14
1993	15	40	37.5	15
1994	16	36	44.4	16
1995	11	26	42.3	11
1996	17	40	42.5	17
1997	17	47	36.2	17
1998	17	42	40.5	17
1999	18	40	45.0	18
2000	13	38	34.2	13
2001	13	41	31.7	13
2002	22	50	44.0	22
2003	9	23	39.1	9
2004	13	31	41.9	13
2005	13	41	31.7	13
2006	13	34	38.2	13
2007	15	36	41.7	15

(*continued*)

* The final few years may be subject to change due to children born in these years applying and/or being admitted after these figures were published: see the right-hand column which gives final/estimated accepted claimant numbers.

[31] Data taken from the 2015 Financial Statements of the Florida Birth-Related Neurological Injury Compensation Association and the 2015 Actuarial reports, available at www.nica.com/news_publications.html.

Table 12.1: (*Continued*)

Birth year	Accepted claims	Reported claims	% of claims accepted	Actual/estimated claims accepted
2008	11	37	42.3	11
2009	16	47	29.7	17
2010	12	35	50.0	14
2011	12	33	33.3	15
2012	9	32	28.1	16
2013	6	12	50.0	18
2014 (9 months)	2	2	100.0	14

Source: sheet 1b of the Actuarial report of 30 September 2014 available at https://www.nica.com/news_publications.html.

This time lag also impacts upon the way the Fund is invested: there is inevitably a difference between the point at which the funds are collected for that year and the time at which they will be paid out to claimants born in that year. Funds are therefore invested from the point that they are paid into the Fund until the point they are paid out. At the point they are invested, there is no certainty about the number of claimants or the value of any awards which will be given, requiring the Fund to take actuarial advice and to consider carefully its investment and contingency positions, this includes taking out reinsurance to spread the risk (the reinsurance policies in place cover births occurring during the years 1999–2001 and 2002–03; there is no insurance in place from 2004 to date). Table 12.2 shows the net assets for the years 2015 and 2014.

Table 12.2: Net assets ($) for the years 2014 and 2015

	2015	2014
Investments	**1,030,522,152**	1,024,478,268
Receivable for securities sold	**7,296,390**	11,836,531
Securities lending collateral	—	—
Reinsurance recoverable	**33,792,777**	33,356,897
Other assets	**6,721,422**	7,218,553
Total assets	**1,078,332,741**	1,076,890,249
Claims reserves	**898,500,000**	878,000,000
Payable for securities purchased	**18,593,388**	28,353,337
Obligations under securities lending	—	—
Other liabilities	**435,358**	840,917
Total liabilities	**917,528,746**	907,194,254
Net investment in capital assets	**563,191**	654,827
Restricted	**160,240,804**	169,041,168
Total net position	**160,803,995**	169,695,995

J. Costs

As of 30 June 2015, the financial position of NICA was as shown in Table 12.3. For the year ending 30 June 2015 the operating revenues and breakdown of expenses are shown in Table 12.4.

Table 12.3: Financial position of NICA as at 30 June 2015

Assets	
Current assets:	
Cash and cash equivalents	$ 157,883
Prepaid expenses and other current assets	53,115
Total current assets	210,998
Receivable for securities sold	7,296,390
Investment income receivable	4,253,883
Assessments receivable	1,693,350
Investments	1,030,522,152
Reinsurance recoverables	33,792,777
Property and equipment, net	563,191
Total assets	$1,078,332,741
Liabilities and net position	
Current liabilities:	
Accounts payable and accrued expenses	$ 152,916
Accrued investment fees	282,442
Total current liabilities	435,358
Assessments refundable	—
Payable for securities purchased	18,593,388
Claims reserves	898,500,000
Total liabilities	917,528,746
Net position:	
Net investment in capital assets	563,191
Restricted	160,240,804
Total net position	160,803,995
Total liabilities and net position	$1,078,332,741

Table 12.4: NICA operating revenues and expenses for the year ending 30 June 2015

Operating revenues	**$**
Hospital assessments	3,225,650
Physician assessments	21,133,807
Total assessments	**24,359,457**
Expenses	
Claims incurred	36,382,730
General and administrative expenses	1,806,154
Depreciation and amortisation	122,213
Commutation expense	91,532
Total expenses	**38,402,629**

Interestingly, general and administrative expenses make up relatively small proportions of NICA's total expenses, 4.7% in 2015. This is an extremely low level of expenditure on administration and it appears that NICA is incredibly efficient when compared to other redress mechanisms such as litigation. However, it should be remembered that legal fees are included within the claims expenses rather than within the general and administrative expenses. By far the largest of the operating expenses is claims expenses, which made up 94.7% of the total operational expenses in 2015.

Legal expenses as a percentage of incremental payments made has varied between 7.4% and 15.2% for injuries occurring between 2002 and 2014. The combined administration expenses and legal expenses for this period falls between 10% and 20% of the total expenses, indicating that the system is efficient.

These figures show there was clearly a relatively small operating loss ($14,043,172) in 2015. However, NICA's financial position is in credit overall due to non-operating revenues ($5,151,172) and starting the year with a positive net position of $169,695,995. The net position at the end of June 2015 was a credit of $160,803,995.

K. Actuarial Soundness of the Fund and Provisions Related to This

As these figures show, the Fund is currently actuarially sound. Should this position change, then the Fund has the ability to generate additional funding without changes to legislation. In the first instance an addition $20 million is to be transferred from the Insurance Regulatory Trust Fund, with provision that if there are still not sufficient funds for NICA to be actuarially sound, then the Department of Financial Services, Office of Insurance Regulation may charge entities licensed to provide casualty insurance in Florida with a levy of no greater than 0.25% of net direct premiums written.

In the event that these additional funds have been added and that the management estimate that the accumulated costs of reported claims (exclusive of family residential or custodial care as defined in section 766.302) equals or exceeds 80% of the funds NICA possesses, then NICA will cease to accept new claims, except for claims expressly authorised by the legislature. However, this restriction on accepting new claims will not apply to injuries occurring 18 months of more prior to time when the restriction was actioned.

L. Breakdown of the Claims Expenses by Category

Table 12.5 shows the incremental payments by major expense group broken down by the year of injury. The largest cost is the cost of nursing, which is not surprising. Nursing costs paid for nursing provided by non-family members accounts for a substantial proportion of the total incremental costs paid, but is dwarfed by the spend on nursing care provided by the family. Prior to the 2012 class action settlement noted below, payment for nursing care provided by families made up a surprisingly small proportion of the incremental payments, which was a source of contention. Since the 2012 settlement, nursing care provided by family and parents has been the largest cost. Medical expenses are consistently low, averaging around 6% over the years. This is potentially surprising, and is likely to be a reflection of the stipulation that NICA is a payer of last resort, so all other options (private and state) must be utilised before NICA funds. Medical expenses are quite likely to be covered under such arrangements, so excluded from NICA. Legal costs have been a relatively small percentage of the total incremental costs, and have dropped in the last couple of years. It will be interesting to see if they stabilise at this level, revert to previous proportions or continue to drop.

Table 12.5: Incremental payments by major expense groups by birth year (payments as of 30 June that year) with percentages

	Birth year 2011	Birth year 2012	Birth year 2013	Birth year 2014 (estimated)
Legal expenses	**1,348,962** 9.41%	**1,472,264** 11.34%	**1,046,543** 7.42%	**1,046,543** 7.42%
Parental award	**1,772,862** 12.37%	**854,442** 6.58%	**1,254,414** 8.89%	**1,254,414** 8.89%
Medical expenses	**795,785** 5.55%	**792,882** 6.11%	**854,458** 5.99%	**854,458** 5.99%
Nursing care—by parents	**5,533,089** 38.61%	**5,409,777** 41.69%	**6,506,101** 46.10%	**6,506,101** 46.10%
Nursing care—by others	**2,555,000** 17.83%	**3,010,739** 23.20%	**2,668,850** 18.91%	**2,668,850** 18.91%
Other	**2,305,512** 16.09%	**1,339,156** 10.32%	**1,717,202** 12.17%	**1,717,202** 12.17%
Family residential or custodial care	— **19,113** 0.13%	**98,021** 0.76%	**73,571** 0.52%	**73,571** 0.52%
Totals	**14,330,324**	**12,977,280**	**14,112,228**	**14,112,228**

M. Class Actions[32]

A class action was launched against NICA, which resulted in an out-of-court settlement with a Final Judgment and Order entered by the Florida Circuit Court on 26 November 2012.[33] The claim was that NICA had failed to provide adequate funding to cover the care provided by parents to NICA-eligible children, as they were statutorily obliged to do. There was no admission of liability from NICA. The settlement concerns parents or guardians of a child born with a 'birth-related neurological injury' in the State of Florida during the time period of 1 January 1989–6 June 2002, who obtained a final order which imposed upon NICA the 'continuing obligation under the provisions of Section 766.31, Florida Statutes, to pay future expenses as incurred', or other substantially similar words to that effect, and after issuance of the final order, provided any 'residential and custodial care and services' to that child without receiving full payment from NICA for such care and services, and/or who intends to provide such care and services to that child in the futures. The settlement specifies the exact requirements on NICA to meet these obligations, both past and future. The estimated impact of this settlement was incorporated into case reserve estimates from the end of 2012 and as at June 2015 the majority of the retroactive settlements associated with the class action have been paid.[34]

N. Conclusions

The Fund operates on a limited scope as it is a payer of last resort, but it serves a very important function, both as a safety net for individual claimants and as a mechanism for reducing the premiums for liability insurance for healthcare providers.

The Fund has a no-fault approach to eligibility, with an inquisitorial investigation of the facts by NICA. The issues of eligibility and quantum are approved at separate stages of the process by an ALJ.

Of particular interest is the interaction between the Fund and other organisations: the Court is required to report to the regulatory bodies, the Division of Medical Quality Assurance and the Agency for Healthcare Administration. This joined-up approach to compensation and regulation should help to avoid the 'lost data' issues seen in other jurisdictions. Too often information on potentially substandard care progresses through a single channel, usually the courts, and does not always reach regulators.

NICA has a low level of administrative costs and the legal costs are low compared to a tortious claim. The levels of compensation paid by NICA and the construction of the statutes in relation to compensation for care provided by parents has been contentious. The class action settlement should resolve this issue. This settlement only applies to a set of children born between 1989 and 2002; it has no impact on payments to children born

[32] See www.nica.com/classaction.html for details.

[33] www.nica.com/10032012/2012%20FINAL%20ORDER.pdf.

[34] Page 9 of the Actuarial report of 30 September 2014, available at www.nica.com/news_publications.html.

after this date. The class action raises as interesting point: as the exclusive remedy for such children, NICA is in a position of responsibility to ensure that the compensation awarded is as the statute intended it to be. However, NICA is not without oversight; the courts have the ability to provide binding guidance to NICA on statutory interpretation and its responsibilities thereunder (this power was not exercised in this class action as a settlement was reached by consent).

NICA is, and has consistently been, actuarially sound, but should it find itself in deficit has the capacity to raise additional funding.

NICA has succeeded in its threefold primary objectives: to encourage physicians to practice obstetrics, to stabilise the market for malpractice insurance and make it affordable, and to provide for the needs of injured children.

13

September 11 Victim Compensation Fund

The terrorist atrocities committed by Al-Qaeda on 11 September 2001 require little introduction. Two hijacked passenger planes (American Airlines Flight 11 and United Airlines Flight 175) were flown into the North and South Towers of the World Trade Center in New York, a third plane (American Airlines Flight 77) was flown into the Pentagon building, and another plane (United Airlines Flight 93) crashed in a field in Shanksville, Pennsylvania after passengers heard of the other attacks and attempted to overcome the hijackers. The attacks killed 2,977 people: 2,753 people in New York, 184 people at the Pentagon and 40 people on Flight 93. Many more people were injured in the attacks and their aftermath.

The September 11 Victim Compensation Fund was set up to provide a no-fault alternative to litigation for individuals who suffered personal injury or death due to the 9/11 acts of terrorism. The initial fund was created by Title IV—Victim Compensation of The Air Transportation Safety and System Stabilization Act of 2002[1] (the Act) (PL 107-42). This fund was open for a limited period of 2 years from the date at which subordinate regulations were enacted.[2] The fund, VCF1, had a final claim submission date of 22 December 2003. The information presented here on VCF1 draws heavily on the Final Report of the Special Master for the September 11th Victim Compensation Fund of 2001.[3]

The Victim Compensation Fund was re-established by the James Zadroga 9/11 Health and Compensation Act of 2010 (PL 111-347) (Zadroga Act).[4]

A. The First Victim Compensation Fund (VCF1)

It is interesting to note the speed with which VCF1 was established. The Act was signed by the President on 22 September 2001, just 11 days after the attacks. This speed was due to concerns that the national civil aviation industry would collapse, with serious implications for the national economy, unless provisions were put in place to protect it.[5] Indeed the

[1] The Victim Compensation of the Air Transportation Safety and System Stabilization Act 2002, available at www.congress.gov/bill/107th-congress/house-bill/2926.

[2] ibid, s 405(a)(3).

[3] Available at www.justice.gov/final_report.pdf.

[4] Available at www.congress.gov/bill/111th-congress/house-bill/847?q=%7B%22search%22%3A%5B%22zadroga%22%5D%7D.

[5] Various Congressional statements reflect these concerns, see 147 Cong Rec S9599 (statement of Sen Leahy) ('the airline industry of this country is in grave danger of collapse'; '[i]f Congress does not pass this legislation

preamble of the Act has a specified aim: 'to preserve the continued viability of the United States air transportation industry'. Part of the solution was to remove the threat of substantial law suits against the airlines by enacting a compensation scheme.

On 26 November 2001 Kenneth R Feinberg was appointed as the Special Master by the Attorney General. The Act required the Department of Justice to issue Regulations within 90 days of enactment of the Act. Merely 14 weeks after the attacks, on 21 December 2001, Interim Final Regulations, containing detailed information on the claims procedures and the determination of economic and non-economic losses, were issued by the Special Master. Simultaneously the Fund set up a free telephone information service, provided an 'Eligibility Form and Application for Advance Benefits', which enabled claimants to seek immediate interim compensation, and opened temporary 'walk-in' offices in New York and Washington DC. On 13 March 2002 Final Regulations were issued by the Special Master and the Fund produced the final Compensation Forms necessary for claimants to pursue a claim for full and final compensation. This is an impressively short time-scale.

Regulations were drawn up after a period of soliciting comments from families, lawyers, employers, government agencies, Congressmen and women, judges, charities, representatives of the emergency services, lawyers, employers, government agencies, members of the judiciary, associations, charities, representatives of the military, fire and police departments, and individuals in state governments. The Regulations pertaining to the VCF1 Claims process are contained in the Final Rule.[6]

A detailed report on VCF1 can be found in the final report of the Special Master for the September 11th Victim Compensation Fund of 2001.[7] Additional information is available from the archived website.[8]

Eligibility

Eligibility was to be determined by a Special Master. The following criteria applied.

Eligible claimants are defined as an individual who:

> (A)(i) was present at the World Trade Center, (New York, New York), the Pentagon (Arlington, Virginia), or the site of the aircraft crash at Shanksville, Pennsylvania at the time, or in the immediate aftermath, of the terrorist-related aircraft crashes of September 11, 2001; and (ii) suffered physical harm or death as a result of such an air crash;
>
> (B) an individual who was a member of the flight crew or a passenger on American Airlines flight 11 or 77 or United Airlines flight 93 or 175, except that an individual identified by the Attorney

today, it is likely that all of our Nation's air carriers would cease service next Wednesday'); see also 147 Cong Rec S9594 (statement of Sen McCain) ('The effect on the airlines of the September 11 terrorist attack put Congress in the unenviable position of having to take immediate action to prevent the collapse of the aviation industry as a result of the federally ordered grounding of all aircraft and the anticipated reduction of air travel.'); 147 Cong Rec S9600 (statement of Sen Byrd) ('[t]he Federal Government cannot allow this industry to fold without seriously disrupting the United States economy').

[6] September 11th Victim Compensation Fund of 2001, Final Rule, 13 March 2002, available at www.justice.gov/archive/victimcompensation/victimcompfedreg.htm.

[7] Available at www.justice.gov/final_report.pdf.

[8] The fund website is now archived at www.justice.gov/archive/victimcompensation/index.html.

> General to have been a participant or conspirator in the terrorist-related aircraft crashes of September 11, 2001, or a representative of such individual shall not be eligible to receive compensation under this title; or
>
> (C) in the case of a decedent who is an individual described in subparagraph (A) or (B), the personal representative of the decedent who files a claim on behalf of the decedent.[9]

To be eligible there must be no more than one claim submitted per individual. By submitting a claim the claimant waived the right to right to file a civil action (or to be a party to an action) in any federal or state court for damages sustained as a result of the terrorist-related aircraft crashes of 11 September 2001. Individuals who were already party to a civil action were prohibited from submitting a claim under the Act unless they withdrew from their civil action within 90 days of the regulations detailing the claims process[10] being enacted.

Physical harm. This was defined by the regulations as:

— a physical injury to the body that was treated by a medical professional within 24 hours of the injury having been sustained, or within 24 hours of rescue, or within 72 hours of injury or rescue for those victims who were unable to realise immediately the extent of their injuries or for whom treatment by a medical professional was not available on 11 September or within such time period as the Special Master may determine for rescue personnel who did not or could not obtain treatment by a medical professional within 72 hours; and

— required hospitalisation as an in-patient for at least 24 hours or caused, either temporarily or permanently, partial or total physical disability, incapacity or disfigurement.

Psychological harm was not covered. There was discretion for the Special Master to compensate for longer latency physical harm which fully materialised within the time period when the fund was operational, such as issues triggered by the inhalation of dust/toxic fumes at the sites. Long-latency diseases which were not fully apparent by the close of the fund were not compensated as the extent of the disease was not known and so the appropriate level of compensation could not be determined.

Crash sites. The victim needed to have been on board a hijacked aircraft, or physically present at the World Trade Center, Pentagon, or Shanksville, Pennsylvania sites. The World Trade Center site was delineated as the NYPD pedestrian exclusions site plus a block on all sides. If there was clear evidence of physical harm, eg from debris falling, occurring outside this zone, then the Special Master had the discretion to award compensation. Definitions of the other sites were more obvious, so a geographical delineation was not necessary.

Immediate aftermath. For a victim who was not a rescue worker, this was defined as the period of time from the crashes until 12 hours after the crashes. If the wictim was a rescue worker who assisted in efforts to search for and recover other victims, this was defined as the period from the crashes until 96 hours after the crashes.

Compensation calculations. One of the interesting questions in this process was whether compensation should be a standardised flat rate or individualised to meet individual

[9] The Victim Compensation of the Air Transportation Safety and System Stabilization Act 2002, s 405(c)(1) and (2).

[10] ibid, s 407 Regulations. The Act provided the overarching legislative framework for the Fund and the limitation on tort claims; the Regulations provided the detail on how the Fund was to function.

claimant's needs. The legislation was silent on this point. Different presumptions were used for deceased and surviving injured claimants.

Deceased victim compensation quantum. Non-economic losses were paid on a fixed basis. Each claim received a uniform non-economic award of $250,000 for the death of the victim and an additional non-economic award of $100,000 for the spouse and each dependant of the victim. The Regulations allowed for deviation from these amounts by the Special Master.

Economic losses were calculated using the presumed methodology. This provided consistency and ensured that claimants of the same age and earning power were awarded the same level of compensation. Briefly, the victim's earning history was supplied and the appropriate compensation level selected. This was usually, but not always, the earnings at the time of death. Any applicable state and federal taxes were subtracted. A deduction was made for the portion of the victim's earnings which would have been self-consumed[11] (and therefore not available to their dependants). The calculation included an average income growth rate over time and an average work-life duration. The formula incorporated a deduction for periods of unemployment based on national averages.

The aim of the presumed methodology was to provide consistent, generous awards to families in a way which was easy to administer and which did not require claimants to supply onerous details.

Under the Regulations, the presumed award methodology only applied to income levels up to the 98th percentile of individual income in the United States. This limitation was dictated both by policy and fact. First, the Act was intended to provide fair and appropriate awards based upon families' individual circumstances, needs and resources. Second, incomes above the 98th percentile are often comprised of a variety of forms of compensation, some of which are variable and volatile and tied to factors other than standard inflation and promotion increases. It was concluded that this rendered the assumptions applied in the presumed methodology inappropriate as they become extremely speculative and conjectural when applied to this type of income. When a victim's income exceeded the 98th percentile ($231,000 was the 98th percentile income level in the year 2000), the Fund calculated a 'presumed' economic loss using $231,000 as the income level. There was provision within the Regulations for the Special Master to deviate from the presumed income. This was often requested by dependants of high-income victims who felt the compensation calculated using $231,000 was insufficient. Unsurprisingly, this deviation was not without controversy as it effectively 'valued' a high-income victim's life as 'more valuable' than, say, that of a rescue worker.

Personal injury victims

(a) *Non-economic loss.* Given the physical injuries were so vastly different and had significantly different long-term effects, individualised payments were deemed appropriate for personal injury. The Regulations directed the Fund to evaluate each individual physical injury claim and establish the non-economic loss according to the extent, nature and permanence of the injury. There was no formula for calculating personal injury non-economic loss, nor a pre-set amount. The Regulations provided that the Special

[11] This was calculated using national data.

Master could rely upon the non-economic loss methodology for deceased victims and adjust the compensation according to the extent of the claimant's physical harm.

(b) *Economic loss.* There was no specific methodology for the calculation of economic loss awards for personal injury victims in the Regulations. The same methodology that was used to calculate deceased victims' economic loss was used for the personal injury claimants, with adjustment for the duration of economic loss made on an individual basis.

The Claims Process

Claimants had to prove eligibility, for which the three requirements in the legislation were physical harm, presence at site, and injury sustained at the time or in the immediate aftermath. These requirements were relatively straightforward for physical injury claimants, who were able to provide sworn affidavits as well as other evidence, though there were issues as to what constituted a physical injury. Proof of eligibility was slightly more difficult for claims for deceased individuals. In these cases, the personal representative was charged with making the claim, and hence proving eligibility.

Personal representatives. In claims for deceased individuals, the statute provided that only one claim could be filed with VCF1 and it had to be filed by the personal representative.[12] There was no further guidance in the Act on identifying the personal representative, obligations to notify potential beneficiaries, obligations to distribute compensation between potential beneficiaries, and the rights of potential beneficiaries to object to decisions of the personal representative. The Regulations stated that the personal representative appointed under the deceased's local state laws should be regarded as the appropriate personal representative for claims against VCF1. There were provisions for the appointment of a personal representative by the Special Master if the deceased's local jurisdiction had not already appointed one. The degree of involvement of the probate courts varied between jurisdictions, with some, such as New York, requesting a greater degree of involvement with VCF1 decisions than others. By utilising the locally recognised personal representative, VCF1 was able to avoid directing resources at the potentially costly process of establishing who had the right to claim.

In order to ensure that potential beneficiaries were notified of their potential claim, all personal representatives had to supply a list of potential beneficiaries to VCF1 and to swear to notify them of the claim. Additionally, VCF1 published a list of the deceased for whom claims were being made online, giving potential beneficiaries another means to discover the claim.

Proof of death for deceased victims. Given the catastrophic nature of the attacks, there were fatalities where there was difficulty in identifying a body. VCF1 took a pragmatic, flexible approach to the issue of proof of death. Various forms of proof of death were permitted, including a death certificate or other evidence, such as an employer affidavit and presence of a victim on an airline manifest. Death certificates based on an actual autopsy or medical examiner confirmation of identity were sufficient without additional corroboration.

[12] Ss 405(c)(2)(C) and 405(c)(3)(A) Air Transportation Safety and System Stabilization Act 2002, available at www.congress.gov/bill/107th-congress/house-bill/2926.

If the claimant could not obtain a death certificate, then other official documentation, such as a court order establishing death, could be used.

Presence at site for deceased victims. Unlike physical injury claimants, deceased victims could not assert their presence by means of affidavits, so VCF1 permitted a range of evidence, including death certificates, records of employment, affidavits from employers and other knowledgeable persons, airline records, DNA test results and other pertinent documentation. In total 45 claims asserted on behalf of deceased victims were denied by VCF1, based on lack of proof of death, lack of proof of presence at the site, or lack of proof that the death was related to the September 11th attacks.

Quantification of Compensation

VCF1 was designed so that it did not put an onerous burden on claimants. Claimants had to provide the victim's age, historical earnings, employment benefits and collateral offset data along with the age and status of members of the victim's household. This information was all relatively simple for claimants to acquire.

Claim Tracks

There were two claim-processing options available to claimants: Track A and Track B. The procedures for these different tracks were as follows (and are shown diagrammatically in Figure 13.1).

Track A: A Claims Evaluator and Special Master's office reviewed the claim to determine eligibility and determine that the claim was substantially complete. The Special Master then sent a letter to the claimant stating that the claim was substantially complete. The Special Master subsequently notified the claimant in writing of the eligibility determination, the amount of the presumed award, and the right to request a hearing before the Special Master or his designee. Once an eligible claimant had been notified of the presumed award, the claimant elected either to accept the presumed compensation determination as the final determination and request payment, or instead to request a hearing before the Special Master or his designee. If a claimant opted for a hearing, the claimant was permitted to make supplemental submissions. The Special Master had the option of altering or modifying the award if the presumed award had been calculated erroneously, or if the claimant demonstrated extraordinary circumstances indicating that the presumed award would not have adequately addressed the claimant's injury. There was no further review or appeal from this determination.

Track B: A Claims Evaluator and Special Master reviewed the claim to determine eligibility. The Special Master then sent a letter to the claimant stating that the claim is substantially complete. Subsequently, the Special Master notified the claimant in writing of the eligibility determination. Upon notification of eligibility, the claimant automatically proceeded to a hearing. Hearings were decided using a variety of methods, such as on the papers, telephonically, or in person. At the hearing, the Special Master or his designee utilised the presumed award methodology, but could modify or vary the award if the claimant presented extraordinary circumstances which were not adequately addressed by the presumed award methodology. There was no review or appeal from this determination.

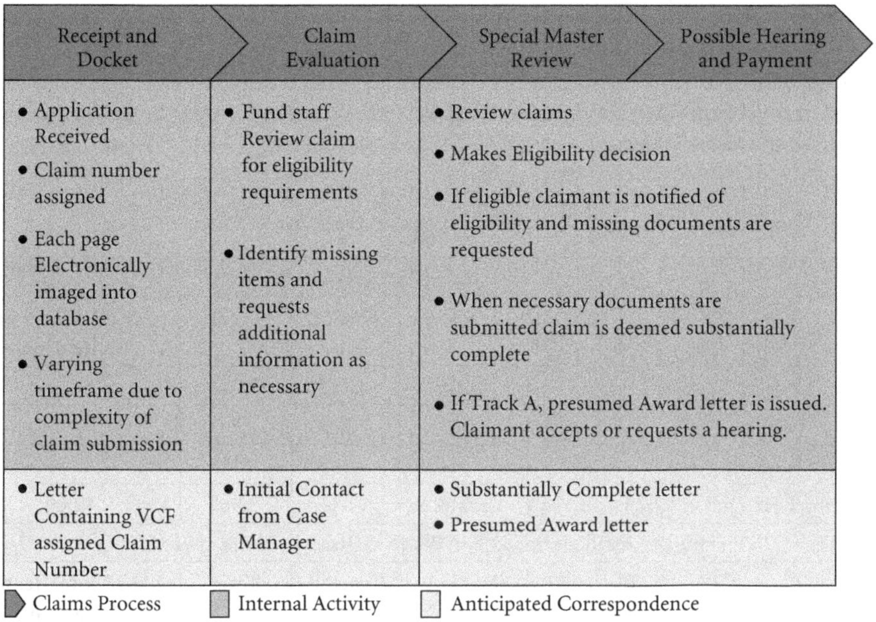

Figure 13.1: VCF1 claims process (from archived website)

Track A provided a more streamlined process without the need for a hearing. Claims were divided between the tracks as follows:

	Track A	Track B
Deceased victims claims	47%	53%
Personal injury claims	89%	11%

There is a clear difference between the choice of track according to the claim type.[13] Claims for deceased victims were nearly equally divided between Track A and Track B: 47% of claims for deceased victims elected for Track A and 53% elected for Track B. However, there were clear differences between the demographics of claimants using the different tracks. Representatives of high-income victims were more likely to have selected Track B. Of claims for victims with income levels in the top 2% (ie over $231,000), over 69% elected for Track B, compared with 48% of victims with the lowest income levels (under $25,000). Personal representatives of higher-income victims often expressed the view that the presumed methodology-calculated sum based on an income of $231,000 was inadequate for these families' needs and individual circumstances.

Similarly, as with claims for deceased victims, physical injury victims with higher incomes preferred Track B: it was selected by over 67% of physical injury claimants with

[13] Data taken from the VCF1 Final Report, ibid.

income levels above $200,000, whilst while 89% of physical injury victims with income below $150,000 elected for Track A.

Hearings. Table 13.1 shows the number of hearings broken down by claim type and track. Hearings on objections and statements of interest are counted as separate hearings, even if they relate to the same claim.

Table 13.1: Numbers of hearings held by claim type and track for VCF1

Hearings—all claims (7,403 claims filed)				
	Track A	Track B	Total	% of all claims filed with hearing(s)
Award hearings	1,302	1,742	3,044	
Eligibility hearings	653	109	762	
Other official hearings	91	65	156	
Total (3,962 hearings on 3,629 claims)	2,046	1,916	3,962	49.00
Hearings—death claims (2,880 death claims with final awards)				
	Track A	Track B	Total	% of total death claims with hearing(s)
Award hearings	518	1,455	1,973	
Eligibility hearings	6	2	8	
Other official hearings	32	61	93	
Total (2,074 hearings on 1,977 claims)	556	1,518	2,074	68.60
Hearings—physical injury claims (2,680 physical injury claims with final awards)				
	Track A	Track B	Total	% of total physical injury claims with hearing(s)
Award hearings	783	283	1,066	
Eligibility hearings	331	62	393	
Other official hearings	30	2	32	
Total (1,490 hearings on 1,265 claims)	1,144	347	1,491	47.20
Claims with multiple hearings—(claims with final awards)				
	Number of claims		% of claims with multiple hearings	
Claims with multiple hearings—all claims with final awards	285		5.13	
Claims with multiple hearings—death claims with final awards	62		2.15	
Claims with multiple hearings—physical injury claims with final awards	223		8.32	

While the number of claims for physical injury and fatality were broadly similar, there were marked differences both in the track and the compensation claimed. Death claims comprised a fixed non-economic sum and economic loss for lifetime loss of earnings. The non-economic elements of physical injury claims were more variable, as was economic loss claimed as the duration of loss of earnings was variable. Table 13.2 shows the physical injury claims broken down by injury type.

Table 13.2: Breakdown of VCF1 physical injury claims by injury type

Injury type	Number of claims	% of claims
Asthma/other respiratory	1,377	51.38
Back injury (disc problem, back pain, etc)	94	3.51
Broken bones/fractures	87	3.25
Bruises/cuts	44	1.64
Burns	40	1.49
Heart attack/other cardiac problems	6	0.22
Neurological problems (stroke, seizure, brain damage, etc)	8	0.30
Other injury	67	2.50
Sensory problems (vision, hearing, etc)	31	1.15
Soft tissue (ligaments and cartilage)	91	3.40
Multiple injuries	835	31.16
Total	2,680	100.00

Claims Statistics and Compensation Awarded

VCF1 awarded a total of $7,049,415,536.64 in compensation, with $5,996,261,002.08 awarded for fatality claims and $1,053,154,534.56 for physical injuries. Tables 13.3, 13.4 and 13.5 give an overall breakdown for claims, for fatality claims, and for physical injury claims respectively.

Table 13.3: Award statistics for all claims

General award statistics for all claims		
Claim data for all claims		
Death	2,968	
Physical Injury	4,435	
Total claims received	**7,403**	
Award details—all eligible claimants		
Total—all claims with awards issued		

(continued)

Table 13.3: ***(Continued)***

Track A 67% (of awarded claims)	3,735	$3,029,856,022.91
Track B 33% (of awarded claims)	1,825	$4,019,559,513.73
Total awards issued	5,560	
Total amount awarded		**$7,049,415,536.64**
Average award		$1,267,880.49
Median award		$855,919.50
Maximum award		$8,597,732.00
Minimum award		$500.00
Minimum offset		$0.00
Maximum offset		$9,875,656.44
Average offset		$524,285.58
Median offset		$236,810.76
Total economic and non-economic awards before offsets		$9,964,443,386.52
Total offsets (all claims)		$2,915,027,849.88
Claims with advance benefits	236	$11,300,000.00
Claimants receiving structures (entirely or partially)	181	$528,589,421.13

Table 13.4: Award statistics for deceased victim claims

General award statistics for deceased victims		
Claim data for deceased victims		
Number of claims received	2,968	
Number of claims denied/withdrawn/abandoned	88	
Total claims with awards issued	**2,880**	
Award details—eligible deceased claims		
Total—all claims with awards issued		
Track A 47%	1,348	$2,187,641,071.92
Track B 53%	1,532	$3,808,619,930.16
Total awards issued	2,880	
Total amount awarded		**$5,996,261,002.08**
Average award		$2,082,035.07
Median award		$1,677,632.54
Maximum award		$7,100,000.00
Minimum award		$250,000.00
Minimum offset		$0.00
Maximum offset		$9,875,656.44

(continued)

Table 13.4: ***(Continued)***

Average offset		$855,826.66
Median offset		$585,657.09
Total economic and non-economic awards before offsets		$8,461,041,778.69
Total offsets (all claims)		$2,464,780,776.61
Claims with advance benefits (216)		$10,800,000.00
Claimants receiving structures (178—entirely or partially)		$519,602,850.13

Table 13.5: Award statistics for physical injury claims

General award statistics for physical injury victims		
Claim data for physical injury victims		
Number of claims received	4,435	
Number of claims denied/withdrawn/abandoned	1,755	
Total claims with awards issued	**2,680**	
Award details—eligible physical injury claims		
Total—all claims with awards issued		
track A 89%	2,387	$842,214,950.99
track B 11%	293	$210,939,583.57
Total awards issued	2,680	
Total amount awarded		**$1,053,154,534.56**
Average award		$392,968.11
Median award		$108,746.50
Maximum award		$8,597,732.00
Minimum award		$500.00
Minimum offset		$0.00
Maximum offset		$2,972,238.00
Average offset		$168,002.64
Median offset		$0.00
Total economic and non-economic awards before offsets		$1,503,401,607.83
Total offsets (all claims)		$450,247,073.27
Claims with advance benefits (20)		$500,000.00
Claimants receiving structures (3—entirely or partially)		$8,986,571.00

The bulk of the compensation awarded was for fatalities, with the average fatal claim attracting an award of over $2 million, compared to an average physical injury claim of just under $393,000. Given the scope of the compensation awarded, this differential is entirely consistent with what would be expected.

Administrative Costs

Insight into the efficiency and success of the Fund can be found by examining the overall costs of administering the programme. VCF1 provided over $7 billion in benefits to over 5,560 eligible claimants. The total administrative costs of VCF1 were approximately $86 million; broken down as shown in Table 13.6.

Table 13.6: Administrative costs of VCF1

PricewaterhouseCoopers (including costs of subcontractors)	$76,511,000
Government attorneys and support assigned to the programme	$3,667,000
Administrative law judges	$679,000
Aspen Systems	$4,674,000
CACI	$862,000
Consultants	$76,312
Feinberg Group professional services	$0
Out-of-pocket-expenses	$404,000
Total	**$86,873,312**

The overall cost of administration was $86,873,312, which constitutes an incredibly low 1.2% of the total dollars awarded to eligible claimants. This is an almost unbelievably efficient compensation distribution mechanism. However, it must be remembered that the awards were high (effectively reducing the ratio of administrative costs to compensation awards) and a variety of individuals provided their time on a pro bono basis. For example, the Feinberg Group, the Special Master and a large number of lawyers did not charge for their time. If these costs were to have been factored in, the ratio of dollars paid for administrative costs to dollars paid in compensations for VCF1 would have been somewhat less impressive. That said, the alternative option, a tort-based court claim, would usually incur substantially higher transactional costs.

The scope and the speed of claims handling by VCF1 was also remarkable. Ninety-seven per cent of eligible families chose to claim from the scheme. Almost 7,500 individualised applications were processed and completed in less than 3 years. Compared to the delays, costs and uncertainties of the civil justice system, the Fund proved to be an efficient and effective alternative.

B. The Second Incarnation of the Victim Compensation Fund (VCF)

The compensation fund was reactivated in 2011. On 2 January 2011, President Obama signed into law the James Zadroga 9/11 Health and Compensation Act (Zadroga Act).[14]

[14] 2010 (PL 111-347).

James Zadroga was a New York Police Department (NYPD) officer who participated in the recovery and rubble clearance operations following the 9/11 attacks. He then developed respiratory problems, which were attributed to dust exposure at Ground Zero. He was awarded compensation from the VCF1 and was retired on the grounds of his respiratory condition. When he died in January 2006, he was the first NYPD officer whose death was attributed to exposure to toxic dust at Ground Zero, though his death has also been attributed to other factors.[15] There followed a campaign to increase the coverage and eligibility for compensation; this gained momentum and resulted in the Zadroga Act.

Title I of the Zadroga Act created the World Trade Center (WTC) Health Program, which is operated by NIOSH. Title II of the Zadroga Act reactivated the September 11th Victim Compensation Fund of 2001 that operated from 2001 to 2004. The reactivated fund is referred to as VCF. It required a Special Master, appointed by the Attorney General, to provide compensation for any individual (or a personal representative of a deceased individual) who suffered physical harm or was killed as a result of the terrorist-related aircraft crashes of 11 September 2001 or the debris removal efforts that took place in the immediate aftermath of those crashes.

On 18 May 2011, Attorney General Eric Holder announced that Sheila L Birnbaum would head the September 11th Victim Compensation Fund Program. On 31 August 2011, the Special Master published the Final Rule for the operation of the VCF in the Federal Register. These rules were largely based on the VCF1 rules, but this was not straight reactivation as some substantial amendments were included. Essentially, VCF differs from VCF1 in having a more expansive definition of 'eligible individuals' and 'physical harm', larger geographical coverage for the crash sites, and wider tolerances of the limits of the 'immediate aftermath'. These parameters are summarised below.

Eligible Individuals. Individuals were defined as having been present at a 9/11 crash site at the time of or in the immediate aftermath of the terrorist-related aircraft crashes and who suffered physical harm as a direct result of the crashes or debris removal. Secondly, they could be the personal representatives of individuals who were present at a 9/11 crash site at the time of or in the immediate aftermath of the terrorist-related aircraft crashes and who died as a direct result of the crashes or debris removal.

Physical Harm. The definition of physical harm was expanded to include certain latent conditions that were presumed to have resulted from exposure at the crash sites. These include: interstitial lung disease; chronic respiratory disorder—fumes/vapours; asthma; reactive airways dysfunction syndrome (RADS); WTC-exacerbated chronic obstructive pulmonary disease (COPD); chronic cough syndrome; upper-airway hyper-reactivity; chronic rhinosinusitis; chronic nasopharyngitis; chronic laryngitis; gastro-oesophageal reflux disease (GERD); sleep apnea exacerbated by or related to the above conditions; lower back pain; carpal tunnel syndrome (CTS); certain other musculoskeletal disorders occurring in responders and defined as 'a chronic or recurrent disorder of the musculoskeletal system caused by heavy lifting or repetitive strain on the joints or musculoskeletal system occurring during' the period between 11 September 2001 and 30 May 2002 or as determined by the Special Master; traumatic injuries; and certain types of cancer. Injuries included in

[15] It was suggested by the New York Chief Medical Officer that James Zadroga's death was not due to dust from the Ground Zero, in contradiction to an earlier and a later autopsy findings, and to his family's beliefs: see www.nytimes.com/2007/10/19/nyregion/19autopsy.html?_r=0.

this list carried a presumption of causality. Other conditions would also be considered for compensation, but the causality presumption did not apply to them.

Crash Sites. Under the Zadroga Act and the final regulations, the 9/11 crash sites include:

(a) The World Trade Center site, the Pentagon site and the Shanksville, Pennsylvania site.
(b) The buildings or portions of buildings that were destroyed as a result of the terrorist-related airplane crashes of September 11, 2001.
(c) The New York City Exposure Zone (shown in Figure 13.2), which consists of:
    (i) The area in Manhattan south of the line that runs along Canal Street from the Hudson River to the intersection of Canal Street and East Broadway, north on East Broadway to Clinton Street, and east on Clinton Street to the East River; *and*
    (ii) Any area related to, or along, routes of debris removal, such as barges and the Fresh Kills landfill site.

In rare circumstances, the Special Master may identify, based on additional evidence, additional areas that presented a demonstrable risk of physical harm resulting from the impact of the aircraft or any subsequent fire, explosions, or collapse of buildings (generally, the immediate area in which the impact occurred, fire occurred, portions of buildings fell, or debris fell upon and injured persons) as 9/11 crash sites.

Figure 13.2: Map showing the New York City Exposure Zone. Source: VCF website

Immediate Aftermath. Immediate Aftermath is defined as any time during the period beginning with the crashes and ending on 30 May 2002. To be eligible, the Zadroga Act requires an individual's presence at a 9/11 crash site prior to 30 May 2002. An individual's eligibility would not be affected by whether he or she continued to be present at a site after that date. Once an individual is deemed to have been eligible based on presence during the relevant time period, it is not necessary for the Fund to determine the precise date that the condition was caused.

Limitation. The Zadroga Act authorises the VCF to accept claims for 5 years, starting in October 2011 and ending in October 2016, with final payments to be made during a 6th year in 2016–17. Strict limits were imposed on fund availability: a fixed sum of $2.775 billion was allocated for VCF. Congress also determined that only $875 million may be paid out in the first 5 years of the VCF Program. The remainder may be paid in the 6th year. Thus, for the first 5 years of the Program, claimants will receive only a portion of the compensation allowed under the rules of the VCF. Depending on the number and type of claims, and in order to ensure that all eligible claimants receive an award, the Zadroga Act's cap on funding means that that claimants' awards will be made in two payments: one pro-rated payment during the first 5 years, and a second payment during the 6th year when the additional money may be spent. Like the original fund, the reactivated fund will pay economic and non-economic losses. However, advance benefits will not be available under VCF. This is unsurprising given the time that has elapsed since 9/11, advance benefits provided prompt relief in the aftermath, and the passage of time means that claims are not so likely to be urgent.

The inclusion of claims for deceased individuals is initially surprising. These deaths are not generally directly caused by the events of 9/11, but are attributable via indirect routes, such as illnesses caused by the inhalation of dust/toxins.

The available funds must cover both compensation awards and administrative costs. In an effort to reduce administrative costs, the reactivated fund only offers a single claims evaluation procedure where VCF1 offered two. Under the new fund, the eligibility and award will be decided based on the claimant's filings. The claimant then has the option either to accept the award or to request a hearing.

The reactivated fund also allows individuals to amend previously submitted claims with relevant information. Both amended and new claims are evaluated on a case-by-case basis.

Claims Process

There is no cost to file a VCF claim. Claims forms are available for online filing or can be sent to VCF as hard copy. Claimants can use the services of an attorney to file a claim should they wish, but the system is designed so that it can be used by claimants representing themselves. It is a two-stage process, see Figure 13.3, with the first stage being to establish eligibility and the second to determine quantum. A claimant has to undertake a three-stage process: firstly, to register online and download the forms; second, to complete and submit the eligibility form; and finally, to submit and complete the compensation form.

The VCF website gives clear options for the different types of claim—physical injury and death claims.[16] The forms require personal details from the claimant, if applicable,

[16] www.vcf.gov/register.html.

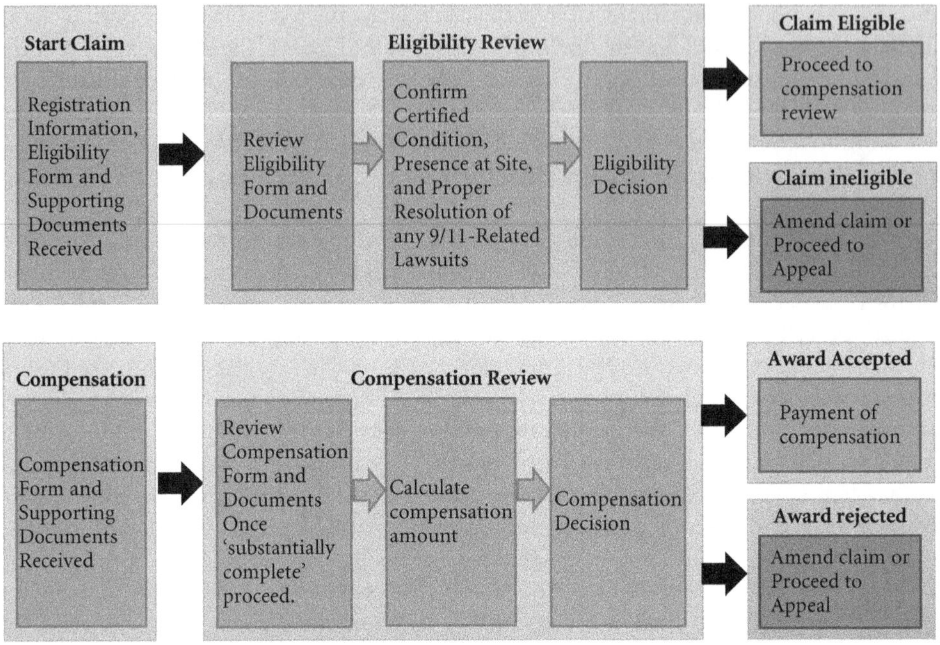

Figure 13.3: Claims process for claims from VCF

details of the deceased, including a death certificate or proof of death, a list of potential beneficiaries, and it gives permission for the fund to access medical records and pension records. There are undertakings for attorneys to state that the fee charged will not exceed 10% of the claim awarded.[17] The website contains clear instructions on the forms, including step-by-step help with filing them in, video tutorials, tip sheets, a map of the New York Exposure Zone, a document checklist of required certificates and information, all of which is designed to enable claimants to easily complete the forms. Should they require any additional help/information, there is a toll-free phone number for claimants, and regular claimant assistance clinics are given on how to make a claim.

Once an application is deemed eligible, it proceeds to the determination of compensation. An application is deemed 'substantially complete' once all the information needed to calculate an award is received and verified. A claimant will be sent a letter stating that the application is 'substantially complete'. Once an application is 'substantially complete' the Special Master has 120 days to make a compensation decision and notify the claimant in writing of the amount awarded. The Special Master then has 20 days from the date of notification of the award amount to authorise payment of the claim. This stage enables a final review of the claim by VCF, including checking the claimant has provided bank details. VCF then submits payment information to the Department of Treasury, who make the payment. The payment is an initial payment, which is 10% of the calculated payment sum.

[17] S 104.81 of the Final Rule.

Limitation

There are two deadlines for VCF claims: a Registration Deadline and a Filing Deadline. The deadlines are detailed in Table 13.7.[18] The Special Master does not have the authority to extend these deadlines. The registration deadline is the date by which a claimant must have registered a claim to be recognised as a claimant, which differs for different categories of claimant. The filing deadline is 3 October 2016 and applies to all claimants. Claimants must have submitted eligibility and compensation forms and all required supporting documentation. This date is absolute and applies even if it is less 2 two years from the date they were diagnosed with their condition or less than 2 years from the date their illness was added to the list of eligible conditions.

Table 13.7: Deadlines for registration and filing VCF claims

To register (file a Registration Form) with the VCF:			
Claimant circumstances	**Applicable registration deadlines**	**What must be filed to meet the registration deadline**	
		Online	**Hard Copy**
Claimant diagnosed *after 3 October 2011* with a 9/11-related condition on the *original list of qualified conditions*	2 years from the date of diagnosis, but no later than *3 October 2016*	Complete Registration Form (receive VCF claim number)	Complete and mail Eligibility Form Part I (even if the remainder of the form is not complete)
Claimant diagnosed *on or before 12 October 2012* with one of the *cancers* added to the list on that date	*12 October 2014*		
Claimant diagnosed *after 12 October 2012* with one of the *cancers* added to the list on that date	2 years from the date of diagnosis, but no later than *3 October 2016*		
Claimant diagnosed with *prostate cancer on or before 21 October 2013*	*21 October 2015*		
Claimant diagnosed with *prostate cancer after 21 October 2013*	2 years from the date of diagnosis, but no later than *3 October 2016*		
Claimant diagnosed *on or before 18 February 2014* with a *rare cancer* that meets the definition as explained by the WTC Health Program on that date	*18 February 2016*		
Claimant diagnosed *after 18 February 2014* with a *rare cancer* that meets the definition as explained by the WTC Health Program on that date	2 years from the date of diagnosis, but no later than *3 October 2016*		

(continued)

[18] Taken from the VCF website, www.vcf.gov/pdf/RegisterChart.pdf.

Table 13.7: ***(Continued)***

To register (file a Registration Form) with the VCF:			
Claimant circumstances	**Applicable registration deadlines**	**What must be filed to meet the registration deadline**	
		Online	**Hard Copy**
Claimant's known condition was not on the original list of qualified conditions or subsequent additions to the list but was added to the list *after 18 February 2014*	2 years from the date the specific condition was added to the list of qualified conditions, but no later than *3 October 2016*		
To file a complete claim (Eligibility and Compensation Forms) with the VCF:			
Claimant circumstances	**Applicable registration deadlines***	**What must be filed to meet the registration deadline**	
		Online	**Hard Copy**
All claimants	*3 October 2016*	All claim forms, including *completed Eligibility and Compensation Forms* with properly executed attestations, certifications and exhibits	

All claimants have the right to a hearing if they do not agree with either the refusal of their claim or the quantum of compensation calculated. Hearings can be by phone or in person, and are transcribed by the court recording service. Transcripts are used in the final decision-making process. The first hearings were held in January 2014. Claimants are notified of the outcome of a hearing in writing as soon as possible.

Claim Statistics, Eligibility and Compensation Awarded

VCF deals with a substantial volume of enquiries. The vast majority of these are via the website, see Table 13.8, and so considerable effort has been made to make the website more user-friendly. User numbers seem to have started slowly in the first year and then stabilised at a higher level, though it is very early to be predicting a pattern.

Table 13.8: Approximate caller and website visitor numbers for VCF's first 3 years

	3 October 2011–30 September 2012	**1 October 2012–31 October 2013**	**1 November 2013–30 September 2014**	**Total**
Website visitors	73,000	282,500	245,749	**601,249**
Inbound calls to helpline	5,600	32,900	33,439	**71,939**

As VCF is in the 3rd year of a 5-year term the available claims statistics are incomplete. Claims statistics are published quarterly. Despite being early in the lifecycle of VCF, some analysis can be performed.

Eligibility

Determining eligibility is the first stage in a VCF claim. Interestingly, the eligibility claims profile differs substantially from claims to VCF1. Table 13.9 shows the breakdown of the status of eligibility forms received as of 31 March 2015. As would be expected, only a small percentage of VCF claims are for deceased individuals, in contrast to approximately half the claims against VCF1.

Table 13.9: Status of eligibility forms received by VCF as of March 2015

Eligibility forms	March 2015		March 2014	
Total eligibility forms	18,895		13,863	
Eligibility claims that can be decided	11,771	62.3%	6,702	48.3%
Personal injury claims	11,646			
Claims for deceased individuals	125			
Eligibility decisions rendered—approved	10,549	89.6%	5,603	83.6%
Personal injury claims	10,467			
Claims for deceased individuals	82			
Eligibility reviews in progress	1,222	10.4%	1,099	16.4%
Eligibility decisions rendered—not eligible	178			
Personal injury claims	174			
Claims for deceased individuals	4			
Eligibility claims that cannot be decided now	6,946	36.8%	7,161	51.7%
Claims missing signed Exhibit A—which allows the VCF to obtain information about eligible conditions	1,005		2,062	
Claims missing requested information regarding presence at a 9/11 site, lawsuit resolution, or additional medical documentation and/or claims pending WTC Health Program review	2,463		1,936	
Claimants who do not appear to have an eligible condition and have not responded to requests for additional medical documentation	3,478		3,163	

When the eligibility statistics are reviewed in detail (Table 13.10), it becomes clear that VCF includes a number of claims for cancer. The original VCF1 fund dealt with immediate injuries, soft tissue, trauma, burns and similar injuries, but did not generally include longer-latency illnesses such as cancer. It is interesting to note that had the decision to reactivate VCF not occurred, then individuals with long-latency illnesses may have gone uncompensated.

Table 13.10: Eligibility statistics 2015: breakdown

Eligibility Forms	Total	Claims with cancer as the only condition	Claims that do not include cancer as a condition	Claims with both cancer and non-cancer conditions
Eligibility claims that can be decided	11,771	1,056	9,376	1,339
Eligibility reviews in progress	1,222	187	870	165
Eligibility decisions rendered—approved	10,549	869	8,506	1,174
Eligibility decisions rendered—not eligible	178	4	57	7

Note: this table does not include the 110 claims found not eligible where the claimant did not prove he or she suffered from any 9/11-related condition.

Compensation

The majority of compensation forms (68.5%) are received once a claimant has been deemed eligible, of which 81.4% have submitted the documentary and supporting evidence required to start reviewing the claim (see Table 13.11).

Table 13.11: Compensation forms 2014 and 2015

Compensation forms	March 2015		March 2014	
Total compensation forms	9,736		3,589	
Total compensation forms from claimants who have been deemed eligible (percentage is the percentage of claimants deemed eligible who have submitted a compensation form)	7,225	68.5%	2,046	36.5%
Compensation forms that can be reviewed—forms with proper signatures/ exhibits that are under review to assess whether there is sufficient information to determine loss	5,881	81.4%	1,220	59.6%
Personal injury claims	5,863			

(continued)

Table 13.11: ***(Continued)***

Compensation forms	**March 2015**		**March 2014**	
Claims for deceased individuals	18			
Compensation decisions rendered	4,415	75.1%	502	41.1%
Personal injury claims	4,409			
Claims for deceased individuals	6			
Total dollar value before proration	$1,058,398,144		$194,608,716	
Compensation reviews in progress	1,466 718			
Compensation forms missing required information	1,344	18.6%	826	40.4%

Table 13.12 outlines the headline compensation figures awarded by VCF to date, as well as a breakdown by self-reported claimant category. It is interesting to note that the mean dollar award figure of $241,000 from VCF is much lower than the average award from VCF1 of $1,267,880.49. It is also considerably lower than the more comparable average physical injury award from VCF1 of $392,968.11. There are several factors which could drive this. It could be due to differences in the injury profiles of claimants. It may be due to differences in the regulation governing the funds or their interpretation. It could be that higher value cases are more complex cases requiring longer to process, so are not yet finalised. When VCF is complete it will be interesting to compare the two schemes to see if further analysis can shed light on these differences.

Table 13.12: Compensation decisions rendered by VCF as of 31 March 2015

Compensation decisions rendered	**Total**	**Dollar value before proration**
Decisions rendered	4,415	$1,058,398,144.44
Highest amount		$4,133,466.00
Lowest amount		$10,000.00
Mean dollar value		$241,038.07
Compensation amounts by self-reported claimant category		
Responder—NYC	3,775	$973,341,889.32
Responder—Pentagon	3	$326,761.68
Responder—Shanksville	1	$50,000.00
Non-responder NYC—cleaning or maintenance work	237	$19,533,941.91
Non-responder NYC—other employment	268	$47,268,819.88
Resident within NYC zone	51	$7,181,592.41
Attend school/childcare/adult care facility	6	$365,250.00
NYC—other capacity	18	$2,550,373.00

(continued)

Table 13.12: ***(Continued)***

Compensation decisions rendered	Total	Dollar value before proration
Non-responder—Pentagon	1	$250,000.00
No response	55	$7,529,516.24

Note: the data includes visitors, tourists and any individual who was in the Exposure Zone during the relevant time period and is not included in another category.

Administrative Costs

As at 30 September 2014, just under $40 million had been spent on administering VCF. This is broken down into three categories of administrative costs, which are detailed in Table 13.13:

— operation costs;
— system development and maintenance costs; and
— labour and contractor support costs.

Table 13.13: Breakdown of VCF's administrative cost up to 30 September 2014

Cost category	Breakdown	Cost at 30 September 2014 (% of admin costs)
Operation costs	Rent of office space in Washington DC (where the claims-processing centre, CPC, is located) and New York. Equipment and services (phone lines, internet, etc)	$4.4 million (11%)
System development and maintenance costs	Creation and ongoing development of and maintenance of VCF's claims-handling system	$13.0 million (33%)
Labour and contractor support costs	Staff necessary to run the CPC, the helpline and to assist claimants with claims. Attorneys who make eligibility and compensation determinations	$22.0 million (56%)

As the administrative costs come from the same capped source as the compensation, every effort is made to minimise them. VCF has a total staff count at September 2014 of 103, up from 75 the previous year. These staff are predominantly at the Claims Processing Centre, with 23 people on the claim-review team, 24 on the document-intake team, and 10 on the helpline team. The Office of the Special Master contains 13 Special Masters Attorneys. To curtail costs, the Special Master and the Deputy Special Master Deborah Greenspan have elected to be uncompensated for their work. This clearly reduces the administrative costs, but must be remembered when comparing VCF cost profiles to other no-fault schemes.

C. Conclusions

The 9/11 funds were a post-hoc response to an event that will hopefully never be repeated.[19] Viewed against the stated aim in the preamble of the Act, namely to preserve the viability of the US airline industry, VCF has been a clear success. However, the funds have not been entirely without criticism.[20]

In his final report, Special Master Feinberg questioned why compensation was limited to the victims of September 11. In contrast, no compensation fund was created for other families and victims of terrorist attacks in Oklahoma City or Kenya, the families and victims of the terrorist attacks on the *USS Cole*, or at the first World Trade Center attack in 1993. He concluded that the scale of both the attacks of 9/11 and of the national response in America justified the creation of VCF1 as a matter of public policy, rather than the attributes of the victims.[21] This runs counter to the standard view of tort-based redress usually seen in the United States and accords more with the European approach of social solidarity seen particularly in the Nordic countries.

The issue of quantum of compensation has also proved troubling. The decision to base compensation on personalised loss of earnings has been questioned: why was the life of a banker worth more than that of a rescue worker? Special Master Feinberg felt that the answer lay in the genesis of the Fund; it was intended as an alternative to a claim in tort through the court system. Congress limited the right to sue, so VCF1 therefore had to be an attractive alternative for claimants. The selection of individualised awards rather than the easier-to-administer flat-rate option pushed VCF1 towards developing a transparent, robust, fair system for determining award value. The presumed methodology provided claimants with a consistent, understandable framework, while enabling VCF1 staff to administer complex individualised awards very swiftly.

It should also be remembered that a flat rate of compensation would almost certainly not have been considered satisfactory by all claimants, and there are difficulties in determining what that rate should be. The impact of an award based on an 'average' wage would obviously be very different depending on the circumstances of the family receiving the award. How much is enough?

There are no easy answers to these questions. The fact that 97% of the eligible families chose to claim from VCF1 indicates that uptake was high, given a lack of alternative litigation that is potentially not surprising. What might be more incisive would be an investigation into claimant satisfaction levels. Once VCF is closed in 2016 a thorough appraisal of both funds would be valuable.

VCF1 did not operate in isolation. Although substantial, it was part of an entire compensation picture. Figure 13.4, taken from a study by RAND,[22] shows that government funding

[19] For an interesting historical review of how VCF1 fitted into the general no-fault landscape, see RL Rabin, 'The Renaissance of Accident Law Plans Revisited' (2005) 64 *Maryland Law Review* 699, http://digitalcommons.law.umaryland.edu/mlr/vol64/iss1/22.

[20] ibid, and the Final Rule of VCF1.

[21] ibid.

[22] Research Brief, *Compensating the Victims of 9/11* (RAND Corporation, 2005), www.rand.org/content/dam/rand/pubs/research_briefs/2005/RAND_RB9087.pdf.

formed only part of the overall compensation awarded. An analysis of the 9/11 quantified benefit by funding source shows that insurers provided the majority of the money spent.

Finally, there are questions as to the effectiveness of the compensation funds. While they have undoubtedly been extremely efficient when compared with any other available compensation distribution option, it must be remembered that some of the administrative efficiency came from the pro bono contributions of a large number of people. That said, the achievements of VCF1 in distributing large amounts of compensation in a very tight time-frame were impressive. The standards set by VCF1 are exceptionally high, over $7 billion in individualised benefits to over 5,560 eligible claimant (a 97% take up) in less than 3 years, with administrative costs of just 1.2%. On many measures, VCF will struggle to match VCF1. For example, VCF will not be able to equal the proportion of administrative costs to compensation seen with VCF1. This is not a reflection on their relative efficiencies, but on the size of compensation awards. It would be premature to evaluate VCF, as this can only be fully undertaken once it is finalised. It is worth remembering that the value of a compensation scheme cannot be reduced to just numbers. VCF1 and VCF were not created solely to fulfil a financial need, though that was an important factor, but also as an expression of social solidarity in the aftermath of a heinous act. Their function is the allocation and distribution of compensation, but their purpose extends beyond the purely functional.

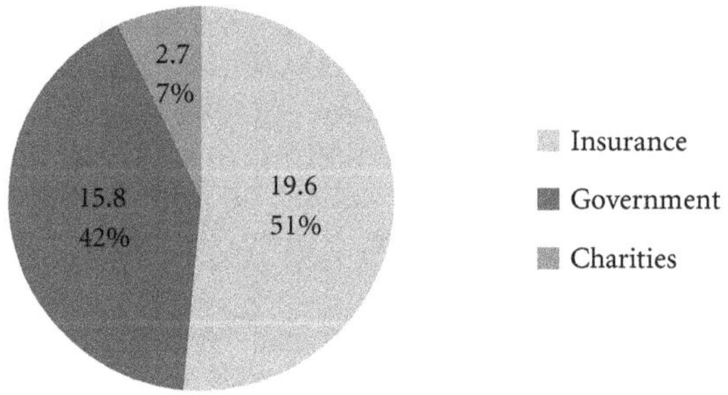

Figure 13.4: Quantified benefits by compensation mechanism

Part V

Vaccines and Pharmaceuticals

14

The Thalidomide Trust

SONIA MACLEOD

A. History

The German firm Chemie Grünenthal applied for a patent for thalidomide (alpha-phthalimido-glutarimide) in West Germany in May 1954.[1] Questions have been raised as to whether this was the first synthesis and testing of thalidomide, but these will not be addressed in this chapter.[2] Early trials showed thalidomide did not function as an anticonvulsant as had been expected, but was an effective sedative and sleeping tablet. It seemed a very attractive sedation option: it was well tolerated, not habit forming and, unlike other products (predominantly barbiturates) on the market, overdoses caused prolonged sleep, not death. At this point in time there were no international standards for pre-marketing testing of pharmaceuticals. Chemie Grünenthal stated that thalidomide was tested in accordance with West German standards; however, this statement is controversial, with some authors indicating that teratogenicity testing was standard practice at the time.[3] Chemie Grünenthal tested thalidomide on animal models (the type and extent of the testing is uncertain as test results were lost),[4] but the testing regime was not sufficiently comprehensive to detect the teratogenic effect. Thalidomide was first marketed over the counter as Contergan in Germany in 1957, and in the UK in April 1958 on prescription as Distaval. Later, compound preparations which combined it with other drugs were marketed for a wide variety of indications, including Asmaval for asthma, Tensival for hypertension and Valgraine for migraine. The promotion of these products stressed the safety of thalidomide, based on the overdose profile.

In 1959 early questions about the safety of thalidomide arose, but these concerned reports of irreversible peripheral neuritis (nerve damage to the hands and feet) usually in the elderly as a side effect, not birth defects. As a result, the package insert was amended to include details of these side effects and in May 1961 Chemie Gruüenthal applied for a change of status to have Contergan upgraded to prescription-only status. The United

[1] Patent number DE00000107584 (process number 5 produces thalidomide) available at https://depatisnet.dpma.de/DepatisNet/depatisnet?action=pdf&docid=DE000001074584A and https://depatisnet.dpma.de/DepatisNet/depatisnet?action=bibdat&docid=DE000001074584B.

[2] For example, see www.bbc.co.uk/news/magazine-15536544.

[3] For a comprehensive review, see R Greek, N Shanks and MJ Rice, 'The History and Implications of Testing Thalidomide on Animals' (2011) 11 *Journal of Philosophy, Science & Law*, http://jpsl.org/archives/history-and-implications-testing-thalidomide-animals/.

[4] Other tests were referred to in the Contergan trial in Germany, but Grünenthal stated that written evidence of these additional tests had been lost.

States had a centralised drug safety assessment procedure, overseen by the Food and Drug Administration (FDA). Safety concerns meant Contergan was not licensed in the United States and it was only available there in clinical trials. Interestingly this was almost certainly due to the vigilance of FDA worker Dr Francis Kelsey, who steadfastly refused to accept that the widespread usage in Europe constituted adequate safety data.

After thalidomide was introduced German paediatricians and geneticists began to see children with gross limb malformations of a novel and unusual pattern. In October 1960 two case studies were presented at a paediatric meeting in Kassel by Kosenow and Pfeiffer;[5] at this time few people present had previously seen such patterns of defects. By 1961 Wiedemann described 13 affected infants who had been referred to him over a period of 10 months,[6] noting that this was a substantial increase. He highlighted a number of associated malformations present in these children, including congenital heart disease, microphthalmos and coloborna, intestinal atresis, renal malformations, abnormal pinnae and facial nacvus.

In a letter dated 16 November 1961 Dr Widukind Lenz suggested to Chemie Grünenthal that these deformities resulted from the mothers having taken thalidomide during pregnancy.[7] By a remarkable coincidence, the same suggestion was made at the same time by Dr William McBride to the distributor of thalidomide in Australia, Distillers.[8] In response to these two independent sources suggesting the same conclusion Chemie Grünenthal withdrew thalidomide from the West German market on 27 November 1961. Distillers, which was also the UK distributor, withdrew thalidomide in the UK in December 1961,[9] while continuing to make it available to hospitals where its administration could be closely supervised.

Confirmation of the causal relationship between thalidomide and congenital malformations came from all parts of the world: the numbers of affected births dramatically declined 9 months after thalidomide was removed from the market. However, there were a very small number of affected births in the UK after September 1962. Thalidomide was in a wide range of medicines and it is thought that these were attributable to mothers using medicines which had been stored for some time and/or which were not clearly labelled as containing thalidomide. The first animal model confirmation of thalidomide as a teratogen came from testing New Zealand White rabbits in 1962.[10] In May 1962 the UK government sent a

[5] W Kosenow and RA Pfeiffer (1961) 109 *Mschr Kinderheilk* 227.

[6] HR Wiedemann, ['Indications of a Current Increase of Hypoplastic and Aplastic Deformities of the Extremities']. (1961) 37 *Med Welt* 1863–66.

[7] Lenz wrote to Grünenthal suggesting a link between thalidomide and birth defects on 16 November 1961. On 18 November 1961 Lenz raised the possibility that thalidomide was a teratogen at a conference: see RA Pfeiffer and W Kosenow ['On the problem of exogenous causes of severe malformations of the extremities']. (1962) 104 *Munch Med Wochenschr* 68–74. These concerns were more widely expressed in a letter to the *Lancet* published 6 January 1962, in vol 279(7219) 45–46 (originally published as vol 1, issue 7219), available at http://ac.els-cdn.com/S014067366292665X/1-s2.0-S014067366292665X-main.pdf?_tid=23f49fbc-5b5a-11e3-a104-00000aab0f6c&acdnat=1385993026_86f7746f0ecf4bf36b3d1ebefb684a1f.

[8] Letter to the *Lancet*, vol 278(7216) 1358 (originally published as vol 2, issue 7216), available at http://ac.els-cdn.com/S0140673661909278/1-s2.0-S0140673661909278-main.pdf?_tid=e0f771d2-5b52-11e3-9d6f-00000aacb35e&acdnat=1385989907_95275eb90cb730827b30a0070cea9e39.

[9] DJ Hayman, Managing Director of Distillers Company (Biochemicals) Ltd, in a letter to the *Lancet*, published vol 279(7219) 45–46 (originally published as vol 1, issue 7219), available at http://ac.els-cdn.com/S014067366292665X/1-s2.0-S014067366292665X-main.pdf?_tid=23f49fbc-5b5a-11e3-a104-00000aab0f6c&acdnat=1385993026_86f7746f0ecf4bf36b3d1ebefb684a1f.

[10] GF Somers, 'Thalidomide and Congenital Abnormalities' (1962) 279(7235) *Lancet* 912–13.

letter to all doctors warning of the dangers of the use of thalidomide, and instructing that all stocks be destroyed.

The restricted availability of the drug in the USA meant there were very few cases there. By contrast, it had been on sale over the counter in Germany, and there were consequently more affected children, approximately 6,000, born there than anywhere else. In the UK, although the drug was available on prescription only, it was used very widely for common symptoms of early pregnancy such as morning sickness, with approximately 500 babies born affected by thalidomide.

World-wide it is thought that approximately 10,000 babies were born with developmental abnormalities due to thalidomide.[11] In addition to limb defects, thalidomide also caused heart and intestinal damage, which was severely live limiting in many cases: 1-year survival figures for live births of thalidomide-affected children suggest a 40% mortality rate.[12] It should be remembered that the number of children born damaged by thalidomide does not include any miscarriages and stillbirths which may have been caused by the drug, but which would not have been attributed to the drug.

The Lady Hoare Trust

A charitable trust funded was established by Lady Hoare, the Lady Mayoress of London, in 1962 following a national appeal for donations from the general public. The aim of the trust was to provide financial and other assistance for those damaged by thalidomide. This trust was a privately funded charitable trust and was not part of any of the official compensation arrangements.

Compensation for Thalidomiders

In order to obtain compensation, the children affected sued Distillers in negligence for personal injuries and associated compensations. The litigation in response to the thalidomide disaster was groundbreaking in many ways: it was one of the first collective actions in England, the first settlement established solely for children, the first settlement where the cause of action was for damages accrued before birth, and eventually the Thalidomide Trust became one of the first structured settlements. Although all of these elements are now an established part of UK legal culture it should be remembered that they were new and unknown when the cases were first brought; some commentators were uncertain whether an action could be brought in negligence for cause of action accrued before birth.

The 1968 Settlement[13]

The 1968 settlement was for the benefit of 65 children and their parents who had individual writs against Distillers. They were offered private settlements of 40% of the value of their

[11] N Vargesson, 'Thalidomide-Induced Limb Defects: Resolving a 50-year-old Puzzle' (2009) 31 *BioEssays* 1327–36; W Lenz, 'A Short History of Thalidomide Embryopathy' (1988) 38 *Teratology* 203–15.

[12] Lenz (n 11).

[13] Reached on 19 February 1968.

claims had they won, with 33% of the award value added on to cover inflation. Individual payments varied between £5,000 and £45,000 depending on the extent of the disability suffered. If accepting effectively a 60% reduction in the award value seems questionable it should be remembered that the success of the action was uncertain; indeed Desmond Ackner QC, who appeared for the children and parents, felt the settlement was a good option for them as had they continued they may not have received anything. Other commentators were not so convinced this was an adequate amount. As the settlement was made for children (62 of whom were still alive) it required the involvement of the High Court. Agreement over liability and the form of the settlement was approved by the High Court on 19 February 1968. Quantum for damages was assessed using two representative cases: RS who had injuries which were in the mid-range in terms of severity; and DJ who was one of the most severely injured of the children.[14]

At the time Distillers indicated that it was intending to provide funds for the remaining 389 thalidomide-affected claimants, and this settlement paved the way for further settlement by establishing a precedent. However, the 1968 settlement only provided for 65 individuals, a small subgroup of affected children; it did not provide anything for any of the other affected children. The parents maintained a consistent and high-profile campaign for a just settlement for all the affected children, and from 1972 onwards they were aided by the *Sunday Times*, beginning a tradition of campaigning and profile-raising which the thalidomiders themselves continue to date.

The 1973 Settlement and Establishment of the Thalidomide Trust

After various offers from Distillers a settlement was finally reached in 1973. At this point there were three lists of children.

List X contained over 300 children who had injuries accepted by all parties as being thalidomide-induced malformations as well as some parents of deceased children whose injuries were agreed to be due to maternal thalidomide ingestion during pregnancy.

List Y, approximately 80 children, had damage which was not confirmed as attributable to thalidomide.

Finally, list Z comprised 4 children who had received compensation as a result of claims against solicitors (eg one case involved a writ for thalidomide damage being issued in a timely manner, but the issuing solicitor failed to file the writ).

Eligibility

To be eligible for compensation in 1973 the child must have been affected by a thalidomide preparation taken by his mother during pregnancy and meet the following criteria specified in the settlement:

(i) … either:—
  (a) that he was born in the United Kingdom, or
  (b) that his mother took the said UK Thalidomide preparation in the United Kingdom, or

[14] *S and Another v Distillers Company (Biochemicals) Ltd J* and *Another v Same* [1970] 1 WLR 114; [1969] 3 All ER 1412; (1969) 113 SJ 672; Times, July 31, 1969.

(c) that he was on 22 February 1973 domiciled or resident within the United Kingdom or any of the countries specified in the First Schedule of the Trust Deed.

The countries specified were Canada, the Channel Islands, Cyprus, Jamaica, Kenya, the Republic of Ireland and the Republic of South Africa. Settlements were paid to children living in these countries at the date of the settlement if they could show that the thalidomide ingested by their mothers had been formulated in the UK.

The 1973 settlement consisted of both individual settlements for the children and their families and also established a trust, the Thalidomide Trust, for the benefit of all children with damage which had been accepted as being due to thalidomide. Acceptance of the settlement required the withdrawing of any claims for negligence by the children or their parents against Distillers and the waiving of any rights to any ongoing or future claims arising from taking thalidomide. Initially, to allow a final settlement to be reached, it was proposed that the settlement would only crystallise if it were accepted by all parents; however, consensus proved difficult to achieve and this requirement was amended from all parents to a majority accepting the settlement.

To achieve a complete and final settlement it was desired that all children were joined in the proceedings. The majority of children and their parents had issued writs which had not been disposed of, but if no writs had been issued the children and their parents were included on the new writ. The new writ and all existing writs were then dismissed by the judge on the terms that the defendant fulfilled the order of the court instructing them to pay the sums listed above into court. The terms of the court order varied according to which list the child was on and were as follows:

Children on the X List and their Parents

Individual awards for these children consisted of:

— a £5,000 award for parents (£10,000 for parents of twins if both twins were affected);
— an award for the individual child of 40% of the value of the claim had it been successful increased by 33% to account for inflation (with quantum as per the 1968 cases).

Children on the Y List and their Parents

The judge was asked to dismiss all claims of children on the Y list (whether contained in existing writs or in the new writ) with provisions to allow determination of whether the child met the X list criteria to be determined at trial. If the child was found to fall within the X list criteria, then they would be eligible for the same benefits as if they have been on the X list.

Following the establishment of the settlement there was considerable discourse on whether certain children belonged on the Y list of the X list, see below.

Children on the List Z and their Parents

It was determined that if it could be demonstrated that these children fulfilled the criteria to be on list X (it was accepted by the defendants that their injuries were attributable to thalidomide and the residence criteria were met), then they should be treated accordingly and would receive all the benefits applicable to X list children, but taking into account any compensation already recovered from actions against solicitors for professional negligence.

Reviews of the Y List Children

The court order made provision for the Y list children to be considered to see whether they could be incorporated into the X list. By early 1978 the Y list had been investigated and approximately 20 children had been added to the X list. This slow rate of processing and the difficulties in obtaining proof for Y list children was of concern. In 1978 the UK government called Sir Alan Marre to investigate and provide a solution. As part of his recommendations approximately a further 20 children were added to the X list. Payments of £10,000 were made to 48 of the children on the list whose injuries were not thought to be due to thalidomide for the anxiety they had suffered. Distillers was required to contribute an additional £3 million to the Trust and the government added £800,000 in tax relief.

Individual Awards

The awards detailed above were paid directly to the recipient. As the recipients were children these awards were almost entirely paid to the children's parents; there were some exceptions where money was paid into the Court of Protection. In some cases the child's award was depleted or very occasionally entirely spent when they reached majority; usually this was due to poor parental investment choices—instances where it was due to parental embezzling were very rare. In these circumstances payments from the Thalidomide Trust went some way towards mitigating the child's loss.

Collective Redress—The Thalidomide Children's Trust

In addition to the individual awards the offer of settlement contained the beginnings of the Thalidomide Children's Trust. The Trust was established on 10 August 1973. It is a charitable trust fund established under court order holding funds for the benefit of the children who met the eligibility criteria for residence and whose congenital disability was (in the opinion of the trustees) caused by their mother taking thalidomide while pregnant. This included all of the children under the 1963 settlement and the 1978 settlement as well as all children subsequently deemed eligible.

The Trust was originally intended to be funded by seven donations of £2,000,000 made by Distillers, with the first being paid on acceptance of the settlement offer, with the remaining six payments being made at annual intervals.

The original intention was for the Trust to spend only the interest and not the capital. However, in the mid-1990s the Trustees realised this was not was not sustainable and the Deed of Trust was amended to allow distribution of the capital. In 1996 Guinness Plc (who had purchased Distillers) pledged to continue funding the Trust by annual contributions totalling £37.5 million until 2009. In 2000 and again in 2008 Diageo (who had purchased Guinness Plc) agreed to make further contributions to fund the Trust until 2022.

Taxation

The original understanding was that the payments from the Thalidomide Trust would be tax free. However, this was not the case and as the UK government did not wish to benefit from the thalidomide tragedy, in 1975 it added £5,000,000 to the Trust fund to offset the tax liability. In 1996 the Trust applied for a tax offset for the new monies contributed by Guinness Plc, but the government declined to continue the existing arrangement, and instead provided a grant so the Trust remained tax neutral.

In 2002 the thalidomiders launched a campaign to change the tax status of the benefits they received from the Trust, which resulted in the removal of income tax from the payments from the Trust.[15]

Allocating Funding to Beneficiaries

The Thalidomide Trust calculates the amount it has available for beneficiaries each year. They have a known number of beneficiaries and they know the severity of the injuries of each beneficiary, ranked according to the 1968 personal injury tariffs. In this way the payments are tiered according to severity of injury.

Government Health Grants

In addition to allocating its own funds the Trust also distributes the government health grants. In 2009 the UK government announced an additional £20 million of funding for the Trust to provide additional health grants, required because thalidomiders faced additional health needs as they aged. In England the government provided an additional £80 million to help the remaining 325 thalidomiders resident in England in December 2012. The devolved regional governments followed suit by providing comparable per-capita funding to the beneficiaries resident in their countries.

New Applications

The majority of beneficiaries under the Trust were known before its inception; however, applicants are still coming forward with claims that they have birth defects which were caused by thalidomide taken by their mothers during early pregnancy, and that they should therefore be accepted as beneficiaries of the Thalidomide Trust. When the Trust was formed in 1973 there were 429 beneficiaries. As at 1 June 2016 102 new beneficiaries have been accepted and 65 have died, leaving the Trust with 466 current beneficiaries.[16] Prior to 2007 evaluating potential beneficiaries was undertaken by the defendant's solicitors; since 2007 the process has been undertaken by the Trust.

Applicants have to pass an initial screen; this is a simple 1-page form, available from the Trust website, asking for personal details and a brief description of their damage (including photographs of external damage) and any evidence of maternal thalidomide ingestion. This information is used to screen out claimants; for example, if they are the wrong age they can be quickly assessed as non-eligible.

Potentially eligible claimants then progress to the second stage of the process: evidence gathering. Two strands are investigated, namely the damage to the claimant, and evidence of thalidomide ingestion by the claimant's mother. On the first, more detailed questionnaires are sent out, and claimants undergo a medical examination and medical/surgical records. To address the ingestion point, claimants can provide prescription records, or

[15] The Thalidomide Children's Trust (Application of Section 329AA of the Income and Corporation Taxes Act 1988), SI 1819 of 2004, available at www.legislation.gov.uk/search?title=&year=2004&number=1819&type=uksi.

[16] Figures obtained from Deborah Jack, Executive Director, The Thalidomide Trust.

statements from doctors if they are available, or they can provide witness statements relating to ingestion.

Stages 1 and 2 are undertaken by Trust staff. Stage 3 involves assessment and recommendation by the Trust Claims Committee. An application summary is produced for the Claims Committee. Thalidomide caused a wide variety of birth defects, none of which was unique to that drug, but experts can detect patterns of injuries associated with the drug. The injuries that the applicant presents with are compared to the known pattern of thalidomide injuries. The Claims Committee considers whether, on the balance of probabilities, it feels the injuries are attributable to thalidomide (if required, the Committee has access to considerable expert medical opinion on this issue). The Claims Committee then decides whether or not to recommend an application for consideration by the full Board of Trustees.

Stage 4 is the final decision making. This is done by the full Board of Trustees. The Board is presented with the recommendation of the Claims Committee and has to decide whether to follow the Claims Committee recommendation, to disregard it or to ask for further information/clarification. The majority of applications recommended by the Claims Committee are accepted.

The percentage of applicants accepted is consistently low—see Table 14.1. It is interesting that new beneficiaries are still coming forward so many years after the triggering events. This can be because the applicants were unaware that their mother took thalidomide, or it is often because their damage is at the less severe end of the damage seen by the Trust, so they have simply 'gotten on with it' but as they age their injuries deteriorate more rapidly, sometimes causing difficulties working and financial hardship as well as pain, etc.

There are some notable peaks in application rates, in 2010 there was a considerable surge in applications, almost certainly due to the introduction of the Department of Health grants and the associated publicity and raised awareness. From provisional numbers it looks like 2016 may also have an elevated number of applications, again thought to be due to raised awareness following coverage of the thalidomide tragedy in the BBC drama series *Call the Midwife*, first screened in early 2016.

Table 14.1: Annual applications and acceptances to the Thalidomide Trust

Year	Applications received	Claims accepted
2007	35	2
2008	53	3
2009	61	3
2010	224	8
2011	39	2
2012	90	3
2013	76	1
2014	63	0
2015	38	0
2016 (as at 30/04/16)	19	0
Total	698	22

If an applicant is accepted, then the injuries he/she present with are evaluated against the comparators to see where the applicant fits within the range of injuries seen within existing beneficiaries. Every beneficiary has what is known as a 6(4)(b) score, a reference to clause 6(4)(b) in the original agreement to give every affected individual an award. These scores reflect the severity of the person's injuries, and are scaled from 1 to 100, with 0 being no disability. Current beneficiaries have scores ranging from 3.5 to 75. These scores are used to allocate funds. The scores from all of the Trust beneficiaries are added together and the available funds are divided by the scores total; this number is then multiplied by the individual's 6(4)(b) score to give the value of the individual's award.

There is currently no review or reconsideration process for refused applications. Applicants are informed that they should reapply if any new evidence comes to light. There have been a handful of legal challenges over the Trust's refusals of potential beneficiaries. In the majority of these cases the original refusal was upheld, in one case the original decision was overturned on the basis of new evidence which had not previously been presented to the Trust.

B. Conclusions

The Thalidomide Trust acts as a unified portal for the administration of payments from both a court settlement (and subsequent additions to this fund) and from relevant UK health departments. The Trust is far more than an administrative agency to distribute funds; it also provides a wide range of health and welfare support, including the provision of multidisciplinary expertise, and the organisation of events of interest and benefit to the thalidomiders (such as their 'Fit for the Future' event, focusing on exercise and nutritional advice tailed to the needs of the beneficiaries) as well as providing information and funding research. Going forward, the intention is to take a more practical approach to assessing the changing needs of individual beneficiaries as they age. A programme of holistic needs assessments, involving visits by Trust staff on a regular basis to individuals in their own homes, is planned to see how the Trust can best meet their needs.

The thalidomiders have a facebook group where useful information can be exchanged, as well as social conversation. Prior to a digitally connected world the Trust provided a forum, initially for families and later for individuals affected by this tragedy, to meet, discuss issues affecting them and coordinate campaigns. The success of these is self-evident in the shape of additional funds from the original defendants and their successors, a more favourable tax position and governmental grants.

One of the areas of contention is over apologies. In 2010 the UK government issued the following statement: 'The Government wish to express their deep sympathy for the injury and suffering endured by all those affected.'[17]

[17] *Hansard*, HC 14 January 2010, vol 503 part 24 col 859, available at http://www.publications.parliament.uk/pa/cm200910/cmhansrd/cm100114/debtext/100114-0006.htm#10011456000006.

In a historic move in 2012 Grünenthal issued an apology for its 50-year silence on the matter and expressed its deep regret for those affected by thalidomide.[18] This has been greeted with anger in some quarters, including by British thalidomiders,[19] as it was not felt to be full and frank. The British thalidomide story highlights that there are situations which require more than just compensation; recognition and acknowledgement of liability and learning from mistakes are also desired by those affected.

[18] www.contergan.grunenthal.info/grt-ctg/GRT-CTG/Stellungnahme/Rede_anlaesslich_Einweihung_des_Contergan-Denkmals/224600963.jsp.

[19] www.bbc.co.uk/news/health-19448046.

15

Vaccine Injury Compensation Schemes

SONIA MACLEOD

Vaccine damage schemes were introduced in many countries around the world from 1961.[1] A key motivation was to improve public health by supporting public confidence in the safety of vaccines through the back-up provision of a state-guaranteed right to compensation. The objective of vaccination is to induce immunity to disease by administering killed or weakened pathogens, producing antibodies that protect against the disease. During the 20th century, vaccination became increasingly available against numerous infectious diseases. Unfortunately, a small number of those who are vaccinated suffer an adverse reaction, which, in some cases, can lead to severe lifelong disability. It is neither scientifically avoidable nor predictable which children will or will not suffer an adverse reaction. But it is an important aspect of national policy to improve and maintain public health that children should be vaccinated against a number of diseases, which are thereby made avoidable. Accordingly, the moral argument arose that a society and state that encourages vaccination should accept responsibility towards those vulnerable members who suffer harm as a result of that policy, where the element of individual choice is restricted because of their young age and of society's general public health policy. Vaccine schemes can be either immunisation specific, such as the National Swine Flu Vaccination scheme in the United States following mass vaccination in 1976 and the proposed Ebola Vaccination Scheme underpinned by the World Bank, or they can be general schemes which cover a variety of immunisations. Examples of inoculation-specific and national schemes (the United States of America and the United Kingdom) are given below. These are by no means the only schemes; such compensation exists across many countries.

A. The US National Swine Flu Immunization Program of 1976

Although this scheme is no longer open (the immunisation programme has long ceased and new claimants would be time-barred) it is worth considering as it provided the template for governmental assumption of liability for vaccine injuries.

[1] Including Germany 1961, France 1964, Japan 1970 (revised 1977), Switzerland 1970, Denmark 1972 (revised 1978), New Zealand 1974, Sweden 1978, United Kingdom 1979, Quebec 1985, United States of America 1988, Taiwan 1988, Italy 1992, Norway 1995.

Background

In January 1976 there was an outbreak of a flu-like illness at an army base in Fort Dix, New Jersey.[2] Testing at the Centers for Disease Control (CDC) in Atlanta confirmed this was an influenza virus, specifically H1N1, a strain of swine flu, and suggested that person-to-person transmission had occurred. This caused serious concern as the 'Spanish flu' epidemic of 1918–19 was also an H1N1 strain of swine flu.[3] Spanish flu infected approximately 2 billion people and conservative estimates suggest it killed at least 20 million people. Given the virulence and deadliness of Spanish flu it is hardly surprising that the discovery of a strain of swine flu would be taken very seriously. President Ford spearheaded a campaign to appropriate emergency funds for a nationwide vaccination programme.[4] $135,064,000 was appropriated by Congress under an Act of 15 April 1976.[5]

Liability under the 1976 Swine Flu Immunization Program

Four manufacturers[6] were tasked with producing the vaccine; contemporary press cuttings show that all of the manufacturers were reluctant to be involved in a mass vaccination programme unless protection from liability was offered.[7] This protection came in the form of the National Swine Flu Immunization Program of 1976.[8] The Act provided that a claim against the state was the exclusive remedy for claims for personal injury arising from vaccination. Ordinarily the US government cannot be held strictly liable under the Federal Tort Claims Act. To counter this, the flu statute contained an exception ensuring that the United States would be liable to the same extent as if an action could have been brought under the Federal Tort Claims Act. The liability of the United States could 'be based on any theory of liability that would govern an action against such program participant[9] under the law of the place where the act or omission occurred, including negligence, strict liability in tort, and breach of warranty'. The protection for programme participants was not unrestricted, and could be removed if a programme participant failed to fulfil their contractual duties owed to the US government or if the injury to the vaccine recipient was caused by the negligence of the programme participant.[10]

[2] For a detailed discussion of the swine flu programme of 1976, see NS Appel, 'Liability in Mass Immunization Programs' [1980] *Brigham Young University Law Review* 69, available at http://digitalcommons.law.byu.edu/lawreview/vol1980/iss1/6/.

[3] https://virus.stanford.edu/uda/ and www.cdc.gov/flu/about/qa/1918flupandemic.htm.

[4] The Gerald R Ford Presidential Library and Museum provides an overview of the 1976 programme, available at www.fordlibrarymuseum.gov/library/exhibits/swineflu/sf.asp.

[5] Act of 15 April 1976, Pub L No 94-266, 90 Stat 363 (1976). Text available at www.gpo.gov/fdsys/pkg/STATUTE-90/pdf/STATUTE-90-Pg362.pdf.

[6] Merrill-National Laboratories; Merck Sharpe & Dohme; Parke, Davis & Co; and Wyeth Laboratories.

[7] This was likely to be due, at least in part, to the findings in the case of *Reyes v Wyeth Laboratories* 498 F2d 1264 (5th Cir 1974), cert denied, 419 US 1096 (1974); for a more detailed review, see AW Reitze, Jr, 'Federal Compensation for Vaccination Induced Injuries' (1986) 13 *Boston College Environmental Affairs Law Review* 169, http://lawdigitalcommons.bc.edu/ealr/vol13/iss2/2.

[8] National Swine Flu Immunization Program of 1976, 42 USC 5 247bG)-(1) (1976), signed into law 12 August 1976, available at www.gpo.gov/fdsys/pkg/STATUTE-90/pdf/STATUTE-90-Pg1113.pdf.

[9] Defined as the manufacturer or distributor of the vaccine, and the public or private agencies, organisations, and medical personnel who provided swine flu inoculations without charge and in compliance with the informed consent form and procedures.

[10] 42 USC § 247b(k)(7) (1976).

The Vaccination Programme and Guillain–Barré Syndrome

Vaccinations began on 1 October 1976, in advance of the main flu season, but only ran for a very short period, ending on 16 December 1976. During this time approximately 40 million inoculations were given. The cessation of the programme was due to a number of cases of Guillain–Barré syndrome appearing shortly after the administration of the vaccine. Guillain–Barré syndrome is a rare and serious condition of the peripheral nervous system.[11] It occurs when the body's immune system attacks part of the nervous system. Symptoms include pain, tingling and numbness, progressive muscle weakness and coordination problems and unsteadiness (sufferers may not be able to walk unaided), and potentially paralysis. The exact aetiology of Guillain–Barré is unknown, but it often occurs (approximately 60% of cases) 2–4 weeks after a viral or bacterial infection or after vaccinations. It is thought infection/inoculation may trigger the immune system to attack nerve roots and peripheral nerves. The majority of patients make a full recovery fairly promptly, but others are more severely affected, and fatalities are possible.

The decision to halt the vaccination programme was aided by the fact that the swine flu outbreak had not been on the scale that had been feared.

The Act contained provisions on the inclusion of a written consent form, which was to be given to every recipient of the vaccine (or a guardian in the case of minors and others lacking capacity and/or competence). The consent form contained general warnings and a paragraph on 'Special Precautions', which read: '[A]s with any vaccine or drug, the possibility of a severe or potentially fatal reaction exists. However, the flu vaccine has rarely been associated with severe or fatal reactions.'[12] There was no mention of Guillain–Barré syndrome on the consent form.

A second form was also provided to participants providing information on the legal redress options available to them.

Claims Process

There were two linked options for potential claimants, an administrative programme and litigation. The Act required the Attorney General to defend all court actions brought against programme participants, their insurers and federal employees. Upon certification, the Attorney General was replaced by the US government and the case was removed to the federal court. There was an indemnification for the United States against programme participants if there was negligence or contractual breach, which was never used. The Act also required claimants to lodge an administrative claim.

There was a 2-year limitation on swine flu claims within the Act, starting from the date of accrual, usually the date of vaccination or shortly thereafter. The programme was suspended in December 1976, meaning the majority of claims had to be lodged by the end of 1978.

[11] www.nhs.uk/Conditions/Guillain-Barre-syndrome/Pages/Symptoms.aspx.

[12] Guillain–Barré syndrome following influenza vaccines remains a concern, though studies suggest that the risk of developing the syndrome is lower following flu vaccinations than after an influenza infection. JC Kwong et al, 'Risk of Guillain-Barré Syndrome after Seasonal Influenza Vaccination and Influenza Health-care Encounters: A Self-controlled Study' (2013) 13 Lancet Infectious Diseases 769–76.

Initially Guillain–Barré claims required the proof of negligence to fall within of the compensation programme, but in June 1978 a change of policy was adopted. In a Memorandum to President Carter, Department of Health, Education and Welfare Secretary Joseph Califano explained that the informed consent given to participants did not warn of Guillain–Barré, consequently expecting claimants to prove negligence seemed unreasonable.[13] There was a legitimate defence open to the government, namely that the risk of Guillain–Barré was unknown, but given that the government had actively encouraged mass vaccination it was considered unfair to effectively bar recovery. At the point this decision was made there were over 400 Guillain–Barré cases pending, claiming in excess of $370 million in damages.

Conclusions

Although long-since concluded, the National Swine Flu Immunization Program provides an early example of an attempt to balance the public good of mass vaccination against the risk of compensation claims arising from such inoculations. It is one of the first times that manufacturers have refused to accept the potential burden of attendant litigation, leaving the government with a stark choice: abandon the mass vaccination (with the public health and reputational risks that runs) or meet the cost of claims arising from it. The fact that the eventual cost of the litigation was approximately equal to the cost of the vaccination programme itself could not have been foreseen. However, it is hardly surprising that manufacturers were not willing to undertake even a very small risk of such a high-value pay-out.

The Guillain–Barré litigation is particularly surprising: the US government voluntarily removed the obstacle of proving negligence from these patients, not because there was a legal imperative to do so, but because it was considered 'unfair'. This illustrates the flexibility of compensation mechanisms, which can incorporate social norms, rather than simply reflecting the established legal position. Although surprising at the time, the removal of negligence from vaccine compensation has now become the norm.

B. The US Vaccine Injury Compensation Scheme

In the USA, the motivation for the introduction of a compensation scheme was the collapse in the number of vaccine manufacturers as a result of product liability litigation. Between 1981 and 1987 the number of product liability claims involving vaccines skyrocketed from a handful to around 800 annually, accompanied by a rise in the wholesale price of the vaccine against diphtheria, pertussis (whooping cough) and tetanus (in triple DPT vaccine) by over 6,000%.[14] Associated with this price increase was a decline in the level of

[13] Memorandum for the President. Weekly Report of HEW Activities, 23 June 1978, see pp 110–11 of www.jimmycarterlibrary.gov/digital_library/sso/148878/82/SSO_148878_082_09.pdf. There is a handwritten note in the margin which reads 'I think this is excessively lenient. Protect US Gov't within bounds of your statement', demonstrating that this was not a universally popular proposition.

[14] This and the following statistics are taken from R Manning, 'Changing Rules in Tort Law and the Market for Childhood Vaccines' (1994) 37 *Journal of Law and Economics* 247–75.

vaccine consumption. In that period, the number of manufacturers fell from 4 to 1, and supply seemed likely to disappear.

In order to safeguard the supply of vaccines, and diversity of supply from a sufficient number of manufacturers, Congress enacted the national Vaccine Injury Compensation Program (VICP), which became operative in 1988.[15] VICP underpinned the USA's mass vaccination programme and created a federal no-fault, non-adversarial alternative to suing vaccine manufacturers and providers in civil court. The law also contained requirements for doctors to (1) give parents information on the benefits and risks of vaccines before vaccination; (2) report serious health problems, injuries, hospitalisations and deaths following vaccination to a centralised, federally operated Vaccine Adverse Event Reporting System (VAERS); (3) write down serious health problems following vaccination in the individual's permanent medical record; and (4) keep a permanent record of vaccinations given, including manufacturer's names and lot numbers.

Reviewing the history of these reforms, one commentator concluded that: 'The tort system is not an efficient insurance mechanism for most losses. Third-party insurance is much better equipped to handle the problems of risk pooling, moral hazards, and adverse selection.'[16]

Scope of the VICP

The VICP is administered jointly by the US Department of Health and Human Services (HSS), the US Court of Federal Claims (located in Washington, DC), and the US Department of Justice (DoJ). The VICP is located in the HRSA Healthcare Systems Bureau. The Advisory Commission on Childhood Vaccines (ACCV) advises and makes recommendations to the Secretary of HHS on issues relating to the operation of the VICP, and its nine voting members provide the oversight of the programme. Covered vaccines and compensable injuries are described on the 'Vaccine Injury Table'.[17]

The system includes a link to the regulatory system that oversees the safety of medicines, so as to monitor the incidence of injuries with individual products and batches. Healthcare providers are required by law to report all suspected vaccine reactions to VAERS, operated by the FDA and the CDC.

The VICP is funded by a surcharge on all doses of vaccines recommended by the CDC for 'universal use' by all children, paid by all vaccine manufacturers on the VICP list. In 2013, all vaccines in the VICP list were taxed at $0.75 per analyte, which is passed on to users in product prices. Trivalent influenza vaccine for example, is taxed $0.75 because it prevents one disease; measles–mumps–rubella vaccine, which prevents three diseases, is taxed $2.25. Private insurance is, therefore, not involved in these arrangements. Monies from the surcharge are placed in a Trust Fund maintained by the government for use to pay vaccine victims. By 2013, $2.24 billion had been awarded to vaccine victims in 14,214 claims, $148.6 million paid in attorneys' fees and costs, and there was a surplus of

[15] The originating legislation was the National Childhood Vaccine Injury Act of 1986 (PL-99-660).

[16] R Manning, 'Is the Insurance Aspect of Producer Liability Valued by Consumers? Liability Changes and Childhood Vaccine Consumption' (1996) 13 *Journal of Risk and Uncertainty* 37–41.

[17] www.cdc.gov/vaccines/pubs/pinkbook/downloads/appendices/appdx-full-f.pdf.

$3.404 billion in the VICP Trust Fund. Roughly two out of three plaintiffs are denied compensation (31%).

The Claims Process

If a vaccine injury or death occurred after 1 October 1988 an application for federal compensation must be made before pursuing a lawsuit against a vaccine manufacturer or vaccine provider in civil court. A vaccinated individual, the parent or guardian of a vaccinated child/adult with disabilities or the personal representative of the deceased can claim providing eligibility conditions are met.[18] A claim for vaccine-related injury or death is filed with a petition for compensation at the Court of Claims, copied to the HHS/Director to the Division of Vaccine Injury Compensation. Claims can be in person or represented by an attorney. Claims must be filed within 24 months of a death and 36 months of an injury. There is a filing fee of $400.

The Secretary of HHS is named as the Respondent. An HHS physician reviews the petition to determine whether it meets the medical and VICP legal criteria for compensation. This recommendation is provided to the court and petitioner through a Respondent's report filed by the DoJ. The HHS position is presented by an attorney from the DoJ in hearings before a Special Master, who makes the decision for compensation under the VICP. A decision may be appealed to the court, then to the Federal Circuit Court of Appeals, and eventually to the US Supreme Court. If a case is found eligible for compensation, the amount of the award is usually negotiated between the DoJ and the petitioner's attorneys. If the attorneys cannot agree, the case is scheduled for a hearing for the Special Master to assess the amount of compensation.

The principal hurdle for a claimant is to prove causation. The VICP Act specifies two ways to prove causation. Firstly, if certain injuries and conditions are found to have occurred within a prescribed period of time after the administration of a vaccine that is listed in the Vaccine Injury Table, there is a rebuttable presumption of causation. This mechanism is designed to minimise petitioners' difficulties in proving injury as a result of a vaccine. Secondly, a petitioner may demonstrate causation under traditional standards applied in tort cases, through scientific studies and expert medical testimony as evidence.

Compensatable claims, and even most claims found to be non-compensatable, are awarded reimbursement for attorney's fees and costs. A petitioner may file a claim in civil court against the vaccine company and/or the vaccine administrator only after first filing a claim under the VICP and then rejecting the decision of the court. It can take 2–10 years to resolve vaccine injury claims in the VICP.

[18] In addition, to be eligible to file a claim, the effects of the person's injury must have: (1) lasted for more than 6 months after the vaccine was given; or (2) resulted in a hospital stay *and* surgery; or (3) resulted in death. The vaccinated individual must be a US citizen. The vaccines must have been covered by the VICP and given in the following circumstances:

— the injured person must have received a vaccine in the US trust territories; or
— if the vaccine was administered outside of the US or its trust territories: (1) the injured person must have been a US citizen serving in the military or a US government employee, or have been a dependant of such a citizen; or (2) the injured person must have received a vaccine manufactured by a vaccine company located in the US and returned to the US within 6 months after the date of vaccination.

If the US Court of Claims awards compensation to the vaccine-injured person:

- The VICP will offer to pay up to $250,000 for a vaccine associated death.
- The VICP will offer to pay for all past and future unreimbursed medical expenses, custodial and nursing home care; and up to $250,000 pain and suffering as well as loss of earned income.

If an individual rejects the award or is denied compensation, a lawsuit may be filed in civil court but with certain restrictions.

The detailed criteria for different vaccines and injuries are shown in Table 15.1.

Statistics

The distribution of claims filed with the VICP for the period 1989–2007 is shown in Figure 15.1.

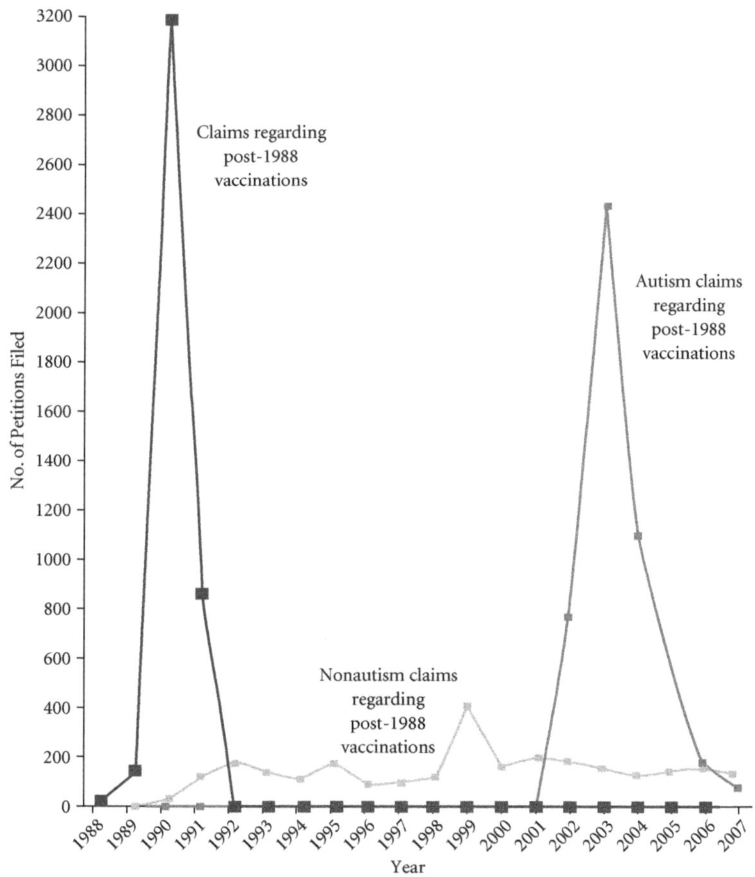

Figure 15.1: Claims filed with the Vaccine Injury Compensation Program 1989–2007. Post-1988 vaccinations are those that occurred on or after 1 October 1988; pre-1988 vaccinations are those that occurred before that date. Source: Health Resources and Services Administration

Table 15.1: Vaccine Injury Table (see: www.hrsa.gov/vaccinecompensation/index.html)

Vaccine	Illness, disability, injury or condition covered[a]	Time interval
I. Vaccines containing tetanus toxoid (eg DTaP, DTP, DT, Td, TT)	A. Anaphylaxis or anaphylactic shock	0–4 hours
	B. Brachial neuritis	2–28 days
	C. Any acute complication or sequela (including death) of above events	NA
II. Pertussis antigen-containing vaccines (eg DTP, Tdap, DTaP, P, DTP-Hib)	A. Anaphylaxis or anaphylactic shock	0–4 hours
	B. Encephalopathy (or encephalitis)	0–72 hours
	C. Any acute complication or sequela (including death) of above events.	NA
III. Measles, mumps and rubella-containing vaccines in any combination (eg MMR, MR, M, R)	A. Anaphylaxis or anaphylactic shock	0–4 hours
	B. Encephalopathy (or encephalitis)	5–15 days (not less than 5 days and not more than 15 days)
	C. Any acute complication or sequela (including death) of above events	NA
IV. Vaccines containing rubella virus (eg MMR, MR, R)	A. Chronic arthritis	7–42 days
	B. Any acute complication or sequela (including death) of above events	Not applicable
V. Vaccines containing measles virus (eg MMR, MR, M)	A. Thrombocytopenic purpura	7–30 days
	B. Vaccine-strain measles viral infection in an immunodeficient recipient	0–6 months
	C. Any acute complication or sequela (including death) of above events	Not applicable
VI. Vaccines containing polio live virus (OPV)	A. Paralytic polio	
	— in a non-immunodeficient recipient	0–30 days
	— in an immunodeficient recipient	0–6 months
	— in a vaccine-associated community case	NA
	B. Vaccine-strain polio viral infection	
	— in a non-immunodeficient recipient	0–30 days
	— in an immunodeficient recipient	0–6 months
	— in a vaccine-associated community case	NA
	C. Any acute complication or sequela (including death) of above events	Not applicable
VII. Vaccines containing polio inactivated (eg IPV)	A. Anaphylaxis or anaphylactic shock	0–4 hours
	B. Any acute complication or sequela (including death) of above events	NA

(*continued*)

Table 15.1: ***(Continued)***

Vaccine	Illness, disability, injury or condition covered[a]	Time interval
VIII. Hepatitis B vaccines	A. Anaphylaxis or anaphylactic shock	0–4 hours
	B. Any acute complication or sequela (including death) of above events	NA
IX. Hemophilus influenzae (type b polysaccharide conjugate vaccines)	A. No condition specified	NA
X. Varicella vaccine	A. No condition specified	NA
XI. Rotavirus vaccine	A. No condition specified	NA
XII. Pneumococcal conjugate vaccines	A. No condition specified	NA
XIII. Hepatitis A vaccines	A. No condition specified	NA
XIV. Trivalent influenza vaccines	A. No condition specified	NA
XV. Meningococcal vaccines	A. No condition specified	NA
XVI. Human papillomavirus (HPV) vaccines	A. No condition specified	NA
XVII. Any new vaccine recommended by the CDC for routine administration to children, after publication by the Secretary of HHS of a notice of coverage [b,c]	A. No condition specified	

NA, not applicable.

[a] Effective date 10 November 2008, proposed updates can be found at www.hrsa.gov/vaccinecompensation/proposedchanges1010.pdf.

[b] As of 1 December 2004 hepatitis A vaccines have been added to the Vaccine Injury Table (Table) under this category. As of 1 July 2005 trivalent influenza vaccines have been added to the Table under this category. Trivalent influenza vaccines are given annually during the flu season either by needle and syringe or in a nasal spray. All influenza vaccines routinely administered in the US are trivalent vaccines covered in this category.

[c] As of 1 February 2007 meningococcal (conjugate and polysaccharide) and human papillomavirus (HPV) vaccines have been added to the Table under this category.[19]

Qualifications and Aids to Interpretation

(1) *Anaphylaxis and anaphylactic shock* mean an acute, severe and potentially lethal systemic allergic reaction. Most cases resolve without sequelae. Signs and symptoms begin minutes to a few hours after exposure. Death, if it occurs, usually results from airway obstruction caused by laryngeal oedema or bronchospasm and may be associated with cardiovascular collapse. Other significant clinical signs and symptoms may include the following: cyanosis, hypotension, bradycardia, tachycardia, arrhythmia, oedema of the pharynx and/or trachea and/or larynx with stridor and dyspnea. Autopsy findings may include acute emphysema which results from lower respiratory tract obstruction, oedema of the hypopharynx, epiglottis, larynx or trachea, and minimal findings of eosinophilia in the liver, spleen and lungs. When death occurs within minutes of exposure and without signs of respiratory distress, there may not be significant pathologic findings.

[19] See News section on the VICP website: www.hrsa.gov/vaccinecompensation.

(2) *Encephalopathy.* For purposes of the Vaccine Injury Table, a vaccine recipient shall be considered to have suffered an encephalopathy only if such recipient manifests, within the applicable period, an injury meeting the description below of an acute encephalopathy, and then a chronic encephalopathy persists in such person for more than 6 months beyond the date of vaccination.

(i) *An acute encephalopathy* is one that is sufficiently severe so as to require hospitalisation (whether or not hospitalisation occurred).

- (A) *For children less than 18 months of age* who present without an associated seizure event, an acute encephalopathy is indicated by a 'significantly decreased level of consciousness' (see 'D' below) lasting for at least 24 hours. Those children less than 18 months of age who present following a seizure shall be viewed as having an acute encephalopathy if their significantly decreased level of consciousness persists beyond 24 hours and cannot be attributed to a postictal state (seizure) or medication.
- (B) *For adults and children 18 months of age or older*, an acute encephalopathy is one that persists for at least 24 hours and characterized by at least two of the following:
  - (1) A significant change in mental status that is not medication related; specifically a confusional state, or a delirium, or a psychosis;
  - (2) A significantly decreased level of consciousness, which is independent of a seizure and cannot be attributed to the effects of medication; and
  - (3) A seizure associated with loss of consciousness.
- (C) Increased intracranial pressure may be a clinical feature of acute encephalopathy in any age group.
- (D) A 'significantly decreased level of consciousness' is indicated by the presence of at least one of the following clinical signs for at least 24 hours or greater (see paragraphs (2)(I)(A) and (2)(I)(B) of this section for applicable timeframes):
  - (1) Decreased or absent response to environment (responds, if at all, only to loud voice or painful stimuli);
  - (2) Decreased or absent eye contact (does not fix gaze upon family members or other individuals); or
  - (3) Inconsistent or absent responses to external stimuli (does not recognize familiar people or things).
- (E) The following clinical features alone, or in combination, do not demonstrate an acute encephalopathy or a significant change in either mental status or level of consciousness as described above: sleepiness, irritability (fussiness), high-pitched and unusual screaming, persistent inconsolable crying, and bulging fontanelle. Seizures in themselves are not sufficient to constitute a diagnosis of encephalopathy. In the absence of other evidence of an acute encephalopathy, seizures shall not be viewed as the first symptom or manifestation of the onset of an acute encephalopathy.

(ii) *Chronic encephalopathy* occurs when a change in mental or neurologic status, first manifested during the applicable time period, persists for a period of at least 6 months from the date of vaccination. Individuals who return to a normal neurologic state after the acute encephalopathy shall not be presumed to have suffered residual neurologic damage from that event; any subsequent chronic encephalopathy shall not be presumed to be a sequela of the acute encephalopathy. If a preponderance of the evidence indicates that a child's chronic encephalopathy is secondary to genetic, prenatal or perinatal factors, that chronic encephalopathy shall not be considered to be a condition set forth in the Table.

(iii) An encephalopathy shall not be considered to be a condition set forth in the Table if in a proceeding on a petition it is shown by a preponderance of the evidence that the encephalopathy was caused by an infection, a toxin, a metabolic disturbance, a structural lesion, a genetic disorder or trauma (without regard to whether the cause of the infection, toxin, trauma, metabolic disturbance, structural lesion or genetic disorder is known). If at the time a decision is made on a petition filed under section 2111(b) of the Act for a vaccine-related injury or death, it is not possible to determine the cause by a preponderance of the evidence of an encephalopathy, the encephalopathy shall be considered to be a condition set forth in the Table.

(iv) In determining whether or not an encephalopathy is a condition set forth in the Table, the Court shall consider the entire medical record.

(3) *Seizure and convulsion.* For purposes of paragraphs (b)(2) of this section, the terms, 'seizure' and 'convulsion' include myoclonic, generalized tonic-clonic (grand mal), and simple and complex partial seizures. Absence (petit mal) seizures shall not be considered to be a condition set forth in the Table. Jerking movements or staring episodes alone are not necessarily an indication of seizure activity.

(4) *Sequela.* The term 'sequela' means a condition or event which was actually caused by a condition listed in the Vaccine Injury Table.

(5) *Chronic arthritis.* For purposes of the Vaccine Injury Table, chronic arthritis may be found in a person with no history in the 3 years prior to vaccination of arthropathy (joint disease) on the basis of:

(A) Medical documentation, recorded within 30 days after the onset, of objective signs of acute arthritis (joint swelling) that occurred between 7 and 42 days after a rubella vaccination;
(B) Medical documentation (recorded within 3 years after the onset of acute arthritis) of the persistence of objective signs of intermittent or continuous arthritis for more than 6 months following vaccination;
(C) Medical documentation of an antibody response to the rubella virus.

For purposes of the Vaccine Injury Table, the following shall not be considered as chronic arthritis: musculoskeletal disorders such as diffuse connective tissue diseases (including but not limited to rheumatoid arthritis, juvenile rheumatoid arthritis, systemic lupus erythematosus, systemic sclerosis, mixed connective tissue disease, polymyositis/dermatomyositis, fibromyalgia, necrotising vasculitis and vasculopathies and Sjogren's syndrome), degenerative joint disease, infectious agents other than rubella (whether by direct invasion or as an immune reaction), metabolic and endocrine diseases, trauma, neoplasms, neuropathic disorders, bone and cartilage disorders and arthritis associated with ankylosing spondylitis, psoriasis, inflammatory bowel disease, Reiter's syndrome, or blood disorders.

Arthralgia (joint pain) or stiffness without joint swelling shall not be viewed as chronic arthritis for purposes of the Vaccine Injury Table.

(6) *Brachial neuritis* is defined as dysfunction limited to the upper extremity nerve plexus (ie its trunks, divisions or cords) without involvement of other peripheral (eg nerve roots or a single peripheral nerve) or central (eg spinal cord) nervous system structures. A deep, steady, often severe aching pain in the shoulder and upper arm usually heralds onset of the condition. The pain is followed in days or weeks by weakness and atrophy in upper extremity muscle groups. Sensory loss may accompany the motor deficits, but is generally a less notable clinical feature. The neuritis, or plexopathy, may be present on the same side as or the opposite side of the injection; it is sometimes bilateral, affecting both upper extremities. Weakness is required before the diagnosis can be made. Motor, sensory and reflex findings on physical examination and the results of nerve conduction and electromyographic studies must be consistent in confirming that dysfunction is attributable to the brachial plexus. The condition should thereby be distinguishable from conditions that may give rise to dysfunction of nerve roots (ie radiculopathies) and peripheral nerves (ie including multiple mononeuropathies), as well as other peripheral and central nervous system structures (eg cranial neuropathies and myelopathies).

(7) *Thrombocytopenic purpura* is defined by a serum platelet count less than 50,000/mm^3. Thrombocytopenic purpura does not include cases of thrombocytopenia associated with other causes such as hypersplenism, autoimmune disorders (including alloantibodies from previous transfusions) myelodysplasias, lymphoproliferative disorders, congenital thrombocytopenia or haemolytic uremic syndrome. This does not include cases of immune (formerly called idiopathic) thrombocytopenic purpura (ITP) that are mediated, for example, by viral or fungal infections, toxins or drugs. Thrombocytopenic purpura does not include cases of thrombocytopenia associated with disseminated intravascular coagulation, as observed with bacterial and viral infections. Viral infections include, for example, those infections secondary to Epstein–Barr virus, cytomegalovirus, hepatitis A and B, rhinovirus, human immunodeficiency virus (HIV), adenovirus and dengue virus. An antecedent viral infection may be demonstrated by clinical signs and symptoms and need not be confirmed by culture or serologic testing. Bone marrow examination, if performed, must reveal a normal or an increased number of megakaryocytes in an otherwise normal marrow.

(8) *Vaccine-strain measles viral infection* is defined as a disease caused by the vaccine-strain that should be determined by vaccine-specific monoclonal antibody or polymerase chain reaction tests.

(9) *Vaccine-strain polio viral infection* is defined as a disease caused by poliovirus that is isolated from the affected tissue and should be determined to be the vaccine strain by oligonucleotide or polymerase chain reaction. Isolation of poliovirus from the stool is not sufficient to establish a tissue specific infection or disease caused by vaccine-strain poliovirus.

This information reflects the current thinking of the US Department of HHS on the topics addressed. This information is not legal advice and does not create or confer any rights for or on any person and does not operate to bind the Department or the public. The ultimate decision about the scope of the statutes authorising the VICP is within the authority of the US Court of Federal Claims, which is responsible for resolving claims for compensation under the VICP.

Compensation Numbers and Amounts[20]

Generally, petitions/claims are not adjudicated in the same fiscal year in which they are filed. On average, it takes 2–3 years to adjudicate a petition/claim after it is filed. Due to the populations receiving vaccines that have been added to the VICP in recent years, the proportion of adults to children seeking compensation has changed. Since influenza vaccines (vaccines administered to large numbers of adults each year) were added to the VICP in 2005, many adult claims related to that vaccine have been filed. Table 15.2 shows the number of claims filed by vaccine as reported by petitioners in claims since the VICP began on 1 October 1988, and how many of those have been compensated or dismissed by the US Court of Federal Claims (Court). Claims can be compensated by a settlement between parties or a decision by the Court

Table 15.2: Numbers of claims filed and compensated or dismissed by vaccine up to 13 August 2013

Vaccine(s)	Filed			Compensated	Dismissed
	Injury	Death	Total		
DT (diphtheria–tetanus)	67	9	76	23	50
DTaP (diphtheria–tetanus–acellular pertussis)	350	76	426	163	189
DTaP–Hep B–IPV	56	23	79	23	23
DTaP–HIB	6	1	7	4	3
DTaP–IPV–HIB	12	8	20	4	4
DTP (diphtheria–tetanus–whole cell pertussis)	3,284	697	3,981	1,269	2,702
DTP–HIB	17	8	25	4	19
Hep A–Hep B	14	0	14	8	1
Hep B–HIB	8	0	8	3	3
Hepatitis A (Hep A)	52	2	54	21	13
Hepatitis B (Hep B)	590	53	643	230	352
HIB (*Haemophilus influenzae* type b)	23	3	26	11	13
HPV (human papillomarvirus)	199	10	209	58	61
Influenza (trivalent)	1080	64	1144	592	112
IPV (inactivated polio)	262	14	276	7	267
Measles	143	19	162	55	107
Meningococcal	31	1	32	17	3
MMR (measles–mumps–rubella)	860	57	917	350	481
MMR–varicella	24	1	25	11	7

(continued)

[20] Taken from www.hrsa.gov/vaccinecompensation/statisticsreports.html#Claims.

Table 15.2: *(Continued)*

Vaccine(s)	Filed			Compensated	Dismissed
	Injury	Death	Total		
MR	15	0	15	6	9
Mumps	10	0	10	1	9
Non-qualified (claims for vaccines not covered by VICP)	76	9	85	0	82
OPV (oral polio)	280	28	308	158	150
Pertussis	4	3	7	2	5
Pneumococcal conjugate	35	5	40	9	23
Rotavirus	55	1	56	34	16
Rubella	189	4	193	70	123
Td (tetanus–diphtheria)	180	3	183	97	63
Tdap	127	0	127	58	6
Tetanus	84	2	86	36	34
Unspecified (insufficient information to make a determination)	5,409	8	5,417	4	4.693
Varicella	73	6	79	39	19
Total	**13,615**	**1,114**	**14,730**	**3,368**	**9,642**

Table 15.3 shows the outlay of the scheme for each financial year from 1989 to 2013.

Table 15.3: Outlay ($) 1989–2013

Fiscal year	Compensated			Dismissed		Interim Fees		Total outlays
	No of awards	Petitioners' award amount ($)	Attorneys' fees/ costs payments	No of payments to attorneys	Attorneys' fees/costs payments	No of payments to attorneys	Attorneys' fees/ costs payments	
1989	6	1,317,654.78	54,107.14	0	0.00	0	0.00	1,371,761.92
1990	88	53,252,510.46	1,379,005.79	4	57,699.48	0	0.00	54,689,215.73
1991	114	95,980,493.16	2,364,758.91	30	496,809.21	0	0.00	98,842,061.28
1992	130	94,538,071.30	3,001,927.97	118	1,212,677.14	0	0.00	98,752,676.41
1993	162	119,693,267.87	3,262,453.06	272	2,447,273.05	0	0.00	125,402,993.98
1994	158	98,151,900.08	3,571,179.67	335	3,166,527.38	0	0.00	104,889,607.13
1995	169	104,085,265.72	3,652,770.57	221	2,276,136.32	0	0.00	110,014,172.61
1996	163	100,425,325.22	3,096,231.96	216	2,364,122.71	0	0.00	105,885,679.89
1997	179	113,620,171.68	3,898,284.77	142	1,879,418.14	0	0.00	119,397,874.59
1998	165	127,546,009.19	4,002,278.55	121	1,938,065.50	0	0.00	133,484,353.24
1999	96	95,917,680.51	2,799,910.85	117	2,306,957.40	0	0.00	101,024,548.76
2000	136	125,945,195.64	4,112,369.02	80	1,724,451.08	0	0.00	131,782,015.74

(continued)

Table 15.3: (***Continued***)

Fiscal year	Compensated			Dismissed		Interim Fees		Total outlays
	No of awards	Petitioners' award amount ($)	Attorneys' fees/ costs payments	No of payments to attorneys	Attorneys' fees/costs payments	No of payments to attorneys	Attorneys' fees/ costs payments	
2001	97	105,878,632.57	3,373,865.88	57	2,066,224.67	0	0.00	111,318,723.12
2002	80	59,799,604.39	2,653,598.89	50	656,244.79	0	0.00	63,109,448.07
2003	65	82,816,240.07	3,147,755.12	69	1,545,654.87	0	0.00	87,509,650.08
2004	57	61,933,764.20	3,079,328.55	69	1,198,615.96	0	0.00	66,211,708.71
2005	64	55,065,797.01	2,694,664.03	71	1,790,587.29	0	0.00	59,551,048.33
2006	68	48,746,162.74	2,441,199.02	54	1,353,632.61	0	0.00	52,540,994.37
2007	82	91,449,433.89	4,034,154.37	61	1,692,020.25	0	0.00	97,175,608.51
2008	141	75,716,552.06	5,270,237.04	72	2,432,847.05	2	117,265.31	83,536,901.46
2009	131	74,142,490.58	5,404,711.98	36	1,557,139.53	28	4,241,362.55	85,345,704.64
2010	173	179,387,341.30	5,961,744.40	56	1,886,239.95	22	1,978,803.88	189,214,129.53
2011	251	216,323,760.31	9,736,216.87	402	5,425,243.19	28	2,001,770.91	233,486,991.28
2012	250	163,511,998.82	9,106,720.30	1,017	8,621,182.32	37	5,420,257.99	186,660,159.43
2013	323	215,023,417.10	11,519,456.02	599	5,687,099.47	46	1,305,646.78	233,535,619.37
Total	3,348	2,560,268,740.65	103,618,930.73	4,269	55,780,869.36	163	15,065,107.42	2,734,733,648.16

'Compensated' claims are claims that have been paid as a result of a settlement between parties or a decision made by the US Court of Federal Claims (Court). The number of awards is the number of petitioner awards paid, including the attorneys' fees/costs payments, if made during a fiscal year. However, petitioners' awards and attorneys' fees/costs are not necessarily paid in the same fiscal year as when the petitions/claims are determined compensatable. 'Dismissed' includes the number of payments to attorneys and the total amount of payments for attorneys' fees/costs per fiscal year. The VICP will pay attorneys' fees/costs related to the claim, whether or not the petition/claim is awarded compensation by the Court, if certain minimal requirements are met. 'Total Outlays' are the total amount of funds expended for compensation and attorneys' fees/costs from the Vaccine Injury Compensation Trust Fund by fiscal year.

C. The UK Vaccine Damage Scheme

Background

The Vaccine Damage Payments (VDP) scheme[21] was created under the Vaccine Damage Payments Act 1979, although operated before that on an ex gratia basis for about a year. The forces that created it can be traced under four broad headings: medical logic, a climate of

[21] See www.gov.uk/vaccine-damage-payment.

concern that legal arrangements were inadequate, social concern fanned through a political campaign, and concern that excessive liability claims could drive vaccine manufacturers out of the market thereby threatening the supply of vaccines.

The legal climate in the United Kingdom—and Europe[22]—was affected by the thalidomide tragedy of the early 1960s, in which fetuses were born deformed as a result of their mothers taking an anti-nausea drug. That catastrophe gave rise to considerable public concern and to two responses within the legal system, namely the introduction of extensive regulatory controls on medicine[23] and review of compensation law. In relation to the latter, deficiencies were identified in the substantive law of negligence and the legal system. The response to concerns over the law of negligence was the introduction, although not until 1988, of strict liability for defective products.[24] The review process of compensating law began with the Pearson Commission, established in 1972.[25]

However, a public campaign was started for compensation for vaccine-damaged children in 1973,[26] led by Rosemary Fox, the parent of a child believed to have suffered such damages, who persuaded Jack Ashley MP to raise the issue in Parliament on behalf of 300 (later 630) families, and founded the Association of Parents of Vaccine Damaged Children (the Association).[27] The Association sought an acknowledgement that vaccination caused the damage to their healthy children and special provision over and above the existing state provision to enable them to provide a better standard of care. It gave evidence to the Pearson Commission urging the need for special provision for vaccine damage. Their arguments were strongly supported by the leading medical associations and Royal Colleges, and the Commission recommended in 1978 that: 'The Government should be strictly liable in tort for severe damage suffered by anyone as a result of vaccination recommended in the interests of the community.'

After the Association complained on behalf of some parents whose children were wrongly immunised that they were not given enough information to make an informed choice, the Health Service Ombudsman issued a report in 1977 confirming that the Department of Health had to 'accept a large measure of responsibility … for not recognising earlier the desirability of alerting parents, as they had now done'. The NHS then produced leaflets for parents that gave details of the vaccines they were being offered for their children, with some reference to possible risk factors.

In advance of the Pearson Commission's solution of amending the legal rights, these various factors led the government in 1977 to accept the principle that there should be a scheme of payments for those who were seriously damaged as a result of vaccination, which was enshrined in the Vaccine Damage Payments Act 1979.[28] The scheme has since been amended a few times.

[22] But not the United States of America, see above.

[23] See CJS Hodges, 'Regulating Medicinal Products and Medical Devices' in A Grubb, J Laing and J McHale (eds), *Principles of Medical Law*, 4th edn (Oxford, Oxford University Press, 2017); P Feldschreiber (ed), *The Law and Regulation of Medicines* (Oxford, Oxford University Press, 2008).

[24] Under the Consumer Protection Act 1987, implementing Directive 85/374/EEC.

[25] *Report of the Royal Commission on the Civil Liability and Compensation for Personal Injury. Chairman: Lord Pearson*. Cmnd 7054-1, 1978.

[26] R Tidley, 'A critical Analysis of the Vaccine Damage Payment Scheme' (2008) 19 *European Business Law Review* 321–63.

[27] J Ashley, *Acts of Defiance* (London, Reinhard Books, 1992) 253.

[28] *Vaccine Damage Payments Scheme—United Kingdom* (Department of Social Security, 2000).

Details of the Scheme

The VDP scheme provides a single, tax-free payment made by the government to a person who has suffered severe mental and/or physical disablement as a result of vaccination against one or more specified diseases. The amount is £120,000.[29] The payment is not regarded as compensation, or as tailored to the financial needs of the individual,[30] 'but is designed to ease the present and future burdens of those suffering vaccine damages, and their families'.[31] The scheme is funded out of general taxation. The payment is not a bar to any civil proceedings for compensation. The VDP can affect other benefits and entitlements.[32] Recipients are advised that the effect the payment will have depends on a number of things, including whether the payment is put into a trust from which periodic payments are made.[33]

VDPs may be claimed by people who are 'severely disabled' where the disability was caused by vaccination against any of a number of the following specified diseases:

— diphtheria;
— tetanus;
— pertussis (whooping cough)
— poliomyelitis;
— measles;
— mumps (from 9 April 1990);
— rubella (German measles);
— tuberculosis (TB);
— *Haemophilus influenzae* type B (HIB);
— meningococcal group C (meningitis C) (from 30 May 2001);
— pneumococcal infection;
— human papillomavirus (from 1 September 2008);[34]
— pandemic influenza A (H1N1) 2009 (swine flu);[35]
— smallpox (up to 1 August 1971).

[29] The Vaccine Damage Payments Act 1979 Statutory Sum Order 2007, available at www.opsi.gov.uk/si/si2007/em/uksiem_20071931_en.pdf (accessed 6 August 2010). See www.direct.gov.uk/en/MoneyTaxAndBenefits/BenefitsTaxCreditsAndOtherSupport/Disabledpeople/DG_10018714 in August 2010. See also: Explanatory Memorandum to the Vaccine Damage Payments Act 1979 Statutory Sum Order 2007 No 1931, available at: www.opsi.gov.uk/si/si2007/em/uksiem_20071931_en.pdf (accessed 8 August 2010). The sum when the scheme as introduced in 1979 was £10,000, increased to £20,000 in 1985, £30,000 from 1991, £40,000 from 1998, £100,000 from 2000.

[30] MA Jones, 'Compensation in the English Health Care Sector' in J Dute, M Faure and H Koziol (eds), *No-Fault Compensation in the Health Care Sector* (New York, Springer, 2004).

[31] *Vaccine Damage Payments Scheme—United Kingdom* (Department of Social Security, 2000).

[32] Such as Income Support, Income-based Jobseeker's Allowance, Working Tax Credit, Child Tax Credit, Pension Credit, Housing Benefit, Council Tax Benefit, Employment and Support Allowance.

[33] www.direct.gov.uk/en/MoneyTaxAndBenefits/BenefitsTaxCreditsAndOtherSupport/Disabledpeople/DG_10018714 (accessed 6 August 2010).

[34] The Vaccine Damage Payments (Specified Disease) Order 2008, 2008 No 2103, available at: www.opsi.gov.uk/si/si2008/em/uksiem_20082103_en.pdf (accessed 6 August 2010).

[35] The Vaccine Damage Payments (Specified Disease) Order 2009, 2009 No 2516, available at: www.legislation.gov.uk/uksi/2009/2516/contents/made and The Vaccine Damage Payments (Specified Disease) (Revocation and Savings) Order 2010, 2010 No 1988, available at: www.legislation.gov.uk/uksi/2010/1988/article/1/made.

Further, a payment may be made to people who are thought to be severely disabled because either:

— the person's mother was vaccinated against one of the diseases in the list while she was pregnant;
— the person has been in close physical contact with someone who had been vaccinated against poliomyelitis with vaccine that was given orally.

The term 'severely disabled' is defined to mean at least 60% disabled, as assessed under the separate industrial injuries' compensation scheme.[36]

Since 1968, all vaccines, like other medicinal products, have been subject to licensing requirements designed to ensure that they satisfy requirements of safety, efficacy and quality.[37] The regulatory system is designed to prevent unsafe medicines from being placed on the market, or from remaining on the market after their risk–benefit balance is seen to be adverse. All relevant safety information, including adverse reaction reports, is required to be collected and reviewed by manufacturers and by the public authorities.[38] Suspected adverse reactions to vaccines, and other medicines, can be reported through the yellow card scheme[39] by doctors, dentists, coroners and pharmacists.

Procedure

Under the Vaccine Damage Payment Act 1979, a claim for a payment must be made to the Secretary of State (1) by or on behalf of the disabled person concerned or, as the case may be, by his personal representatives; (2) in the prescribed manner;[40] and (3) within the period of 6 years beginning on the latest of the following dates, namely the date of the vaccination to which the claim relates, the date on which the disabled person attained the age of 2 and 9 May 1978.

A claim form must be submitted to Vaccine Damage Payments Unit,[41] which will obtain medical evidence from the doctors or hospitals involved in the applicant's treatment. The payment will be made if it is satisfied that the applicant's disablement was caused by vaccination, and the level of disablement that can be directly attributed to the vaccination is

[36] Vaccine Damage Payments Act 1979, s 1(4), amended by the Regulatory Reform (Vaccine Damage Payments Act 1979) Order 2002/1592 for claims after 16 June 2002. Prior to that date the requirement was 80% disablement.

[37] Art 8 of Directive (EC) 2001/83.

[38] The UK competent authority is the Medicines and Healthcare Products Regulatory Authority (MHRA), see www.mhra.gov.uk.

[39] http:/yellowcard.mhra.gov.uk and www.mhra.gov.uk/Safetyinformation/Howwemonitorthesafetyofproducts/Medicines/TheYellowCardScheme/UsingYellowCarddatatoprotectpublichealth/ImpactoftheYellowCard/index.htm.

[40] Vaccine Damage Payment Regulations 1979, SI 1979 No 43 2, reg 2.

[41] www.dwp.gov.uk/advisers/claimforms/vadla_print.pdf. Vaccine Damage Payments Unit, Palatine House, Lancaster Road, Preston PR1 1HB, tel: 01772 899 944, textphone: 0845 60 45 312, www.direct.gov.uk/en/MoneyTaxAndBenefits/BenefitsTaxCreditsAndOtherSupport/Disabledpeople/DG_10018714.

at least 60%. The result of the claim will be notified to the applicant in writing. An applicant who is turned down can ask for the decision to be looked at again by the Vaccine Damage Payments Unit, or can appeal to an independent First-tier Tribunal. If the applicant disagrees with the decision of the First-tier Tribunal, he or she can ask for the decision to be looked at again by the Vaccine Damage Payment Unit or can request an appeal to the Upper Tribunal. Unlike the US system, legal representation is very rare during the initial stages: of the 1551 claims registered between 1 April 2000 and 1 August 2013 only 2 are noted as having legal representation.[42]

Various reforms were introduced in 2002.[43] Firstly, the 6-year time limit within which to bring a claim was extended to whichever is the later of the date on which the disabled person attains the age of 21, or, where he has died, would have attained that age, and the end of a period of 6 years beginning on the date of the vaccination against the disease to which the claim relates. Secondly, the disability threshold was reduced from 80% to 60%. In 2005, further amendments extended the scheme to vaccinations given abroad to serving members of the armed forces and their families.[44]

Claims

Table 15.4 lists the number of claims made, awards grated and total amounts paid between 1977/78 and 1999/2000, and Table 15.5 shows the awards granted and amounts paid between 1999/2000 and 2011/2012. Table 15.6 lists the reasons for refusal of claims up until 2013, with a more detailed breakdown of refusal reasons from 2005 until 2013 given in Table 15.7. It will be seen that the number of accepted claims peaked around 1979–83, since when numbers have fallen to very low levels. Proof of causation is usually the problem, as is indicated by the difference between the incidence of claims and number of awards.

Table 15.4: Total number of claims, awards and sums paid between 1977 and 2000

Year	Claims	Awards	Total amount paid (£)
1977/78	6	0	
1978/79	1,712	85	940,000
1979/80	874	272	2,720,000
1980/81	121	255	2,550,000
1981/82	72	68	680,000
1982/83	98	37	370,000
1983/84	101	40	400,000

(continued)

[42] Freedom of Information request 4051, dated 23 August 2013.

[43] The Regulatory Reform (Vaccine Damage Payments Act 1979) Order 2002, available at www.opsi.gov.uk/si/si2002/uksi_20021592_en.pdf.

[44] The Vaccine Damage Payments (Amendment) Regulations 2005, available at: www.opsi.gov.uk/si/si2005/20053070.htm (accessed 6 August 2010).

Table 15.4: ***(Continued)***

Year	Claims	Awards	Total amount paid (£)
1984/85	111	36	360,000
1985/86	64	24	250,000
1986/87	76	17	210,000
1987/88	66	10	160,000
1988/89	38	3	40,000
1989/90	51	3	60,000
1990/91	34	1	10,000
1991/92	43	6	100,000
1992/93	49	11	260,000
1993/94	54	3	50,000
1994/95	78	7	160,000
1995/96	70	4	110,000
1996/97	69	5	130,000
1997/98	202	3	90,000
1998/99	135	0	0
1999/2000	202	5	110,000
Total	4326	895	9760000

Table 15.5: Awards made between 1999/2000 and 2011/2012

Year	Awards	Amount of award
1999/2000	4	**£359,000** (1 @ £78,000; 2 @ £91,500; 1 @ £98,000)
2000/2001	2	**£176,000** (1 @ £78,000; 1 @ £98,000)
2001/2002	3	**£291,500** (1 @ £91,500; 2 @ £100,000)
2002/2003	5	**£500,000** (5 @ £100,000)
2003/2004	5	**£478,000** (1 @ £78,000; 4 @ £100,000)
2004/2005	4	**£400,000** (4 @ £100,000)
2005/2006	5	**£500,000** (5 @ £100,000)
2006/2007	4	**£378,000** (1 @ £78,000; 3 @ £100,000)
2007/2008	2	**£200,000** (2 @ £100,000)
2008/2009	0	
2009/2010	3	**£291,500** (1 @ £91,500; 2 @ £100,000)
2010/2011	1	**£91,500** (1 @ £91,500)
2011/2012	0	

Table 15.6: Totals of the reasons for refusal up to August 2013

Reasons for refusal	Number
Out of time	551
Under age 2 at date of claim	16
Died before 9 May 1978 or under age 2	15
Over 18 when vaccinated	64
Disease not covered by VDPA	76
Vaccinated outside UK/Isle of Man	15
Vaccinated before 5 July 1948	57
Smallpox vaccination after 1 August 1971	3
Claim withdrawn	12
Invalid claim	24
Payment already authorised	2
Causation not accepted	4403
Causation accepted, but not severely disabled	113
Total	5333

Table 15.7: Breakdown of reasons for refusal of claims from 1 April 2005 to 23 August 2013. Source: Freedom of Information request 4051 dated 23 August 2013

Year	Claim out of time	Child under age 2	Died prior to 9 May 1978/prior to age 2	Over age 18 when vaccinated	Disease not included in the Act	Not vaccinated in UK/Isle of Man	Vaccinated prior to 1948	Smallpox vaccination after 1971	Claim withdrawn	Invalid Claim	Payment already authorised	Causation not accepted	Causation accepted, not severely disabled	Total
2006/07	13	0	1	1	5	0	0	0	0	1	0	46	1	68
2007/08	9	1	0	2	0	0	0	0	0	0	1	32	1	46
2008/09	1	1	0	0	0	1	0	0	1	2	0	46	0	52
2009/10	5	0	0	3	0	0	0	0	0	1	0	55	5	69
2010/11	9	0	0	2	2	0	0	0	0	1	0	65	1	80
2011/12	11	0	0	1	8	0	0	0	0	4	0	68	0	92
2012/13	13	1	0	1	3	2	0	0	0	1	0	60	1	82
2013/14	2	0	1	0	1	0	0	0	0	0	0	14	0	18

Attribution to a Vaccine

Most claims are made for children who have received a number of different vaccinations, all of which are included in the claim. When the claim is considered, the doctors who advise the Secretary of State on whether to award a payment or not are not required to say which vaccine caused the disability. Therefore, it is not possible to tell which vaccine was responsible for the severe disability in each case. That is why the number of vaccinations listed is greater than the number of awards made (see Table 15.8). A change in policy at the Vaccine Damages Payments Unit means that these awards granted after 2001 cannot be attributed to specific vaccine or vaccines.[45] Between 1 January 2000 and 31 December 2010 the Vaccine Damage Payments Unit received 1,483 claims, of these, 26 resulted in an award (see Table 15.8). The total amount paid out for those awards was £2,600,000.

Table 15.8: Breakdown of claims by vaccine 1978–2001. Source: www.gov.uk/government/publications/vaccine-damage-awards-2000-to-2011-foi-45822012

			Vaccine									
Year	**Claims**	**Awards**	**Diphtheria**	**Measles**	**MMR**	**Pertussis**	**Polio**	**Rubella**	**Smallpox**	**Tetanus**	**Triple**	**TB**
1978/79	1,711	4	0	0	0	0	1	1	2	0	0	0
1979/80	871	259	15	12	0	20	49	1	40	2	173	0
1980/81	121	254	13	12	0	43	77	1	13	2	172	1
1981/82	72	68	7	3	0	6	22	0	5	3	49	0
1982/83	98	37	1	3	0	5	13	0	2	1	22	0
1983/84	101	38	1	3	0	13	18	0	0	1	18	0
1984/85	111	36	2	3	0	16	11	0	1	3	11	0
1985/86	64	24	2	0	0	7	7	0	0	1	13	0
1986/87	76	17	1	0	0	3	8	0	0	0	13	0
1987/88	65	10	0	0	0	5	3	0	0	0	3	0
1988/89	38	3	0	0	0	0	2	0	0	0	1	0
1989/90	51	3	0	2	0	0	1	0	0	0	0	0
1990/91	34	1	0	1	0	0	0	0	0	0	0	0
1991/92	43	7	0	2	0	1	3	0	0	0	4	0
1992/93	49	11	5	1	2	6	8	0	0	5	2	0
1993/94	54	3	1	1	1	1	2	0	0	1	2	0
1994/95	78	7	3	2	0	3	4	1	1	3	0	1
1995/96	70	3	1	1	2	1	3	0	0	1	2	0
1996/97	69	5	1	0	2	1	4	0	0	1	3	0
1997/98	201	3	0	0	0	0	3	0	0	0	2	0
98/99	135	0	0	0	0	0	0	0	0	0	0	0
1999/2000	88	3			0	0	3				2	
2000/01	39	3			1	1	1				2	

Conclusions

Although broadly similar and shaped by the same rationale, there are clear differences between the UK and US vaccine compensation schemes. The US scheme has a substantially higher level of lawyer involvement, including the payment of considerable sums in

[45] www.gov.uk/government/uploads/system/uploads/attachment_data/file/223291/foi_4582_2012.pdf and www.gov.uk/government/uploads/system/uploads/attachment_data/file/223234/foi_3426_2012.pdf.

lawyers' fees. The most striking differential for victims is the contrasting success rates of applications: up to 2012 approximately 35% of US applications have been compensated; in the UK between 1977 and 2000 just under 21% of claims were compensated, but for the period of 2006/07–2011/12 a total of 38 awards were made and 407 were rejected, a success rate of 8.5%.

D. Ebola

Background

Ebola virus[46] disease (previously known as Ebola haemorrhagic fever) is a serious, acute disease. It was first discovered in two simultaneous outbreaks in 1976, one in a village near the Ebola River (from which the disease takes its name) in Yambuku, Democratic Republic of Congo and the second in Nzara, Sudan. The disease spreads from wild animals to humans (possibly through the consumption of bush meat) and then via human-to-human transmission. Transmission occurs through direct contact (through broken skin or mucous membranes) with the blood, secretions, organs or other bodily fluids of infected people, and with contaminated surfaces and materials (eg bedding, clothing).

Ebola has an incubation period of 2–21 days. Patients are not infectious until they become symptomatic. Initial symptoms include the sudden onset of fever fatigue, muscle pain, headache and sore throat. This is followed by vomiting, diarrhoea, rash, symptoms of impaired kidney and liver function, and in some cases, both internal and external bleeding (eg oozing from the gums, blood in the stools). Laboratory findings include low white blood cell and platelet counts and elevated liver enzymes.

Ebola is a serious illness, with an average fatality rate of 50%, though rates vary considerably from 20% to 90%. Early treatment, consisting of rehydration and symptom reduction, dramatically improves survival, but there is no licensed treatment. Containment and community surveillance are the only available prevention. Two vaccines are currently undergoing human safety trials.

2014 Outbreak

The outbreak in 2014 was the largest outbreak since the discovery of Ebola, and affected Guinea, Sierra Leone and Liberia. This outbreak was declared an international emergency by WHO in August 2014. Outbreaks have previously tended to affect rural, relatively sparsely populated areas; this outbreak differed as it infiltrated urban areas. The outbreak was reported in Guinea by the WHO's regional team on 23 March 2014. By 29 March a case had been confirmed in Liberia; following that, the first case was confirmed by the Ministry of Health in Sierra Leone on 25 May 2014. The three graphs in Figure 15.2—taken from the WHO Ebola Situation Report dated 4 March 2015[47]—show the progress of case numbers

[46] For general information, see WHO factsheet on Ebola, available at http://www.who.int/mediacentre/factsheets/fs103/en/.

[47] Available at http://apps.who.int/ebola/current-situation/ebola-situation-report-4-march-2015.

of Ebola in these three countries up until March 2015. This is undoubtedly one of the most serious public health issues of recent years.

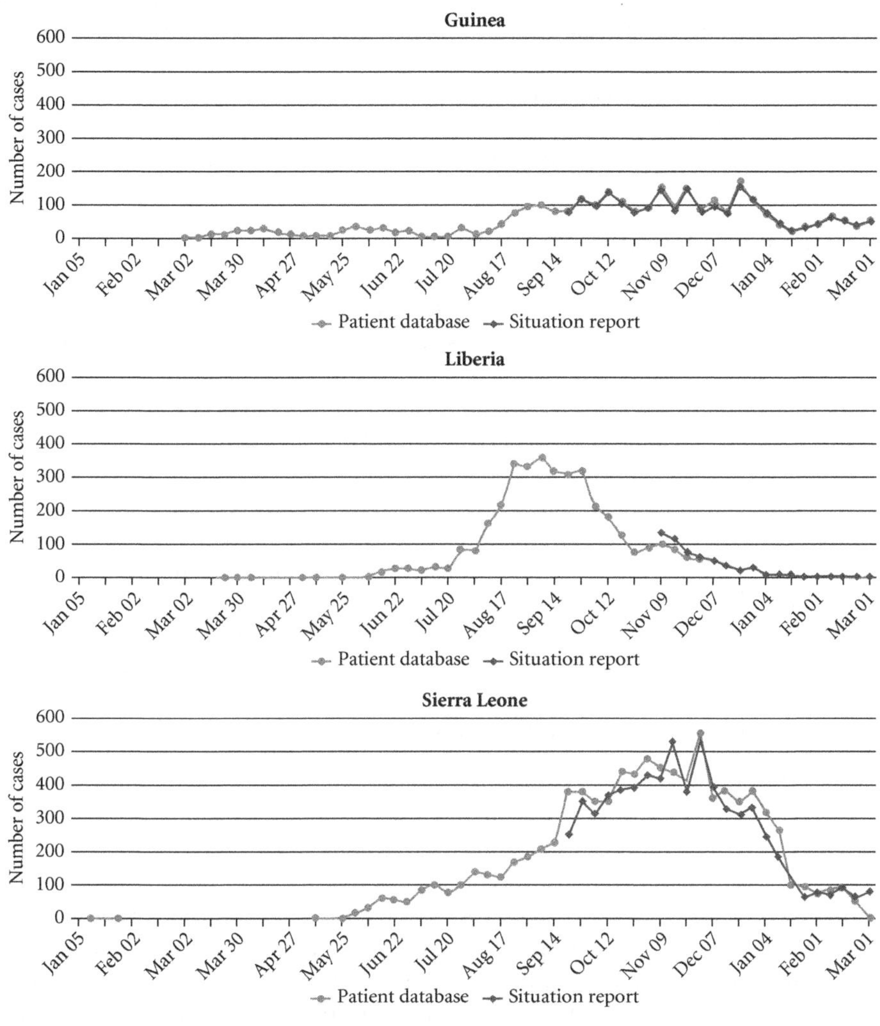

Figure 15.2: Distribution of Ebola cases in Guinea, Liberia and Sierra Leone 2005–March 2014

Vaccine Development

Two vaccine manufacturers, GlaxoSmithKline (GSK)[48] and NewLink Genetics,[49] have vaccines currently being tested in West Africa. Johnson & Johnson (J&J) and Bavarian

[48] The ChAd3-ZEBOV vaccine: www.gsk.com/en-gb/our-stories/health-for-all/our-contribution-to-the-fight-against-ebola/.

[49] The rVSV-ZEBOV vaccine: www.newlinkgenetics.com/development-pipeline.

Nordic have a 2-dose vaccine (with separate vaccines in the first and second doses), which began phase I trials in early in 2015.[50] Although cooperation between manufacturers is not what would normally be expected, this is what has been seen in the face of such enormous humanitarian need. All manufacturers are accelerating their production and testing in response to overwhelming need rather than profit. Other companies have also offered help; for example, Pfizer offered to share specialised production facilities to help increase capacity. Additionally worldwide regulators have been looking at ways to remove obstacles from the development of an Ebola vaccine, without compromising safety.

Liability for Ebola Vaccines

The United States has added the Ebola vaccines to the Public Readiness and Emergency Preparation (PREP) Act,[51] which provides US government-backed immunity under USlaw against legal claims related to the manufacturing, testing, development, distribution and administration of three vaccines for Ebola virus disease. It does not, generally, provide immunity for a claim brought in a court outside the United States.

Given that the Ebola outbreak is focused in countries that do not have developed social security or the capacity to fund large-scale compensation programmes in the event of a serious adverse reaction to vaccinations, it is essential that the issues of liability and compensation are addressed. Discussions of a compensation fund underpinned by the World Bank are ongoing.[52] Liability is complex due to the severity of the disease making mass vaccination desirable within a shorter time-frame than would usually be expected. Thus the accelerated development of a vaccine would make it harder for a manufacturer to obtain insurance against adverse events. While it is too early to say what form any compensation fund may take, it is worth noting the sacrifice of tortious liability in the face of humanitarian need.

Conclusions

The Ebola epidemic was contained and the impetus to rush a vaccine through was therefore reduced. This diminished the concerns over testing and accelerated time-scales. However, the spread of Ebola triggered a unified compassion-driven response, and the potential underpinning of a compensation scheme by the World Bank illustrates the function of compensation funds in providing access to vaccines.

Reports of fresh Ebola cases in 2017 may lead to further attempts to fast-track a vaccine.

[50] Johnson & Johnson, in association with Bavarian Nordic, are developing 2-dose vaccination approaches for Ebola using different vaccines for the first and second doses. This approach is known as heterologous prime-boost. The two vaccine candidates are known as Ad26-EBOV and MVA-EBOV. See www.jnj.com/news/all/Johnson-Johnson-Announces-Start-of-Phase-1-Clinical-Trial-of-Ebola-Vaccine-Regimen.

[51] www.hhs.gov/news/press/2014pres/12/20141209a.html.

[52] www.who.int/mediacentre/news/ebola/23-october-2014/en/ and www.bloomberg.com/news/articles/2014-11-13/world-bank-may-help-indemnify-ebola-vaccine-makers-who-says.

16

The Japanese Pharmaceutical Injury Scheme

SONIA MACLEOD AND CHRISTOPHER HODGES

A. General Description

A compensation scheme for pharmaceutical injuries was established in Japan in 1980 under the Adverse Drug Reaction Sufferings Relief Fund Law of 1 October 1979. The law was passed as a response to an epidemic of injuries in the 1960s attributed to subacute myelo-optic neuropathy (SMON), an iatrogenic disease of the nervous system leading to a disabling paralysis, blindness and even death, after the Tokyo District Court ruled in 1978 that SMON had been caused by Ciba-Geigy's product Clioquinol, which had been marketed as a prophylaxis for gastroenteritis.

The legislation authorised the Minister for Health and Welfare to create a fund to provide compensation in specific cases. The fund is administered by the national pharmaceutical regulatory agency, the Pharmaceuticals and Medical Devices Agency (PMDA), which was created in 2004. The PMDA has three main areas of focus, constituting a so-called 'safety triangle' system: product safety review and regulation; post-marketing safety measures; and relief services for adverse health effects. In relation to the third area, PMDA operates five adverse health effects relief services:[1]

1. Relief Systems for Adverse Drug Reaction: relief benefits relating to health damage such as diseases and disabilities requiring hospitalisation that were caused by adverse reactions to drugs prescribed at hospitals or clinics as well as drugs purchased at pharmacies, etc, even if such drugs were properly used.
2. Relief Systems for Infections derived from Biological Products (since April 2004).
3. Relief for HIV-positive and AIDS patients infected through contaminated blood products.
4. Relief for SMON patients.
5. Relief for Individuals Affected by Hepatitis C through specified Fibrinogen Products and Specified Blood Coagulation Factor IX Products[2] (since January 2008).

[1] PMDA, *Profile of Services 2011–2012*, 12.

[2] Act on Special Measures concerning the Payment of Benefits to Assist Individuals Affected by Hepatitis C through Specified Contaminated by Hepatitis C Virus (Act No 2 of 2008) (Specified Relief Service).

The following cases are not eligible for relief benefits from either the Adverse Drug Reaction Relief System or the Relief System for Infections Derived from Biological Products.

— Cases of adverse health effects resulting from statutory vaccinations (a different public relief system is available for such cases). Cases of adverse health effects resulting from voluntary vaccinations are eligible for relief benefits.
— When the liability for damage by the marketing authorisation holder of the pharmaceutical or biological product is clear.
— Cases where it was necessary to use the pharmaceutical or biological product in an amount exceeding the regular dosage for the purpose of saving the patient's life, even if it was acknowledged beforehand that adverse health effects may occur.
— Cases of infections, etc, resulting from adverse reactions to pharmaceutical and biological products, where the extent of adverse health effects is minor, or where the period for claiming relief benefits has passed.
— Cases where the pharmaceutical or biological product was not properly used.
— Cases of adverse health effects caused by pharmaceuticals not applicable to relief benefits (eg anticancer and immunosuppressant drugs) (applies only to the Adverse Drug Reaction Relief System).

B. Funding

The single fund in fact contains five separate compensation schemes, each of which is funded from distinct sources, so there is no overall risk of pooling or cross-subsidy.[3] PMDA is commissioned to distribute funds to claimants under the SMON, HIV and hepatitis C schemes. The SMON compensation is funded by a court settlement with some additional governmental funding. PMDA works under the commission of the Yu-ai Welfare Foundation to make payments for healthcare expenses for HIV-positive and AIDS patients, which are funded from different sources depending on the category of claimant. The hepatitis C compensation schemes are funded by central government. The Adverse Reactions to Drugs and Biologicals-Acquired Infections schemes have hybrid funding, partly from government and partly from industry.

C. SMON Claims

PMDA is commissioned by the government and pharmaceutical companies to provide healthcare allowances and nursing care expenses to SMON patients. The size of the class

[3] The Pharmaceuticals and Medical Devices Agency Annual Report FY 2014, available at www.pmda.go.jp/english/about/.

of claimants (approximately 7,500 claimants) was known when the court settlement was established, and is steadily declining. PMDA pays healthcare allowances and nursing care expenses to SMON patients. At the end of 2014 the number of patients in receipt of these allowances was 1,533 and the total paid in 2014 was 1,082 million yen—see Table 16.1 for a breakdown.

Table 16.1: SMON claimants and payments 2008–14

	2008	**2009**	**2010**	**2011**	**2012**	**2013**	**2014**
Number of recipients	2,180	2,075	1,960	1,855	1,748	1,639	1,533
Healthcare allowances (from companies) (yen)	1,140,517	1,089,491	1,031,376	975,567	924,669	864,462	811,727
Allowances for nursing care expenses (from companies) (yen)	284,981	268,749	250,946	241,890	233,050	219,630	201,919
Allowances for nursing care expenses (from government) (yen)	106,247	99,485	93,300	888,72	83,650	76,902	69,346
Total paid (yen)	**1,531,745**	**1,457,724**	**1,375,622**	**1,306,329**	**1,241,368**	**1,160,994**	**1,082,992**

D. Payments for HIV and AIDS

Payments for HIV-positive patients infected through contaminated blood products are mainly funded by the government and are provided under three streams:

(a) Payment of healthcare allowances for HIV-positive patients (who have not developed AIDS) as part of investigational research.

(b) Payment of healthcare allowances for AIDS patients for whom a settlement has been reached in court.

(c) Payment of special allowances, etc, for AIDS patients for whom a settlement has not been reached in court.

As would be expected, the number of claimants under these schemes is slowly declining year on year. Payments under these schemes are detailed in Table 16.2.

Table 16.2: Benefits paid to HIV and AIDS positive claimants by funding stream 2010–14 (amounts paid in 000s yen)

	2010		2011		2012		2013		2014	
	Number of recipients	Amount paid	Number of recipients	Amount paid	Number of recipients	Amount paid	Number of recipients	Amount paid	Number of recipients	Amount paid
Investigational research	562	309,355	547	302,763	540	297,790	529	292,349	524	288,736
Healthcare support	116	206,100	115	210,000	112	199,500	112	119,650	110	197,400
Special allowances	2	6,300	2	6,276	3	6,362	2	6,232	2	6,190
Total	**680**	**521,755**	**664**	**519,039**	**655**	**503,652**	**643**	**498,230**	**636**	**492,325**

E. Hepatitis C Payments

PMDA makes payments to individuals infected by hepatitis C under the Act on Special Measures concerning the Payment of Benefits to Assist Individuals Affected by Hepatitis C through Specified Contaminated by Hepatitis C Virus (Act No 2 of 2008) (Specified Relief Service). The Act was amended on 14 September 2012 to extend time-frames for claiming by 5 years (until 30 January 2018). Payments under this scheme may be eligible for 2 streams of compensation: standard and additional payments. Table 16.3 details the benefits by stream.

Table 16.3: Payment values and recipient numbers for hepatitis C relief 2008–14

	2008	2009	2010	2011	2012	2013	2014
Total number of recipients (additional payments recipients)	660 (4)	661 (22)	305 (20)	220 (20)	129 (28)	133 (18)	95 (20)
Amount paid (value of additional payments) (yen millions)	13,632 (68)	13,748 (272)	6,293 (324)	4,732 (268)	2,624 (488)	2,888 (332)	2,100 (368)
Number of consultations (cases)	3,607	894	1,286	674	982	473	660

F. Adverse Drug Reactions

This accounts for the vast majority of the relief fund work undertaken by PMDA. Half of the administrative fees for the Adverse Drug Reaction Relief System and the Relief System for Infections Derived from Biological Products are subsidised by the government. The remainder is funded mainly from contributions from industry. PMDA collects adverse drug reaction contributions from marketing authorisation holders (MAHs) of approved drugs to cover the scheme costs (but not compensation costs)—see Table 16.4. In financial year 2014, the contribution rate applied to drug MAHs was 0.27/1000 of sales, and 3,852 million yen was collected. PMDA also commissions the Japan Pharmaceutical Agency to collect contributions from the MAHs of pharmacy-compounded drugs.

Table 16.4: Contributions by marketing authorisation holders to fund the Adverse Drug Reaction Relief System by financial year 2008–14

Fiscal year	2008	2009	2010	2011	2012	2013	2014
MAHs of drugs (no of MAHs)	3,722 (752)	3,783 (742)	3,984 (716)	4,330 (713)	4,548 (688)	3,590 (688)	3,852 (692)
MAHs of pharmacy-compounded drugs (no of MAHs)	8 (8,015)	8 (7,598)	7 (7,082)	7 (6,694)	6 (6,186)	6 (5,866)	6 (5,658)
Total number of drugs	3,730	3,790	3,991	4,337	4,554	3,596	3,857
Contribution rate per million yen	0.35/1000	0.35/1000	0.35/1000	0.35/1000	0.35/1000	0.27/1000	0.27/1000

The rate of contribution for drugs has remained relatively stable for the last decade, while the contribution income has increased because more drugs have been marketed (see Figure 16.1).

PMDA collects adverse drug reaction contributions from MAHs of approved biologicals (see Table 16.5). In financial year 2014, the contribution rate applied to MAHs of biologicals was 0.1/1000 and 93 million yen was the collected. This was a substantial decrease in the contribution rate of 1/1000, which had been the stable contribution rate for some years beforehand.

If an MAH is found to be the 'cause' of an injury, such as a manufacturing or labelling defect, then PMDA can claim an additional contribution towards the compensation of 25%.

PMDA has accumulated a liability fund (mostly accrued from government funding) to cover the estimated costs of relief benefits due to be paid to eligible claimants in the future. Figure 16.2 shows the trends of the liability reserves, from 1981 to date; at the end of the 2014 fiscal year the liability reserve was 20,141 million yen.

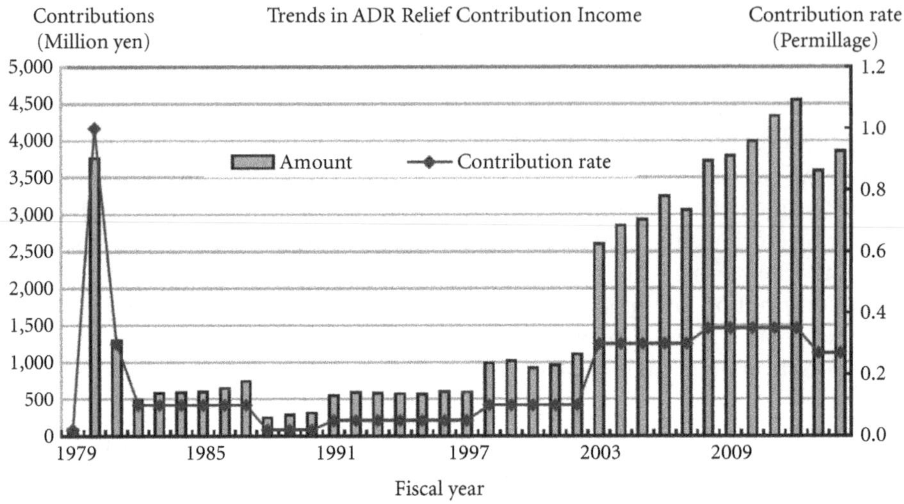

Figure 16.1: The rate and amount (in millions of yen) of contributions from the inception of the fund to 2014

Table 16.5: Contributions for biologicals 2009–14

Fiscal year	2009	2010	2011	2012	2013	2014
MAHs of approved biological products (no of MAHs)	631 (97)	693 (93)	785 (92)	866 (92)	869 (94)	93 (92)
Contribution rate (million yen)	1/1000	1/1000	1/1000	1/1000	1/1000	0.1/1000

G. The Claims Process

Claims for relief benefits are submitted direct to the PMDA Relief System Consultation Service by the patient or by his or her bereaved family. Claimants can contact the scheme by phone on a toll-free number or via the website. Table 16.6 shows the ways claimants choose to contact the Relief System Consultation Service: website access has grown significantly in recent years. The rule is that the claimant must prove the pathogenesis and symptoms/progress of the adverse drug reaction and infection, and that these resulted from use of pharmaceuticals or the relevant type of product. Accordingly, the prescribed claim form must be accompanied by a medical certificate from the doctor who conducted treatment for the adverse reaction or infection, and proof of prescription.[4] In fact, the process is that the claimant is not required to provide further proof than the doctor's certification.

[4] PMDA, *Profile of Services*, 2008, 14.

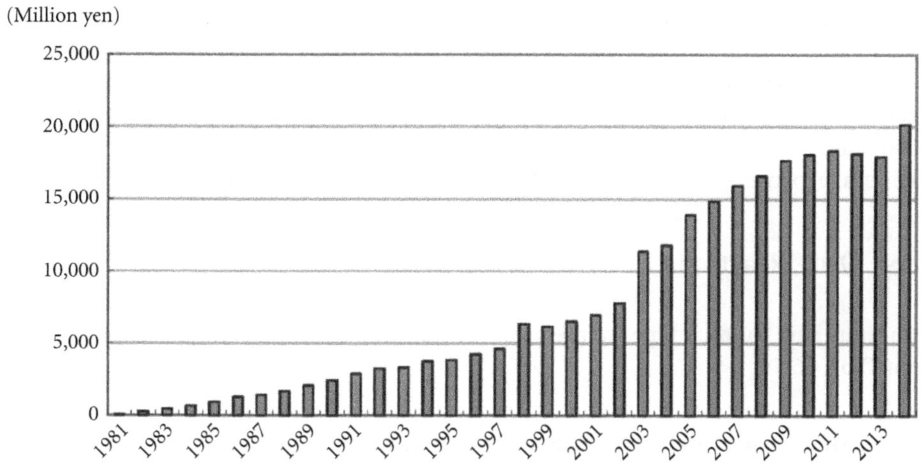

Figure 16.2: Trends of the liability reserves from 1981 to 2014 in millions of yen

Table 16.6: Ways in which claimants access the Relief System Consultation Service 2008–14

Fiscal year	**2008**	**2009**	**2010**	**2011**	**2012**	**2013**	**2014**
Number of phone consultations	17,296	34,586	16,123	21,577	22,324	21,843	21,300
Number of website accesses	67,711	87,109	89,500	72,688	113,182	151,925	137,359

For Adverse Drug Reactions and Reactions to Biologicals the amount paid is fixed by the type of benefit, regardless of the extent of injury. Seven types of relief are paid,[5] and claims can be made simultaneously for more than one type of benefit:

— Medical expenses—compensation will reflect the actual costs of treatment borne by the patient.
— Medical allowances—relief is provided for costs other than medical costs for disease treatment.
— Disability pension—relief is provided for living costs of patients over 18 years old, who suffer from a certain degree of disability.
— Pension for raising handicapped children—relief is provided for those who are responsible for raising patients under 18 years old who suffer from a certain degree of disability

[5] ibid.

— Bereaved family pension—relief is provided for bereaved families in rebuilding their life following death of their main provider.
— Lump-sum benefit for bereaved family—relief is provided for bereaved families as a gesture of sympathy following the death of their family member who is not the main provider.
— Funeral expenses—relief is provided for the costs of holding a funeral.

Adverse Drug Reaction Claims and Payments

The amounts paid for each type (to the nearest thousand yen) of benefit per fiscal year are listed in Table 16.7

Table 16.7: Annual payments in thousands of yen by type of benefit for fiscal years 2008–14

	2008	2009	2010	2011	2012	2013	2014
Medical expenses	75,339	86,666	87,475	93,284	97,905	95,025	123,987
Medical allowances	62,055	70,963	71,142	75,198	75,326	82,730	95,457
Disability pension	747,362	804,251	853,854	881,885	861,595	905,233	943,939
Pension for raising handicapped children	40,127	50,804	44,210	49,606	43,744	40,785	38,965
Bereaved family pension	523,455	545,843	583,501	614,318	602,068	603,130	585,626
Lump-sum benefit for bereaved family	335,977	215,342	214,081	328,093	227,696	220,032	310,806
Funeral expenses	14,391	9,914	12,927	16,006	12,438	12,249	14,507
Total	**1,798,706**	**1,783,783**	**1,867,190**	**2,058,389**	**1,920,771**	**1,959,184**	**2,113,286**

Table 16.8: Claims filed for adverse drug reactions by type of benefit for fiscal years 2008–14

	2008	2009	2010	2011	2012	2013	2014
Total number of adverse drug reaction claims	**926**	**1,052**	**1,018**	**1,075**	**1,280**	**1,371**	**1,412**
Medical expenses claims	769	902	854	909	1,101	1,200	1,221
Medical allowances claims	824	943	911	964	1,168	1,252	1,290

(continued)

Table 16.8: ***(Continued)***

	2008	2009	2010	2011	2012	2013	2014
Disability pension claims	79	71	74	77	83	88	95
Pension for raising handicapped children claims	7	11	4	4	1	7	12
Bereaved family pension claims	26	36	46	47	46	49	41
Lump-sum benefit for bereaved families claims	49	50	54	63	53	54	65
Funeral expenses	78	83	100	107	98	105	103

The success rate for claims varies by the type of benefit sought as shown in Table 16.8. Table 16.9 details the number of claims which were made and which were successful for 2008–14. As claims can span more than one fiscal year, in some years the number of cases approved exceeds the number filed. There appears to be a trend for moderate increases in claim filings and approvals.

Table 16.9: Claims filed and approved for adverse drug reactions for fiscal years 2008–12

	Fiscal year	2008		2009		2010		2011		2012		2013		2014	
		Filed	Approved	Filed	Approved	Filed	Approved	Filed	Approved	Filed	Approved	Filed	Approved	Filed	Approved
Types of benefit	Total claims	926		1052		1018		1075		1280		1371		1412	
	Medical expenses	769	659	902	763	854	803	909	836	1101	892	1200	886	1221	1108
	Medical allowances	824	711	943	813	911	837	964	895	1168	947	1252	945	1290	1151
	Disability pensions	79	27	71	26	74	38	77	28	83	28	88	39	95	37
	Pension for raising handicapped children	7	7	11	7	4	5	4	6	1	0	7	3	12	2
	Bereaved family pension	26	22	36	18	46	31	47	35	46	32	49	31	41	31
	Lump-sum benefit for bereaved families	49	47	50	30	54	29	63	47	53	32	54	32	65	45
	Funeral expenses	78	72	83	46	100	63	107	80	98	62	105	59	103	72

Once a claim is filed it will then be processed and 'judged' by the schemes' administrators. Judged cases are generally either approved or rejected, with a small number withdrawn. The proportions accepted and rejected appear to be fairly stable. In 2014 61.9% of cases were judged within 6 months of filing, with a median processing time of 5.7 months. These case processing speeds appear to be either consistent or slightly improving for the past 7 years (see Table 16.10).

Table 16.10: The number and percentages of adverse drug reaction cases in progress and judged, broken down by result and processing times for fiscal years 2008–14

		2008	2009	2010	2011	2012	2013	2014
Number of claims		926	1,052	1,018	1,075	1,280	1,371	1,412
Judged cases	Total number of judged cases	919	990	1,021	1,103	1,216	1,240	1,400
	Number of judged cases approved (% of total number of Judged cases)	782 (85.1)	861 (87.0)	897 (87.9)	959 (86.9)	997 (82.0)	1,007 (81.2)	1,204 (86.0)
	Number of judged cases rejected (% of total number of Judged cases)	136 (14.8)	127 (12.8)	122 (11.9)	143 (13.0)	215 (17.7)	232 (18.7)	192 (13.7)
	Number of judged cases withdrawn (% of total number of Judged cases)	1 (0.1)	2 (0.2)	2 (0.2)	1 (0.1)	4 (0.3)	1 (0.1)	4 (0.3)
	Number judged within 8 months (% judged within 8 months)[a]	683 (74.3)	733 (74.0)	765 (74.9)	809 (73.3)	923 (75.9)		
	Number judged within 6 months (% judged within 6 months)[b]	355 (38.6)	360 (36.4)	434 (42.5)	534 (48.4)	553 (45.5)	754 (60.8)	867 (61.9)
Cases in progress[c]		684	746	743	715	779	910	922
Median processing time (months)		6.5	6.8	6.4	6.1	6.2	5.8	5.7

[a] The percentage of cases judged within 8 months out of the total number of cases judged during the financial year.
[b] The percentage of cases judged within 6 months out of the total number of cases judged during the financial year.
[c] The number of cases in progress at the end of each fiscal year.

The performance of the adverse drug reactions component of the scheme is summarised in Figure 16.3. The vast majority of claims are accepted.

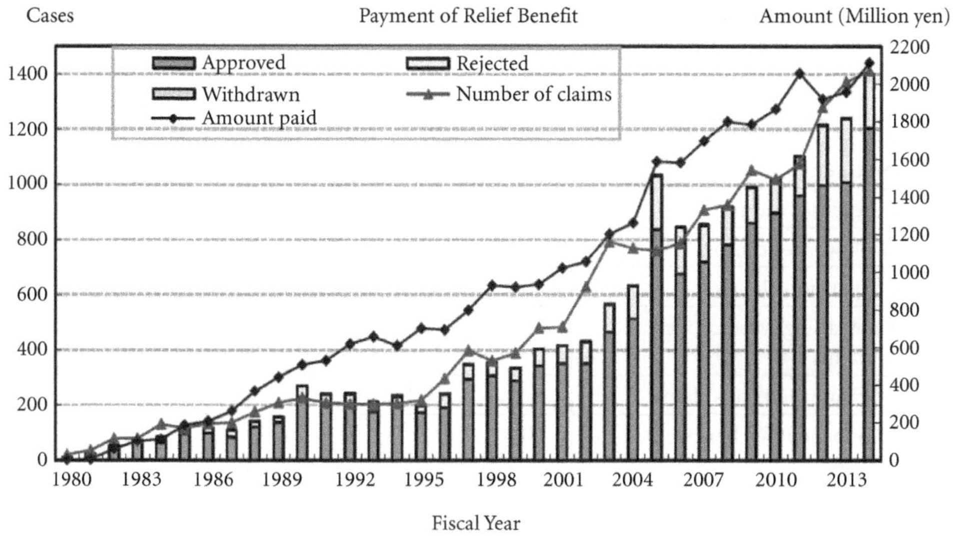

Figure 16.3: Summary of the relief benefits for adverse drug reactions 1980–2014

Claims for Infections Acquired through Biological Products

In relation to biological products, the amounts paid for each type (to the nearest thousand yen) of benefit per fiscal year are listed in Table 16.11.

Table 16.11: Annual payment for biologicals infections in thousands of yen by type of benefit for fiscal years 2008–14

	2008	2009	2010	2011	2012	2013	2014
Medical expenses	204	375	425	213	83	258	336
Medical allowances	386	567	384	282	282	356	566
Disability pension	–	–	–	–	–	–	–
Pension for raising handicapped children	–	–	–	–	–	–	–
Bereaved family payment	2,378	2,378	2,378	2,370	2,362	2,353	2,338
Lump-sum benefit for bereaved family	–	–	–	–	–	–	–
Funeral expenses	–	–	–	–	–	–	–
Total	**10,302**	**3,320**	**10,540**	**2,865**	**2,726**	**2,967**	**3,239**

It is noteworthy that the funds paid from the biologicals acquired infections component is substantially lower than the adverse drug reactions payments. The biologicals infection scheme receives far fewer claims (see Table 16.12).

Table 16.12: Claims filed for biologicals acquired infections by type of benefit for fiscal years 2008–14

	2008	2009	2010	2011	2012	2013	2014
Total number of biologicals acquired infection claims	**13**	**6**	**6**	**9**	**4**	**7**	**3**
Medical expenses claims	11	5	5	6	2	6	2
Medical allowances claims	13	6	5	8	4	7	3
Disability pension claims	–	–	1	–	–	–	–
Pensios for raising handicapped children claims	–	–	–	1	–	–	–
Bereaved family pensions claims	–	–	–	–	–	–	1
Lump-sum benefit for bereaved families claims	1	–	1	–	–	1	1
Funeral expenses	1	–	1	–	–	1	2

Table 16.13 details the number of claims which were made for biologicals acquired infections and which were successful for 2008–14. As claims can span more than 1 fiscal year in some years the number of cases approved exceeds the number filed. As with adverse drug reaction claims, once a claim is filed it will then be processed and judged.

Table 16.13: Claims filed and approved for biologicals acquired infections 2008–14

	Fiscal year	**2008**		**2009**		**2010**		**2011**		**2012**		**2013**		**2014**	
		Filed	**Approved**	**Filed**	**Approved**	**Filed**	**Approved**	**Filed**	**Approved**	**Filed**	**Approved**	**Filed**	**Approved**	**Filed**	**Approved**
Types of benefit	**Total claims**	13		6		6		9		4		7		3	
	Medical expenses	11	5	5	6	5	5	6	3	2	2	6	3	2	3
	Medical allowances	13	6	6	8	5	5	8	3	4	4	7	7	3	7
	Disability pensions	0	0	0	0	1	0	0	0	0	0	0	0	0	0
	Pension for raising handicapped children	0	0	0	0	0	0	1	0	0	0	0	0	0	0
	Bereaved family pension	0	0	0	0	0	0	0	0	0	0	0	0	1	0
	Lump-sum benefits for bereaved families	1	1	0	0	1	1	0	0	0	0	1	0	1	0
	Funeral expenses	1	1	0	0	1	1	0	0	0	0	1	0	2	0

Appeals

Decisions on payment of compensation are taken by the Minister of Health, Labour and Welfare, acting on the advice of the Central Council for Pharmaceutical Affairs. An applicant who is not satisfied with the judgment on eligibility for relief benefits may request consideration by the Minister within 2 months from the day after being informed about the judgment.[6]

H. Criteria

For the HIV, AIDS and hepatitis C claims the criterion is the existence of the virus and use of contaminated product. For SMON, the class of claimants was established by the court. Entitlement for adverse drug and biologicals reactions compensation requires proof of prescription of product and causation.

I. Conclusions

The Japanese Pharmaceutical Injury Scheme is stable and successful. Its stability is shown by the fact that it has a reasonably steady-state profile of claims and contribution rates, and has been able to build up a reserve fund.

A single administrative system covers claims against five different product types, which each have different sources of funding, spread across public and business, which do not cross-subsidise. The arrangement provides a single point of contact for claimants regardless of the source of the funding they are claiming from, and a seamless and consistent process. The public regulatory body for medicines, PMDA, evaluates claims and allocates and distributes funds as appropriate from the correct source.

[6] ibid, 15.

17

Injuries in Clinical Trials on Medicines and Clinical Evaluation of Medical Devices

CHRISTOPHER HODGES AND SHUNA MASON

A. Medicinal Products

A person who suffers personal injury as a result of being involved in clinical research on a medicinal product may, as an alternative to bringing a claim in the courts for breach of contract, negligence or strict product liability,[1] claim under one of the UK's two pharmaceutical industry voluntary 'no-fault' compensation schemes. The specific purpose of these schemes is to enable a claimant to overcome the legal difficulties (proving liability, causation, quantum and an absence of limitation), cost and delay that arise when seeking to establish a legal right. Such schemes were introduced contemporaneously with the development of ethical rules by the world medical community on the circumstances in which research in human subjects is or is not acceptable, and requirements for care, compensation and insurance for injuries to such subjects.[2] They were established specifically to assist the speedy delivery of compensation, so as to found public confidence in clinical research in the UK. From the perspectives of industry and of medical researchers and the National Health Service (NHS), the schemes reduce the reputational risk that is associated with a long and uncertain legal battle, and enable compensation to be delivered swiftly and without fuss. Hence, volunteers and patients should be encouraged to assist with research into innovative products.

Negligence and strict liability can be difficult and costly for claimants to prove, especially in medical situations. Standards of testing, manufacture, labelling, supply and safety are generally high in the pharmaceutical sector, and extensively controlled and documented, so it should be rare that the conduct of manufacturers or suppliers falls below the required

[1] Under the Consumer Protection Act 1987, Part I, based on Directive 85/374/EEC.

[2] *Declaration of Helsinki: Ethical Principles for Medical Research Involving Human Subjects*, World Medical Association, 1964, revised Tokyo 1975, Venice 1983, Hong Kong 1989, Somerset West 1996, Edinburgh 2000, Washington 2002, Toyko 2004, Seoul 2008 and Fortaleza 2013, available at: www.wma.net/wp-content/uploads/2016/11/DoH-Oct2013-JAMA.pdf (accessed 28 April 2017). Directive 2005/28/EC laying down principles and detailed guidelines for good clinical practice as regards investigational medical products for human use, as well as the requirements for authorisation of the manufacturing or importation of such products requires clinical trials to be conducted in accordance with Declaration of Helsinki 1996, Art 3, available at: http://eur-lex.europa.eu/legal-content/EN/TXT/PDF/?uri=CELEX:32005L0028&qid=1493371067598&from=EN (accessed 28 April 2017).

level of reasonableness. Liability will be clearer if a product contains a contaminant as a result of a substandard manufacturing process. The area that causes most problems is where an adverse drug reaction (ADR) occurs that was not listed in the safety labelling: the argument will be whether the risk should have been identified earlier or more clearly, or whether the doctor knew but did not pass the information on (the 'learned intermediary defence'), or the patient knew and accepted the risk.

The testing of unlicensed medicinal products on humans is subject to prior legal and ethical approval. Legal requirements govern the manufacture and labelling of the medicinal substance, and the need for prior ethical approval of the trial.[3] In the UK, research protocols involving NHS personnel and facilities must be authorised by the Medicines and Healthcare products Regulatory Agency (MHRA) and must additionally follow the Health Research Authority's (HRA) centralised approval process 'HRA Approval', which includes an assessment by an independent Research Ethics Committee (REC), undertaken by the UK research ethics service.[4] A clinical trial may only be undertaken, inter alia, if the foreseeable risks and inconveniences have been weighed against the anticipated benefit for the individual trial subject and other present and future patients [and] if the Ethics Committee and/or [product licensing] authority comes to the conclusion that the anticipated therapeutic and public health benefits justify the risks.[5]

Ethical evaluation of a study automatically involves consideration being given to the provision made for the payment of compensation and the basis upon which it may be payable to a subject injured by participation in proposed research. The European Clinical Trials Directive makes it a prerequisite for conducting a trial that provision for insurance or indemnity has been made.[6] This Directive is due to be replaced by a Regulation expected to become applicable in 2018.[7] Under the new Regulation, Member States have an obligation to ensure that the systems for compensation (in the form of insurance, a guarantee, or a similar arrangement) for damage suffered by participants are appropriate to the nature and extent of the risk involved in the trial.[8]

[3] The Medicines for Human Use Clinical Trial Regulations 2004, SI 2004/1031, as amended, which implements Directive 2001/20/EC on the approximation of the laws, regulations and administrative provisions of the Member States relating to the implementation of good clinical practice in the conduct of clinical trials on medicinal products for human use.

[4] HRA guidance on the 'HRA Approval' process, available at: www.hra.nhs.uk/research-community/hra-approval-the-new-process-for-the-nhs-in-england/ (accessed 28 April 2017). The Research Governance Framework for Health and Social Care (2005), sets out the NHS RECs responsibilities and requirements, clause 3.12, available at: www.gov.uk/government/uploads/system/uploads/attachment_data/file/139565/dh_4122427.pdf (accessed 28 April 2017).

[5] Directive 2001/20/EC on the approximation of the laws, regulations and administrative provisions of the Member States relating to the implementation of good clinical practice in the conduct of clinical trials on medicinal products for human use, Art 3(2), available at: http://eur-lex.europa.eu/legal-content/EN/TXT/HTML/?uri=CELEX:02001L0020-20090807&qid=1493905478720&from=EN (accessed on 4 May 2017). Directive 2001/20/EC, Art 3(2).

[6] Directive 2001/20/EC, Art 3.2(f) The Research Governance Framework for Health and Social Care (2005) also details that it is a Sponsor's responsibility to ensure that an agreement has been reached about compensation in the event of harm to research participants, clause 3.8.7.

[7] Regulation (EU) No 536/2014 of the European Parliament and of the Council of 16 April 2014 on clinical trials on medicinal products for human use, and repealing Directive 2001/20/EC.

[8] Regulation (EU) No 536/2014 of the European Parliament and of the Council of 16 April 2014 on clinical trials on medicinal products for human use, and repealing Directive 2001/20/EC, Recital 62 and Article 76(1). Directive 2001/20/EC, Art 3.2(f), available at: www.eortc.be/Services/Doc/clinical-EU-directive-04-April-01.pdf (accessed 17 August 2010).

Unlike the position in relation to normal NHS care with licensed medicines, see above, particular 'no-fault' arrangements apply in the UK to compensation of research subjects who are injured as a result of taking part in research. UK ethics committees will require confirmation of the intention of the sponsoring company to apply the Association of the British Pharmaceutical Industry (ABPI) Guidelines,[9] even from companies who are not members of ABPI. This common practice is reflected by the inclusion of the ABPI Guidelines, first published in 1991, in the model clinical trial agreement templates.[10] The templates were developed jointly by the ABPI and the Department of Health.

Two UK industry compensation schemes exist for drug injuries, one for research involving healthy volunteers and patients not expected to receive a therapeutic benefit from the trial (known as Phase I research) and the other for research involving patients (known as Phases II, III and IV). Both schemes were drawn up by ABPI, with Phase I introduced in 1970 and Phases II to IV in 1983.[11] Revisions for Phase I studies were introduced in 1977[12] and 1988, with sets of guidelines for procedures[13] and for facilities[14] and again in 2007. In 1986, the Royal College of Physicians also published a report on research in healthy volunteers.[15] Since the two sets of guidelines were originally published, ABPI has conducted periodic reviews of these and amendments have been adopted.[16] Taking account of the fact that Phase I studies may involve patients as well as healthy subjects, the ABPI produced composite guidance, which came into effect from January 2015.[17] Under this guidance, patients who are not expected to gain a therapeutic benefit from the trial will benefit from the Phase I compensation arrangements. Should a patient in a Phase I trial be expected to receive a benefit from the trial, the Phase II-IV compensation arrangements will apply to that patient.

Before confirming a favourable opinion on any research, the REC must assure itself that the sponsors and investigators will have appropriate insurance or indemnity cover in place.[18] Companies that sponsor Phase I research are required as a matter of ethical practice to enter into a written contract with each healthy volunteer research subject, which includes compensation terms should the research subject suffer injury as a result of taking part in the trial.[19] Where products are being investigated for unlicensed indications in Phases II,

[9] *Clinical Trial Compensation Guidelines* (ABPI 2014), available at www.abpi.org.uk/our-work/library/guidelines/Documents/compensation_guidelines_2014.pdf (accessed 28 April 2017). *Guidelines for Compensation in Clinical Trials* (ABPI, 1991); *The Use of Healthy Volunteers in Research* (ABPI, 1988) as amended.

[10] Model Clinical Trial Agreement, England (NHS, ABPI & BIA 2011), clause 3.3 and Appendix 3.

[11] The schemes were drafted by the leading pharmaceutical sector lawyer Ian Dodds-Smith, then a partner in city law firm McKenna & Co, now of Arnold & Porter LLP. See C Hodges, 'Harmonization of European Controls over Research: Ethics Committees, Consent, Compensation and Indemnity' in A Goldberg and I Dodds-Smith (eds), *Pharmaceutical Medicine and the Law* (London, Royal College of Physicians and Faculty of Pharmaceutical Medicine of the Royal Colleges of Physicians of the UK, 1991).

[12] *Guidelines for Preclinical and Clinical Testing of New Medicinal Products* (ABPI, 1977).

[13] *Guidelines for Medical Experiments in Non-Patient Volunteers* (ABPI, 1988).

[14] *Guidelines for Facilities for Non-Patient Volunteer Studies* (ABPI, 1988).

[15] *Research on Healthy Volunteers* (Royal College of Physicians, 1986).

[16] *Guidelines for Phase 1 Clinical Trials* (ABPI, 2012), available at www.abpi.org.uk/our-work/library/guidelines/Documents/guidelines_phase1_clinical_trials.pdf (accessed 28 April 2017).

[17] *Clinical Trial Compensation Guidelines* (ABPI, 2014).

[18] HRA *Standard Operating Procedures for Research Ethics Committees* (2017), clause 3.48, available at www.hra.nhs.uk/documents/2017/01/standard-operating-procedures-version-7-2.pdf (accessed 3 May 2017).

[19] *Clinical Trial Compensation Guidelines—Phase 1 Clinical Trial Compensation Guidelines* (ABPI, 2014), clause 1 and *Insurance and Compensation in the Event of Injury in Phase 1 Clinical Trials* (ABPI, the BioIndustry Association and the Clinical Contract Research Association, 2012), clause 3.1.

III or IV the sponsoring company is ethically required to provide undertakings, verified as a part of the ethical review process, to provide compensation. In all phases the amounts of compensation are to be the same as would be awarded by a court.

Under the model clinical trial agreement, the sponsor indemnifies the NHS (in accordance with the appended ABPI Compensation Guidelines) against all claims and proceedings brought by or on behalf of the clinical trial subjects.[20]

In the absence of a 'no-fault' scheme, the participant who suffers injury and believes it to be trial-related would have to rely on her or his ability to claim compensation through the courts often a difficult and lengthy process. Under the new EU Regulation, a participant (or their legal representative) must give written, dated and signed 'informed consent' after having received information about the trial including the applicable damage compensation system.[21] Although under the Directive it is not an express requirement that participants be given information about the compensation available, most ethical guidelines require a study participant to be told in advance what provision has been made—if any.

The MHRA, the body which authorises clinical trials of pharmaceuticals in the UK, reported that it received 842 commercial clinical trial applications in 2015 and in relation to the National Institute for Health Research (NIHR) network commercial contract studies alone, almost 35,000 participants were recruited between 2015 and 2016.[22] Nonetheless, despite these high participant numbers, injuries in trials on medicinal products have been extremely rare over the past 50 years. One student in Cardiff had a heart problem after he did not report that he had been involved in more than one trial. There have only been two serious cases of injury in clinical trials in around forty years of extensive drug research in Europe.[23] The first occurred in a private research unit at Northwick Park Hospital in London in 2006 during a Phase I trial of an anti-inflammatory drug, which was being developed for the treatment of arthritis and leukaemia.

The product owner was a very small German company called Te Generis, and the drug was its only product. The healthy volunteer study was being run by the larger clinical research organisation (CRO) Parexel. It involved 6 subjects, 3 of whom were controls and 3 who took the active ingredient; the latter were seriously affected, suffering multiple organ failure.[24] The damages claims exceeded the £2 million insurance cover held by the trial sponsor, because of the serious and long-term nature of the injuries. Te Generis went into administration. Claims were then made directly against the insurers. The problem was that Te Generis only had insurance for around £500,000. The insurers paid out the policy, but that left an unrecovered shortfall for the claimants. The claimants then sued Parexel. That claim was settled, on confidential terms, so it is not known what, if any, money was paid. The trial was unblinded, to reveal which subjects had taken the active or a safe

[20] Model Clinical Trial Agreement, England (NHS, ABPI & BIA 2011), Appendix 3.

[21] Regulation (EU) No 536/2014 of the European Parliament and of the Council of 16 April 2014 on clinical trials on medicinal products for human use, and repealing Directive 2001/20/EC, Art 29.

[22] *Research Delivery in the NHS 2015–2016* (NIHR), available at www.nihr.ac.uk/about-us/documents/CRN%20Key%20statistics%202015-16.pdf (accessed 3 May 2017).

[23] See J McHale, 'Clinical Research' in A Grubb, J Laing and J McHale (eds), *Principles of Medical Law*, 3rd edn (Oxford, Oxford University Press, 2011) para 13.02, for a report on a single volunteer in the USA in 2001.

[24] See the report of the expert scientific group established by the Secretary of State, *The Expert Group on Phase One Clinical Trials: Final Report* (2006).

substance. Claims were also made by the two volunteers who had taken placebo, who had no physical injury but suffered anxiety.

The second case occurred in Rennes, in France in January 2016 in a Phase I clinical trial conducted by a company called Biotrial. The trial was testing a compound called BIA 10-2474, which had been developed by the Portuguese pharma company Bial, as a candidate drug for a number of diseases. One man died and five others suffered serious adverse effects.

After the Northwick Park hospital trial, the Department of Health was concerned that ministers could be criticised if other cases of uninsured and uncompensated injury to research subjects were to occur, and also that the UK would not be an attractive venue for clinical research. The Department raised with the ABPI the need to ensure that insurance should be held ((the previous guidelines had been unclear about whether external insurance was needed)) and that it should cover all foreseeable circumstances; in other words have much higher cover. As outlined above, adequate insurance or indemnity cover must be obtained before a favourable opinion for a trial will be given for first-into-man studies. Presently, in relation to Phase I trials, RECs determine whether insurance cover represents adequate protection for volunteers on a case-by-case basis, having regard to all the circumstances of the case.[25] For pharmaceutical trials generally it is not considered adequate for a sponsor to undertake to 'self-insure' against any potential liability from its own funds and insurance or indemnity must be provided by another legal entity although this may be another company within the same corporate group.[26] As at January 2017, guidance to RECs states that: 'It is recommended that sponsors purchase insurance that provides a minimum of £5m indemnity cover for any first-in-human Phase 1 trial, or £2.5m for other Phase 1 trials.'[27] Those levels are massively more than the historical evidence would justify. There is a concern about imposing excessive burdens on a comparative basis. £2.5 million is high compared with other EU schemes.

B. Medical Devices

Conformity of medical devices with the essential requirements required by law must be based on clinical data before being approved for placing on the market.[28] This will remain the position under the revised EU legislation due to take mandatory effect in 2020.[29] Such evaluation may involve a clinical investigation of a device.[30] Clinical investigations must be

[25] *Insurance and Compensation in the Event of Injury in Phase 1 Clinical Trials* (ABPI, the BioIndustry Association and the Clinical Contract Research Association 2012), clause 5.1 and HRA *Standard Operating Procedures for Research Ethics Committees* (2017), Annex G, clause 2.

[26] HRA *Standard Operating Procedures for Research Ethics Committees* (2017), Annex G, clause 6.

[27] HRA *Standard Operating Procedures for Research Ethics Committees* (2017), Annex G, clause 23(vii).

[28] Directive 93/42/EEC on medical devices, Art 15 and Annex X; Directive 90/385/EEC on active implantable medical devices, Art 10 and Annex 7. See CJS Hodges, 'Regulating Medicinal Products and Medical Devices' in Grubb, Laing and McHale (n 15).

[29] Regulation 2017/745/EU, Art 61 and Annex XIV.

[30] See *Clinical Investigation of Medical Devices—Guidance for Manufacturers* (MHRA, March 2016).

approved in advance by the competent authority[31] and by the relevant ethics committee(s). The NHS ethics committees that scrutinise such investigations are those referred to above in relation to medicinal products.

After the above regulatory requirements were introduced in the mid-1990s, the UK medical device manufacturers proposed to the NHS to adopt the same mechanism for compensating anyone injured in a clinical investigation of a medical device as the then well-established mechanism for compensating injuries caused by medicinal products, namely to agree to abide by a professional code of indemnity. The ABHI Compensation Guidelines were therefore adopted by the medical device manufacturers' trade association, the Association of the British Healthcare Industries (ABHI).[32] Allowing for differences in terminology, these Guidelines provide substantively the same level and type of redress as do the ABPI Compensation Guidelines discussed above in relation to medical device studies involving patients. The ABHI Compensation Guidelines apply in circumstances where the medical device is used in the course of a clinical investigation where the device in question does not bear the CE Marking in relation to the aspect(s) under investigation. Like the ABPI Compensation Guidelines, the ABHI Compensation Guidelines have now been incorporated into a model Clinical Investigation Agreement agreed between the Department of Health and the ABHI,[33] which NHS bodies are encouraged to use.

Where a clinical investigation of a medical device is to be performed, the MHRA assesses issues relating to the safety of the study and the technical specification and performance of the device while the REC is responsible for addressing other ethical issues arising from clinical investigations, such as recruitment, informed consent, confidentiality, indemnity and compensation, incentives and payments, follow-up treatment at the end of the study, suitability of sites, registration or publication of study results. Unlike current pharmaceutical legislation, the medical devices legislation will not contain mandatory provisions regarding insurance and indemnity in respect of injuries to study participants until the new legislation comes into mandatory application in 2020.[34] That legislation will require EU member states to establish compensation systems for participants in medical devices studies in the form of insurance, guarantees or similar arrangements—in other words the medical devices legislation will adopt the approach set out in the pharmaceutical legislation discussed above. At present, for medical devices research conducted in relation to NHS patients, the REC is required to assure itself that the sponsor and investigators will have appropriate insurance or indemnity cover for the potential legal liability arising from the research, and to consider provision in proportion to the risk for compensation or treatment in the event of injury, disability or death attributable to participation.[35]

[31] In UK the MHRA, see above.

[32] They were drafted by C Hodges, who based them on the ABPI Compensation Guidelines drafted by I Dodds-Smith.

[33] Model Clinical Investigation Agreement, England (NHS and ABHI 2011), Appendix 3.

[34] Regulation 2017/745/EU, Art 69 and Annex XV.

[35] HRA, *Standard Operating Procedures for Research Ethics Committees* (2017) s 3.48.

Part VI

Patient Injuries

18

France: The ONIAM Scheme

CLAIRE BRIGHT AND CHRISTOPHER HODGES

A. Overview

Following the Law of 4 March 2002,[1] new Articles L 1142-1 to L 1142-29 were added to the French Public Health Code (Code de la santé publique), creating an out-of-court settlement scheme for victims of medical accidents.[2] Three administrative bodies are in charge of operating the scheme:

— The National Office for Compensation for Medical Accidents, Iatrogenic Disorders and Nosocomial Infections (Office National d'Indemnisation des Accidents Médicaux, des affections iathrogènes et des infections nosocomiales: ONIAM),[3] which operates under the supervision of the Ministry of Health.[4] ONIAM offers no-fault compensation to victims of serious medical accidents from a state fund and also manages a variety of other funds that have been set up for the compensation of victims of particular categories of medical accident. ONIAM also covers the operating costs of the Conciliation and Compensation Commissions and provides them with administrative and technical support.

— The Conciliation and Compensation Commissions (Commissions de Conciliation et d'Indemnisation: CCI) have the objective of facilitating the amicable settlement of disputes related to medical accidents, iatrogenic disorders and nosocomial infections as well as disputes between healthcare professionals, institutions which provide healthcare, or producers of health products and users of health services. The CCIs assess the merits of the claims brought before them. There is at least one CCI per region.

— The National Commission for Medical Accidents (Commission Nationale des Accidents Médicaux: CNAMed), whose role is to draw up a list of independent experts who assess cases that come before ONIAM or the CCIs, to draft guidelines for these experts, and to ensure the harmonised application of the scheme across the CCIs as well as to evaluate its functioning through an annual report.

[1] Loi no 2002-303 of 4 March 2002 relative aux droits des malades et à la qualité du système de santé [Law on the rights of patients and on the quality of the health system].

[2] See BA Koch (ed), *Medical Liability in Europe: A Comparison of Selected Jurisdictions* (Berlin, De Gruyter, 2011) 217.

[3] Art L 1142-22 of the French Public Health Code.

[4] See R Cascão and R Hendrickx, 'Shifts in the Compensation of Medical Adverse Events' in WH van Boom and M Faure (eds), *Shifts in Compensation between Private and Public Systems* (New York, Springer, 2007).

B. Background

The introduction of ONIAM in 2002 was part of a wider package of reforms to the health system under the 'Kouchner Law', named after the then Minister of Health, Bernard Kouchner (a former doctor and co-founder of Médecins sans Frontières).[5] The overarching aim of the law was a greater recognition of patients' rights in the health system, placing patients in more of an equal position with the doctors treating them.

A variety of factors drove the introduction of the law. Firstly, the generally increasing education of patients made them increasingly less satisfied in accepting uncritically advice from doctors (the development of patients' associations also contributed in this respect). Secondly, and related to this, the continuing advance of medical science had greatly increased both the volume and complexity of medical interventions available to patients, resulting of course in an increased incidence of medical accidents.[6] Thirdly, there was widespread dissatisfaction with existing systems of redress in the cases of these accidents, with victims struggling to prove fault in a complex system with varying liability rules in administrative and civil courts.[7] Lastly, there was a perceived loss of trust in the health system, which stemmed largely from widely publicised scandals such as the distribution of infected blood to patients in the mid-1980s, which it was argued was done knowingly by the doctors involved.[8]

The creation of ONIAM must therefore be placed in this context—as a way of restoring public trust in the health system and improving patients' rights, in particular by giving patients who had suffered as a result of their treatment easier access to redress. ONIAM offered several improvements in order to achieve this aim. Most important among these was the establishment of a scheme providing full compensation for those injured in serious medical accidents, regardless of whether any fault could be established. This is consistent with a French philosophy of a collectivised vision of risk-spreading through civil liability within the community,[9] and an 'ideology of compensation'.[10]

Also important was the simplification of the claims procedure. French medical law is relatively unusual, in that the health system combines both public and private hospitals or practices, which in practice means that the type of tribunals competent, or the type of rules applicable, vary according to context: administrative tribunals apply administrative law for public hospitals or practices; civil tribunals apply private law for private ones. Although less common, criminal investigations and prosecutions are also part of the landscape where fault has caused death or personal injury, and offer cost-free opportunities for injured persons to claim damages as civil parties.[11] The creation of ONIAM provided a means of

[5] C Bouillard, P-A Lecocq, S Lecocq and J Saison-Demars, 'L'indemnisation amiable des accidents médicaux, Bilan du traitement de 15000 dossiers par les CRCI et l'Oniam et perspectives d'avenir' (2009) 13 *Revue Générale du Droit Médical* 11.

[6] S Taylor, *Medical Accident Liability and Redress in English and French Law* (Cambridge, Cambridge University Press, 2015) 24.

[7] Y. Lambert-Faivre, 'Lindemnisation du préjudice des victimes d'accidents médicaux. N'est-ce pas temps d'adopter un système d'indemnisation cohérent et stable?' [2001] *Gazette du Palais* 13.

[8] See www.cairn.info/revue-les-tribunes-de-la-sante-2004-4-page-39.htm.

[9] Taylor (n 6) 63, 75.

[10] L Cadiet, 'Sur les faits et les méfaits de l'idéologie de la réparation', *Mélanges offerts à Pierre Drai: Le juge entre deux millénaires* (Paris, Dalloz, 2000) 495.

[11] See P Mistretta, *Droit penal medical* (Paris, Cujas, 2013); M Kazarian, D Griggiths and M Brazier, 'Criminal Responsibility for Medical Malpractice in France' (2011) 27 *Journal of Professional Negligence* 188. There appear to be perhaps under 100 criminal investigations into doctors and under 10 convictions a year: see Taylor (n 6) 52.

unifying and simplifying compensation procedures, regardless of whether the claimant was treated in the public or private sector.[12] Finally, the use of alternative dispute resolution techniques has meant that the system is more accessible, including to those on low incomes, and provides decisions more rapidly.

Before the introduction of the ONIAM scheme, French law continued to reply on the fault principle for liability in both civil and administrative courts, but a number of exceptions had evolved[13] that alleviated the burden on healthcare professionals and insurers, whilst presenting some barriers for patients. The 2002 Kouchner Law clarified that medical professionals and healthcare establishments would only be liable in the case of a fault.[14] Two exceptions were provided where strict liability applies. The first one relates to the harm suffered as a result of a defective medical product or device. The second one concerns, in certain cases,[15] the situation in which the harm has been caused by a nosocomial infection (Article 6.6111-6 of the Public Health Code defines a nosocomial infection as 'any infection linked to medical care which is contracted in a health establishment').

The law also provides that patients who suffer medical accidents can choose whether to claim through traditional civil, administrative or criminal liability regimes or claim through the new scheme. They can also pursue parallel claims through the scheme and the courts.

C. Organisational Structure

ONIAM is managed by a Board of Directors, which consists of a President and 22 members representing different elements of the state, the public and private health systems, and patients' and users' groups. There are in total 11 members drawn from state organisations (the Director-General for Health, the Director of National Social Security, the Director-General for Medical Care, the Director-General for Social Cohesion, the Secretary-General of the Social Affairs Ministry, the Budget Director, the Director-General of the Treasury, the Director of Civil Affairs and Justice, the Director-General for Forestry and Rural Affairs, the Director-General for Industry, Information Technologies and Postal Services, and the Director-General for Labour, or their representatives). A further 9 members are appointed by the Health Minister for 3-year renewable terms: 2 experts in medical liability and compensation for health risks, 2 representatives put forward by associations of patients and users of the health system, 1 representative of public hospitals and 1 of private hospitals, 1 representative of the National Health Insurance Fund for Employees, and 1 representative of self-employed health professionals and one of publicly employed institutions). Finally, 2 ONIAM staff representatives also sit on the board.[16]

The CCI are presided over by a magistrate of the administrative or judiciary court system. Membership of the CCI comprises representatives of patients and users of the

[12] P Frati and M Gulino, 'European Legislative and Juridical Overview', in SD Ferrara, R Boscolo-Berto and G Viel (eds), *Malpractice and Medical Liability—European State of the Art and Guidelines* (Berlin, Springer, 2013) 84–85.

[13] S Taylor, 'Clinical Negligence Reform: Lessons from France?' (2003) 52 *International and Comparative Law Quarterly* 737.

[14] Art L 1142-1 of the French Public Health Code.

[15] Under Art L 1142-1-1 of the French Public Health Code, if either death or permanent incapacity exceeding 25% occurs, the state will take on the burden of compensation, regardless of fault.

[16] Art L 1142-43 of the French Public Health Code.

health system, healthcare professionals, liability insurers, as well as heads of health institutions and representatives of ONIAM. The operational costs incurred by the CCIs are borne by ONIAM, which also provides appropriate technical, administrative and where necessary staff support.[17] ONIAM itself is provided for by the state.

D. Compensation Scheme: General Overview

The scheme reflects the importance of the principle of social solidarity in a civil liability tradition which is victim orientated.[18] The aim of the scheme is to facilitate access to compensation for the victims of medical accidents.

Under the scheme, injured patients, their representatives or beneficiaries may submit a claim for injuries caused by:

- medical accidents, 'iatrogenic' disorders (resulting from medical examination or treatment), 'nosocomial' (hospital-acquired) infections and persons who suffer damage in medical research;
- mandatory vaccinations;
- vaccination against influenza A (H1N1) 09;
- contamination with human immunodeficiency virus (HIV) from transfusion of blood or blood products;
- contamination with the hepatitis C virus from transfusion of blood or blood products;
- contamination with Creutzfeldt–Jacob disease from extractive growth hormone;
- the administration of Benfluorex (Mediator, Benfluorex Mylan and Benfluorex Qualimed).

The compensation procedure varies depending on the type of harm. However, in all cases, access to the scheme is free of cost for the claimants. In order to have access to the scheme, claimants must complete a paper-based compensation form and send it to the relevant authority (either ONIAM or the regional CCI depending on the cause of injury). The authority in question will carry out an investigation based on the medical records of the claimant, and, if it deems necessary, order a medical examination. The costs of the examination are also borne by ONIAM.

A reasoned decision must be handed out within 6 months upon receipt of a complete compensation claim. Where the evidence justifies an award, an offer is made to the claimant indicating the assessment made for each item of loss and the applicable amount of damages, taking into account any benefits he or she has already received or may receive from other third-party payers. Damages are calculated by reference to a standardised assessment scale weighing the different types of harm, with victims receiving full compensation for their harm. If the victim rejects the offer, it is deemed to be withdrawn, and the victim retains the right to sue in court. A claim can be filed in court concurrently to a claim with ONIAM, but both the court and ONIAM must be informed.

[17] Art L 1142-6 of the French Public Health Code.

[18] Taylor (n 6) 173.

E. Compensation Scheme: Specific Procedures

The procedure for the general scheme for compensation for medical accidents varies according to whether or not there has been a fault on the part of the medical service provider. Where fault is established, the liability insurer of the medical service provider takes on the burden of compensation (and such insurance is of course mandatory). Where no fault is established, a state fund will compensate the victim.[19]

As highlighted above, compensation for specific categories of medical accident victims is also assumed by ONIAM regardless of fault.

The General Scheme

The general scheme concerns serious harms caused by medical accidents, or other injuries suffered as a result of medical research,[20] iatrogenic disorders (side effects linked to a medical treatment) and nosocomial infections (infections contracted in a health establishment).[21]

Applicants submit their claim to the competent CCI (which is the one responsible for the region where the incident took place). The procedure requires applicants to complete a compensation form accompanied by their relevant medical records, including proof of the damage and the cause it is attributed to. This form is sent by recorded delivery to the CCI.

The CCI then verifies the circumstances, causes, nature and extent of the injury. The CCI may reject the claim at the initial stage if it considers that it does not meet the requirement of seriousness.[22] When the CCI is satisfied that the seriousness threshold has been met, it requests a medical examination from an expert or board of experts who will file a report. This will then be followed by a fuller examination before the Commission. All the involved parties must be present during the investigation, where claimants may be assisted by a person of their choice. The report of the examination is then sent to both the parties, who can make written observations and can ask to be heard by the Commission before a decision is made. The CCI is required to hand out a decision on the circumstances, the causes, the nature and extent of the damage and the compensation regime applicable within 6 months of the submission of a complete claim.[23] The decision is then communicated to all the interested parties, ie the victim, the insurer and ONIAM.[24] Where the CCI is satisfied that an award is justified, it decides whether the injury was due, on the one hand, to negligent medical examination or treatment or, on the other hand, resulted from a serious non-negligent medically adverse event or hospital-acquired infection.

[19] ibid, 53.

[20] Art L 1121-10 of the French Public Health Code.

[21] Art L 1142-1 of the French Public Health Code.

[22] The medical act underlying the accident must have been carried out from 5 September 2001, and must have resulted in serious damage. Damage is considered serious if (a) the adverse effects on physical or mental well-being are (permanently) higher than 24%, or (b) it resulted in a temporary suspension of professional activities for at least 6 consecutive months or 6 non-consecutive months over a period of 12 months, or (c) on an exceptional basis, (i) when the victim is declared permanently unable to perform the professional activity he/she performed before the medical accident, the iatrogenic disorder or the nosocomial infection; (ii) when the medical accident, the iatrogenic disorder or the nosocomial infection brings about particularly serious problems, including of an economic nature, for his/her living conditions.

[23] Art L 1142-8 of the French Public Health Code.

[24] Koch (n 2) 218.

In the case of fault, notification is sent to the insurer of the relevant healthcare provider, who is required to make a satisfactory settlement proposal for full compensation of damages within 4 months of the receipt of the decision.[25] Should the insurer refuse to compensate the claimant, the claimant may sue in court or ask ONIAM to pay compensation; ONIAM then has a right of subrogation against the insurer to recover the sums paid out on its behalf. If ONIAM wins, it is entitled in addition to a punitive fee of up to 15% of the compensation award.[26] If the claimant rejects the offer made by the insurer or ONIAM, he/she may bring a standard legal claim, which will be in an administrative court for treatment arising from a public hospital, or a civil court for an independent professional, private hospital or healthcare provider.

Where there is no fault, the compensation recommendation of the CCI is sent to ONIAM, which has 4 months to decide on whether or not to follow the recommendation and make a compensation offer to the claimant.[27] Compensation is then paid by a state fund managed by ONIAM. The claimant can either accept the decision taken by ONIAM or challenge it in court. However, the acceptance rate of the offers made by ONIAM to the victim is very high: 99%.[28] In case of acceptance, payment is to be made to the claimant within a month.

For Injuries Caused by Mandatory Vaccinations,[29] Vaccination against Influenza A (H1N1) 09,[30] Contamination with HIV from Transfusion of Blood or Blood Products,[31] Contamination with the Hepatitis C Virus from Transfusion of Blood or Blood Products[32]

A variety of other compensation funds have also been established for particular types of medical accident. In these cases, compensation is always assumed by ONIAM, regardless of fault. Claimants must fill a compensation form accompanied by their relevant medical records including proofs of the damage and the cause it is attributed to. This form must be sent to ONIAM by recorded delivery.

ONIAM then carries out an investigation and may request, if it deems it necessary, a medical examination that is carried out by a doctor appointed on the basis of his/her expertise in the field, selected from a national list of judicial experts and specialists in medical accidents.[33] The claimant may be assisted or represented by a person of his/her choice during the procedure, but, unlike the procedure described above, will have to bear the costs incurred. The expert's report must be sent to the claimant, who has 15 days to make potential observations. The expert's final report, including his/her response to the claimant's observations, must be sent to ONIAM within 3 months of the expert's

[25] Art L 1142-14 of the French Public Health Code.

[26] Art L 1142-15 of the French Public Health Code.

[27] Art L 1142-17 of the French Public Health Code. ONIAM followed the advice of the CRCI in 91.5% of cases in 2015. See: www.oniam.fr/indemnisation-accidents-medicaux/rapport-d-activite.

[28] www.oniam.fr/indemnisation-accidents-medicaux/rapport-d-activite.

[29] Art L 3111-9 of the French Public Health Code.

[30] Art L 3131 of the French Public Health Code.

[31] Art L 3122 of the French Public Health Code.

[32] Art L 1221-14 of the French Public Health Code.

[33] Art L 1142-12 of the French Public Health Code.

appointment. ONIAM sends it to the claimant, who has 15 days to make potential further observations.

A reasoned decision on the compensation claim must be handed out by ONIAM within 6 months of receipt of the claim. After verifying the injury and its cause, if it is satisfied that the claim meets the requirements for compensation, ONIAM sends by recorded delivery an offer for full compensation to the claimant. The offer indicates the assessment made by ONIAM for each type of injury and the applicable amount of damages, taking into account any benefits he or she has already received or may receive from other third-party payers. The claimant shall inform ONIAM by recorded delivery if he/she accept the offer. In case of acceptance, payment is to be made to the claimant within a month. If the claimant rejects the offer made by ONIAM, he/she may challenge it in an administrative court.

For Injuries Caused by the Administration of Benfluorex (Mediator)[34]

As an alternative to bringing a claim in court, an amiable compensation procedure for victims of injuries caused by the administration of Benfluorex was put in place on 1 September 2011.

Claimants can submit a claim to ONIAM by filling out a compensation form accompanied by their relevant medical records and sending it by recorded delivery. At ONIAM, a panel of experts reviews each individual claim based on pre-established guidelines. The claimant may be assisted or represented by a person of his/her choice during the procedure, but will have to bear the costs incurred. The panel of experts may either decide on the injury and its cause on the basis of the written file or request further examination. The panel of experts must hand out its decision within 6 months of receipt of the claim and send a notification to the parties by recorded delivery.

Where the panel of experts is satisfied that the injury has been caused by the administration of Benfluorex, it makes a ruling on the circumstances, causes, nature and extent of the injury as well as a recommendation on the liability regime. If the panel of experts considers that an award is justified, the persons that it deems liable or their insurer are required to make a satisfactory settlement proposal which fully compensates the damage within 3 months of receipt of the decision.

In case of silence, refusal or an inadequate offer from the liable party, the claimant may ask ONIAM to pay compensation. ONIAM will then be subrogated to the claimant's rights against the liable party or its insurer. The claimant can furthermore always challenge the decision of the panel of experts, the offer made by the liable party or the decision of ONIAM in court.

F. Statistics on the CCIs

The number of compensation claims received by the CCIs and ONIAM has increased steadily since the scheme was created over 10 years ago, as shown in Table 18.1 and

[34] Art L 1142-24 of the French Public Health Code.

Figure 18.1.[35] The popularity of the scheme is shown by the fact that it has received over 4,000 claims in each of the last 5 years.

Table 18.1: CCI and ONIAM claims 2003–15

Year	Claims received	Monthly average	Claims rejected		Claims accepted
			Before expert report	After expert report	
2003	1,719		232	40	48
2004	3,453	288	668	576	511
2005	2,671	223	874	1,001	947
2006	2,719	227	659	977	1,147
2007	3,439	287	845	1,140	1,383
2008	3,556	296	1,192	1,308	1,316
2009	3,611	301	1,071	1,409	1,327
2010	4,113	343	1,019	1,411	1,293
2011	4,281	357	1,342	1,621	1,410
2012	4,255	355	1,348	1,874	1,527
2013	4,394	366	1,616	1,779	1,504
2014	4,479	373	1,598	1,734	1,482
2015	4,319	360	1,506	1,664	1,582
Total	47,009		13,970	16,534	15,477

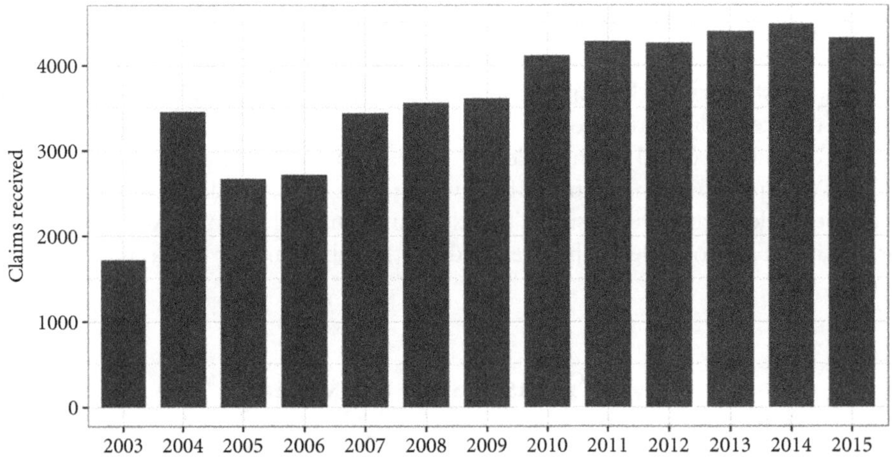

Figure 18.1: Number of claims received 2003–15 across all CCIs

[35] See ONIAM's 2015 activity report: www.oniam.fr/indemnisation-accidents-medicaux/rapport-d-activite.

In total, since 2003, the CCI has received over 47,000 compensation claims. As shown in Table 18.1, approximately one-third of these claims have been accepted, with two-thirds rejected either before or after a medical examination. These proportions have remained roughly constant throughout the period that ONIAM has been in operation. The reasons for rejecting applications mainly lie in the fact that the seriousness threshold of the harm had not been met, or that a causal link between the medical act and the injury could not be established.[36]

In total it takes on average approximately 8 months from initially receiving a claim to the notification of a decision by a CCI (see Table 18.2). It is worth noting, therefore, that the scheme has thus far not been able to fulfil the requirement of reaching a decision within 6 months.

Table 18.2: Average time for claim processing per year 2012–15

Year	2012	2013	2014	2015
Average time (in months) from complete file submission to notification of decision by a CCI	8.9	8.2	7.6	8.1

G. Performance Statistics: Medical Accidents

The number of claims per year received by ONIAM in the case of medical accidents has remained relatively stable (approximately 750 per year), whilst the average payout for successful claims has fluctuated between roughly €80,000 and €99,000 (see Table 18.3).

Table 18.3: Number of claims received by ONIAM in the case of medical accidents and average payouts

	Files received from CCIs (in the case of medical accidents)	Files closed	Average payout in the case of a successful claim (€)
2006	658	292	61,547
2007	758	513	86,934
2008	813	561	98,688
2009	786	748	75,173
2010	732	458	79,392
2011	669	384	86,341
2012	769	464	86,298
2013	806	529	98,915
2014	718	362	95,105
2015	816	655	85,927

[36] Commission Nationale des Accidents Médicaux (CNAMed), *Rapport au Parlement et au Gouvernement*, Année 2013, 21.

H. Conclusion

The ONIAM scheme is considered to have been a success in France, providing a simple, swift, cheap and non-adversarial pathway for patients who suffer medical accidents to obtain compensation. The regime exists alongside the three court pathways, and seems to be the preferred mechanism. The scheme has been financially sustainable, and commentators ascribe this to advantages of cost, speed and user-friendliness,[37] whilst potentially more generous damages rules established by the courts and the legislature only apply to restricted categories of victim.[38]

Taylor notes that the most important issues that have arisen with the ONIAM scheme are uncertainty over definitions ('medical accident', 'seriousness'), the potential for inconsistency in decision-making between CCIs, and whether claims are accurately attributed to the different categories, each of which has its own liability and redress principles.[39]

[37] Applicants will be given a hearing once past the initial filtering stage. The relative informality of hearings has been said to have an 'appeasing' effect: P-A Lecocq, 'L'indemnisation amiable des accidents médicaux. Bilan du traitement de 15 000 dossiers par les CRCI et l'ONIAM: perspectives d'avenir' [2009] *Revue Générale de Droit Médical* (special issue) 29.

[38] Taylor (n 6), 6.

[39] ibid, 104, 115.

19

Polish No-Fault Medical Compensation Scheme

MAGDALENA TULIBACKA

A. Overview

The 2011 amendment to the 2008 Act on patients' rights and the Patients' Rights Ombudsman[1] introduced a new scheme for seeking compensation for injuries or death caused by 'medical incidents'[2] in hospitals. Regional Committees for Adjudicating on Medical Incidents (the Committees—16 in total) have been in operation since 1 January 2012. They are financed in a large part by the state, although, following the loser-pays principle, one of the parties ultimately bears the costs of the proceedings including the filing fee and disbursements for experts and witnesses.[3]

Initially it was established that the core element of the scheme would be an insurance obligation for hospitals: the Act on Healthcare Activity of 18 March 2011 required all healthcare bodies in charge of hospitals to take out an additional insurance policy.[4] This insurance policy specifically covered liability for medical incidents. Under the scheme, injuries and death caused by medical incidents must be compensated by the insurer, or the hospital if there is no insurance coverage. Initially the insurance obligation was to come into force in 2014, but it was postponed twice due to complaints from hospitals concerning the high costs of policies and the lack of competition in the insurance market.[5] In 2015, out of the total of about 700 public hospitals and 1,000 private hospitals operating in Poland, only 7 had the additional insurance policy.[6] In April 2016 the Ministry of Health started working on an amendment to the Act on Healthcare Activity making insurance

[1] Act of 6 November 2008 on patients' rights and the Patients' Rights Ombudsman ('the Act on patients' rights' or 'the Act'); amendment of 28 April 2011. The text of the whole Act including the amendment (and certain other amendments) was published in *Dziennik Ustaw* (*Journal of Laws*) of 2012, item 159. (www.bpp.gov.pl/dok/ustawa/ustawa06112008-pr.p.pdf). The amendment introduced a new ch 13A to the Act on patients' rights.

[2] See below for definition.

[3] The Act on patients' rights specifies remuneration for Committee members: up to 430 PLN for each attended session (Art 67h.1). The remuneration is paid for by the budget of the region of Poland in which the Committee is based.

[4] Published in *Dziennik Ustaw* of 2011, no 112, item 654. Art 17(1)4.

[5] It was to finally come into force on 1 January 2017.

[6] Source: www.rynekzdrowia.pl/Finanse-i-zarzadzanie/Tylko-siedem-szpitali-ma-ubezpieczenie-od-zdarzen-medycznych,155064,1.html (accessed 5 January 2017).

against liability for medical incidents entirely voluntary. Thus, while the scheme remains unchanged, hospitals are to have a choice whether to insure against this type of liability or whether to cover the compensation from their own resources.[7] As a result of the amendment, in force since 1 January 2017, the Act on Healthcare Activity no longer requires hospitals to obtain an insurance policy against medical incidents.[8] The assumption is that most of them will not buy coverage and will cover the compensation under the scheme from their own resources.

The model is similar to the French ONIAM model, albeit somewhat simpler and with influences of civil litigation procedure. The Explanatory Note to the Act also mentions the Nordic (especially Swedish) schemes as an inspiration for the Polish scheme.[9]

Decisions made by the Committees are based on evidence produced by the claimant of whether there was a 'medical incident'. The subjective fault of a particular doctor or hospital does not seem to be a relevant factor in these decisions. Nevertheless, the definition of a 'medical incident' contained in the Act requires the Committees to establish the exact nature of the incident, which must have taken place in a manner 'contrary to current medical knowledge'.[10] Thus, according to academic commentators, this scheme is not a significant detraction from the general, fault-based liability for medical malpractice as developed by Polish courts.[11] On the other hand, certain high state officials (eg the former General Prosecutor, or the former Speaker of the Sejm—the lower chamber of Parliament) believed that it was a no-fault liability system.[12] An Explanatory Note produced for the benefit of potential claimants and their families by the Patients Rights' Ombudsman expressly states that Committees do not decide on fault of medical personnel, but only confirm that a medical incident took place.[13] There certainly is scope for more clarity with regard to this fundamental issue. The key question seems to pertain to the very nature of the new scheme: is it a civil liability scheme, or an insurance-like compensation scheme? The Committees appear to be suffering from an identity crisis.

The decisions of the Committees are binding on the insurer and hospital, but not on the claimant. The decision confirming a medical incident triggers a duty by the insurer (or the hospital) to make an offer of payment within 30 days, which the claimant has 7 days to accept. Compensation levels are capped by a Ministerial Regulation, although there is no lower threshold on offers and the Committees have no power to control the amounts

[7] www.politykazdrowotna.com/7878,szpitale-nie-musza-sie-ubezpieczac-od-zdarzen-medycznych (accessed 7 December 2016).

[8] The latest version of the Act (as amended by the Act of 10 June 2016, published in *Dziennik Ustaw* of 2016, no 960) can be found at http://isap.sejm.gov.pl/DetailsServlet?id=WDU20111120654.

[9] Official Explanatory Note to the Draft Act amending the Act on patients' rights, available on the website of the Polish Parliament: http://orka.sejm.gov.pl/Druki6ka.nsf (note no 3488). Here referred to as the 'Explanatory Note to the Act'.

[10] Art 67a.1 of the Act on patients' rights.

[11] E Baginska, 'The New Extra-Judicial Compensation System for Victims of Medical Malpractice and Accidents in Poland' (2012) 3 *Journal of Tort Law* 101, 105. See below for further analysis.

[12] The opinions of the Prosecutor and of the Speaker were presented during proceedings before the Constitutional Tribunal, hearing a claim by the General Medical Council concerning constitutionality of certain provisions of the Act on patients' rights regulating the new scheme. The case is discussed below (Decision of 11 March 2014, K 6/13).

[13] The note can be found on the website of the Ombudsman: www.bpp.gov.pl/gfx/bpp/userfiles/_public/edukacja_-_poznaj_swoje_prawa/ulotka_wkom.pdf.

offered.[14] Acceptance of the offer by the claimant bars taking an ordinary court action, whether to obtain compensation for the injury or death which was the subject of the Committee decision, or to obtain further compensation to supplement the one awarded by the Committee. The claimant has no obligation to accept the offer and may refer the case to an ordinary court instead.

Claims are easy and cheap to bring (no legal aid is possible), and need to be dealt with within 4 months. Proceedings are quite informal. The Committees are extra-judicial, quasi-administrative bodies,[15] and the procedure before them was described by the Explanatory Note to the Act on patients' rights as a sui generis settlement or mediation procedure.[16] It is certainly different from an ordinary mediation procedure because one of the parties (the insurer or hospital) has no right to object to taking part in it.[17]

Ever since their introduction, and even before—during parliamentary deliberations and consultations concerning the shape of the new mechanism—the Committees attracted a considerable amount of criticism. Legal scholars express mostly negative views about the scheme.[18] Problems noted by them focus on the substantive legal elements, such as the definition of 'medical incident', which often requires a thorough examination with the help of an expert. They also encompass procedural issues, such as the fact that no judicial oversight has been envisaged over the Committees' decisions or the compensation offers to which they lead. They also concern the limited scope of the system, in only covering incidents occurring in hospitals, and not at outpatient clinics or private individual practices.[19]

It is very clear that the scheme is not satisfactory, both from the point of view of the victims and their families, and the doctors and hospitals involved. However, its present form and scope are unlikely to be the ultimate arrangements. The Explanatory Note to the Act amending the Act on patients' rights gives examples of schemes in other countries (eg Sweden) which evolved, were extended and improved with time, and appears to indicate the same future for the Polish scheme which for now remains a sort of test. The Minister of Health has been considering a thorough revision of the scheme, but no concrete proposals have yet appeared.[20]

[14] Rozporzadzenie Ministra Zdrowia (Regulation of the Health Minister), 10 February 2012, w sprawie szczegolowego zakresu oraz warunkow ustalania wysokosci swiadczenia w przypadku zdarzenia medycznego [on the scope and conditions for establishing the level of compensation for medical incidents], published in *Dziennik Ustaw* of 23 February 2012, item 207.

[15] Baginska (n 11) 103.

[16] Explanatory Note to the Act, 7.

[17] E Gmurzyńska, 'Mediacja jako preferencyjna metoda rozwiązywania sporów dotyczących zdarzeń medycznych' ['Mediation as the Preferred Method of Resolving Disputes Concerning Medical Incidents'] (2016) 10(4) *Medium* 4.

[18] M Nesterowicz and M Walachowska, 'Liability for Damage caused by Medical Treatment in Connection with the New Extra-Judicial Compensation System in Poland' in E Kowalewski (ed) *Kompensacja szkod wyniklych ze zdarzen medycznych. Problematyka cywilnoprawna i ubezpieczeniowa [Compensation of damage caused by Medical Incidents, Civil Law and Insurance Aspects]* (Torun, 2011) 24.

[19] Baginska (n 11) 104; see also 'Komisyjny niewypal: czyli zdarzenia medyczne raczej przed sadem' ['The Committee Failure: Medical Incidents Rather Before Courts'], *Rynek Zdrowia*, 24 August 2012, www.rynekzdrowia.pl/Prawo/Komisyjny-niewypal-czyli-zdarzenia-medyczne-raczej-przed-sadem,122572,2.html.

[20] The 2015 Report of the Patients' Rights Ombudsman refers to the Ministry's work in this area: www.bpp.gov.pl/gfx/bpp/userfiles/_public/bip/sprawozdania_roczne/spr_za_2015.pdf, 51.

B. Context and History

The Polish healthcare system has undergone significant transformations over the past 30 years (after the end of socialism). Legislative and regulatory measures were introduced to improve the management, operation and financial standing of healthcare entities, and to reorganise their insurance coverage. These reforms notwithstanding, it remains a state-funded and state-run, heavily regulated system. In 2008 and 2011 a series of legislative acts were introduced, aimed at improving the quality of medical services, their accessibility and stability of prices.[21] A key element of this reform was the 2008 Act on patients' rights and the establishment of a Patients' Rights Ombudsman. The Act introduced a list of patients' rights, which can be enforced before courts and are being safeguarded by the Ombudsman.[22] In 2011, among other legislative measures, the Act amending the Act on patients' rights (establishing the medical incident compensation scheme) was introduced.

Liability for medical incidents has been a highly discussed topic in Poland for some time, in the wider context of the liability of medical practitioners. The number of claims concerning breaches of patients' rights, as well as injuries and deaths caused by medical treatment or diagnosis, that the patients or their families put through various redress paths is steadily increasing. Indeed, at between 10% and 30% per year it is above average among EU Member States.[23] Before the introduction of the new compensation scheme, the avenues of redress for patients whose rights were infringed included ordinary courts, the medical courts[24] and the Patients' Rights Ombudsman. Only the ordinary courts could award compensation for injuries or death caused by medical treatment. The fault-based provisions of the Civil Code were the sole basis for liability.[25] According to a report prepared by the Patients' Association, Primum Non Noncere, the most common types of problems with medical treatment and diagnoses, and complaints resulting therefrom, include: problems during childbirth, hospital-originated infections, failures to recognise signs of a heart attack, injuries while operating on a thyroid, and leaving a medical instrument or a part of it inside a human body during a procedure.[26] Statistical data gathered by the Ministry

[21] For analysis see: A Mrozek and A Dabek, 'Rzadowy Projekt Pakietu Ustaw Zdrowotnych—Kolejny Etap Reformy Ochrony Zdrowia w Polsce' ['The Government's Package of Healthcare Bills—The Next Stage of Healthcare Reform in Poland'] [2011] *Polityka Zdrowotna* 35.

[22] Patients' rights as set out by the Act on patients' rights include: the right to information, the right to privacy, the right to healthcare services complying with the state of current medical knowledge, the right to notify appropriate bodies of side effects of medical products, the right to consent to medical treatment, the right to intimacy and dignity, the right of access to medical documentation, the right to contest the opinions and decisions of medical practitioners, the right of respect for family and private life, the right to pastoral care, and the right to having valuables stored safely.

[23] T Kaczmarek and JT Marcinkowski, 'Odszkodowania za niepowodzenia lecznicze' ['Compensation for medical incidents'] (2011) 8(2) *Orzecznictwo Lekarskie* 79–90.

[24] The self-governance body for doctors, Naczelna Izba Lekarska (General Medical Chamber), and the regional chambers, operate medical courts: the General Medical Court and regional medical courts.

[25] Before 2004, medical liability was often seen as state liability: medical professionals were state functionaries. This was liability based on the principle of equity, the basis of which was the fault of a doctor or other medical professional (Baginska (n 11) 105). After 2004, ordinary contractual or tortious liability is applied to medical professionals. It is based on fault, but modified by courts to focus on objective factors such as conduct contrary to current medical knowledge. See below for a further elaboration.

[26] Report available from www.sppnn.org.pl (quoted in the Explanatory Note to the Act, 4).

of Justice for 2001–09 indicated that cases before ordinary courts could take as long as 10 years from lodging the claim to conclusion.[27] They also tend to be expensive: the court fee is 5% of the value of the claim.[28] A study of medical liability litigation between 2011 and 2013[29] demonstrated that these cases continue to be complex and difficult to win. Further, on average only 40% of civil litigation concerning medical liability is won by claimants. This is striking when contrasted with the total average of 60% of all civil claims.[30]

As long ago as 2003 the Special Assembly of Medical Professionals had called for an extra-judicial, no-fault scheme for medical injuries similar to the Nordic medical compensation systems.[31] Scholars dealing with medical liability have argued for some type of out-of-court, no-fault system for the past 20 years.[32] The new compensation scheme builds on these proposals, followed by a series of consultations and discussions. The Act introducing the scheme, passed in April 2011, inserted a new Chapter (13a) to the Act on patients' rights. As mentioned above, the Explanatory Note to the Act indicates that, in common with other countries where such schemes have appeared and evolved for many years, the Polish scheme may well be extended, amended and improved in future.[33]

C. Medical Incident

Claimants are required to provide evidence demonstrating a 'probability' that a medical incident occurred.[34] A medical incident is defined as contamination by a biological agent, injury or death of a patient caused by one of the following types of medical activities that are contrary to current medical knowledge:

- an incorrect or delayed diagnosis (resulting in incorrect or delayed treatment);
- treatment (including surgeries); or
- use of a medical product or device.[35]

The incident must have been the result of receiving medical treatment or other medical services in a hospital.[36] The actual state of current medical knowledge can be determined either by the Committee itself or, if expert knowledge is required, with the assistance of an expert. Because expert knowledge is often required to determine the current state of

[27] Data mentioned in the Explanatory Note, 2.

[28] ibid.

[29] *'Poszkodowany pacjent w sądzie i przed komisją' Raport z badania spraw sądowych o błędy medycznei i z badania orzecznictwa Wojewódzkich Komisji do Spraw Orzekania o Zdarzeniach Medycznych* [*Injured Patients in Courts and before Committees. Research Report of Civil Litigation of Medical Accidents and of Decisions of Regional Committees for Adjudicating on Medical Incidents*], Kancelaria BFP (J Budzowska), Krakow, 2015 ('BFP Report 2015').

[30] ibid, quoting data provided by the Ministry of Justice.

[31] Explanatory Note to the Act.

[32] Baginska (n 11) 101, also K Baczyk-Rozwadowska, *Odpowiedzialnosc cywilna za szkody wyrzadzone przy leczeniu* [*Civil Liability for Medical Malpractice*] (Torun, 2007) 355.

[33] Explanatory Note, 5.

[34] Article 67d.2(1) of the Act on patients' rights.

[35] Here the scheme covers situations where an incident was caused by the improper use by a medical practitioner, and not the fault/defect in the product or device itself. The latter is covered by the product liability rules (mostly, Art 449.1 et seq of the Civil Code).

[36] Art 67a.1 of the Act.

knowledge with regard to specific medical fields, procedures, or products, experts are used quite often.[37]

D. No-Fault Liability?

The precondition for payment of compensation under the scheme is the determination, by one of the Committees, that a medical incident took place. The Act does not mention fault as another precondition. Views concerning the standard of liability in this new compensation scheme are divided. Scholars are almost unanimous in their judgement that the Act does not offer anything new in the way Polish civil law already approaches medical liability. Ordinary courts dealing with medical malpractice cases use fault-based provisions of the Civil Code on contractual and tortious liability, albeit within the scope of these provisions they introduce very strict duties for medical professionals. The position confirmed by numerous judicial decisions is that conduct not complying with the current state of medical knowledge satisfies the definition of fault, irrespective of any subjective factors related to conduct or approach of a particular person or medical team.[38] The commonalities with the definition of 'medical incident' in the Act on patients' rights are clear. The categorical statement by scholars concluding that the new compensation scheme is therefore fault based appear, however, to be going too far. While both the general Civil Code-based system developed by courts and the new compensation scheme are grounded in the objectively determined duties of medical professionals, they detract from the strict idea of fault. Standards of conduct are of an objective nature. On the other hand, it is also very clear that liability in both systems can only be established if medical professionals did not fulfil their legal duties, the scope of which is determined by the notion of the current state of medical knowledge. Mere accidents or some unexpected side effects of medical procedures or medical products will not result in liability.

The Patients' Rights Ombudsman, in a leaflet for potential claimants, states that fault of doctors is not the subject of the Committees' decisions. The Constitutional Tribunal recently agreed with the General Prosecutor and the Speaker of the Sejm that liability is detached from any individual doctor or a group of doctors performing medical services. These views were presented during the proceedings before the Tribunal in a case brought by the General Medical Council testing the constitutionality of the new scheme. The case is analysed further below.[39]

However one looks at this issue, there is certainly need for a much clearer approach to liability: if the Act established no-fault liability, why refer to a medical incident as an event which took place against the current state of medical knowledge? Such an expression will always invoke an analysis of what exactly was done, and what ought to be done, by the

[37] The evidence with regard to the use of experts is only anecdotal for the moment. No data is available concerning the exact number of cases where experts were used by the Committees.

[38] Baginska (n 11) 105. See also: S Kalus, 'Liability for Medical Damage in Polish Law' in B Heiderdorff and G Zmij (eds) *Tort Law in Poland, Germany and Europe* (Munich, Sellier, 2007) 47; M Nesterowicz, E Baginska and A den Exter, *Medical Law in Poland* (Alphen aan den Rijn, Wolters Kluwer, 2011) 47.

[39] Decision of 11 March 2014, K 6/13, published on the website of the Constitutional Tribunal: http://trybunal.gov.pl/rozprawy/wyroki/art/6751-postepowanie-przed-wojewodzkimi-komisjami-do-spraw-orzekania-o-zdarzeniach-medycznych-popelnionyc/.

doctors, and may lead to serious evidentiary problems. If the aim is swift and efficient resolution of claims and provision of compensation, requiring such a thorough analysis of the incident which led to injury or death seems unnecessary. Perhaps a good option would be to eliminate the phrase 'contrary to the current state of medical knowledge' from the Act, thus focusing the definition of a medical incident on its detrimental consequences.

This brings the discussion to the very nature of the new scheme. It is not entirely clear for now whether it is a pure compensation scheme (closer to an insurance system), or perhaps a civil liability scheme grounded in civil law concepts of culpability and responsibility. The (official) sentiments around the Act on patients' rights appear to indicate that it should be the former, but the text of the Act and the academics interpreting it suggest a more complex position. Further, a preliminary analysis of the Committee decisions issued so far indicates that they rather follow the civil liability scheme path. The majority of decisions deny occurrence of a medical incident.[40] Often the decision is negative because it was not possible to demonstrate that something contrary to the current medical knowledge took place at the hospital in question.

E. Causation

The Act on patients' rights does not expressly mention causation, although the definition of medical incident focuses on injuries and deaths *caused by* practices contrary to current medical knowledge. It is not absolutely clear to what extent claimants need to prove causation. Normally, the burden of proving causation rests on the claimant. In Polish civil procedure causation does not need to be proven with absolute certainty, but rather only to a high degree of probability.[41]

It is possible that the Committees may further simplify proof of causation in order to facilitate enforcement of compensation claims.[42] On the other hand, if the scheme starts moving away from the current framework of 'medical incident' towards a no-fault approach, causation may well become a very prominent element in the analysis. A good example of how significant a filter to claims it can become is the UK Vaccine Damage Payments system, where the majority of claims are rejected because 'causation due to vaccination has not been accepted'.[43]

F. Membership and Term of Office

Membership of the Committees is carefully regulated in an attempt to balance the interests of patients and medical practitioners. Each Committee has 16 members, of which half

[40] BFP Report, 2015, fn 25. See below for the statistical data.

[41] Baginska (n 11) 106.

[42] ibid.

[43] Data provided in response to a request by Dr Duncan Fairgrieve of the British Institute of International and Comparative Law in 2010. Quoted by R Goldberg, *Medicinal Product Liability and Regulation* (Oxford, Hart Publishing, 2013) 11.

must have medical degrees and 5 years of professional experience in the medical profession, and the other half must have law degrees and 5 years of experience in the legal profession. Fourteen members are nominated by the Head of the Region (Wojewoda), of whom 4 are chosen from people nominated by lawyers' self-governing bodies, 4 from people nominated by doctors, dentists and nurses' self-governing bodies, and 6 from people nominated by social organisations concerned with patients' rights. Two additional Committee members are appointed by the Health Minister and the Patients' Rights Ombudsman, respectively. The term of appointment is 6 years.[44]

G. Time Limitations, Costs and Fees

The victim or the victim's heirs must bring a claim within 1 year from finding out about the injury or from the death of the patient, but not later than 3 years from the incident that caused it.[45] The claim should be brought before the Committee responsible for the region where the hospital in which the injury or death occurred is located.

The fee for lodging the claim is 200 PLN (about €50).[46] This fee becomes part of the total fee for the case, which is covered by the claimant only if he or she loses the case (see below). Other costs of the proceedings include reimbursement of travel, accommodation costs and reimbursement of lost earnings for persons requested to appear at Committee meetings, and the cost of an expert opinion if one is required. The Regulation of the Minister of Health of 23 December 2011 set out the tariffs for these other costs.[47] Persons requested to appear before a committee receive 83.58 PLN for each 100 km travelled, and 34.50 PLN to cover overnight accommodation. They also receive 153.57 PLN for each day of lost income. Experts receive 300 PLN for their opinions. This amount increases if the expert has an academic title: by 150 PLN for a professor, 100 PLN for a habilitated doctor and 60 PLN for a doctor (PhD).

The costs are assessed by the final decision of the Committee. They are to be covered by the claimant if he or she loses the case, by the hospital if the claimant wins, and by insurer if the latter refuses to make a formal offer of compensation after receiving the final decision of the Committee.[48]

H. Procedure

After receiving the claim, which states the amount of compensation the claimant wishes to obtain and provides evidence of a 'medical incident', the Chair of the Committee reviews

[44] Art 67e of the Act.

[45] Art 67c.2.

[46] Art 67d.3.

[47] Rozporzadzenie Ministra Zdrowia w sprawie zryczaltowanej wysokosci kosztow w postepowaniu przed wojewodzka komisja do orzekania o zdarzeniach medycznych [Regulation of the Minister of Health on the tariff system for costs of proceedings before a regional committee for adjudicating on medical incidents], published in *Dziennik Ustaw* of 2011, No 294, item 1740.

[48] Art 67l.3.

it for compliance with formal requirements and evidence of payment of the 200 PLN fee (claims with formal deficiencies are returned). The reviewed and accepted claims are sent to the hospital and its insurer, both of whom must respond within 30 days. If they fail to respond within that time it is considered that the hospital and insurer accept the claim and the amount.[49]

The Committee considers the claim in smaller chambers of 4 members (half of them lawyers, the other half medical professionals). The meetings of the Committee can be attended by the claimant and by the representatives of the hospital and its insurer.[50] In many respects the Committee is expected to follow the same rules as an ordinary court, especially with regard to types of admissible evidence, presumptions, service of documents, or suspension of proceedings.[51]

The Committee should complete the proceedings within 4 months from receiving the complaint. This is not always possible and many Committees are reporting delays[52] caused by the following factors:

- complex cases requiring more than the usual number of sessions;
- the need for an expert opinion;
- one of the parties, or a witness, does not attend a session;
- resignation of a Committee member;
- the need to ask the hospital for medical documentation and delays in receiving a reply.[53]

As mentioned above, the Committee makes decisions based on evidence presented by the parties as well as, when necessary, an expert opinion. Experts must be doctors who are members of the Physicians' Commission (a consultative body established by the Patients' Rights Ombudsman).[54] The complexity of the notion 'medical incident' often requires an expert opinion. The Polish civil justice system suffers from shortages of experts, and long waiting times for their opinions are one of the major causes of judicial delays. It has already been confirmed that having to obtain an expert opinion is one of the key reasons why the Committees take more than 4 months to complete a case.[55]

I. Decisions, Appeals and Compensation

In its final decision, the Committee declares whether an event that caused the claimant's injury or death was, or was not, a medical incident. The amount of compensation is not

[49] Art 67d.6.

[50] Doctors are not entitled to be present at the sessions of the Committee: this was one of the key complaints in the case before the Constitutional Tribunal brought by the National Medical Council—case K 6/13. Analysed below.

[51] Art 67o of the Act.

[52] According to the BFP Report 2015, the average time for completion of proceedings is 5 months, although in only 52% of cases were proceedings completed within the recommended 4 months.

[53] Rzecznik Praw Pacjenta [Patients Rights Ombudsman] KB Kozlowska, *Sprawozdanie dotyczace przestrzegania praw pacjenta na terytorium Rzeczypospolitej Polskiej* (1 January 2012–31 December 2012), available on the website of the Ombudsman: www.bpp.gov.pl/gfx/bpp/userfiles/_public/bip/sprawozdania_roczne/sprawozdanie2012.pdf. Here referred to as the Patients' Rights Ombudsman 2012 Report.

[54] Baginska (n 11) 107. Arts 32.2 and 67i.7 of the Act on patients' rights.

[55] Patients' Rights Ombudsman 2012 Report, 133.

determined at this stage. Upon receiving the decision and its written justification, the claimant, the hospital or its insurer have 14 days to lodge a request that the claim be considered again by different Committee members. There is also a possibility to make a formal complaint that the Committee breached its rules of procedure in considering the claim. This complaint can be made to the same Committee, within 30 days from the final decision. It is considered by 6 members of the Committee. No other redress or appeal procedure was envisaged for Committee decisions.

The Act specifies that the Committees do not exercise public authority.[56] Consequently, the participants in the proceedings have no possibility of using state liability remedies against their decisions and other activities. For instance, the participants have no power to claim compensation for unlawful conduct by the Committees (based on Article 77(1) of the Constitution). Further, Article 77(2) of the Constitution, providing access to a court in order to obtain a remedy for infringement of rights or freedoms by public authorities, also does not apply. This evasion of the key constitutional provisions led some academics to conclude that this particular provision of the Act is contrary to the Constitution.[57]

Within 30 days of receiving the final decision of the Committee, the insurer of the medical establishment must propose an amount of compensation to the claimant, which the latter has 7 days to accept. As explained above, if the hospital has no insurance policy against medical incidents, it must offer compensation.

The compensation offered cannot exceed the maximum amount established by the Act on patients' rights: 100,000 PLN in cases of contamination or injury, and 300,000 PLN in cases of death. There is also an annual limit for any given insurance policy of 1,200,000 PLN per patient. The Minister of Health was required by the Act to set up tariffs for compensation payable for different types of contamination and injuries, and for death. The ministerial regulation to this effect was adopted on 23 February 2012.[58] The regulation sets out detailed criteria for establishing levels of compensation for pecuniary and non-pecuniary losses in cases of contamination or injury, depending on the type of injury and the level of incapacity. The following elements are to be part of the final calculation of the award: the level of health deterioration or injury caused by the incident (60%, ie 60,000 PLN maximum), deterioration in the quality of life (including the need for care by third persons (10%), and loss of ability to work or study (5%)), as well as the inconvenience of treatment and other non-pecuniary losses caused by the incident (25%).[59] There are further, very specific factors to be taken into account when assessing the exact amount to be paid in each case. The Regulation lists a large number of examples of inconveniences of treatment and other non-pecuniary losses (eg isolation in a hospital for at least 7 days, an additional surgery, or hospital stay exceeding 30 days) and attributes a percentage value to each. In general, the maximum amount of 100,000 PLN seems too low to cover all the possible costs and non-pecuniary losses that can result from an injury or contamination.[60]

[56] Art 67e.2 of the Act. Baginska (n 11) 213 does not agree.

[57] Baginska (n 11) 213.

[58] Rozporzadzenie Ministra Zdrowia [Regulation of the Minister of Health], 23 February 2012, published in *Dziennik Ustaw* of 2012, item 207.

[59] Para 2 of the Regulation.

[60] Baginska (n 11) 107, and M Nesterowicz and M Walachowska, 'Liability for Damage caused by Medical Treatment in Connection with the New Extra-Judicial Compensation System in Poland' in E Kowalewski (ed), *Kompensacja szkod wyniklych ze zdarzen medycznych. Problematyka cywilnoprawna i ubezpieczeniowa* (Torun, 2011) 24.

Unfortunately, no details are provided as regards pecuniary and non-pecuniary damage in cases of death of a family member. The Regulation sets the upper limit for pecuniary damage as 200,000 PLN, and for non-pecuniary damage as 100,000 PLN. This amount, on the other hand, may appear too high.[61]

The Act and the Regulation do not introduce minimum amounts of compensation that can be offered by the insurer or hospital. The Committees have no power to influence the compensation offers by insurers and hospitals. This proves to be a problem in many cases. Insurers are reported to be offering very low amounts. Sometimes the offers are 10 times lower than what the victims or their families requested. Representatives of the Committees across Poland are reporting cases where such compensation offers are obviously too low and the claimants reject them in order to bring the claim before an ordinary court instead. In the region of Lodz it was reported that an insurer offered 5,000 PLN for the death of a patient.[62] The Patients' Rights Ombudsman reported on a number of cases where insurers who do not agree with a Committee decision have offered a symbolic 1 PLN.[63] While offering an amount that is lower than what the claimant requested is not necessarily a problem, there is scope for some form of regulatory or judicial oversight of compensation offers. Any possible systemic abuses can thus be avoided. It is not clear why the Committees themselves were not given powers of oversight over the compensation offers.

Acceptance of a compensation offer must always be accompanied by a written statement that no further claims will be made with respect of injuries or death covered by the decision, in courts or otherwise.[64] However, nothing prevents the claimant from bringing a claim in court if he or she does not accept the compensation offer.

The Act expressly states that the offer of compensation by the insurer or hospital is not deemed to be an admission of liability in any further civil lawsuit.[65]

The compensation offer that was accepted becomes an execution title without having to be endorsed by a court.[66] This means that the patient can take enforcement action without additional formalities on the basis of it, as if it were a court's judgment. If the insurer or hospital refuses to make an offer within 30 days, the Committee issues a document specifying the amount requested by the claimant as the amount to be recovered. This document becomes an execution title.[67]

J. Insurance—The Original Core Element and One of the Main Problems of the Scheme

As mentioned above, initially the scheme was designed around a single core element: all entities managing hospitals were to be obliged to take out an insurance policy to cover

[61] ibid.

[62] Data from M Sygut, 'Komisje orzekaja o zdarzeniach medycznych—pacjenci nie przyjmuja odszkodowan' ['The Committees Confirm a Medical Incident—The Patients Reject Compensation Offers'], 16 July 2013, available at: www.rynekzdrowia.pl/Prawo/Komisje-orzekaja-o-zdarzeniach-medycznych-pacjenci-nie-przyjmuja-odszkodowan,132306,2.html.

[63] 2015 Report, www.bpp.gov.pl/gfx/bpp/userfiles/_public/bip/sprawozdania_roczne/spr_za_2015.pdf, 51.

[64] Art 67k.6 of the Act on patients' rights.

[65] Art 67k.9.

[66] Art 67k.8.

[67] Art 67k.4.

liability for medical incidents. Very quickly it transpired that the costs of obtaining coverage for a hospital were very high and there were not many insurance companies offering it. In fact, insurers started denying liability coverage for medical incidents, and others significantly increased premiums.[68] It is difficult to assess why these problems arose. After all, it should be relatively straightforward to calculate liability exposure under this scheme: the Act on patients' rights introduced maximum thresholds of compensation, annually and per incident. Due to difficulties with obtaining insurance coverage, the deadline for all hospitals was initially moved a number of times, the latest to 1 January 2017.[69] As a result of the recent amendment of the law, the obligation to take out the insurance policy against medical incidents disappeared from the law governing the scheme on 1 January 2017, making it entirely voluntary.

Overall, the obligations to compensate under the scheme fall on the hospital if the total annual limit has been exceeded or if there is no insurance coverage. There is no secondary insurance (reinsurance) body specified for the scheme. Considering that the compensation that can be offered under the scheme is never very high, the potential liability risk for hospitals is not significant. This is especially the case because, without statutory lower limits to compensation, the hospitals may offer amounts that are much lower than what would be obtained during a court case. On the other hand, however, were such minimum limits to be established in the future (which the author of this chapter recommends) and the Committees to gain in popularity, the insurance problem will resurface again. Scholars have suggested that a guarantee fund similar to the traffic liability insurance fund ought to be set up.[70]

K. No Participation of Doctors in the Procedure

The scheme does not allow for the participation of a doctor or another medical professional whose conduct is under review in the proceedings before the Committees (unless they are called as witnesses). Further, doctors cannot bring either of the possible two recourse actions against the Committee decisions mentioned above. These issues were the subject of a complaint by the General Medical Council to the Constitutional Tribunal against the Act. The unconstitutionality was, in the view of the Council, manifest in a breach of the con-

[68] A Sekura, 'Nowe zasady ustalania odszkodowania za zdarzenia medyczne' ['New Principles for Awarding Compensation for Medical Accidents'], February 2012, available online at http://biznes.onet.pl/nowe-zasady-ustalania-odszkodowania-za-zdarzenia-m,18651,5043112,analizy-detal (accessed 24 October 2012).

[69] Art 11 of the Act of 14 June 2012 o zmianie ustawy o działalności leczniczej oraz niektórych innych ustaw [On the amendment of the act on medical activity and some other acts], published in *Dziennik Ustaw* of 2012, item 742, amended by the Act of 11 October 2013 o zmianie ustawy o świadczeniach opieki zdrowotnej finansowanych ze środków publicznych oraz niektórych innych ustaw [On the amendment of the act on health benefits financed by public means and of some other acts], published in *Dziennik Ustaw* of 2013, item 1290.

[70] Baginska (n 11) 112, and E Kowalewski, 'Patient Insurance against Medical Incidents—Lights and Shadows' in E Kowalewski (ed), *Kompensacja szkod wyniklych ze zdarzen medycznych. Problematyka cywilnoprawna i ubezpieczeniowa* (Torun, 2011) 93, 109.

stitutional provisions concerning the right of access to justice (Article 45) and the right to bring an appeal action or another recourse action in proceedings pertaining to one's rights and obligations (Article 78). The Tribunal agreed with the arguments presented by the Speaker of the Parliament and the General Prosecutor that liability for medical incidents is not linked to liability of a specific doctor or medical professional.[71] It is rather liability of an insurer or a hospital. As doctors are not directly legally affected by the action before a Committee or its decision, the Tribunal did not see an infringement of the right of access to justice or the right to recourse against a decision.

L. Statistics

The Report of the Patients' Rights Ombudsman for 2015 notes that between 1 January and 31 December 2012, 432 claims were brought before the Committees across Poland, 28 of which were requests to reconsider a previous decision.[72] The number of claims brought before the Committees was initially increasing: in only the first 4 months of 2013, 448 claims were brought.[73] For the next 3 years, however, the numbers were as follows: 2013, 1,310 claims; 2014, 1,197 claims; 2015, 1,030 claims. While it is still too early to make any firm assessment of the work of the Committees, it is clear so far that the number of claims has been steadily decreasing.

As mentioned above, it has been reported that the majority of the Committee decisions deny the occurrence of a medical incident. For example, in the Wrocław region, between January 2012 and July 2013, 87 claims were brought: 28 decisions were issued, of which only 8 confirmed a medical incident. Interestingly, at least 4 compensations offers made following these decisions were rejected by the claimants.[74] In the Western Pomerania region (Zachodniopomorskie), 64 claims were brought: 8 decisions confirmed a medical incident.[75] In the Łódź region, 69 claims were brought: out of 46 decisions, 19 confirmed a medical incident.[76] The Patients' Rights Ombudsman reported that in 2015 in only 33% of cases did the Committees confirm that a medical incident took place.

The Committees are aware of the fact that many claimants reject compensation offers, although they have not always been able to check whether these claimants decided to pursue their cases in court.

[71] Decision of 11 March 2014, K 6/13.

[72] The Report is available on the Ombudsman's website: www.bpp.gov.pl/gfx/bpp/userfiles/_public/bip/sprawozdania_roczne/sprawozdanie_rpp_2015.pdf.

[73] ibid, 134. A significant number (110) were returned because of various formal deficiencies (the incident did not take place in a hospital, it took place before the Act came into force, the compensation amount was not indicated, or the claim fee was not paid). The Patients' Rights Ombudsman 2012 Report, 132, 133.

[74] Sygut (n 62).

[75] ibid.

[76] ibid.

M. The Role of the Patients' Rights Ombudsman

The Patients' Rights Ombudsman was established by the 2008 Act on patients' rights and the Patients' Rights Ombudsman.[77] The Ombudsman is a public body safeguarding rights of patients. These rights are listed in the Act on patients' rights and include, among others, the right to healthcare services complying with the state of current medical knowledge.[78] The Ombudsman's powers range from administrative to civil. They include bringing patients' complaints to the hospital or an institution in a supervisory function over the hospital, making recommendations to those institutions, and requesting that disciplinary action be taken. The institutions are not obliged to follow the Ombudsman's recommendations, but his/her position is further strengthened by the power to take cases to court and join ongoing litigation. These powers extend over mass rights violations as well as individual rights violations. There are no compensatory powers. The Ombudsman also cannot bring cases before the Committees examined in this chapter. Further, the Ombudsman has no power to appeal or in other ways revise the Committees' decisions or the insurers' and hospitals' compensation offers. On the other hand, if the Ombudsman were to consider that one of the patients' rights was infringed upon during the proceedings or in the compensation offer, there is a potential for action, either in administrative proceedings or in civil action before court. It is, for instance, possible to envisage an action against a hospital or an insurer who persistently offer very low compensation. No such action was taken so far.

N. Evaluation

As mentioned above, the model is similar to the French ONIAM model. The Polish scheme omits the French requirement that the insurer's offer must be reasonable in amount, otherwise significant consequences will follow. This has already shown to be a significant problem with the scheme. The Patients' Rights Ombudsman currently considers the Committees ineffective in providing redress to victims of medical incidents.[79] According to the Ombudsman's assessment, the Committees are not a viable alternative to judicial proceedings. The main concern is that they do not control the compensation offers made by the insurers and hospitals. The Ombudsman reported that in many cases the offer was a mere 1 PLN. This happens, according to the Ombudsman, where the insurer or hospital do not agree with the Committee's assessment, and are expecting the claimant to take his or her case to court.[80] While no data have been collected so far, there is anecdotal evidence that because of low compensation offers claimants are forced to reject them and to take the judicial route in order to obtain what they see as adequate redress. This calls into question the very existence of the Committees, and this situation should be reconsidered.

[77] See n 1.
[78] See n 24 for the complete list of rights.
[79] See also Gmurzyńska (n 17) 5.
[80] 2015 Report, 51.

The aim of lightening the load of overburdened courts and providing quick and efficient compensation has not yet been achieved. The scheme in its present form has significant problems and weaknesses that need to be revisited. One of the most significant issues, apart from lack of control over compensation offers, is the confusion as to the standard of liability for this new scheme: is it no-fault liability as some presume, or is it no different from the ordinary fault-based liability, albeit modified by courts? As discussed in this chapter, the very essence of this debate is the following question: is the system an insurance-like compensation system, or is it grounded in civil liability notions of responsibility and culpability? When observing the approach of the Committees themselves, as well as noting the problems with obtaining insurance coverage for this new scheme, one may draw the following conclusion: Poland may not be ready for a true no-fault compensation system for medical injuries and death.

The Committees have not been in operation for long, and while they are rapidly gaining in popularity (they are talked about in the media and analysed by scholars), the number of cases brought before them is gradually decreasing. Because ordinary civil procedure in Poland can be a long and costly process, the Committees have the potential to be a very attractive mechanism for obtaining redress. The level of compensation is limited by law and not very high, but it should be high enough to satisfy many ordinary medical claims. While the future may be bleak for the Committees in their present shape, I remain optimistic that after a number of well-thought reforms the system is capable of performing its desired function.

20

Ireland: The Injuries Board

CHRISTOPHER HODGES

The Injuries Board ('the Board'), previously known as the Personal Injuries Assessment Board,[1] is an independent statutory body, established in 2004 with the following aims:

— to assess how much compensation is due to an injured party;
— to reduce costs and fees involved in the administration of personal injury claims; and
— to reduce the amount of time it takes to finalise a claim for compensation.

The Board covers civil actions for personal injuries arising from:

(a) a civil action by an employee against his or her employer for negligence or breach of duty arising in the course of the employee's employment with that employer,
(b) a civil action by a person against another arising out of that other's ownership, driving or use of a mechanically propelled vehicle,
(c) a civil action by a person against another arising out of that other's use or occupation of land or any structure or building,
(d) A civil action not falling within any of the preceding paragraphs (other than one arising out of the provision of any health service to a person, the carrying out of a medical or surgical procedure in relation to a person or the provision of any medical advice or treatment to a person).[2]

It was originally intended that medical negligence cases should be included, but indemnity issues over state and private insurers were not resolved in time for that category of claims to be included in the legislation. There has been ongoing debate over whether to include medical negligence. One consideration is that medical cases typically give rise to complex issues of causation, allocation of fault, the impact on defendants' reputations and catastrophic injuries. However, such issues would not appear to justify the exclusion of all medical negligence claims from assessment by the Board, which is solely involved in the determination of the compensation due to a patient/claimant. The issues of causation, fault and liability are determined by the parties outside of the Board's process and before the respondent (the person(s) against whom the claim is made) decides if he or she wants the compensation payable assessed by the Board.

The Board assesses the amount of damages a claimant is entitled to in respect of a claim on the assumption that a respondent is fully liable.[3] It awards the same level of compensation

[1] Personal Injuries Assessment Board (PIAB) Act 2003 ('PIAB Act').

[2] PIAB Act, s 3.

[3] PIAB Act, s 19. There is no legal aid for personal injury cases in Ireland (only criminal cases), so people have to fund claims themselves. There are no 'claims farmers' or claims management companies, and advertising by lawyers is prohibited, although would be permitted under the 2015 Legal Services Bill.

as the courts[4] but within a faster time frame and without associated litigation costs. As required under its founding legislation, the Board published a Book of Quantum, which set out for the first time in comprehensive form general guidelines on awards for specified types of injury.[5] This function of comprehensive collation of damages awards assists courts and parties in taking a consistent approach to awards. The courts are required under the Civil Liability and Courts Act 2004 to have regard to the Book of Quantum. The position is different from the UK where the *Guidelines for the Assessment of General Damages in Personal Injury Cases* are produced by the judiciary. The amounts listed in the Book of Quantum are, however, not always endorsed in the courts, and it would be preferable if the judiciary were involved in some form in approving its contents.[6]

Anyone seeking compensation for a personal injury must first apply to the Board: there is a statutory bar on bringing proceedings unless such an application has been made and the proceedings are authorised.[7] A claimant fills out a claim application form online or sends it to the Board by post with a medical report from their treating doctor and a €45 processing fee.[8] The report from a treating doctor, for which a downloadable template is available,[9] must give brief details of the accident, injuries sustained (including diagnostic information), relevant medical history, clinical description of illness or disablement, anticipated treatment required, and opinion on prognosis. The Board notifies the respondent of the claim with a Formal Notice including a copy of the application form and medical report, thereby ensuring full transparency in relation to the claim.[10] The respondent or its insurance company have 90 days from the date of the Formal Notice to confirm in writing to the Board whether or not it consents to the Board assessing the claim. The parties might settle a case at that stage, or the respondent/insurer can pay the fixed fee for the next assessment stage.

If the respondent consents, it pays a fee.[11] The assessors at the Board will then make an assessment of damages, if necessary with the assistance of external retained experts,[12] on the basis of the records or documents available, but without an oral hearing.[13] An assessor may request such additional information or documents as he or she considers necessary, or request the claimant to attend an independent medical examination.[14] Where a claimant fails to provide information or documents, or submit to an examination, the assessors

[4] PIAB, s 20(4).

[5] PIAB Act, s 54(b).

[6] Judicial approval is intrinsic in England and Wales, under the *Guidelines for the Assessment of General Damages in Personal Injury Cases* (Judicial College, 2015). In 2015 the upper jurisdictional limit of the Circuit Court was raised from €38 to €60, after which far fewer Injuries Board cases came under the jurisdiction of the High Court, and also the introduction of a Court of Appeal created a new tier in the judicial system. After these changes, a number of new judges were appointed, and there was some inconsistency in subsequent judicial damages awards.

[7] PIAB Act, s 12.

[8] PIAB Act, ss 11 and 22 in relation to the Minister's power to specify charges.

[9] www.injuriesboard.ie/eng/Forms-Guidelines/Resources-for-Doctors/.

[10] PIAB Act, s 13.

[11] The fee has reduced from an initial €1,050 to €850 from 1 February 2011, and €600 from 2013. These changes do not appear to have sparked significant increases in claim numbers, since the trend has generally been upward since inception.

[12] PIAB Act, s 20.

[13] PIAB Act, s 21.

[14] PIAB Act, s 24.

shall in general proceed to make the assessment 'as best they may'.[15] An assessor may also request information from any person or public body, and such requestee is under a duty to comply.[16]

If the respondent does not consent to an Injuries Board assessment, it must advise the Board in writing within 90 days. The Board will then issue the claimant with an Authorisation which allows him or her pursue the claim through the court system if so wished. If the Board does not receive a reply in writing within the 90 days, the respondent will be deemed by default to have consented to the assessment of the claim and become liable for payment of the statutory fee.

Following a 2008 Supreme Court case, the Board now sends to represented claimants copies of key communications, to ensure full transparency for all parties.[17] The Board assesses damages on the same basis and by reference to the same principles as damages measured by the courts, based on the Book of Quantum, a general guide as to the amounts that may be awarded or assessed in respect of specified types of injury. Damages comprise general damages (pain and suffering) and special damages (loss of earnings, medical expenses, out-of-pocket expenses, vehicle damage costs, and in serious cases future loss of earnings, future expenses, etc). The assessment is based on the medical report(s) submitted by the claimant's treating doctor and the report(s) of any independent medical examiner(s) that it may seek, the cost of which is borne by the respondent.

The Board issues its Notice of Assessment to both claimant and respondent.[18] Where both parties accept the assessment, it is binding on both parties[19] and the Board issues an Order to Pay, which has the same effect in relation to enforcement as a court order.[20] If one of the parties does not accept the assessment, the Board issues an Authorisation which allows the claimant to pursue the claim through the courts.[21]

The Board's website contains an 'Estimator' for people to estimate the quantum of their claims,[22] and a downloadable smartphone app is also now available. It also gives a number of illustrative case studies, covering motor accidents, employer liability and public liability claims.

The Board is required to complete its assessment of damages 'as expeditiously as may be', and within 9 months of the respondent consenting to the claim, with some permitted exceptions.[23] The average time from consent to award has consistently averaged less than 7 months, with 93% completed within 9 months. This is compared with an average 36 months under the previous litigation system.

The Board has not witnessed a spate of whiplash claims, as has occurred in England. The process of certification by the treating doctor provides a filter, and the ability to refer cases to a panel of independent doctors, plus giving all reports to both sides, is thought to be effective against most inflated claims. Panel doctors are encouraged by the Board to give

[15] PIAB Act, s 25.

[16] PIAB Act, s 26.

[17] *O'Brien*, 19 December 2008.

[18] PIAB Act, s 30.

[19] PIAB Act, s 33.

[20] PIAB Act, ss 38–43. Note that periodic payments are not permitted in Ireland, despite lengthy debate over their introduction.

[21] PIAB Act, s 32.

[22] www.injuriesboard.ie/eng/How-to-make-a-claim/Estimate-your-claim/.

[23] PIAB Act, s 49.

a clear view on prognosis and not 'sit on the fence'. They are indemnified by the Board. If an insurer suspects fraud, such as a staged motor crash, it will not give consent for a Board assessment, so that the case, if the claimant wishes to proceed further, must go to court.

The Board has a discretionary power to include in an award an amount in respect of fees or expenses reasonably and necessarily incurred by a claimant in complying with the provisions of the Act.[24] However, a number of years ago, the Board noted an increasing number of cases where assessments were accepted by respondents but seemingly unnecessarily declined by claimants, only to be later concluded at the same compensation figure after initiation of legal proceedings but with additional avoidable legal costs being incurred. As a result, the Personal Injuries Assessment Board (Amendment) Act 2007 provided that where a claimant does not receive more than the amount of the original Board award in subsequent proceedings, he/she will not recover his/her legal costs and risks being held responsible for costs incurred by the respondent/defendant. The Amendment Act also provides that legal costs incurred in dealing with the Board process are not recoverable in the litigation process. It is reported that only about 2% of all personal injury claims are decided by a court.

A 2010 judgment removes the possibility of legal fees having to be paid in respect of all claims, an occurrence that would have increased insurance premiums in Ireland.[25] The judgment clarified that claimants who opt to use a solicitor must pay their own legal costs from the injury compensation awarded by the Board, unless the Board discerns a need for legal assistance/costs on specific cases.

Over 90% of claims received by the Board are from solicitors. 'Claims farmer' intermediaries have been established far less in Ireland than in England. Although individuals can 'do it themselves', it appears to be the general perception in Ireland that there is no real advantage to not using a lawyer other than the claimant having to fund the legal costs out of their final settlement.

All insurance companies have established sections that now aim to investigate and settle cases themselves before the end of the initial 90-day period that they have to elect whether to agree to a Board assessment. The insurance industry is therefore constantly engaged in settling cases swiftly where this can be done.

Prior to the establishment of the Board, between 30,000 and 35,000 personal injury proceedings were initiated in the courts each year. The number of cases involving such litigation has reduced to approximately 15,000 per annum.[26] This includes cases outside the remit of the Board (notably, medical negligence cases are currently excluded) and cases released by the Board where consent is not received to assess the claim (eg where liability is disputed).

The Board stated in 2012 that it 'has proven to be robust yet capable of adjustment to ensure policy goals are met' and '[b]y year end 2011, the Board had assessed over 50,000 claims and made awards of over €1 billion, resulting in direct savings of approximately €250 million with similar indirect savings from early settlements facilitated by the new model with its strict timelines and transparent administrative procedures'.[27]

[24] PIAB Act, s 44.

[25] *Plewa* and *Giniewicz* cases, October 2010.

[26] Annual Report 2013.

[27] *Personal Injuries Assessment Board: Strategic Plan 2012–2016* (July 2012).

Claims to the Board reached a full flow of activity in 2007. The Board currently receives about 30,000 applications (see Table 20.1) and made 12,420 assessments in 2014. The balance of the claims were either outside the remit of the Board, settled or contested by respondents/insurers. A 2011 inspection of personal injury settlements by insurance companies found that around 70% of cases are either settled directly between the parties or through the formal awards made by the Board before the commencement of legal proceedings.[28]

The acceptance rate of awards has been reasonably consistent, with 6,476 of 10,656 awards accepted in 2013, resulting in €142.57 million of compensation being agreed, and 7,519 of 12,420 awards accepted in 2014, resulting in €165.97 million of compensation being agreed. Amongst the reasons why awards are not accepted are that the parties may for their own reasons consider the award too high or low or questions arise as to the veracity of the claimant: such claims would be appropriate to progress for resolution by a court. Even so, most claims clearly settle, because court statistics show that cases do not appear in court.

Table 20.1: Personal injury applications to the Injuries Board 2014, awards and accepted awards 2006–14

Year	Motor liability	Public liability	Employer's liability	Total applications	Total awards	Total accepted awards
2006	11,590	5,554	3,782	20,926	5,573	
2007	13,060	6,050	4,235	23,345	8,208	
2008	14,012	6,320	4,390	24,722	8,845	5,670
2009	15,079	6,721	4,119	25,919	8,643	5,387
2010	15,971	7,251	3,742	26,964	8,380	5,038
2011	16,351	7,452	3,866	27,669	9,833	5,875
2012	17,443	8,332	3,828	29,603	10,136	6,124
2013	18,877	8,394	4,040	31,311	10,656	6,476
2014	18,994	8,214	4,368	31,576	12,420	7,519

The breakdown by claim category (see Table 20.2) has been consistent, with approximately 70% of awards relating to motor liability (75% in 2013 and 2014), and the lowest claims volumes being employer's liability (8% in 2013 and 2014). The value of the vast majority of cases was under €38,000, which is consistent with cases heard in the courts, where the vast majority are in the Circuit Court, where the jurisdictional limit is €38,000. In 2013, 61% of all awards were under €20,000 (59% in 2014) and 90% (same again in 2014) under €38,000. The highest award in 2013 was €1,390,043 (2014: €972,898) and the lowest €80 (2014: €70).

[28] Inspection by the Central Bank of Ireland, see Injuries Board Annual Report 2013, 16.

Table 20.2: Breakdown of total number of awards in 2014 (by value and claim category)

Average value category (€ 000s)	Motor liability	Employer's liability	Public liability	Total
<20	5,930	351	1,073	7,354
20–38	2,679	363	730	3,772
38–100	664	208	338	1,210
>100	55	19	10	84
All	9,328	941	2,151	12,420

The overhead costs associated with delivering compensation (€10.86 million in 2011; €11.17 million in 2014) have been reduced from 46% to less than 7% in the case of formal awards. Savings delivered by the Board feed directly into the cost of insurance.

The Board is self-financing from application fees, currently €600 per case. Initial funding of €6.9 million was provided by the state, and repaid by the end of 2011, since when the Board has been self-funding from fees charged. The Board employed 68 staff in 2014. Delivery costs have represented between 7% and 9% of claim values. This compares with 46% under the previous court system.

Functionally, therefore, the system has provided a non-adversarial administrative process to deliver compensation to claimants largely without the need for legal proceedings. The intention is to assist swifter settlement of claims, thereby saving court time, and enabling swifter payment of compensation to victims. Furthermore, the assessment of damages, including much of the investigation element and the entire quantification element, has been moved from the courts to the Board. This has resulted in savings in litigation costs of approximately 50%.[29] Indirect benefits realised include removal of worry from claimants about their claim, swifter rehabilitation and return to work, so benefiting claimants, employers and the economy, and reduction in the burden for doctors of writing reports for litigation, repeat examinations and court attendance. The process is administrative and non-contentious. It assists in the settlement of 70% of all personal injury without the need for legal proceedings.

[29] Annual Report 2013.

21

Ireland: Health Services and Redress for Women Formerly Resident in Magdalen Laundries

RAYMOND BYRNE[1]

A. The Magdalen Laundries: Revealing the Story

There is a long history in a number of countries of establishing institutions to house women and young girls primarily where they had become pregnant outside marriage or were otherwise considered to be at risk. After the establishment of the Irish state in 1922 until the last 'Magdalen laundry' closed in 1996, 10,012 women were admitted to 10 such institutions in Ireland. Separately, thousands of young children, mostly boys, who were deemed to be 'delinquent' were also sent to residential 'industrial schools' between the 1930s and 1970s. By the early 1990s, it began to emerge that conditions in the Magdalen laundries and in the industrial schools had included forced labour as well as physical, emotional and sexual abuse.

After a mass grave was discovered in Dublin in 1993 on the site of a former convent that had also housed a Magdalen laundry, there was extensive public concern.[2] Revelations were made in the 1999 groundbreaking TV documentary series *States of Fear*, produced for Radio Telefis Éireann (RTÉ, the Irish public service broadcaster) by Mary Raftery, which first identified forced labour as hallmarks of the Magdalen laundry life and the industrial schools. This was followed by the publication of Raftery and O'Sullivan's book *Suffer the Little Children*,[3] which also documented physical and sexual abuse of boys in the industrial school system. Peter Mullen's 2002 film *The Magdalene Sisters* also gained wide publicity.

As a direct result of the revelations in *States of Fear*, in May 1999 the then Taoiseach (Irish Prime Minister) made a public statement which included 'a sincere and long overdue apology to the victims of childhood abuse for our collective failure to intervene, to detect

[1] Some portions of this discussion, in particular concerning the Redress for Women Resident in Certain Institutions Act 2015, have previously appeared in R Byrne and W Binchy, *Annual Review of Irish Law 2015* (Dublin, Round Hall Thomson Reuters, 2016) 449–53.

[2] See RL McCarthy, *Origins of the Magdalene Laundries: An Analytical History* (Jefferson, NC, McFarland, 2010).

[3] M Raftery and E O'Sullivan, *Suffer the Little Children: The Inside Story of Ireland's Industrial Schools* (Dublin, New Island, 1999).

their pain, to come to their rescue'. The statement also indicated that a range of policy and legislative initiatives would be put in train to address the historical abuse and to prevent any future occurrence. In terms of the historical abuse, this included the establishment of a statutory commission to determine in detail the level of abuse that had occurred in the industrial schools;[4] the 2009 *Report of the Commission to Inquire into Child Abuse* (CICA) concluded that in the network of state-administered and church-run industrial schools there had been widespread physical, emotional and sexual abuse of children between the 1930s and the 1970s.[5]

The 1999 statement also included the establishment of a statutory Residential Institutions Redress Board (RIRB),[6] which provided for a redress scheme for residents of the industrial schools, a small percentage of whom had also been in the Magdalen laundries. Although this redress scheme did not require applicants to establish negligence or other fault, the scheme operated in practice in an adversarial manner and applicants were almost invariably represented by solicitor and counsel. Nonetheless, between 2003 and 2014 the RIRB had dealt with over 16,000 applications for redress and had made over 15,000 awards totalling €967.6 million, ie an average award of €62,245.[7]

B. The McAleese Committee Report on the Magdalen Laundries

Some of the applicants who had made submissions to the CICA and who had received awards from the RIRB had also worked in Magdalen laundries, but neither the CICA nor the RIRB had dealt with the Magdalen institutions. Indeed, when the revelations concerning the Magdalen laundries emerged initially in the 1990s, it was suggested that the state had no involvement in the referral of the women to the institutions and this view was maintained for many years. Representative groups of the Magdalen women disputed this view

[4] Established under the Commission to Inquire into Child Abuse Act 2000.

[5] *Report of the Commission to Inquire into Child Abuse* (2009) (chaired by Mr Justice Seán Ryan, and often referred to as 'the Ryan Report'), available at www.childabusecommission.ie.

[6] Established under the Residential Institutions Redress Board Act 2002.

[7] In its annual report for 2014, by which time the RIRB had virtually completed its work, it reported that between 2003 and 2014 it had dealt with 16,617 cases, of which: 11,988 offers/awards were made following settlement; 3,000 awards were made following hearings (of which 16 awards were rejected by applicants); 557 awards were made following review by an independent review committee chaired by a former Supreme Court Judge; and 1,072 applications were withdrawn, refused or resulted in a nil or no award. The total awards made by the end of 2014 amounted to €967.6 million, of which the average value of awards was €62,245, the largest award being €300,500. Of the total number of awards: 7,506 awards (48.29%) were between €50,000 and €100,000; 5,641 awards (36.29%) were under €50,000; 2,070 awards (13.32%) were between €100,000 and €150,000; 280 awards (1.8%) were between €150,000 and €200,000; and 48 awards (0.31%) were between €200,000 and €300,000. The RIRB reported that 97.7% of applicants were represented by solicitors (over 1,000 solicitors' firms were involved), and that legal costs totalling €186.98 million had been paid between 2003 and 2014 in respect of 15,064 applications. Of this total, €174.38 million had been paid in respect of applications to the RIRB and in 2,601 of these applications a further €12.6 million had been paid in respect of the costs of associated court proceedings. These costs were net of any payments made by the RIRB for medical reports received prior to the completion of an application. The average costs and expenses paid to applicants' solicitors, including payments made for medical reports, to the end of 2014 in respect of applications amounted to €12,035 per application, or 19.3% of the award. Source: *Annual Report of the Residential Institutions Redress Board 2014*, available at www.rirb.ie.

and the government belatedly established, in 2012, a non-statutory committee chaired by Senator Martin McAleese to investigate the extent of state involvement with the Magdalen laundries. The McAleese Committee published its Report in 2013.

The McAleese Report[8] found that just over 10,000 women had spent time in a Magdalen laundry since the foundation of the state, of whom between 800 and 1,000 were alive in 2013. The manner in which the women concerned found themselves in Magdalen laundries varied greatly. There were a large number of referrals by the women themselves. There were also many referrals to Magdalen laundries made by families, priests or clergy, as well as the Society of St Vincent de Paul, the Red Cross and Samaritans. In addition, 27% of referrals were made by or facilitated by the state. The McAleese Report commented that, judged by contemporary standards, it was difficult to understand the circumstances which could in the past have led to daughters, sisters and mothers being placed in the laundries.

The McAleese Report also stated that it would be unfair to judge these cases or the people concerned by applying contemporary standards and societal norms. It pointed out that, because of life expectancy at the time, it was common for one or more parents of the girls or women to be deceased. Thus, many of the women ended up in these institutions simply because they did not have any person in a position to look after them. The Magdalen laundries were, in many cases, used as a temporary measure in the absence of more appropriate services which simply were not available at the time. They were described in the McAleese Report as 'often the only door open to these women'.

Many of the women worked in the laundries for long periods, about half being in the laundries for periods in excess of 3 years. The referrals made by state agencies came about in a variety of ways. There were referrals from industrial and reform schools, as well as health and social services, from the then existing county homes and from psychiatric hospitals. There were also referrals from the criminal justice system, including: persons on remand; referrals as a condition of probation; other less formalised referrals facilitated by the courts; referrals from prison; and referrals made by An Garda Síochána (the Irish police service). Owing to the lack of more appropriate services, many of the case studies demonstrated widespread acceptance of practices that would be unacceptable in the early 21st century. The case studies often illustrated the fact that alternative social care arrangements for the women concerned were not available at the time.

The McAleese Report also noted that there appeared to be a mistaken public perception that women referred to these institutions were predominantly lone parents. This perception did not reflect the reality in most cases. The report found that placements of girls or women in Magdalen laundries were made for a wide range of reasons. Some referrals were seen as a means of providing for women with physical disabilities, or with a mental or psychiatric illness. Some had intellectual disabilities and special needs, while others were referred because of advanced age. In other cases women were placed in Magdalen laundries by their families following family disputes or as a result of abuse or neglect in the home. Only in a minority of cases did they relate to girls or women rejected by their families for having a child.

Arrangements for compensation were not forthcoming from the religious or other authorities who owned and ran the various laundries. Since the state was involved in

[8] *Report of the Inter-Departmental Committee to Establish the Facts of State Involvement with the Magdalen Laundries* (February 2013) (the McAleese Report), available at www.justice.ie.

oversight of the running and operation of the Magdalen laundries in various ways, the McAleese Report recommended that an ex gratia scheme should be established for the benefit of the women. The McAleese Report noted the harsh and physically demanding work required of the women and the traumatic, ongoing effects which their incarceration and misery within the laundries had upon their security, confidence and self-esteem.

C. The Magdalen Commission's Design of an Ex Gratia Scheme

Following the recommendation in the McAleese Report, the government established a further non-statutory Commission, the Magdalen Commission, to advise on the details of the ex gratia scheme for the Magdalen women.[9] The Magdalen Commission formally comprised one member, Mr Justice Quirke (President of the Irish Law Reform Commission and a former High Court judge), and it was also assisted by a small advisory group and a further group of over 40 women barristers who volunteered to interview the women who had been sent to Magdalen laundries. The work of this group of volunteer barristers was recognised in 2014 as the largest single exercise *in pro bono* legal work in Ireland.[10]

The Magdalen Commission was specifically instructed that the approach was to contribute to a healing and reconciliation process. Accordingly, the Magdalen Commission decided that

> Since the proposed Scheme ... must operate on a non-adversarial basis, lengthy and mutually antagonistic roles and positions leading to invasive and painful inquiry and interrogation cannot be entertained and must be specifically avoided. It follows that the 'traditional' methods and principles which apply to the awarding and assessment of damages cannot appropriately and effectively be applied to the function and administration of the proposed Scheme and must be replaced in order to meet its requirements.[11]

The Magdalen Commission therefore decided that the principles of restorative justice and the methods applicable to and used in alternative dispute resolution (ADR), notably the needs, interests and underlying requirements of the parties rather than their differences, should be adopted and applied to the design of the proposed scheme.[12] In consequence, the Commission undertook an 'interest-based dispute resolution process' which acknowledged the 'blameless' status of the women and focused upon their present and future needs, interests and underlying requirements, rather than the parties' respective differences. It undertook a swift but extensive consultation process with the women who wished to be heard.

[9] *The Magdalen Commission Report: Report of Mr Justice John Quirke on the Establishment of an Ex Gratia Scheme and Related Matters for the Benefit of those Women who were Admitted to and Worked in the Magdalen Laundries* (May 2013) (Magdalen Report), available at www.justice.ie.

[10] See www.irishlawawards.ie.

[11] Magdalen Report, para 3.08.

[12] This approach expressly drew on the analysis in the Law Reform Commission of Ireland's *Report on Alternative Dispute Resolution: Mediation and Conciliation* (LRC 98-2010) para 2.22, cited in the Magdalen Report, para 3.09.

In particular, the Magdalen Commission noted that the 'traditional' methods and principles which apply to the awarding and assessment of damages for personal injury are adversarial in nature and require individual assessment based on sworn and documentary evidence of liability, cause, injury and consequence. Individual assessments of the 520 women who, by 23 May 2013, had expressed an interest in the scheme would probably take several years to complete and might be further delayed by appeal processes.

These considerations led to the decision that a flexible and consensus-orientated approach applying the principles of restorative justice was the appropriate means of achieving the objectives envisaged by the Terms of Reference.

> Those principles place an emphasis on underlying interests rather than divisions and seek a creative resolution based upon those factors.
>
> The process adopted was a form of 'alternative dispute resolution' ('ADR') which is an umbrella term for a variety of alternative and flexible processes for resolving disputes. The primary focus of these processes is not the parties' respective differences, but the needs, interests and underlying requirements of the parties.
>
> The consultation process chosen was intended to give the participants significant control over its conduct and outcome.[13]

The Magdalen Commission was also required by its terms of reference to consider the approach to be taken in circumstances in which a payment had already been made by the statutory RIRB, discussed above, by way of redress to a former resident of an industrial school, where such payment included a sum specifically due to the direct transfer of that person from an industrial school to a relevant laundry and their time or part of their time spent in a laundry. The Magdalen Report concluded that it would be highly unlikely that any examination of RIRB orders, transcripts or other documents would identify any part of an RIRB award that had been made specifically due to the direct transfer of a person from an industrial school to a Magdalen laundry. As a result, the Magdalen Commission recommended that the ex gratia scheme it was recommending should not take into account any award made by the RIRB.[14]

The Magdalen Commission recommended that the ex gratia scheme should have the following characteristics:

(i) to exclude mutually antagonistic roles and positions and avoid invasive and painful inquiry and interrogation;
(ii) not to require the individual assessment of any of the Magdalen women and
(iii) to be a speedy procedure as part of a final process of healing, reconciliation and closure and, in consequence,
(iv) to reflect the expressed wishes of an overwhelming majority of the 337 Magdalen women who actively participated in a consultation process with the Commission.[15]

The Commission noted Australian academic criticism of Australian ex gratia redress programmes as not correlating with the principle of corrective justice (since no liability

[13] Magdalen Report, para 2.03.
[14] ibid, ch 7, 11th recommendation.
[15] ibid, para 2.01.

is admitted and payments are made voluntarily), and having more to do with the healing process and hence with the principle of restorative justice.[16]

The criteria and factors taken into account in the design of the scheme were:

(a) The needs and interests of the Magdalen women,
(b) The avoidance of 'positional' or confrontational advocacy,
(c) The nature and extent of the work undertaken by the Magdalen women,
(d) The wish of the contemporary Irish community to reduce the hurt and pain suffered by the Magdalen women by providing them with monetary and other benefits;
(e) The needs and interests of that community including its capacity to provide such benefits in a manner which will be (i) fair, just and equitable to all of those, who will be affected by the proposed Scheme and (ii) as proportionate as is reasonably possible having regard to the assessment and calculation of other similar monetary awards made within this and other jurisdictions and;
(f) The need to facilitate and encourage any reasonable measures which may be required to accommodate consensual healing and reconciliation between parties having an interest in the Scheme.[17]

In considering the supports and benefits, the Magdalen Commission found the following considerations significant:

> The following important information and recurring trends and themes emerged from the consultation process namely: (a) an age profile; (b) diverse and sometimes serious health concerns and problems; (c) concerns relating to the vulnerability and capacity of some women, (d) a significant education deficit and (e) some unsatisfactory living conditions.
>
> The age profile of the Magdalen women was predominantly elderly, with some of advanced age: of 386 women who provided their age, the average age was 68 years, 3 percent were under 50 years and 14% were over 80 years.[18]

The ex gratia scheme recommended in the Magdalen Report included a significant element of monetary compensation element, but this was based on restorative principles rather than the criteria applied in awarding damages in civil claims. The Report explained the restorative aspects as follows:

> Firstly, the monetary payments are intended to assist the Magdalen women who have endured suffering through no fault of theirs. The payments therefore form an important and sincere part of the recognition of past State failures. Secondly, by avoiding any adversarial process, the needs and interests of the Magdalen women and of Irish society generally are also met, because the voluntary nature of the offer of payments contributes to the process of on-going reconciliation to which the other aspects of the Scheme also contribute.[19]

In addition to these monetary elements, the scheme recommended in the Magdalen Report also reflected ADR principles because it recommended that the general scheme for Magdalen women would include free access to healthcare for them and the establishment of a dedicated unit to assist them to obtain the health, monetary and other benefits to which

[16] S Winter, 'Australia's *Ex Gratia* Redress (2009) 13(1) *Australian Industrial Law Review* 49, cited in the Magdalen Report, para 5.07.
[17] Magdalen Report, para 3.11.
[18] ibid, para 2.05.
[19] ibid, para 5.07.

they would be entitled. This unit would also be required to acquire, maintain and administer 'any garden, museum or other form of memorial' which the women had requested. The Magdalen Report also recommended that this unit would be established only after the scheme's administrator 'has first consulted with and received written submissions from an advisory body or committee representing the needs and interests of the Magdalen women'.[20]

A number of the women had been deprived of an education because they were admitted to the laundries at an early age and received no additional education. Many of the women stated that they found it hard to make ends meet and to pay household bills. Others indicated that they were 'just about' able to pay their bills. While many of the women were in receipt of social welfare and related payments, a very small proportion of the women received the State Pension (Contributory). Since all of the women who worked within the designated laundries worked without pay, some for very long periods of time, it was recommended that the scheme should include provisions enabling all of the women who were admitted to and worked in the designated laundries upon reaching pension age to receive (from an administrative scheme created for that purpose) the equivalent of the State Pension (Contributory) for the remainder of their lives.

The Magdalen Commission noted the traumatic effect of admission and incarceration within the laundries had a significant adverse effect upon the education, perceived reputation, subsequent health, life, lifestyle, confidence and self-esteem of the Magdalen women. It recommended that:

(a) those who were admitted to a designated laundry and worked there for less than 3 months should be paid the sum of €10,000;
(b) payment should be supplemented to reflect the 'work undertaken' by the women within the laundries for up to 3 months who should be paid a further €1,500, so that women who have spent less than 3 months in the laundries will receive €11,500 (€10,000 as a general payment and €1,500 as an additional payment for work done);
(c) if they worked for longer than 3 months, €1,000 for each month in the laundries up to 5 years and 3 months, and thereafter €500 per month up to a maximum total payment of €100,000.

The Commission found that a significant number of the Magdalen women were frail and some were very vulnerable. A concern was to protect the resources which they would acquire when they received monetary payments arising out of the scheme. Accordingly, their needs and interests would be best addressed by making any ex gratia payments in excess of €50,000 payable as tax-free weekly income for the remainder of their lives. Those who had not yet reached the state retirement age should be granted a minimum tax-free weekly income of not less than €100 (€5,200 annually), to be paid until the payments equivalent to the State Pension (Contributory) commence. This should overlap with any ex gratia payments in excess of the €50,000 which would also be converted into an income set at a minimum of €100 per week (€5,200 annually) which would be paid until state retirement age has been reached or to increase their income if they have already attained state retirement age. Similarly, for women who had reached the state retirement age, they should be provided with payments equivalent to the State Pension (Contributory), which

[20] ibid, para 5.21.

should also overlap with any ex gratia payments in excess of the €50,000 which would also be converted into a life income.

D. Implementation of the Magdalen Laundries Scheme

The Magdalen Commission's recommendations were implemented primarily on a non-statutory basis but with some statutory support in the form of the Redress for Women Resident in Certain Institutions Act 2015 ('the 2015 Act'). Speaking during the Oireachtas (parliamentary) debates on the 2015 Act, the then Minister of State in the Department of Justice and Equality noted that the Irish government had accepted in principle all the recommendations in the Magdalen Report in June 2013, ie 1 month after the Report was completed. Following this, the Department of Justice and Equality then immediately set up a non-statutory ex gratia scheme which was publicised widely both in Ireland and abroad.

The state established a dedicated unit within the Department of Justice and Equality charged with providing the following services for eligible Magdalen women:

- A helpline accessible daily by the women to assist them to obtain the health, monetary and other benefits to which they were now entitled.
- Investigative and other help and assistance in obtaining such sheltered or other housing as they may be entitled to.
- Investigative and other help and assistance in obtaining such educational assistance as they may be entitled to.
- Practical and, if necessary professional, assistance to enable those women who wish to do so to meet with those members of the religious orders who have similar wishes to meet and interact.
- Similar practical assistance to meet and interact with other Magdalen women.
- The acquisition, maintenance and administration of any garden, museum or other form of memorial which the scheme's administrator, after consultation with the advisory body or committee referred to below, has decided to construct or establish.

Applicants under the ex gratia scheme do not have to prove that they suffered any abuse or damages, nor is the manner of their referral to the institution considered relevant. All that has to be established is that a woman was admitted to and worked in a relevant institution. The processing of all applications begins with the premise that the testimony of the applicant is correct and the officers processing the application then seek to verify the application by checking appropriate records. Where the records of institutions are incomplete, as they are in a small number of cases, the Department officials in the dedicated unit arrange to meet the applicants affected and the Department is generally able to verify the information provided by the applicant by corroborating the information with the institutions or third parties.

The Magdalen Report recommended that the scheme's administrator should apply a fair and robust eligibility or qualification process, giving eligible applicants access to institutional and other relevant records and affording such additional and other cooperation and assistance from state and other agencies as they may require in order to enable them to properly record and verify the work which the women have done and the periods(s) of time which they have spent within the laundries. Those women who applied for, were included within and obtained the benefits of the scheme are not permitted to make any further claim against the state or any of its agencies in respect of their admission to and time spent within the designated laundries.

Under the scheme's provisions, qualifying women are eligible for a payment of between €11,500 and €100,000, depending on the length of stay. The balance of lump sums in excess of €50,000 is paid in weekly instalments to provide an income spread over a longer period. During the Oireachtas debates on the 2015 Act the then Minister of State stated that, by mid-2015, decisions had been made on almost 90% of applications out of 776 received to that point, and a total of €18 million had been paid out to the women. It was expected that a small number of further applications would be received.

As the Magdalen Report recommended, the ex gratia payments also include top-up pension payments of up to €100 for those under the age of 66 years and up to €230.30 for those aged 66 or over. These payments are made by the Department of Social Protection on a non-statutory basis. The 2015 Act provided that the payments being made under the non-statutory ex gratia scheme, including top-up pension payments, are not included in the financial assessment of means under the Nursing Homes Support Scheme Act 2009, a statutory means-tested scheme for persons who receive long-stay care in a residential nursing home.

The Magdalen Report recommended that Magdalen women should be provided with access to a comprehensive suite of health services, and that the women receive medical services equivalent to those provided to the holder of a HAA card, provided to persons with hepatitis C under the Health (Amendment) Act 1996. The Report added that 'not all of the services described in the [Irish Health Service] Guide [to the HAA card] may be directly relevant to the Magdalen women and any comparable Guide for the Magdalen women would require suitable adaptation.' In the Oireachtas debates on the 2015 Act the then Minister of State commented that the services provided for in section 2 of the 2015 Act were 'precisely' those recommended in the Magdalen Report: GP services; prescribed drugs, medicines, aids and appliances; dental services; ophthalmic services; aural services; home support; home nursing; counselling services; and a range of other services, to include chiropody–podiatry and physiotherapy. With regard to the question of alternative therapies, the Minister of State commented that the Minister for Health had

> reservations about such therapies being provided and funded through the health service and, for that reason, they were not included in the [2015 Act], just as they are not included in the Health (Amendment) Act 1996, or in [the Magdalen Commission's] recommendations. However, the Minister for Justice and Equality is considering options for provision of such services on an administrative basis outside of legislation.

The Minister of State also referred to whether specific provision was required for women entitled to these supports who do not have capacity to make the necessary applications or arrangements. The Minister pointed out that the women could have a care representative appointed for them under section 21 of the Nursing Homes Support Scheme Act 2009. The Minister also noted that the Assisted Decision-Making (Capacity) Act 2015 (which contains a rights-based approach to financial and health decision-making, comparable to adult guardianship legislation in many jurisdictions) was, at the time, going through its final stages in the Oireachtas; and it was, indeed, enacted by the end of 2015. Because of this, the Minister of State considered that it would 'not make sense' to introduce a special scheme for the Magdalen women who lack capacity in the light of the enactment of the Assisted Decision-Making (Capacity) Act 2015. It needs to be acknowledged that, at the time of writing (January 2017) the main elements of the Assisted Decision-Making (Capacity) Act 2015 have not yet been brought into force, and thus were never applicable to the Magdalen women. Nonetheless, it is also worth noting that the provisions on appointing a care representative under section 21 of the 2009 Act apply a functional test to determine capacity, which anticipated the more generally applicable functional test in the Assisted Decision-Making (Capacity) Act 2015. The Minister of State indicated that around 40 Magdalen women might not have had full decision-making capacity. In this respect, section 2 of the 2015 Act provides that in some cases a family member or a carer, rather than the woman herself, is empowered to make decisions about GP or other services.

E. From Adversarial Redress Scheme to Transitional Justice

In 2017, the government of Ireland declined to extend the adversarial model used in the Residential Institutions Redress Scheme, discussed above, to an entirely new group of potential applicants. Instead, the government decided to adopt a transitional justice approach to past wrongs.

The government was responding to the *Second Interim Report of the Commission of Investigation into Mother and Baby Homes*, completed in September 2016 and published in April 2017.[21] The Commission of Investigation, a statutory judicial investigation,[22] was established to investigate conditions in mother and baby homes and county homes in Ireland in the period 1922–98. The Commission's terms of reference were extremely wide ranging, and included investigating general living conditions in the institutions, whether unauthorised vaccine trials were carried out in them, whether children born there were illegally adopted from them, to what extent women who gave birth were required to carry out unpaid work, mortality rates in the institutions, whether all deaths were recorded and the burial arrangements in place. The immediate catalyst to the Commission's establishment was the discovery of a mass, unmarked, grave beside one of the homes, in Tuam, County Galway, and the revelation that a local historian, Catherine Corless, had identified that 796 babies and children (the youngest 10 minutes old, the oldest 9 years of age) had died in the institution between 1922 and 1998. A TV documentary broadcast in 2015 brought this to the attention of the public, and the Commission was established in its wake, mirroring the background to the establishment of the McAleese Committee (see above).

The *Second Interim Report* suggested that the government should consider extending the Residential Institutions Redress Scheme to residents of the mother and baby homes.

[21] Available at www.mbhcoi.ie/MBH.nsf/page/LPRN-ALCFND1238712-en/$File/MBHCOI%202nd%20Interim %20Report.pdf.

[22] Established under the *Commissions of Investigation Act 2004*: see *Commission of Investigation (Mother and Baby Homes and Certain Related Matters) Order 2015* (SI No 57, 2015).

The government declined, citing a number of reasons.[23] First, the government stated that the Commission of Investigation had made no findings regarding abuse or neglect of women or children in the mother and baby homes, and that therefore it would not be appropriate to deal with the question of redress in advance of any conclusions on this issue by the Commission. Second, the government noted that the Residential Institutions Redress Scheme 'was complex to administer and often difficult for applicants', a reference to the adversarial nature of the Redress Scheme. Third, the government had regard to the potential cost of such an extension, bearing in mind that the Residential Institutions Redress Scheme had originally been estimated to cost €250 million but had cost over €2.5 billion, 10 times the original estimate.[24] While not completely ruling out a redress scheme, this response appears to suggest a declining appetite for such an approach to past events.

Instead, it is notable that the government appointed a person to prepare a potential model of transitional justice as a means of giving voice to former residents of mother and baby homes and county homes.[25] The government noted that the United Nations defines transitional justice as the set of approaches a society uses 'to try to come to terms with a range of large scale past abuses'. The government added that transitional justice 'puts survivors and victims at the heart of the process. It commits to pursuing justice through truth. It aims to achieve not only individual justice, but a wider societal transition from more repressive times, to move from one era to another.' The government referred in this respect to the need to 'learn from international best practice in transitional justice, such as the Museums of Memory in Argentina and Chile, for example'. In many respect, this analysis reflects the approach in the 2013 *Magdalen Commission Report*, discussed above.

[23] www.dcya.gov.ie/viewdoc.asp?fn=%2Fdocuments%2FMotherandBabyInfoPage.htm.

[24] Comptroller and Auditor General, *Special Report 96: Cost of Child Abuse Inquiry and Redress* (2016), available at www.audgen.gov.ie/reports/Cost_of_Child_Abuse_Inquiry_and_Redress.pdf.

[25] The government's appointee was chosen because of his experience in issues of transitional justice: see eg J Gallen, 'Jesus Wept: The Roman Catholic Church, Child Sexual Abuse and Transitional Justice' (2016) 10 *International Journal of Transitional Justice* 332.

22

Germany

HERBERT WOOPEN AND CHRISTOPHER HODGES

This chapter explains the ADR-based approaches that have developed in Germany to resolve medical claims and medical insurance claims, outside the court system. It begins by explaining the context of liability and regulatory changes as a result of the shock of the thalidomide tragedy, including the development of a particular insurance arrangement for pharmaceutical manufacturers, the Pharmapool.

A. The Pharmapool

The 1960s Contergan (thalidomide) tragedy, in which the children of mothers who had taken this medicinal product as sleeping pills or to control morning sickness were born with tragic deformed limbs (phocomelia), provoked three consequences that are relevant to the current analysis.

Firstly, the company contributed hesitatingly to compensation arrangements for children in some countries, and was 'supported' in that in Germany by the legislator and the federal budget. Secondly, legislation was enacted that regulated the marketing of medicinal products. Thirdly, insurance was made compulsory and the industry arranged high-level pooled insurance cover.[1] These developments are explained in further detail below.

The producer, Grünenthal, started in 1961 to pay compensation of DM 200–20,000 to patients with damaged nervous systems, and took the drug out of sale only under heavy pressure from the press and politicians in November 1961. Of 4,000 children born with phocomelia between 1958 and 1962 in Germany alone, 2,800 survived. Under the pressure of the penal trial that started in 1968, in which 400 victims joined the case as claimants, the company offered the sum of DM 100 million as compensation if all those entitled to compensation declared not to claim any further compensation. This settlement never became effective as the Bundestag issued a law creating a specific foundation for handicapped children into which the money was paid by Grünenthal, matched by the same sum paid from the federal budget. The federal contributions were increased in 1976 and in 1980. The funds were exhausted in 1987, and since then pensions for the victims are entirely paid from

[1] Stellpflug, Meier and Hildebrandt, *Handbuch Medizinrecht*, update 12/2014, 8 Arzneimittelhaftung, no 172–86.

the federal budget, depending on the degree of the handicap. The foundation, based in Cologne, is now called Conterganstiftung für behinderte Menschen, and since 2009 victims receive bonus payments on top of a reduplication of their pensions in that year. Grünenthal paid another DM 50 million voluntarily into the foundation in 2002, topped up by the federal government with another DM 50 million. The bonus payments will go on until this fund is exhausted.[2]

As a consequence of this tragedy, the German legislator created as early as in May 1961 the first version of the Medicinal Products Act (Arzneimittelgesetz, AMG),[3] a law regulating admission of pharmacological products for use in Germany. This provided for an extensive administrative procedure under which medicinal products have to prove their quality, efficacy and safety. These three criteria have to be demonstrated in toxicological tests and randomised, placebo-checked and double-blind studies, which have to be executed in three phases with an increasing number of patients (Phase I = a few, Phase II = 100–200, Phase III ≥ 200). Germany was the first European country to do this. As a reaction to the same scandal, the provisions laid down by law, regulation or administrative action relating to proprietary medicinal products were harmonised in 1965 by EEC Directive 65/65, requiring involvement of the respective national authority for admission of medicines to each national market (Art 3) before providing for mutual recognition, requiring submission of test data including those of clinical trials (Art 4 no 8), of course without dismissing the manufacturer from liability by the mere granting of permissions to market the product (Art 9). Germany's AMG was unusual in enacting a special liability provision for pharmaceutical manufacturers, and by doing so in its regulatory legislation. Under section 84 of the AMG, strict, 'absolute' liability applies for manufacturers of medicinal products if the product is used in accordance with its intended purpose and its harmful effects go beyond the limits considered tolerable in the light of current medical knowledge or if the damage resulted of insufficient instructions for use. Section 84 subsection 2 then offers the presumption that the damage has been caused by the medicinal product in question.[4] If required, the injured party can, pursuant to section 84a AMG, request information from the pharmaceutical entrepreneur on similar facts known to the entrepreneur which could be of significance in assessing the justifiability of harmful effects—unless statutory provisions require that the data remain secret or when non-disclosure is justified by an overriding interest of the pharmaceutical entrepreneur or a third party. In practice, this 'discovery' provision does not play a decisive role.[5]

In 1976, a duty to insure persons participating in the various stages of testing was introduced, as well as a duty to insure for potential liability (sections 88 and 94 AMG)—this prescribed cover per single medicinal product by law, without any possibility of limiting this by contract (sections 88 and 92 AMG), to the tune of minimum amounts of currently €600,000 or an annuity of up to €36,000 per year for a single person, and a capital amount of €120 million or an annuity of at least €7.2 million per year for several persons.

[2] See the sequence of events described at www1.wdr.de/themen/archiv/sp_contergan/contergan176.html (as of 5 May 2015).

[3] Current version available in English at www.gesetze-im-internet.de/english_amg/index.html.

[4] Arzneimittelgesetz (AMG). This regime was exempted from the subsequent EU Product Liability Directive 85/374/EEC under Art 13, which exempted pre-existing special liability regimes.

[5] E Deutsch, 'Die Methodenlehre des Medizinrechts' [2013] *MedR* 708.

Behind this scene, a reinsurance pool (the Pharmapool) was created in 1978 among the primary insurers and reinsurers. Initially, this group comprised 120 members; with the passage of time and increasing concentration in the German insurance sector, the number has shrunk to 54 members as of 2013, mainly due to takeovers.[6] The Pharmapool operates vis-à-vis injured persons as follows: section 94 AMG requires a pharmaceutical company to justify its provision either by a bank guarantee or by an insurance policy issued by a primary insurer licensed for business in Germany. As most primary insurers cannot bear the risk of €120 million alone, this risk provision is divided into two parts. If an insured event occurs, a first tranche of €6 million is covered completely by the respective primary insurer issuing the policy; a further €114 million as a second tranche is covered by the Pharmapool. Externally, the primary insurers take the full risk of €120 million, but internally, the second tranche is covered by the consortium. Legally speaking, the primary insurer concludes a reinsurance contract for €120 million with the Pharmapool and agrees not be reimbursed for the first €6 million (retained amount). Therefore, the existence of the Pharmapool is not in practice visible to the insured pharmaceutical companies, which conduct their business directly with their primary insurers. As several primary and reinsurers are parties to the Pharmapool and carry the risks jointly, the primary insurers participating in the consortium are also actively engaged in the reinsurance business.

It is quite remarkable that the statutes of the Pharmapool provide for a deficiency guarantee (an indemnity bond) with respect to the member companies of the Pharmapool. Each member company undertakes, to the extent of its participation in the pool, to pay for the debt of any other member company should it fail.

Notwithstanding European harmonisation, the development in Germany is considered to be particular in that it drew not only on the Contergan case and European legislation in order to transform liability for medicinal products into a liability by 'mere creation of a danger', but stiffened that liability further as a consequence of the HIV cases in the second law on amending liability provisions in 2002. Liability therefore is now based on more objective scientific medical standards and heavily influenced by notions of evidence-based medicine.[7]

B. Claims against Doctors[8]

Claims for medical negligence against physicians in Germany would normally have to be brought in court, but in practice are for a major part handled by 12 specialist settlement schemes, which are organised on a regional basis at the *Länder* level, although some schemes cover more than a single *Land*.

[6] S Materne and T Seemayer, in Lüer and Schwepcke, *Rückversicherungsrecht* (Munich, Beck, 2013) 699, also for the ensuing details.

[7] Rieger, Dahm and Katzenmeier, *Arztrecht, Krankenhausrecht, Medizinrecht* (so-called Heidelberger Kommentar—2010), Arzneimittelhaftung, Entwicklung, § 243 no 5.

[8] This section is updated from C Hodges, I Benöhr and N Creutzfeldt-Banda, *Consumer ADR in Europe* (Oxford, Hart Publishing, 2012).

The medical ADR bodies, *Schlichtungsstellen*, were created as a response to a widespread perception that valid claims in the courts were succeeding too slowly, were cumbersome, complex and untransparent, and had a stigmatising effect on the doctors involved. The 'contradictory' procedure in court has a tendency to exacerbate conflict, puts parties in conflict with each other and can therefore finally destroy their relationship. Further, a feeling was widespread that medical experts appointed in litigation were reluctant to criticise their colleagues, and patient confidence was falling. The network of federal and regional medical associations called *Aerztekammern* (the chambers of which all practising doctors are by law compulsory members) wished to create a separate, more satisfactory and quicker dispute-resolution service. However, the different regional *Gutachterkommissionen und Schlichtungsstellen* operate under different rules, established by their local medical chambers, and the result is confusing and makes comparisons challenging. However, it should be kept in mind that lawyers and other professionals are organised regionally as well. Some attempts have been made to work towards a single national standard, bearing in mind that lawyers operate across *Land* borders and bring claims in different *Länder*, under the same laws and rules before state courts, but under different procedures and rules before the various *Schlichtungsstellen.*

The schemes all have in common that they work on the basis of three principles: participation is *voluntary*, *non-binding* and *free of charge* as such. Costs outside of the procedure, particularly those of lawyers, are to be borne by the parties.[9]

Mediation as a technique has been found to be unsuited to medical claims in the *Schlichtungsstellen*, since it aims to bring the parties together and restore trust, whereas patients are generally unwilling to approach a medical problem in this way when the primary available outcome is limited to financial compensation.[10] Once the question of liability as such has been decided, mediation can prove helpful in determining the amount and the items to be included in future payments of compensation.[11] While asymmetry of expertise between doctors and patients may hinder reaching consensus, it is often the doctor's fear of losing authority which prevents him from exhibiting compassion and engaging with the patient. To the extent that the conflict concerns the level of the relationship rather than the level of facts, the mere fact of starting communication (in form of mediation) may be able to resolve it.

The general aims of the *Gutachterstellen und Schlichtungskommissionen* are to improve the relationship between doctor and patient, and so keep the number of court cases low. The aims over time have expanded to encompass preventing future mistakes in treatment and improving patient safety, by publishing their recommendations and feeding them into a centralised information system called MERS (see below). Over the period 2012–15, the 12 schemes jointly handled approximately 12,000 cases each year, closing approximately the same number per year, of which almost 8,000 were by a decision on substance, about 90% of which were accepted by both parties without recourse to the court system.[12]

[9] See C Katzenmeier, in A Laufs, C Katzenmeier and V Lipp, *Arztrecht*, 7th edn (Munich, Beck, 2015) X F I, 383–85.

[10] See BK Hattemer, *Mediation bei Störungen im Arzt-Patient-Verhältnis* (Berlin, Springer, 2011). The book provides a detailed analysis of the possibilities and limitations of mediation in a variety of problems in the doctor–patient relationship.

[11] Katzenmeier (n 9) X F II, 385–86.

[12] Statistics 2015: www.bundesaerztekammer.de/patienten/gutachterkommissionen-schlichtungsstellen/behandlungsfehler-statistik-2015/ (accessed 26 March 2017); charts for 2015: www.bundesaerztekammer.de/patienten/gutachterkommissionen-schlichtungsstellen/behandlungsfehler-statistik-2015/praesentation.

It is difficult to determine the number of cases which are treated in court without having been introduced into the *Schlichtungsstelle* beforehand. Various sources, however, report a figure of 40,000 cases per year claimed via the courts, *Schlichtungsstellen*, public health insurers or private liability insurers,[13] therefore indicating that roughly one-third of all cases are covered by the *Schlichtungsstellen*. For 2008 the precise figures comparing court and *Schlichtungsstellen* are 9,214 cases resolved in court and 10,498 by *Gutachterkommissionen und Schlichtungsstellen* (amount solved directly with insurers and public health service without involvement of courts or *Schlichtungsstellen* not known).[14] For 2014, the figures rose to 10,824 cases in court[15] and 12,053 for the *Schlichtungsstellen*.[16] It should also be noted that within the system of compulsory health insurance (*Gesetzliche Krankenversicherungen*) there is also a possibility to have medical treatment reviewed. The medical service of compulsory health insurances (*Medizinischer Dienst der Krankenkassen*) has a group of experts who, on application by patients and/or by insurances, can review treatment just like the *Gutachterkommissionen*. The number of cases dealt with under this system is even slightly higher than the overall number of cases dealt with by the *Gutachterkommissionen*.

Two regional medical ADR arrangements are outlined below.

North Germany: Schlichtungsstelle für Arzthaftpflichtfragen der norddeutschen Aerztekammern

The scheme for North Germany, *Schlichtungsstelle für Arzthaftpflichtfragen der norddeutschen Aerztekammern*[17] (out-of-court settlement for medical liability disputes), was established in 1976 and has handled a steadily increasing number of claims, totalling around 85,000 claims between 1976 and 2010; in 2012 it dealt with around 4,300 cases[18] (see Figure 22.1).[19]

The scheme is organisationally independent of the 10 medical chambers that it covers.[20] The chambers finance the scheme but have no influence on the it. The chambers have the right to change the procedures but not to be involved in individual cases. The scheme only

[13] H-D Laum (at the time President of Gutachterkommission Nordrhein), 'Außergerichtliche Streitbeilegung in Arzthaftungssachen—Bilanz, Analysen, Prognosen der Gutachterkkommission Nordrhein', in *Medizinrecht* (2011) 230; Report in the Journal of the Ärtzekammer Nordrhein, November 2010, www.aekno.de/page.asp?pageID=8586&noredir=True.

[14] Laum (n 13) 230.

[15] Sum of numbers for medical liability cases (Arzthaftungssachen) mentioned on pp 18 and 48: www.destatis.de/DE/Publikationen/Thematisch/Rechtspflege/GerichtePersonal/Zivilgerichte2100210147004.pdf?__blob=publicationFile (accessed 26 March 2017).

[16] See n 12, quoted as previous year (accessed 26 March 2017).

[17] www.norddeutsche-schlichtungsstelle.de—webpages also provided in English, see eg on the success rates and the MERS system: www.norddeutsche-schlichtungsstelle.de/fehlerprophylaxe/patient-safety/.

[18] See eg 4,230 cases in 2015; www.bundesaerztekammer.de/fileadmin/user_upload/downloads/pdf-Ordner/Behandlungsfehler/Statement_Schaffartzik.pdf, 2 with a map of the territories belonging to the various *Schlichtungsstellen*.

[19] This chart includes from 3 October 1990 4 of the 5 new *Länder* (Brandenburg, Mecklenburg-Vorpommern, Sachsen-Anhalt und Thüringen); and from 1 January 2014 also includes Saarland. See www.norddeutsche-schlichtungsstelle.de/fakten-und-zahlen/antragsentwicklung/ (accessed 24 February 2017).

[20] The following information was supplied by Johann Neu and Christine Wohlers of *Schlichtungsstelle für* Arzthaftpflichtfragen *der norddeutschen Aerztekammern*.

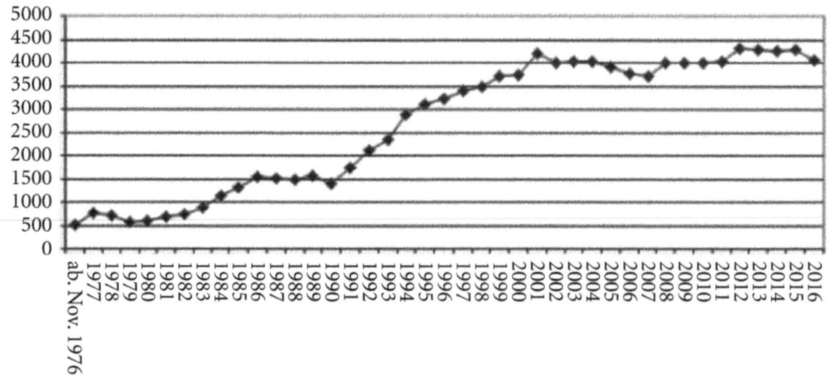

Figure 22.1: Claims lodged at the Norddeutsche Schlichtungsstelle 1976–2016

has the obligation to report back to the chambers on financial matters. The advantage of being independent is that at least in most systems patient and doctor can be sure that their dispute will not be 'known' by the medical chambers: no 'name and shame' policy operates. Hamburg, however, seems to be an exception to this rule, as the administration of the medical chamber, as a rule, is informed of the outcome of a case.

The north German scheme has been described in a previous book in the current series[21] and will not be repeated here.

The 2016 evaluation of cases from 2011 shows that of those cases that were introduced at the *Schlichtungsstelle* (ie beyond those that were dealt with directly by court, insurance companies or the *Medizinischer Dienst der Kranklenkassen* under the system of compulsory health insurance, see introductory remarks above), 85.4% were handled by the *Schlichtungsstelle* without further recourse to a court—a ratio termed the 'court avoidance ratio'.[22] Since final results can only be analysed after a few years when several levels of courts have been involved, the *Schlichtungsstelle* asked the respective liability insurers of the doctors/clinics held liable about the development of the 2,520 cases closed by the *Schlichtungsstelle* in 2011. Replies were received for 1,586 cases and showed the distribution in Table 22.1. The data shows that the court reached a different decision from the *Schlichtungstelle* in 41.8% of the cases that were brought to court after having been closed by the *Schlichtungstelle*—rather a high 'revision rate'.

Table 22.1: Follow-up on 2011 cases in north Germany

2,520 cases decided by out-of-court settlement in Hanover in 2011, replies received for 1,586 cases, of which:			
Court avoidance ratio	**Court follow-up**	**Court deviation from *Schlichtungsstelle***	
1,354 cases: 85.4%	232 cases: 14.6%	97 cases: 3.57% of all cases	97 cases/232 court cases: 41.8%

[21] Hodges, Benöhr and Creutzfeldt-Banda (n 8) 94–96.

[22] www.norddeutsche-schlichtungsstelle.de/fakten-und-zahlen/evaluation (accessed 24 February 2017).

Since 2000, anonymised data from cases has been entered into a Medical Error Reporting System (MERS), which the federal *Ärztekammer* decided in 2006 to operate throughout Germany.[23] Beyond that system, interesting commercial solutions are gaining ground in Germany about which a separate section below gives more detail.

North Rhine: Gutachterkommission Ärztekammer Nordrhein

The *Ärztekammer Nordrhein* is the Regional Chamber of Doctors based in Düsseldorf. It operates an ADR scheme, the *Gutachterkommission Ärztekammer Nordrhein* (Evaluators for Personal Injury Claims against Doctors).[24] The procedure is voluntary and free to participants. The body comprises around 100 doctors and 8 lawyers, former high-ranking civil law judges. Considering the hierarchical culture still dominating the medical profession, hierarchy matters for the doctors concerned more than profound knowledge of medical malpractice law.

The procedure is that patients may lodge a claim with the *Gutachterkommission* instead of, or before, going to court. The process is all in writing, so there is no way to resolve conflicts of evidence by oral evidence or examination. This issue is partly addressed by the standard rules of German law, which require doctors to keep documentation of all that they do. That leads to a presumption that if something is recorded in the documentation it occurred, and vice versa. There is also a rule under German law of civil procedure in medical malpractice cases that where there is a grave mistake, there is an assumption of causality (ie the gross negligence caused the damage),[25] a rule which has been developed by the courts for tort law and recently been codified for contract law in section 630h subsection 5 sentence 1 BGB (Civil Code). A case is sent out to a medical expert, who normally does not examine a patient, and writes a draft opinion. The draft is reviewed by one of the lawyers. If the opinion is clear, the medical expert opinion is sent out with a letter from the President explaining the position. If the position is debatable, a formal decision is sent out, signed by the president, adopting the reasoning of the medical expert. Claimants are told that they can appeal to the *Gesamtkommission*: as of summer 2015 this still comprised 5 doctors and 1 judge, but its constitution was changed in December 2015, when the *Gesamtkommission* was replaced by teams of 1 medical and 1 legal member of the Commission, which are assisted in their review by either recommendations given by specialised medical members during the Plenary meetings, or by setting up special consultation meetings in small groups involving specialised doctors.[26] Another new rule is that objections in straightforward cases (such as standard appeals motivated by doctors' insurers so as to avoid a finding of negligence) can be reviewed and disposed of quickly by the original medical expert, and with a discretion to involve others in the review.

The *Gutachterkommission* makes a determination on whether there has been a mistake (ie fault/negligence liability, which is the legal standard that applies in court proceedings) and whether the mistake has had negative consequences for the health of the patient

[23] www.norddeutsche-schlichtungsstelle.de/fehlerprophylaxe/patient-safety.

[24] www.aekno.de/page.asp?pageID=64 (accessed 26 February 2017).

[25] Katzenmeier (n 9) 436–52.

[26] www.aekno.de/downloads/aekno/gak-statut-2015.pdf and www.aekno.de/page.asp?pageID=5214 (accessed 26 February 2017).

(causality) but not on the amount of damages. In about 30% of the cases brought before the commission the finding is that there has been a mistake. Cases are subsequently dealt with by negotiated settlement between insurers and patients, or by the courts, but only a very small percentage proceeds to court.

The *Gutachterkommission* issues an annual report and publishes decisions it considers of general importance, mainly with a view of supporting continuing education of medical practitioners. It maintains a database of all cases and has made them accessible as a PDF and ebook of 255 pages.[27] The annual report to September 2016 is shown in Table 22.2.[28]

Table 22.2: Claims data from the *Gutachterkommission Ärztekammer Nordrhein* 2015–16

Statistical overview			
	Reported period (1 October 2015–30 September 2016)	**Previously reported period**	**Total number (since 1 December 1975)**
I.			
1. Number of *Applications*	2,183	2,141	53,401
2. Number of *cases closed* of which	2,385	2,114	51,813
2.1. *By mere medical opinion*	1,839	1,597	38,615
2.2. *Formal notices* by the President (eg procedural obstacles), (of which	296	271	5,358
a) challengeable	(56)	271	
b) not challengeable)	(240)		
2.3 *Other settlements* (withdrawal, lack of 'jurisdiction')	250	246	7,840
3. Still pending applications	**1,588**	**1,790**	
(of 2.1): Number of confirmed cases of *medical malpractice*	550	453	12,244
(in %)	(29.91%)	(29.08%)	(31.71%)
II. Applications and settlements in second instance			
1. Number of *applications*	490	262	8,801
(in per cent of first instance I.2.1 + 2.2a)	(25.86 %)	(29.08 %)	(20.01 %)

(continued)

[27] www.aekno.de/downloads/aekno/gak-entscheidungen-2000-2015.pdf and http://epaper-archiv.de/aekno/2015_GAK_entscheidungen/index.html#96 (accessed 27 February 2017).

[28] www.aekno.de/page.asp?pageID=5214.

Table 22.2: ***(Continued)***

Statistical overview			
	Reported period (1 October 2015–30 September 2016)	**Previously reported period**	**Total number (since 1 December 1975)**
2. Number of			
2.1 *Final medical opinions* (of which	318	293	8,185
a) Decisions by the commission (§ 10 a. F.)	73		
b) Final medical opinion (§ 11))	245		
(number of deviations from the result of the first instance)	(31)	(19)	(541)
2.2 *Other settlements* (withdrawals, cessations)	15	11	315
3. Still pending	**301**	**144**	

It is noteworthy that the President of the *Gutachterkommission Nordrhein* also reported repeated invitations to Russia, where there is great interest in out-of-court settlement of cases of medical malpractice.

C. Claims against Insurers: Ombudsman for Private Health Insurers (PKV-Verband)

Citizens in Germany may opt to purchase private healthcare insurance, so-called substitutive health insurance, as an alternative to public healthcare. Citizens eligible for this substitutional cover must earn, in 2017, more than €57,600 p.a., and those who do opt for that switch, obtain from their employer the same monthly amount as they would for public healthcare membership. Public healthcare is paid for by employers' and employees' contributions roughly half-half depending on the amount of the respective employee's salary, with a cap on the relevant income of €52,200 in 2017 (*Beitragsbemessungsgrenze*). Public healthcare is organised through several dozen regional, professional or company-affiliated public bodies called *Gesetzliche Krankenkassen* or *Gesetzliche Ersatzkassen*. Citizens insured with these public bodies and not eligible for full private insurance cover may still purchase extra cover to supplement public cover.

In 2001, the Association of Private Healthcare Insurers (*PKV-Verband*) established its own PKV-Ombudsman to resolve claims under private substitutive and complementary policies, at the same time as general insurers established the Insurance Ombudsman.[29] Those two ombudsmen are separate from each other, both being based in Berlin.

[29] Hodges, Benöhr and Creutzfeldt-Banda (n 8) 90–94.

Like the general Insurance Ombudsman, a former President of the highest Federal Civil Court, the PKV-Ombudsman is a distinguished figure, appointed by the Board of the *PKV-Verband* upon proposal by the Executive Directors of this Association.[30] His staff comprises 7 lawyers and 5 secretaries. The PKV-Ombudsman is recognised as an arbitration board by the Ministry of Justice, under section 114 of the Insurance Contract Act (Versicherungs-Vertrags-Gesetz, VVG). The PKV-Ombudsman is a member of FIN-Net, although it has never had to handle any cross-border private insurance healthcare claims.

The PKV-Ombudsman is funded by the insurers, and its services are free unless a patient chooses to use a lawyer to raise his claim—these costs have to be borne by the patient himself. Claims are received in writing only, including by fax or via the Ombudsman's website. The Ombudsman staff can also be reached by phone for general information or enquiry purposes—they receive approximately 30 calls per day.[31] Claims can only be made within 1 year after the decision of the insurer which is being challenged. The same holds true for claims which are already being treated in court. Proscription of the claim in state courts is suspended for as long as the Ombudsman declares the procedure as unfinished.

Case managers build up the necessary documentation on a file, with the help of a checklist that is kept on the front of the file. The managers solve some cases quickly, for example with a phone call. Case managers spend a great deal of effort in explaining to patients how to properly understand which therapies and medical services are covered by the policies they had chosen to buy. Patients can drop complaints as a result of receiving a sympathetic explanation. Case managers usually find compromises that satisfy the parties. Companies often wish to avoid formal decisions being issued against them.

Ultimately, the case managers prepare a draft decision letter, which is signed by the Ombudsman personally: it is considered important that patients should see that the Ombudsman has considered their case personally. Difficult points can be discussed between case managers and the Ombudsman at any stage of a case.

Decisions are not legally binding, and either side may take a case to court. However, adherence by insurers to Ombudsman decisions is over 99.9%. The Ombudsman considers it to be important that decisions are non-binding. Relevant factors are the uncertainties that arise with medical science, and the advantage of having some flexibility to make proposals that are fair to both sides. The Ombudsman aims to apply the law, but some cases are best resolved by reaching results that are fair rather than strictly legal.

There is a lower threshold for claims of €50 and no upper limit. Cases involve a wide range of sums of money: some, such as cancer treatments, can involve very large sums. Cases generally take 3–4 months to resolve. Some are quicker, and some longer.

There are around 42 million health insurance contracts in Germany. The number of claims has been declining from an all-time high of 6,511 in 2011 to 5,984 in 2013 and 5,770 in 2015. In 2016 it went up again to 6,084. Based on the figure of 42 million private health insurance contracts, these figures amount to only 0.014–0.013% of the existing contracts. While 4,577 of the complaints received in 2016 (2013 4,418; 2014 4,015; 2015 4,328) were admitted for further scrutiny (75.2%), only roughly a quarter were successful (25.8% in 2013, 27% in 2014 and 2015, 25,3 in 2016). The Ombudsman considers this to be a consequence of an essentially functioning community of health insurers.

[30] www.pkv-ombudsmann.de/statut/, s 1.1.

[31] Tätigkeitsbericht 2013, 24, at www.pkv-ombudsmann.de/taetigkeitsbericht/2013.pdf.

Quite a number of complaints refer to the medical necessity of treatments and to the amount to be reimbursed in a particular tariff. Government scales, which have a degree of flexibility, apply to doctors' fees, and their application can be subject to different interpretations. Courts themselves can reach inconsistent decisions on them.

Problems sometimes arise not with the insurer but with a doctor or dentist, such as overuse of a new treatment for cancer. Patients need to be informed what is recognised therapy, and what is a new therapy that has not been proved. Problems about range of the coverage arise also when tariffs are changed by the insured against the following background: particularly the substitutive tariffs are a lifelong contract without a right for the insurer to give notice of the contract; insurers can, however, increase the price they charge, under the control of an independent trustee supervised by the Federal Insurance Supervisory body (BaFin). Particularly when prices rise, policy holders try to lower their financial burden by reducing the treatments covered and the quality level of the insurers' obligations, which may create uncertainties for the policy holder about the treatments covered. Another reason for changes in the range of insurance cover is that a health insurer is also forced by statutory law to accept tariff switching of their clients to new tariffs when these are created. New tariffs may seem easier but what may appear to be a cheaper tariff can raise issues of treatments that are not covered.

The PKV-Ombudsman publishes an annual report, and holds an annual meeting to explain his views. There are biannual meetings with the insurance companies which cover topics that have proved problematic over previous 2 years. The Ombudsman can request insurers to change their practice, and can do this in bilateral meetings arranged at any time. He has done both of these things.

The current 2016 figure of 25.3% of successful claims has fallen from around 38% when the PKV-Ombudsman was established. Of the 74.7% that are unsuccessful, only a handful of cases are subsequently pursued in court. In 2014, one formal recommendation was made against an insurance broker.

D. Recent Developments Targeting Reduction of Medical Liability Cases and Further Outlook

Since the German legal system provides that liability of doctors and hospitals is dependent on incorrect behaviour/practice, a change of culture towards prioritising the avoidance of future mistakes rather than a defensive attitude on the part of insurers and those held liable appears a strategically difficult task to accomplish. Such systemic change, however, may be facilitated by new technological developments as these gain ground in a number of particularly progressive German hospitals. The insurance broker Marsh has built a database that now contains 20 million cases in a real-time online system named CRAB™.[32] It is used by the UK NHS in order to provide more transparency down to the level of individual hospitals and doctors.[33] It is based on results gathered by another proprietary system,

[32] General page: www.marsh.com/us/industries/healthcare/clinical-healthcare-consulting.html. The acronym CRAB is derived from the surname of the developer, Prof Copeland: 'Copeland's Risk Adjustment Barometer'.

[33] See eg pp 10 and 11 of www.whh.nhs.uk/_store/documents/whhqualityaccounts0910final.pdf; reports published later show the structure, but not any more the name of the application, eg www.nhs.uk/Services/Trusts/Overview/DefaultView.aspx?id=866. Quality Report 2015–16 of that hospital pp 81 and 82: www.nhs.uk/Services/UserControls/UploadHandlers/MediaServerHandler.ashx?id=368&t=636264741543768414

ClearSight[34] (formerly Stars), which is integrated into the IT of a hospital and collects all cases of actual or closely avoided problems. It is accessible by all medical personnel and administrative staff and links together all concerned, whereas traditionally the issue of medical malpractice is easily pushed aside as not convenient and painful for those involved. In particular, the link between quality control units and the surgeons operationally in charge tends to be concentrated in a limited number of meetings per year, whereas the systems provided by Marsh ensure immediate and real-time attention, ensuring that measures recognised as appropriate in order to avoid a repetition of problems encountered are swiftly implemented into practice. They also have the merit of drawing on information about similar events based on the experience collected in hundreds of other hospitals worldwide. It remains to be seen whether such technological change will lead to a change of mindset across the board in German hospitals; several neighbouring countries seem to have embraced this development more actively.

In this context it may be worthwhile mentioning that a few scholars have already aired ideas about switching the protection of patients from a fault-oriented system towards a 'patient insurance' system as exists in Sweden and New Zealand.[35] This is supposed to increase the protection of the patient and to improve the relationship with the doctor treating the patient as compensation would no longer rely on a victory of the patient over the doctor but would be based on the reasonably determined needs of the affected patient. This idea was taken up by the SPD group in the German Bundestag not with a view to replacing medical liability, but rather to provide some basic security to affected patients by creating a fund for victims of medical malpractice, similar to the solution existing in Austria under section 27a paragraph 6 KAKuG.[36] Such a complement to the current fault-oriented system appears, however, to be unacceptable to the majority view of German scholars on the grounds that it does not yield better results for patients, and in particular because it would eliminate one of the central tasks of liability law: preventing damages due to fear of the need also to compensate for any non-material damage suffered—there only is a legal basis in law to allocate payments of money to compensate pain and suffering if it is a consequence of faulty behaviour, and these can amount to considerable sums of money. Furthermore, the responsibility of the doctor to compensate damages is being seen as a corollary of the constitutional legitimacy of the freedom of the medical profession.[37]

This raises the question whether a limitation of the liability of hospitals and/or doctors is possible either by contract or by law. Several US states have introduced 'capping statutes' for damages for pain and suffering and for punitive damages, often at a sum of $250,000. This has led to a reduction in medical liability claims by more than 50%, and insurance premiums have been lowered by 20% on average. Katzenmeier considers that contractual limitations of liability might be acceptable to the extent that they are usually only denied

[34] www.marshclearsight.com and www.marsh.com/us/media/stars-becomes-marsh-clearsight.html (accessed 27 February 2017).

[35] C Katzenmeier, ‚Arzthaftpflicht in der Krise—Entwicklungen, Perspektiven, Alternativen', in *Medizinrecht* 2011, 201/208.

[36] ibid, fn 126 and: www.ris.bka.gv.at/GeltendeFassung.wxe?Abfrage=Bundesnormen&Gesetzesnummer=10010285 (accessed 27 February 2017). €0.73 per day per patient for a maximum of 28 days levied in order to feed a fund which compensates persons having suffered damage without any clear responsibility of the hospital/doctors involved.

[37] ibid, 209.

based on their insurability; so, if insurance becomes excessively expensive, contractual limitations might be lawful, as is the case for lawyers and patent lawyers who can limit their liability to the minimum amount that must be insured by law—which also was a response to the increasing liability risks these professions faced. Another argument in favour of a maximum liability sum for doctors could be that various court decisions and developments in the law of liability can only be explained by the fact that insurance cover exists.[38] Against this background, various discussions have taken place in the political arena though without any tangible results so far; the most acceptable idea could prove to be the creation of a complementary fund in order to cover peak damages, as a corollary to liability limitations by contract or law.[39]

E. Conclusions

The paradigm means of resolution of civil disputes in Germany has long been the remit of courts. German civil courts pride themselves on an efficient procedure, which involves a more inquisitorial model than the common law paradigm, the former being in general cheaper for participants and amenable to the application of fixed fees, which in turn supports the widespread use of legal expenses insurance and thus a significant amount of litigation. However, ADR and ombudsman schemes have been spreading, led by doctors and insurers—both groups that wish to retain good relations with their 'customers'. The various insurance ombudsmen, established around 2000, operate highly effectively. Medical ADR *Schlichtungsstellen* were established as far back as the 1970s, and likewise operate highly effectively.

However, all of these bodies operate under the shadow of the court system and its paradigm model, notably in making decisions by applying the law rather than notions of fairness, and within a similar procedural format. The insurance ombudsmen are able to utilise mediation as a technique and to promote compromises that are agreed by the parties, whereas the experience in Germany is that mediation and compromise are far more difficult to achieve as between doctors and patients. The reasons for this resistance to compromise and mediation in the medical context appear to be related to the somewhat inflexible authoritarian position of doctors rather than to an unwillingness by patients to compromise, and to the fact that apologies might be felt to imply admission of fault or lack of professionalism. In addition, doctors as insured persons are bound by the way their insurer decides to handle a claim, and this quite often seems to be the cause for a certain reluctance to compromise.

[38] ibid, 211 et seq.

[39] C Katzenmeier, 'Patientenentschädigungsfonds—rechtspolitische Forderungen und rechtsdogmatische Erwägungen', in: *Versicherungsrecht* 2014, 405, 412—referring also to the French systems of Fonds de garantie des dommages consécutifs à des actes de prévention, de diagnostic ou de soins dispensés par des professionnels de santé fed by a reinsurance fund supervised by the state (Caisse centrale de réassurance) which levies a small fee from all members of the medical professions (between €15 and €25).

Part VII

UK Schemes

23

The Armed Forces Compensation Scheme

SONIA MACLEOD

A. Introduction

The Armed Forces Compensation Scheme (AFCS) is a no-fault scheme to compensate injuries sustained as a result of serving in the UK armed forces, including reservists. The scheme covers injuries sustained from 6 April 2005 regardless of how they are sustained, and replaces the War Pension Scheme. No distinction is made between injuries sustained on operations, and those incurred during training, service approved sport, or while exercising to maintain fitness. Payments are available for injuries to service personnel and for their dependants if the injury proved fatal.

The AFCS rules are set out in legislation in the Armed Forces and Reserve Forces (Compensation Scheme) Order 2011[1] and the Scheme is administered by Veterans UK, a part of the Ministry of Defence with responsibility for administering armed forces pensions and compensation for those injured during service.

The AFCS was designed with a set of fundamental underlying principles at its core. These principles were examined and further clarified in the report of the former Chief of Defence Staff, Admiral the Lord Boyce, in his 2010 review of the Scheme.[2] As amended, the principles are to:

— be fair;
— be understandable, accessible and transparent;
— be contemporary and joined up;
— provide security;
— encourage employability;
— be compatible with human rights and fairness at work;
— be sustainable.

[1] The Armed Forces and Reserve Forces (Compensation Scheme) Order 2011 SI 2011/517, Armed Forces Compensation Scheme Statement of Policy. With details on application set out in Ministry of Defence JSP 765, updated December 2015, available at www.gov.uk/government/uploads/system/uploads/attachment_data/file/485419/20151203_-_JSP765_FINAL-Official.pdf.

[2] *Review of the Armed Forces Compensation Scheme 2010*, ISBN 9780101779821, Cmd 7798, available at www.gov.uk/government/uploads/system/uploads/attachment_data/file/228893/7798.pdf.

B. Eligibility

Causation of Injuries 'Caused by Service'

Compensation is paid for personal injury (physical or mental) to members of the armed forces and reservists where the injury was caused wholly or partially by the individual's service. Injuries are classified according to a 15-level tariff system, with level 1 being the most severe injuries.[3] A claimant must show that the injury is more likely than not to have been caused by service (balance of probabilities test). Where an injury is partly caused by service, compensation will be awarded if the service is the predominant cause of the injury.

A number of defined contexts are specifically excluded from the Scheme as such injuries are not considered to have been caused by service. These include injuries caused by or occurring during: social events, travel to and from work, sport which is not formally recognised or not had prior approval, and the majority of slips and trips.[4]

There is no automatic entitlement to obtaining a payment under the Scheme. When an injury is caused in circumstances that are not excluded by the Scheme, an assessment will be made as to whether the injury or injuries sustained are caused by service. Each case is assessed, using all the facts of the case, to establish whether the injury was caused (wholly or partly) by service or not. Where the case is borderline the decision-maker may need to engage with the Ministry of Defence (MoD) policy team. Factors that may mean an injury was more likely to be caused by service include, but are not confined to: acting on orders, responding to a service-related emergency, being on operations or exercise, participation in service-approved and -recognised sport, being in a hazardous environment as a result of service, or performing an activity specified in the individual's job description.

Factors that may mean an injury was less likely to be caused by service include, but are not confined to: disobeying orders, being at a social event, using alcohol or drugs, or participation in a sporting activity that has not received prior approval from the Defence Council. Causation of Injuries: 'Injury or Illness Made Worse by Service'

There are limited provisions for the payment of an illness or injury which is worsened by service. The specific circumstances when this applies are when an injury:

- was sustained before entry into the services, and was recorded in the individual's medical reports on entry to service;
- was sustained before entry into the services but about which the individual was not aware; and
- the injury was not found at medical examination on entry; or
- arose during service but was not caused by service.

These injuries have specific timing and discharge requirements.[5]

[3] See The Armed Forces and Reserve Forces (Compensation Scheme) Order 2011 2011 No.517

[4] Armed Forces Compensation Scheme Statement of Policy (n 1) paras 2.4 and 2.21–2.33.

[5] ibid, paras 2.17–2.20.

Time Limits

For the majority of cases a 7-year time limit is enforced.[6] Time starts to run from the earliest of the following: the date of the injury was sustained/made worse by service; the day the claimant's service ends; or the date the claimant first sought medical attention in relation to an illness. If the illness started within the 7-year period but was not diagnosed until less than a year before the end of the 7-year limitation, then the time limit may be extended for a further 3 years to allow the applicant sufficient time to bring a claim.

Late-onset illnesses (including mental illnesses) are not subject to the 7-year limitation, but have a 3-year time limit which starts from the date of diagnosis.

When an individual dies a 3-year time limit exists; should the individual have died while in service a claim for bereavement compensation will automatically be considered. If an individual dies within 7 years of leaving the service from an injury or illness caused by service, then there is a 3-year limitation period for claiming bereavement compensation which starts on the date of death. This 7-year limitation can be extended in the case of death caused by late-onset illnesses or death due to an injury in which an AFCS award has been made where the lump sum was within levels 1–9 of the tariff.

Potential Applicants

The injured individual is eligible to apply. If the injuries are of such a degree that the injured party is unable to manage his or her own affairs, then the claim may be managed by another person. The representative can either have legal authority to manage the injured individual's affairs or, if such as person does not exist, Veterans UK can appoint a representative.

Eligible partners (spouses or civil partners of more than 6 months duration prior to the injury) are automatically able to apply for compensation for bereavement. If there is no spouse or civil partner, then an individual will have to show he/she is eligible to be able to claim. To be eligible the individual must have lived with the deceased in a substantial and exclusive relationship, where they were not prevented from marrying or forming a civil partnership, and must have been either financially dependent upon the deceased or financially interdependent.

In the event of a death, then eligible children may receive child payments. Eligible children are birth children, adopted children or other children who were financially dependent upon the deceased. For the purposes of child payments children are either under 18 years old or, if in full-time education or vocational training, under 23.

C. Application Process

The standard of causation required is that the injury was on the balance of probability due to service. While the responsibility to show that the injury was caused by service rests

[6] ibid, paras 5.4–5.22.

with the claimant, the process is designed not to be onerous. The determination of claims is inquisitorial and not adversarial, with Veterans UK undertaking the vast majority of evidence gathering on the claimant's behalf.

There are two types of claim which will be considered automatically without the need for a claim form:

— When an individual has been medically discharged and has been considered for an ill-health pension, then Veterans UK will automatically consider whether the injury was caused by service and warrants an AFCS award (assuming that an AFCS award has not already been given for that injury).
— When an individual dies in service and leaves an eligible partner and/or eligible children, then bereavement compensation will automatically be considered.

Additionally, if an application to the AFCS has been rejected because the injury listed as the principal medical condition was not considered to have been caused by service, the application will be reconsidered, without the need for a further claim form, to determine if the principal medical condition qualifies as a pre-existing injury which was worsened by service.

All other injuries require an application forms; these are available online to complete, print, sign and send to Veterans UK.

D. Awards

Lump Sum Awards

All AFCS awards include a lump sum payment for the pain and suffering caused by the injury. There is a 15-level tariff system, with level 1 being the most severe injuries. The Scheme's aim is to ensure that the compensation awarded reflects the level of injury sustained; to that end the compensation for a number of less debilitating injuries will be lower than that awarded for a single, more profound injury.

Injuries which do not impact on future earning potential, but may have a significant impact on an individual's self-image, confidence and self-worth, attract an additional payment, a supplementary award. For example, genital injuries resulting in infertility do not impact on future earning potential but do impact on lifestyle so would attract a supplementary award.

Tariff 1 lump sum awards (including supplementary awards) are capped at £570,000 per incident.

Guaranteed Income Payment

If an individual has sustained an injury that will impact on their earning potential, then a Guaranteed Income Payment (GIP) may be provided. This is a lifelong, tax-free, inflation-proof income that covers both salary and pension provision. There is no cap on GIP, and it is calculated on a fixed basis taking into account factors such as age, salary and future earning potential.

GIP is banded, see Table 23.1. Only one GIP is paid, regardless of the number of injuries sustained. If injuries are sustained in more than one incident, then GIP will be paid at the rate of the highest tariff injury. If more than one injury is sustained in a single incident, then GIP will either be paid at the band corresponding to the tariff level of the most severe injury if the injuries fall into different tariffs, or if both injuries fall within a single tariff level, then the GIP will be paid at the banding immediately above the band into which the injuries fall.

GIP is adjusted in the light of Armed Forces Pensions Payments, the Gurkha Pension Scheme and the Armed Forces Early Departure Payment Scheme, where applicable.[7]

Table 23.1: GIP bandings by injury tariff level

Band	Tariff levels	% of future earnings replaced by GIP
A	1–4	100
B	5–6	75
C	7–8	50
D	9–11	30
Unbanded	12–15	0 (as the injury is not considered to have significant permanent effects on an individual's earning potential)

Paired Body Structures

Both the lump sum and GIP payments recognise that damage to constituents of paired body structures[8] has more than double the impact on an individual than equivalent injuries to half of the pairing. If one foot is lost, then a lump sum of £60,000 is paid along with 50% GIP, but if both feet are lost, then the lump sum is £175,000 with 75% GIP.

Multiple Injuries and Body Zoning

Where a single incident results in multiple injuries there are rules[9] to ensure that:

— those most seriously injured receive the most compensation;
— those with a large number of minor injuries receive a lower award than those with fewer, more serious injuries;
— every injury is compensated.

[7] See ibid, paras 3.19–3.21.
[8] Paired body structures to which these rules apply are arms or parts of an arm, feet, hands, kidneys, legs or parts of a leg, total loss of sight in both eyes and total loss of hearing in both ears. See ibid, paras 3.22–3.29.
[9] Detailed at ibid, paras 3.38–3.47.

The overall impact of the injury may be assessed by considering the impact of the injury in relation to five body zones:

A. the head and neck;
B. the torso;
C. the upper and lower limbs;
D. the senses;
E. mental health.

The more zones affected, the greater the impact of the injury on the individual.

Armed Forces Independence Payment[10]

Where an individual has been awarded a GIP of 50% or more they are eligible for an Armed Forces Independence Payment (AFIP). This is to cover any extra costs incurred as a result of the service-related injury. It is paid by the Department for Work and Pensions, but is administered by Veterans UK. Application is by form, submitted to Veterans UK. There is no time limit on submitting the form. Payment is currently set at £134.40 per week, rising with annual uprating.

Fast Payment[11]

This is intended to provide the most seriously injured with the option to claim an upfront AFCS payment without the need to wait for the full AFCS claims process. The aim is to provide financial assistance as quickly as possible after the injury. Fast payment applies when the injury occurred within service after 9 May 2011 and it is beyond doubt that it was caused by service and it is beyond doubt that it is a tariff 1–8 injury. Fast payment has a separate, shorter form.[12]

Temporary Awards

If an injury occurs for which there is no descriptor on the tariff and the assessor feels the injury is sufficiently serious to warrant compensation, then a temporary award will be granted.[13] Within a year of the grant of this temporary award the tariff will be amended to include an appropriate descriptor for the injury and if the temporary award was lower than the new tariff award, then the difference will be paid to the applicant.

[10] ibid, paras 3.50–3.53 and www.gov.uk/government/uploads/system/uploads/attachment_data/file/414586/20130322_afip_faq_8april_comms_combined.pdf.
[11] Armed Forces Compensation Scheme Statement of Policy (n 1), paras 3.54–3.58.
[12] See www.mod.uk/afcs to access the fast payment form.
[13] Armed Forces Compensation Scheme Statement of Policy (n 1) paras 3.30–3.33.

Interim Awards

If the extend of an injury is uncertain, then an interim award may be granted until it is possible to ascertain the extent of the injury.[14] A first interim award will be made for a maximum period of 2 years. This can be extended only once for another 2 years. After this the award will be finalised.

Overseas Medical Expenses

AFCS may pay medical expenses for those who are treated overseas in some circumstances.[15] As an applicant's needs may change, any application with an ongoing element will be reviewed.

Review

An individual's healthcare needs will be reviewed once every 24 months at the minimum.[16]

Bereavement Payments

Bereavement payments are available to eligible partners and eligible children. These can take three forms: Survivors' Guaranteed Income Payments (SGIP), Child Payments and Bereavement Grants.

Survivors' Guaranteed Income Payments

The SGIP is an ongoing income stream paid to an eligible partner to compensate for the loss of the deceased partner's earnings. It is calculated according to a set formula similar to GIP and is then multiplied by a factor of 0.6 to give the SGIP payment levels.[17]

Child Payments

A Child Payment is an income stream paid to eligible children to compensate for the loss of the deceased's income upon which they were financially reliant. It is paid up to the point when it would be reasonable to expect the child to be financially independent.[18] The amount of child payment is calculated on a set formula and depends upon whether there is an eligible partner and the number of eligible children.

[14] ibid, paras 3.34–3.36.
[15] Detailed at ibid, 3.59–3.73.
[16] ibid, paras 3.74–33.75.
[17] Detailed at ibid, paras 4.15–4.16.
[18] Detailed at ibid, paras 4.17–4.27.

Bereavement Grants

A Bereavement Grant is paid to an eligible partner; in the event there is no eligible partner, then it will be divided between the eligible children. It is a tax-free lump sum paid in certain circumstances. It is intended to offset the difference in death-in-service lump sum payments available from the two Armed Forces Pension Schemes.[19]

E. Reviews, Reconsiderations and Appeals

If an individual is unhappy with the award or his/her circumstances have changed, then there are options to reconsider the applicant, to review the award or to appeal.[20]

Reconsideration

Applicants can explain in writing on a set form why they feel their case should be reconsidered. Veterans UK will then look at the case in the light of this input and will reconsider the case. This can result in an award being either increased or remaining the same; it cannot be decreased.

Reviews

There are five types of review:

- review of an AFCS award when service ends;
- review of an AFCS award in exceptional circumstances;
- review of an AFCS award due to ignorance of fact or mistake;
- final review of an AFCS award;
- review of an AFCS award to take account of an award of damages.

These reviews are intended to re-examine the case. All four types of review can be instigated by the claimant; a review due to ignorance of fact or mistake can also be instigated by the MOD. The only reviews that can result in a reduction in compensation are the where there is ignorance of fact or mistake or a review to take account of an award of damages.

Appeals

When an individual is unhappy with the decision of AFCS he or she has the right to appeal to an independent tribunal.[21] If the case has already been reconsidered, then it will

[19] Detailed at ibid, paras 4.27–4.35.
[20] ibid, ch 8.
[21] Further information available at www.justice.gov.uk/tribunals/war-pensions-and-armed-forces-compensation/appeals.

proceed straight to appeal; if not a reconsideration will automatically be undertaken by AFCS. This may result in resolution and remove the need for an appeal. Appeals must be made within 12 months of the original decision or reconsideration decision. Appeals must be made in writing (usually on the standardised appeal form) to Veterans UK and must contain certain mandated information.[22] The tribunal will consider the case, often with oral evidence from the claimant, and will make a decision. The appeal can result in an award remaining the same, being increased or being decreased.

If either party does not agree with the decision of the tribunal, each has the right of a further (and final) appeal. The further appeal must be lodged within 6 weeks of the original appeal decision.

F. Statistics[23]

Claim Numbers and Values

In the decade since its inception in 2005 the AFCS has had a total of 60,885 claims made by 46,110 people. The vast majority of these, 60,030, have been injury claims, with just 855 survivors' claims being made (see Table 23.2).

Table 23.2: AFCS claims numbers (revised) from 6 April 2005 to 30 September 2015

	Claims registered during										
	2005–06	2006–07	2007–08	2008–09	2009–10	2010–11	2011–12	2012–13	2013–14	2014–15	2015–16[a]
Total number of claims	360	1,595	2,935	4,570	5,505	6,710	7,495	7,985	8,990	9,535	5,215
Total number of people	355	1,545	2,730	4,255	4,965	5,925	6,785	7,170	7,760	8,175	4,625
Injury claims numbers	330	1,475	2,800	4,455	5,370	6,600	7,420	7,925	8,945	9,505	5,205
In-service	200	735	1,630	2,960	4,095	5,120	5,950	6,160	6,710	7,215	4,065
Medical discharge	115	605	630	730	300	320	390	330	355	295	145
Post-service	10	130	515	685	910	1,090	975	1,320	1,690	1,825	905
Additional claim	0	–	30	80	65	70	105	115	190	170	90

(continued)

[22] Detailed at Armed Forces Compensation Scheme Statement of Policy (n 1) paras 8.23–8.33.

[23] Taken from www.gov.uk/government/collections/armed-forces-compensation-scheme-statistics-index

Table 23.2: ***(Continued)***

	Claims registered during										
	2005–06	2006–07	2007–08	2008–09	2009–10	2010–11	2011–12	2012–13	2013–14	2014–15	2015–16[a]
Survivors' claims numbers[b]	**25**	**120**	**130**	**115**	**135**	**110**	**75**	**60**	**45**	**30**	**10**
Death in-service	25	115	130	105	135	105	70	60	45	25	10
Death post-service	0	0	–	–	–	–	–	0	0	–	0
Additional child	0	–	0	–	–	–	–	–	–	0	0

Source: Armed Forces Compensation Scheme Biannual Statistics, 4 December 2015, available at www.gov.uk/government/statistics/armed-forces-compensation-scheme-statistics-financial-year-201516.

[a] Only includes figures to 30 September 2015 and all figures for this period are provisional.

[b] A single survivors' claim may result in an award that gives entitlement to one or more SGIP

The number of injury claims made each year has been steadily rising, and provisional figures indicate that this trend continued in 2015/16. In contrast, the number of survivor claims was fairly constant from 2006 to 2010/11 and has decreased thereafter; this corresponds to the periods of highest activity of British Army deployments in Afghanistan.

The value of compensation awards paid out by AFCS is shown in Figure 23.1. Between 2005 and 2015 over £422 million was paid out as lump sum awards under the scheme, and over £53 million was paid out as GIPs for injury and illness claims. Survivors' GIPs payments over the same time frame totalled over £44 million.

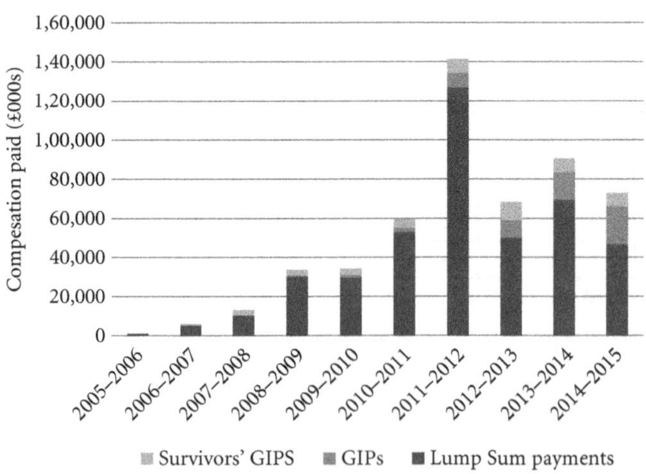

Figure 23.1: AFCS Compensation payment amounts by award type 2005–15

Claim Outcomes

Of the injury claims made, there are startling differences in the acceptance and rejection rates of claims when broken down by category, with medical discharge claims having a far higher rejection rate, 62%, than other claim types (see Table 23.3).

Table 23.3: Injury claims from 6 April 2005 to 30 September 2015 by claim type and outcome

	In-service	Medical discharge	Post-service	Additional claims	Total
Claims made	42,240	4,200	9,245	815	**56,495**
Accepted—lump sum plus GIP	1,110 (3%)	100 (2%)	70 (1%)	295 (36%)	**1,575 (3%)**
Accepted—lump sum only award	26,565 (63%)	1,145 (27%)	3,680 (40%)	440 (54%)	**31,830 (56%)**
Accepted—no award	5,180 (12%)	350 (8%)	1,485 (16%)	50 (6%)	**7,065 (13%)**
Rejected	9,390 (22%)	2,600 (62%)	4,010 (43%)	30 (4%)	**16,030 (28%)**

For survivors' claims the picture is equally variable by claim category. All claims for death post-service have been rejected so far, but as these are very low claim numbers it would be difficult to draw concrete conclusions from this data (see Table 23.4).

Table 23.4: Survivors' claims from 6 April 2005 to 30 September 2015 by claim type and outcome

	Death in-service	Death post-service	Additional child claims	Total
Claims made	815	10	15	**840**
Awarded	315 (39%)	0 (0%)	15 (93%)	**330 (39%)**
Rejected	500 (61%)	10 (100%)	~ (7%)	**510 (61%)**

Claim Clearance Times

As well as differences in outcome, injury/illness claims and survivors' claims also differ in their processing times (see Figures 23.2 and 23.3). The median clearance time frame for survivors' claims between 6 April and 30 September 2015 is 50 working days, and for an illness/injury claim it was 75 working days. These figures are very prompt when compared with a negligence claim.

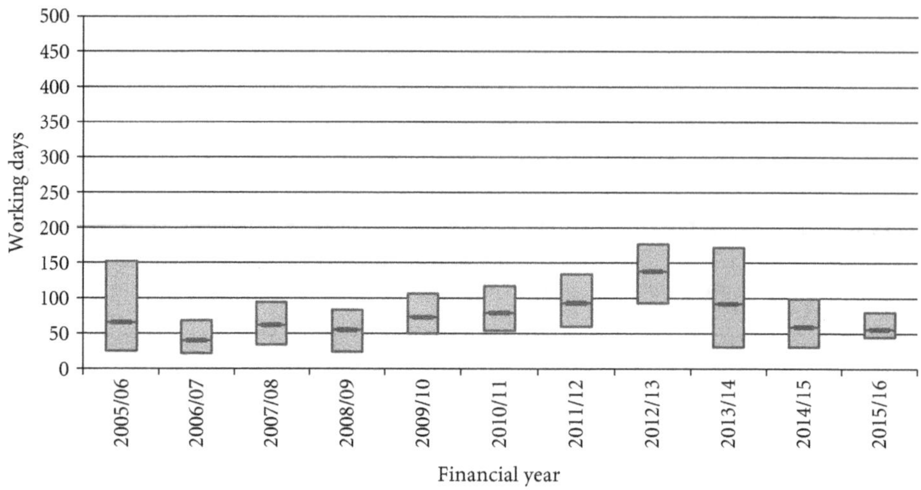

Figure 23.2: The median average clearance times for injury/illness claims. The graph shows the median average and the interquartile range (IQR) for clearance times, in working days, by financial year

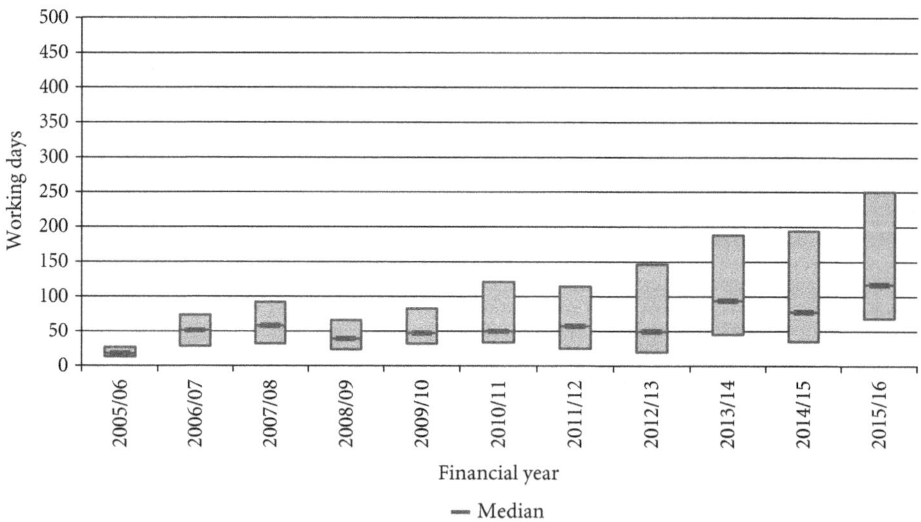

Figure 23.3: The median average clearance times for survivors' claims. The graph shows the median average and the interquartile range (IQR) for clearance times, in working days, by financial year

Reconsiderations and Appeals

It is difficult to directly calculate the rate of appeals as claims may not be appealed in the year in which they were registered. However, overall figures of reconsiderations and appeals per year are valuable. Table 23.5 indicates that the number of reconsiderations and appeals has been rising each year.

Table 23.5: Numbers of reconsiderations and appeals registered per year 2005/06–2015/16

	Claims registered during										
	2005–06	2006–07	2007–08	2008–09	2009–10	2010–11	2011–12	2012–13	2013–14	2014–15	2015–16[a]
Number of people seeking reconsideration	0	125	250	600	755	940	1,210	1,145	1,770	1,880	360
Total number of reconsiderations	0	125	255	625	805	995	1,330	1,230	1,940	2,075	395
Total number of people appealing	0	40	110	270	300	430	570	600	925	1,140	530
Total number of appeals made	0	40	115	275	315	455	630	650	1,010	1,275	585

[a] Provisional figures, likely to increase.

Interestingly, a fact which is also illustrated by Figure 23.4, this increase does not appear to be out of proportion with the year-on-year increases in claim numbers, suggesting a consistent level of reconsiderations and appeals.

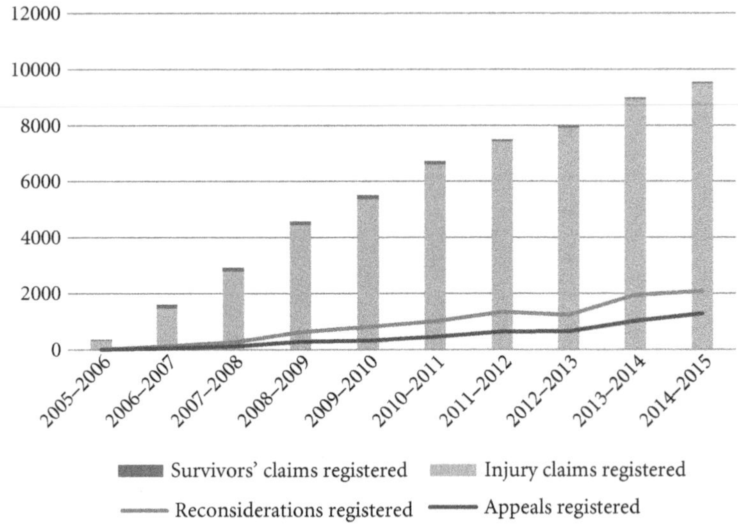

Figure 23.4: Numbers of claims, reconsiderations and appeals registered 2005/06–2014/15

The outcome of appeals and reconsiderations is a key indicator of the Scheme's decision making. Table 23.6 shows the number of reconsiderations requested from 6 April 2005 to 30 September 2015 and the outcomes of these reconsiderations as a percentage of requests. These figures show very low levels of decreasing awards and of granting new awards upon reconsideration, which indicates that the eligibility decisions of the Scheme are consistent with those applied on reconsideration. The quantification is slightly less consistent: the majority of awards remain at the same level, but almost a quarter of awards to live claimants are increased. The number of survivors' reconsiderations is very low, so it is difficult to drawn meaningful conclusions, but a similar pattern is seen between living claimants' and survivors' awards.

Table 23.6: AFCS reconsiderations from 2005 to 2015 claims by outcome

	Injury/illness claims				Survivors' claims		
	In-service	Medical discharge	Post-service	Injury totals	Death in-service	Death post-service	Additional child
Reconsiderations requested	7,620	600	1,260	9,480	80	—	—
Outcomes							
New award	60 (0.8%)	10 (1.7%)	5 (0.4%)	75 (0.8%)	—	0	0

(continued)

Table 23.6: *(Continued)*

	Injury/illness claims				Survivors' claims		
	In-service	Medical discharge	Post-service	Injury totals	Death in-service	Death post-service	Additional child
Increased award	2,025 (26.6%)	95 (15.8%)	215 (17.1%)	2,335 (24.6%)	5 (6.3%)	0	0
Maintained award	5,500 (72.2%)	495 (82.5%)	1,035 (82.1%)	7,030 (74.2%)	70 (87.5%)	—	—
Reduced award	40 (0.5%)	5 (0.8%)	—	45 (0.5%)	0	0	0

The figures for appeals (Table 23.7) also show a very low rate of disallowed and out-of-jurisdiction appeals. Very few appeals reduce the award value. A substantial number of claims, just over 40%, are maintained at the same level. Approximately 30% of appellants receive a favourable reconsideration. This is interesting as these ought to have been settled before the appeal was lodged. Another quarter of appeals result in either a new award or an increased award. This suggests scope for improvement in both the initial eligibility and quantification and in the reconsideration processes.

Table 23.7: AFCS claims appeals from 2005 to 2015 by outcome

	Illness/injury claims				Survivors' claims		
	In service	Medical discharge	Post-service	Total	Death in service	Death post-service	Additional child
Appeals lodged	2,675	275	490	3,440	30	~	~
				Outcomes			
New award	240 (9.0%)	40 (14.5%)	55 (11.2%)	335 (9.7%)	—	0	0
Increased award	465 (17.4%)	45 (14.5%)	60 (12.2%)	570 (16.6%)	0	0	0
Maintained award	1,095 (40.9%)	115 (41.8%)	230 (46.9%)	1,440 (41.9%)	20 (66.7%)	—	—
Reduced award	30 (1.1%)	—	5 (1.0%)	35 (1.0%)	0	0	0
Favourable Reconsideration	800 (29.9%)	65 (23.6%)	135 (27.6%)	1,000 (29.1%)	—	0	0
Disallowed—late appeal	20 (0.7%)	—	—	20 (0.6%)	0	0	0
Overturned by upper tier tribunal	—	0 (0%)	0 (0%)	0 (0%)	0	0	0
Out of jurisdiction	20 (0.7%)	5 (1.8%)	—	25 (0.7%)	—	0	0

G. Conclusions

The AFCS provides a welcome alternative to litigation for those injured during service. The difficulties of the very high thresholds presented by negligence claims are well documented. For servicemen and women the no-fault approach provides a pragmatic solution where it would be difficult for a claimant to succeed in negligence—for example injuries due to improvised explosive devices often could not have been prevented, so proving negligence, and hence obtaining compensation, would have been very difficult.

One of the major criticisms of the AFCS is the level at which injuries are quantified; the amounts are substantially lower than would be obtained in tort. Claimants can still litigate should they so wish, but the difficulties outlined above mean that the AFCS is often the only realistic option for compensation. Given that these injuries occur during service to the country it is easy to see why the low-values awarded have been considered inadequate by some.

In practical terms, the AFCS seems to work fairly well, though the rate of successful appeals suggests that there is room for improvement. Anecdotal evidence from claimants suggests that they would like more a more consistent point of contact.

The idea of having a rules-based scheme underpinned by a set of fundamental principles is interesting. The principles provide a guide for the interpretation of the rules, and also enable amendment of the rules if they are inconsistent with the principles. This provides an effective mechanism for enabling both externally and internally driven changes in the rules in accordance with the overarching principles.

24

Criminal Injuries Compensation Scheme

SONIA MACLEOD

A. Background

A scheme to compensate blameless victims of crime has been in existence in Great Britain[1] since 1964, with considerable alterations and amendments. Prior to 1 April 1996 the scheme was 'common law' based and cases were assessed in the same way as a civil case in the courts would have been. These payments were administered by the Criminal Injuries Compensation Board and were governed by ministerial guidelines rather than having a statutory basis. In 1988 a statutory scheme was enacted,[2] but these provisions were never brought into force. The scheme was placed on a statutory footing by the Secretary of State in 1996 using powers granted to him by the Criminal Injuries Compensation Act 1995. This framework applied to all injuries from 1 April 1996 when the Criminal Injuries Compensation Authority (CICA) replaced the Criminal Injuries Compensation Board[3] and a tariff of injuries was introduced. Various incarnations of the scheme have existed with major amendments made in 2001, 2008 and to its current form in 2012, but throughout these changes, injury tariffs have remained. Since 27 November 2012 the Criminal Injuries Compensation Scheme 2012 has been in force.[4]

B. Compensation Available

CICA states that it is a government organisation that can pay money (an award) to people who have been physically or mentally injured because they were the blameless victim of a violent crime. Claims can be made under the following heads of damage:

— personal injury following a single incident;
— personal injury following a period of abuse;

[1] The Scheme covers England, Wales and Scotland. Northern Ireland has its own scheme.
[2] Criminal Justice Act 1988, ss 108–17 and schedules 6 and 7.
[3] www.justice.gov.uk/victims-and-witnesses/cica.
[4] Criminal Injuries Compensation Scheme 2012, available at www.justice.gov.uk/downloads/victims-and-witnesses/cic-a/am-i-eligible/criminal-injuries-comp-scheme-2012.pdf.

- loss of earnings;
- special expenses payments (to cover special injury-related requirements that are not available free of charge from any other source);
- fatal injuries, including loss of parental services and financial dependency; and
- funeral payments.

Awards are capped; a claim must be for at least £1,000 to be eligible under this scheme and there is a maximum limit of £250,000 per person under the tariffs. In severe cases non-tariff expenses, such as loss of earnings and medical expenses, may also be awarded, but there is a total cap of £500,000 which includes both tariff and non-tariff claim elements.

C. Eligibility[5]

With the exception of terrorism the criminal act to which the injury is attributable must occur in Great Britain. There are various ways to establish eligibility:

- As a primary victim when the injury (physical or mental) resulted from a criminal act; or
- When a criminal injury resulted from the taking of an exceptional and justified risk for the purpose of apprehending an offender or suspected offender, preventing a crime, containing or remedying the consequences of a crime, or assisting a constable who is acting for one or more of the purposes described previously.
- Secondary victims can also establish eligibility if they sustained a mental injury which is directly attributable to being present at and witnessing a criminal incident, or the immediate aftermath of an incident, in which a loved one was the primary victim of a criminal injury. A loved one is defined as a person with whom the applicant had a close relationship of love and affection at the time of the incident; and if the loved one is alive at the date of the application, continues to have such a relationship.

There are requirements on residency, that the victim was ordinarily resident on the day of the injury and most commonly that they are a British or EU citizen or close family member of a British or EU citizen. Other more unusual ways of qualifying such as being a victim of people trafficking are outlined in detail in the scheme.[6]

A criminal conviction and/or identification of the assailant are not necessary to establish that a criminal injury has occurred; however, the crime which lead to the injury must have been reported (as soon as reasonably practicable). If a conviction has not occurred, then the incident will be judged on the facts using the civil standard of proof, 'the balance of probabilities', rather than the criminal standard of 'beyond reasonable doubt'.

[5] Details can be found in the Criminal Injuries Scheme 2012.

[6] S 2 and Appendix 1 of *A Guide to the Criminal Injuries Compensation Scheme 2012* (Criminal Injuries Compensation Authority, 28 June 2013) available at www.justice.gov.uk/downloads/victims-and-witnesses/cic-a/how-to-apply/cica-guide.pdf.

Multiple Injuries

If a claimant has more than 1 injury then he/she will be paid 100% of the tariff payment for the highest-value injury, 30% of the tariff payment for an injury with equal or second highest value and 15% of the tariff payment for an injury with an equal or third highest value. The scheme only pays for 3 injuries. However, additional tariffs are available if the injury has led to the claimant becoming pregnant, losing a fetus or contracting a sexually transmitted disease.

Deductions from Awards

The scheme is intended to assist 'blameless' victims; awards are therefore refused or deductions are made where it is considered that a claimant has contributed to his/her own injuries or is in some other way blameworthy. Factors that can influence the making and level of a payment include not cooperating with the police, criminal justice system or CICA, engaging in aggressive conduct (excessive consumption of alcohol or use of illicit drugs will not excuse aggressive behaviour) and having unspent criminal convictions[7] and/or evidence of bad character on the part of the claimant.

The scheme is intended as a last-resort option when all other compensation avenues have been exhausted. The award offered will be reduced by the amount of compensation received from other sources, civil claims, etc.

Limitation

For those over 18 years old the claim should be registered within 2 years of the incident. There is a narrow exception to this rule, namely if the claimant's injuries are such that it would not have been reasonable to have expected him to submit a claim within the 2-year timeframe and he supplies sufficient evidence in his application to enable a decision to be made without further extensive enquiries. Claimants have to supply evidence to justify the extension of the limitation period.

If an incident took place before the victim was 18, the preferred option is for an adult to apply on behalf of the child victim as this enables evidence of the crime to be obtained more easily. If compensation was not applied for and the crime was reported before the victim turned 18, then the 2-year limitation period starts to run when the victim turns 18. If compensation was not applied for and the crime was not reported before the victim turned 18, then the 2-year limitation period will start to run from the date when the crime was first reported.

The scheme is a last-resort option, but the limitation periods mean that claims should be lodged with CICA within the 2-year limitation period and CICA will defer its proceedings until any other outstanding claims have been finalised.

[7] Deductions are explained at ibid, Appendix 2.

D. Application Process, Application Numbers and Refusal Numbers

The onus is on the applicant to demonstrate that he or she is eligible for an award. Applications can be made online or by telephone. The forms are designed to be user-friendly and quick: providing the applicant has all the relevant information to hand, an online application takes approximately 30 minutes and a telephone application approximately 20 minutes. Over 90% of applications are now made online. Applications can also be made via Victim Support, a national charity which helps the victims of crime.

Once an application is submitted, CICA checks the applicant meets the eligibility criteria, including contacting the police to cross-refer the application with the reported crime. Further investigations are only carried out once eligibility has been established. Claims may be refused on various grounds as detailed above. Reasons for refusing claims from 2009 to 2013 are listed in Table 24.1.[8] Since 2013 the reporting of reasons for refusing claims has altered; Table 24.2 gives data for 2013/14–2014/15.

Table 24.1: Reasons for refusing CIC claims from 2009 to 2013

Scheme paragraph	Criterion	2013–12	2011–12	2010–11	2009–10
6	Injury sustained before 1 August 1964	13	6	6	15
7A	Previous claim for same injury	749	611	543	447
7B	Injury sustained in family setting before 1 October 1979	76	67	78	136
8A	Mainly, injury did not result from crime of violence	3,050	3,201	3,566	4,183
9	Application did not meet restrictions in paragraph	246	265	380	322
11	Motor vehicle cases: mainly, vehicle not used as a weapon to injure	255	312	305	346
12	Accidental injury sustained in law enforcement: risk not justifiable	78	103	115	123
13A	Failure to report without delay	1,378	1,533	1,610	2,147
13B	Failure to cooperate with police in bringing assailant to justice	3,690	4,381	4,983	5,290
13C	Failure to cooperate with the authority	2,140	1,996	2,126	3,685
13D	Conduct before, during or after the incident	3,110	2,967	3,200	3,898

(continued)

[8] Some claims are refused for more than one reasons, so the number of refusals listed may be greater than the total number of refusals. Figures obtained from CICA Annual Reports and returns for 2009 onwards.

Table 24.1: ***(Continued)***

Scheme paragraph	Criterion	2013–12	2011–12	2010–11	2009–10
13E	Applicants criminal record/character	5,235	4,485	3,864	4,929
16A	Assailant would have benefited from award	10	35	102	108
16B	Award would have been against a minor's interests	0	1	5	9
17A	Assailant in family violence/abuse claim not prosecuted	140	81	2	3
17B	Violence between adults in the same family	16	34	58	79
18	Claim not submitted within 2 years of incident	1,550	1,804	1,196	1,371
25	Injury not serious enough to qualify for minimum award of £1000	7,406	7,839	8,862	10,879
26	Pre-existing medical condition	447	498	580	576
Total rejections		25,455	30,219	31,581	38,546

Table 24.2: Reasons for refusing CIC claims from 2013 to 2015

Criterion	2014–15	2013–14
Injury does not qualify for compensation	7,584	8,444
Failure to cooperate with the authority	3,328	2,408
Injury did not result from crime of violence	2,246	2,907
Applicant's criminal record/character	2,017	4,635
Failure to cooperate with police in bringing assailant to justice	1,910	2,846
Conduct before, during or after the incident	1,359	2,564
Failure to report without delay	1,119	1,429
Claim not submitted within time limit	754	1,048
Other	1,624	2,026

The application process and timeframe depends upon the complexity of the case and the type of award sought. Medical information may be sought and it may be necessary to wait until an applicant's medical prognosis is clear before reaching a final decision. During this process CICA will write to the applicant, keeping him/her informed of progress. Figure 24.1 shows the time taken to reach a first decision in 2014–15. The majority of applications are decided within 10 months. The average time to reach a first decision in 2014–15 was 10.9 months, a slight increase from 10.4 months in 2013–14.[9]

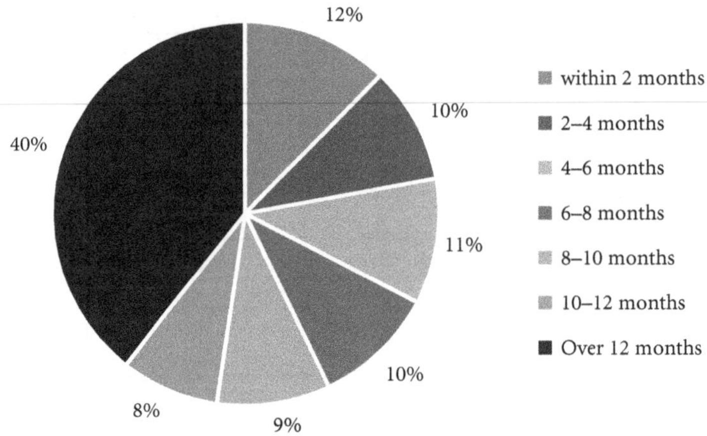

Figure 24.1: Percentage of claims decided by time period for 2014–15

Once an application is decided, CICA will write to the applicant. If an award is offered the applicant has 56 days in which to complete and return an acceptance form. Failure to do so may result in the award being withdrawn.

The preferred option is a final award, which will be given when the losses are clear and established. If the investigations are ongoing and cannot fix a figure for financial or other losses and the claimant is clearly eligible, then an interim payment may be given; this amount will be deducted from any final payment. Payments are made electronically into the applicant's bank or building society account. If an adult applicant cannot manage his/her own affairs due to lack of capacity, then payment will be made to a person with authority over the applicant's affairs to be held for the applicant's benefit. In the case of child applicants the award will generally be held in an interest-bearing account until the child reaches 18, when it will be released to them, though other arrangements, such as payment into a Child Trust Fund or Child ISA, may be permitted.

E. Claims Settled

As can be seen in Figure 24.2 the number of claims settled per year has been decreasing year on year; the compensation paid peaked in 2011/12 and has been decreasing since then. The live caseload of CICA has also been decreasing year on year during this time period.

[9] CICA Annual Report 2014–2015.

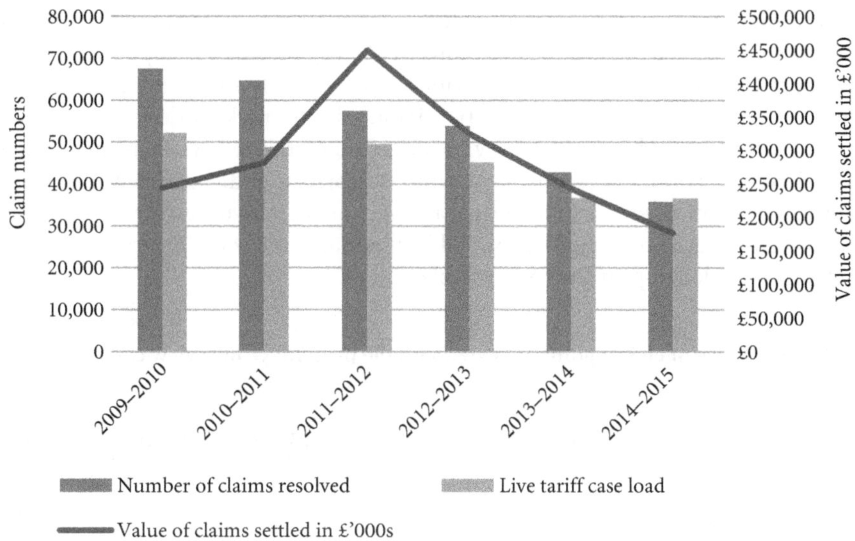

Figure 24.2: Claims resolved, settlement values and live tariff caseload numbers 2009–15

F. Reopening Cases, Reviewing Decisions and Appeals

A case can be reopened at CICA's discretion or at the request of a First-tier Tribunal (Criminal Injuries Compensation). This mechanism is used to allow an additional payment where either a person has previously accepted a payment and then subsequently died as a result of the original criminal injury or when there has been such a change in the applicant's medical condition that an injustice would occur if the case were not reopened. Cases are not normally reopened unless it is within 2 years of the final decision. Requests to reopen a case on medical grounds more than 2 years after a decision will only considered if the applicant provides enough evidence to make a decision on the case without the need for further extensive enquiries.

If an applicant is unhappy with the decision made by CICA he or she can request a review of the case by a different assessor. The request must be in writing and within 56 days of the original decision. Additional information to assist with the review may be included. If it will take longer than 56 days to gather such information, then the applicant can request an extension of the time limit by up to another 56 days. Reviews can result in more favourable, less favourable or unaltered decisions, and awards offered can be either increased or decreased. Review decisions will be posted to the applicant along with an appeal application form.

If an applicant is still unhappy after a review he or she can appeal to the First-tier Tribunal (Criminal Injuries Compensation) in accordance with Tribunal Procedure Rules.[10]

[10] www.justice.gov.uk/tribunals/criminal-injuries-compensation/rules-andlegislation.

The application form for an appeal needs to be completed and lodged with the Tribunal within 90 days of the review decision, along with any additional supporting material. The members and staff of the First-tier Tribunal are entirely independent of CICA and will consider the whole claim. The First-tier Tribunal may make a decision that is more favourable or less favourable than the review decision, or the review decision can stay the same. The First-tier Tribunal decision is final, but is subject to judicial review if appropriate. The number of claims made and the percentage of claims going to review and the percentage of review cases going to appeal since 2008 is shown in Table 24.3. The percentages being reviewed and appealed appear to remain stable.

Table 24.3: Percentage of claims going to review and the percentage of review cases going to appeal for 2008–13

	2014–15	2013–14	2012–13	2011–12	2010–11	2009–10
Claims made	35,801	42,859	53,821	57,480	64,768	67,597
% of claims going to review	20.5	19.7	18.8	19.1	18.2	20.1
% overall appeal rate	4.0	4.5	4.5	4.3	4.1	4.6

The average time to complete an internal review was 5.8 months in 2014/15, down from 6.7 months in 2013/14. Over 95% of cases are resolved without the need for an appeal. Of those decisions that are appealed, only a very low percentage result in overturning the CICA decision, just 2.43% in 2014/15, down from 2.97% in 2013/14.

The low rates of appeal, and the low success rates of these appeals, suggest that the combined decisions and review process makes decisions that are appropriate to the framework underpinning the Criminal Injuries Compensation scheme. The scheme also has high rates of customer satisfaction, often a good indicator of efficiency. Overall the scheme seems to function well.

G. Conclusions

One of the more interesting points about the Criminal Injuries Compensation scheme is the inclusion of blamelessness and behaviour as eligibility criteria. Behavioural 'fault' is used as an eligibility criteria; this clearly delineates the scheme from being 'no-fault' in the usually understood term of reference. This scheme is an alternative non-adversarial, administratively based provision for situations in which there is often limited hope of recourse through the traditional tort mechanisms.

25

Industrial Injuries Disablement Benefit

SONIA MACLEOD

Workers' compensation schemes first appeared in Europe in the 1880s, beginning with the inception of the German scheme in 1884.[1] This movement started primarily in civil law jurisdictions, but spread to common law jurisdictions. The largest common law jurisdiction to have adopted workers' compensation is the USA, but the UK, Australia, Canada and many other common law jurisdictions have also incorporated workers' compensation into their legal frameworks.

A formalised industrial injuries scheme has existed in Britain in various forms since 1897, with informal local or trade-specific arrangements pre-dating this. The earliest reported English scheme was of an English pirate crew compensation scheme to compensate injured or disabled crew members, which pre-dated the official statute of 1897.[2]

The major source of compensation for injured workers in Britain is Employers' Liability Insurance. All employers are required by law to carry Employers' Liability Insurance of at least £5 million from an authorised insurer.[3] This covers the compensation in the event of illness or injury to their employees and associated legal fees, though policies vary from provider to provider. The insurance is provided by private insurance companies and claims are processed by the companies. There are also some statutory benefits and schemes to cover specific work injuries. Industrial Injuries Disablement Benefit (IIDB) is a benefit available to eligible individuals who have an injury accrued during the course of employment. Several types of industrial injury are dealt with under separate specific schemes. These include provisions for mesothelioma, pneumoconiosis, miners' compensation (for chronic obstructive pulmonary disease (COPD)—respiratory disease (chronic bronchitis and emphysema)

[1] Starting with Germany in 1871 and 1884, Austria in 1887, Norway in 1894, Finland in 1895, Denmark, France and Italy in 1898, Greece in 1901, Belgium and Russia in 1903; taken from SB Horovitz, 'Worldwide Workmen's Compensation Trends' (1970) 59 *Kentucky Law Journal* 37, 40–41.

[2] S Talty, *Empire of Blue Water: Captain Morgan's Great Pirate Army, the Epic Battle for the Americas, and the Catastrophe that ended the Outlaws' Bloody Reign*, reprint edn (New York, Three Rivers Press, 2008).

[3] The Employers' Liability (Compulsory Insurance) Act 1969, the Employers' Liability (Compulsory Insurance) Regulations 1998 (SI 1998/2573), the Employers' Liability (Compulsory Insurance) (Amendment) Regulations 2004 (SI 2004/2882) and the Employers' Liability (Compulsory Insurance) (Amendment) Regulations 2008 (SI 2008/1765).

and vibration white finger (VWF)). These will be described further in the following two chapters.

IIDB is a benefit available to eligible individuals. To be eligible an injury must have accrued during the course of employment or a recognised training scheme either as a result of an accident at work (industrial accident) or a prescribed disease, with the injury-causing incident occurring in England, Scotland or Wales. IIDB offers compensation to employees injured during the course of employment regardless of fault.[4] The objective of the scheme is to provide:[5]

1. no-fault contributory benefits to people disabled by injuries or diseases through employed earners employment; *and*
2. assistance with the extra costs of care needed because of the industrial disablement.

The original principles behind the scheme were that:

— many industries vital to the community were especially dangerous and because of the essential nature of the work, people would be encouraged to take up work if they were assured of compensation in the event of injury;
— a person disabled in the course of work was under orders to carry out his duties; and
— employer's liability at common law could be limited to cases of negligence only if no-fault provisions were made.

The industrial scheme is funded by the employers' and employees contributions to the National Insurance (NI) fund, which is collected through the taxation system.

Two conditions need to be met to claim under this scheme and to be entitled to disablement benefit. First, the person making a claim must be an 'employed earner' at the time of the accident. Second, the injury is a result of an 'accident arising out of and in the course of employment' or resulted from an occupational disease prescribed in relation to that employment.[6] The scheme does not cover certain groups of people, eg the self-employed, members of the armed services[7] or some trainees.

An industrial accident is an incident or series of discrete identifiable incidents that result in personal injury. A prescribed disease is an industrial disease 'prescribed' by regulations to be included in the IIDB scheme. There are currently about 70 prescribed diseases. Diseases are included in the list of prescribed diseases by Order of the Secretary of State for Work and Pensions, who acts on the basis of advice from the independent Industrial Injuries Advisory Council.[8,9]

[4] Social Security Contributions and Benefits Act 1992, para V, subs 94–111. See D Bonner, I Hooker, R White, *Social Security Legislation 2003*, vol 1: *Non-Means Tested Benefits* (2003) 190–232.

[5] DWP, *Background to the Industrial Injuries Scheme 2007*, available at: www.dwp.gov.uk/docs/dmgch64.pdf.

[6] More about the scheme: www.gov.uk/industrial-injuries-disablement-benefit/eligibility.

[7] See chapter 23 above on armed forces compensation for more details on this scheme.

[8] The Social Security Administration Act 1992, s 171, www.dwp.gov.uk/healthcare-professional/benefits-and-services/industrial-injuries-disablement/;http://www.dwp.gov.uk/publications/specialist-guides/technical-guidance/db1-a-guide-to-industrial-injuries/appendix/appendix-1/.

[9] http://iiac.independent.gov.uk/.

To get the basic benefit the claimant needs a medical assessment of the degree of his/her disability caused by the injury or disease. This is expressed as a percentage. Since October 1986, an injured employee must reach a threshold of 14% disablement, unless the disability is caused by pneumoconiosis, byssinosis or diffuse mesothelioma.[10] If a person claims benefit for more than one industrial accident or disease the assessments may be added together and benefit awarded on the total. A person who has the threshold level of disablement receives a weekly disablement pension. How much he gets depends on the percentage of his disability. If he is assessed as being 14–19% disabled, he is paid at the 20% rate. If he is assessed as over 20% disabled, his assessments are rounded up or down to the nearest 10%. For example, 34% would be rounded down to 30%, and 35% would be rounded up to 40%. Benefit can only be paid in respect of occupational deafness if the assessment is 20% or more. A person deemed to have 20% disablement currently receives £33.60 a week, increasing in stages (30% = £50.40; 40% = £67.20; etc) up to £168.00 per week for 100% disablement.[11] With some exceptions, IIDB is not payable until 90 days after the accident or date of onset of a prescribed disease. The claimant's occupation or any loss of earnings are not taken into account in the assessment of their disability, but he may get allowances added to his basic benefit. If the claimant is entitled to other Social Security benefits, these will be paid as well as IIDB.

The instructions on how to make a claim for IIDB are shown in Figure 25.1.[12] The claims forms (one for accidents and one for diseases) are available online for printing and sending to the claims office.[13] Non-medical decision-makers collect evidence to decide on claims and make decisions about entitlement. In a claim for an industrial accident, the decision-maker collects evidence to establish if there has been an incident which is an accident for entitlement purposes. In a claim for a prescribed disease, the decision-maker will establish whether the person has worked in a prescribed occupation. If the decision-maker is satisfied that the person has suffered an industrial accident or has worked in a prescribed occupation, he/she usually seeks advice from a healthcare professional working for Atos Healthcare before making a decision on the claim. To receive benefit, a person's disablement has to be 1% or more for certain respiratory diseases, 20% or more for occupational deafness and 14% or more in all other cases.[14]

The IIDB quarterly statistics are shown in Table 25.1.

[10] M Jones, 'Compensation in the English Health Care Sector' in J Dute, M Faure and H Koziol (eds) *No-Fault Compensation in the Health Care sector*, Tort and Insurance Law, vol 8 (Berlin, Springer, 2004) 32.

[11] Social Security Contributions and Benefits Act 1992, Sch 4, available at: www.legislation.gov.uk/ukpga/1992/4/enacted.

[12] www.gov.uk/industrial-injuries-disablement-benefit/how-to-claim, www.gov.uk/industrial-injuries-disablement-benefit/further-information. The claim form is at www.dwp.gov.uk/advisers/claimforms/bi100a_print.pdf.

[13] www.gov.uk/government/publications/industrial-injuries-disablement-benefit-claim-forms.

[14] www.dwp.gov.uk/healthcare-professional/benefits-and-services/industrial-injuries-disablement/.

Reporting the accident to your employer or trainer

If you have an accident at work, tell your employer, trainer or someone else in authority at once, even if the accident does not seem serious at the time.

Most employers and trainers have an accident book. Record the details of the accident in the book as soon as you can. If there is no accident book, tell your employer or trainer about the accident by:

— speaking to them, or
— writing to them.

You must tell them:

— your name and address
— your job
— the cause and nature of your injury, and
— the date, time and place of the accident.

If you cannot report the accident yourself, ask someone else to do it for you.

Am I entitled to Industrial Injuries Disablement Benefit?

You may be entitled to the benefit if you are disabled because of an accident at work or whilst on an approved or registered training scheme or course.

The accident must have happened because of your work, and it must usually have happened in Great Britain.

We will not pay Industrial Injuries Disablement Benefit if you were self-employed when the accident happened.

The amount you get depends on how seriously you are disabled. We usually need you to go to a medical examination.

The doctor who examines you will tell us about how seriously you are disabled and how long you are expected to be disabled.

When to claim

If you become disabled because of an accident, claim Industrial Injuries Disablement Benefit straight away. But usually you will not be entitled to benefit for 90 days after the date of the accident. If you delay you may lose some benefit.

Signing the form for someone else

The claim form should only be signed by someone else if:

— the person who is making the claim is not mentally able to act on their own behalf, **and**
— someone is willing to act on their behalf in all social security matters, including telling us about any change in their circumstances and collecting money for them.

Someone accepted by the Department to act on a person's behalf is known as the *appointee*.

A person who is physically disabled but mentally able will not normally need someone to act for them.

If someone applies to act on a person's behalf we will:

- arrange a visit to:
  - the person who is making the claim, and
  - the person who is applying to act on their behalf, **and**
- decide if the person needs someone to act on their behalf, **and**
- explain the responsibilities that the appointee would be taking on.

We will not pay any benefit until this process is complete.
If someone has power of attorney or legal authority to act on behalf of the person making the claim, then the person with power of attorney or legal authority must:

- sign the claim form, **and**
- send us a copy of the legal authority with the claim form.

The copy of the legal authority must be certified and signed by a solicitor as a true copy.

What happens after we get your claim for benefit

We will write to you and tell you that we have got your claim.

A decision maker will look at your claim.

Decision makers are people who decide

- if the law says your accident is an industrial accident
- if the law says you are entitled to benefit or not
- how much benefit the law says you are entitled to.

To help the decision maker decide on your claim, we may need to get more information. We may need to ask people about your claim. For example, we may write to any employer, doctor or hospital that you have told us about in the form.

If the decision maker decides that your accident was an industrial accident, we may ask you to go to a medical examination. We will write and tell you where and when to go for the examination. If you are not fit to travel or you are in hospital, we may ask the doctor to come to you.

Please tell us any date and times that you may not be able to go to a medical examination in the next six months. For example, holidays and hospital appointments. Please give us full details.

If you can get Industrial Injuries Disablement Benefit, we may write and tell you:

- how much money you can get
- more about the benefit.

If you cannot get Industrial Injuries Disablement Benefit, we will write and tell you the reason and what to do if you disagree.

Figure 25.1: How to make a claim for IIDB

Table 25.1: Industrial Injuries Disablement Benefits quarterly statistics. Source: www.gov.uk/government/collections/industrial-injuries-disablement-benefit-quarterly-statistics#history

Industrial Injuries Disablement Benefit and Reduced Earnings Allowance in payment by type of assessments payable in £000						
		Total[1]	Industrial Injuries Disablement Benefit only	Reduced Earnings Allowance only	Industrial Injuries Disablement Benefit with Reduced Earnings Allowance	Not known[2]
1998	March[3]	**343.0**	194.3	81.0	67.8	.
1999	March	**353.4**	204.8	82.0	66.7	.
2000	March	**357.5**	208.4	82.9	66.2	.
2001	March	**358.0**	209.1	82.6	66.3	.
2002	March	**354.7**	209.3	81.0	64.4	.
						Number
	June	**341,620**	197,650	76,490	67,460	10
	September	**342,370**	198,540	76,500	67,300	20
	December	**342,390**	198,980	76,360	67,030	20
2003	March	**343,370**	200,330	76,220	66,800	20
	June	**342,960**	200,770	75,870	66,310	20
	September	**342,340**	200,770	75,560	66,000	20
	December	**341,910**	201,130	75,250	65,510	20
2004	March	**341,310**	201,330	74,810	65,150	20
	June	**340,410**	201,490	74,220	64,690	20
	September	**340,720**	202,320	73,900	64,480	20
	December	**341,010**	203,260	73,550	64,190	10
2005	March	**340,300**	203,410	73,150	63,710	20
	June	**339,690**	203,450	72,860	63,370	10
	September	**339,390**	203,800	72,500	63,070	20
	December	**338,970**	204,230	72,030	62,680	20
2006	March	**337,840**	204,220	71,380	62,230	20
	June	**336,610**	203,940	70,880	61,770	20
	September	**336,020**	204,220	70,410	61,380	10
	December	**335,420**	204,540	69,930	60,940	10
2007	March	**334,240**	204,480	69,360	60,400	10
	June	**332,790**	204,130	68,750	59,890	10
	September	**332,070**	204,220	68,330	59,510	10
	December [4]	..	..	..	..	..

(continued)

Table 25.1: ***(Continued)***

Industrial Injuries Disablement Benefit and Reduced Earnings Allowance in payment by type of assessments payable in £000						
		Total[1]	Industrial Injuries Disablement Benefit only	Reduced Earnings Allowance only	Industrial Injuries Disablement Benefit with Reduced Earnings Allowance	Not known[2]
2008	March	**329,930**	204,300	67,190	58,430	10
	June	**328,730**	204,020	66,720	57,980	10
	September	**327,940**	204,330	66,140	57,460	10
	December	**327,270**	204,610	65,670	56,980	10
2009	March	**325,680**	204,070	65,130	56,480	10
	June	**325,020**	204,250	64,730	56,040	10
	September	**324,360**	204,380	64,170	55,810	10
	December	**324,120**	205,200	62,840	56,080	10
2010	March	**325,880**	208,030	60,100	57,740	10
	June	**326,400**	209,420	58,730	58,240	10
	September	**327,240**	211,070	57,670	58,500	10
	December	**326,860**	211,600	56,760	58,490	10
2011	March	**326,430**	212,080	56,000	58,340	10
	June	**325,330**	211,910	55,400	58,010	10
	September	**324,270**	211,720	54,870	57,670	10
	December	**323,250**	211,710	54,320	57,210	10
2012	March	**321,500**	211,100	53,660	56,730	10
	June	**319,790**	210,680	53,000	56,100	10
	September	**318,760**	210,530	52,580	55,640	10
	December	**317,460**	210,290	51,990	55,170	10
2013	March	**315,840**	209,730	51,460	54,640	10
	June	**314,420**	209,360	50,910	54,150	10
	September	**313,420**	209,250	50,400	53,760	10
	December	**312,840**	209,560	49,940	53,330	10
2014	March	**311,360**	209,230	49,380	52,750	10
	June	**309,520**	208,400	48,850	52,260	10
	September	**307,030**	206,970	48,310	51,740	10
	December	**305,870**	206,780	47,850	51,230	10

Figures are subject to a high degree of sampling error and should only be used as a guide.

- Nil or Negligible.

. Not applicable.

.. Not available.

Table 25.1: ***(Continued)***

(1) Figures up to and including March 2002 are based on a 10% sample and include recipients resident overseas. Figures from June 2002 onwards are based on 100% data and may include a few recipients not resident in Great Britain and may also include some temporarily suspended cases. Latest estimates suggest there are around 6,500 overseas cases.
(2) Not known—cases where the types of assessment cannot be accurately identified at the extract date.
(3) Figure for March 1998 has been revised due to an amendment in the calculation.
(4) Due to DWP embargo there was no December 2007 data.

Industrial Injury assessments in payment by prescribed disease in £000											
	Quarter[1]	All prescribed diseases	Synovial inflammation	Occupational deafness	Vibration white finger	Carpal tunnel syndrome	Pneumo-coniosis	Diffuse mesothelioma	Unilateral or bilateral diffuse pleural thickening	Bronchitis and emphysema	
			(A8)	(A10)	(A11)	(A12)	(D1)	(D3)	(D9)	(D12)	Others
1998	March	**57.5**	2.3	14.7	5.2	0.6	11.4	0.7	1.4	12.0	9.4
1999	March	**61.4**	2.5	14.5	6.2	0.8	11.2	0.9	1.7	13.6	10.0
2000	March	**61.5**	2.5	14.3	6.8	0.9	11.0	1.0	2.0	13.2	9.8
2001	March	**61.3**	2.5	13.7	7.4	1.1	11.1	1.0	2.2	12.5	9.8
2002	March	**60.6**	2.5	13.2	8.1	1.4	11.0	0.8	2.4	11.5	9.7
2003	March	**59.9**	2.6	12.6	8.3	1.7	11.5	0.7	2.4	10.7	9.5
2004	March	**59.6**	2.4	12.3	8.3	1.9	11.8	0.8	2.7	9.9	9.5
2005	March	**59.4**	2.5	11.9	8.0	2.0	12.3	1.0	3.1	9.2	9.5
2006	March	**59.3**	2.5	11.5	8.2	2.1	12.5	1.2	3.4	8.4	9.5
2007	March	**57.9**	2.5	11.1	8.1	2.1	12.2	1.3	3.7	7.6	9.5
2008	March	**57.1**	2.5	10.7	8.0	2.2	12.0	1.5	3.8	6.9	9.5
2009	March	**57.3**	2.5	10.4	8.0	2.3	11.9	1.5	3.9	6.3	10.6
2010	March[2]	..	..	..	..	..	..	..	..	..	..
2011	March[2]	..	..	..	..	..	..	..	..	..	..
2012	March[2]	..	..	..	..	..	..	..	..	..	..

Figures are subject to a high degree of sampling error and should only be used as a guide.

- Nil or negligible.
. Not applicable.
.. Not available.

(1) Figures are based on a 10% sample and include recipients resident overseas. Latest estimates suggest there are around 6,500 overseas cases.
(2) Figures not available yet.

Table 25.1: ***(Continued)***

	New Industrial Injury claims in the quarter by type of claim in £000					
		Claim type				
Year	**Quarter**[1]	**All new claims**	**Prescribed diseases**	**Accidents**	**Reduced Earnings Allowance**	**Accident declarations**
1998	March	**22.5**	14.6	7.9	..	..
	June	**18.4**	11.3	7.1		..
	September	**19.0**	10.7	8.2	..	..
	December	**15.7**	8.9	6.8	..	..
1999	March	**20.1**	12.0	8.2	..	..
	June	**18.2**	11.2	7.1	..	..
	September	**16.2**	9.1	7.1	..	..
	December	**16.0**	9.1	7.0	..	..
2000	March	**18.3**	11.1	7.3	..	..
	June	**17.7**	11.1	6.6	..	..
	September	**22.1**	15.5	6.6	..	..
	December	**17.6**	11.1	6.5	..	..
2001	March	**18.0**	11.0	7.1	..	..
	June	**17.0**	9.7	7.3	..	..
	September	**14.8**	8.8	6.0	..	..
	December	**15.2**	9.5	5.7	..	..
2002	March	**16.4**	10.5	5.9	..	..
	June	**20,450**	10,400	5,490	1,560	3,000
	September	**21,590**	10,750	5,680	1,470	3,690
	December	**18,980**	9,410	5,190	1,300	3,080
2003	March	**19,180**	9,320	5,570	1,020	3,270
	June	**19,540**	10,160	5,300	1,120	2,970
	September	**19,360**	10,100	5,080	980	3,210
	December	**15,820**	7,870	4,680	860	2,410
2004	March	**17,650**	8,890	5,500	690	2,570
	June	**14,980**	7,080	5,050	800	2,060
	September	**14,310**	6,810	4,900	640	1,960
	December	**12,210**	5,480	4,390	710	1,640
2005	March	**13,030**	5,640	4,700	660	2,030
	June	**12,040**	5,240	4,380	580	1,840
	September	**12,530**	5,620	4,640	580	1,700
	December	**11,440**	4,800	4,570	510	1,570

(continued)

Table 25.1: ***(Continued)***

	New Industrial Injury claims in the quarter by type of claim in £000					
		Claim type				
Year	**Quarter**[1]	**All new claims**	**Prescribed diseases**	**Accidents**	**Reduced Earnings Allowance**	**Accident declarations**
2006	March	**12,990**	5,610	5,180	450	1,750
	June	**11,960**	5,030	5,030	340	1,560
	September	**11,940**	4,690	5,380	400	1,470
	December	**11,980**	4,750	5,390	430	1,410
2007	March	**11,650**	5,000	4,950	320	1,380
	June	**10,800**	4,440	4,900	360	1,100
	September	**11,260**	4,540	5,230	350	1,140
	December [3 4]	**11,090**	4,800	4,950	200	1,130
2008	March [3 4]	**11,200**	4,710	4,960	260	1,260
	June [3 4]	**10,320**	4,080	4,830	300	1,120
	September [3 4]	**10,160**	3,960	4,900	300	1,010
	December	**8,870**	3,680	3,940	280	980
2009	March	**8,310**	3,640	3,310	330	1,040
	June	**8,700**	4,080	3,380	340	900
	September [5]	**35,160**	30,390	3,660	360	760
	December	**16,420**	12,060	3,330	330	700
2010	March	**14,000**	9,140	3,690	330	830
	June	**12,060**	7,400	3,510	370	780
	September	**10,740**	6,150	3,550	310	730
	December	**9,230**	5,090	3,210	320	610
2011	March	**9,330**	4,920	3,270	350	790
	June	**8,420**	4,390	3,050	300	680
	September	**8,150**	4,090	3,190	270	600
	December	**7,470**	3,630	3,070	210	570
2012	March [5]	**8,250**	4,000	3,340	250	670
	June	**7,100**	3,530	2,890	180	510
	September [6 7]	**6,860**	3,330	2,860	170	500
	December	**7,320**	3,530	3,070	230	500
2013	March [8]	**6,720**	3,390	3,160	170	-
	June	**6,350**	3,250	2,950	160	-
	September	**6,010**	3,000	2,840	170	-
	December	**5,900**	3,120	2,560	220	-

(continued)

Table 25.1: ***(Continued)***

	New Industrial Injury claims in the quarter by type of claim in £000					
		Claim type				
Year	**Quarter[1]**	**All new claims**	**Prescribed diseases**	**Accidents**	**Reduced Earnings Allowance**	**Accident declarations**
2014	March	**6,280**	3,240	2,840	200	-
	June	**5,820**	3,000	2,630	190	-
	September	**6,590**	3,530	2,790	270	-
	December	**6,340**	3,440	2,630	270	-

Figures are subject to a high degree of sampling error and should only be used as a guide.

- Nil or negligible.
. Not applicable.
.. Not available.

(1) Figures up to and including March 2002 are based on a 10% sample. Figures from June 2002 onwards are based on 100% data. Claims from people resident overseas are included in all quarters.
(2) Figures are provisional and are subject to revision.
(3) 'All new claims' include a small number of cases where the claim type is not known. For this reason the individual columns may not add to the total.
(4) Includes claims to Prescribed Disease A11 under new regulations from 1 October 2007.
(5) Prescribed Disease A14 was introduced from 13 July 2009. New regulations introduced from 30 March 2012.
(6) Prescribed Disease D10 was introduced from 1 April 1987. New regulations introduced from 1 August 2012.
(7) Includes claims for Prescribed Disease C31 which was introduced from 18 July 2011.
(8) The Accident Declaration procedure was abolished from 5.12.12 so no new applications were received from this date.

A. Conclusions

It is interesting to note that the IIDB scheme persists in the UK despite the fact that Employers' Liability insurance has been compulsory for many years. Other jurisdictions, such as the USA, rely exclusively on mandatory employers' insurance. The scheme is a social security payment, and as such is intended as a safety net, rather than as a first port of call. Despite this, a considerable number of claims are made each year.

26

UK Mesothelioma Compulsory Insurance Fund and Compensation Scheme

SONIA MACLEOD

A. Mesothelioma

Mesothelioma is a cancer of the lining of the internal organs, almost always arising from exposure to asbestos. It is a progressive and fatal disorder, which takes a long time to develop. There are several routes to compensation for mesothelioma suffers, dependent upon their exposure, their employment status, whether they can trace a liable employer and their date of diagnosis. Figure 26.1 shows a schematic of possible claims routes for a mesothelioma sufferer, though this will depend on diagnosis dates and other factors.

B. Industrial Injury Disablement Benefit

If eligible, then Industrial Injury Disablement Benefit (IIDB) is payable to the mesothelioma sufferer, as detailed above. However, there are other sources of support available to mesothelioma sufferers.

Standard social security payments are available to disabled persons, including sufferers of pneumoconiosis, byssinosis and diffuse mesothelioma, where IIDB is payable to him in respect of the disease, every relevant employer of his has ceased to carry on business, and he has not brought any action, or compromised any claim, for damages in respect of the disablement.[1]

C. Pneumoconiosis Act 1979 Lump Sum Claim

A lump sum payment is also potentially available under the Mesothelioma Lump Sum Payments (Conditions and Amounts) (Amendment) Regulations 2013. A claim must be

[1] The Pneumoconiosis etc (Workers' Compensation) Act 1979; and the Child Maintenance and Other Payments Act 2008, ss 46(3) and 47.

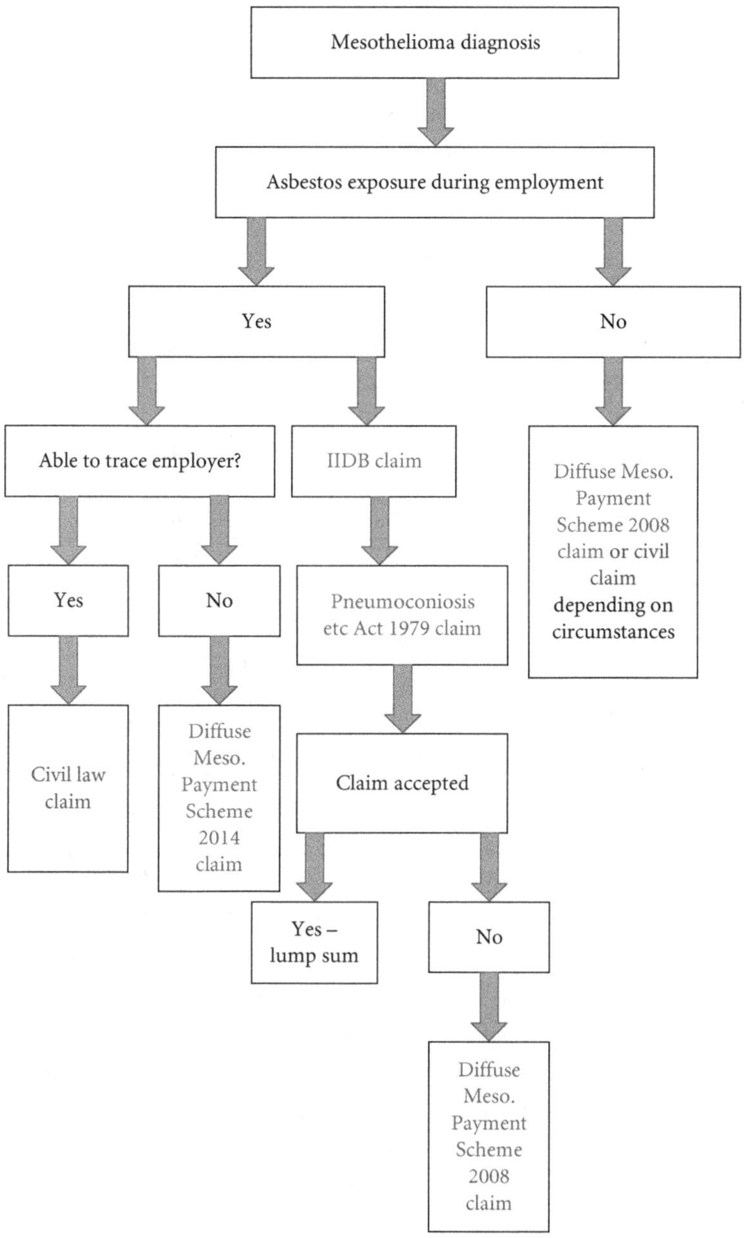

Figure 26.1: Potential claim routes for mesothelioma sufferers. State payments highlighted in black, private payments highlighted in grey

brought within 12 months of the decision awarding IIDB. Payments are also made to the dependants of those who suffered from the relevant conditions before they died, for claims made within 12 months of the death of the sufferer.[2]

The amounts payable are specified in Regulations, and regularly increased.[3] The lump sum payments applicable for persons with diffuse mesothelioma and their dependants respectively are shown at Tables 26.1 and 26.2.[4]

Table 26.1: Lump sum values (£) for a person diagnosed with mesothelioma on or after 1 April 2013

Age of disabled person	Percentage assessment for the relevant period									
	10% or under	11–20%	21–30%	31–40%	41–50%	51–60%	61–70%	71–80%	81–90%	91–100%
37 and under	33,977	60,678	71,193	73,217	75,239	76,856	78,474	80,093	81,710	83,330
38	33,007	58,248	69,048	71,520	73,618	75,239	76,856	78,474	80,093	81,710
39	32,035	55,821	66,905	69,818	72,005	73,405	75,239	76,856	78,474	80,093
40	31,067	53,391	64,761	68,117	70,381	72,003	73,618	75,239	76,856	78,476
41	30,095	50,968	62,619	66,415	68,768	70,381	72,003	73,618	75,239	76,856
42	29,118	48,543	60,473	64,724	67,148	68,768	70,381	72,003	73,618	75,239
43	27,669	45,708	58,330	63,427	66,018	67,960	69,575	71,193	72,812	74,431
44	26,210	42,877	56,183	62,134	64,884	67,148	68,768	70,381	72,003	73,618
45	24,758	40,046	54,040	60,837	63,753	66,336	67,960	69,575	71,193	72,812
46	23,300	37,216	51,899	59,545	62,619	65,533	67,148	68,768	70,381	72,003
47	21,846	34,383	49,753	58,248	61,488	64,724	66,336	67,960	69,575	71,193
48	20,590	33,250	48,058	55,661	59,545	62,453	64,075	65,692	67,306	68,931
49	19,336	32,118	46,357	53,071	57,605	60,189	61,809	63,427	65,047	66,666
50	18,082	30,986	44,690	50,484	55,661	57,924	59,545	61,161	62,777	64,397
51	16,828	29,853	42,957	47,893	53,717	55,661	57,294	58,893	60,515	62,134
52	15,574	28,720	41,260	45,304	51,775	53,391	55,016	56,631	58,248	59,864
53	14,399	26,700	38,668	43,041	49,834	51,775	53,391	55,016	56,631	58,248
54	13,227	24,678	36,080	40,779	47,893	50,162	51,775	53,391	55,016	56,631
55	12,056	22,655	33,497	38,507	45,949	48,543	50,162	51,775	53,391	55,016

(continued)

[2] ibid.

[3] See the Pneumoconiosis etc (Workers' Compensation) (Payment of Claims) Regulations 1988/668, as amended, eg by the Pneumoconiosis etc (Workers' Compensation) (Payment of Claims) (Amendment) Regulations 2013/ 690; and the Mesothelioma Lump Sum Payments (Conditions and Amounts) Regulations 2008/1963, as amended, eg by the Mesothelioma Lump Sum Payments (Conditions and Amounts) (Amendment) Regulations 2013.

[4] Taken from Schedule 1 of The Pneumoconiosis etc (Workers' Compensation) (Payment of Claims) (Amendment) Regulations 2013/ 690.

Table 26.1: *(Continued)*

Age of disabled person	Percentage assessment for the relevant period									
	10% or under	11–20%	21–30%	31–40%	41–50%	51–60%	61–70%	71–80%	81–90%	91–100%
56	10,880	20,630	30,905	36,244	44,012	46,921	48,543	50,162	51,775	53,391
57	9,708	18,606	28,317	33,977	42,076	45,305	46,921	48,543	50,162	51,775
58	8,939	16,909	25,285	30,496	37,861	40,856	42,595	44,297	45,950	47,570
59	8,169	15,208	22,247	27,022	33,653	36,404	38,265	40,046	41,745	43,360
60	7,406	13,509	19,213	23,541	29,452	31,957	33,933	35,800	37,540	39,159
61	6,632	11,810	16,183	20,062	25,241	27,506	29,608	31,552	33,332	34,949
62	5,865	10,113	13,147	16,584	21,035	23,056	25,285	27,306	29,118	30,743
63	5,461	9,141	11,810	14,828	18,712	20,708	22,815	24,758	26,534	28,152
64	5,065	8,169	10,476	13,064	16,381	18,366	20,349	22,208	23,944	25,562
65	4,653	7,200	9,141	11,309	14,059	16,019	17,879	19,661	21,360	22,978
66	4,250	6,231	7,806	9,544	11,732	13,671	15,412	17,108	18,770	20,388
67	3,845	5,258	6,469	7,787	9,406	11,329	12,945	14,561	16,183	17,799
68	3,739	5,092	6,249	7,465	9,084	10,880	12,439	14,118	15,672	17,272
69	3,641	4,936	6,030	7,141	8,761	10,437	11,936	13,671	15,171	16,744
70	3,538	4,776	5,803	6,817	8,435	9,988	11,429	13,227	14,662	16,222
71	3,440	4,609	5,582	6,491	8,110	9,544	10,924	12,782	14,160	15,697
72	3,338	4,450	5,359	6,170	7,787	9,104	10,415	12,337	13,649	15,171
73	3,254	4,371	5,280	6,048	7,664	8,900	10,214	12,016	13,249	14,724
74	3,176	4,287	5,198	5,927	7,547	8,695	10,015	11,693	12,844	14,269
75	3,097	4,210	5,119	5,803	7,421	8,495	9,808	11,367	12,439	13,832
76	3,014	4,123	5,035	5,684	7,300	8,296	9,610	11,044	12,030	13,392
77 and over	2,936	4,044	4,957	5,560	7,180	8,089	9,406	10,725	11,631	12,945

Table 26.2: Lump sum payment (£) for a dependant of a deceased mesothelioma sufferer who died on or after 1 April 2013

Age of disabled person on their last birthday before their death	Payment
37 and under	43,366
38	42,433
39	41,503
40	40,573
41	39,643
42	38,713

(continued)

Table 26.2: ***(Continued)***

Age of disabled person on their last birthday before their death	Payment
43	37,823
44	36,924
45	36,040
46	35,151
47	34,263
48	33,169
49	32,076
50	30,985
51	29,897
52	28,806
53	27,910
54	27,023
55	26,133
56	25,238
57	24,351
58	21,888
59	19,417
60	16,951
61	14,482
62	12,012
63	11,307
64	10,607
65	9,889
66	9,184
67 and over	7,180

If a lump sum claim under the 1979 Act is not successful, the claim is automatically passed to the Diffuse Mesothelioma Payment Scheme 2008 or 2014 depending upon the date of diagnosis.

D. Employers' Liability Claims and Schemes

In order to obtain further compensation there are two routes available: a private civil law claim and compensation under a government-created scheme. There are two government-created schemes, the Diffuse Mesothelioma Scheme 2008 ('the 2008 Scheme') and the

Diffuse Mesothelioma Payment Scheme (DMPS). These schemes have different eligibility criteria so cover different groups of claimants. These schemes are essentially safety nets to cover individuals who are not able to obtain compensation from other sources.

Employers' Liability Insurance Claims

Civil compensation claims may also made against employers for disease due to unsafe working conditions. The insurance industry currently pays £200 million a year under employers' liability (EL) policies to sufferers of mesothelioma. Claims can be made by the individual or if the mesothelioma sufferer has died, then claims can be brought by dependants.

In order to make a civil law claim an eligible defendant has to be identified. By the time it has been diagnosed, some employers are no longer in existence and tracing relevant employers' insurance cover can be difficult. Median life expectancy from diagnosis is 12–21 months.[5] This provides a very short time period in which to claim, and claimants are often very ill during this period. In order to speed up the process of tracing relevant insurance, the Association of British Insurers established the Employers' Liability Tracing Office (ELTO)[6] in April 2011 to run an electronic database of EL policies, to which 99% of EL insurers provide data. This is funded by the insurance providers. In February 2011, the Financial Standards Agency (subsequently the Financial Conduct Authority) introduced new regulations which make it compulsory for all companies that provide EL insurance to produce searchable information on these policies;[7] the majority of EL insurance companies fulfil this requirement by joining ELTO.

Following a 10-week consultation period (which ended 2 October 2013), the government shelved plans for a fixed-fee online portal for mesothelioma.[8] This portal was to have covered claims by claimants who could trace a liable employer as well as those who could not. It was intended to have enacted a dedicated pre-action protocol specifically designed for mesothelioma claims against a known employer, with fixed fees for lawyers making the claims.

The consultation also incorporated the review required by section 48 of the Legal Aid, Sentencing and Punishment of Offenders (LASPO) Act 2012 regarding the application of conditional fee agreement reforms[9] to mesothelioma claims. In response to the review the government argued that the case had not been made for treating mesothelioma cases differently. Therefore sections 44 and 46 of the LASPO Act apply to mesothelioma cases, as for all other personal injury cases.[10] The changes in claims handling for cases where a

[5] www.mesothelioma.com/mesothelioma/prognosis/life-expectancy.htm.

[6] www.elto.org.uk/.

[7] www.fsa.gov.uk/pubs/policy/ps11_04.pdf.

[8] 'Mesothelioma Claims', written Ministerial Statement, Wednesday, 4 December 2012, Ministry of Justice, available at www.parliament.uk/documents/commons-vote-office/December%202013/3%20December/4th-December-2013/9.JUSTICE-Mesothlioma-claims.pdf.

[9] Part 2 of the Legal Aid, Sentencing and Punishment of Offenders (LASPO) Act 2012 which means claimants can no longer recover success fees and after-the-event insurance premiums from losing defendants, as per other non-mesothelioma personal injury claims.

[10] This review remains the subject of ongoing controversy; see www.publications.parliament.uk/pa/cm201415/cmselect/cmjust/308/308.pdf.

liable insurer is known started from July 2014, at the same time as the compulsory payment scheme for victims of mesothelioma came into force under the Mesothelioma Bill.

There is a specialised mesothelioma 'fast-track' process. Under the Mesothelioma Practice Direction (PDD3) defendants have to 'show cause' at an early hearing, showing why judgment should not be entered in favour of the mesothelioma claimant.[11] This essentially reverses the burden of proof in mesothelioma claims, as the defendant is required to provide evidence that is 'more than fanciful' in order to be able to deny liability.

Mesothelioma civil claims are afforded some differences from other personal injury claims. Under the Compensation Act 2006, claimants in mesothelioma cases who have had more than one potentially liable employer are not required to trace them all: a claim for the full amount of damages need only be brought against one employer, who is then responsible for tracing other liable employers to obtain contributions from them.[12]

There is considerable disquiet with the way in which mesothelioma cases progress through the courts, but until the impact of the Mesothelioma Act 2014 can be analysed it seems unlikely that further legislative and/or procedural changes will be implemented.

Claims under the 2008 Scheme and the DMPS

There are some claimants who were not exposed to asbestos by an employer, for example the self-employed or relatives who were exposed when washing contaminated clothing, etc. These individuals cannot make a civil law claim as there is no liable defendant. Other individuals who were exposed to asbestos while employed may not be able to trace their employer (eg the employer may have gone bankrupt), and so may not be able to formulate a civil claim. The 2008 Scheme and the DMPS were created to provide these individuals with compensation.

E. The 2008 Scheme[13]

The 2008 Scheme entitles claimants to a one-off payment. It is designed to cover those who are ineligible for a payment under the Pneumoconiosis Act 1979 or a Ministry of Defence scheme and who have received a payment for the disease from an employer, a civil claim or other sources. To be eligible the exposure to asbestos must have happened in the United Kingdom. This not limited to exposure during employment and included, for example:

- coming into contact with asbestos via a relative, eg by washing their clothes;
- exposure to asbestos in the environment, eg by living near a factory using asbestos;
- exposure to asbestos while self-employed;
- exposure to asbestos cannot be specified but it occurred in the United Kingdom.

Claims must be made within 12 months of diagnosis.

[11] www.justice.gov.uk/courts/procedure-rules/civil/rules/part03/pd_part03d.
[12] S 3 Compensation Act 2006.
[13] www.gov.uk/diffuse-mesothelioma-payment/overview.

The scheme includes payments for dependants of a sufferer who has died. Again, claims must be made within 12 months of their death.

Application is by completing form PWC1, which can be obtained by telephoning the Barrow Benefits Office. The same form is used for claims under the 1979 Act, and if a claim is rejected under the 1979 Act it will automatically be considered for payment under the 2008 Scheme. Medical evidence is required, but the claimant will not usually be examined. There is no stipulated format, though form BI 100PN(A) has been specially designed for this purpose. Medical evidence is acceptable from any appropriate healthcare professional providing it states that the person is suffering from diffuse mesothelioma and gives the date the disease was diagnosed. A statement from a hospital consultant, the claimant's GP (usually after the disease has been diagnosed by the hospital) or from the claimant's lung cancer or Macmillan nurse is equally acceptable. Detailed hospital reports are not required, just a statement as above.

Claims under the 2008 Scheme and for lump sum payments under the 1979 Act are both processed at the same Benefits Centre in Barrow. If it is clear from the PWC1 form that the claimant is entitled to IIDB, then a request for payment under the 2008 Scheme will automatically be considered as a request under the 1979 Act. The sums payable are related to age at diagnosis, as shown in Table 26.3.

Table 26.3: Sums payable under the 2008 Scheme

Age of person with mesothelioma at diagnosis, or if unknown at date of claim	2008 Scheme payment (£)[a]
37 and under	85,580
38	83,916
39	82,256
40	80,595
41	78,931
42	77,270
43	76,441
44	75,606
45	74,778
46	73,947
47	73,115
48	70,792
49	68,466
50	66,136
51	63,812
52	61,480
53	59,821
54	58,160

(continued)

Table 26.3: ***(Continued)***

Age of person with mesothelioma at diagnosis, or if unknown at date of claim	2008 Scheme payment (£)[a]
55	56,501
56	54,833
57	53,173
58	48,854
59	44,531
60	40,216
61	35,893
62	31,573
63	28,912
64	26,252
65	23,589
66	20,938
67	18,280
68	17,738
69	17,196
70	16,660
71	16,121
72	15,581
73	15,122
74	14,654
75	14,205
76	13,754
77 and over	13,295

[a] Values taken from Regulation 2 of The Mesothelioma Lump Sum Payments (Conditions and Amounts) (Amendment) Regulations 2014 SI 2014/868.

F. DMPS

Around 300 sufferers of mesothelioma a year were exposed to asbestos during employment, but cannot trace a liable employer. Therefore, they are unable to claim compensation under their employer's EL insurance. The total existing number of mesothelioma sufferers who have been or are anticipated to be unable to claim is 3,000.

In order to improve the situation for these claimants, on 25 July 2012 the UK Minister for Welfare and the Association of British Insurers announced the creation of the DMPS.

The scheme was created by the Mesothelioma Act 2014 and subordinate legislation.[14] DMPS is funded by compulsory contributions from all EL insurers,[15] totalling around £25–35 million a year.[16] Originally the Scheme regulations specified that successful applicants would receive the equivalent of 80% of the average payment that would have been obtained from a successful civil courts claim. In February 2015, the scheme's tariff payment was increased to achieve parity with average equivalent civil compensation payments (see Table 26.4). Values are regularly updated.

Table 26.4: Payment amounts available from DMPS

Age of person with mesothelioma at diagnosis, or if unknown at date of claim	Payment (£)	
	Diagnosis between 25 July 2012 and 9 February 2015[a]	Diagnosis from 10 February 2015[b]
40 and under	216,896	271,120
41	213,951	267,439
42	211,006	263,758
43	208,062	260,078
44	205,117	256,396
45	202,172	252,715
46	199,227	249,034
47	196,282	245,353
48	193,337	241,671
49	190,392	237,990
50	187,447	234,309
51	184,502	230,628
52	181,557	226,946
53	178,612	223,265
54	175,667	219,584
55	172,722	215,903
56	169,777	212,221
57	166,832	208,540
58	163,887	204,859
59	160,943	201,179

(continued)

[14] The Mesothelioma Act 2014 and the Diffuse Mesothelioma Payment Scheme Regulations 2014 (2014/916).

[15] The Mesothelioma Act 2014 and the Diffuse Mesothelioma Payment Scheme (Levy) Regulations 2014 (2014/2904).

[16] In the first year of the Scheme's operation (April 2014–March 2015) the levy was £32 million. The levy for the financial year 2015/16 was £23.2 million. Taken from Diffuse Mesothelioma Payment Scheme Official Statistics, April 2014–September 2015, produced by Department for Work and Pensions. Available at www.gov.uk/government/uploads/system/uploads/attachment_data/file/495389/diffuse-mesothelioma-statistics-apr-2014-sep-2015.pdf.

Table 26.4: ***(Continued)***

Age of person with mesothelioma at diagnosis, or if unknown at date of claim	Payment (£)	
	Diagnosis between 25 July 2012 and 9 February 2015[a]	Diagnosis from 10 February 2015[b]
60	157,998	197,498
61	155,053	193,816
62	152,108	190,135
63	149,163	186,454
64	146,218	182,773
65	143,273	179,091
66	140,328	175,410
67	137,383	171,729
68	134,438	168,048
69	131,493	164,366
70	128,548	160,685
71	125,603	157,004
72	122,658	153,323
73	119,713	149,641
74	116,768	145,960
75	113,823	142,279
76	110,879	138,599
77	107,934	134,918
78	104,989	131,236
79	102,044	127,555
80	99,099	123,874
81	96,154	120,193
82	93,209	116,511
83	90,264	112,830
84	87,319	109,149
85	84,374	105,468
86	81,429	101,786
87	78,484	98,105
88	75,539	94,424
89	72,594	90,743
90 and over	69,649	87,061

[a] As per Diffuse Mesothelioma Payment Scheme (Amendments) Regulations 2014 (2014/917).
[b] As per The Diffuse Mesothelioma Payment Scheme (Amendments) Regulations 2015 (2015/367).

DMPS was launched on 6 April 2014, and is administered by Gallagher Bassett.[17] Claims are made by submitting a claim form available from the DMPS website; this must include:

- National Insurance number of the claimant;
- full employment history;
- evidence of employment history (can be obtained from HMRC or can be P60s/payslips);
- evidence of unsuccessful attempts to trace the claimant's employer/his insurer (evidence of a search from the Employers Liability Tracing Office, ELTO);
- date and age at diagnosis;
- evidence of diagnosis (a letter from medical practitioner, GP, etc);
- details of any previous claim;
- witness statement (if a claim has been made use the witness statement from that, otherwise it should contain details of where the individual was working, how the exposure occurred, and details of any witnesses and other evidence that would support the case).

If an applicant cannot supply the necessary paperwork electronically they may post it. If an application is incomplete it will be put on hold and Gallagher Bassett will inform the applicant of this and will detail what is missing. If a claimant is struggling to obtain information from a third party Gallagher Bassett will assist if it can.

The intention is not to make the application process overly onerous, and the claimant should be able to formulate a claim without the assistance of a legal representative (though solicitors, etc, can help if the claimant prefers). Successful applicants are paid a fixed fee of £7,000, out of which legal fees can be paid. If the legal costs are lower than £7,000 the applicant can keep the difference; if higher, then the applicant is liable for the difference.

Payment should be received within 6 weeks of a completed application.

If a claimant is dissatisfied with the decision there is a right to request a review by Gallagher Bassett. If an internal review does not resolve the issue there is a right of appeal to a First-Tier Tribunal.[18]

G. Statistics

DMPS has only been running for a relatively short time, so there are no reliable statistics on trends. A snapshot of the data is provided by the DMPS statistics published by the government.[19] Table 26.5 shows the numbers and proportions of DPMS claims by outcome from April 2014 to September 2015. Of the 610 applications received, 575(94%) are from men and only 35 (6%) are from women.

[17] www.gallagherbassett.co.uk/.

[18] Ss 21–25 of the Diffuse Mesothelioma Payment Scheme Regulations 2014 (2014/916).

[19] www.gov.uk/government/statistics/diffuse-mesothelioma-payment-scheme-statistics.

Table 26.5: Numbers and proportions of DMPS claims by outcome from April 2014 to September 2015

Outcome	Number	Percentage
Successful	410	67
Unsuccessful	135	22
Pending	40	7
Withdrawn	25	4
Applications received	610	

As would be expected with a long-latency disease, applicants are generally in the older age brackets at diagnosis (see Figure 26.2).

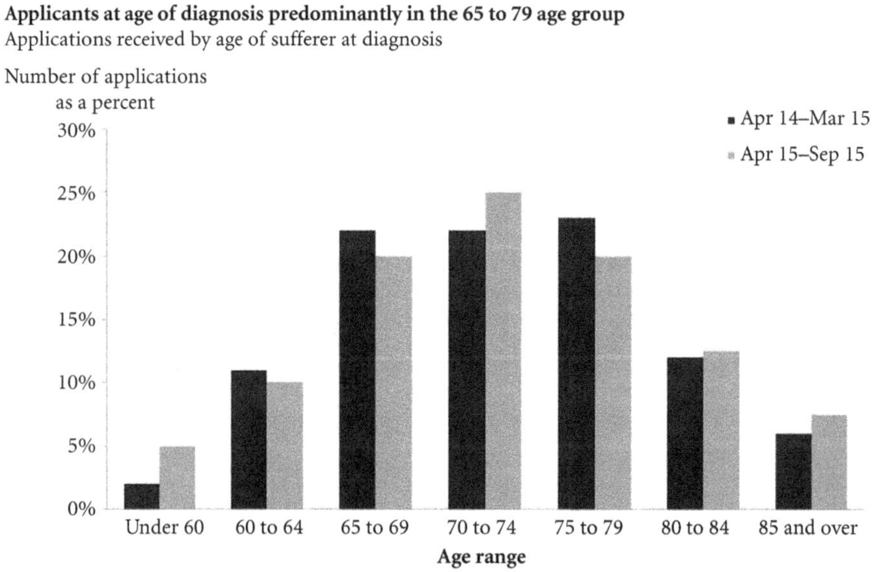

Figure 26.2: DMPS applicants at age of diagnosis

From inception until September 2015 £47.4 million had been paid out (approximately £3.2 million per month), with an average payment of £125,000.

Table 26.6 shows the figures for reviews and appeals. Relatively few cases are reviewed. In the first year of operation 15 out of the 40 (37.5%) reviewed cases resulted in a different outcome, which is concerning. This may be because the scheme was new, but this figure should be kept under review.

Table 26.6: Outcome of claims 2014–15

	April 2014–March 2015	April 2015–September 15	Total
Unsuccessful applications	100	35	135
Review requested	40	10	50
(i) same or pending outcome	25	10	35
(ii) different outcome	15	–	15
Referred to a First-Tier Tribunal	15	0	15
Complaints	0	0	0

H. Compensation Recovery

If an individual who has received IIDB and/or a lump sum under either the 1979 Act or the 2008 Scheme subsequently receives compensation from a civil law claim, then the civil award may be subject to compensation recovery. Essentially this is intended to prevent double compensation for the same injury; the compensator may be required to reimburse the government the value of any of the above benefit types which have already been paid to the claimant.

I. Conclusions

If an individual has a diagnosis of mesothelioma he usually has a very limited life expectancy. The large number of potential routes to compensation is, while understandable, not straightforward for an ill claimant to have to navigate. Recent improvements, such as the Employers Liability Tracing Office and the fast-track Mesothelioma Practice Direction, have improved the process for claimants. However, the shelving of the Mesothelioma Portal appears a missed opportunity to improve the position for claimants.

27

Miners' Compensation Schemes

SONIA MACLEOD

A variety of compensations schemes exist for miners, including schemes for pneumoconiosis, chronic obstructive pulmonary disease (COPD) and vibration white finger (VWF). Of these, only the pneumoconiosis scheme remains open; the COPD and VWF schemes are closed to new applicants.

A. The Coal Workers' Pneumoconiosis Scheme 1974

The Coal Workers' Pneumoconiosis Scheme[1] was set up as a no-fault scheme in 1974. It was created by agreement between the National Coal Board, the National Union of Mineworkers, the National Association of Colliery Overmen, Deputies and Shotfirers, and the British Association of Colliery Management. The Department of Trade and Industry assumed liability on 1 January 1998. The scheme was originally administered by Capita Insurance Services; since November 2011 Nabarro Claims Solutions has administered the scheme.

The scheme provided compensation to mineworkers (or their families) who had certain respiratory diseases, principally pneumoconiosis, though it is also possible to claim under the scheme for bilateral pleural thickening, primary carcinoma of the lung with accompanying asbestosis, and primary carcinoma of the lung with accompanying bilateral pleural thickening.

A claim is in lieu of a common law claim, with three forms of compensation available:

(a) Lump Sum Payments (LSPs): These are paid calculated based upon date of onset/development of pneumoconiosis as determined for Industrial Injuries Disablement Benefit (IIDB) and a miner's age at that date.
(b) Loss of Earnings Allowance Payments (LEAPs): These are paid to men up to the age of 62 who have been certified as being incapable for work for 28 successive weeks or more as a result of pneumoconiosis.
(c) Progression Payments (PPs): These are paid to men who previously received an LSP and whose disability has subsequently been certified as having increased.

[1] *Coal Health Compensation Schemes. Report by the Comptroller and Auditor General*, HC 608 Session 2006–07, 18 July 2007, available at: www.nao.org.uk/wp-content/uploads/2007/07/0607608es.pdf.

Eligibility depended upon:

(a) 10 years employment in a British Coal Corporation mine; and
(b) the claimant should be in receipt of IIDB for pneumoconiosis; or
(c) pneumoconiosis should appear on the death certificate.

Since 1974 there have been some 90,000 claims, involving a total of approximately £150 million in lump sum payments and some £45 million loss-of-earnings payments.

At its peak the scheme dealt with a considerable number of claims, over 100 per month on average. Claim numbers have been dropping off: Figure 27.1 shows the numbers of pneumoconiosis cases assessed for IIBD by year from 2003 to 2014, and a clear decline can be seen. Since November 2011 an average of 20 claims have been made per month and that figure has remained fairly consistent. Of these claims, the following payments have been awarded, 582 LSPs, 52 PPs and 2 LEAPs.

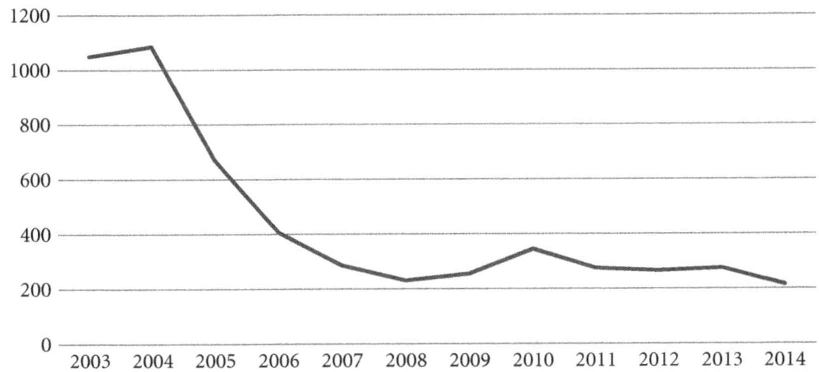

Figure 27.1: British Coal pneumoconiosis cases assessed for IIBD 2003–14. (Source: Health and Safety Executive Statistics on Pneumoconiosis, available at www.hse.gov.uk/statistics/causdis/pneumoconiosis/)

B. The Coal Health Compensation Schemes

The Coal Health Compensation Schemes were the largest personal injury schemes in British legal history.[2] They were negotiated to compensate miners and their families for two mining-related health problems caused by working underground in British coalmines. The two medical conditions that were compensatable were:

— COPD—respiratory disease (chronic bronchitis and emphysema) resulting from the dusty conditions; and
— VWF—a disease caused by using vibrating tools.

[2] Documents relating to the schemes including the Claims Handling Agreements, statistics and summaries of the respective Orders can be found at www.gov.uk/government/collections/coal-health-claims.

British Coal was held liable for negligence in litigation group actions in 1997 in relation to VWF and in 1998 in relation to lung diseases.[3] Test cases were brought for each disease, and found for the claimants. Faced with a large number of litigants, negotiations were undertaken with the claimants' solicitors. In January 1998 the Department of Trade and Industry (DTI)[4] took over responsibility for the accumulated personal injury liabilities of the British Coal Corporation.[5] DTI, in negotiation with the claimants' solicitors' groups,[6] and subject to the approval of the High Court, introduced two compensation schemes for former miners, one for COPD (including respiratory conditions such as chronic bronchitis, small airways disease, asthma, chronic airflow limitation and emphysema) and one for VWF. The VWF Agreement was signed in January 1999, and the COPD Agreement was signed in September 1999.

A Claims Handling Agreement (CHA) for each scheme defined how claims would be dealt with.[7] Claimants could make applications for compensation through their legal representative to the claims handlers appointed by the government. The CHAs specified the Medical Assessment Process, and set out the compensation a claimant would have expected to get if he had pursued his claim under common law through the courts. The government met the cost of the claimant's legal representation where these claims were successful. These arrangements were intended to mirror court procedure but to avoid taking a claim through the courts since the large number of miners and widows that were expected to lodge claims would have clogged the courts up for many years and taken perhaps up to 15–20 years to resolve.

The CHAs were approved by the judges concerned who remained in overall charge of the process of paying out the compensation. Any matters upon which the claimants and the DTI could not agree were referred to the court to determine. The DTI and the miners' solicitors reported back regularly to the judge on progress (3–4 times a year for each disease). Claimants also retained the right to pursue a common law claim should they so wish.

The schemes were administered by IRISC, a claims management firm owned by Capita.[8] Claimants were expected to lodge a completed application pack through their solicitors. A dispute resolution procedure was provided for each scheme.[9]

[3] *Armstrong and Others v British Coal Corporation*: CA 28 1996; [1996] EWCA Civ 1049, *Armstrong and Others v British Coal Corporation (2)*: CA 31 July 1998; see chs 29–30 in C Hodges, *Multi Party Actions* (Oxford, Oxford University Press, 2001).

[4] Restructured and renamed in June 2007 the Department for Business, Enterprise and Regulatory Reform (BERR), and in 2008 the Department for Business Innovation and Skills (BIS).

[5] For a summary of the history of the schemes, see the National Audit Office Report, *Coal Health Schemes* (12 July 2007), available at www.nao.org.uk/wp-content/uploads/2007/07/0607608.pdf.

[6] The claimants' solicitors' groups were steering groups each led by the same three firms of solicitors, whose role was to represent the interests of claimants.

[7] www.gov.uk/government/publications/chronic-obstructive-pulmonary-disease-scheme-claims-handling-agreement and https://www.gov.uk/government/publications/vibration-white-finger-compensation-scheme-claims-handling-agreement.

[8] The function was initially performed by Aon-IRISC.

[9] www.publications.parliament.uk/pa/cm200405/cmselect/cmtrdind/375/37505.htm.

The COPD Scheme

COPD is an umbrella term for set of a medical conditions that affect the lungs. Narrowing of the airways means that blood is not sufficiently oxygenated, reducing the supply of oxygen to the muscles, and limiting normal exertion. The main symptoms are breathlessness, a phlegmy cough and frequent chest infections. At its mildest COPD manifests as chronic bronchitis, which is non-disabling and reversible. Emphysema, a permanent narrowing of the airways, is the severest form and is life-shortening.

The most usual cause of COPD in the UK is smoking; smokers are 6 times more likely to die from COPD than non-smokers, but exposure to fumes, dust and air pollution can also cause it. All of these substances irritate and inflame the lungs, eventually causing scarring. Over many years, the inflammation leads to permanent changes in the lung. The walls of the airways thicken and more mucus is produced. Smaller airways become scarred and narrowed. Damage to the delicate walls of the air sacs in the lungs causes emphysema and the lungs lose their normal elasticity. These changes cause the symptoms of breathlessness, cough and phlegm associated with COPD.

A 1998 High Court judgment found British Coal liable for COPD caused by dust exposure while working underground. Exposure to dust on the surface was not covered by the judgment. The judgment also specifically excluded liability for injury attributable to a claimant's smoking. A matrix and formula were developed by the claimants' primary medical expert in order to apportion injury between smoking and dust exposure (see Table 27.1).

Table 27.1: The ratio and formula to apportion injury between smoking and dust in COPD cases

	High dust exposure	**Moderate dust**
Light smoking	0.5	1
Average smoking	1.0	2
Heavy smoking	1.5	3

$$\text{Proportion due to job} = \frac{\text{years of underground work}}{\text{years of underground work} + (\text{ratio} \times \text{years of smoking})}$$

Smoking could substantially reduce the claimant's entitlement to damages. Non-smoking miners received the full damages amount, or 100% of the damages. For example, a miner who had been exposed to moderate dust for 20 years and smoked lightly for 20 years would receive ½ or 50% of the full damages award. A miner who had been exposed to moderate dust for 20 years and smoked heavily for 20 years would receive ¼ or 25% of the full damages award.

COPD Claims Process

Two types of damages were available under the COPD scheme:

— general damages—to compensate for pain, suffering and loss of amenity; and

— special damages—to cover financial losses due to the injury, including, for example, loss of earnings, loss of benefits in kind, reductions in redundancy and pension benefits as well as the cost of nursing care.

The application pack included details for health conditions, work history, a smoking record and permission to access the claimant's medical records. A schematic of the claims process is shown in Figure 27.2. The claims handlers requested British Coal employment and other records from Iron Mountain, a document storage company employed by the DTI to store these data.

There were four claim pathways: the main scheme, the expedited option and the Optional Risk Offer Scheme, and finally the Entry of Judgment/Unaccepted Offer Process.

All schemes required the collection of employment records, and for deceased claimants the collection of post-mortem records, for live claimants a spirometer reading. As COPD ranges from mild to severe the scheme had to have diagnostic tools to evaluate the degree of lung damage a claimant was suffering from and the severity of the consequent disability. A spirometer reading is used to measure lung damage. It measures the volume of air expelled in a single forced exhalation following a deep breath in. Two main measures are taken, the forced vital capacity (FVC) and the forced expired volume in 1 second (FEV1). FVC is the maximum volume of air that can be exhaled during a forced manoeuvre. FEV1 is the volume expired in the first second of maximal expiration after a maximal inspiration. This is a measure of how quickly the lungs can be emptied. FEV1 expressed as a percentage of the FVC (FEV1/FVC) gives a clinically useful index of airflow limitation. In a normal adult FEV1/FVC is between 70% and 80%, a recording of 60–80% indicates mild COPD, 40–60% moderate and less than 40% is severe.

For live claimants an initial screening spirometer, combined with age and health information, was used to prioritise claimants for the Medical Assessment Process.

Under the main scheme claimants went through the entire process involving a medical assessment, and received a full offer of compensation for the entire value of the claim. There were two types of compensation for claimants who completed the full medical process.

The Medical Assessment Process provided a nation-wide standardised process. The Medical Assessment Process for respiratory disease was dealt with by Atos Origin. This stage included more complex lung function tests and, on the same day, a consultation with a respiratory consultant. The doctor, who had the claimant's medical records and a copy of his claim, produced a detailed medical report (known as the 'MAP') that rated the injury in gradations of 10%. The doctor also assessed, by talking to the man and looking at his records, how his disability had developed in the past and was likely to progress in the future. Diagnosing COPD and making these judgements about disease progression was difficult, and therefore required specialists, ie doctors of consultant status.

The medical assessment procedures for deceased men were based on medical records.

The COPD scheme also included risk offer options, where the offer was made on the basis of the initial spirometer reading and a qualifying period of employment. These risk offers were made without an examination of medical records or the involvement of a doctor. Although 'full and final' offers, they were quantified at significantly lower values than if the claim had gone through the Medical Assessment Process, but they provided faster payment.

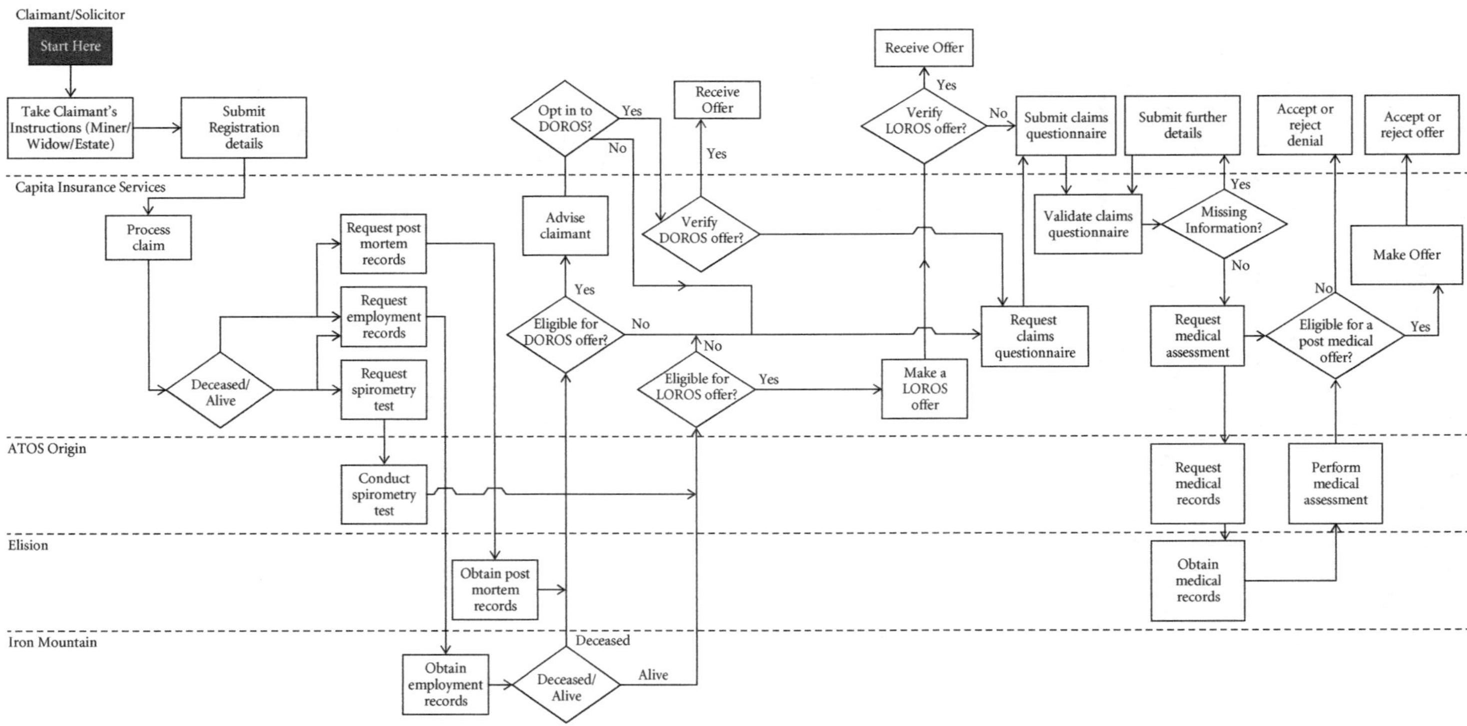

Figure 27.2: Schematic of the COPD claims process

Notes: (1) All claimants who were eligible for Deceased Optional Risk Offers (DOROS) were advised, through their solicitors, in September 2005. The solicitors were required to indicate before February 2007 whether or not their claimant accepted the offer. (2) Eligibility for a Live Optional Risk Offer (LOROS) was determined on the basis of both a claimant's employment history and their spirometry test result.

From the outset an expedited scheme option was available to claimants. Under this option the offer of compensation was made on the basis of the spirometry test plus a minimum period of employment; no medical assessment was undertaken. The compensation offered was individualised, but was lower than the equivalent main scheme offer; however, the process was faster. This was an opt-in scheme and was not widely used.

From 2005 the Optional Risk Offer Schemes (live and deceased, LOROS and DOROS, respectively) were implemented. These were essentially similar to the expedited payment option, but claimants were automatically opted in and the awards were based on past awards rather than being individualised. Claimants could opt out and undergo the full medical assessment procedure if they felt they would be better off taking that path.

Entry of Judgment/Unaccepted Offer Process were brought in to help finalise and close the scheme. These were imposed upon claimants where cases were stalling or unresponsive.

COPD Statistics[10]

The COPD scheme closing date for registering new claims was 31 March 2004. The total number of claims received under the COPD scheme was 591,768. Final statistics were drawn up on 19 June 2012 when the main claims-handling contract ended.[11] There were several claims-handling options available to claimants; the breakdown of claims by claimant category and settlement mechanism is shown in Table 27.2. The majority of claims were for deceased workers (44% of claims were by estates, 15% by widows) with live claimants making up 41% of the claims.

Table 27.2: Breakdown of total CODP claims by claimant category and settlement mechanism

	Number of claims			
	Live	**Widow**	**Estate**	**Total**
Total	**240,376**	**89,136**	**262,256**	**591,768**
Settled by main scheme payment	100,406	54,114	113,049	267,569
Settled by expedited	22,132	820	1,231	24,183
Settled by Optional Risk Offer Scheme (OROS)	72,112	16,114	74,501	162,727
Settled by Entry of Judgment/Unaccepted Offer Process (EoJ/UOP)	1,332	191	628	2,151
Struck out	36,202	14,005	51,542	101,749
Duplicate denials	8,191	3,882	21,299	33,372
Outstanding	1	10	6	17

[10] Taken from www.gov.uk/government/uploads/system/uploads/attachment_data/file/48284/4489-copd-detailed-stats.pdf.

[11] Ibid.

Expenditure for the scheme using all settlement mechanisms is shown in Table 27.3; total other costs includes medical costs, scheme administration, etc. Payments to live claimants were 33% to the total expenditure, with widows receiving 25% and estates receiving 42%. Under the main scheme, costs (including solicitors costs) as a percentage of compensation were 43.6% and the total scheme expenditure was £3,471,209,360.

Table 27.3: Total COPD scheme expenditure

	Total expenditure				
	Total damages (£)	**Total solicitors' costs (£)**	**Total other costs (excluding compensation recovery unit payments) (£)**	**Total compensation recovery unit payments (£)**	**Total expenditure**
Live	722,934,191	369,343,337	1,634,550	51,084,625	1,144,996,702
Widow	687,715,056	184,138,340	654,239	9,832,983	882,340,618
Estate	958,839,380	475,634,718	1,090,020	8,307,923	1,443,872,040
Total	**2,369,488,627**	**1,029,116,394**	**3,378,809**	**69,225,531**	**3,471,209,360**

The use of different settlement mechanisms changed over time, as shown in Figure 27.3.

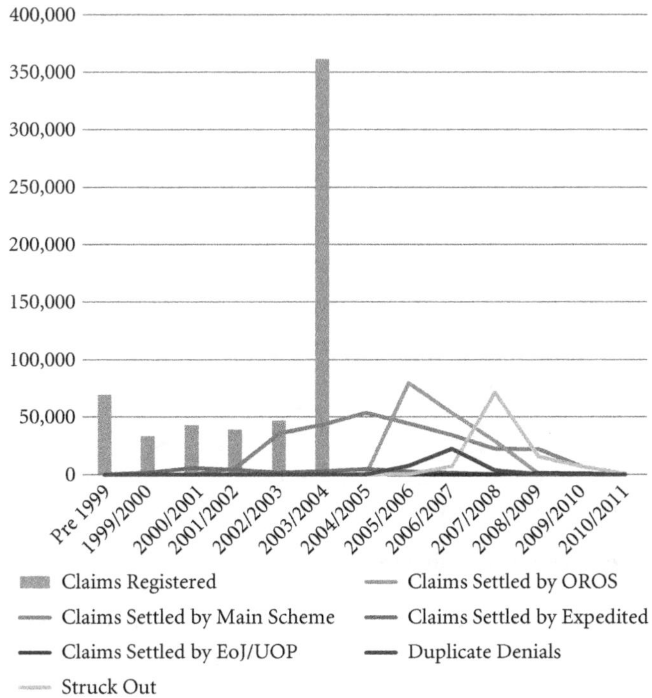

Figure 27.3: COPD scheme claim numbers and settlement means

The majority of the claims under the COPD scheme were for relatively low-value sums (see Figure 27.4).

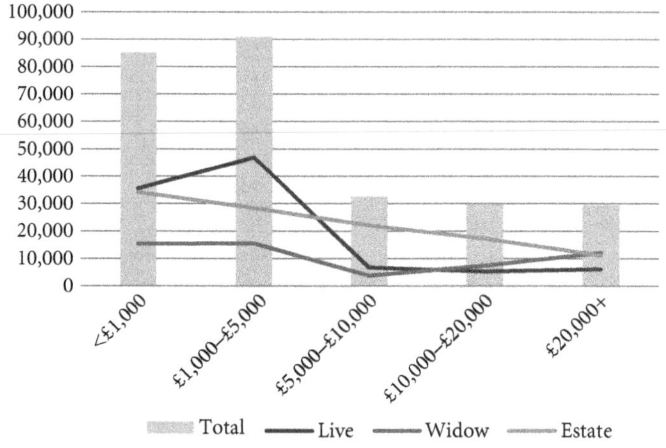

Figure 27.4: Number of claims be settlement banding

Despite being on average low value, these claims took a substantial time to reach settlement, as shown in Figure 27.5. The median claim-processing duration from registration to settlement was 2–3 years, but there were a reasonable number of claims which took over 5 years and over 8,000 claims which took more than a decade to settle. This is not impressive.

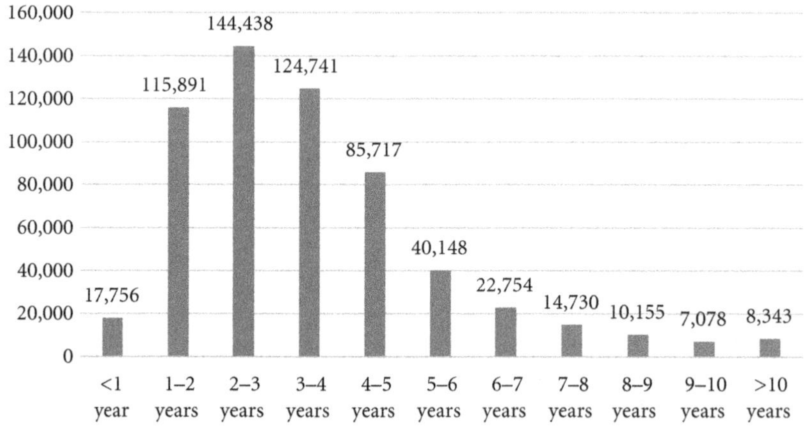

Figure 27.5: Claim numbers by duration bandings

A detailed breakdown of the various settlement options, the Main Scheme, the Expedited process, the Optional Risk Offer Scheme, and Entry of Judgment/Unaccepted Offer Process follows.

The Main Scheme

The Main Scheme proceeded according to the CHA rules. Expenditure under the Main Scheme is shown in Table 27.4, where 'total other costs' includes medical costs, scheme administration, etc. The Main Scheme settlements were 45% of the total settlements made. Of the Main Scheme, expenditure payments to live claimants made up 27%, with widows receiving 30% and estates receiving 43%. Under the Main Scheme, costs (including solicitors' costs) as a percentage of compensation was 37.8% and the total expenditure was £2,780,911,882.

Table 27.4: Expenditure under the Main Scheme

	Total damages (£)	Total solicitors' costs (£)	Total other costs (excluding compensation recovery unit payments) (£)	Total compensation recovery unit payments (£)	Total expenditure
Live	454,846,415	245,428,333	1,461,226	41,353,234	743,089,209
Widow	659,028,399	156,381,457	607,699	9,127,211	825,144,766
Estate	862,713,674	342,008,190	854,866	7,101,178	1,212,677,908
Total	**1,976,588,488**	**743,817,981**	**2,923,791**	**57,581,623**	**2,780,911,882**

The Expedited Scheme

The Expedited Scheme was designed for those claimants with a short life expectancy and with high-value claims. It was a voluntary opt-in scheme, with personalised compensation awards, and it was available from the start of the scheme. It was not widely used—see Figure 27.3.

Expenditure under the Expedited Scheme is shown in Table 27.5. The Expedited Scheme settlements were 4% of the total settlements made. As would be expected, the recipient profile of the Expedited Scheme was very different to that of the Main Scheme. Of the Expedited Scheme expenditure, payments to live claimants made up 88%, with widows receiving 5% and estates receiving 7%. Under the Expedited Scheme, costs (including solicitors' costs) as a percentage of compensation were 24.1% and the total expenditure was £163,639,055.

Table 27.5: Expenditure under the Expedited Scheme

	Total damages (£)	Total solicitors' costs (£)	Total other costs (excluding compensation recovery unit payments) (£)	Total compensation recovery unit payments (£)	Total expenditure
Live	109,625,921	26,970,835	67,835	8,057,829	144,722,421
Widow	5,983,286	1,172,090	4,891	602,289	7,762,556
Estate	8,449,415	1,685,253	6,878	1,012,531	11,154,078
Total	**124,058,622**	**29,828,178**	**79,605**	**9,672,650**	**163,639,055**

The Optional Risk Offer Scheme

In an effort to fast-track claims the Optional Risk Offer Scheme (OROS) was introduced at the end of February 2005. This was aimed at cases likely to attract smaller amounts of compensation, and had separate procedures for live (LOROS) and estate (DOROS) claims. OROS eligibility criteria were less stringent than the Expedited Scheme route and the compensation offered reflected the amounts paid to claimants in a similar position (by this point over 100,000 COPD compensation payments had been made, enabling comparisons based on experience, medical evidence, etc). From March 2005 all living COPD applicants who had not yet attended a full medical examination were automatically added to the OROS; they could then opt-out of OROS to join the Main Scheme (with its full assessment requirements).

Expenditure under the OROS is shown in Table 27.6. The OROS settlements were 28% of the total settlements made. Of the OROS expenditure, payments to live claimants made up 49%, with widows receiving 9% and estates receiving 42%. Under OROS costs (including solicitors' costs) as a percentage of compensation were an utterly staggeringly high 94.8% and the total expenditure was £514,899,503.

Table 27.6: Expenditure under the Optional Risk Offer Scheme (OROS)

	Total damages (£)	Total solicitors' costs (£)	Total other costs (excluding compensation recovery unit payments) (£)	Total compensation recovery unit payments (£)	Total expenditure
Live	156,551,793	93,401,603	98,534	1,205,911	251,257,842
Widow	21,801,925	26,041,926	40,164	61,741	47,945,757
Estate	85,231,382	130,168,377	223,418	72,727	215,695,905
Total	**263,585,100**	**249,611,907**	**362,116**	**1,340,379**	**514,899,503**

Entry of Judgment and Unaccepted Offer Process

Towards the latter stages of the scheme some claims were settled using the Entry of Judgment (EoJ) and Unaccepted Offer Processes (UOP)—see Table 27.7. These mechanisms assisted with the settling of claims in advance of the closing of the schemes. In total only 2,162 claims out of the 591,768 total claims were dealt with using these procedures.

Expenditure under the EoJ Scheme is shown in Table 27.7. The EoJ settlements were 0.4% of the total settlements made. Of the EoJ expenditure, payments to live claimants made up 50%, with widows receiving 13% and estates receiving 37%. Under the EoJ scheme, costs (including solicitors' costs) as a percentage of compensation were shockingly high at 111.7% and the total expenditure was £11,758,920.

Table 27.7: Expenditure under the Entry of Judgment (EoJ) scheme

	Total damages (£)	Total solicitors' costs (£)	Total other costs (excluding compensation recovery unit payments) (£)	Total compensation recovery unit payments (£)	Total expenditure[a]
Live	1,910,062	3,542,565	6,954	467,650	5,927,231
Widow	901,446	542,866	1,485	41,742	1,487,539
Estate	2,444,908	1,772,897	4,857	121,487	4,344,149
Total	**5,256,416**	**5,858,328**	**13,297**	**630,879**	**11,758,920**

[a] As before, figures are taken from www.gov.uk/government/uploads/system/uploads/attachment_data/file/48284/4489-copd-detailed-stats.pdf; however, the total expenditure given in the EoJ Scheme expenditure, £11,758,920, was not the sum of the breakdown of the total expenditure (£135,208 for live claims, £126,911 for widow claims, £31,622 for estate claims). These figures are therefore amended to correct this.

Summary of the COPD Scheme

The National Audit Office reviewed the schemes as at 31 March 2007 and found the following.[12] COPD compensation awards accepted by claimant miners up to that date ranged from 50p to £375,000, with a median award of £1,500. Calculations based on the final statistics show that the average solicitors' fee in the COPD cases was £1,739, out of a total £3,200 cost of administering a COPD claim.[13] The total solicitors' costs under the COPD scheme were £743,817,981 and total damages £1,976,588,488. For every £2.66 spend on compensation to former miners, the DTI spend £1.00 on solicitors. Top-earning law firms, including Thompsons (£156 million) and Beresfords (£144 million), made substantial sums from the coal health compensation schemes. Interesting there was considerable variation in the average COPD settlement received, as shown in Table 27.8.[14] These figures for average settlements are from the 10 claimants' representatives who had submitted the largest volume of COPD coal health claims as at 21 June 2009. As can be clearly seen, Beresfords, who handled the largest volume of COPD claims, had an average damages settlement figure of £2,559, less than a third of the average damages settlement of £9,202 paid to Thompsons. For a scheme that was intended to ensure all miners would receive an equitable settlement, this differential could be indicative of serious flaws in the scheme, or it may be due to differences in the claimant profiles of the different firms.

[12] National Audit Office Report, *Coal Health Compensation Schemes* (July 2007) HC 608 Session 2006–07.

[13] The total cost of administering a case is taken from the National Audit Office Report, ibid; as it involves apportioning costs, there were higher estimates of solicitors' costs: see J Dean, 'Controversy Continues over Miners' Claims' *Law Society Gazette*, 30 July 2009, at www.lawgazette.co.uk/news/news-focus-miners-compensation.

[14] Hansard HC 26 June 2009, vol 494 col 1201W, www.publications.parliament.uk/pa/cm200809/cmhansrd/cm090626/text/90626w0013.htm; see also Dean (n 13).

Table 27.8: Average settlements from the 10 claimants' representatives who had submitted the largest volume of COPD coal health claims as at 21 June 2009

Solicitor	Total number of claims	Total number of claims settled by payment	Average damages paid on claims settled by payment (£)
Thompsons	58,133	45,600	9,202
Mark Gilbert Morse	25,721	20,641	7,535
Hugh James	56,811	42,214	7,261
Browell Smith and Co	32,998	24,962	6,538
Raleys	48,914	35,591	5,990
Watson Burton LLP	14,072	12,482	4,346
Union of Democratic Mineworkers	16,586	13,794	3,097
Barber and Co	14,090	11,494	2,832
Beresfords	80,817	62,425	2,559
Avalon	32,413	19,354	2,375

It was reported in 2007 that as at 31 March 2007, around 69% of all claimants who had paid compensation had received less than the average cost of administering the claims (£3,200 per claim). Probably the most disparate example was a case where compensation of 50p was awarded, with associated solicitors' costs of £1,974 (note this figure only includes legal costs, and is not the total cost of administering the claim, which was presumably higher).[15] There were clear issues with the relative values of the compensation awarded compared to the administration (and particularly the legal) costs of the claims.

Severe criticism was levelled at the DTI for underestimating the number of claims and for agreeing to a fee structure which contained annual inflationary increases, with no option to review. Once agreed, the Department was tied to the agreement for the duration of the scheme. The DTI had initially assumed in March 1998 that the volume of claims for COPD would be around 173,500. In actuality almost 592,000 claims were made, over 3 times higher. The Department had not sought any actuarial advice prior to the inception of the scheme, but had relied upon figures provided by British Coal, which turned out to be incorrect. Additionally, it appears that the DTI had not appreciated that, should an entitled miner die, any entitlement to compensation could be passed to their estate and to subsequent estates if unclaimed. Given that 44% of COPD claims were estate claims, this was a serious omission.

The DTI stated that the fact that the courts were already seised of the cases effectively bound its hands, preventing it from putting in place a statutory scheme which extinguished the right to litigate. It therefore had to 'accept' the court-imposed scheme rules. This is highly questionable on many grounds. It should be remembered that the claims-handling agreements were drawn up in consultations between the steering groups (claimant lawyers)

[15] *Coal Health Compensation Schemes* (n 1); House of Commons Committees of Public Accounts Report, *Coal Health Compensation Schemes*, HC 350 2007–08.

and the DTI. The agreements included a flat fee according to the type of case, irrespective of the quantum of compensation awarded. Clearly this had the effect that solicitors were paid even when they obtained very minimal levels of compensation for their clients. From November 2006 a minimum payment scheme was introduced. Under the minimum payment scheme if a client was awarded less than £500 in compensation, then the solicitors' firms had to increase the award to £500 using their own funds. While this was progress, the DTI acknowledged that this change had taken far too long to implement.

There was considerable disquiet at the level of costs payable under the schemes CHA. An assessment was done in 2007 by a Senior Costs Judge, comparing the cost of an assessment by a Senior Costs Judge and the costs payable under the CHA; the results are shown in Table 27.9. It is clear that the costs paid under the CHA were generous to say the least, and it has been estimated that the DTI could have saved £295 million had the costs been curtailed at the level of a Senior Cost Judge's analysis.

Table 27.9: Comparison of costs under the CHA with the cost of an assessment by a Senior Cost Judge, taken from [2007] EWHC (QB) case no 960177

Category of case	Assessment by Senior Cost Judge	Cost payable under the CHA
Claim following full medical	£922	£2,023
Widow and estates claim following full medical	£985	£2,023
Expedited settlement	£725	£1,061

When the scheme's CHA had been negotiated it was on the basis that each claim was valued in isolation; no account was taken over the economies of scale that could be (and subsequently were) achieved by firms handling large numbers of claims. High-volume processing enabled firms with large numbers of claims to reduce their transactional cost per claim, providing them with even greater profit margins.

There was considerable controversy over the level of fees attached to the OROS claims. Initially these were paid at the same level as claims under the main scheme. The DTI felt that this was unjustified given that the level of work required per claim was lower. The claimants' solicitors' groups contested this and initially the scheme judge ruled against the DTI. This was appealed to the High Court, which ruled in the Department's favour: costs were reduced and solicitors were asked to pay back overpayments.[16] In total, this repayment amounted to £80.6 million.

C. Vibration White Finger

VWF is a medical condition affecting the hands, mainly the fingers; it has now been renamed to hand arm vibration syndrome to reflect more accurately the condition, but VWF will be used here. VFW is caused by damage to nerves and blood vessels in the hand and arm.

[16] [2007] EWHC (QB) 960177.

Damage to nerve endings results in changes in sensory perception, such as numbness and tingling. Damage to the blood vessels in the hand can causes parts of the fingers to suffer from reduced flow of blood, turning them white. Damage to the blood flow is known as secondary Raynaud's disease. The severity of the disorder varies. Mild cases exhibit temporary numbness and tingling. This can cause problems with fine dexterity movements, such as difficulty in doing up buttons or holding small objects, such as coins. Exposure to the cold may result in restricted blood flow which can be uncomfortable or painful. As the disease becomes more serious, there is a progressive loss of feeling, and therefore of dexterity and more frequent bouts of restricted blood flow. In the most extreme cases amputation of the digits can be required. There is no cure for the disease, which is irreversible, but without further exposure the damage usually stabilises.

In 1968 a scale for the classification of VWF was created by Dr Pelmear, a leading expert, and Professor Taylor, an industrial hygienist. Initially this scale only incorporated vascular damage as demonstrated by whitening of parts of the fingers; in 1986 it was refined to include nerve damage, and became known as the Stockholm scale. Under the Stockholm scale, each hand and each of the three parts of each finger is classified separately.

VWF Claims Process

After considerable denials of liability, the hearing of lead cases and various appeals by British Coal, the Court of Appeal ruled on the threshold for negligent exposure and the levels of awards available for VWF in July 1998. After this the VFW CHA was drawn up.

The claims process for VWF is outlined in Figure 27.6. The application form required personal information including employment history, symptoms, and permission to contact the claimant's doctor.

The VWF scheme did not have an expedited option as each case required a medical assessment to ensure a fair assessment of the claim; however, there were arrangements for claimants to receive interim payments.

Three types of damages were available; these were:

- pain, suffering and loss of amenity (general damages);
- 'Smith and Manchester damages' for handicap on the labour market (general damages); and
- services for assistance with gardening, window cleaning, DIY, decorating, car washing and car maintenance (special damages).

Some claims were limited to general damages; others included special damages as well. The nature of damages claimed determined the way in which the claim progressed (see Figure 27.6).

There were three categories of VWF claims:

A. claims where court proceedings commenced on or before 22 January 1999.

B. claims:

    (i) where court proceedings commenced after 22 January 1999 and a medical report was served with those proceedings;

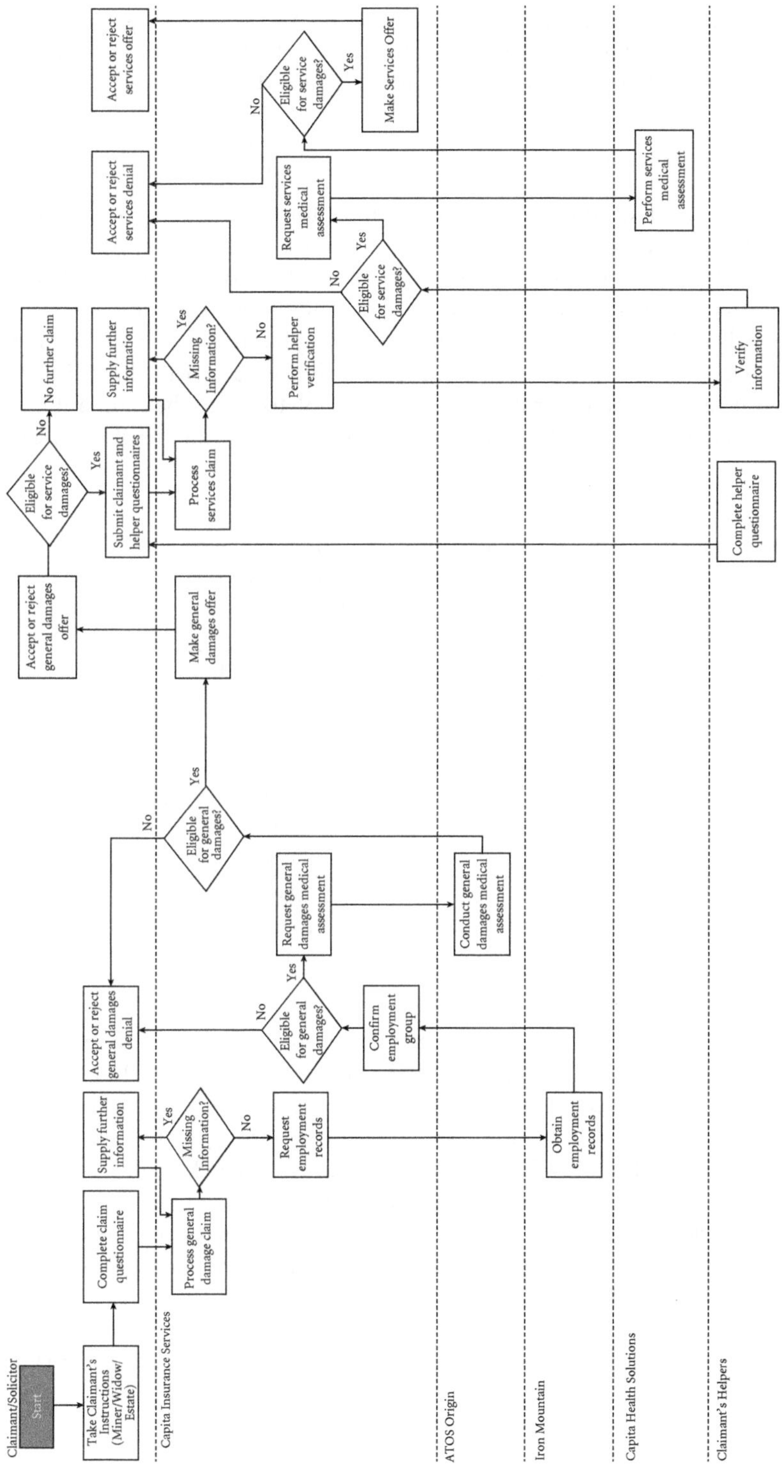

Figure 27.6: Schematic of the VWF claims process

(ii) which had not resulted in the commencement of court proceedings but where a medical report has been served on the claimant's behalf or had been commissioned prior to 25 March 1999 and subsequently served; and

C. claims which had not resulted in the commencement of court proceedings and where a medical report had not been served on the claimant's behalf.

There were slightly different requirements and timeframes for the handling of category A claims. Category A claims had shorter processing times and did not have to demonstrate an occupational link. In contrast, to be eligible under the VWF CHA, category B and C claimants had to demonstrate an occupational link. There were three defined occupation groups,[17] which had different requirements to be eligible under the VWF CHA (see Table 27.10). For groups 1 and 2 there was a presumption in favour of the claimant (providing the requisite employment duration had been completed) that the claimant had been exposed to damaging vibration; those in group 3 had to show evidence of exposure to vibration.

Table 27.10: Eligibility under the VWF CHA by occupational group

Group	Qualifying requirements	Occupations
1	12 months employment after 1 January 1975	Occupations where the use of hand-held vibratory or percussive tools was recognised as a substantial part of the occupation (examples from defined list: face production team, roof bolter).
2	24 months employment after 1 January 1975	Occupations where the use of vibratory or percussive tools was not necessarily a substantial part of the job, but use by individuals within this group may well have been significant (example from defined list: general underground labourer)
3	Evidence of tortious exposure to vibration	Any occupations not shown in the defined lists of groups 1 and 2 (eg canteen staff, crane driver, and locomotive driver and guard (underground and surface)).

Once eligibility on the basis of occupation had been established, the case progressed to a general damages Medical Assessment Process. This Medical Assessment Process was based on the Stockholm scale. However, the application of the scale was not straightforward. Measurement depended not only upon the accurate reporting of the symptoms by the claimant to the doctor, but also the doctor's evaluation of the claimant's description of his symptoms. Clearly there was the potential for variation and lack of consistency. In an effort to develop a consistent assessment methodology that could be applied on a large volume of claimants the DTI established the Independent Medical Advisory Group. The Group devised an Medical Assessment Process that comprised 3 main tests (see Table 27.11), which could be delivered by a trained technician within a controlled environment (eg ambient temperatures and a period of acclimatisation for the claimant) in combination with a consultation with a specialist doctor.

[17] Schedule 2(2) of the VWF CHA.

Table 27.11: Tests used in the Medical Assessment Process to assess general damages

Name	Area tested	Test variable	Injury under test
Cold provocation test (discontinued in 2001 as unreliable)	Both hands held in 15°C water for 5 minutes	Rewarming speed	Vascular damage
Vibrotactile threshold test	Forefinger and little finger of each hand	Ability to detect vibration	Neurological injury
Thermal aesthesiometry test	Forefinger and little finger of each hand	Ability to detect changes in temperature	Neurological injury

The claimant was then seen by a doctor who was specifically trained to interpret the test results and make an assessment of the overall injury due to VWF, discounting other conditions which can produce similar test results or symptoms. The doctor could also conduct further specified tests if it was deemed appropriate.

A services damages medical assessment was conducted to see if the VWF was responsible for the inability to carry out the tasks or if there were other co-morbid conditions which could be causal. This assessment was questionnaire based and carried out by a doctor.

VFW Statistics[18]

The VWF scheme closing date for registering new claims was 31 October 2002 for live claims and January 2003 for deceased claims. The breakdown of claims is shown in Table 27.12. The total number of claims received under the COPD scheme was 169,611. Final statistics were drawn up on 19 June 2012 when the main claims-handling contract ended.[19] The majority of claims (85%) were for live claimants, with deceased claimants making up just 15% of the claims.

Table 27.12: Breakdown of total VWF claims by claimant category and settlement mechanism

	Live	Deceased	Total
Total	**144,888**	**24,723**	**169,611**
Settled by payment	116,122	17,089	133,211
Settled by denial/withdrawn	28,739	7,634	36,373
Outstanding	27	0	27

Expenditure for the scheme using all settlement mechanisms is shown in Table 27.13; 'total other costs' includes medical costs, scheme administration, etc. Payments to live claimants

[18] Taken from www.gov.uk/government/uploads/system/uploads/attachment_data/file/48283/4490-vwf-detailed-statistics.pdf.

[19] ibid.

were 33% of the total expenditure, with widows receiving 25% and estates receiving 42%. Under the main scheme, costs (including solicitors' costs) as a percentage of compensation were 43.6%, and the total scheme expenditure was £3,471,209,360. Under the VWF scheme, costs (including solicitors' costs, other costs (excluding compensation recovery unit payments) and defendant costs) as a percentage of compensation were 11.4% and the total expenditure was £2,054,022,280.

Table 27.13: Total VWF Scheme expenditure

	Total damages (£)	Total solicitors' costs (£)	Total compensation recovery unit payments (£)	Total other costs (excluding compensation recovery unit payments) (£)	Defendant costs (£)	Total expenditure (£)
Live	1,566,863,461	25,166,018	150,177,019	1,983,138	3,203,771	1,747,393,408
Deceased	134,694,234	162,282,784	8,796,461	573,362	282,031	306,628,873
Total	1,701,557,696	187,448,802	158,973,480	2,556,500	3,485,802	2,054,022,280

The use of different settlement mechanisms changed over time, with the highest number of claims paid in 2007/08, 4 years after the close of the scheme (see Figure 27.7).

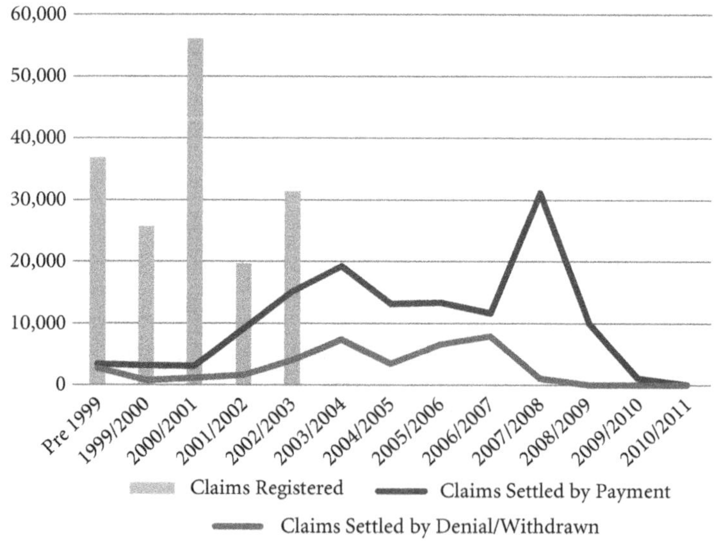

Figure 27.7: VWF Scheme claim numbers and settlement means

Claims under the VWF Scheme were for higher-value sums, with the majority being between £5,000 and £20,000 (see Figure 27.8).

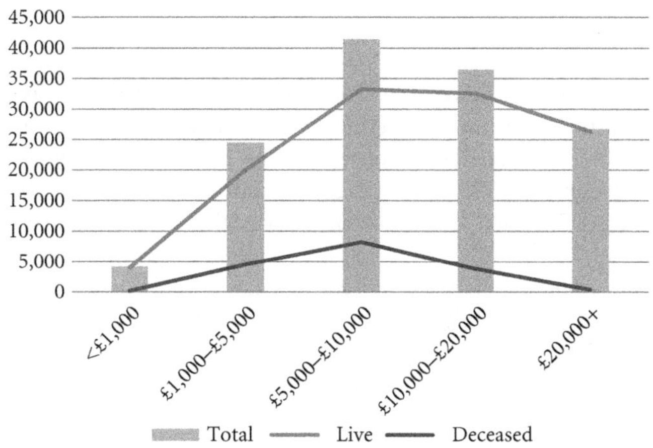

Figure 27.8: Number of claims by settlement banding

These claims took a substantial time to reach settlement (see Figure 27.9). Given that each claim had to involve a medical assessment, very fast processing would have been surprising. The median claim-processing duration from registration to settlement was 5–6 years, but there were a substantial number of claims which took over 5 years and almost 5,000 claims which took more than a decade to settle. This is not impressive, but it should be remembered that these figures include claims where service claims were made (approximately a quarter of VWF claims included a service claim) and this added additional time. Claims without a service claim element had a median processing time of 20 months.[20]

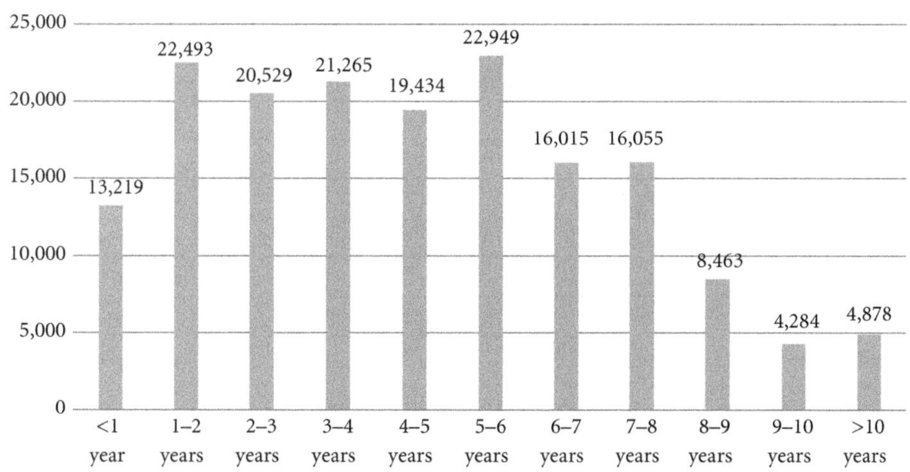

Figure 27.9: VWF claim numbers by duration bandings

[20] National Audit Office Report (n 12).

Summary of the VWF Scheme

The National Audit Office reviewed the schemes as at 31 March 2007 and found the following.[21] VFW compensation awards accepted by claimant miners up to that date had a median award of £8,300. Calculations based on the final statistics figures show that the average solicitors' fee in the VWF cases was £1,105, out of the total £2,600 cost of administering a VWF claim.[22] The total solicitor's costs under the VWF scheme were £187,448,802 and total damages £1,701,557,696. For every £9.07 spend on compensation to former miners the DTI spend £1.00 on solicitors. Top-earning law firms, including Thompsons (£156 million) and Beresfords (£144 million), made substantial sums from the coal health compensation schemes.[23] Interestingly, there was considerable variation in the average VWF settlement received (see Table 27.14).[24] These figures for average settlements are from the 10 claimants' representatives who had submitted the largest volume of VWF coal health claims as at 21 June 2009. As can be clearly seen, Thompsons handled the largest volume of VWF claims and had an average damages settlement figure of £16,379. The highest average damages settlement figure, £17,106, was paid to Watson Burton LLP's claimants, compared to £10,399 for Beresfords' clients, on average £6,707 less. For a scheme that was intended to ensure all miners would receive an equitable settlement this differential could be indicative of serious flaws in the scheme, or it may be due to differences in the claimant profiles of the different firms. Interestingly there are difference in the relative successful claim rates of the different firms: 62.4% of Beresfords' claims were settled by payment; by contrast, 88.4% of Raley's claims resulted in settlement by payment.

Table 27.14: Average settlements from the 10 claimants' representatives who had submitted the largest volume of VWF coal health claims as at 21 June 2009

Solicitor	Total number of claims	Total number of claims settled by payment	Average damages paid on claims settled by payment (£)
Watson Burton LLP	5,935	5,066	17,106
Thompsons	21,231	16,997	16,379
Graysons	8,700	7,201	15,537
Towells	5,368	4,219	12,468
Raleys	11,279	9,968	12,405
Hugh James	10,637	8,552	11,319
Browell Smith and Co	16,462	11,957	10,854

(continued)

[21] ibid.
[22] Dean (n 13).
[23] Mann (n 14).
[24] Hansard HC, 26 June 2009, vol 494 col 1201W, www.publications.parliament.uk/pa/cm200809/cmhansrd/cm090626/text/90626w0013.htm; see also Dean (n 13).

Table 27.14: ***(Continued)***

Solicitor	Total number of claims	Total number of claims settled by payment	Average damages paid on claims settled by payment (£)
Union of Democratic Mineworkers	11,520	9,946	10,607
Moss	7,398	5,031	10,514
Beresfords	11,619	7,251	10,399

It was reported in 2007 that as at 31 March 2007, around 7% of all claimants who had been paid compensation had received less than the average cost of administering the claims (£2,600 per claim). This compares favourably to the COPD scheme.[25]

Severe criticism was levied at the DTI for underestimating the number of claims and for agreeing to a fee structure that contained annual inflationary increases, with no option to review. Once agreed, the DTI was tied to the agreement for the duration of the scheme. The Department had initially assumed in March 1998 that the volume of claims for VWF would be around 45,000. In actuality almost 169,000 claims were made, over 3 times higher.

The VWF scheme was far more efficient than the COPD scheme, but some of the same criticisms were levelled: the lack of actuarial advice prior to the inception of the scheme; the flat-fee agreements; the lack of any option to renegotiate fees, etc; the lack of resource dedicated to the claims by the DTI; the fact that economies of scale for large-volume claims handling firms were not considered, etc.

D. Subsequent Claims against Solicitors

The compensation schemes produced unacceptable behaviour by a sequence of lawyers and other claims handlers. In February 2005, there was a national outcry when an investigation into the Coal Health Compensation Scheme revealed that law firms across the country had wrongly taken a share of their clients' compensation. It has been estimated that 20% of the 750,000 beneficiaries had been improperly charged by solicitors, who made deductions to compensation awards despite a fixed-fee agreement with the government.[26] It appeared to be widespread practice that solicitors failed to advise their clients that the government was funding all the costs of the scheme, and charged fees or deducted them from the compensation payments. Alternately, some solicitors gave money to third-party claims handlers

[25] *Coal Health Compensation Schemes* (n 1); House of Commons Committees of Public Accounts Report: Coal Health Compensation Schemes, HC 350 2007–08.

[26] *Review of Coal Health Compensation Schemes*, http://webarchive.nationalarchives.gov.uk/+/http://www.berr.gov.uk/files/file16630.pdf.

and others deducted the money for the miners' trade union without making clear that the deduction was voluntary.[27]

A number of solicitors were disciplined by the Law Society, whose Legal Complaints Service (LCS) dealt with complaints about deductions. The Solicitors Regulatory Authority (SRA) also aided the LCS with the work on deductions and investigated the impropriety by solicitors.[28] The LCS and SRA worked with 22 law firms to write to 91,000 clients to tell them about the deductions. It reported that solicitors repaid nearly £10 million to clients as a result of this intervention.

The LCS was replaced by the Legal Ombudsman[29] in late 2010. A decision was taken not to include coal health claims within the new Legal Ombudsman Scheme Rules. Therefore, prior to its closure the LCS made a concerted effort to contact potential claimants to inform them that they might have a claim about their legal services provider, and the need to act.

The SRA also carried out 67 separate onsite inspections of firms involved in miners' compensation claims: 28 of these resulted in successful prosecutions before the Solicitors Disciplinary Tribunal (SDT), and reprimands were made against another 22 firms.[30]

In 2009 11 law firms were sued for allegedly failing to obtain the full amount of compensation owed to some miners under their original compensation claims.[31] Somewhat ironically, given that a number of complaints related to improper payments to claims handlers, this was followed by a number of claims management companies and law firms setting up businesses that specialised in harvesting potentially undersettled claims.

[27] N Rose, 'Miners' Compensation: Why Did Some Solicitors Take Too Big a Slice of the Pie' *The Guardian*, 8 March 2011, www.theguardian.com/law/2011/mar/08/sick-miners-compensation-solicitors; A Hirsh, 'Lawyers Made Millions from Sick Miners' *The Guardian*, 12 December 2008, http://www.theguardian.com/uk/2008/dec/12/beresfords-misconduct-solicitors-miners?guni=Article:in%20body%20link.

[28] For a review, see Legal Services Complaints Commissioner, *Special Report: Investigation into the handling of Coal Health Service Complaints by the Legal Complaints Service and the Solicitors Regulatory Authority.* January 2008.

[29] Established under the Legal Services Act 2007 c 29.

[30] SRA Update Issue, 17 March 2011, www.sra.org.uk/sra/news/sra-update/issue-17-miners-compensation.page.

[31] J Dean, 'Beresfords Pays Out to Miners', *Law Society Gazette*, 29 April 2010, available at: www.lawgazette.co.uk/news.

28

Compensation Scheme for Hepatitis C and HIV

SONIA MACLEOD

A. Background[1]

During the 1970s and 1980s, an estimated 32,718 people were exposed to hepatitis C[2] as a result of being treated by the NHS with contaminated blood products.[3] Slightly fewer than 6,000 of these individuals have been identified. This was before heat-treatment of blood products began in the mid-1980s to kill viruses present in donated blood that was used in transfusions or blood clotting factor. Those injured were mainly haemophiliacs,[4] who were given a product to help their blood to clot. Between 1978 and 1985 more than 1,500

[1] This chapter draws heavily on information taken from the *Inquiry into the Current Support for those affected by the Contaminated Blood Scandal in the UK*, All-Party Parliamentary Group (APPG) on Haemophilia and Contaminated Blood (January 2015), available at www.haemophilia.org.uk/what_we_do/influencing_advocacy/appg_hcb_fr.pdf. In particular, the information on diseases and the historical contexts of the five compensation schemes is taken from this report.

[2] Hepatitis C virus (HCV) is a member of the hepatitis virus family, and cases inflammation of the liver and loss of function in liver cells. It is usually transmitted through the blood, but can also be sexually transmitted. HCV infection falls into three main stages. Acute HCV is usually mild or asymptomatic; 15–25% of people with acute HCV infections naturally clear the virus from their bodies. Those who do not naturally clear the HCV go on to develop chronic HCV or stage 1 HCV. The symptoms of stage 1 HCV include jaundice, moderate to severe drowsiness, and it can cause death. With drug treatment approximately 55% of individuals clear HCV from their bodies at this stage. There are a variety of complications associated with stage 1 HCV, with the government-funded compensation arrangements only recognising 5 of these. Stage 2compensation occurs if the HVC infection has led to cirrhosis, non-Hodgkin's lymphoma or primary liver cancer.

[3] Estimate of total number of hepatitis C-infected individuals in UK infected over the period 1970–91, Annex 2, Table 1, of Department of Health (2011) *Review of the Support Available to Individuals Infected with Hepatitis C and/or HIV by NHS-supplied Blood Transfusions or Blood Products and their Dependants*, available at www.gov.uk/government/uploads/system/uploads/attachment_data/file/215828/dh_125977.pdf.

[4] Haemophilia is a bleeding disorder, almost exclusively found in men (von Willebrand disease affects both sexes equally), which is caused by the lack of a protein required for blood clotting. Haemophilia A is a factor VIII deficiency, haemophilia B is a Factor IX deficiency and von Willebrand disease is due to a lack of von Willebrand factor (vWF). From the 1960s onwards haemophiliacs were treated with concentrated solutions of blood-clotting factors obtained from donated blood; those with von Willebrand disease could be treated with either blood-derived products or a non-blood-derived alternative DDAVP (desmopressin acetate). Unfortunately during the 1970s and early 1980 there was no way of screening donated blood for viral infections and the pooling of donations to produce clotting factors increased the risk of infection for recipients. During this time many haemophiliacs were treated with blood products containing HIV and HCV.

individuals were infected with human immunodeficiency virus (HIV).[5] Almost 2,000 of those people have since died.

For the purposes of this chapter, primary infectees are those who were infected through receiving contaminated blood products; secondary infectees were infected by personal contact with a primary infectee.[6] Primary beneficiaries are those who are recipients of support from any of the 5 schemes because they are infected with hepatitis C and/or HIV and meet the relevant eligibility criteria. Secondary beneficiaries are widows/widowers, carers or dependants who are eligible to register with any of the 5 schemes because of their relationship with a beneficiary of that scheme.

The UK government has recently held a consultation on the future of funding for contaminated blood recipients.[7] The final report highlights the difficulties with the current system and proposes considerable reforms to it. This chapter will briefly consider the current arrangements, and will then detail the reforms suggested in the report. However, until these reforms are implemented they should be regarded as intention rather than actuality.

B. Overview of the Current Arrangements

There are 5 separate sources of support for different groups of beneficiaries, listed in Table 28.1 These were developed at different times and have different remits.

The two private companies are:

— the Skipton Fund Ltd (established 2004);[8] and
— the MacFarlane and Eileen Trust Ltd (MFET Ltd) (established 2010).

They give non-discretionary one-off lump sum and ongoing payments. The companies making these payments have no discretion in allocating or quantifying the payments; they have to pay them at the set values when the eligibility criteria are met.

[5] HIV is a virus which inhabits T cells, part of the immune system, which usually fight bacterial infections. T cells are hijacked by the HIV, which uses the T cell to replicate itself many times, and then splits the T cell, releasing the new HIV viruses into the body and killing the T cell in the process. Initial infection often produces a short flue-like illness, followed by an asymptomatic period, which can last for years. Eventually if left untreated, the number of T cells in the body drops so low that the immune system does not function, and opportunistic infections take hold; this is known as late-stage HIV or acquired immunodeficiency syndrome (AIDS). Treatment with a combination of antiretroviral drug therapy prevents the HIV from replicating, and helps to allow the immune system to recover. This is generally more effective if taken earlier, but can also be effective even after a diagnosis of AIDS. Treatment also reduces the likelihood of passing the disease on to a sexual partner or from mother to child. HIV is predominantly transmitted through unprotected sex, with lower levels of transmission occurring from mother to baby during pregnancy, birth or breastfeeding, sharing needles and less commonly by blood contamination, such as needle-stick injuries by healthcare workers. HIV is not currently considered curable and there are very, very limited reports of spontaneous remission, where individuals who were infected show an undetectable viral load. With treatment the life expectancy of someone with HIV does not differ from that of the general population.

[6] Sometimes known as 'infected intimates'.

[7] Infectious Diseases and Blood Policy, *Infected Blood: Government Response to Consultation on Reform of Financial and Other Support*, 13 July 2016, available at www.gov.uk/government/uploads/system/uploads/attachment_data/file/539993/Consultation_response_acc.pdf.

[8] http://www.skiptonfund.org

The three charities are:

— the Caxton Foundation (established 2011);[9]
— the MacFarlane Trust (established 1988);[10] and
— the Eileen Trust (established 1993).

These give discretionary support in various forms (lump sums, grants and ongoing payments), and the MacFarlane Trust and the Eileen Trust offer means-tested ongoing payments to top up non-discretionary payments. The fact that these payments are discretionary means that the charities can chose whether or not to pay claimants, and can alter eligibility requirements and can alter the quantum of payments.

The eligibility criteria for the 5 schemes are shown in Table 28.1. It is not the most straightforward of situations for potential claimants, who have to establish which of the 5 schemes they can claim under. In particular, the fact that the Eileen Trust does not even have a website makes obtaining information about them tricky.

Table 28.1: Eligibility criteria for the five contaminated blood compensation schemes

	Eligibility criteria		
	HIV/HCV infection status	**Haemophiliac status**	**Claimant type**
The Skipton Fund Ltd	HVC positive[a]	Haemophiliacs and non-haemophiliacs[c]	Primary infectees and specified secondary infectees[c]
The MacFarlane and Eileen Trust Ltd (MFET Ltd)	HIV positive[b]	Haemophiliacs and non-haemophiliacs[c]	Primary infectees and specified secondary infectees[c]
The Caxton Foundation	HVC positive, HIV negative	Haemophiliacs and non-haemophiliacs and their families	Primary and secondary infectees Uninfected secondary beneficiaries may be eligible Infectees must be registered with the Skipton Fund Ltd
The MacFarlane Trust	HIV positive and either HVC negative or HVC positive	Only haemophiliacs and their families	Primary and secondary infectees Uninfected secondary beneficiaries may be eligible
The Eileen Trust	HIV positive and either HVC negative or HVC positive	Only non-haemophiliacs and their families	Primary and secondary infectees Uninfected secondary beneficiaries may be eligible

[a] Can also claim from the MFET Ltd.
[b] Can also claim from the Skipton Fund Ltd.
[c] There are limited circumstances in which family members may be able to claim previously unclaimed lump sum payments, but generally benefits are not available to secondary beneficiaries.

[9] http://www.caxtonfoundation.org.uk
[10] http://www.macfarlane.org.uk

The schemes do not offer the same levels of support; Table 28.2 outlines the differences between the support available according to the scheme. The support available from the three charities is discretionary, whereas the two companies offer non-discretionary lump sums and ongoing payments. Obtaining support from the charities is dependent upon the infected individual already being registered with the appropriate company.

Table 28.2: Types of support available from the five contaminated blood compensation schemes

	Available types of support		
	Recipients	**Financial support**	**Non-pecuniary support**
The Skipton Fund Ltd	People infected with HCV Stage 1: chronic hepatitis C Stage 2: cirrhosis, primary liver cancer, B cell non-Hodgkin lymphoma, liver transplant	— Lump sum at stage 1 and stage 2 — Annual payments for Stage 2 (£14,749 in 2015/16) — Reimbursement for prescription prepayment certificate	
The MacFarlane and Eileen Trust Ltd (MFET Ltd)	People infected with HIV through blood transfusions	— Variable lump sum on entry — Annual payments (£14,749 in 2015/16) — Reimbursement for prescription prepayment certificate (PPC)	
The Caxton Foundation	People infected with only HCV and their families, including bereaved family members and dependants	— Means-tested income top ups — One-off grants — Winter payments	— Benefits advice — Other support, such as debt advice and counselling
The MacFarlane Trust	Haemophiliacs infected with HIV (and those co-infected with HCV) and their families, including bereaved family members and dependants	— Means-tested income top ups — One-off grants — Winter payments	— Benefits advice — Other support, such as debt advice and counselling
The Eileen Trust	Non-haemophiliacs infected with HIV via blood transfusions (and those co-infected with HCV) and their families, including bereaved family members and dependants	— Means-tested income regular payments — One-off grants — Winter payments	— Benefits advice — Beneficiary events — Other support

C. Hepatitis C

A UK-wide ex gratia government payment scheme was established on 25 March 2004,[11] called the Skipton Fund,[12] to provide support for those affected.[13] The Skipton Fund makes payments to defined people who were infected with hepatitis C through treatment with NHS blood or blood products prior to September 1991 and other persons eligible for payment in accordance with the scheme's provisions. The Caxton Foundation provides additional discretionary support for those eligible for support under the Skipton Fund Ltd.

Application Procedures

Eligibility: Skipton Fund Ltd

Determining eligibility for the Skipton Fund Ltd is a two-stage process. A potential claimant must first register, and the Skipton Fund Ltd makes a determination on whether to accept the claim. Interestingly claims are only rejected if an applicant dies; all other claims are either 'accepted' or 'deferred'. Financial support is only available to individuals whose claim has been registered and accepted.

When a potential claimant registers a potential claim they are then send the first-stage application form and guidance notes.[14] The onus is on the claimant to prove that he is eligible. To do this he must prove that he meets the criteria for being a primary or secondary infectee *and* has stage 1 hepatitis C.

The second-stage claim can only be paid subsequent to a successful application for a first-stage payment. If a claimant has second-stage HCV he must ask the Skipton Fund Ltd for a second-stage application form once his first stage application has succeeded.

— *Primary infectees.* To be eligible an individual must show that he has chronic hepatitis C virus, and that infection was, on the balance of probabilities, due to receiving contaminated NHS blood or blood products prior to September 1991. To demonstrate this the primary infectee must produce NHS medical records showing treatment with potentially infected blood products. The applicant has to show that it is a probability, not just a possibility, that the infection with hepatitis C occurred from contaminated blood treatment delivered by the NHS.

[11] It has never been established that anyone was negligent or at fault: the establishment of the scheme ended attempts to sue the government and the Blood Products Laboratory.

[12] www.skiptonfund.org/.

[13] The Skipton Fund was established by the Department of Health (England), acting for and on behalf of the Secretary of State for Health, Scottish Ministers, the National Assembly for Wales and the Department of Health, Social Services and Public Safety (Northern Ireland), to administer the scheme and make payments to relevant claimants on behalf of each of the UK health administrations. Scottish Ministers adopted the Skipton Fund as a scheme under s 28 (Infection with Hepatitis C as a Result of NHS Treatment etc) of the Smoking, Health and Social Care (Scotland) Act 2005.

[14] Sample application forms are available on the Skipton Fund website, but these cannot be submitted to the Fund as only original hard copies will be considered.

- *Secondary infectees.* To be eligible as a secondary infectee the claimant must be able to show their relationship with a primary infectee falls into one of the following categories:
  - spouse or civil partner;
  - person who was living with the person from whom the virus was transmitted as husband or wife or in a relationship which had the characteristics of the relationship between civil partners (or if the person from whom the virus was transmitted was in hospital immediately before death, had been so living when that person was admitted to hospital);
  - other cohabitant (whether or not of the same sex as the person from whom the virus was transmitted);
  - where the person from whom the virus was transmitted was a mother, a son or daughter of the mother;
- *and* that, on the balance of probabilities, that the relationship with the primary infectee was the cause of their infection. *And* the primary infectee *must* have already made a successful first-stage claim to the Skipton Fund Ltd. Determining if the primary infectee was the source of infection will also be dealt with on the balance of probabilities.

In order for a claim to be accepted the applicant must provide evidence of infection with either stage 1 or stage 2 hepatitis C virus. This evidence must come from a GMC-registered doctor; usually it comes from the claimant's hepatologist, or if the claimant has a bleeding disorder their haematologist. The gold standard is evidence from a liver biopsy. However, liver biopsies are costly, and potentially fatal for individuals with bleeding disorders, so alternative proof can be submitted for them (see Table 28.3).

Table 28.3: Skipton Fund Criteria

	Preferred evidence	**Alternative evidence**
Stage 1: chronic hepatitis C virus	Liver biopsy[a]	Must show at least 6 months of HCV infection in the blood, usually achieved by polymerase chain reaction tests showing viral presence in the blood over this time or by ongoing medical treatment for HCV
Stage 2: either — cirrhosis, or he/she is more likely than not to progress to have it, *or* — B-cell non-Hodgkin's lymphoma or primary liver cancer	Liver biopsy[a]	Blood test results 3 months apart showing AST, ALT and platelet levels for cirrhosis, or a variety of other forms of evidence including a fibroscan to measure levels of fibrosis in the liver or proof the claimant is undergoing a liver transplant

[a] The Skipton Fund Ltd explicitly disqualifies haemophiliacs from registering if they have had a liver biopsy solely for the purpose of proving stage 1 or stage 2 HCV infection.

Completed applications are then considered by the 5 directors of the Skipton Fund. Of the 5 directors, 2 are hepatologists, the other 3 are not. Unclear or complex applications are considered by the 2 hepatologist directors.

Although the onus is on the claimant to prove his illness he cannot submit the relevant information; this has to come from his doctor. This situation can be potentially difficult, and there have also been problems with hospital trusts not being able to find patient records dating from the 1980s and early 1990s.

Until 2011, anyone with chronic hepatitis C contracted from contaminated blood received a one-off payment of £20,000. A further £25,000 was paid if the person's condition progressed to more severe disease, such as cirrhosis or liver cancer.

The level of payments has been criticised as being insufficient, but successive governments rejected calls for an independent inquiry and additional compensation. A non-statutory inquiry funded from private donations, the Archer Inquiry, reported in February 2009.[15] Amongst its recommendations was a call for reform of support for those affected, in line with a scheme used in the Republic of Ireland. The Irish scheme offers substantially higher payments than the UK schemes, but successive UK governments rejected comparability on the basis that the Irish scheme was established to compensate victims for wrongdoing by a government agency but that no similar wrongdoing occurred in the UK. An April 2010 judicial review of the previous government's response to the Archer Inquiry questioned that assertion,[16] and the then government responded with a proposal to review aspects of support, while restating a rejection of the Irish model.[17]

On 10 January 2011 the Coalition government announced an increase in payments to some of those infected with hepatitis C, removed a provision preventing payments for hepatitis C patients who died before 29 August 2003, and offered additional medical and psychological support for those with hepatitis C and/or HIV.[18] These new measures apply to patients in England and are estimated to be worth an additional £100–130 million. While the measures were welcomed as a step in the right direction by some contaminated blood activists and their supporters, remaining concerns included the level of payments relative to the Republic of Ireland, and whether the provisions will extend to the rest of the UK.

The new measure outlined by the Coalition government for those with severe infection included:

— Introducing an annual payment of £12,800 for those with the most serious hepatitis C-related disease from NHS blood transfusions and blood products, in line with that received by those infected with HIV.
— Increasing the one-off payment already made to those with the most serious hepatitis C-related disease from £25,000 to £50,000.
— Making discretionary payments for those most in need.
— Increasing the annual payments for both hepatitis C and HIV patients in line with the Consumer Price Index.

[15] Lord Archer of Sandwell, *Independent Public Inquiry Report on NHS Supplied Contaminated Blood and Blood Products* (Archer Inquiry, 2009), available at: www.archercbbp.com.

[16] *The Queen on the application of March v Secretary of State for Health* (2010) 116 BMLR 57, [2010] EWHC 765 (Admin), [2010] Med LR 271.

[17] G Colthard, *HIV and Hepatitis C Infection from Contaminated Blood and Blood Products*. House of Commons SN/SC/5698, 13 January 2011, available at: www.parliament.uk/briefingpapers/commons/lib/research/briefings/snsc-05698.pdf.

[18] *Review of the Support Available to Individuals Infected with Hep C and/or HIV by NHS-supplied Blood Transfusions or Products and their Dependants* (Department of Health, 10 January 2010), available at: www.gov.uk/government/uploads/system/uploads/attachment_data/file/215828/dh_125977.pdf.

- Providing £100,000 per annum to selected third-sector organisations over the following 3 years to provide additional access to counselling for individuals infected with HIV and/or hepatitis C by NHS-supplied blood transfusions and blood products.
- Permitting a posthumous claim on behalf of those who died before 29 August 2003, and payments to the individual's estate.

The government recognised that some infected individuals might be uninsurable for some risks, but decided that a state-run insurance scheme was not considered to represent value for money. Individuals for whom insurance is available have freedom to use the ex gratia payments that they receive to help pay for the premiums.

The annual payments by the Skipton Fund are shown in Table 28.4.[19]

Table 28.4: Skipton Fund annual payments 2005–15

Year ending 31 March	Stage 1 living or died after 29 August 2003	Died before 29 August 2003	Stage 2 living or died after 29 August 2003	Died before 29 August 2003	'Top-up' payments	Receiving annual payments at 31 March	PPC England only	Cumulative proportion of stage 2 to stage 1
2005	3,034	NA	294	NA	NA	NA	NA	10
2006	433	NA	188	NA	NA	NA	NA	14
2007	245	NA	101	NA	NA	NA	NA	16
2008	204	NA	101	NA	NA	NA	NA	17
2009	130	NA	76	NA	NA	NA	NA	19
2010	162	NA	78	NA	NA	NA	NA	20
2011	116	0	82	0	0	0	0	21
2012	148	436	148	169	697	548	104	26
2013	111	19	116	24	14	615	101	27
2014	86	8	79	40	152	709	110	29
2015	93	9	70	11	2	730	240	30

The Caxton Foundation

Becoming eligible to obtain funding from the Caxton Foundation is a two-stage process, involving registration followed by checking of eligibility with the Skipton Fund Ltd. Applicants must first register with the Caxton Foundation, which can be done online. To be eligible for support from the Caxton Foundation the infected individual must be, or have been if claiming in relation to a deceased individual, registered with the Skipton Fund Ltd; this information is checked by the Foundation after first registration.

[19] *Skipton Fund Ltd Directors Report and Financial Statement for the Year ended March 2013*, available at www.skiptonfund.org/scripts/editor/uploaded/SFL%20Annual%20Financial%20Accounts%202013.pdf.

Application is by application form, which is available to download from the charity's website. The form requires identification details, a description of the grant sought with an outline of the reasons it is required, and details of income, expenditure, savings, capital and investments. As a charity the Caxton Foundation will only support individuals where there is a demonstrable 'charitable need'; this means that if an individual is able to afford support from his own income the charity will not provide any benefits to him. This is the reason that details of income, expenditure, savings, etcm are requested when grant applications are made.

The following items should be sent:

- application form;
- proof of income;
- if the application relates to financial assistance while undergoing treatment or for respite breaks or home adaptations, then a supporting letter from the clinician should be provided;
- two quotes for the item requested.

There has been considerable dissatisfaction expressed at the way in which support from the Caxton Foundation is accessed, with recipients describing the process as having to 'beg' for support.

D. HIV

The access to support for individuals infected by HIV from contaminated NHS blood products is more complex. Whether a claimant was treated for a bleeding disorder or not will determine where they can apply for funding. Since 2010 support for HIV-infected individuals has been available from the MacFarlane/Eileen Trust Ltd.

Discretionary support is also available from the two charities: the MacFarlane Trust and the Eileen Trust. The MacFarlane Trust, set up in 1998, provides discretionary support for haemophiliacs and other bleeding disorder sufferers, who were infected with HIV from NHS blood products. Claimants who are co-infected with both HIV and HVC can claim under the MacFarlane Trust. Non-haemophiliacs can access additional discretionary support from the Eileen Trust, set up in 1993, if they were infected with HIV (including co-infection with HVC) by contaminated NHS blood products.

It is interesting to note that the discretionary support from the two charities appears to pre-date the fixed sum support available from the MacFarlane/Eileen Trust Ltd. This is because the MacFarlane/Eileen Trust replaced two other schemes, which historically had made one-off lump sum payments to individuals infected with HIV. The Macfarlane Special Payments Trust (MSPT) provided a one-off payment of £20,000 to each individual with a bleeding disorder who had been infected with HIV. The Macfarlane Special Payments Trust No 2 (MSPT No 2) was an out-of-court settlement to the same group of people, and their families, and payments ranged from £2,000 to £60,500. Both of these schemes are now closed.

MacFarlane/Eileen Trust Ltd

This is a source of non-discretionary support in the form of lump sum payments, an ongoing annual payment (£14,749 in 2015/16) and reimbursement for a prescription pre-payment certificate.

To receive the payments, an individual has to prove:

— that have HIV; and
— that they underwent treatment with NHS-supplied blood products before 1985.

In order to obtain support from the MacFarlane/Eileen Trust, the individual must provide proof directly to the Department of Health. This has the same problems detailed above, namely that an individual may struggle to obtain medical records from the 1970s and 1980s. If an individual is successful with registering for non-discretionary support, he will then be paid lump sum and some ongoing payments; the precise level of these payments differs according to the recipient's individual circumstances.

Lump Sums

All MacFarlane Trust beneficiaries received £20,000 in 1991.[20] In 1992 a further £21,500–60,500 was paid, varying dependent on individual family circumstances.[21] Initially these payments were only available to those with bleeding disorders. Since 1993, however, those without bleeding disorders have been eligible to claim the same lump sum payments upon registration. Lump sum payments are not index linked and do not increase annually.

Ongoing Payments

These are index-linked and increase annually. For 2015/16 the payment value was set at £14,749.

Payments are discounted for tax purposes and are not counted as income or savings when deciding eligibility for government welfare assistance. If an individual transfers the savings accrued from these payments over to a partner, they are not considered savings for his/her partner's tax and benefits purposes, and nor is any capital, such as interest, gained from these payments; but money transferred over to a dependant is not disregarded in these ways.

The MacFarlane/Eileen Trust Ltd will make payments to families of an infected individual who would have been eligible under for payments, but who died before making a claim. The families of deceased HIV infectees who died before claiming under the HIV

[20] Written Parliamentary Question 211081, Diana Johnson MP, answered 27 October 2014. These payments were made by predecessor organisations, not by MacFarlane/Eileen Trust Ltd, which was not formed until much later.

[21] The intention behind the variance at the time was that those who faced higher day-to-day living costs to support dependants should receive greater payments to make up for this. Because of the short life expectancy assumed at the time, no appreciation was given to the fact that those who subsequently married and/or had children could not claim further lump sums and so lost out on payments. The categories for the 1991 lump sum are: under-18, £21,500; single adult, £23,500; infected by partner, £23,500; with a partner, £32,000; with a partner and dependent children, £60,500.

compensation scheme are entitled to claim £43,500 from the Trust at any point, although no such payments have been made since the Trust's inception in 2010.[22] These lump sums are the same value as the infectee would have obtained had they applied during their lifetimes.

Prescription Pre-payment Certificates

Two forms are required for this:[23] the MacFarlane/Eileen Trust Ltd prescription pre-payment certificate application form along with an FP95 form. These are completed and returned to the joint office of the MacFarlane/Eileen Trust Ltd and the MacFarlane Trust. On receipt of the completed forms, applications will be sent for processing and the prescription pre-payment certificate is sent directly to the claimant by the NHS Business Services Authority.

The MacFarlane Trust

The MacFarlane Trust offers several types of discretionary financial assistance (see Table 28.5).[24] Primary beneficiaries can receive regular payments, grants, winter fuel payments and prescription pre-payment certificates. Non-financial assistance including benefits advice, advice on debt and money management, and counselling services are also available. Secondary beneficiaries, such as children and spouses/partners and widowers, may also be eligible for some awards.

Regular Payments

Each year trust beneficiaries are asked to complete a 'census' form for the Trust, detailing their income, and to return it with proof of income. Primary beneficiaries whose household income is below £37,900 qualify for regular top-up payments. These payments range from as little as £756 per annum for households earning £30,001–37,900 per year up to £5,676 for those earning below £7600 per annum, with three additional thresholds in between.

Discretionary Payments

These are also made to bereaved spouses/partners who have an annual household income below £19,000.

Should circumstances change, then beneficiaries can contact the Trust to have their situation re-evaluated.

[22] Written Parliamentary Question 214996, Diana Johnson MP, answered 24 November 2014. It should be noted that before 2010, two predecessor companies—MSPT 1 and MSPT 2—existed before MacFarlane/Eileen Trust Ltd. The APPG did not enquire as to the arrangements in place for the families of deceased infectees to obtain lump sum payments from MSPT 1 and MSPT 2, and it is not known what arrangements, if any, were in place for the families of deceased HIV infectees to make MSPT 1 and MSPT 2 claims, and how many such registrants made claims.

[23] The forms are available from the MacFarlane Trust website: www.macfarlane.org.uk/support-for-beneficiaries.php/.

[24] www.macfarlane.org.uk/home.php.

Grants

Grants are made by application form available from the Trust website. Grants will only be made where a 'charitable need' can be demonstrated, so details of income and expenditure are included in the application. Some grant applications require that the applicant includes any supporting evidence, such as medical letters; this is specified on the application form. If the application involves the purchase of equipment or the completion of work, then two quotes for the cost will be needed.

Since May 2014 some grants can be processed and approved by Trust staff without the need for consideration by the Grants Committee. The claimant will be notified if his/her application is being sent to the Grant Committee for consideration. The Grant Committee meets every 6 weeks. A complete application and any supporting documentation needs to be in the office at least 10 working days before a Grant Committee meeting to be considered at that meeting. The dates of the Grant Committee meetings are on the Trust website. The applicant will be notified of the Grants Committee decision in writing within 5 working days of the meeting. Table 28.5 lists the types of funding available from the Trust.

Table 28.5: Grant funding available from the MacFarlane Trust

Grant area	Examples
Accommodation	House repairs and adaptation related to health and mobility issues Costs associated with setting up home for the first time
Health	Mobility equipment Counselling Respite breaks following treatment for hepatitis C
Education	The cost of a vocational training course Further education expenses
Bereavement	Funeral costs, or funeral plans Costs of hospital travel for parents/dependants/cares/spouse to visit ill beneficiaries
Empowerment (bereaved spouse/partners only)	Grants from the Honeycombe Legacy Fund for bereaved spouses or partners for personal development and to develop economic independence, eg academic development, skills training, driving lessons

Because of the requirement to demonstrate 'charitable need' the Trust is unlikely to consider grant applications for retrospective items, even if the item would normally fall into an area that would fund. Retrospective requests will be considered by the Grants Committee on a case-by-case basis, and applications will need to demonstrate why an application could not be made in advance.

Appeals

All grant application decisions can be appealed. Appeals must be made in writing within 1 month of the decision date and must include new evidence in support of the appeal. The appeal will be considered at the next Grants Committee meeting. If the Grants Committee upholds its original decision, but the applicant is still not satisfied, a second appeal may be made, again in writing and within 1 month of the decision date. The second appeal will be heard by the full board at its next scheduled meeting.

Winter Fuel Payments

In recent years MFT has made a winter fuel payment to primary beneficiary households who are in receipt of a discretionary top-up payment. A total of £100,250 was awarded for this purpose in 2015/16. This is means tested, with those earning up to £15,200 receiving £500, and those earning £15,201–37,900 receiving £250 per annum.

Prescription Pre-payment Certificates

The application process is exactly the same as for MacFarlane/Eileen Trust Ltd, detailed above.

Awards for Children of Infected Parents

The Trust makes monthly payments to the children of infected parents when the parent has died; £3,000 per annum is awarded to the first child, with £1,200 per annum to subsequent children. The Trust also makes monthly payments to children of living infected parents of £1,200 per child. An annual payment of £250 is made to children of infected parents (regardless of whether the parent is alive or dead) every summer. A one-off lump sum payment of £1,200 is made to children of infected parents (regardless of whether the parent is alive or dead).

Awards for Partners of Infected Beneficiaries

When a beneficiary dies the Trust provides financial support to widowed partners whose household income falls below £19,000 per annum.

Eileen Trust

The Eileen Trust offers several types of discretionary financial assistance. Primary beneficiaries can receive regular payments, grants, winter fuel payments and prescription prepayment certificates. Non-financial assistance including benefits advice, advice on debt and money management, and counselling services are also available. Secondary beneficiaries, such as children and spouses/partners and widowers may also be eligible for some awards.

Essentially the benefits available from the Eileen Fund are the same as those to those available from the MacFarlane Trust, with very limited differences.

Regular Payments

The Eileen Trust does not have a fixed threshold for making these; payments are on the basis that beneficiary households are judged to have insufficient incomes. Only a small number of such payments are made per year.

E. Proposed Reforms

One of the major criticisms of the current arrangements is that the discretionary nature of the charities' funding forces claimants to 'beg' and means that there is little long-term certainty for these claimants as the benefits can be changed at any time—for example, the Caxton Foundation reduced its winter fuel payments from £500 per annum to £350 with very little warning to counter a funding short-fall. The discretionary nature of funding has led to claimants being uncertain as to the funding criteria, and there have been calls for a clearer, more transparent awards policy.

In addition, the 5 different organisations have different criteria for claimants, leading to discrepancies between the available funding for claimants. For example, the HIV charities provide payment to the children of infected beneficiaries, but no such provision exists for children of HVC-infected parents.

Another factor is the inefficiency of having 5 overlapping organisations, rather than a single unified organisation, which would be easier for potential claimants to navigate and, presumably, less expensive for government to run.

The devolved governments of the UK act independently in providing compensation for infected blood cases. Scotland has taken the lead, and undertook a consultation on financial support available to those people in Scotland who were infected with hepatitis C (including those co-infected with HIV).[25] This followed the Penrose Inquiry in 2015, a statutory investigation into the manner in which contaminated blood was used by the NHS.

The key elements of the Scottish reforms are:

— Annual payments for those with HIV and advanced hepatitis C increased from £15,000 a year to £27,000 a year, to reflect average earnings.
— Those with both HIV and hepatitis C had annual payments increased from £30,000 to £37,000 to reflect additional health needs.
— When a recipient dies, his/her spouse or civil partner will continue to receive 75% of his/her annual payment.
— Those infected with chronic hepatitis C will receive a £50,000 lump sum payment (previously £20,000), meaning an additional £30,000 for those who have already received the lower payment.
— A new Support and Assistance Grants scheme will be established in Scotland, to administer and provide more flexible grants to cover additional needs. Scottish government funding for this scheme will be increased from £300,000 to £1 million per year.

[25] http://news.scotland.gov.uk/News/Extra-20-million-for-infected-blood-support-2418.aspx and https://haemophilia.scot/advocacy/infected-blood-scottish-financial-support-review-group/.

England has followed, by issuing a consultation. It is expected that Wales and Northern Ireland will mirror any changes made in England. The main elements of the proposals for reform in England are:[26]

— to replace the current five schemes with one operated by a single body;
— to keep eligibility for the reformed scheme broadly the same as it is for the current schemes;
— to offer some access to new hepatitis C treatments for those considered clinically appropriate on the basis of a treatment assessment;
— to introduce individual assessments for those with hepatitis C stage 1 and, for all new entrants to the scheme, to determine the amount of a new annual payment, the highest level of which will be the same as payments that will be received by those with hepatitis C stage 2;
— to retain annual payments (for HIV and/or hepatitis C stage 2) for those who currently receive them;
— to seek views on whether to provide newly bereaved partners/spouses with a final payment equivalent to one further annual payment at the level his/her partner was receiving at the time of his/her death or to provide access to a discretionary fund or a choice of either;
— to seek views on the future arrangements for those already bereaved, and whether that should be through a one-off lump sum or through continuation of a means-tested discretionary fund, or a choice of either.

It will be interesting to see how these proposals are maintained in the transition to the new arrangements.[27]

F. Conclusions

The HIV and hepatitis C compensation arrangements form a considerable ex gratia settlement. The historic developments of the 5 individual funds have led to a piecemeal, and not particularly satisfactory compensation landscape. The UK government's responses to the consultation on funding suggest that long-overdue improvements may be on the horizon. It is interesting to note that the dissatisfaction expressed by claimants is largely directed at the charitable funds, with the non-discretionary corporate funds provoking far less criticism. It will be interesting to observe the form of the promised new entity, and to see whether lessons are learnt from the consultation.

[26] Infectious Diseases and Blood Policy, *Infected Blood* (n 7) ch 2.

[27] In February 2017 an HCV infectee (Alex Smith) was granted permission to bring a judicial review of a decision to make a lower award to him than if he had been infected with HIV, despite the government's own medical advice that contracting HVC was at least as serious as contracting HIV.

29

Creutzfeldt–Jakob Disease

SONIA MACLEOD

A. Background

Creutzfeldt–Jakob disease (CJD) is one of a group of prion diseases and is progressive and fatal. There is currently one treatment option, pentosan polysulphate, which prolongs life, but is not curative. CJD was first observed in 1920 and occurs in several different forms. In 1996 a new form of CJD was observed,[1] variant CJD (vCJD), which was linked to bovine spongiform encephalopathy (BSE), a prion disease seen in cattle. Between 1996 and 2012 vCJD accounted for an average of 9.7% of the total UK CJD cases, reaching a peak in 2000 when 34% of UK CJD cases were vCJD. Of non-variant CJD cases the vast majority (approximately 86%) of CJD cases are sporadic CJD, which causes between 1 and 2 deaths per million people per year. The remainder of non-variant CJD cases are be due to a genetic mutation (approximately 9% of CJD cases) or caused by infection from medical or surgical treatment, so-called iatrogenic CJD (4.5% of cases). (See Figure 29.1.)[2]

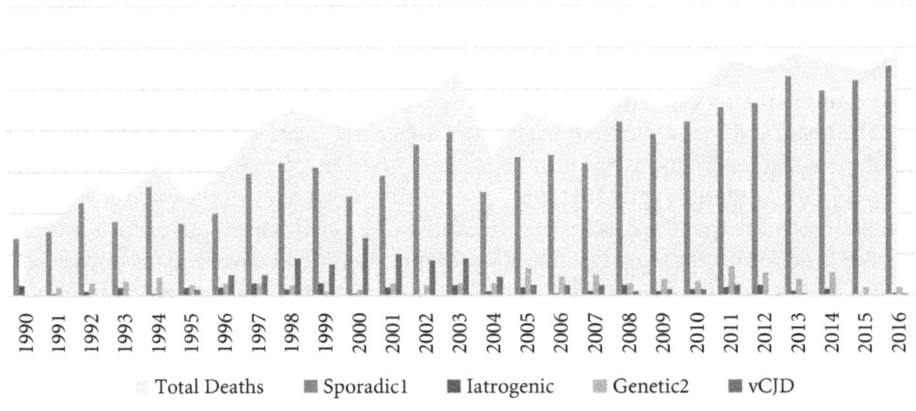

Figure 29.1: Breakdown of UK CJD referrals and deaths 1990–2016. Data taken from the National CJD Research and Surveillance Unit Monthly Statistics

[1] (1996) 347 *Lancet* 921–25, www.thelancet.com/journals/lancet/article/PIIS0140-6736(96)91412-9/abstract.
[2] See www.cjd.ed.ac.uk/data-and-reports/latest-cjd-figures.

The 2001 Report of the BSE Inquiry chaired by Lord Phillips[3] concluded that BSE probably arose from a single source, such as a genetic mutation in a single animal. The practice of rendering cattle to produce meat and bone meal (MBM) and using MBM in cattle feed resulted in widespread distribution of the BSE prion and the BSE epidemic. The Spongiform Encephalopathy Advisory Committee considered that exposure to the BSE agent was the most likely explanation for vCJD.

Human Growth Hormone Programme and Related Litigation

Between 1959 and 1985, almost 2,000 children in the UK were given human growth hormone derived from donated human pituitary glands prepared by the Wilhelmi method. The Medical Research Council (MRC) ran the programme as a clinical trial until 1977, when the Department of Health assumed full control. The programme was suspended in 1985 when the link to vCJD was established and from that point onwards synthetic growth hormone was used.

Between 1996 and 1998 a tranche of litigation known as the Creutzfeldt–Jakob disease litigation followed, largely handled by Mr Justice Morland. Victims, those affected by vCJD (group A) and those claiming psychiatric injury as a result of knowing there was a chance they had been infected with the prion and were at risk of developing vCJD (group B), claimed against the MRC.[4] The eventual outcome of these cases was that victims who had received at least half of their treatment before 1 July 1977 (when the MRC were warned of the danger that the growth hormone could be contaminated with a prion infective agent) could succeed in negligence for both physical and psychological injury. At this stage there was no intention of creating a no-fault compensation scheme, on 25 November 1998 Ms Tessa Jowell, then the Minister for Public Health stated:

> The Government's policy—like that of the previous Government—is that compensation will be paid only when negligence has been proved. We believe that paying compensation on a no-fault basis would be enormously complicated, would not necessarily solve the problems that are being encountered, would be expensive and would have ramifications beyond the health service.[5]

[3] The BSE Inquiry: The Report, at http://webarchive.nationalarchives.gov.uk/20060715141954/http://bseinquiry.gov.uk/report/index.htm.

[4] *Creutzfeldt–Jacob Disease Litigation*, unreported, 24 May 1996; QBD: *Creutzfeldt–Jacob Disease Litigation (No 1), Re* (2000) 54 MLR 1; QBD: *Creutzfeldt–Jacob Disease Litigation Claimants v United Kingdom Research Council*, unreported, 19 July 1996; QBD: *Newman v Secretary of State for Health (No 1)* [1996] 7 Med LR 309, (2000) 54 BMLR 8 (overturned), *Re Creutzfeldt–Jacob Disease Litigation (No 3)* (2000) 54 BMLR 79; QBD: *Newman v Secretary of State for Health (No 1)* (2000) 54 BMLR 85; Times 20 December 1997 CA (Civ Div); *Newman v Secretary of State for Health (No 2)* (2000) 54 BMLR 95; QBD.

[5] Hansard, November 1998, vol 321, para 187, http://hansard.millbanksystems.com/commons/1998/nov/25/cjd-and-human-growth-hormone.

B. The vCJD Trust: A No-Fault Compensation Scheme[6]

Over the next 3 years there was a change in policy regarding a no-fault compensation scheme for vCJD. In 2001 the UK government, without an admission of liability, agreed a compensation scheme for vCJD sufferers and carers following the Report of the BSE Inquiry chaired by Lord Phillips.[7] Initially £67.5 million was allocated for the first 250 cases, with the intention to review if the number of cases rose above 250.

Eligibility for Compensation

The two eligibility requirements as set out in the Trust Deed are:

— the victim suffered vCJD on the legal test of the balance of probabilities; and
— that he or she was resident in the UK for at least 5 years between 1982 and 1996.[8]

To establish that a victim probably suffered from vCJD a letter must be obtained confirming that this is the case from the National CJD Surveillance Unit, the University of Edinburgh. Confirmation of the residence requirement can either come from the National CJD Surveillance Unit (after taking a statement from the victim/qualifying person). Alternatively, an Authority can be completed and attached sent to Field Fisher Waterhouse Solicitors (the Trust solicitors) who will obtain the relevant information from the National CJD Surveillance Unit.

Who is Entitled to Compensation

Payments under the Scheme can be made to the victim during his or her lifetime, and to 'Qualifiers'. The definition of 'Qualifiers' in the Trust Deed is very wide, and includes all of the victim's immediate and more remote family. It also includes the family of any spouse or (in some circumstances) partner. However, more remote family members will only receive compensation if the Trustees think it appropriate, which will usually require there to have been contact with the victim. The following may be considered as 'Qualifiers':

— Spouse, or a partner of the victim who was living in the same household for at least 2 years as husband or wife or as a partner in a hetero- or homosexual relationship.
— Ancestor or descendant of the victim.
— Children of the victim.[9]

[6] www.vcjdtrust.co.uk/.
[7] BSE Inquiry (n 3).
[8] This can be waived at the Trustee's discretion. This requires a directive from the Secretary of State that that person can be included in the scheme.
[9] Including any child treated by the victim as their own child, eg stepchildren.

— Brothers, sisters, aunts or uncles, half-brothers, half-sisters, brothers-in-law, sisters-in-law, or their children.
— Dependants of the victim, being those whom the victim financially maintained.
— Carers who do not fall within any of the above, but who were significantly involved in caring for the victim.

The Trust Funds

To facilitate quick payments to victims two interim trusts, the First and Second Interim Trusts, were set up while the details for the Main Trust were being agreed. The majority of families received interim payments of £25,000 from the Interim Trusts. The first interim payments to claimants were paid in 2001from First Interim Trust (dated 10 April 2001).[10] The First Interim Trust continued until 11 December 2001, at which point the Second Interim Trust replaced it. The Second Interim Trust continued until the Main Trust deed was signed on 15 March 2002. The two interim trusts were closed, but money paid out under them is taken into account when awarding payments from the Main Trust. In April 2001 legislation was amended so that payments from the vCJD trusts were disregarded when calculating income-related and other benefits.[11]

The Main Trust consists of two funds, the Main Fund and the Discretionary Fund. In 2010 an agreement was reached to simplify the procedure for claiming compensation, and an amended Trust deed was signed on 12 February 2010.[12] Under this, two versions of the scheme apply, one for those diagnosed on or before 30 March 2010 and one for those diagnosed on or after 31 March 2010. The benefits available are summarised in Table 29.1.

Table 29.1: Benefits

MAIN FUND		
	Diagnosis on or prior to 30 March 2010	**Diagnosis on or after 31 March 2010**
Basic sum (clause 3.1.1)	£125,000 if diagnosis was suspected before 26 October 2000 £120,000 if diagnosis was suspected after 26 October 2000	£120,000, rising to £122,400 for diagnosis on or after 31 March 2012 and increasing by 2% or whatever sum the Trustees see fit annually thereafter
Experience of family	£10,000 to be shared by victim's family if diagnosis was suspected before 26 October 2000	£40,000 to be shared by victim's family

(continued)

[10] JM Williams QC 'Setting up the vCJD Trust—Operations and Lessons for the Future', www.vcjdtrust.co.uk/articles/setting-up-the-vcjd-trust-operations-and-lessons-for-the-future/.
[11] Social Security Amendment (Capital Disregards and Recovery of Benefits) Regulations 2001 SI 1118 www.legislation.gov.uk/uksi/2001/1118/contents/made.
[12] www.vcjdtrust.co.uk/the-trust-deed/.

Table 29.1: ***(Continued)***

MAIN FUND		
	Diagnosis on or prior to 30 March 2010	**Diagnosis on or after 31 March 2010**
(clauses 4.1 and 4.2. Clause 4.3 in cases of psychiatric injury to relatives with diagnosis before 30 March 2010)	£5,000 to be shared by victim's family if diagnosis was suspected after 26 October 2000	
	£5,000 to be shared between carers who have been significantly involved in care	£5,000 to be shared between carers who have been significantly involved in care
	£5,000 for psychiatric injury to qualifying carer/family member (one who has suffered an identifiable psychiatric injury lasting more than a month caused or aggravated by the victim's contracting vCJD)	
Costs incurred as a direct result of a victim suffering vCJD (clauses 5.1 and 5.2)	Up to £10,000 (more in exceptional circumstances) in respect of expenditure by the victim and/or his qualifiers on real or personal property or on its repair, alteration or improvement	Up to £10,000 (more in exceptional circumstances) in respect of expenditure by the victim and/or his qualifiers on real or personal property or on its repair, alteration or improvement
Dependency claims (clause 6)	Dependency sum as defined and specified in Schedule 2 of the Trust Deed	Dependency sum as defined and specified in Schedule 2 of the Trust Deed

The Discretionary Fund has more limited funds than the Main Fund, and is able to meet costs under various heads at the discretion of the trustees. All hardship claims can only be made in cases where diagnosis was made on or before 30 March 2010. These include the cost of care provided before 31 March 2001 or the implementation of a care package for that victim, whichever is the sooner (clause 5.3), carer's loss of earnings where a qualifying carer has suffered 'such loss of earnings as to cause them particular hardship' (clause 5.4), victim's loss of earnings, where a victim has suffered 'such loss of earnings as to cause them particular hardship' (clause 5.5), additional payments to dependants (clauses 6.2 and 6.3), and payments where the victim has suffered particular hardship obtaining life insurance or mortgage protection. In the original Trust Deeds there was also provision for claims arising from emotional or financial hardship due to psychiatric injury (clause 4.3.2), but this has been removed and replaced with an increased award for families.

Making a Claim

Applicants must fill in an application form, available from www.vcjdtrust.co.uk, which is processed by solicitors Field Fisher Waterhouse. All claims are sent to Field Fisher

Waterhouse, and usually take 4–9 months before they are ready to be considered by the Trustees. It usually takes 4–5 weeks after a Trustees' Meeting to release compensation funds. Claims can be either for interim payments or for full compensation; urgent interim claims, usually where a victim is still alive, can be dealt with expeditiously (payments can be as fast as 48 hours after receiving a completed application form).

There are two different procedures for making claims depending upon the date of diagnosis. Claims on or before 30 March 2010 fall under Scheme 1,[13] whereas claims for diagnosis after this date fall under Scheme 2.[14] The application forms vary slightly between the 2 schemes.

If a claim is made by an individual for their own loss it is allocated to them. If a claim is for a sum where there are a number of Qualifiers the Trustees are able to apportion funds at their discretion. Usually the family will agree an apportionment that they collectively feel is correct and the Trustees will approve this. The Trustees need to be sure, as far as possible, that all Qualifiers who are entitled to claim have been informed of their right to claim. If individual Qualifiers are not listed in the apportionment, then a disclaimer will usually be obtained from them to say they do not wish to claim.

If the family are not in agreement as to apportionment, then Field Fisher Waterhouse will ask for witness statements as to views on apportionment and the Trustees will consider these and apportion as they see fit.

Claim Data

As at the 58th meeting on 25 October 2013[15] there were 177 victims on the National CJD Surveillance Unit website, but Field Fisher Waterhouse were aware of 183 victims. As at 25 October 2013 the Trustees had considered 181 of the claims submitted to date. These included:

— 406 claims for £5,000 for psychiatric injury;
— 216 claims for particular emotional hardship; and
— 45 claims for particular financial hardship.

By this date the following sums had been paid out:

— £41,078,281.89 in compensation
— £14,494,181.98 in costs, expenses and taxes.

This provides an administrative-costs-to-compensation ratio of 35%, which is lower than would be expected in litigation, but considerably higher than some other schemes.

CJD Care Package

The Department of Health provides additional funding to assist local authorities with the care of patients with all forms of CJD in the form of the National Care Package (NCP).

[13] www.vcjdtrust.co.uk/scheme1/.
[14] www.vcjdtrust.co.uk/scheme2/.
[15] See meeting minutes, available at www.fieldfisher.com/pdf/vcjd-abbreviated-minutes-of-58th-meeting-25-october-2013.pdf.

The NCP was established in October 2000 to ensure that the health and social care needs of patients with CJD are met quickly and effectively. The NCP is accessed via the National CJD Care Team, based within the National CJD Research and Surveillance Unit. It provides funding for local care and equipment, but local organisations are responsible for the actual provision, for example for specific equipment or specialist nursing staff. The care package is individually assessed and is intended to be flexible and adaptable should the patient's needs change. In 2011 NCP payments amounted to £199,622.58.[16]

Court Cases and Settlement

The Trust specifically does not limit an individual's right to seek redress through the courts, but if he/she choses to do so, then any funds paid by the Trust will be taken into account in any damages award The compensation levels paid by the Trust are equivalent to what would be awarded at common law. Interestingly some of the claims considered by the Trust in the past have been more generous than those awarded at common law, for example claims for psychiatric injury of qualifying carers would normally have failed at common law as the sufferers would have been considered secondary victims.

[16] S 6 of the Twelfth Annual Report 2011, 'Creutzfeldt–Jakob Disease Surveillance in the UK', www.cjd.ed.ac.uk/documents/report20_000.pdf.

Part VIII

Corporate Schemes for Medical Devices

SONIA MACLEOD

A number of compensation schemes have been created by manufacturers for people claiming for injury caused by their products. Some schemes have been created by pharmaceutical manufacturers, such as Imperial Chemical Industries in the UK in relation to Eraldin, but it appears that more recent schemes have mainly been created by manufacturers of medical device implants, such as Dow Corning in relation to breast implants and Johnson & Johnson for the ASR hip implant.

Conclusions on Private Schemes

Corporate schemes arise in similar situations to governmental and court settlement schemes; the factors that determine whether compensation is provided privately or by the state are a mix of the economic, legal and cultural. For example, an unexpected adverse drug reaction could prompt the setting up of a corporate compensation scheme or one jointly funded from court settlements and governmental funding, for example the UK Thalidomide Trust or the PMDA stream for SMON patients in Japan. Adversarial legal systems disincentivise the admission of errors, which may explain the relatively scarcity of corporate compensation schemes. Were a more open no-fault culture combined with high expectations of corporate responsibility to exist, then this might lead to an increase in privately funded compensation. The ASR reimbursement scheme is an interesting example of an equitable corporate culture; it is not a compensation programme, and does not limit claimants' right to redress through the courts—the objective is to ensure ASR recipients' financial resources do not have any ASR-related health impacts.

The Trilucent example highlights that only if the compensation scheme is an attractive option for the claimant can litigation be all but avoided. Obviously financial elements of the compensation package are essential in making a scheme attractive to potential claimants, but other important elements, such as ease of use, clear scheme parameters, admissions (if appropriate) of liability and apologies, and the knowledge that data garnered by the complaint will be used to improve the service being complained about, should not be overlooked.

30

Dow Corning Breast Implant Compensation

A. History

Dow Corning was the first US manufacturer of silicone breast implants, which were released to market in 1964.[1] They were then sold both within the USA and abroad, with approximately half of Dow Corning's total breast implant sales occurring outside the USA.[2] The implants were marketed both for reconstruction and cosmetic augmentation. Although injury claims were an international issue, this chapter will primarily focus on the US situation as this is where the settlement fund was established.

In 1976 changes were made to the regulation of medical devices in the USA. The Federal Food, Drug and Cosmetic Act was amended by Congress, extending the powers of the Food and Drug Administration (FDA) to review and approve the safety and effectiveness of medical devices. At this point, silicone breast implants had been available for over a decade and were considered to be moderate risk, so were classified as class II devices. Those implants already on the market (including Dow Corning's silicone implants) were 'grandfathered in' and subsequently all new-to-the-market implants went through the pre-market notification or 501(k) process.[3]

In 1988 the FDA reclassified silicone breast implants as class III devices. This required that all new-to-market implants undergo pre-market approval (PMA). Implants that were already on the market continued to be reviewed under the 501(k) process until the FDA made a final call for PMA applications in April 1991, which required manufacturers to demonstrate that their implants were safe and effective.

In November 1991 the FDA's General and Plastic Surgery Devices Panel found that the Dow Corning had failed to supply sufficient information in its PMAs to demonstrate that its implants were safe. However, the Panel returned a unanimous vote advising the FDA that silicone implants (including those made by Dow Corning) should be available to the public, as implants fulfilled a public health need, while the manufacturers collected further data. On 6 January 1992 a voluntary 45-day moratorium on the use of silicone implants was

[1] *In Re Dow Corning Corp*, 211 BR, 550.

[2] *In Re Silicone Gel Breast Implant Product Liability Litigation Multi-District Litigation*, No 926 Case No CV 92-P-10000-S.

[3] For a more detailed timeline, see 'Regulatory History of Breast Implants in the US', US Food and Drug Administration, at www.fda.gov/MedicalDevices/ProductsandMedicalProcedures/ImplantsandProsthetics/BreastImplants/ucm064461.htm.

implemented until the Panel received further information on their safety. On 20 February 1992 a second Panel meeting recommended restrictions on the marketing of silicone implants until safety and efficacy data had been supplied, but again did not recommend total removal from the market as silicone breast implants fulfilled a public health need.

On 16 April 1992, the FDA retained the moratorium on augmentation, restricting the use of silicone implants, which would only be available '[u]nder controlled clinical studies for reconstruction after mastectomy, correction of congenital deformities, or replacement for ruptured silicone-gel filled implants for augmentation; silicone breast implants would no longer be available for augmentation patients'. The FDA also concluded that none of the PMAs submitted for silicone implants that had been on the market prior to the moratorium contained sufficient information for them to be approved. Since then only a handful of implants have been deemed fit to be placed on the market, none of which were manufactured by Dow Corning, which left the breast implant market early in January 1992 when the moratorium commenced.[4] The moratorium on silicone breast implants for augmentation was reversed in 2006.

Litigation

Concerns about silicone breast implant safety began to be raised in the later 1970s. The first known successful case was won by attorney Richard Mithoff in 1977 for a woman who had had one implant rupture. The defective product claim was against Dow Corning as the manufacturer and V Mueller & Company, Inc, the distributor, and the complaint was that the rupture necessitated further surgery, causing financial loss and additional pain and suffering. There was no claim for systemic or autoimmune symptoms. A jury awarded the woman $170,000, which was upheld on appeal.[5] Until the 1990s, there was a relative hiatus on litigation against Dow Corning, with only a handful of cases actually going to trial.[6] By the 1990s, questions regarding the safety of silicone breast implants were starting to be raised in the media, in particular by a 1990 TV programme *Face to Face with Connie Chung*,[7] leading to a rapid increase in the number of claims.

Under US tort law the burden of proving causation falls on the claimant. The initial claims were medically relatively simple and localised, involving rupture, removal and replacement, with economic loss and pain and suffering. Later cases were far more complex, with claims for systemic body diseases such as scleroderma, systemic lupus, erythymatosis, and other symptoms including chronic fatigue, joint pain and muscle pain. These claims led to conflicting and complex expert testimony, with issues of causation often being the keystone of a case. Claimants usually preferred trial by jury in these cases, with some states perceived as being more sympathetic. Some of the awards granted by juries were of substantial

[4] For a detailed timeline, see ch 11 'Dow Corning Corporation's Silicone Gel-Filled Breast Implants: An Ethical Analysis' in O Mascarenhas, *Responsible Marketing: Concepts, Theories, Models, Strategies and Cases.* (North Richland Hills, TX: Roval Publishing, 2007).

[5] *V Mueller & Co v Corley*, 570 SW2d 140 (1978) Court of Civil Appeals of Texas, Houston (1st Dist) 27 July 1978.

[6] The only successful claim against Dow Corning between 1977 and 1989 was that of Maria Stern in 1984 who was awarded $211,000 in compensatory damages and $1.5 million in punitive damages: *Stern v Dow-Corning Corp*, No C83-2348 MHP (MDCal. 15 November 1984).

[7] CBS, originally shown on 10 December 1990.

value: in 1991 an award was made of $7.34 million,[8] which was dwarfed by later awards such as a 1992 award of $25 million against manufacturer Bristol-Myers Squibb.[9]

Multiple Claims

In 1992 a consolidation of multiple claims was made via a multi-district litigation suit filed against Dow Corning in Cincinnati, Ohio.[10] This litigation included foreign as well as US claimants. The size of litigation was such that in 1994 Dow Corning claimed to have spent over $200 million on defence costs.[11]

Creation of the Dow Corning Settlement Trust

On 1 September 1994, the global action against the manufacturers of breast implants was settled[12] and the Dow Corning Settlement Trust (DCST) was created. Thought had been given to the ability of Dow Corning and others to meet the liabilities arising from the DCST, and this point was specifically referred to in Judge Pointer's Opinion approving the settlement.[13] At this time the liabilities due to the entire settlement were valued at $4,225,070,000.

Chapter 11 Bankruptcy and Reorganisation

However, on 15 May 1995, facing 20,000 lawsuits (some with multiple plaintiffs) and with over 400,000 claims already filed in the global settlement, Dow Corning voluntarily filed for Chapter 11 bankruptcy protection as it decided it could not fulfil its liabilities under the settlement.[14] The bankruptcy halted all litigation,[15] thereby protecting the company from further suits.

This was followed by a corporate reorganisation under the US Chapter 11 bankruptcy law. In 1998 Dow Corning filed for bankruptcy reorganisation, which included the previously agreed-to settlement funds. The Amended Joint Plan of Reorganisation[16] (the Plan) was submitted on 4 February 1999 and was confirmed by the Bankruptcy Court on

[8] *Hopkins v Dow-Corning Corp*, No 92-16132, 1992 WL 176560.

[9] *Johnson v Bristol-Myers Squibb Corp et al*, CN 91-21770 (TX Dist Ct, 125 Jud Dist 1992.

[10] *MDL 926 Breast Implant Litigation*, at www.fjc.gov/BREIMLIT/mdl926.htm.

[11] *In Re Dow Corning Corp*, 211 BR, 552.

[12] See the Settlement Notice and Orders 15-20 of *In re Silicone Gel Breast Implant Products Liability Litigation* (MDL 926) Civil Action No CV 94-P-11558-S, available at http://www.tortcomm.org/mdl_orders/1994-09-01%20OPINION%20GLOBAL%20SETTLEMENT%20CORRECTED.pdf.

[13] See fn 2 of Opinion (Corrected) (Approval of Settlement) from *In re Silicone Gel Breast Implant*, ibid.

[14] Filed at the Eastern District of Michigan, Case No 95-20512; further information available at www.mied.uscourts.gov/Information/Dow/.

[15] Litigation continues regarding the Settlement Trust, mainly over interest payable on the period between entering Chapter 11 bankruptcy and the reorganisation. For further details, see *In Re Settlement Facility Dow Corning Trust*, No 11-2632; *In re Settlement Facility Dow Corning Trust. Dow Corning Corporation*, Interested Party-Appellant v Claimants' Advisory Committee, Interested Party-Appellee, US Court of Appeals, Sixth Circuit. Filed 18 April 2013.

[16] Available at www.sfdct.com/_sfdct/resources/Final%20Plan1.pdf.

30 November 1999. A considerable number of appeals and other litigation followed, delaying the Plan taking of effect until 1 June 2004.[17]

Options for Potential Claimants

Under the Plan, claimants could either accept the settlement or opt out of it. Claimants who opted out to pursue litigation filed a lawsuit in the District Court in Michigan against DCC Litigation Facility Inc, which was created to defend lawsuits filed by opt-out claimants. Given the level of payments awarded in the early breast implant litigation, opting out might have seemed an attractive option. However, claimants who opted out were not permitted to recover punitive damages, reducing the potential value of litigation settlements. Other disincentives included the fact that if a claimant chose to opt-out she permanently lost all eligibility to claim from the DCST. Litigation was slower than the DCST and was public, whereas settlement under the Trust was confidential. In addition, a claimant had the burden of proving that her breast implant caused her disease or other problems, which DCC Litigation Facility, Inc would contest. Any lawsuit was required to follow the procedures and deadlines established in the Case Management Order and relevant MDL Orders.[18]

B. Claiming from the Dow Corning Settlement Trust

As part of the Joint Reorganisation Plan, a resource was established to provide advice for applicants,[19] known as the Claimants Advisory Committee (CAC). The CAC provided online and paper guidance, as well as having a phoneline for potential claimants. In order to be eligible for compensation, claimants were required to meet certain threshold criteria, including residence, proof of implant manufacturer, time limitations and proof of illness; additional requirements were that:[20]

1. They had not released their claim against Dow Corning or its shareholders (or had such a claim resolved by final judgment, dismissal or order); and
2. They did not opt out of the Settlement Facility; and
3. They filed a Proof of Claim form on or before 30 November 1999 or were a 'Notice of Intent' (NOI)[21] claimant; and
4. They had one of the acceptable types of proof that Dow Corning made their implants.

Once a claimant had submitted a Proof of Claim form they were allocated to 1 of the 5 categories of claimant. They were sent the appropriate claims form package for that claimant category, or could obtain them online.[22] The 5 categories of claimant were defined as shown in Table 30.1.

[17] Available at www.sfdct.com/_sfdct/resources/Final%20SFA1.pdf.
[18] The MDL orders are located at www.fjc.gov/BREIMLIT/mdl926.htm.
[19] www.tortcomm.org/.
[20] Detailed in ss 5.01–5.02 of Annex A to the Settlement Facility and Fund Distribution Agreement, available at www.sfdct.com/_sfdct/index.cfm/am-i-eligible-to-file-a-claim-for-benefits/.
[21] For further details, see http://www.sfdct.com/_sfdct/index.cfm/am-i-eligible-to-file-a-claim-for-benefits/information-on-notice-of-intent-claimants-noi/.
[22] www.sfdct.com/_sfdct/index.cfm/how-to-file-a-claim-for-benefits/claim-forms-by-class/.

Table 30.1: Categories of DCST claimants

Class	Description	Category characteristics	Subcategory
5	**Domestic Dow Corning breast implant claimants**	Implanted with a Dow Corning breast implant and either a US citizen or resident alien, or Had Dow Corning implant(s) implanted in the USA	NA
6	**Foreign Dow Corning breast implant claimants**	Implanted with a Dow Corning breast implant, which was implanted outside of the USA, and They are not a citizen of the USA or a resident alien within the USA, Puerto Rico, or the territories and possessions of the USA	Class 6.1—resident of 1 of the countries listed in Categories 1 or 2 on the chart located below Class 6.2—resident of 1 of the countries listed in Categories 3 or 4 on the chart located below Class 6A—a member of the class of plaintiffs in a class action filed in the province of Quebec Class 6B—a member of the Class of plaintiffs in a class action filed in the province of Ontario Class 6C—a member of the class of plaintiffs in a class action filed in the province of British Columbia. The class includes resident claimants in British Columbia who did not opt out of the class action as well as those claimants who are residents of any province of Canada other than British Columbia, Quebec and Ontario who timely elected to be bound by the British Columbia class action Class 6D—resident of Australia or received your implants in Australia, and elected to participate in the Australia Breast Implant Settlement Option on your ballot on the Amended Joint Plan in 1999 Classes 6A–D are governed by specific definitions contained in the class action settlements and judgments relating to these Classes. Membership in these Classes is based on residence at specific periods of time. Members of 1 of these classes should receive a separate notice

(continued)

Table 30.1: *(Continued)*

Class	Description	Category characteristics	Subcategory
7	**Silicone material claimants/ foreign gel claimants**	Implanted with a silicone gel breast implant after 1 January 1976 and before 1 January 1992 from either Baxter, Bioplasty, Bristol, Cox-Uphoff, Mentor, Koken, Silimed, Societe Prometel or Medasil Surgical, and have never had a Dow Corning implant They are members of Class 7 regardless of their country of residence or citizenship	
9	**Domestic Dow Corning other products claimants**	Implanted with an eligible Dow Corning implant (other than a breast implant) listed in Table 1, Part II,[a] and are a US citizen or resident alien, or if the eligible Dow Corning implant was implanted in the USA	
10	**Foreign Dow Corning other products claimants**	Were implanted with an eligible Dow Corning implant (other than a breast implant) listed in Table 1, Part II,[a] and it was implanted outside of the USA, and are not a citizen of the USA or a resident alien within the USA, Puerto Rico, or the territories and possessions of the USA	Class 10.1—resident of 1 of the countries listed in Categories 1 or 2 on the chart located below Class 10.2—resident of 1 of the countries listed in Categories 3 or 4 on the chart located below

Category 1 countries	Category 2 countries	Category 3 countries	Category 4 countries
Australia, Canada, New Zealand, United Kingdom	Austria, Bahamas, Belgium, Bermuda, Cayman Islands, Denmark, Finland, France including: French Polynesia & New Caledonia, Germany, Greece, Hong Kong, Iceland, Ireland, Italy, Japan, Kuwait, Liechtenstein, Luxembourg, Monaco, Netherlands, Norway, Portugal, Singapore, Spain, Sweden, Switzerland, United Arab Emirates	Argentina, Barbados, British Virgin Islands, Chile, Cyprus, Czech Republic, Israel including: Gaza Strip & West Bank, Korea, Malaysia, Malta, Mauritius, Qatar, Saudi Arabia, Taiwan	Algeria, Belize, Bolivia, Botswana, Brazil, Bulgaria, Cambodia, Central African Republic, China, Colombia, Cook Islands, Costa Rica, Cote d'Ivoire (Ivory Coast), Croatia, Cuba, Dominican Republic, Ecuador, Egypt, Estonia, Fiji, Ghana, Grenada, Guatemala, Guyana, Haiti, Honduras, Hungary, India, Indonesia, Jamaica, Jordan, Kenya, Lebanon, Lithuania, Mali, Mexico, Morocco, Namibia, New Guinea, Nicaragua, Nigeria, Oman, Pakistan, Panama, Paraguay, Peru, Philippines, Poland, Saint Kitts and Nevis, Senegal, South Africa, Thailand, Tonga, Turkey, Uruguay, Venezuela, Vietnam, Zambia, Zimbabwe

[a] See Table I, Part II of www.tortcomm.org/downloads/CIG_ENG_9.pdf for a full listing.

Certain claimants could be members of more than one category, though some categories were mutually exclusive.[23] Categories 9 and 10 cover implants other than breast implant and will not be discussed further in this chapter.

Settlement Options for Breast Implant Recipients

The settlement options varied by category of claimant. (All awards are in US dollars.)

Class 5 and 6.1 Claimants

These claimants were offered three settlement options (payment could be received for all three options). The domestic payment rates are detailed below. International claimants received reduced rates of compensation depending on where they resided; 60% of the domestic rate for category 1 and 2 countries and 35% of domestic rates for category 3 and 4 claimants.

1. *$5,000 Explant Payment*: to receive the $5,000 Explant Payment, claimants were required to submit medical records that showed that their Dow Corning breast implant was removed after 31 December 1990 and on or before 10 years after the Effective Date. Claimants were not eligible if they received silicone gel breast implants after their Dow Corning breast implants were removed.
2. *$25,000 Rupture Payment*: to receive the $25,000 Rupture Payment (including a Premium Payment), claimants were required to submit medical records on or before 2 years after the Effective Date that showed that their Dow Corning silicone gel breast implant was removed and was ruptured.
3. *(A) $2,000 Expedited Release Payment*: to receive a $2,000 Expedited Release Payment, claimants were required to submit:
    - A Proof of Manufacturer Form and medical records or documents that showed that they were implanted with a Dow Corning breast implant.
    - An Expedited Release Payment Claim Form on or before 3 years after the Effective Date.

    or (B) Disease Payment: to receive a Disease Payment ranging from $12,000 to $300,000 (including a Premium Payment), claimants were required to submit medical records that showed that they had 1 of the eligible diseases or conditions and that they had an eligible disability or met the severity criteria for that disease or condition. If a claimant became more ill in the future, there was provision for them to apply for additional payment from the Increased Severity Fund depending on the disease option that they had been were approved for.

[23] Details available at www.tortcomm.org.

Class 6.2 Claimants

Claimants who were in Class 6.2 and elected Option 1 had 3 available settlement benefit options. Payment could be recovered from all three options, detailed below:

1. *(A) Explant Payment $1,750*: to receive the $1,750 Explant Payment claimants must have submitted medical records to show that their Dow Corning breast implant was removed after 31 December 1990 and on or before 10 years after the Effective Date. Claimants who received silicone gel breast implants after their Dow Corning breast implant was removed are not eligible for this;

   or

   (B) Increased Explant Payment $3,000: to receive the $3,000 Increased Explant Payment claimants must have submitted the records described above for the $1,750 Explant Payment. To obtain the Increased Explant Payment of $3,000 claimants must have waived their right to the Premium Payment on their Rupture Claim (the Rupture Payment of $7,000 is still available for qualifying claimants);
2. *Rupture Payment $8,750*: To receive the $8,750 Rupture Payment (including a Premium Payment), claimants must have submitted medical records on or before 2 years after the Effective Date that show that their Dow Corning silicone gel breast implant was removed and was ruptured;
3. *(A) Expedited Release Payment $700*: to receive a $700 Expedited Release Payment, claimants must have submitted the Proof of Manufacturer Form and medical records or documents to show that they were implanted with a Dow Corning breast implant and the Expedited Release Payment Claim Form on or before 3 years after the Effective Date

   or

   (B) Disease Payment: to receive a Disease Payment ranging from $4,200 to $105,000 (including a Premium Payment), claimants must have submitted medical records showing that they had 1 of the eligible diseases or conditions and that they had a disability or met the severity criteria for that disease or condition.[24] If a claimant became more ill in the future, there was provision to apply for additional payment from the Increased Severity Fund depending on the Disease Payment Option that they were approved for.

The values of the various claim options are detailed in Table 30.2.

Table 30.2: Values of settlement options for category 6.2 claimants

Class 6.2, Option 1 settlement options	Base premium ($)	Premium payment ($)	Total ($)
Explant Payment Option *or* Increased Explant Payment Option	1,750 *or* 3,000	N/A	1,750 *or* 3,000
Rupture Payment Option	7,000	1,750	8,750
Expedited Release Payment Option	700	N/A	700

(continued)

[24] Guidance on disease payment options 1 and 2 and the conditions covered and severity/disability levels is available at table 5 of the Dow Corning Foreign Breast Implant Claimants (Class 6.2), at www.tortcomm.org/downloads/CIG_ENG_6.2.pdf.

Table 30.2: ***(Continued)***

Class 6.2, Option 1 settlement options	**Base premium ($)**	**Premium payment ($)**	**Total ($)**
Disease Payment Option 1: Disability C or D	3,500	700	4,200
Disease Payment Option 1: Disability B	7,000	1,400	8,400
Disease Payment Option 1: Disability A	17,500	3,500	21,000
Disease Payment Option 2: General Connective Tissue Symptoms (GCTS) Severity B	26,250	5,250	31,500
Disease Payment Option 2: GCTS Severity A or Polymyositis (PM)/ Dermomyositis (DM)	38,500	7,700	46,200
Disease Payment Option 2: Systemic Sclerosis or Lupus Severity C	52,500	10,500	63,000
Disease Payment Option 2: Systemic Sclerosis or Lupus Severity B	70,000	14,000	84,000
Disease Payment Option 2: Systemic Sclerosis or Lupus Severity A	87,500	17,500	105,000

Class 7 Claimants

These claimants were offered two settlement options (payment could be received for only 1 option)

A. *Expedited Release Payment*: claimants were required to show that they were implanted with a silicone gel breast implant at any time after 1 January 1976 and before 1 January 1992 from 1 of the following manufacturers: Bioplasty, Baxter, Bristol, Cox-Uphoff (CUI) or Mentor.

   or

B. *Disease Payment*: claims could be made for payment for 1 of the 9 eligible diseases and conditions. Claims required the submission of medical records and documents that show that the claimant had 1 of the diseases or conditions listed below and a related disability or met the severity criteria for that disease or condition:
   - atypical connective tissue disease (ACTD)
   - atypical neurological disease syndrome (ANDS)
   - primary Sjogren's syndrome (PSS)
   - mixed connective tissue disease (MCTD)/overlap syndrome
   - systemic sclerosis/scleroderma (SS)
   - systemic lupus erythematosus (SLE)
   - polymyositis (PM)
   - dermatomyositis (DM)
   - general connective tissue symptoms (GCTS).

The values of the various claim options are detailed in Table 30.3.

Table 30.3: Values of settlement options for Class 7 claimants

Class 7 settlement options	**Total ($)**
Expedited Release Payment Option	To be determined by Claims Administrator
Disease Payment Option 1: Disability C or D	Up to 4,000
Disease Payment Option 1: Disability B	Up to 8,000
Disease Payment Option 1: Disability A	Up to 20,000
Disease Payment Option 2: GCTS Severity B	Up to 30,000
Disease Payment Option 2: GCTS Severity A or PM/DM	Up to 44,000
Disease Payment Option 2: Systemic Sclerosis or Lupus Severity C	Up to 60,000
Disease Payment Option 2: Systemic Sclerosis or Lupus Severity B	Up to 80,000
Disease Payment Option 2: Systemic Sclerosis or Lupus Severity A	Up to 100,000

Deadlines for Filing Claims

The deadlines for filing are listed below: the majority have now passed.[25]

1. Deadlines to 'Register' a Claim with the Bankruptcy Court

During the bankruptcy proceedings, claimants were required to file a Proof of Claim form. The deadline for 'timely' claims was set as 30 November 1999. Section 2.02(b)(i) of Annex A[26] to the Settlement Facility and Fund Distribution Agreement also established a process where claimants who had not filed a timely Proof of Claim, but on whose behalf such a form was filed, could file a 'Notice of Intent' form with the court. Section 2.02(b)(iii) set the deadline as 90 days after the Effective Date of the Plan, which was 1 June 2004. Therefore, the deadline to file a Notice of Intent form was 30 August 2004.

2. Deadlines to File a Claim for Benefits with the Settlement Facility

All deadlines for timely claimants to file a claim for benefits were linked to the Effective Date of the Plan—1 June 2004—and are detailed in Table 30.4.

3. Deadlines to Cure Deficiencies in Claim Submissions

Annex A established time limits for claimants to fix problems in their claims after the Settlement Facility notified them of the problems in a 'Notification of Status' letter: see Table 30.5.

25 Information taken from www.sfdct.com/_sfdct/index.cfm/deadlines/.
26 Available at www.sfdct.com/_sfdct/resources/ANNEX%20A%20to%20SFA%20-%20FINAL.5-26-04.pdf.

Table 30.4: Deadline for filing claims with the Settlement Facility

Deadline	Benefit	Annex A reference
1 June 2006	Deadline to submit Rupture claim	Section 6.02(a)(iii)
1 June 2006	Deadline to submit Class 7 claim	Section 6.04(h)(ii)
1 June 2006	Deadline to submit Class 9/10 claim	Section 6.03(c)
2 June 2014	Deadline to submit Explant claim	Section 6.02(a)(i)
3 June 2019	Deadline to submit Disease or Expedited Release claim	Section 6.02(a)(ii)

Table 30.5: Deadline for curing deficiencies in claims submissions with the Settlement Facility

Type of claim	Deadline	Annex A reference
Explant	6 months from the date of the NOS letter	Section 7.09(a)(iii)
Rupture	6 months from the date of the NOS letter	Section 7.09(c)(ii)
Disease	1 year from the date of the NOS letter	Section 7.09(b)(ii)(1)
Class 7 Disease	1 year from the date of the NOS letter	Section 7.09(b)(ii)(1)
Class 9, 10	6 months from the date of the NOS letter	Section 6.03(f)

4. Deadlines that Apply only to Notice of Intent Claimants

Explant and Rupture claims had a deadline of 20 October 2008 set in the NOI Consent order.[27]

5. Deadlines that Apply only to 'Eligible Late Claimants'

Certain late claimants were allowed to join the settlement on a more limited basis. They were named in Appendices 1–3 of the Late Claimant Order.[28] Class 7, 9 and 10 Eligible late claimants had to file their claims by 21 July 2008. Class 5, 6.1 and 6.2 claimants had to file their claims by 21 January 2009.

Evidential Requirements for Claims

In addition to the appropriate claim form, certain evidential requirements were common to all breast implant claims from the DCST. All claims required:

— proof of manufacturer;
— dated medical records to confirm implantation and/or explantation (with or without rupture).

If a disease payment was claimed then appropriate medical evidence had to be supplied to confirm the diagnosis and/or severity level.

[27] Available at www.sfdct.com/tasks/sites/_sfdct/assets/File/NOI_Consent_Order(1).pdf.
[28] Available at www.sfdct.com/tasks/sites/_sfdct/assets/File/Late_Claimant_Order.pdf.

One of the major criticisms of the fund has been the difficulty encountered by some claimants in providing proof of manufacturer and/or medical records. This information is not usually held by the claimant, but by healthcare professionals, the majority of whom practice in the private sector. Having to go to a third party to obtain information that is critical to the claim has proved a substantial obstacle for some claimants. The time delay between implantation and making a claim potentially further exacerbates the problem of obtaining records—for example, surgeons may have moved or died in the intervening years. There are some procedures to try to facilitate struggling claimants—for example, there are experts who can identify the brand of an explanted implants and there is a list of surgeons who have sworn affidavits stating they predominantly used one brand in a given time frame.[29]

Requests for Re-Review, Error Correction and Appeals

Included within the Settlement Plan was a process for reviewing claims and an appeals process. A claimant who disagreed with the decision of the DCST may seek review of the claim through the error correction and appeal process.[30]

Error correction[31] is an administrative request whereby the claimant informs the Claims Administrator of the error in the Claims Office records, and if the Claims Administrator determines that it is an error, it will be corrected.

Appeals to a Claims Administrator[32] were only possible after all corrections had been made to the claimant's records. If the claimant was unsuccessful in her efforts to cure any deficiencies in the Claims Office records, or if the claimant was dissatisfied with the allowed benefits, an appeal could be made to the Claims Administrator. Before a claimant could file an appeal to a Claims Administrator all records, including additional documentation that a claimant wishes to have considered for the appeal, must have been submitted for review by the Settlement Facility for the Dow Corning Trust. An appeal is not automatic: it must be requested in writing. The appeal was limited to the benefit status contained in the Notification of Status letter. Because there could be multiple Notification of Status letters for each of the settlement options, claimants could appeal each of the rulings.

Once an appeal had been filed, the Claims Administrator conducted a de novo review and promptly issued a ruling in writing to the claimant and/or her counsel. The Claims Administrator could request further submissions from the Claimant, or seek further information from the claimant's physicians in deciding the appeal. In the event that the Claims Office determined that the Qualified Medical Doctor or records submitted in support of the claim were unreliable, the Notification of Status letter advised the claimant of such determination and identified the particular records or statements that are deemed unreliable.

Appeals to an Appeals Judge[33] could be made by claimants who disagreed with the ruling of the Claims Administrator, by submitting a written statement outlining the claimant's

[29] www.tortcomm.org/prod_ident.shtml.

[30] Art 8 of Annex A of the Settlement Facility and Funds Distribution Agreement, available at www.sfdct.com/_sfdct/resources/ANNEX%20A%20to%20SFA%20-%20FINAL.5-26-04.pdf.

[31] ibid, Art 8.03 of Annex A.

[32] ibid, Art 8.04 of Annex A.

[33] ibid, Art 8.04 of Annex A.

position and a statement as to why the claimant believed the Claims Office and Claims Administrator had erred. The Appeals Judge reviewed the appeal record and claim file in deciding the appeal. The Appeals Judge had to apply the guidelines and protocols established in Annex A to the Settlement Facility Agreement, including the provisions of the Revised Settlement Program as adopted by Annex A, and the appeals process was required not to result in any modification of substantive eligibility criteria. The Appeals Judge issued a determination on the appeal in writing. The decision of the Appeals Judge was final and binding on the claimant. The decisions of the Appeals Judge were served on the claimant (and her counsel), the debtor's representatives and the CAC.

C. Claims Data from the Dow Corning Settlement Fund

As of 31 May 2013,[34] the DCSF had received 79,695 Proof of Manufacturer claim forms in Classes 5, 6.1 and 6.2 (Dow Corning Breast Implant Claims), see Table 30.6. Of this, 69,634 (87.38%) had acceptable proof of at least one Dow Corning breast implant. Of the 69,634 Dow Corning breast implant claimants with acceptable Proof of Manufacturer, almost 32% (21,911 women) had been approved for a Rupture claim for their Dow Corning silicone gel breast implant.

Table 30.6: Dow Corning breast implant claims data for Class 5, 6.1 and 6.2 claims (exclusive of NOI) payments to 31 May 2013

Claim category	Number of claims paid to 31 May 2013	% of Class 5 claims approved	% of Class 6 claims approved	Total paid per category to 31 May 2013 ($)
Rupture	21,911	71	40	421,674,223
Explant	28,713	99	96	138,975,935
Explant Assistance	619	92	83	3,058,641
Increased Explant	293	NA	92 (6.2 only)	659,700
Expedited Release	24,067	92	88	44,653,006
Disease Option 1	28,380	84	70	418,326,891
Disease Option 2	1,447	30	22	185,724,209
Total	105,430	NA	NA	**1,213,072,606**

[34] Taken from www.tortcomm.org/claimsdata.shtml.

D. Conclusions

The DCST has been widely criticised for being slow and difficult to claim from. Delays have been attributable to ongoing litigation concerning the bankruptcy of the company and the status of claimants. In particular, there have been complaints that the requirement for third-party medical documentation to support a claim has proved difficult for claimants to fulfil.

31

Trilucent Breast Implants

A. Background

Trilucent breast implants were a textured silicone elastomer shell implant filled with Trilipid $Z^{6\,TM}$ (triglyceride) filler. They were available from 1993 to 1999. The intention was to produce a radiolucent breast implant which enabled better detection of early cancers in women who had had breast implants.

Corporate History

In January 1996 Collagen Corporation purchased Lipomatrix Inc (founded in 1992 by Dr T R Kanpp, one of the founders of Collagen Corporation) as a subsidiary business with the intention of producing and testing what became Trilucent breast implants. In November 1998 LipoMatrix Inc was sold to Sierra Medical Technologies Inc. Collagen Corporation became Collagen Aesthetics (with a subsidiary AEI Inc), and in November 1999 Collagen was purchased by Inamed (previously known as McGhan Medical Corp). Inamed was acquired by Allergan in 2006.[1]

Pre-market and Marketing of Trilucent Implants

Tilucent breast implants were round, non-anatomically shaped implants made in a variety of sizes with volumes from 120 to 510 cc. Clinical testing of these implants began in Europe in 1993 and they were CE marked (the notified body was TüV Munich) in 1994. The US FDA also approved a clinical trial of Trilucent implants in the USA, which began in December 1994. Trilucent implants were launched onto the European market in 1995. Sales in 9 European countries by the end of 1996 accounted for 7% of the market share of breast implants, with relatively high volumes of sales in the UK and Germany, and smaller sales volumes in Spain, Portugal, Switzerland, Benelux, Scandinavia and Italy.[2] In the UK approximately 4,500 women were implanted with Trilucent breast implants.

[1] See http://agn.client.shareholder.com/releases.cfm?ReleasesType=&Year=2005 and US Securities and Exchange Commission, Washington, DC, Inamed Corporation (IRS Employer Identification No 59-0920629) Form 10-K Annual Report pursuant to s 13 or 15(d) of the Securities Exchange Act of 1934 for the fiscal year ended 31 December 2001, available at www.sec.gov/Archives/edgar/data/109831/000091205702012689/a2073866z10-k.htm.

[2] Trilucent Breast Implants, *Sales Training Manual* (LipoMatrix Inc).

The Trilucent were marketed as a safer alternative to saline and silicone implants as they had a better radiolucency profile. Human fat and Trilipid Z^{6} ™ have a Z number (a logarithmic indicator of opacity to X-rays: the higher the number, the lower visual transparency on an X-ray) of 6, compared to 10.4 for silicone gel and 7.6 for saline. Silicone and saline breast implants presented difficulties when the breast was radiographically imaged. These difficulties could be minimised by imaging using, for example, the Eklund technique, where the breast tissue is pulled forward of the implant. This allows for better imaging of the breast tissue; however, approximately 10% of the breast tissue will still be obscured by the implant.[3] Trilucent implants were developed and marketed to provide an implant with a better imaging profile.

Adverse Events and Traceability of Trilucent Implants

There are regulatory requirements that cover the reporting of reporting adverse events to the UK regulatory agency (the Medical Devices Agency (MDA), later the Medicines and Healthcare Products Regulatory Agency (MHRA)). LipoMatrix had in place internal standard operating procedures for the reporting of adverse events to the regulator, with detailed protocols provided to sales staff on how complaints should be made and describing how explants should be treated and returned to them for testing.[4]

In addition, all Trilucent implants contained a passive transponder providing a unique Vigilance Identification Device System™. Each implant had an individual alphanumeric code, which could be read by a handheld reader that activated the transponder with a low-frequency radio signal through any tissue, including bone. This enabled traceability of each individual implant down to the exact batch of raw materials used. There were criticisms that the transponder only responded to proprietary transducers and introduced a non-radiolucent element to the implant.

As well as the transponder, two cards were included with the implant on delivery: a registry card was to be completed by the surgeon and a warranty card by the patient, and in both cases then sent back to LipoMatrix. These cards contained a unique number that corresponded to the number on the transponder. The information on these cards enabled LipoMatrix to build a very detailed database. The UK Breast Implant Registry was also operational throughout the period when Trilucent implants were on the market, but the data collected on this resource was somewhat patchy, because patients failed to return cards.

The first warning sign that there was an issue with Trilucent implants was the higher number of capsular contraction seen with these implants. Capsular contractions are a known risk with all breast implants; they occur when the immune system responds to the implant by forming a tight capsule of scar tissue around the implant, which can cause the breast to become hard and/or misshapen. There was initially a question over whether insertion impacted on device performance. Trilucent implants were designed with an open-cell texture to stick in place, so an incision of at least 6 cm was recommended when inserting

[3] www.cancerscreening.nhs.uk/breastscreen/publications/nhsbsp53.pdf.

[4] Trilucent Breast Implants (n 2) ch XI.

them,[5] which was more difficult to achieve with transaxillary incisions, which were often used as they generally resulted in less visible scarring.

Withdrawal of Trilucent from the Market

In March 1999 the MDA requested a voluntary withdrawal of Trilucent implants from the market in the UK.[6] A programme of research and further testing followed. At the company's request shells were examined by the Rubber and Plastics Research Association (RAPRA) and the oil filler was sent to the Department of Chemistry, London University for analysis, where the filler was found to be genotoxic. The problems with the implants were found to be due to oxidation of the filling material and the way this interacted with the shell.[7]

During this period UK court proceedings were issued by several women who had had Trilucent implants removed after complications occurred.[8] On 6 June 2000, the MDA issued a Hazard Notice which recommended removal of Trilucent implants.[9] This also recommended not becoming pregnant or breastfeeding until implants had been removed.

B. Care Package and Compensation Scheme

At the same time, the MDA announced that an agreement had been made with AEI Inc[10] that the latter would bear the cost of removal and replacement of these implants from this point forward. The Trilucent Care Centre was set up in the UK, which administered a care package. The first point of contact was a trained nurse, followed by referral to a consultant, who undertook a full examination including ultrasound and magnetic resonance imaging and explantation. If desired, replacement with non-Trilucent implants was undertaken. There were complaints that this system did not function as it should, with surgeons refusing to accept the fee scales offered by AEI for removal and replacement surgery as adequate.[11]

The compensation scheme was structured in two parts: a fixed-cost settlement for more straightforward cases; and a non-fixed option for women with more complex problems, either physical, psychological or both.

[5] ibid, ch VIII.

[6] www.mhra.gov.uk/Safetyinformation/Safetywarningsalertsandrecalls/MedicalDeviceAlerts/Advicenotices/CON2025118.

[7] For more detailed analysis, see WN Kirkpatrick and BM Jones, 'The History of Trilucent Implants, and a Chemical Analysis of the Triglyceride Filler in 51 Consecutively Removed Trilucent Breast Prostheses' (2002) 55 *British Journal of Plastic Surgery* 479–89 and S Monstrey, A Chrisophe, J Delanghe, S De Vriese, M Hamdi, K Van Landuyt and P Blondeel, 'What Exactly Was Wrong with the Trilucent Breast Implants? A Unifying Hypothesis' (2004) 113 *Plastic and Reconstructive Surgery* 847–56.

[8] Personal communication, Paul Balen, consultant at Freeth Cartwright LLP (www.freethcartwright.co.uk), who represented many of the claimants.

[9] www.mhra.gov.uk/Safetyinformation/Safetywarningsalertsandrecalls/MedicalDeviceAlerts/Hazardnotices/CON008727.

[10] Negotiations had previously been undertaken with Paul Balen (see n 8 above).

[11] See s 9 of Minutes of Evidence taken before the Health Committee, Thursday, 22 March 2001, at www.publications.parliament.uk/pa/cm200001/cmselect/cmhealth/308/1032205.htm#n43.

Under the fixed-cost option, affected women were only required to provide their name, address and National Insurance number (so the Department for Work and Pensions could be informed of the compensation). There were three levels of payment for removal and replacement:

— removal and replacement in the same operation (£5,250);
— removal without replacement (£7,500); and
— removal and replacement in two different operations with an interval between them (£8,750).

In addition, claimants could claim a fixed sum for legal expenses of £750, and £500 for personal expenses, which were paid automatically. If they claimed expenses over £500, every item had to be detailed and proved. The vast majority of UK claimants used the fixed-cost scheme. As the compensation scheme ran for some years the costs were periodically increased.

For more complex needs, a non-fixed-cost option was established. In addition to the information listed above, claimants had to provide a witness statement detailing their claim, a schedule of loss and medical records. These were then scrutinised by AEI's lawyers, who decided if further medical evidence was needed. If additional medical evidence was sought, a relevant expert would be jointly instructed (at AEI's expense) and a report produced. If a settlement could not be negotiated even with this additional medical report, then a single barrister was selected from an agreed pool to handle the case. The figure reached by the barrister became the 'Part 36 offer' for AEI,[12] should the claimant decide to go to court. Claimants were free to go to court, but to succeed they would need to recover more than the Part 36 offer. It is notable that no one chose to litigate in the UK.

Initially, the compensation scheme covered women who had had their implants removed in response to the MDA announcement of 6 June 2000. It was later expanded to cover retrospectively women who had had their Trilucent implants removed before this date.

A final Medical Device Alert was issued by the MHRA in 2004, which reiterated the need for removal of Trilucent implants and highlighted that the Trilucent Care Centre would cease to exist at the end of 2004.[13] The compensation scheme was closed to new claims on 15 September 2005.

Table 31.1 shows the proportion of women who received an explantation after they had requested it from 2001 to 2006. This illustrates that once a woman had requested an explantation the process worked very efficiently in the UK, though not so well for non-UK residents. However, these figures are for women who came forward to an initial consultation, and there may have been women who did not respond to, or were not aware of, the recommendations for removal. A comprehensive recipient database could potentially have helped with traceability.

[12] A Part 36 offer was an offer of settlement under the Civil Procedure Rules that had implications for costs if it was not subsequently beaten by the level of the court's award: see www.justice.gov.uk/courts/procedure-rules/civil/rules/part36.

[13] MDA/2004/047, at www.mhra.gov.uk/home/groups/dts-bs/documents/medicaldevicealert/con008502.pdf.

Table 31.1: Proportion (%) of patients by year who had a Trilucent explantation after an initial consultation where they requested explantation

	2001	2002	2003	2004	2005	2006
UK residents	90	90	Over 90	Over 90	Virtually all	Virtually all
Non-UK residents	75	Almost 60	Almost 67	Over 67	Over 95	Over 95

Source: Data taken from Inamed's annual 10-K filings, available at www.sec.gov/cgi-bin/browse-edgar?action=getcompany&CIK=0000109831&type=10-k&dateb=&owner=exclude&count=40.

Scheme Costs

The scheme was partly self-funded and partly funded by insurance. The UK scheme was only a part of the global liabilities, the costs of which follow. It was agreed at the sale of LipoMatrix to Sierra Medical Technologies Inc in November 1998 that Collagen would retain liabilities for Trilucent implants sold before 6 November 1998. On 30 June 1999 Collagen increased its provision for LipoMatrix by $11.5 million. When Inamed purchased Collagen it took on Collagen's liabilities. Inamed funded these liabilities in three ways: using existing insurance policies, by purchasing a medical expense reimbursement insurance policy and by reserving money.

In 2001 Inamed entered into a 'Trilucent Claims Settlement Protocol Interim Funding and Non-Waiver Agreement' under which (subject to a full reservation of rights by all parties) Inamed's carriers agreed to fund up to $20 million of settlement protocol payments, on condition that Inamed make pro-rata co-payments of up to $2 million and satisfy a $0.4 million self-insured retention (the SIR). Inamed satisfied the SIR, and both Inamed and the carriers funded the settlement protocol as agreed. In 2002, Inamed entered final settlements with each of its two insurers for product liability claims arising from the Trilucent implant.[14]

Under the first settlement with MEDMARC Casualty Insurance Company, MEDMARC paid $6.0 million in cash to Inamed in January 2003, $1.5 million in cash in May 2003, and, effective 16 November 2002, agreed to make a policy with a limit of $10.0 million available to Inamed for defence and indemnification of Trilucent-related bodily injury claims worldwide. The policy only covered claims filed on or before 7 November 2005.

Under the second settlement, AISLIC, an AIG company, agreed to make an excess policy with a limit of $10 million available to Inamed for the indemnification of non-US Trilucent claims. Under the settlement with AIG, subject to certain conditions, Inamed were required to pay up to $1.5 million to AIG on or before November 2005.

As a result of these settlements, Inamed released both MEDMARC and AISLIC from any other actual or potential financial claims under the policies and dismissed all litigation

[14] Inamed Corporation (n 1) Annual Report For the fiscal year ended 31 December 2002, available at www.sec.gov/Archives/edgar/data/109831/000104746903011036/a2106833z10-k.htm.

related to them with prejudice. Table 31.2 shows the amounts remaining on these policy commitments at the end of the December in each year from 2002 to 2005 in millions of dollars.[15]

Inamed and later Allergan consistently maintained that they felt they had sufficient to cover their liabilities, but that this could not be guaranteed. The possibility of claims for late-onset illnesses, such as claimants going on to develop cancer many years after explantation, meant that the potential for future liabilities was difficult to calculate.

Table 31.2: Sums remaining (in millions of dollars) on Trilucent companies' insurance policy commitments at the end of the December for the years 2002–05

	Initial value	Value at 31 December 2002	Value at 31 December 2003	Value at 31 December 2004	Value at 31 December 2005
MEDMARC policy	10	9.6	3.3	0 (spent by 30 June 2004)	0
AISLIC policy	10	7.9	6.2	4.2	1.6

C. Conclusions

After the creation of the care package, no new court cases proceeded through to judgment in UK courts,[16] so the package would appear to have been a reasonably well-functioning alternative to litigation. The two-tier system appeared to provide an efficient and appropriate means to claim. Some higher-value, more complex cases have been very slow to finalise, though this may be due to the nature of the injuries to the claimant, such as those claims that included psychological injuries, rather than procedural inadequacies. There was a wide uptake of the scheme, but it was not complete. Some women were not reached, which highlights the need for better traceability and coverage. The UK Breast Implant Registry was in place and potentially could have provided better coverage, but the uptake was low, so it was of little assistance. The NHS announced the reopening of the Breast and Cosmetic Implant Registry on 10 October 2016 in response to the PIP scandal (in which the French company Poly Implant Prothèse was found to be manufacturing implants using substandard non-medical-grade silicone). However, this is a voluntary registry so may be subject to the same uptake issues as the previous UK Breast Implant Registry.

One factor which improved identification of claimants was the inclusion of a transponder within the implant. This allowed for a clear and non-invasive identification of

[15] Taken from Inamed's annual 10-K filings available at www.sec.gov/cgi-bin/browse-edgar?action=getcompany&CIK=0000109831&type=10-k&dateb=&owner=exclude&count=40.

[16] Three cases were brought by British women in the federal Trilucent actions in the District Court in San Francisco, California, but were declared *forum non conveniens* as the British courts offered an avenue for the women to seek redress.

relevant devices, saving women time and worry. Obviously this is advantageous, but the only entity that had all the data on these implants was the company. In this case there was a voluntary product withdrawal, a settlement scheme was established and it was prepared to release data when asked by claimants through their lawyers.[17] However, if a manufacturer were not cooperative or if a PIP-type situation involving fraud were to arise again, this type of settlement scheme would be unlikely to function efficiently.

[17] In this case, such representatives could have been an ombudsman, or claimants themselves, if such a mechanism had existed.

32

ASR Hip Reimbursement Programme

A. Background

Although not technically a compensation scheme, the reimbursement scheme offered by DePuy/Johnson & Johnson merits consideration in this volume. This scheme reimburses costs associated with investigation and, if necessary replacement, of articular surface replacement (ASR) hip joints following their recall from the market. These devices were implanted into just under 100,000 people worldwide. It is a voluntary scheme offered by the manufacturers with no specific statutory or court-determined underpinnings, so offers an example of an internally driven corrective action scheme undertaken by the corporation itself.

ASR Hip Joints

Hip joints are formed of a 'ball and socket', the ball being the femoral head, a rounded protrusion formed by the end of the femur (thigh bone), and the socket being the acetabulum, a concave structure within the pelvic bone. Total hip replacements and resurfacing hip replacements are used when osteoarthritis, injury or other conditions cause the hip joint to lose mobility and become painful. The ASR hip systems were manufactured by DePuy (a Johnson & Johnson company). There were two products in this range: the ASR XL Acetabular System and the ASR Hip Resurfacing System. The ASR XL Acetabular System was a total hip replacement: the natural bone femoral head was removed and replaced with a metal femoral head (or ball) fixed using a stem inserted into the femur. The metal head fitted within a one-piece metal cup which lined the acetabulum, thereby forming a 'metal-on-metal' bearing. The ASR Hip Resurfacing System comprised two components, a metal cap which was fitted over the natural bone femoral head and a one-piece metal cup which lined the acetabulum.

The very early hip replacements of the 1960s included metal-on-metal articulations, but by the 1980s and 1990s units comprising a metal ball fitted into a polyethylene cup were more commonly used. These polyethylene cups were prone to wearing, necessitating replacement surgery. Plastic wear is also associated with osteolysis, which can lead to bone resorption and implant loosening.

Subsequently, large-diameter metal-on-metal hip replacements such as the ASR systems were developed. The aim of these replacements was to provide a longer-lasting joint (through the use of materials harder than polyethylene) with a lower dislocation rate (through the use of a large-diameter head). These implants would be particularly suitable for younger, fitter recipients.

The components of the ASR systems were made of cobalt–chromium–molybdenum alloy (CoCrMo), which is much harder wearing than either ceramic or plastic. Calculations demonstrated that, providing the surface on the hip joint components was smooth enough, a layer for fluid should be trapped between the ball and the cup, minimizing actual contact and resultant surface wear. Further calculations demonstrated that the more an individual moved, the more viscous the trapped fluid became, reducing wear levels even further in active recipients.[1,2]

Pre-market Testing and Marketing Approval

DePuy began work on metal-on-metal hip systems in the late 1990s, with the characteristic large head appearing in the 2000s. The products went through a rigorous testing process, including fretting tests, wear tests, deflection tests, materials tests, clearance tests, hip simulation tests, torque testing, lever-out testing, scientific calculations, computerised finite-element analysis, cadaver tests and Sawbones tests. In accordance with international standards, DePuy ran several 5–6-month-long tests of the ASR XL in hip simulators that ran 5 million cycles of motion, representing 2.5 years of use by a patient. In order to better understand the long-term survival of the hip, DePuy also ran additional tests up to and including 15 million cycle tests, exceeding international standards. Data from these tests was published[3] and formed part of the applications for regulatory approval.

Europe

Within the European regulatory system,[4] medical devices are classified into 1 of 4 categories (I, IIa, IIb and III). Class I are the lowest-risk devices, such as plasters. A medical implant[5] is always categorised as class IIb or III. ASR hip replacements were categorised as class IIb in 2003.

To be placed on the market in Europe, a conformity assessment must be undertaken to determine whether a device meets the regulatory requirements and, when used correctly, works properly and is acceptably safe.[6] Conformity assessments can be based on clinical

[1] SL Smith, D Dowson and AA Goldsmith, 'The Effect of Femoral Head Diameter upon Lubrication and Wear of Metal-on-Metal Total Hip Replacements' (2001) 215 *Proceedings of the Institution of Mechanical Engineers [H]* 161–70.

[2] SL Smith, D Dowson and AA Goldsmith, 'The Effect of Diametral Clearances, Motion and Loading Cycles upon Lubrication of Metal-on-Metal Hip Replacements' (2001) 215 *Proceedings of the Institution of Mechanical Engineers [C]* 1–5.

[3] D Dowson et al, 'A Hip Joint Simulator Study of the Performance of Metal-on-Metal Joints, Part II: Design' (2004) 19 *Journal of Arthroplasty* Suppl 3.

[4] Pursuant to Directive 93/42/EC on Medical devices. See C Hodges, 'The Regulation of Medicines and Medical Devices' in A Grubb, J Laing, J McHale and I Kennedy (eds), *Principles of Medical Law*, 3rd edn (Oxford, Oxford University Press, 2011).

[5] A device which is intended to be either totally introduced into the body or to be partially introduced into the body through surgery and to remain there for at least 30 days: Directive 93/42/EEC at Ev7, para A3.

[6] Included within the conformity assessment is a requirement for manufacturers to provide the notified body with evidence of a quality system to ensure the product and the manufacturing methods continue to conform to the CE standards. This must include arrangements to monitor the marketing phase by obtaining, recording and reviewing data, including conducting risk analyses and creating and implementing plans for corrective action should issues arise with the device as per 3.1 of Annex II of Directive 93/42/EEC.

trials data or on equivalence, data which shows the device is 'equivalent' to an already approved device. Once a device has passed the conformity assessment a CE (Conformité Européenne) mark can be applied and the product can then be marketed in all EU countries.

In order to place a class II or III device on the market in Europe it has to be certified by one of the notified bodies. Notified bodies are third-party private-sector organisations.[7] The conformity assessment for ASR implants was undertaken by BSI, who approved the ASR XL Acetabular System and the ASR Hip Resurfacing System on the basis of equivalence. Having completed the certification process, DePuy launched the ASR Hip Resurfacing System in Europe in July 2003 and the ASR XL Acetabular System in March 2004.

USA

The Food and Drug Administration (FDA) approves medical devices in the United States. There are several different routes to approval depending upon the class of device and what data is relied upon, which can derive from clinical trial results, equivalence data or long-standing use. The FDA has three classifications of device, I, II and III, with I being the lowest risk and III the highest. Hip replacements are class III.

The ASR XL Acetabular System was cleared for marketing by the FDA in 2005.[8] It was approved under the 510(k) procedure, whereby pre-marketing notification is supplied to the FDA, which then determines if the device is substantially equivalent to a device already on the market. Substantial equivalence reduces/removes the need for clinical trials data. The ASR XL System was deemed substantially equivalent to devices already on the market so marketing approval was granted.[9]

The ASR Hip Resurfacing System was never marketed in the USA. Hip resurfacing was a relatively new technique at the time and so the equivalence route would have been more difficult to establish, necessitating more extensive trials/data gathering. An application for approval was started and then withdrawn after the market for metal-on-metal hips declined. Surgeons can legitimately use a device that is not approved by the FDA 'off-label', but the number of cases will usually be small in comparison to the usage of approved alternatives.

Australia

There is a mutual recognition agreement between Australia and the European Union, so that a device evaluation undertaken in the EU can be used for regulatory approval in Australia.[10] Many medical devices available on the Australian market have undergone initial assessment by a European assessment body that is approved and monitored by European regulators. In the case of the ASR systems, the BSI assessment was relied upon to put the devices approved onto the Australian Register of Therapeutic Goods (ARTG). The ASR Hip

[7] There are six notified bodies in the UK: Amtac Certification Services Ltd, BSI Healthcare, Intertek Testing and Certification Ltd, Lloyd's Register Quality Assurance Ltd, SGS United Kingdom Ltd and UL International (UK) Ltd.

[8] No K040627, decision date 5 August 2005.

[9] www.accessdata.fda.gov/cdrh_docs/pdf4/K040627.pdf.

[10] For details see www.tga.gov.au/industry/devices-argmd.htm#.Us_3Ny1FBeU.

Resurfacing System prosthesis was first approved for use in Australia in October 2004. The ASR XL Acetabular System femoral head prosthesis was first approved for use in Australia in 2005.[11]

Usage and Countries

Approximately 93,000 people were implanted with ASR hip systems before the worldwide recall in 2010. The breakdown varies by country, as shown in Table 32.1.

Table 32.1: Numbers of ASR hip implants by country

Country	Number of implants
UK	9,996
Ireland	3,500
South Africa	3,388
USA	36,845
Australia	5,590

Reported Complications

There has been some controversy over who knew, or should have known, that there was an elevated failure rate with ASR hips at what time-point, and what was known, or should have been known, about the cause of the elevated revision rates. This chapter focuses on the reimbursement scheme, and a detailed exploration of the events leading to the withdrawal from the market is outside the scope of this volume. A brief summary of the major safety signals that led to the recall will be outlined, followed by an outline of the major actions taken in Australia, the UK and the USA.

Data on Revision Rates and Type of Issues with ASR Hips

Although information from individual surgeons can be extremely valuable in highlighting potential safety issues, the data provided by an individual is usually anecdotal and not statistically significant. Some individual surgeons did raise concerns about ASR hips; with hindsight these can be seen as early warning signs, but at the time it would have been difficult to evaluate effectively the concerns raised by individuals.[12] Larger-scale data sources were needed to validate the findings reported by individual surgeons. Registries can provide fertile data sources for post-marketing safety surveillance, particularly for products such as hip replacements which have a long in vivo lifespan. At the same time, national

[11] For a brief history see www.tga.gov.au/behind-news/recall-depuy-orthopaedics-asr-hip-replacement-device.
[12] D Cohen, 'Out of Joint: The Story of the ASR' (2011) 342 *BMJ* d2905.

joint replacement registry data is best considered 'a signalling device' which identifies situations that may require further investigation, and which may or may not actually involve a safety issue. Two major sources of data about the performance of ASR (and other brand) hip replacements were the Australian Orthopaedic Association National Joint Replacement Registry (AOANJRR)[13] and the UK National Joint Registry.[14]

In the UK, the National Institute for Health and Care Excellence (NICE) develops and publishes guidance on the selection of prostheses. NICE assesses the quality and value for money offered by treatments and is not involved in the marketing approval of healthcare products. At the time, in its guidance on the selection of prostheses for total hip replacement (NICE does not have a standard for resurfacing hip prostheses), NICE stated that:

> Using the most recent available evidence of clinical effectiveness, the best prostheses (using long term viability as the determinant) demonstrate a revision rate (the rate at which they need to be replaced) of 10% or less at 10 years. This should be regarded as the current 'benchmark' in the selection of prostheses for primary Total Hip Replacement (THR).[15]

Statistically significant indications of possible issues with the ASR hips first came from the AOANJRR Annual Reports. The 2007[16] report indicated that the ASR Hip Resurfacing System had twice the revision rate of all other resurfacing procedures (3.0 revisions per 100 observed component years (hazard ratio = 2.18, 95% CI (1.5, 3.2), $P < 0.001$); the cumulative percentage revision at 2 years for the ASR is 5.16%). This trend for the ASR hip to have a higher than expected revision rate was seen in the AOANJRR Annual Reports, with a 3-year cumulative revision rate of 6% in the 2008 Annual Report,[17] and a 5-year cumulative revision rate of 8.7% in the 2009 Annual Report.[18] These figures were concerning, but it was uncertain whether they were attributable to device failure, or to surgical technique, or were artefacts due to differential recording of revision events. At the same time, the market for metal-on-metal hip systems had declined, and DePuy had decided to discontinue or 'rationalise' the ASR Hip Resurfacing System worldwide because of declining demand by surgeons. Because rationalisation was a business decision, and not based on any safety concerns, the timing to discontinue sales varied from country to country. After discussions with the Therapeutic Goods Administration (TGA),[19] Hip Resurfacing Systems were voluntarily removed from the Australian market by DePuy on 31 December 2009.

In the UK ASR remained on the market until 2010. The UK National Joint Registry, did not have sufficient data to compare other brands of resurfacing (including ASR) to the Birmingham Hip Resurfacing (BHR) procedure in 2006/07.[20] Again in 2007/08 the only resurfacing brands with sufficient market share to enable meaningful comparison were the

[13] https://aoanjrr.dmac.adelaide.edu.au/home.

[14] www.njrcentre.org.uk/njrcentre/Home/tabid/36/Default.aspx.

[15] 'Guidance on the Selection of Prostheses for Primary Total Hip Replacement', NICE Technology Appraisal Guidance 2 (April 2000) available at www.nice.org.uk/nicemedia/live/11386/32002/32002.pdf.

[16] AOANJJR 2007 Annual Report, 64, available at https://aoanjrr.dmac.adelaide.edu.au/documents/10180/42612/Annual%20Report%202007?version=1.1&t=1349406300550.

[17] AOANJJR 2008 Annual Report, 57, available at https://aoanjrr.dmac.adelaide.edu.au/documents/10180/42662/Annual%20Report%202008?version=1.1&t=1349406277970.

[18] AOANJJR 2009 Annual Report, 49, available at https://aoanjrr.dmac.adelaide.edu.au/documents/10180/42728/Annual%20Report%202009?version=1.1&t=1349406243327.

[19] TGA are the state regulators of Therapeutic Goods, see www.tga.gov.au/about-tga.

[20] National Joint Registry 4th Annual Report (April 2006–March 2007) s 4.2.2. ISSN 1753-9382, available at www.njrcentre.org.uk/NjrCentre/Portals/0/Documents/England/Reports/NJR_AR_4.pdf.

BHR and the Cormet 2000.[21] By 2008/09, 3-year revision data was available for 3 brands of hip resurfacing systems, with ASR having a revision rate of 7.5%.[22] In the 2009/10 Annual Report the 5-year revision rate for ASR resurfacing was 12% (compared to 4.3% for the BHR). The National Joint Registry then notified the regulator, the Medicines and Healthcare Products Regulatory Agency (MHRA). In the light of the data from the National Joint Registry, De Puy voluntarily withdrew all ASR products worldwide[23] on 24 August 2010.

Notices Concerning ASR

A series of notices was issued in relation to the recall, as shown in Table 32.2.

Table 32.2: Notices issued concerning the ASR recall

Date (dd/mm/yy)	Notice type	Issuer/country	Outcome
31/12/09	Withdrawal from market	Australia	Voluntary withdrawal of ASR products from the Australian market
08/03/10	Field Safety Notice	DePuy/worldwide[a]	— Care placing the implant — Not using in women of child-bearing age — Caution implanting a head <50 mm diameter
22/04/10	Medical Device Alert MDA/2010/033	MHRA/UK	Advice to clinicians on the management of patients with MoM hips
25/05/10	Medical Devices Alert MDA/2010/044	MHRA/UK[b]	— follow-up patients with ASR acetabular cups — ensure ASR acetabular cups are used in accordance with the manufacturer's instructions
24/08/10	Field Safety Notice	DePuy/worldwide[c]	*Voluntary recall all ASR devices*
07/09/10	Medical Device Alert MDA/2010/069	MHRA/UK[d]	Recalling all ASR hips and recommending monitoring
24/03/11	Class 2 Recall	FDA/USA[e]	Recall on all ASR devices used in USA

(continued)

[21] of the National Joint Registry 5th Annual Report (April 2007–March 2008) table 4.18. ISSN1753-9374, available at www.njrcentre.org.uk/NjrCentre/Portals/0/Documents/England/Reports/5th%20Annual.pdf.
[22] of the National Joint Registry 6th Annual Report (April 2008–March 2009) table 3.4. ISSN1753-9374, available at www.njrcentre.org.uk/NjrCentre/Portals/0/Sixth%20annual%20NJR%20report.pdf.
[23] www.mhra.gov.uk/home/groups/dts-bi/documents/fieldsafetynotice/con093710.pdf.

Table 32.2: ***(Continued)***

28/02/12	Medical Device Alert MDA/2012/008	MHRA/UK	advisory to professionals and patients on the use of MoM hips and the monitoring of MoM-implanted patients
09/05/12	Public Health Communication[f]	Health Canada	advisory to professionals and patients on the use of MoM hips and the monitoring of Mo-implanted patients
25/06/12	Medical Device Alert MDA/2012/036	MHRA/UK[g]	monitoring of all brands of MoM hips

[a] www.mhra.gov.uk/home/groups/dts-bi/documents/fieldsafetynotice/con076187.pdf.

[b] http://www.mhra.gov.uk/Publications/Safetywarnings/MedicalDeviceAlerts/CON082089.

[c] www.mhra.gov.uk/home/groups/dts-bi/documents/fieldsafetynotice/con093710.pdf.

[d] www.mhra.gov.uk/Publications/Safetywarnings/MedicalDeviceAlerts/CON093789?tabName=Problem.

[e] www.accessdata.fda.gov/scripts/cdrh/cfdocs/cfres/res.cfm?start_search=1&event_id=57177&productdescriptiontxt=&centerclassificationtypetext=&recallnumber=&postdatefrom=&postdateto=&productshortreasontxt=&firmlegalnam=&knumber=&pnumber=&PAGENUM=100.

[f] http://healthycanadians.gc.ca/recall-alert-rappel-avis/hc-sc/2012/15835a-eng.php and http://healthycanadians.gc.ca/recall-alert-rappel-avis/hc-sc/2012/14120a-eng.php.

[g] www.mhra.gov.uk/home/groups/dts-bs/documents/medicaldevicealert/con155767.pdf.

B. Reimbursment Programme

Along with the worldwide recall DePuy announced that it would set up a reimbursement programme.[24] The reimbursement programme is self-funded by DePuy (with funds set aside and reserved for this purpose)[25] and is administered by Broadspire.[26] It is designed to ensure that implanted patients are not prevented by financial constraints from seeking testing and treatment (if required). As tissue damage from ASR hips can remain asymptomatic in the initial stages of joint degradation, the concern was that patients may not seek testing and treatment until joint damage had already occurred. This was thought to be particularly likely in countries where patients have to pay for their own healthcare.

Losses Covered

The scheme covers the cost of testing and revision (if required), along with associated economic losses, such as loss of earnings and travel cost. There is no compensation element

[24] www.asrhipinfo.com/sites/default/files/pdf/Broadspire.pdf.

[25] Exhibit 13 of Johnson & Johnson's 10-k filing for 2013, available at www.sec.gov/Archives/edgar/data/200406/000020040613000038/ex13-form10xk20121230.htm.

[26] Broadspire are a third-party claims administration company; for details see http://www.accesscrawford.com/BroadspireTPA/Home.

for pain, suffering, loss of amenity, etc, but there is also no bar on claimants paid by the reimbursement scheme seeking redress for these elements through the courts.

Making a Claim

When the recall was announced, surgeons were advised to contact patients implanted with ASR hips. In countries where there is good medical record-keeping and high patient traceability, the majority of claimants should have been contacted by letter explaining that they have an ASR hip and detailing the steps to be taken, including details of how to contact the reimbursement programme and what it would provide. Rates of contact will vary from country to country and potentially from doctor to doctor, depending upon standards of record-keeping and patient traceability (including whether centralised regional/national databases exist and the extent of health tourism).

Claims to the programme are initiated by a claimant contacting Broadspire on a local phone number or via the internet. Details are collected from the claimant.

Uptake of the Reimbursement Programme

There are considerable differences in the percentage uptake of the reimbursement programme in different countries. These can be partially explained by variations in patient traceability, but other elements, including, but not limited to, cultural attitudes towards litigation and availability of healthcare, will also affect the uptake.

Claim Processing

Once a claimant registers with Broadspire, he or she will be asked for details about his/her implanted ASR hip. Once it is established that the claimant has an ASR hip, the patient is then entered into the reimbursement programme. Eligible claimants are then assigned a reference number, which should be provided at hospital visits, along with informing the hospital that the patient is part of the recall programme. This reference number enables the hospital to submit claims either directly to the reimbursement programme for uninsured patients, or to insurers, as normal, if the patient is insured. Insurers have established channels with the reimbursement programme to such handle claims. This should provide a simple process by which the patient can claim reimbursement expenses, and aims to ensure that patients do not have to provide any payment for included medical treatment. Any out-of-pocket arrangements can be claimed by the patient via the ASR help line.

Appeals

If a claimant is unhappy with the outcome of his/her reimbursement claim, he or she can contact Broadspire, who will review the claim. If still unhappy at the end of the internal review, the claimant can resort to legal action.

Time Limits on the Reimbursement Programme

The programme is currently open and will terminate at some point in the future.

Litigation

It will be interesting to see what the impact of the reimbursement programme is on litigation levels and on the overall amount of damages paid. It may be that once patients have had their treatment under the reimbursement programme, they do not then engage with litigation. Free health monitoring and treatment, plus payment of out-of -pocket expenses, may be sufficient redress for some claimants. It is too early to tell if offering this programme will reduce litigant numbers.

The reimbursement programme may impact on the average amount claimed per litigant but should not affect the aggregate amount payable to any affected patient. Claimants who have claimed for testing and treatment under the reimbursement programme will not include these elements in any litigation, reducing the per capita claim. Claimants may find the reimbursement programme preferable to pursuing litigation as they do not have to pay upfront costs and there should be no delay in receiving treatment. By contrast, under a court order, the litigant has to wait for the legal process to complete before payment is made, usually necessitating either upfront payment or a delay in treatment until completion of the process. These factors may make the reimbursement programme the first option for claimants, with litigation as a second string. Whether a claimant claims testing and treatment costs via the reimbursement programme or via a court order will make little difference to DePuy as the amounts claimed are quantifiable economic losses and the only differential will be the 'processing costs'. If, as would be expected, claimants do prefer to use the reimbursement scheme in addition to litigation, then a reduction in court settlement values would be expected. Therefore, court settlement 'headline figures' should not be used to determine the amount of compensation paid out by the manufacturer: more detailed analysis of both court and reimbursement programme awards would be needed for meaningful conclusions.

C. Conclusions

In some quarters, the reimbursement programme has been greeted with suspicion: it has even been suggested that it is an effort to 'buy off' patients cheaply.[27] It should be remembered that personal injury lawyers, who constitute some of the most vocal opponents, are well placed to understand the reimbursement programme, but would also potentially benefit from claims being processed by the courts. Interestingly, the reimbursement programme

[27] T Clarke, 'Insight: DePuy's Handling of Hip Recall Sparks Questions' *Reuters*, 22 August 2011, www.reuters.com/article/2011/08/22/us-depuy-idUSTRE77L35H20110822.

website also provides details of the US court settlement.[28] The fact that there is a clear presentation of the options open to patients provided within a single website, and there is no attempt to prevent patients claiming through the courts, seemingly indicates that the reimbursement programme is a genuine attempt to provide an effective, equitable service for patients.

It would be premature to comment on the impact of the ASR reimbursement programme on how the medical device manufacturers respond to recalls, but it is interesting to note that some other manufacturers, such as Stryker, have put similar reimbursement programmes in place after hip joint recalls.[29] Whether such programmes will be restricted to hip or joint replacements or will expand to incorporate other types of devices remains to be seen. One of the issues with ASR hips was the potential for latent damage, and it may be that such reimbursement schemes are used in these situations. The ASR reimbursement programme was field-leading, and may have set a new industry-wide standard of care following hip recalls.

[28] www.asrhipinfo.com/content/reimbursement-program. The court settlement will not be discussed in this volume.

[29] http://strykerhipreplacementhelpline.com/stryker-hip-replacement-reimbursement-program-offered-by-stryker/.

Part IX

Conclusions

SONIA MACLEOD AND CHRISTOPHER HODGES

33

Conclusions

This chapter sets out the conclusions of this review of personal injury compensation schemes. We first summarise key aspects of the various schemes described above, before noting some important features, similarities and lessons. We then note that a compensation scheme approach towards personal injuries may offer significant advantages, and chimes with the current strong move towards alternative dispute resolution (ADR) in Europe.

We draw on the findings of the current research to suggest best practice in relation to key features of how a contemporary compensation scheme should be designed, and what features may give rise to difficulties or are best avoided. These conclusions should found decisions on how to design a recommended model for personal injury compensation.

One of the key issues is terminology. What is meant by 'no-fault'? Is it simply that these are schemes where the decisions are not founded on 'fault'-based negligence criteria, such as is seen with workers' compensation or motor vehicle 'no-fault' policies? The term 'no-fault' has certainly been used (perhaps incorrectly) to cover far wider issues, such as the intention to introduce a 'no-fault' scheme for birth injuries in England under the NHS Redress Act 2006, with compensation based on an existing liability in tort. No-fault has been variously used to describe a system using a wider trigger, non-tort, or a different process and non-court schemes. In actuality there has never been a truly no-fault compensation scheme, as all schemes have sought to limit eligibility for compensation in some way, so retaining an element of 'fault'. As such it is a confusing term and not one we would recommend. The term 'no-blame' provides a clearer option.

A. Origins and Motivations of the Schemes

The compensation schemes described here are surprisingly diverse in scope, subject-matter and the details of their procedural operation. Hence, it is interesting to consider why and how they were created, as this goes some way to explaining the differences between them, because of what they are trying to improve, remedy and achieve. It also prompts thoughts on how some might be improved.

We start this overview analysis of the schemes studied in this book, therefore, by noting the reasons why the compensation schemes were created. After that we will be able to proceed to look at objectives and mechanisms. Our analysis is that some schemes were created to avoid litigation—whether part-way through litigation so as to provide an efficient short-cut to resolution, or as against the likelihood of future litigation, or more generally as a generically different means of resolving disputes that would otherwise end up in court. But the more important shift, evidenced in some major and comprehensive schemes, was not so much with the process (litigation or some other form of dispute resolution) as with a change away from tort liability to a wider basis on which to pay compensation—generically, a no-blame basis.

Schemes Created to Avoid Continuation of Litigation

Some schemes were created *after* legal liability had been established in litigation, or was accepted, so as to provide an administratively efficient procedure for the subsequent distribution of compensation to those entitled. Examples of this are Japan's Adverse Drug Reaction Scheme after a Tokyo court's ruling on liability for SMON injuries in 1978 (although the Japanese Adverse Drug Reaction Scheme was put in place as a retrospective measure, it now functions in a pre-emptive manner), and the UK's COPD and VWF schemes for miners. Some schemes are subsequent to court action but *without* a formal admission or acceptance of liability, for example the Skipton Fund, the HIV scheme and the vCJD Trust in the UK. It is worth remembering that the Thalidomide Trust, although arising from litigation, was established as a result of public pressure, in circumstances where a legal claim based on negligence was widely thought not to have any chance of success.

It is interesting to note that the German medical injury panels were created by the medical profession, with the support of the insurance industry, to counter the phenomenon in the 1970s that valid claims were not succeeding in court because medical experts were reluctant to criticise their peers in court. This phenomenon had reached the point that the professional ethical standing of doctors in the country was being undermined, and a resultant crisis in public confidence in the profession had to be addressed. An effective response had to be created, and the solution was to create a specialist medical ADR body that would have greater professional expertise in investigating a resolving medical liability issues, after which insurers could propose settlements with patients.

Schemes Created for Processing Specific Disputes

There are various examples of schemes that were created to process particular disputes, especially as a better way of doing so from the perspectives of both potential defendants and claimants than through court proceedings.

The phenomenon of manufacturers of products establishing voluntary redress schemes may have been increasing. The schemes are often established before liability has been tested in court. Major examples are the schemes from the medicines sector for Eraldin (1970s) and the medical devices sector for Trilucent breast implants and the ASR hip implant reimbursement programmes. Under these schemes, manufacturers may either admit liability or choose not to require the claimant to prove it. In some cases, it is questionable whether claimants would have been successful had they had to prove liability in court; the legal advice in the Eraldin case, for example, was that ICI would not have been found liable. The terms of the scheme can, by definition, be set by the manufacturer, and such terms may be criticised by claimants or their representatives. Indeed, although some litigation was associated with the Trilucent and ASR products, it is notable that the manufacturers were proactive and swift in creating the schemes. A comparison of the schemes described in this book indicates a clear shift away from the output of the scheme being to provide just money (as was the case 3 decades or more ago) to it being a focus on the needs of patients, and to become far less bureaucratic and much speedier in delivering patient access to expert medical assistance and in providing a swift and focused care package for patients, including advice, assessment, removal of an implant, replacement by another implant if desired (revision), and general care. Part of the rationale behind the ASR scheme was that latent damage from the implanted joints was hard to detect until it had reached clinically significant levels. The close monitoring of patients enabled earlier detection of damage occurring, and therefore enabled timely clinical interventions and reduced the clinical harm suffered by these patients. That very functional approach contrasts well with the ex post facto approach that is inherent in litigation, in which the remedy cannot by definition focus on providing care or service to the person who needs it but can only provide financial compensation after the event.

The same motivation to prevent litigation and deliver compensation can be seen in the UK industry arrangements (which exist in various other countries in similar form) for injuries sustained by healthy volunteers and patients involved in trials of new medicines, and by patients involved in clinical evaluation of new medical devices. The indemnity arrangements set out in industry guidelines, adherence to which has become required as part of the approval process, exceed the requirements of liability law, and are designed to make patients' recourse to legal procedures unnecessary. In practice, at least healthy volunteers in drug trials should automatically have the benefit of instant medical attention if they become ill during the trial, when they are closely monitored. It would be worth the industry reviewing whether developments could be made in provision of swift medical investigation, advice and treatment in other types of trials, to go beyond the mere provision of compensation.

It should be noted that although UK manufacturers have put in place voluntary compensation arrangements for clinical research and evaluation injuries, they continue to rely on tort-based litigation for claims over injuries arising post-marketing. In contrast, the traumatic Japanese experience with SMON prompted that nation to create an Adverse Drug Reaction Scheme for all post-marketing injuries of licensed products, as in the Nordic states and the French ONIAM model.

Schemes Created to Protect an Industry

In addition to providing compensation to individuals there are some injury compensation schemes which were created to protect an industry for the public good. The US vaccine compensation schemes limit the liability of vaccine manufacturers, which enabled the continued production of vaccines, and the obvious health benefits that accompanied mass vaccination. Similarly, the two US birth injury compensation schemes arose from a situation where insurance was becoming unaffordable for healthcare professionals. This created the risk of the healthcare professionals either withdrawing from practice, or working without insurance. Neither outcome was desirable for the public, so schemes were created that enabled doctors and midwives to continue their work, without unaffordable insurance bills. Both birth injury schemes were also hugely beneficial to claimants, providing prompt payment based on clear eligibility criteria, and removing the delays and difficulties associated with the tortious requirement to prove negligent fault.

The 9/11 schemes are the most recent example of a scheme created to rescue an industry from potentially crippling liabilities. Unusually for a personal injury compensation scheme, the 9/11 fund was created to support the airline industry, rather than a healthcare-related industry. Interestingly the beneficiaries of keeping the US airline industry flying were a very wide and diverse group, and those who received compensation from the 9/11 fund were not necessarily direct users of the US airline industry. The 9/11 fund provided compensation to individuals who were hurt or killed by the terrorist attack, a situation that would usually be excluded from standard insurance policies, leaving the victims with little chance of compensation from insurance. The creation of the 9/11 schemes both protected the airline industries and also provided an extremely efficient compensation mechanism in circumstances where the entitlement to compensation was unlikely to be clear cut.

Schemes Created to Reduce the Burden on the Courts

In looking for motivations for the creation of injury compensation schemes, it should be remembered that there are circumstances where schemes have been created to reduce the potential burden on the litigation system. The UK coal health claims were of such magnitude that it was thought that they would swamp the courts. An alternative scheme was proposed with the aim of removing these claims from the litigation system. This was perceived to have the dual advantages of removing a mass of litigation that would have clogged up the courts for years, and an administrative scheme was thought to provide a faster, more efficient system for claimants. The overriding motivation was not to make an ideological move away from a tort-based decision-making framework, but to move the claims out of the tort system. As a result, the scheme was designed by tort lawyers. Although the scheme achieved its objective of keeping these cases out of the courts it is arguably one of the less successful of the schemes detailed here.

The Irish Injuries Board is another scheme that was intended to reduce the burden on the courts. Interestingly, by reversing the usual order and dealing with loss quantification before attributing liability, the Irish Injuries Board has reduced the number of cases ending in the courts. Once the quantum is established the parties to the case appear to be more willing to settle rather than to go on to fight over liability. While this is a surprising model it seems to meet the brief of reducing case numbers.

Schemes Created to Replace Tort-Based Liability

In looking for motivations for the creation of injury compensation schemes, focusing too much on the court/litigation process appears to be misleading. By far the more powerful motivation has been the desire to switch away from tort as the trigger of liability to a system that has a wider and less onerous gateway and provides a less adversarial environment for the parties. For example, historically dissatisfaction with the scant opportunities for recompense for injured workers led to the widespread development of workers' compensation schemes. In common law jurisdictions the 'unholy trinity' of employer defences often prevented workers from recovering damages in tort.[1] The resulting dissatisfaction with the tortious arrangements[2] culminated in the introduction of workers' compensation. Interestingly the current workers' compensation arrangements in Australia are administrative and non-adversarial and more akin to the civil law tradition, in contrast to the traditional adversarial common law approach in which they evolved. The same is obviously true for the arrangements in New Zealand.

This move away from a narrow adversarial tort system clearly drove the shift in all the Nordic states to create a no-blame patient insurance (after which the smaller no-blame drug schemes also followed). The medical professionals who administer the patient schemes state strongly that it is vital for the safety of their national healthcare systems that healthcare professionals and patients should operate in an honest and open relationship, without blame. Such an open culture, as operates in civil aviation safety,[3] means that information is shared without a second thought and never hidden because of fear of liability or other adverse criticism. The point was made strongly that looking for scapegoats through a legal system based on fault is 'a very efficient way of killing more patients'.[4] A no-blame culture is viewed as vital for success in being able to learn and improve professional practice in healthcare. The results quoted here, such as in the reduction in neonatal neuropathy, are simply stunning.

The Nordic chapters also show that those states have a consistent holistic approach based on social solidarity, which extends laterally across different injury types; a no-blame approach applies in the related systems for workers' compensation, and for motor vehicle and medical injuries. The same ethos of social equality drives the New Zealand scheme. The original scheme covered earners, non-earners and motor vehicle injuries; the inclusion of non-earners highlighted the scheme's social equality credentials—inclusion was not based on payment by the individual and/or employer. Subsequently the ACC's remit has increased to include treatment injuries—initially on a fault basis (medical misadventure), then on a no-blame basis (treatment injuries). This provides New Zealanders (and visitors to New Zealand) with a broad spectrum of coverage for the majority of injuries. For many years, it may have been fashionable for outsiders to dismiss the no-blame compensation arrangements in New Zealand and the Nordic states, but they continue to perform their functions and to work well.

The social solidarity rationale was clearly a major idea in the creation by France of the ONIAM, later copied in Poland. The same idea runs behind providing compensation for military personnel who are injured, or for victims of criminal assaults, or for victims of various other particular traumatic events (HIV, hepatitis C, vCJD in the UK).

[1] The defences of contributory negligence, voluntary assumption of risk and the fellow-servant defence (that an employer could not be found liable for damages arising from the actions of a fellow servant) reduced the chances of a successful claim following an injury at work. See PV Fishback and SE Kantor, *A Prelude to the Welfare State: The Origins of Workers' Compensation* (Chicago, University of Chicago Press, 2000) and KS Abraham, *The Liability Century: Insurance and Tort Law from the Progressive Era to 9/11* (Cambridge, MA, Harvard University Press, 2008) 44–45.

[2] Abraham, ibid, 23.

[3] RL Helmreich, 'Building Safety on the Three Cultures of Aviation' in *Proceedings if the IATA Human Factors Seminar* (Bangkok, 1999) 39–43; S Dekker, *Just Culture. Balancing Safety and Accountability* (Aldershot, Ashgate, 2007); D McCune, C Lewis and D Arendt, 'Safety Culture in Your Safety Management System' in AJ Stolzer, CD Halford and JJ Goglia (eds), *Implementing Safety Management Systems in Aviation* (Farnham, Ashgate, 2011).

[4] Statement by members of the Swedish Patient Insurance to C Hodges in 2012.

The history of a succession of vaccine schemes shows consistently that a principal motivation was to create them so as to avoid litigation against the manufacturers, since the potential litigation cost risk in the USA was both high and uncertain. That commercial risk, manifested in the unavailability of insurance and a drastic fall in the number of manufacturers who were willing to make vaccines, threatened the continued supply of vaccine after the number of manufacturers had fallen dramatically as a result of litigation. Further, considerations of compensating patients *post harm* through the litigation system was not thought to be adequate to overcome families' and parents' concerns over the risk of being vaccinated against very unpleasant diseases. That fear, which also applied elsewhere (illustrated by the UK), created a serious risk to the safeguarding of public health should the vaccination rate fall. Creating a compensation scheme was a means of both limiting the liability exposure of manufacturers and maintaining public confidence in vaccination programmes. In short, the real consideration driving the vaccine schemes was to maintain public health and confidence in the safety of vaccines in order to maintain a high vaccination rate by the general population. Many countries followed the US precedent and created national compensation schemes—and in many of those states, the litigation risk was nowhere near as high as it was in America.

The main point behind the Japanese drug ADR scheme was to maintain the confidence of the population in the safety of medicines and the regulatory system. Likewise, the pharmaceutical and medical technology industry's concern to support public belief in the safety of clinical research, and hence a pipeline of beneficial medicines, albeit also with advantageous commercial implications, drives their spontaneous introduction of voluntary compensation arrangements for participants in clinical trials on medicines and clinical evaluation of medical devices.

Another clear example is the 9/11 Fund, which was created partly with widespread popular support, but expressly in order to rescue the US airline industry—the impending commercial consequences of litigation would have been catastrophic, particularly since, at that time, people had completely lost confidence in flying, reducing industry income.

Conclusion: A Shift from Tort to Providing Compensation on a Wider Social Basis

The main conclusion of this analysis, therefore, is to note the strong and widely experienced shift towards no-blame, liability systems, ie compensation arrangements that left behind the need to establish fault in order to trigger eligibility for compensation for injury. It is well known that this shift occurred in New Zealand in the 1960s and 1970s. It may be less known how pervasive the no-blame approach has also been throughout the Nordic states, encompassing not just patient injuries but also injuries to workers and road traffic injuries within an integrated holistic system. What we have also shown is that no-blame has a hold right across the world in virtually all vaccine injuries, and across Australia and the USA in relation to workers' compensation and road traffic injuries, where schemes have been evaluated on their successful operation and design, rather than on ideological grounds. The only relevance of fault in the French ONIAM scheme is to identify which insurer will pay: the coverage of the scheme is ideologically based on social solidarity and wider coverage than tort.

In the UK, the number of individual schemes that exist for particular injuries has grown considerably: industrial injury disablement, armed forces compensation, criminal injuries compensation, clinical research, thalidomide, coal workers' pneumoconiosis, COPD and VWF, hepatitis C and HIV, and vCJD. It is a long list, and individual schemes have been created piecemeal since the 1960s, almost unnoticed, and they are all different. This leads to confusion over their differing individual scopes, procedures and entitlements, and even a lack of knowledge of their existence. It can also lead to overlaps. Scholars have long understood that tort plays no practical role in the resolution of injury claims (motor vehicle and workers) in the UK because the reality is an administrative practice by insurers that operates largely on causation of injury and does not evaluate whether there was negligent conduct.[5]

B. Structures of the Schemes

First we note some basic points about the structural design of the various schemes. To some extent, this analysis repeats that of the first section in this chapter, where the focus was on the *origins* of the schemes, but it is important also to analyse, as we do here, the context of how *in fact* they operate structurally.

We see a spectrum of models. At one end, there is the paradigm of an adversarial process and binding decision by a judge—which we take for granted will be familiar to readers, so will not analyse it here. There are then various forms of ADR, in which the decision-making tribunal is no longer a normal court, and may be a new specialist tribunal.

Unlike many manifestations of ADR in the context of other types of disputes,[6] we have found almost no reliance on mediation in relation to the resolution of disputes involving personal injury. We found one specific reference to this expressed by German experts, who felt that the imbalance of power undermined the ability of the parties to trust each other, and in any event was irrelevant to the only outcome that their system could deliver, which was money.

Again negotiation, while a common tool in ADR (in practice this is usually a negotiation between the claimant, or the claimant's lawyer, and the defendant's insurer) is not really seen in the schemes we have examined. There may be a variety of factors behind this. Firstly, it is likely to be due to the fact that compensation is often tariff-based and less discretionary. Secondly, schemes are usually independent of the payer, so are unlikely to try to minimise the compensation paid out, removing the need for negotiation.

There are, however, specific examples of arrangements for negotiation, conciliation and arbitration of claims, such as in the Australian workers' compensation architecture. Australian workers' compensation is one of the oldest schemes in existence, which may explain the more traditional framework. Interestingly conciliation and negotiation only

[5] R Lewis, 'How Important Are Insurers in Compensating Claims for Personal Injury in the UK?' (2006) 31 *Geneva Papers* 323; R Lewis and A Morris, 'Tort Law Culture: Image and Reality' (2012) 39 *Journal of Law and Society* 562.

[6] S Roberts and M Palmer, *Dispute Processes: ADR and the Primary Forms of Decision-Making* (Cambridge, Cambridge University Press, 2005); C Hodges, I Benöhr and N Creutzfeldt-Banda, *Consumer ADR in Europe* (Oxford, Hart Publishing, 2012).

applied to certain claim elements. The compensation tariff is fixed by statute, so is not negotiable. The decision on medical causation is governed by external medical panels or specialists, and I therefore not subject to negotiation or conciliation. The only elements which are negotiated or subject to conciliation proceedings were return-to-work timings and arrangements which need to be put in place to facilitate a return to work. Hence, although they are present within the scheme, the scope of negotiation and conciliation is limited.

It is interesting, and somewhat counterintuitive, to note that conventional ADR does not seem to apply to these schemes. However, ADR is a broad term, with two major components. Firstly, ADR techniques, namely the strategies, such as mediation and negotiation, used to resolve disputes. Secondly, the shift away from using litigation and courts to resolve disputes by creating ADR mechanisms or mechanics. A more nuanced analysis of these schemes indicates that ADR techniques are not required within an ADR mechanic. ADR techniques evolved to provide a less adversarial, more remedy focused option within the litigation framework. Given that the entire procedure within an ADR mechanic is non-adversarial and remedy focused, it appears that ADR techniques have little to contribute. Other ADR mechanics, such as the Legal and Financial Ombudsman in the UK, do not rely on ADR techniques, indicating this is not specific to injury redress schemes.

In models where the context has largely shifted away from the courts into its own separate world, such as an administrative world, many different structural variations may occur. Schemes take a plethora of different legal forms including private (not-for-profit) companies, charities, trust funds, governmental departments and arm's-length governmental bodies, which can have implications for the way in which the scheme functions.

The Main Structural Contexts

Reforms that Maintain the Context of the Courts

Some schemes are designed to operate alongside the courts, or even to be supervised by a court. The Florida Birth Injury Fund requires that a claim be lodged in the Administrative Court; the eligibility decisions and compensation quantum are then decided administratively and outside the court system by the Florida Neurological Injury Compensation Association, and the Administrative Court then approves (or amends) the proposed settlement offer.

Sometimes, the scheme merely operates by changing the context from a normal courtroom to a special court or tribunal, such as with the US Vaccine Court. An advantage of this approach may be that the procedure or atmosphere might be less formal and intimidating for a claimant; interestingly the US Vaccine Court makes little attempt to dispense with lawyers (and specifically allows them to be reimbursed for their involvement). One of the major reasons why their administration-to-compensation ratio appears favourable despite the inclusion of substantial legal fees is that compensation payments are often very high value (the majority of claimants are children, often with lifelong claims). The US vaccine scheme has been criticised for being quasi-judicial in process and limited in scope, but despite this it remains essentially an adjudication-based model.

A similar shift is to transfer a dispute from civil procedure to an arbitration-like procedure. An example of this is the German medical injury tribunals, which cover the entire country and handle many medical negligence claims. As has occurred in many court-based litigation processes, ADR was introduced as a preliminary stage, in the form of the specialist medical tribunal. Roughly the same number of claims are sent to the medical injury tribunal as to the courts. However, it appears that, in at least one region, 15% of tribunal decisions are not accepted and return to the courts, of which an alarmingly high 42% are decided differently. The tribunal model may provide less public embarrassment to doctors, but it still applies the law.

Ireland adopted a different approach to modifying the litigation process, by creating a specialist state body to assess the quantum of a claim—assuming liability—as a result of which insurers have been considerably swifter to negotiate settlement than before, when both liability and quantum were in issue. Although it may seem a somewhat illogical step to resolve quantum first, the most difficult parts of disputes to resolve were shown to be about the uncertainties of agreeing quantum. This mechanism addressed the local issue, though it is questionable how transferable this solution is.

Australian workers' compensation has shifted the procedure out of the courts and into specialist agencies. This applies across all the different states and territories studied. One of the major issues with federal rather than national statutes is that there are some regional discrepancies in eligibility and coverage. The procedure varies slightly according to the region, but each one is essentially a specialist ADR mechanic operating under and applying a statutory framework.

Insurers and Insurance Intermediaries

In standard insurance contracts the insurers operate theoretically under the continuing influence of the courts, since the fallback position from failure of the parties to reach agreement is for a case to be able to be brought in the civil procedure system.

The fact that insurers settle claims directly with those who have suffered losses, or with their lawyers, is far from a new phenomenon. However, it deserves some fresh thought in the context of considering who the intermediaries are that have power to settle disputes. Traditionally there are opposing insurers. This raises two types of questions over the role of insurers and loss adjusters: which mechanism is more efficient overall (is it cheaper to have a single independent intermediary or two or more?) and are the intermediaries sufficiently independent (or are they essentially partisan)?

However, in the context of no-blame schemes the insurers do not fulfil the traditional tort-based adversarial role. For example, Australian workers' Compensation schemes are underpinned by statutes, which clearly define the extent of coverage, the benefits available and the process to be applied. Tort-based litigation is not the fallback position for the insurers; instead the insurers are governed by statute. This provides a very clear, unambiguous framework without discretion. In New South Wales, for example, the claim is administered according to the statutes by icare NSW. icare NSW can serve as either the insurer (if the worker is insured by it since March 2017) or as an intermediary (if the insurer was insured by another insurance provider prior to March 2107). For claims under older policies the combination of a strict non-adversarial statutory framework and an independent decision-maker ensures that the system is impartial and non-partisan. For claims under more recent policies offered by icare the strict non-adversarial statutory framework and the clearly defined benefits should remove discretion in the way in which the scheme is administered and in the available benefits.

Interestingly, claims to the Swedish, Norwegian and Finnish insurances for medical/pharmaceutical injuries are also handled directly by the insurers, without the use of an intermediary. Several factors may influence why this appears to function well in these countries. The schemes have very detailed rules, some statutory, some voluntary, but all very clear and widely understood. The compensation is tariff-based; the value of compensation is solely governed by the tariff, not by circumstances. There is a strong culture of settling reasonable claims. Finally, these schemes act to top-up social security payments; in a conventional insurance setting any reduction in compensation is a saving for the insurer. In the Nordic states, where the insurer only pays out a low percentage of the overall compensation package, then any saving is a reduction of a low percentage of the total. In a conventional insurance situation if the total compensation is £100,000, then saving 10% of this would be a gain of £10,000 for the insurer, but if the insurer only pays the last 20% of the compensation, then a 10% overall saving produces a far more modest £2,000 saving for it.

Administrative Schemes: Reforms 'Outside the Box' of the Litigation Framework

The creation of administrative compensation schemes constitutes a major reform 'outside the box' of the litigation framework. Whilst courts may have some limited ability to review procedure, or the ultimate unreasonableness of an administrative decision, the administrative entities essentially exist in their own new universe, largely separate from courts. For example, in Australia workers' compensation arrangements are largely laid out in statute, but are dealt with administratively outside the court system, but with recourse to appeal to the courts if necessary. The same is true of the birth injury scheme in Virginia, which is administered by the Virginia Workers' Compensation Commission.

In New Zealand, no-blame compensation began as an attempt to provide a cheaper, swifter and more effective means of providing compensation for injuries than through the court system based on proof of negligence under tort law, or related legal liability regimes. Its origin in 1974, after the 1967 Woodehouse Report, was at a time when the European response to thalidomide was moving in the different direction of introducing strict liability for defective products.[7] Indeed, the Accident Compensation Scheme has essentially replaced litigation for personal injury compensation in New Zealand.

Nordic states introduced a triple-pack of compensation schemes for road traffic, medical injuries and drug injuries. The three elements were always established in that sequence, with road traffic first and drugs last. The existence of a pre-existing model, and of a general Nordic approach to social solidarity that is evidenced by high levels of taxation that support social security of various kinds, was a prerequisite for the establishment of the medical and then drug insurance schemes. Equally, all the schemes fit into a holistic national and Nordic model, even if there are some differences of detail.

The first initiatives in the Nordic states often occurred in Sweden, which the other states largely copied, usually quickly. The Swedish Drug Compensation Scheme was introduced voluntarily in 1978 in response to a government threat to introduce a statutory scheme. The shape of the Danish Drug Scheme of 1995 was influenced by views on how EU law (the Product Liability Directive) affected it, and resulted in all compensation being funded by the state.

[7] Harmonised by Directive 85/374/EEC.

Japan's scheme arose from 1978 litigation over SMON, but has expanded to encompass post-hoc compensation for contaminated blood and, more recently, to incorporate pre-emptive compensation for injuries caused by pharmaceuticals and biologicals. The combination of the horrendous injuries in the SMON case and the widespread scandal caused by the use of contaminated non-heat-treated blood well into the 1980s, when other countries were warning of the dangers of using such blood, led to public dissatisfaction. As a result PMDA administered post-hoc compensation for these groups, and then made the move towards a more general pre-emptive scheme.

The introduction of the ONIAM scheme in France in 2005 was a revolutionary change for that country. Its origins lay in the congruence of number of different factors, notably a 'patients' revolt' against the traditional autocracy of doctors, fuelled by dissatisfaction at the lack of candour (or worse) in a series of health scandals such as HIV/AIDS, and the desire to support a new culture of sharing of information and decision-making in treatment decisions, supported by a 'social solidarity' approach to treatment injuries and redress. Another strong feature was dissatisfaction with victims having to struggle to prove fault in a complex system with varying liability rules in administrative and civil courts. It is important to note the French reliance on their policy of social solidarity, in providing access to compensation (note: not access to justice) for the victims of medical accidents.

In Poland, a major driver for the new 2011 scheme was simply that court cases can take 10 years, and the success rate for medical claims is lower than the average for all claims (40% as opposed to 60%), indicating the classic difficulties in obtaining evidence to substantiate the existence of fault.

C. Comparative Statistics

We set out here some of the core statistics: see Tables 33.1–33.5. As the schemes are different in their operation and the way in which they classify costs, these statistics should not be used to make direct comparisons between schemes, but they serve as a useful indicator of the performance of a scheme.

To put these figures in context it should be remembered that these schemes generally replace tort proceedings. For example, in the UK the largest defender of clinical negligence claims is the NHSLA, the National Health Service Litigation Authority. NHSLA received 10,965 new clinical negligence claims in 2015/16 and for claims valued at under £100,000 average claimant legal costs as a percentage of the claim value are 55%.[8]

Where a scheme is well established and numbers are consistent from year to year, such as the Nordic schemes and the New Zealand ACC, data from the most recent available year has been used. As ONIAM has a shorter history, and earlier years included start-up phases, data from 2012 has been used. Publically available English-language information on the administrative costs of the Japanese PDMA is limited, so costs have been calculated using actual compensation payments from 2009 to 2012 and estimates of administrative costs take from the strategic plan for 2009–13. PMDA is outperforming the strategic plan, so the

[8] NHSLA Annual Report and Accounts 2015/16.

latter figures will be conservative. The UK Vaccine Damages Payment Unit handles very small numbers of claims, so an average of 5 years has been used.

Table 33.1: Nordic adverse drug reaction schemes

	Sweden	Finland	Norway	Denmark
Average number of annual claims	600	220	50	200
% of claims per head of population	0.01	0.00	0.00	0.00
% of successful claims	30	50	40	30
Number of claims paid per year	200	125	20	50
% of claims paid per head of population	0.00	0.00	0.00	0.00
Average compensation per claim (€)	18,000	10,000	50,000	30,000
Total cost of compensation (excluding admin costs) paid per year (€ million)	11.87	0.89	1.33	1.9
Total contributions as % of national sales turnover	0.4	0.1	0.3	—
Cost of administration of the scheme per year (€ million)	1.5	0.15	0.15	0.265
Cost of total compensation and administration per head of population per year (€)	1.21	0.3	0.85	0.41
Administration costs as a percentage of payments	11.22	14.42	10.14	12.24

Table 33.2: New Zealand ACC

	2016
Number of annual claims	1,933,629
% of population compensated	30.8
% of successful claims	
Number of claims paid per year	
% of claims paid per head of population	
Average compensation per claim (NZ$)	
Total cost of compensation (excluding admin costs) paid per year (NZ$ million)	6,119
Total non-governmental contributions (NZ$ million)	2,851
Cost of administration of the scheme per year (NZ$ million)	494
Cost of total compensation and administration per head of population per year (NZ$)	
Administration costs as a percentage of payments	8.1

Table 33.3: ONIAM

Year	2012[a]
Number of annual claims	4,255
% of claims per head of population	0.01
% of successful claims	35.89
Number of claims paid per year	1,527
% of claims paid per head of population	0.00
Average compensation per claim (€)	72,986.25
Total cost of compensation (excluding admin costs) paid per year (€ million)	111.45
Total non-governmental contributions (€ million)	
Cost of administration of the scheme per year (€ million)	21.24
Cost of total compensation and administration per head of population per year (€)	132.69
Administration costs as a percentage of payments	16.01

[a] 2012 figures have been used as prior to this ONIAM was starting up and so the figures from earlier years would not be very representative.

In comparison, the claim numbers for the Polish scheme were 432 in 2012, 1310 in 2013, 1197 in 2014, and 1030 in 2015.

Table 33.4: PDMA adverse drug reaction account

Number of annual claims	1,106
% of claims per head of population	8.66151×10^{-6}
% of successful claims	84
Number of claims paid per year	929
% of claims paid per head of population	7.26979×10^{-6}
Average compensation per claim (¥)	2,054,424.61
Total cost of compensation (excluding admin costs) paid per year (¥ million)	2323.8
Total non-governmental contributions (¥ million)	4082
Cost of administration of the scheme per year (¥ million)	740.2
Cost of total compensation and administration per head of population per year (¥)	4899.4
Administration costs as a percentage of payments	15.11

Actual figures for 2009–12.
Taken from the Mid-Term Plan 2009–13.

Table 33.5: US and UK vaccine schemes

	US VCIP 2010–13	**UK VDPU 2006–12**
Number of annual claims	396.4	69.50
% of claims per head of population	1.24618×10^{-6}	0.00
% of successful claims	59.54	2.40
Number of claims paid per year	236	1.67
% of claims paid per head of population	0.000124618	0.00
Average compensation per claim (US$ or £)	747,767.82	96,100.00
Total cost of compensation (excluding admin costs) paid per year: (US$ or £ million)	179.91	1.60
Total non-governmental contributions (US$ or £ million)	60.78	0
Cost of administration of the scheme per year (US$ or £ million)	16.62	?
Cost of total compensation and administration per head of population per year (US$ or £)	0.62	?
Administration costs as a percentage of payments	9.24	?

D. Analysis of Major Common Features and Findings

Principal Findings

This section gives an overview of the principal findings of the schemes considered above. It focuses first on the broad general picture, and then in separate sections on the origins of schemes, claim durations, what payment triggers are used, whether they are public or private bodies, what procedures are adopted, and how they are funded. We end by noting some conclusions and arguments on no-blame schemes.

Firstly, more 'no-blame' or administrative compensation schemes exist than is widely recognised. They have been created by both states and private enterprises, both permanently and to handle particular classes of injuries or event.

Secondly, there is a *huge variation in the number of claims* handled by different schemes. For example, the UK vaccine scheme pays 1 or 2 claims a year, the Florida birth scheme a couple of hundred, the Nordic drug schemes vary between 50 and 600 a year by country, whereas the New Zealand comprehensive accident scheme pays 1.7 million claims a year.

Thirdly, the administrative costs of all compensation schemes are vastly lower than the tort litigation system. Tort litigation usually has legal fees of around 55% of total compensation value awarded, often higher. Although administrative schemes are variable in their administrative costs, as a percentage of redress awarded they are hugely more efficient than tort. The larger schemes, such as the Nordic patient injury schemes, run in the low teens;

the New Zealand scheme is similar. In the USA the administrative costs of the Florida and Virginia birth injury schemes are around 5% each, while the 9/11 fund was phenomenally low at just 1.3%.

Fourthly, administration costs are affected by the extent to which work has to be done on assessing applications. In some schemes, a large proportion of applications are rejected. In other schemes, the extent of scrutiny is less, such as where coverage is automatically wider, eg New Zealand, so less time is spent in assessing whether criteria, such as causation, are met.

Fifthly, *economies of scale* and fewer demarcation problems exist where schemes cover several injury types—such as road traffic, workplace, and medical treatment injuries—within the same broad administrative structure.

Sixthly, wherever different injury types are processed within the same administrative structure, different *financial pools* are usually maintained, so that discrete groups of funders are pooled together, thereby limiting cross-subsidisation, enabling pricing to be more closely linked to individual risk groups.

Seventhly, schemes apply different approaches to 'triggers', which can significantly affect the extent to which they accept or reject applications. This variation, including additional filtering, can clearly be seen in the Nordic schemes: the underlying trend in adverse effects ought to be similar between those four states, but there is a variation in the extent to which each country accepts claims. In other words, schemes can control the flow.

Eighthly, the impact on applications received by each scheme is harder to base on any single factor—awareness, cultural expectation, alternative options, etc, can all play a part. But given that vaccines would presumably have adverse events at similar rates, it is interesting to compare the UK and US vaccine schemes. The UK levels are substantially lower, probably due in part to the gating conditions, which signal to applicants, medics and lawyers that there is no point applying for a scheme that has a low acceptance rate or that you are not eligible for.

Claim Duration

The use of administrative schemes provides relatively quick resolution, as can be seen from the illustrative examples below. For a more detailed description of any of the schemes, see the individual scheme chapters.

Almost all of the statutory schemes have defined claims-processing time limits. For example, NICA, the Florida Neurological Injury Compensation Association,[9] requires that a hearing must be arranged between 60 and 120 days from the filing of a claim,[10] and a determination on eligibility under the programme and levels of compensation due follows rapidly. The Virginia Birth Related Injury Fund[11] determines claims within 6 months of them being filed with the Virginia Workers' Compensation Commission. It is interesting to compare these US schemes to litigation in England. For maternity claims brought

[9] www.nica.com.

[10] S 766.305 Florida Statutes.

[11] www.vabirthinjury.com.

in England, the average time from incident to resolution is 8.57 years for cases valued at over £1 million; for lower-value claims it is 4.32 years.[12] This is painfully slow. It has been argued that the complexity and high value of maternity claims necessitates a long timeframe, but the US examples show that resolution can be achieved more rapidly using a different framework.

Under the largest scheme studied, the New Zealand scheme, the average time to make a coverage decision in 2014/15 was 1.2 days, average time to commence payment was 9.2 days, and the 'return to work' rate was 67.3% after 70 days and 93.2% after 273 days.

Claims under most schemes are promptly processed—for example, the investigation process under ONIAM should not exceed 6 months, and most claims are then settled quickly by insurers. However, caution must be used as statutory timeframes are not always followed. For example, the Polish Committee must make a decision within 4 months of receiving the complaint (although in practice delays are reported), and a hospital or insurer must propose an amount of compensation to the claimant within 30 days, which the claimant then has 7 days to accept. While this should lead to prompt resolution it seems that other elements of the Polish system are not sufficiently strong in supporting the making of sensible settlement offers by hospitals and insurers, and these timeframes are not actually realised. While the situation in Poland is not ideal, the scheme is a replacement for litigation which was taking the best part of decade; so, while by no means perfect, the scheme does seem to be an improvement.

Some schemes are much slower; claims to the vCJD Fund in the UK take between 5 and 10 months to process, using a law firm as secretariat to collect a letter from a university for submission to a trustees' meeting.

Public or Private Administration Body?

Examples exist of both public and private compensation administration bodies. The largest of the schemes considered here, ACC in New Zealand, is a Crown organisation, governed by statute, and is therefore a state entity.

The Nordic states have a mixture of models, detailed in Table 33.6. It can be seen that the private model is overwhelmingly used in the Nordic states. The private bodies there are not-for-profit, operating under established criteria (maybe set by law). There seems to be an absence of objection to the fact that the entities are private not public. The conclusion is that the most important criterion is to maximise not just the independence of the body but also the perception of independence.

Although a private company, it should be remembered that LÖF in Sweden is a mutual insurer, so is in fact owned by the regional health authorities. It is neither entirely private nor entirely public.

[12] 'Ten Years of Maternity Claims. An Analysis of NHS Litigation Authority Data' (NHS Litigation Authority, October 2012) 11.

Table 33.6: Public and private compensation administrative bodies in the Nordic states

Public bodies

- The Norwegian Patient Compensation System (NPE) is a public body under the Ministry of Health and Care Services that handles injuries caused by both healthcare and medicines. Payments for public healthcare are made by public providers as contributions on an annual pay-as-you-go basis; private insurers pay for private healthcare; the NPE informs the Norwegian Drug Liability Association what it has to pay.

Private bodies

- The Swedish Patient Association (LÖF is a mutual insurer), subject to statutory requirements.
- The Swedish Pharmaceutical Insurance Association (LFF) acts through a captive insurance company (SLF).
- The Danish Patient Compensation Association (Patienterstatningen) operates on an administrative inquiry basis and is required to be established jointly by the public regional authorities and the private health insurance companies under statutory requirements, even though compensation and administration are funded by the state; payments are made on a single lump sum basis.
- The Finish Patient Compensation is a private body subject to statutory requirements and funded by public and private entities.
- The Finish Mutual Insurance Company for Pharmaceutical Injury Indemnities is a private captive owned by the Finnish Cooperative for Indemnification of Medicine-Related Injuries, which outsources claims handling to the British loss adjuster Crawford.

A number of schemes are public/private hybrids, such as icare in New South Wales, which is a public financial corporation governed by an independent board of directors. Likewise, the Skipton Fund Ltd (which provides support for people infected by hepatitis C virus from contaminated blood in England) is a private company limited by guarantee (meaning there are no shareholders), but the board is populated by the Department of Health.

Interestingly, the function of a scheme does not necessarily determine the scheme structure. In the UK compensation for HIV and hepatitis C virus infection from contaminated blood is administered by both not-for-profit limited by guarantee private companies and by charities. While the function of a compensations scheme clearly does not determine its form, the form does nevertheless impact on the way in which compensation is delivered, with support from charities being entirely discretionary.

A number of the schemes set up to respond to a specific need, such as the vCJD Trust, are trust funds, while others are charities, such as the MacFarlane Trust for HIV-infected haemophiliacs in the UK. Charities and trust funds are seen only as a post hoc response to the needs of a defined class of beneficiaries (though the identity of the beneficiaries may not be fully known at the time of the charity/trust fund creation).

Procedure

Decisions by Experts

Being able to have a decision on causation, eligibility or quantification taken by involving appropriate experts in the process in the most efficient manner is one of the clear benefits of an administrative scheme. However, not all schemes make these determinations in-house; some schemes outsource these decisions to external experts.

The single largest scheme considered here, ACC in New Zealand, takes a very different approach to this compared with other schemes. ACC deals with high claims volumes, with lower levels of scrutiny for individual claims. Essentially the causation and eligibility requirements have been largely stripped out of the New Zealand system; in the majority of circumstances, the question is simply one of injury quantification. Thus, ACC does not need to make detailed assessments on causation and eligibility in most cases; but in cases where these are required they are done by ACC itself.

The Nordic schemes usually have an individual investigator who assesses the merits of the claim. These schemes tend to have an element of avoidable harm, so require a determination on this. Thus, the individual investigator will examine if the injury type is eligible (eligibility) and then if the injury was avoidable (in effect causation). Quantification is determined by statute. The Nordic schemes have access to in-house medical expertise, as well as to a pool of external specialists when they need them. These schemes tend to make all of the decisions on causation, eligibility and quantum within the scheme.

Interestingly, some of the schemes studied do not routinely make determinations on any of these issues. The use of external expertise minimises the requirement for the scheme itself to determine these issues. For example, in Western Australia the claimant must attach a First Medical Certificate from his treating practitioner. This must outline the diagnosis of the injury and the likely impact of the worker's injury on his ability to work, and plans for future appointments and treatment. Essentially the scheme, WorkCover WA, has outsourced the information gathering to an external expert, the treating medical practitioner. If at any point in the process there is a question of the extent of the worker's injury or his capacity to work, then a Medical Assessment Panel can be convened to determine this—again an outsourcing of the decision on the extent and consequences of the injury. The decision on liability then rests with the insurer, who has to decide whether to accept or dispute liability. Again the insurer is an expert in liability, albeit one with an interest in the outcome of the liability decision. In cases where liability is accepted the process is prompt and the scheme itself does not make a judgment on causation, liability or quantum (quantum is outlined in the statutes). It is only where liability is disputed that WorkCover WA becomes involved in determining a claim. This use of external expertise minimises the requirement for the scheme itself to make determinations on these issues in the majority of cases.

Other schemes use external expertise to determine eligibility. For example, the Virginia Birth-Related Neurological Injury Compensation Program has statutory authority to convene a Medical School Panel, which is tasked with determining whether the claim 'satisfies the requirements of the legislation'. Effectively the causation and eligibility recommendations rest outside the scheme administering the payment.

The schemes in France and Poland can be used to highlight how care is needed if the decision on quantum is seated outside of the administrative scheme. There is one highly illuminating difference between the French and Polish patient schemes. Both schemes require the insurer to make an offer within a certain time. But in the former, the insurer is required to make a reasonable offer, and failure to do that should lead to quite serious consequences for the insurer. That attempt to regulate quantum is absent from the Polish system. Not surprisingly, it appears that the Polish system suffers from offers being made at an undervalue by insurers. That leaves the patient—whose claim has already been found to be valid—needing possibly to seek legal advice and then to fund and institute legal proceedings. This is clearly unsatisfactory and could potentially be resolved by providing an appeal stage to a specialist tribunal, at no or limited cost to the claimant. The information on the French system appears to be that it works well. It compels or encourages insurers to make accurate and fair offers, and to do so quickly. Indeed, insurers take a close interest in cases during the earlier stages, so should normally be in a position to make a fair offer quickly. The difference between the functionality of these two very similar schemes indicates that where quantification is outsourced, care must be taken to ensure that the claimant's interests are adequately protected. However, innovations that aid the process are, first, if the decision on causation (if required) and entitlement can be made by an expert third party to the claim. This third party can be within the compensation scheme or independent of it. If the decision rests within the scheme, then it is beneficial if the same administrator who is handling the case (or an adjacent colleague) can take the decision. This decision may require specialist input, for example from medically qualified staff. From a scheme perspective, it is likely to be beneficial if these decisions rest within the scheme as this allows the scheme to control claim acceptance and ensure consistency of decisions.

Second, if as much as possible of the compensation award can be made by reference to an agreed or official tariff, this minimises any arguments over quantification and provides certainty for all parties. Also, it is key that the scheme is able to provide oversight on the decision on quantum if there is discretion in the level of award available. The situation in Poland demonstrates the potential pitfalls of allowing the payer to determine the value of any award. Various schemes have put mechanisms in place to minimise this—for example, the French system of penalties for not making a reasonable offer, the use of agreed tariffs, appeal mechanisms, etc. The simplest, and arguably best, solution is that the decision on quantum should be taken by an independent party, and not by the payer. This removes the possibility of the payer attempting to devalue the award offered.

Innovation and Efficiency

The influence of traditional civil (court) procedure can be seen in some schemes, where innovations have built on the familiar civil procedure model, and taken relatively minor steps away from it. Such basic changes might be to change either the independent decision-maker, or the procedure and format.

The most obvious change of independent decision-maker is to switch from a judge to an arbitrator, or to substitute a committee. Thus, the French and Polish schemes create committees. The advantage of a committee is that it might contain experts (in medicine and law), which should possibly bring efficiency into the process by removing the need for a judge to obtain external expertise from one or more medical experts. Thus, the French and Polish schemes have medics and lawyers as committee members.

However, the steps in this direction have not always produced transformative results. Taking the further shift to a body of administrators, instead of a committee with a precise number of members and with fixed specialisations, is more flexible. The New Zealand and Nordic schemes demonstrate the point clearly. The administrators can either make swift decisions themselves, on the basis of their own medico-legal expertise, or can obtain expertise by asking relevant specialists (who may be members of an expert panel), perhaps referring difficult decisions to a committee, or having all decisions scrutinised by a supervisory committee. The administrative model therefore provides considerable flexibility, and hence efficiency in resources and cost, and increased speed. Those schemes that rely on committees give the impression either of over-engineering the structure, such that it will have difficulty evolving, or of a lack of trust in administrative bodies.

Changes in Procedure and Format

One interesting change is to reverse the order of the normal steps that occur. This is the Irish approach. The normal sequence would be for a decision to be taken first on entitlement (liability, in court terms), followed by a decision on quantum (the sums for which the claimant is entitled). The Irish approach, in the context of a litigation mode, has been to require claimants to obtain a statement of the quantum of damages from the independent State Board before being permitted to start court proceedings. Somewhat counterintuitively, this approach has had a clear success in its context. It has reduced the throughput of cases in the court system, and speeded up their rate of settlement either before or during that court process. But it is a system designed to increase efficiency in the court process rather than in relation to resolving patient injury claims as such. Interesting as that reversed approach is, there seems to be no reason to take such an approach in relation to administrative compensation schemes. Indeed, the evidence is that the 'normal' approach of deciding qualification first and then financial entitlements second works well in administrative schemes.

The Florida NICA scheme retains elements of a traditional administrative court hearing, with the final decision on whether to accept or reject the settlement proposed by NICA resting with a judge. However, the basis of the decision has switched from negligence and tort to a statutory no-blame basis, and both the procedure and the format have changed considerably. The case is not about establishing if a healthcare provider is at fault, but about establishing if the child is eligible to join the NICA programme. Essentially the healthcare provider (or its insurer) is no longer the defendant. Instead of having an adversarial process involving opposing experts, the format has been changed such that the expert medical input on eligibility comes from experts instructed by the scheme.

Conclusions on Procedure and Structure

A wide variety of scheme structures are observed and they can function in a surprising range of ways. At one end of the spectrum is the 'light-touch' Western Australian workers' compensation scheme, where the scheme effectively outsources decisions on causation (such as are required), eligibility and quantum, and is a framework to facilitate interactions between the parties. The other extreme are the Nordic patient injury schemes which investigate and carry out their own determinations on causation, eligibility and quantum. Although there is considerable diversity within the schemes studied, some clear points emerge.

Firstly, the decision-maker does not necessarily need to be within the scheme. The most important factor for causation, eligibility and quantum decisions is that the decision-maker is perceived as credible and sufficiently expert. Where there is discretion on a decision it is essential that the decision-making process is clear and transparent. For example, in the UK compensation for infection with HIV and hepatitis C from infected blood is awarded by both charities and companies; the charitable awards are discretionary and claimants have expressed considerable dissatisfaction about the lack of clarity about the support available and the lack of transparency on how award decisions are reached. It seems prudent for any scheme to have an oversight mechanism, but this is particularly important if the scheme is also the payer. Decisions that are taken externally are usually advisory rather than binding on the scheme, providing a natural oversight.

Secondly, there is no necessity to have a committee to make decisions, and there is a strong argument that committee decision-making is less efficient.

Thirdly, there are potential difficulties associated with allowing the payer to determine causation, eligibility or quantum. These can be reduced by oversight mechanisms, but separation of function entirely removes any risks associated with conflicts of interest.

The Payment Trigger: Fault Liability, 'No-Fault', Criteria and Filters

Every jurisdiction covered in this review maintains a regime of law that gives rise to liability for damage caused by breach of contract or (save mainly for New Zealand) by negligence/fault, and most also have strict liability for defective products. Those laws are well known and it is not the purpose of this book to examine them. With the exception of Ireland, such regimes operate as a parallel universe to that of the compensation scheme regimes analysed here.

It is a major finding of this study that the criteria subject to which an injured person becomes qualified for payment have evolved away from 'fault' liability on the part of the person who is held to have legally caused the injury towards a set of criteria focused on the condition of the person injured. In fact, two important shifts have occurred. First, the move from a judicial procedure focusing on fault liability to the various no-blame, and predominantly administrative, procedures noted in this study, has shifted the focus of the criteria that trigger payment from *actions by the injurer* towards *the condition of the injured and the circumstances of the injury*. The schemes noted in this study indicate that the traditional conception of a 'liable' party providing the compensation is not an essential component of a functioning compensation scheme.

Second, within the various no-blame administrative systems, the triggering criteria have shifted from an assessment of whether the actions or omissions of the payer satisfy a legal standard of *negligence*, based on a judicial opinion on the quality of the healthcare provider's conduct, towards a set of objective *factual criteria* about the condition of the injured person. Instead of 'fault liability' as the determinant of whether compensation is due, a series of alternative 'triggers' has been adopted for providing compensation.

There is no such thing as 'no-fault' liability. Liability arises on the satisfaction of defined legal criteria. The historical paradigm criteria are fault, causation and injury. The point about all of the schemes considered in this book is that they have moved to some other criteria. All of the criteria adopted differ between schemes, so it is almost impossible to generalise, save by exclusion to describe them as 'not fault', but that does not tell one much about what the particular criteria are. Although compensation schemes have traditionally been known as 'no-fault' schemes, that label is inaccurate, as it is not correct to describe any system noted above as being 'no-fault'. Although assessments of fault or blame have been removed, other criteria have been substituted instead of criteria about 'fault'.

A summary of the most significant triggers that are applied is given in Table 33.7. The detailed criteria are all set out in detail in the national chapters above, and this summary is consciously simplified so as to enable an easier overview and comparison to be undertaken.

Table 33.7: Payment triggers

Country	Scheme	Definition of trigger (excluding residence requirements)
UK	NHS treatment (except vaccinations)	Negligence
Germany	Medical negligence	Tort liability (without impact on the doctor's insurance premium)
Germany	Pharmapool	Fault
Ireland	Irish Injuries Board: assessment of quantum for types of personal injuries (other than arising during the course of medical treatment)	Legal liability, eg for negligence or as an occupier of land
France	ONIAM	Injury from: 1. Negligent medical examination or treatment (insurance responds) 2. Serious non-negligent medical adverse event or hospital-acquired infection (insurer responds) 3. Specified vaccines, HIV, etc (ONIAM responds)
Poland	Medical accident	Contamination by a biological agent, injury or death of a patient caused by a diagnosis, treatment, or use of a medical product or device which is contrary to current medical knowledge, in a hospital
New Zealand/ ACC	Accident	Physical injury and/or some mental injuries resulting from physical injuries (including treatment injuries and injuries with contributory negligence elements)
Sweden	Patient	1. the injury could have been avoided by other treatment 2. defect in, or defective use of, a product 3. incorrect diagnosis 4. infection injury 5. injury caused by an accident occurring in connection with examination, care, treatment or any similar procedure 6. medication injury

(continued)

Table 33.7: ***(Continued)***

Country	Scheme	Definition of trigger (excluding residence requirements)
Denmark	Patient	on a preponderance of the evidence: 1. an experienced specialist in the field would have acted differently 2. malfunction or failure of technical apparatus 3. the injury might have been avoided using another available treatment technique or treatment method that would have been just as effective 4. injury occurring as a result of examination or treatment that is more extensive than the patient should reasonably have to endure
Nordic states	Road traffic injuries	Coverage applies to injuries arising from unknown or uninsured drivers
Nordic schemes	Pharmaceutical injuries	unforeseeable injuries, either not warned about or more severe than expected
Japan/PDMA	Pharmaceutical injuries	Variable according to claim: — Adverse drug reaction: health damage caused by a prescribed drug, even when properly used — Infections from biologicals: health damage from an infection caused by a biological — SMON injury: injuries caused by taking SMON as per the court order — HIV/HCV infection: infection with HIV/HCV from contaminated treatments
USA and UK	Vaccine-related injury	Specified injuries occurring as a consequence of vaccination
UK	Armed Forces Compensation Scheme	Discretionary: caused wholly or partially by the individual's service
UK	Industrial Injuries Disablement Benefit	injury accrued during the course of employment (triggers benefits or payments towards extra care)
UK	Criminal Injuries Compensation Scheme	Blameless victim of violent crime
UK	Thalidomide Trust	Phocomelia caused by mother's ingestion of Thalidomide

(continued)

Table 33.7: ***(Continued)***

Country	Scheme	Definition of trigger (excluding residence requirements)
UK	Mesothelioma	Mesothelioma where employer untraceable, and claim within 12 months of award of IIDB (triggers single payment)
UK	Coal Workers' Pneumoconiosis Scheme	Specific conditions to British Coal workers (triggers lump sum payment and some disability benefits)
UK	Coal Health Compensation Schemes	COPD or VWF in miners and their families
UK	Skipton Fund	HCV or HIV infection caused by NHS-supplied blood products
UK	Creutzfeldt–Jakob disease	vCJD on the legal test of the balance of probabilities from human growth hormone treatment
UK	ABPI/ABHI clinical trial schemes	Involvement in a trial or device evaluation (no fault)
UK	Dental Complaints Service	Legal liability (contract or tort)
UK	Parliamentary and Health Service Ombudsman	Maladministration
Proposed schemes (not yet confirmed or in force)		
UK	Proposed neonatal injuries	Avoidable injury for defined births
Scotland	Patient	Avoidable injury

Defining Criteria

The underlying technical advance that has taken place is to define the criteria for triggering payment in terms that are more factually based and can be applied more efficiently and accurately than the more generalised and impressionistic (or judgemental) criterion that an individual was at fault. A good example of this is seen in the NICA birth injury eligibility in Florida, where eligibility is determined by weight rather than gestational age as the former is objective and easily measurable.

It is essential to apply the appropriate trigger to the particular type of injury involved. This goes to the heart of the function of a compensation scheme. The purpose should be to compensate those 'worthy' claimants who have suffered a personal injury, while not compensating 'non-worthy' claimants.

The trigger can sometimes be *variable*, or even applied with some discretion (as in the Nordic schemes). Triggers can change over time to adapt to the needs of the scheme and/or the claimant population. In New Zealand at the inception of the ACC the trigger was an 'accident', but the definition of 'accident' proved difficult to administer, so was subsequently amended to the current position of 'eligible injury' as detailed above. Likewise, until 1 July 2005 ACC used 'medical misadventure' as the trigger for clinical injuries, but since then the trigger has been redefined to 'treatment injuries'. Indeed, it follows that different triggers may be applied not just over time within an individual scheme, but also at the same time across different related schemes.

This finding has considerable significance in designing possible future models. Any new scheme would need to take into account procedural fairness and the perception of the scheme—to succeed the scheme has to provide clear, transparent processing to be seen as fair and just.

It is readily apparent that different schemes use different triggers that fulfil the needs of a particular set of circumstances. Indeed, different triggers can be used within the same scheme for different types of injuries, such as in Sweden and New Zealand. The ACC example indicates that a trigger does not have to be static and can be adjusted as a scheme evolves. Schemes can have triggers that are not injury related—for example, all the industrial schemes have additional eligibility requirements, working for particular bodies, and the UK Criminal Injuries Compensation is only available to 'blameless' victims of crime.

In addition to initial triggers, schemes may also use filters, which may differ for different types of disease or claimants. For example, the initial trigger in the Trilucent breast implant compensation scheme was having received Trilucent breast implants. Once this was established, claimants were then filtered to allow allocation to different claim pathways. Those who simply wanted their implants removed and replaced were allocated to one of three fixed-cost compensation options depending on whether they had removal; removal and replacement in one operation; or removal and replacement in two separate operations. Claimants with more complex cases, either physical or those with psychological damage, were filtered to a specialist track which enabled their needs to be fully assessed and appropriately compensated. This sort of filtering enables schemes to provide more appropriate tailored solutions for claimants.

The essential question for each scheme (including any new schemes) is how to determine the appropriate trigger for claimant needs. There are schemes where the claimant inclusion criteria are fairly self-evident—for example, the vCJD Trust fund requires a diagnosis of vCJD. Where causation is demonstrable it is usually relatively easy to set eligibility parameters, but where causation is in doubt there will be more difficulty. For example, in the recent PIP breast implant situation, the current scientific evidence suggests that PIP implants are not thought to be unsafe, but they are not as described/as licensed. This is essentially an issue over quality rather than safety, albeit probably requiring earlier revision than might have otherwise been the case. A no-fault compensation scheme would need to decide what thresholds to apply in such a case: should it cover the cost of checking, the cost of revision, and pain and suffering?

There are various reasons why fault has been superseded. When a person suffers an adverse reaction to a drug, it is questionable whether the notion of 'fault' is relevant, irrespective of whether the adverse reaction is already known or unforeseen. Under the Nordic Drug Compensation Scheme, the latter will attract compensation, whereas the former will not unless the reaction was unusually severe. In the medical treatment context, it has been found that operating a regime based on fault has a significantly damaging effect on building a culture of trust, openness and learning.

Interestingly, in the UK the move away from fault is not just seen in personal injury; UK ombudsmen use a trigger of 'fair and reasonable' rather than relying upon the relevant legal test. It will be interesting to see which tests any future administrative compensation schemes, such as those proposed by the National Maternity Review and the Scottish government, will choose to apply.

Conclusion on Triggers

Schemes are simpler to administer when there are clear and unambiguous triggers—for example, the two US birth injury schemes both use birth weight, which is a clear objective measure, rather than gestational age, which is based on the opinion of an expert, and is therefore potentially arguable. However, having clarity about the triggers does not mean that triggers need to remain rigidly fixed, as the developments in New Zealand clearly demonstrate. When a scheme is pre-emptive and designed to meet ongoing need, triggers can, and do, change and evolve. This is hardly surprising and is not limited to compensation schemes. A similar effect can be seen in negligence litigation when advances in scientific knowledge impact on the liability threshold. For example, the Copper7 and Gravigard litigation was dropped in the US and UK, respectively, when it became clear that the symptoms suffered by these women were due to infection with chlamydia (which had only just been identified) and were not due to their use of these contraception devices. This example demonstrates why it is desirable to have some flexibility in the trigger criteria.

Triggers enable the scheme to control the numbers of compensation payments. Therefore, triggers for payment are often thought of as the cornerstone of the financial stability of an ongoing compensation scheme, but this is something of a misunderstanding of their primary function. The aim of triggers is to select the worthy claimants from the less deserving and to make payments only to the former. The decision of what a makes a claim worthy of compensation is complex and is influenced by a wide range of factors, from societal norms to prevailing financial conditions. Schemes have to take the earliest opportunity to define the coverage item and the trigger that is relevant to each, and to keep these under review.

It is salutary to remember that many children who suffer catastrophic injuries at birth are *not* compensated as negligence cannot be proved (that is not to say it did not occur), and a strict liability claim in relation to thalidomide, for example, would not succeed.

Financial Aspects

There exist real opportunities to save costs by the use of administrative schemes. Firstly, administrative schemes based on an investigational approach are far cheaper than adversarial litigation.

Secondly, the administrative overheads of various different schemes can be reduced if their administration is combined. Thus, multiple individual schemes should be merged, without necessarily adopting the same rules or procedures for different types of injuries or compensation arrangements. In the event of mergers, care should be taken to ensure the traceability and transparency of both the incomings and outgoings of each of the individual elements are preserved, particularly in cases where the incoming funding arises from several different sources. While merging may provide a cost-saving benefit the case studies have shown that some schemes, eg the Thalidomide Trust and the vCJD Trust, provide specific sources of expertise on particular issues affecting claimants. Merging risks diluting this expertise, and careful consideration should be given before mergers are undertaken.

The advent of the internet, particularly social media, may assist with information-sharing and campaigning. Certain groups, eg HIV and HCV infectees, may not utilise this media or may feel some stigma about discussing the reasons for their claim, so this may not apply equally to all classes of claimant. One potential solution is to have specialist panels under a generic application portal—for example, ONIAM in France have specialist panels for Benfluorex, and PDMA in Japan have a single application process for various claimant types, including patients afflicted by the SMON drug issues.

Some schemes have in-built cost-recovery mechanisms. Denmark considers on-charging in every drug injury case, on the basis of a subrogated claim strict liability under the EU Product Liability Directive, but has never made a claim.

This is an indication of how difficult it is for patients to succeed under this legal basis, and prompts the question of whether some other basis may be relevant, although it would have to maintain a balance between the interests of injured patients and manufacturers/insurers.

Administrative Costs

Compensation schemes have substantially lower administrative costs compared to adversarial processes, so although the per-claimant costs of a scheme may be higher during the start-up period they are still likely to be considerably lower than the per capita cost of adversarial options. The way a scheme is funded can have an impact on administrative costs—for example, the Irish Injuries Board charges a small fee to users, and has become self-funding within 5 years.

Compensation Award Costs

One very interesting development is the inclusion of non-financial redress. Tort litigation offers very limited types of redress; for personal injury claims the only redress a court can order is financial compensation. This contrasts with a shift of approach which has been seen in non-court-based dispute resolution in the UK—for example, the Legal Ombudsman now recommends a range of redress options from apologies and explanations through to compensation. Interestingly, some of the schemes also employ non-financial redress: in the early days of the Thalidomide Trust, for example, the Trust purchased a holiday home for the affected children, which gave them a place to holiday and also contact with other similarly affected families. While financial compensation may be the mainstay of many schemes it should not be forgotten that they often have greater freedom over the type and form of redress on offer than is seen in the courts.

Individual schemes take quite different approaches to quantum: some apply a tariff, some match the level of compensation a court would award, some offer less than a court would, and some are more generous than the settlement a claimant would have reached had the issue gone to trial. Schemes can also compensate under different heads of damages to those seen in tort. There is no universal formula for assessing quantum.

The levels of compensation awarded by a scheme partially depend upon the relationship between all available sources of financial support. This is clearly shown by the Nordic systems, where the medical and drug schemes act as secondary 'top-up' sources for the primary social security sources. Each scheme needs a clear rationale for the level of compensation paid: is it to be restorative, as in the UK courts, a 'top up' as in the Nordic states, or a hybrid as in New Zealand, where it provides not full compensation, but 'fair compensation'? However, several factors need to be considered when determining the appropriate level of compensation for claimants under compensation schemes, such as the following:

- Problems arise and people find it unsatisfactory where there are parallel streams which compensate at different levels. For example, the Armed Forces Compensation Scheme and the Criminal Injuries Compensation Scheme attract criticism as they pay at lower levels than a court award for an equivalent injury, and it is often deemed unfair that circumstances decide compensation levels. These schemes are schemes of last resort, and it is questionable whether claimants would opt for them if there were an alternative available.
- If you opt for 'fair' compensation, who sets it and how is it quantified? The objective for compensation under a tortious system is to restore the individual to the position that they would have been in had the injury not occurred. However, there is evidence that some individuals do not spend the compensation that they received for this purpose and that claimants feel that the awards given are not adequate.[13] How should this impact on the quantum awarded?
- Some sources of funding for compensation are finite; this can make assessing quantum an issue of apportionment rather than restoration. For example, each year the Thalidomide Trust apportions the sum available to it on the basis of injury severity; it does not have the funds to be able to work back from individual needs.
- Where the sources of funding are from the public purse the issues of public goods versus private rights become more apparent. In the UK under current legislation[14] anyone making a claim for personal injury, including claims against public healthcare providers, has the right to request funding for private treatment even if such treatment is available on the NHS. This effectively removes money out of the public purse into private companies (assuming it is spent on treatment). While this appears to be contrary to the public good, from a claimant's perspective the situation may well differ: if you were injured by the NHS would you really want to give it a second bite of the cherry by letting it treat you again?

Assessing the quantum for compensation under a scheme is multifaceted and is not straightforward. Compensation under a scheme may not directly mirror court-awarded compensation awards.

As well as awarding different levels of award, schemes will also compensate under different heads of damage. In a tort-based system claims for compensation can only be made under the recognised heads of damage; compensation schemes, however, can be more flexible and accommodating. It is notable that some of the schemes in this volume do *not* include payment of an equivalent of pain and suffering damages (moral damages, in civil

[13] H Genn, 'Personal Injury Compensation: How Much Is Enough?' (1994) Law Com No 225, available at www.ucl.ac.uk/laws/judicial-institute/docs/Personal%20Injury%20Compensation%20how%20much%20is%20enough.pdf.

[14] S 2(4) of the Law Reform (Personal Injuries) Act 1948 states that 'In an action for damages for personal injuries ... there shall be disregarded in determining the reasonableness of any expenses the possibility of avoiding those expenses or part of them by taking advantage of facilities available under the NHS.'

law systems). For example, the British Dental Council rarely awards such compensation and the ASR hip reimbursement scheme does not offer it. Initially this seems unjustifiable, but as the dental example and other research[15] shows, many claimants do not specifically desire such damages for themselves. Non-monetary remedies are often highly significant to claimants.[16] It may be that financial compensation has a symbolic value as a public symbol of responsibility for error or blame.[17] But would monetary awards be desired to the same extent if the compensation system was based on a 'no-blame' principle, and the behaviour of professionals and commercial entities was clearly accountable through other mechanisms, such as through regulatory systems?

Assuming that a tortious liability-based system is necessary to hold back the tide of potential litigants is an oversimplification. It assumes that claims are more likely to apply and/or to succeed under a compensation scheme and/or that they are likely to be compensated at a comparable or greater level. None of these assumptions are automatically true: the way in which a scheme is structured, the thresholds and eligibility criteria, the quantum and heads of damage compensated will all have considerable impact on the sums awarded to successful claimants.

There is not a simple correlation between increased claim numbers and increased costs: whether costs increase, decrease or remain the same will depend upon the circumstances of the individual scheme. Cost is by no means the only consideration when considering no-blame compensation schemes.

Conclusions on Costs

Many claim types are shifting to new architectures and triggers. The large number of schemes detailed here show it is possible successfully to shift injury claims to a compensation scheme. There are several models from which to learn, and the exact form of any new scheme will be determined by the scheme's purpose and function. Theoretically new schemes should be able to offer simplicity and reduced costs and shortened timeframes, but the examples above show that not all schemes manage this, notably the miners' COPD compensation schemes in the UK. To flourish, schemes need clear parameters, transparent accountable processes, and a reliable and sufficient funding source to provide financial stability. Compensation schemes are not an automatic panacea: costs may be higher than expected; trigger definitions may need updating and amending; coverage may not be sufficiently wide and/or integrated. While potentially problematic, it is interesting to note that these factors are rarely fatal to a scheme. There are plenty of examples of schemes adapting and altering, but we have not found a single example of a scheme which makes ongoing payments collapsing. Some US motor insurance statutes have been withdrawn, but as these

[15] EA Dauer and LJ Marcus, 'Adapting Mediation to Link Resolution of Medical Malpractice Disputes with Health Care Quality Improvement' (1997) 60 *Law and Contemporary Problems* 185–218.

[16] CS Hyman and CB Schechter, 'Mediating Medical Malpractice Lawsuits Against Hospitals: New York City's Pilot Project' (2006) 25 *Health Affairs* 1395–99, doi: 10.1377/hlthaff.25.5.1394; CB Liebman and C Stern Hyman, 'A Mediation Skills Model to Manage Disclosure of Errors and Adverse Events to Patients' (2004) 23 *Health Affairs* 22–32, doi: 10.1377/hlthaff.23.4.22.

[17] WLF Felstiner, RL Abel and A Sarat, 'The Emergence and Transformation of Disputes: Naming, Blaming, Claiming' (1989) 81 15 *Law and Society Review* 634.

were statutory frameworks that governed insurance policies the effect of the sunset clauses was simply to restrict an individual's ability to purchase a particular type of policy.

Summary of Arguments for a Compensation Scheme

The evidence suggests the following. Complainants seek a system that is user-friendly, readily understandable, efficient and swift, provides a sympathetic opportunity to be heard, operates fairly, and produces outcomes that are fair and useful. People do not always seek compensation; they may simply wish to be listened to, taken seriously and believed, or to be given an explanation and apology, and to try to prevent the same thing happening again to others. Health professionals wish to be allowed to operate in an environment that supports learning rather than blame, maximising reporting of adverse event and clinical practice so that feedback such as clinical outcome reporting and self-learning can occur, and problems can be addressed through non-adversarial processes. Manufactures and insurers are prepared to pay liability claims provided the resolution processes operate within fair systems and deliver fair outcomes, with business stability though predictable costs rather than unexpected and especially unjustified spikes, and efficient systems that have low transactional costs. Those responsible for systemic safety wish to see maximised and early capture of reports of problems and adverse events, through free reporting and protection of whistleblowers.

Any insurance mechanism is intended to provide some protection against unexpected and potentially destabilising shocks through the mechanism of spreading the risk of adverse financial costs. The relevance of this for product manufactures can be illustrated by the experience with vaccines. The US Vaccine Injury Compensation Program was introduced in 1988 in order to respond to a mass of non-autism claims against manufacturers. As Figure 33.1 shows (repeated from chapter 15 above), after those claims had been resolved, the number of successful claims fell to a low and relatively stable level (as occurred under the UK scheme). But another spike in claims occurred around 2001, all involving allegations that the MMR vaccine caused autism. This mass of claims emanated from public panic, fuelled by extensive media reports, but was scientifically without foundation. All of the claims were rejected efficiently by the US scheme, whereas extensive litigation in England and Wales took far longer to reach the same conclusion.[18]

Alternatives to Liability: Opening the Floodgates

One of the widely cited arguments against 'no-fault' compensation schemes is that the removal of the liability threshold would lead to an opening of the floodgates and a huge increase in claimants seeking compensation. The number of claimants impacts upon the collective value of the claims, but the legitimacy of an individual claim is not altered by the number of other similar claims made. A large increase in the volume of claims can be an issue for any redress mechanism. An initial screening using appropriate eligibility criteria

[18] See C Hodges, *Multi-Party Actions* (Oxford, Oxford University Press, 2001).

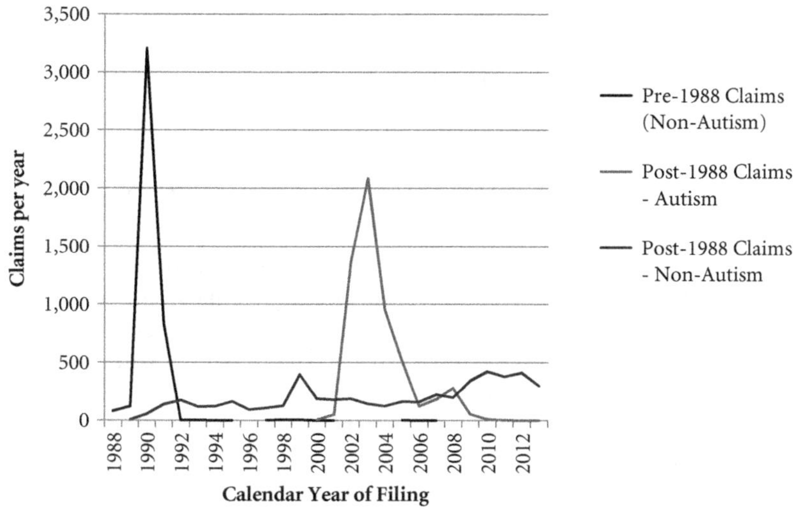

Figure 33.1: VCIP claim numbers by year of filing for pre-1988 claims and autism and non-autism claims since 1988

should minimise the risk of a system being overloaded by the volume of claims. Compensation schemes tend to have lower administrative costs than litigation, enabling them to triage claims more cheaply. The objective of a compensation system is to compensate deserving claimants; if a larger claimant cohort can be assessed in an efficient and cost-effective manner, then it is less likely that deserving individuals will go uncompensated.

One of the potential issues with large claim volumes is that they can assist in the detection of safety signals. This concern only pertains to forms of personal injury redress where there is a feedback mechanism. The data generated by compensation schemes can be used to feedback on safety concerns. There are numerous examples of the way in which awareness of issues can lead to an increase in claimant numbers, as was seen with the US silicone breast implant litigation and UK cases involving MMR vaccine and benzodiazepines. These claims can be driven by public opinion rather than by expert scientific opinion. In some situations the type of claim can be used to gather information; Figure 33.1 shows the increase in autism-related claims to the VICP in the USA. The spike seen can be differentiated from the non-autism claims, enabling separate analysis of the two claimant populations. It is easy to see how changes in claimant behaviour could raise safety concerns which further investigation did not substantiate. This should be balanced against the fact that safety signals can only be investigated once there is some indication that there is a problem, and data from compensation schemes could provide such indications.

A major part of the floodgates argument is the increased cost of a larger number of claimants. There are two elements to this, the administrative cost of processing claims, and the cost of the compensation awarded.

There are examples of an increase in claimant numbers at the start of a scheme—for example, the number of claims under the VICP in the USA was initially very high (as would be expected as retrospective claims were permitted) and then settled to a lower baseline.

Costs for the administration of a new scheme are likely to be higher in the first few years of operation due to set-up costs. Although there may be more claimants under a new scheme (particularly if retrospective claims are allowed), the per capita cost of administration is likely to be lower.

No-Blame Compensation Schemes: Achieving Positive Change

All the jurisdictions discussed here are back-stopped by functioning litigation systems (except New Zealand which has chosen largely to replace personal injury litigation). Despite this, these very diverse countries have all chosen to create no-blame schemes for some types of personal injury. That indicates widespread belief that there is a better way to deal with some types of personal injury claims than tort litigation: if it ain't broke, don't fix it. In actuality there are a large number of drivers for replacing tort with administrative schemes. Some of these are personal: potential claimants want a speedy, user-friendly, easy-to-understand mechanism to resolve their complaint. Some are economic: in general the administrative costs of compensation schemes are far lower than those associated with litigation, making them look a far more attractive option. Some are far larger public policy reasons related to the perception of fairness, or lack of fairness, associated with proving negligence.

A telling rationale for no-blame compensation schemes is illustrated by the comment from Sweden that mistakes happen, and that looking for scapegoats through a legal system for medical injuries based on fault is 'a very efficient way of killing more patients'. This maxim can be seen mirrored in the UK's attempts to introduce an open, no-blame, complaints-reporting system but at the same time maintaining a fault- and litigation-based system for injury compensation. The maintenance of a blame-based litigation system effectively prevents healthcare professionals from openly reporting problems as they occur. In contrast, the Nordic culture is to be open and acknowledge problems immediately, and to assist patients who have suffered unexpected injuries to lodge a claim with the no-blame compensation scheme, which responds irrespective of whether causation involved fault or professional error.

If reporting of every safety issue is important as a matter of institutional learning and ensuring safety, then the incidence of reporting is considerably enhanced by a 'no-blame' culture. Hence it is logical to separate the delivery of compensation from the safety management system. And to make it dependent on a trigger that does not involve blame either.

There are currently two no-blame schemes under consideration in the UK: a scheme for clinical injuries in Scotland and a scheme for birth-related brain injuries in England. Compared to the current position of reliance upon tort litigation, a new scheme could produce many potential benefits including a faster, fairer resolution for those affected, potentially shifting administration costs from the public sector (courts and NHS) to a private body, supporting a positive NHS culture, and creating harm reduction strategies by using scheme data to support learning and improvement. However, there are examples in this book of schemes not working as predicted, and it would be easy to get this wrong: scheme design aspects are absolutely critical.

34

Parameters and Outline of a Modern Compensation Scheme

This chapter suggests the parameters and outline for a contemporary injury compensation scheme based on the findings of the research set out above.

It should be remembered that a scheme has to operate in the context of a specific national, local and organisational landscape, so the model that we set out here is to some extent a generalised and unspecific model. The function and lifespan of a scheme will have substantial impacts upon its design; short-term, post-hoc, single-issue schemes often do not require (or have the financial wherewithal to sustain) the more expensive independent oversight and thorough investigation of claims. By contrast, long-lived schemes and/or high-value claims will usually require far greater stringency in assessing claims and a more complex scheme structure with an oversight mechanism built in.

One of the major advantages that a no-blame compensation scheme offers is the ability quickly to collate information that can be used in harm reduction. The speed of the decision-making process, combined with the open disclosure afforded by a no-blame environment, is unique to no-blame administrative systems. One of the more disappointing findings of this book is that this valuable resource is often not fully utilised. Both existing and future schemes should carefully examine how to maximise the use of their injury data for harm reduction.

A. The Parameters for a Principled Scheme

The design of a compensation scheme model clearly depends on the political and policy objectives, and on decisions driven by economic considerations. There is, therefore, a range of possible design choices. Taking that 'chicken and egg' issue into account, the following design aspects would seem to be suggested from the findings of the research set out in this book on the range of compensation schemes covered.

The essential constitutional principles to be observed are: fairness, stability, simplicity, efficiency. We suggest that the parameters that should govern the design of a compensation scheme in the context of a contemporary European democracy are as follows.

The compensation scheme should be:

1. unbiased and objective;
2. easily identifiable;
3. have expertise and authority;
4. efficient;

5. transparent;
6. fair;
7. support maintenance of and improvement in performance.

It is relatively straightforward to design an efficient compensation scheme model, drawing on best practice and lessons from a significant volume of evidence from different contexts around the world. The model would have a unified investigation portal, operate on a swift inquisitorial process, decide on eligibility, and notify units/insurers of payments to be made.

It would make eligibility decisions on sets of criteria for defined injury types, none of which would involve fault.

B. Outline of a Model for Personal Injury Compensation Schemes

1. The scheme should command widespread trust, including transparency of operation and of outcomes, and independent governance, without political or other partial operational interference.
2. The processes should observe the principle of fairness (due process) in providing predictable, impartial and just procedures.
3. The scheme should have procedural and financial stability.
4. The scheme should be easily accessible, user-friendly and efficient.
5. The general position should be that claimants do not have to pay a fee to access the scheme.[1]
6. The scheme should operate on an administrative investigative/inquisitorial-like basis, rather than on an adversarial basis.
7. The design should incorporate processes for lodging, investigation, deciding and paying, or ordering the payment of, valid claims.
8. The scheme should consist of the following basic stages:
   (a) Lodging claims: direct (ideally online) or through doctors (with verification of injuries and initial attribution of causation).
   (b) Triage: initial scrutiny to screen out invalid claims, refer relevant claims to any other relevant pathway, and allocate valid claims to appropriate internal processing tracks.
   (c) Assessment: confirmation of causation and legal satisfaction of the rules of entitlement. The procedure may differ depending on the type of injury and circumstances, but the various tracks should be specified in advance. For simple injuries or recognised types, paper evaluation may be sufficient. Some cases—but possibly not all—may require: medical examination of a claimant; expert opinion on causation by an external expert or panel of experts; technical input from manufacturers or insurers on causation.

[1] This is the current position with many EU ADR consumer-to-business contract dispute schemes.

(d) Payment: if a claim satisfies the rules, it should be paid as soon as possible. It is preferable that the scheme administration itself holds adequate finance to pay immediately (the New Zealand and Nordic models). Alternatively, the payment instruction should be sent to the relevant insurer (the French model but with a simplified system, not dependent on settlement offers and potential for bargaining).

9. The trigger for acceptance of a claim will always include establishment of causation of injury that is attributable to a defined source (eg a traffic accident, medical treatment, use of a product).
10. Exactly how the trigger is defined will vary depending on the particular source of injury. For example, an option for medical treatment in an unexpectedly serious adverse event[2] but the option for vaccine injuries could be inoculation followed by defined side effects.
11. The trigger should be based on as many simple factual criteria as possible, in order to simplify and speed determination of eligibility.
12. The trigger should not be based on, or include, the concept of fault.
13. Redress should include financial and non-financial elements. For financial payments the basis and quantum for payment of awards made should be defined in advance, and be based on predictable tariffs wherever possible.
14. The relationship between benefits under the scheme and payments and entitlements from any other source, such as social security, state-provided healthcare or first-party insurance, should be defined.
15. It should be possible to amend the operational rules, and levy finance in unexpected circumstances, without undue delay, whilst satisfying requirements of law, predictability, due consultation and fairness.
16. In order to protect European Convention on Human Rights Article 6 rights, claimants should be able to bring claims in court, but only *after* making a claim on the scheme and either having had the claim rejected or having themselves rejected an award.
17. Awards should either be binding on the payer if the claimant accepts the offer,[3] or deductible from any subsequent litigation settlement
18. There should be an administrative review mechanism against adverse decisions.
19. It should be possible to review the scheme, via judicial review if it is classified as a public body, on the usual limited grounds of procedural illegality and manifest unreasonableness: this would not constitute an appeal process.
20. It is preferable, but not essential, that the scheme should cover administration of compensation to as many different types of personal injury as possible, such as medical (public and private), medicines, medical devices, cosmetics, workplace and road traffic injuries. Conversely, fragmentation of schemes is not advisable.
21. If a scheme covers many different injury categories—motor vehicles, workers' compensation, healthcare injuries, etc—then the scheme should operate an appropriate number of category-specific payment pools for the types of claim that it administers.

[2] As with the Nordic schemes.

[3] This is the current position with many EU ADR consumer-to-business contract dispute schemes. See C Hodges, I Benöhr and N Creutzfeldt-Banda, *Consumer ADR in Europe* (Oxford, Hart Publishing, 2012).

Each pool should be funded by discrete homogeneous groups of payers, whose funds are ring-fenced and not generally used for cross-subsidising other pools.

22. The extent to which some or all funding is provided by the state (from general taxation or other levies such as fuel taxes or levies on commercial or professional operators) and by private groups is a political decision, but is encouraged at least for certain types of universal risks that are not attributable to any particular group of actors.[4]
23. The basis on which members of a risk group provide funding should be agreed in advance.[5]
24. The fund should have the option to purchase insurance at the end of each financial year against new claims for that year that may arise some years later, so as to mitigate risk and reduce major budgeting issues that have to be funded by other companies.[6]
25. It should be possible for individual or a group of actors to join the administrative arrangements on an ad hoc basis so as to process a particular group of claims.
26. If the scheme holds its own funds, then it should be permitted to accumulate a reserve that may be allocated to pay a group of unusually large claims.
27. The scheme should be able to administer pooled-sector compensation funds that are funded by a defined group as insurance against the event that one of its members becomes insolvent.[7]
28. The scheme administration should collate all data and feed it back in appropriate form (usually anonymously aggregated) to members, regulators, users and the public to assist learning about improving practices, and identifying new or emerging risks.[8]

While we have laid out some thoughts on scheme design, there is one final caveat: where scheme rules are too restrictive or prescriptive they can impinge on the functioning of a scheme. Rules are essential for providing a framework, but the rules must aid, rather than prevent, the efficient functioning of a scheme.

[4] An example of comprehensive state funding is Denmark, which provides a right of recourse to reclaim sums paid in an individual case against a particular party who can be shown to have liability, such as a manufacturer whose product is in breach of the Product Liability Directive (such a claim has never occurred).

[5] The default position will usually be that all members pay the same level of annual premium (mutually insuring). Particular pools may agree to operate on the basis of individual risk-based premiums, or a combination of an annual overhead charge and individual case fees. For example, the Japanese pharma scheme can charge a manufacturer who has 'caused' an injury a 25% surcharge as a contribution. This may provide some form of incentive to maximise safety by reducing payment costs, but 'deterrence' is only recognised to have a limited effect at best, and other means are adopted to promote appropriate behaviour.

[6] See the Florida birth injuries scheme.

[7] This is similar to a number of schemes such as Directive 97/9/EC on investor compensation schemes. It would cover the insolvency of PIP, the manufacturer of breast implants.

[8] See the arrangements for cooperation between consumer ADR entities and regulators under Directive 2013/11/EU, Art 17.

INDEX

Note: **see main entries for individual countries/schemes for information relating to that country**